P9-EMP-244

INTERNATIONAL UNION OF CRYSTALLOGRAPHY
CRYSTALLOGRAPHIC SYMPOSIA

Interpretation of $|F|^2$ series

$$A(\mathbf{u}) = \iiint_{\text{cell}} \rho(\mathbf{x})\,\rho(\mathbf{x}+\mathbf{u})\,dx_1\,dx_2\,dx_3$$

$$= \frac{1}{V^2}\sum\sum\sum |F(\mathbf{h})|^2\, e^{-2\pi i(\mathbf{h}\cdot\mathbf{u})}$$

$$|F(\mathbf{h})|^2 = \sum_{i=1}^{N} f_i^{\,2} + \sum_{i,j=1}^{N}\sum{}' f_i f_j\, e^{2\pi i\,\mathbf{h}\cdot(\mathbf{x}_i-\mathbf{x}_j)}$$

Lindo Patterson and the Patterson function in his handwriting.

Patterson and Pattersons

FIFTY YEARS OF THE PATTERSON FUNCTION

Proceedings of a Symposium held at the Institute for Cancer Research, the Fox Chase Cancer Center, Philadelphia, PA, USA, November 13–15 1984

EDITED BY

JENNY P. GLUSKER

BETTY K. PATTERSON

AND

MIRIAM ROSSI

INTERNATIONAL UNION OF CRYSTALLOGRAPHY

OXFORD UNIVERSITY PRESS

1987

Oxford University Press, Walton Street, Oxford OX2 6DP
Oxford New York Toronto
Delhi Bombay Calcutta Madras Karachi
Petaling Jaya Singapore Hong Kong Tokyo
Nairobi Dar es Salaam Cape Town
Melbourne Auckland
and associated companies in
Beirut Berlin Ibadan Nicosia

Oxford is a trademark of Oxford University Press

Published in the United States
by Oxford University Press, New York

British Library Cataloguing in Publication Data
Patterson and Pattersons: fifty years of the Patterson function: proceedings of a symposium held at the Institute for Cancer Research, the Fox Chase Cancer Center, Philadelphia, PA, USA, November 13–15, 1984.—(Crystallographic symposia; 1)
1. Patterson functions
I. Glusker, Jenny P. II. Patterson, Betty K.
III. Rossi, Miriam IV. Series
548'.81 QD945
ISBN 0–19–855230–0

Library of Congress Cataloging in Publication Data
Patterson and Pattersons.
'First volume in the crystallographic book series'—Foreword.
Bibliography: P.
Includes index.
1. Patterson function (Crystallography)—Congresses.
2. Crystallography, Mathematical—Congresses.
3. Patterson, A. Lindo—Congresses. I. Glusker, Jenny Pickworth. II. Patterson, Betty K. II. Rossi, Miriam. IV. Institute for Cancer Research (Fox Chase Cancer Center)
QD911.P318 1987 548'.7 87–1701
ISBN 0–19–855230–0

Printed by St Edmundsbury Press
Bury St Edmunds, Suffolk

Preface

This volume is dedicated to the memory of A. Lindo Patterson, who, in 1934, proposed a method for determining crystal structures that forms the basis of our ability to understand three-dimensional structure in chemistry, biochemistry, geology and many other branches of science. To commemorate the fiftieth anniversary of the publication of the article describing this method a symposium was held at the Institute for Cancer Research, The Fox Chase Cancer Center. The lectures given at that meeting, which stressed advances in various scientific fields that have been made possible by the Patterson function, and the detailed discussions that followed them, together with a series of historical and personal reminiscences by his friends and colleagues form the core of this volume. Introductory biographies are included for those who attended the meeting. We were delighted that many others wanted to contribute articles for this volume and their chapters are included here and help to show the broader aspects of the use of the Patterson function.

This meeting, which was the brain-child of Betty Travaglini of the Institute for Cancer Research, would not have been possible without generous financial and other support from the Fox Chase Cancer Center. We thank all of those who helped ensure a successful meeting, including the Board of Associates, the Board of Trustees, the President of the Fox Chase Cancer Center (Dr. John Durant) and the Director of the Institute for Cancer Research (Dr. Jesse Summers). Others who worked hard to contribute to the smooth running of the symposium and publication of this volume include Richard Bomze, Jane Bosley, Joan Cantwell, H. L. Carrell, Jan Feigus, Pat King, Marcia Neeley, Ann Nista, Eileen Pytko, Kimberley Williams and many in addition. The entire meeting was taped and our especial thanks go out to Richard M. Peck and Eileen Pytko who helped the editors by typing the proceedings into the computer and Ann Glusker and Harold Greenwald who helped with proof-reading and Lisa Galati for corrections. Further financial support was provided by NIH grant CA-10925. This volume was typeset using the program TEX at Vassar College and we thank Vassar College for computing time. We also thank Frank

Manion at the Fox Chase Cancer Center for his help in setting up the program at Fox Chase.

Our thanks are also due to Dr. Linus Pauling for permission to reprint his letters to Lindo Patterson; to Dr. Raymond Pepinsky for permission to reprint the article in *Computing Methods and the Phase Problem in X-ray Crystal Structure Analysis*; the Albright-Knox Art Gallery, Buffalo, NY for permission to reproduce *Dynamism of a Dog on a Leash* by Gaicomo Balla; and Cordon Art, Baarn, Holland, for use of the Escher picture *Reptiles.* And finally we especially want to thank Dr Dorothy Hodgkin for permitting us to use the title of her talk, *Patterson and Pattersons*, as the title of this volume.

Jenny P. Glusker
Betty K. Patterson
Miriam Rossi

Philadelphia, PA
March, 1986

Foreword

It was a milestone in the development of the young science, crystallography, when in 1934 A. L. Patterson's first paper on the $|F|^2$ series appeared. The 'Patterson Function' has opened entirely new avenues to the solution of crystal structures and continues to be a widely used tool. To celebrate the fiftieth anniversary of Patterson's discovery, a Symposium, 'Patterson and Pattersons. Fifty Years of the Patterson Function' was held on November 13–15, 1984 at the Fox Chase Cancer Center in Philadelphia, Pennsylvania.

The title of the Symposium and its Proceedings portrays the spirit of the meeting. In addition to the lectures and contributions which deal with the theory and the applications of the Patterson Function, as well as other modern aspects of crystallography, the volume contains many personal reminiscences of Lindo Patterson by his widow and by contemporaries and friends. The intimate intertwining of scientific facts and personal memories gives the book its particular charm. Thanks are due to the editors of this work.

It is thus the hope of the International Union of Crystallography that the Proceedings of the Patterson Symposium will find wide interest, both as a source for the Patterson Function and its use and as a contribution to the history of crystallography.

Theo Hahn
President
International Union
of Crystallography

Aachen
March, 1986

INTERNATIONAL UNION OF CRYSTALLOGRAPHY CRYSTALLOGRAPHIC SYMPOSIA

This work is the first volume in the Crystallographic Symposia series, recently established by the International Union of Crystallography and Oxford University Press. Future volumes will contain proceedings of schools and symposia organized under the auspices of the International Union of Crystallography. In addition, a series of monographs and textbooks on crystallography is forthcoming.

Contents

PART III: CONTRIBUTED PAPERS

A. METHODS

C. HOMOMETRICS

PART IV. PATTERSON — THE MAN AND THE SCIENTIST

Contributors

S. C. ABRAHAMS, AT&T Bell Laboratories, Murray Hill, NJ 07974, USA

W. F. ANDERSON, Medical Research Council Group on Protein Structure and Function, Department of Biochemistry, University of Alberta, Edmonton, Alberta T6G 2H7, Canada

C. ARNOLD BEEVERS, 128 Blackford Avenue, Edinburgh, EH9 3HH, UK

GEZINA BEURSKENS, Crystallography Laboratory, Toernooiveld, 6525 ED Nijmegen, The Netherlands

PAUL T. BEURSKENS, Crystallography Laboratory, Toernooiveld, 6525 ED Nijmegen, The Netherlands

DAVID M. BLOW, Blackett Laboratory, Imperial College, London SW7 2BZ, UK

R. F. BOEHME, IBM Thomas J. Watson Research Center, P.O. Box 218, Yorktown Heights, NY 10598, USA

J. T. BOLIN, Purdue University, Department of Biological Science, Lilly Hall, West Lafayette, IN 47907, USA

BRADFORD C. BRADEN, Thomas C. Jenkins Department of Biophysics, The Johns Hopkins University, Baltimore, MD 21218, USA

MARTIN J. BUERGER, Lincoln Center, MA 01773, USA

RUFUS W. BURLINGAME, Thomas C. Jenkins Department of Biophysics, The Johns Hopkins University, Baltimore, MD 21218, USA

H. L. CARRELL, The Institute for Cancer Research, The Fox Chase Cancer Center, Philadelphia, PA 19111, USA

CHUNG CHIEH, Guelph-Waterloo Centre for Graduate Work in Chemistry, University of Waterloo, Waterloo, Ontario N2L 3G1, Canada

JOAN R. CLARK, Los Altos, CA 94022, USA

RICHARD E. DICKERSON, Molecular Biology Institute, University of California, Los Angeles, CA 90024, USA

GABRIELLE DONNAY, Department of Geological Sciences, McGill University, Montreal, PQ, H3A 2A7, Canada

J. D. H. DONNAY, Department of Geological Sciences, McGill University, Montreal, PQ, H3A 2A7, Canada

JERRY DONOHUE, Department of Chemistry and Laboratory for Research on the Structure of Matter, University of Pennsylvania, Philadelphia, PA 19104, USA

JACK DUNITZ, Laboratorium für Organische Chemie, ETH-Zentrum, CH-8092 Zürich, Switzerland

JOHN DURANT, The Institute for Cancer Research, The Fox Chase Cancer Center, Philadelphia, PA 19111, USA

EMIL ECKLE, Institut für Organische Chemie, Biochemie und Isotopenforschung, Universität Stuttgart, Pfaffenwaldring 55, D-7000 Stuttgart 80, Federal Republic of Germany

Present address: Schering AG, PCI/SPQ, Mullerstrasse 170 - 178, 1000 Berlin 65

ERNST EGERT, Institut für Anorganische Chemie, Universität Göttingen, Tammannstrasse 4, D-3400 Göttingen, Federal Republic of Germany

JONAH ERLEBACHER, The Institute for Cancer Research, The Fox Chase Cancer Center, Philadelphia, PA 19111, USA

JENNY P. GLUSKER, The Institute for Cancer Research, The Fox Chase Cancer Center, Philadelphia, PA 19111, USA

DAVID HARKER, Medical Foundation of Buffalo, 73 High Street, Buffalo, NY 14203, USA

DICK VAN DER HELM, Department of Chemistry, University of Oklahoma, Norman, OK 73019, USA

YEN-SEN HO, SmithKline and French Laboratories, 1500 Spring Garden Street, Philadelphia, PA 19101, USA

DOROTHY CROWFOOT HODGKIN, Somerville College, University of Oxford, Oxford, UK

M. BILAYET HOSSAIN, Department of Chemistry, Oklahoma University, Norman, OK 73019, USA

EDWARD W. HUGHES, Senior Research Associate Emeritus, Division of Chemistry and Chemical Engineering, California Institute of Technology, Pasadena, CA 91125, USA

TOSHIAKI ISOBE, Department of Chemistry, Tokyo Metropolitan University, Setagaya-ku, Tokyo, Japan

ROBERT A. JACOBSON, Department of Chemistry and Ames Laboratory-USDOE, Iowa State University, Ames, IA 50011, USA

G. A. JEFFREY, Department of Crystallography, University of Pittsburgh, Pittsburgh, PA 15260, USA

ISABELLA L. KARLE, Naval Research Laboratory, Laboratory for the Structure of Matter, Washington, DC 20375-5000, USA

JEROME KARLE, Naval Research Laboratory, Laboratory for the Structure of Matter, Washington, DC 20375-5000, USA

JOHN S. KASPER, General Electric Company, Schenectady, NY 12345, USA

HENRY KATZ, The Institute for Cancer Research, The Fox Chase Cancer Center, Philadelphia, PA 19111, USA

THOMAS J. KING, Vincent T. Lombardi Cancer Research Center, Georgetown University Medical Center, 3800 Reservoir Road, N.W., Washington, DC 20007, USA

MARY L. KOPKA, Molecular Biology Institute, University of California, Los Angeles, CA 90024, USA

Z. R. KORSZUN, University City Science Center/ISFS, 3401 Market Street, Room 320, Philadelphia, PA 19104, USA

Present address: Department of Chemistry, Biomedical Research Institute, University of Wisconsin-Parkside, Box 2000, Kenosha, WI 53141, USA.

THOMAS F. KUMOSINSKI, Eastern Regional Research Center, Agricultural Research Service, U.S. Department of Agriculture, Philadelphia, PA 19118, USA

S. J. LA PLACA, IBM Thomas J. Watson Research Center, P.O. Box 218, Yorktown Heights, NY 10598, USA

ROBERT LANGRIDGE, Computer Graphics Laboratory, University of California, San Francisco, CA 94143, USA

MITCHELL LEWIS, SmithKline and French Laboratories, 1500 Spring Garden Street, Philadelphia, PA 19101, USA

HENRY LIPSON, The University of Manchester Institute of Science and Technology, PO Box 88, Manchester M60 1QD, UK

CECILY DARWIN LITTLETON, Wynnewood, PA 19096, USA

D. B. LITVIN, Department of Physics, The Pennsylvania State University, The Berks Campus, P.O. Box 2150, Reading, PA 19608, USA

WARNER E. LOVE, Thomas C. Jenkins Department of Biophysics, The Johns Hopkins University, Baltimore, MD 21218, USA

MARTHA L. LUDWIG, Biophysics Research Division, The University of Michigan, Ann Arbor, MI 48109, USA

B. W. MATTHEWS, Institute of Molecular Biology and Department of Physics, University of Oregon, Eugene, OR 97403, USA

CAROLINE H. MACGILLAVRY, Amsterdam, The Netherlands

L. J. MASSA, Chemistry Department, Graduate School and Hunter College of the City University of New York, 695 Park Avenue, New York, NY 10021, USA

Visiting Scientist: IBM Thomas J. Watson Research Center, Yorktown Heights, New York 10598, USA

JEAN A. MINKIN, U.S. Geological Survey, Reston, VA 22092, USA

LEONARD MULDAWER, Department of Physics, Temple University, Philadelphia, PA 19122, USA

PETER MURRAY-RUST, Glaxo Group Research, Ltd., Greenford Road, Greenford, Middlesex UB6 OHE, UK

CHRISTER E. NORDMAN, Department of Chemistry, University of Michigan, Ann Arbor, MI 48109, USA

D. H. OHLENDORF, Institute of Molecular Biology and Department of Physics, University of Oregon, Eugene, OR 97403, USA

Present address: E. I. du Pont de Nemours & Co. Inc., Experimental Station, Wilmington, DE 19898, USA

JOHN C. OXTOBY, Department of Mathematics, Bryn Mawr College, Bryn Mawr, PA 19010, USA

BETTY K. PATTERSON, The Institute for Cancer Research, The Fox Chase Cancer Center, Philadelphia, PA 19111, USA

KATHERINE A. PATTRIDGE, Biophysics Research Division, The University of Michigan, Ann Arbor, MI 48109, USA

LINUS PAULING, The Linus Pauling Institute of Science and Medicine, Palo Alto, CA 94306, USA

RAYMOND PEPINSKY, Department of Physics, University of Florida, Gainesville, FL 32611, USA

HELMUT PESSEN, Eastern Regional Research Center, Agricultural Research Service, U.S. Department of Agriculture, Philadelphia, PA 19118, USA

SIR DAVID PHILLIPS, Laboratory of Molecular Biophysics, University of Oxford, Rex Richards Building, South Parks Road, Oxford OX1 3QU, UK

JAMES W. RICHARDSON, JR., Department of Chemistry and Ames Laboratory-USDOE, Iowa State University, Ames, IA 50011, USA

JOHN H. ROBERTSON, School of Chemistry, University of Leeds, Leeds, LS2 9JT, UK

MARTIN ROSENBERG, SmithKline and French Laboratories, 1500 Spring Garden Street, Philadelphia, PA 19101, USA

MIRIAM ROSSI, Department of Chemistry, Vassar College, Poughkeepsie, NY 12601, USA

MICHAEL G. ROSSMANN, Department of Biological Sciences, Purdue University, West Lafayette, IN 47907, USA

WILLIAM E. ROYER, JR., Thomas C. Jenkins Department of Biophysics, The Johns Hopkins University, Baltimore, MD 21218, USA

Present address: Biochemistry Department, College of Physicians and Surgeons, Columbia University, New York, NY 10032, USA

JOHN S. SACK, Thomas C. Jenkins Department of Biophysics, The Johns Hopkins University, Baltimore, MD 21218, USA

Present address: Biochemistry Department, Rice University, Houston, TX 77251, USA

DAVID SAYRE, IBM Thomas J. Watson Research Center, Yorktown Heights, NY 10598, USA

VERNER SCHOMAKER, Department of Chemistry, University of Washington, Seattle, WA 98195, USA

DAVID P. SHOEMAKER, 3453 NW Hayes Avenue, Corvallis, OR 97330, USA

CLARA B. SHOEMAKER, 3453 NW Hayes Avenue, Corvallis, OR 97330, USA

WILLIAM C. STALLINGS, Biophysics Research Division, The University of Michigan, Ann Arbor, MI 48109, USA

HOWARD M. STEINMAN, Department of Biochemistry, Albert Einstein College of Medicine, Bronx, NY 10461, USA

JOHN J. STEZOWSKI, Institut für Organische Chemie, Biochemie und Isotopenforschung, Universität Stuttgart, Pfaffenwaldring 55, D-7000 Stuttgart 80, Federal Republic of Germany

ROLAND K. STRONG, Biophysics Research Division, The University of Michigan, Ann Arbor, MI 48109, USA

MARIANNA STRUMPEL, Institut für Kristallographie, Freie Universität, Takustrasse 6, Berlin 31, West Germany

GERALD STUBBS, Department of Molecular Biology, Vanderbilt University, Nashville, TN 37235, USA

Y. TAKEDA, Chemistry Department, University of Maryland, Baltimore County, Catonsville, MD 21228

TIMOTHY R. TALBOT, The Institute for Cancer Research, The Fox Chase Cancer Center, Philadelphia, PA 19111, USA

MAX R. TAYLOR, School of Physical Sciences, Flinders University of South Australia, South Australia 5042, Australia

A. TULINSKY, Department of Chemistry, Michigan State University, East Lansing, MI 48824, USA

GEORGE TUNELL, Professor Emeritus of Geochemistry, Univeristy of California, Santa Barbara, Santa Barbara, CA 93106, USA

BERTRAM WARREN, Massachusetts Institute of Technology, Cambridge, MA, USA

ALICE WELDON, 67 Third Street, Newport, RI 02840, USA

ELIZABETH WOOD, Red Bank, NJ 07701, USA

C. R. WORTHINGTON, Carnegie-Mellon University, 4400 Fifth Avenue, Pittsburgh, PA 15213, USA

FUMIYUKI YAMAKURA, Department of Chemistry, Juntendo University, Narashino, Chiba 275, Japan

Part I
Introduction

1. Patterson functions

1.1. Introduction to Patterson methods

Jenny P. Glusker and Miriam Rossi

The methods used to solve crystal structures initially involved geometrical principles and trial-and-error methods; it was in this way that the structures of sodium chloride,[1] diamond,[2] graphite,[3] hexamethylbenzene[4] and hexachlorobenzene[5] were determined. However, it was not practical to apply these methods to much larger molecules; isomorphous replacement methods were tried,[6] but very few compounds crystallized as isomorphous pairs. Thus, at this time, only small crystal structures that involved few parameters were possible candidates for a structure determination. This caused great frustration, because many very large structures were found to give beautiful diffraction patterns. Patterson was very conscious of this problem and became obsessed with the idea that more could be learned from Fourier theory about how to solve this problem. Fourier had derived mathematical expressions that could be used to describe regularly repeating (periodic) functions. This was done by forming the sum of a set of cosine waves (regularly repeating functions) of appropriately differing amplitudes, frequencies and relative phases. Such expressions were suggested by W. H. Bragg for use in X-ray crystallography.[7] A two-dimensional Fourier summation was prepared by his son, W. L. Bragg, using the magnitudes and signs of the structure factors from the known structure of diopside.[8] They are still used today in the calculation of electron density maps, since the electron density in crystals repeats regularly from unit cell to unit cell.

Patterson, working at MIT, consulted the experts in Fourier theory, such as Norbert Wiener. He also studied the work of Zernike and Prins[9] on liquids and of Warren and Gingrich[10] on powdered solids. Zernike and Prins showed that the radial distribution of atoms in a liquid could be determined (as distances between atoms) by an analysis, using Fourier theory, of the experimental X-ray diffraction pattern of the liquid. Warren and Gingrich then showed that the same method would give the radial distribution of atoms in a solid; only the intensity distribution of the X-ray powder pattern of the solid was needed. Such analyses gave information on distances between atoms in a crystal structure but no information on the directionality of such distances.

A major step forward in the possibility of determining crystal structures was provided by A. L. Patterson in 1934, over 50 years ago. Patterson realized that he could generate, from the summation of a Fourier series, a three-dimensional function that used the squares of the structure amplitudes of the diffracted beams as coefficients.[11–13] He called this an '$|F|^2$ series', although it is now generally known as 'the Patterson function.'The Patterson function $P(uvw)$ is expressed in terms of the coordinate system u, v, w. The important feature of the Patterson series is that it does not depend on any knowledge of the relative phases of the diffracted beams, information that is lost in the diffraction experiment. Only the intensities of the diffracted beams are required, and these are experimentally measurable.

This mathematical function permits a vector map of a crystal structure to be computed directly from the intensities of the spots obtained by beams diffracted from crystals by X-rays. A vector between two atoms is the distance between them in the correct direction. Every vector in the Patterson map is found with one end on the origin and the other end on a peak, the height of which is proportional to the product of the atomic numbers of the atoms on either end. Thus the heavier the atoms, the higher the peak. This is why it became useful to introduce heavy atoms into structures, especially those of complex molecules such as proteins and nucleic acids. This technique is still in use. The coefficients are obtained directly from the measured intensities and no phase information is required. The Patterson function contains peaks at the positions (relative to the origin) of all possible interatomic vectors between the atoms of the crystal under study. Not only are the distances between any two atoms revealed, as for liquids and powdered solids, but the direction of the vector between them is also specified by this three-dimensional function.

The Patterson function $P(uvw)$ is expressed in terms of a coordinate system u, v, w, as opposed to the system x, y, z for the coordinates of atoms in a crystal. The Patterson function at a point u, v, w may be thought of as a convolution of the electron density at all points x, y, z with the electron density at points $x+u$, $y+v$, $z+w$. Such a convolution is computed by multiplying together the values of the function described by x, y, z and the function described by $x+u$, $y+v$, $z+w$ for each set of possible values of x, y, z and summing all these products. To find the convolution as a general function of u, v, w it is then necessary to perform this multiplication and summation for all desired values of u, v, w.[14] Thus if any two atoms are separated by a vector (u, v, w), then there will be a peak (of height proportional to the product of the atomic numbers of the two atoms at the ends of the vector) at the position u, v, w in the Patterson map.[15] Lindo Patterson called this function the F^2-series.

(1) Explanation:

$$P(uvw) = \left(\frac{1}{V}\right) \int\int\int \rho(x, y, z)\rho(x+u, y+v, z+w)\, dxdydz$$

(2) Calculation:

$$P(uvw) = \sum_{all\ hkl} |F_{hkl}|^2 \cos 2\pi(hu + kv + lw)$$

where V is the unit cell volume, h, k, and l are diffracted beam indices, $\rho\ (x, y, z)$ is the electron density and $|F|$ is the structure amplitude with no phase implied.

Lindo described this in terms of the 'Faltung' of $\rho(x)$ as:

$$\begin{aligned} A(u) &= \left(\frac{1}{a}\right) \int_0^a \rho(x)\rho(x+u)dx \\ &= \sum_{h=-\infty}^{\infty} |F(h)|^2 \exp(2\pi i \cdot hu/a). \end{aligned}$$

Consider a distribution of electron density, $\rho(x)$. Any element dx at a distance x from the origin can be expressed as a function of a parameter u in the form $\rho(x+u)$. This distribution is then weighted by the quantity $\rho(x)dx$, which is the total amount of scattering matter in the element dx. Next the weighted average distribution about any element dx, when x is allowed to assume all values within the period, is computed; the result is the first part of the equation for $A(u)$ (above). This integral is called the 'Faltung' of $\rho(x)$, and can be converted to the summation in the second part of the equation. This summation, Lindo showed, can be computed from the measured intensities. He also pointed out that the principal contributions to $A(u)$ will be made when $\rho(x)$ and $\rho(x+u)$ have large values, *i.e.*, if there is a peak in $A(u)$ for $u = u_1$, then there must be two peaks in $\rho(x)$ separated by a distance u_1.

If there are N atoms in the structure, there are then N^2 interatomic vectors, hence N^2 peaks in the Patterson map. Of these, N are at the origin each representing a vector from an atom to itself. The height can be predicted from the atomic numbers of the two atoms at the ends of the vector. Thus, as N increases, *e.g.*, to 10^3 for a protein, the number of vectors increases to 10^6 in the same volume. Therefore Patterson maps of proteins cannot be interpreted directly. However, they are useful for the location of heavy atoms and for revealing extra symmetry in the packing of the structure. Lindo also showed how the use of atomic form factors, f_i, in an F^2/f_i^2-series, gave a 'sharpened' map which could take care of some of the fall-off in intensity with increasing sin θ/λ (where $90° - \theta$ is the angle between a diffracted ray and the normal to a set of crystal lattice planes and λ is the wavelength of the radiation used, in Å). A temperature factor correction, to allow for the reduced scattering (particularly at higher values of θ) as a result of vibration of atoms, may also be made. On the other hand, if there is a peak in the Patterson map at a position u, v, w, then at least one position of that particular vector within the crystal structure must have both ends on atomic positions. Judson, in *The Eighth Day of Creation*, [16] described a vector map using the analogy of a party. Everyone has their shoes nailed to the floor and everyone wants to meet everyone else. The vectors are the direction each person has to turn to shake the hand of another guest and how far that person has to extend his arms. The strength of the handshake is analogous to the atomic numbers of the two atoms at the ends of the vector. This analysis of the location of the guests at the party is as if, for some reason, the only information you had about the party was the distance and angle of the handshake of each person when he met everyone else. With this complicated knowledge it should then be possible to find where everyone stood in the room relative to each other.

Unfortunately, when crystals diffract X-rays certain information is lost so that this diffraction effect cannot be used as a supermicroscope. Because of the Patterson function, interatomic vector maps (that is, all vectors or directions and distances of handshakes) become available, providing a method of analysis far more powerful than any prior method. It allowed for the first time the determinations of quite complex structures. If one atom is very heavy then vectors from it will dominate the Patterson map. Therefore it became useful to introduce heavy atoms into structures that might be difficult to determine. This concept was then applied to the study of crystals of macromolecules, such as proteins. In proteins the positions of heavy atoms, introduced for phase determination by isomorphous replacement, are still generally found from Patterson maps.

Harker[17] extended the interpretation of Patterson maps, showing that certain portions of these maps, depending on the space group of the crystal under study, contain a large proportion of the readily interpretable structural information. This was particularly useful in the days when any computation was tedious. It is still used for protein data.

There are several methods for solving a Patterson map that involve transcribing the map upon itself with different relative origins (minimum function vector superposition map) by rotating the map if it is suspected that there are two or more identical groups in the crystal (rotation function) or by comparing the map with a vector map calculated for known molecular fragments of the molecule under study. In a minimum-function map,[18,19] the origin of the Patterson map is put in turn on each of the symmetry-related positions of a known atom (such as a heavy atom). The *smaller* value at each point in the superposed functions is then recorded; this minimum-function map will contain only peaks obtained from all maps used in the superpositions and will not be affected by a high peak that is found in only one map. The rotation function, described by Rossmann and Blow,[20] measures the degree of coincidence over a given volume when one Patterson function is rotated on another identical one. In the case of proteins such a rotation of a Patterson map upon itself may indicate some symmetry in the macromolecular structure under study. In the third method, when part of the structure of a molecule is known, one can try to find the entire arrangement in the crystal by seeking a fit between the experimental Patterson function and the computed vector map of a known fragment of the molecule.[21] A match between the two maps will indicate the orientation of the known fragment of the structure in the unit cell.

As an extension of his work on vector maps, Patterson became interested in the theory of homometric structures.[22] These are different structures that give identical vector maps and hence give the same diffraction pattern. This work has been continued by Buerger, Franklin, Chieh and others.[23–25] The first set of homometric structures was that reported for bixbyite $(Mn,Fe)_2O_3$.[26] Two possible arrangements of metal positions were found to give identical calculated diffraction patterns. Only when the positions of the oxygen atoms were considered could the correct structure be determined.

Examples: (a) Potassium dihydrogen phosphate

In 1934 Patterson first tested his method on this structure[11,12] which had been solved by J. West in 1930.[27] He used West's published intensities. The structure had earlier been investigated by Hendricks,[28] but it was West who showed that the phosphate groups are tetrahedral. Lindo and his fiancée Betty did the calculations with Patterson–Tunell strips.[29,30]

This compound crystallizes in the tetragonal space group $I\bar{4}2d$ (V_d^{12}). Sixteen oxygen atoms occupy the sixteen-fold general positions. The unit cell, with dimensions $a = b = 7.43$, $c =$

6.97 Å contains four KH_2PO_4. Equivalent positions of the space group are:

x	y	z	$\bar{x}$	$\frac{1}{2}+y$	$\frac{1}{4}-z$
$\bar{x}$	$\bar{y}$	z	x	$\frac{1}{2}-y$	$\frac{1}{4}-z$
$\bar{y}$	x	$\bar{z}$	y	$\frac{1}{2}+x$	$\frac{1}{4}+z$
y	$\bar{x}$	$\bar{z}$	$\bar{y}$	$\frac{1}{2}-x$	$\frac{1}{4}+z$

and each set of these parameters translated $(\frac{1}{2}, \frac{1}{2}, \frac{1}{2})$. There is a four-fold axis parallel to the unique $\vec{c}$ -axis. The four potassium ions and the four phosphorus atoms must lie on four-fold points, of which two sets are available.

4K at	0	0	$\frac{1}{2}$	4P at	0	0	0
	$\frac{1}{2}$	0	$\frac{3}{4}$		$\frac{1}{2}$	0	$\frac{1}{4}$
	$\frac{1}{2}$	$\frac{1}{2}$	0		$\frac{1}{2}$	$\frac{1}{2}$	$\frac{1}{2}$
	0	$\frac{1}{2}$	$\frac{1}{4}$		0	$\frac{1}{2}$	$\frac{3}{4}$

There are only three parameters to be found; the x, y and z of the oxygen atoms. It was found that the oxygen atoms lie in general positions at $x = 0.0828$, $y = 0.1486$, $z = 0.1261$ Note that there are two orientations of the phosphate groups in the crystal structure (shown in Figs. 1.1.1(a and b)) and the Patterson map (shown in Fig. 1.1.1(c)) where two sets of four peaks are around the origin peak in the center of the diagram. In the plane projection shown in the figures, the potassium ions and the phosphorus atoms fall on top of each other. By an examination of the $(hk0)$ and $(h0l)$ Patterson projections, Lindo was able to account for the peaks in the Patterson map. Thus the new method could have been used to establish the structure.

Examples: (b) Hexachlorobenzene

Kathleen Lonsdale studied hexamethylbenzene in 1928[4] and she was the first to show that the benzene ring is a symmetrical hexagon and does not have alternating single and double bonds

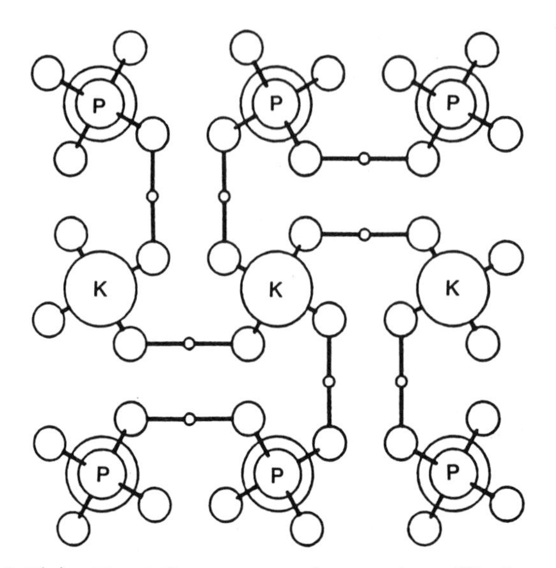

FIG. 1.1.1(a). Crystal structure of potassium dihydrogen phosphate. Small circles represent hydrogen atoms in short symmetrical hydrogen bonds.

FIG. 1.1.1(c). Patterson map of potasium dihydrogen phosphate showing the vectors representing the two possible orientations of the phosphate group.

as drawn by Kekulé [31] (this has been verified by neutron[32] and spectroscopic data[33]).

The crystal structure of hexachlorobenzene was determined by Kathleen Lonsdale in 1931.[5] The crystals are monoclinic. The unit cell dimensions are $a = 8.07$, $b = 3.84$, $c = 16.61$ Å, $\beta = 116°52'$, $Z = 2$ and the space group is $P2_1/c$ (C_{2h}^5). Thus the asymmetric unit consists of half a molecule, related to the other half by a center of symmetry. In this space group the equivalent positions are:

$$\begin{array}{ccc} x & y & z \\ \bar{x} & \frac{1}{2}+y & \frac{1}{2}-z \\ x & \frac{1}{2}-y & \frac{1}{2}+z \\ \bar{x} & \bar{y} & \bar{z} \end{array}$$

The atomic coordinates determined (in projection) by Lonsdale are given below.

	x	z
C(1)	0.181	0.0265
C(2)	0.118	0.0872
C(3)	0.083	-0.0623
Cl(1)	0.412	0.0704
Cl(2)	0.278	0.205
Cl(3)	0.133	-0.138

Better coordinates and unit cell dimensions are now available:[34] $a = 8.08(2)$, $b = 3.87(1)$, $c = 16.65(3)$ Å $\beta = 117.0°$:

	x	y	z
C(1)	0.189	0.072	0.0305
C(2)	0.128	-0.072	0.0945
C(3)	-0.058	-0.128	0.0640
Cl(1)	0.403	0.153	0.0655
Cl(2)	0.278	-0.153	0.2040
Cl(3)	-0.128	-0.275	0.1405

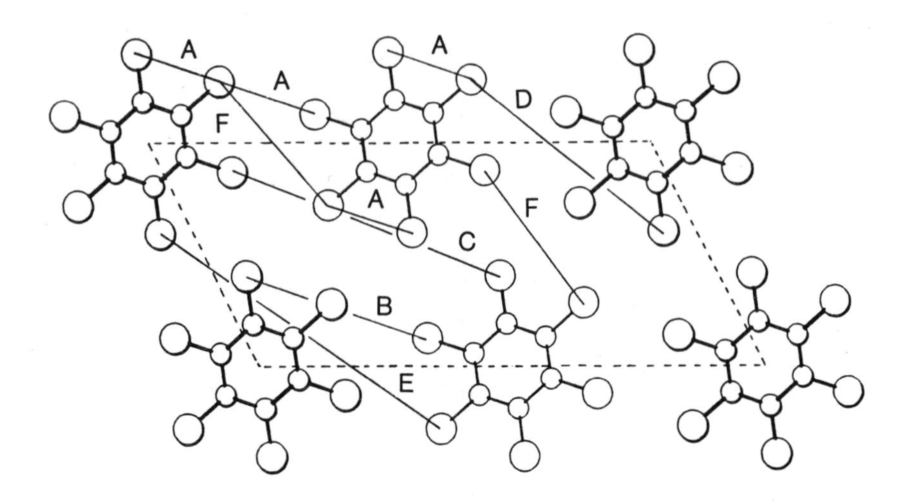

FIG. 1.1.2(a). The crystal structure of hexachlorobenzene viewed down the **b** axis. Note that the horizontally adjacent molecules are tilted differently. Vectors A to F between chlorine atoms are shown.

A view of the structure down the **b** axis is shown in Figs. 1.1.2(a and b). This is the second structure for which Lindo calculated a Patterson function by a consideration of the $(h0l)$ data. Since the heights of peaks in Patterson maps are related to the product of the atomic numbers at the two ends of a vector, it was expected by Lindo that the major peaks in the Patterson map would be vectors between chlorine atoms. The Cl ··· Cl peaks would have a relative height of $17 \times 17 = 289$ versus C ··· Cl peaks of $6 \times 17 = 102$. An analysis is shown.

u	v	w
0	0	0
0	$\frac{1}{2} - 2y$	$\frac{1}{2}$
$2x$	$2y$	$2z$
$2x$	$\frac{1}{2}$	$2z - \frac{1}{2}$

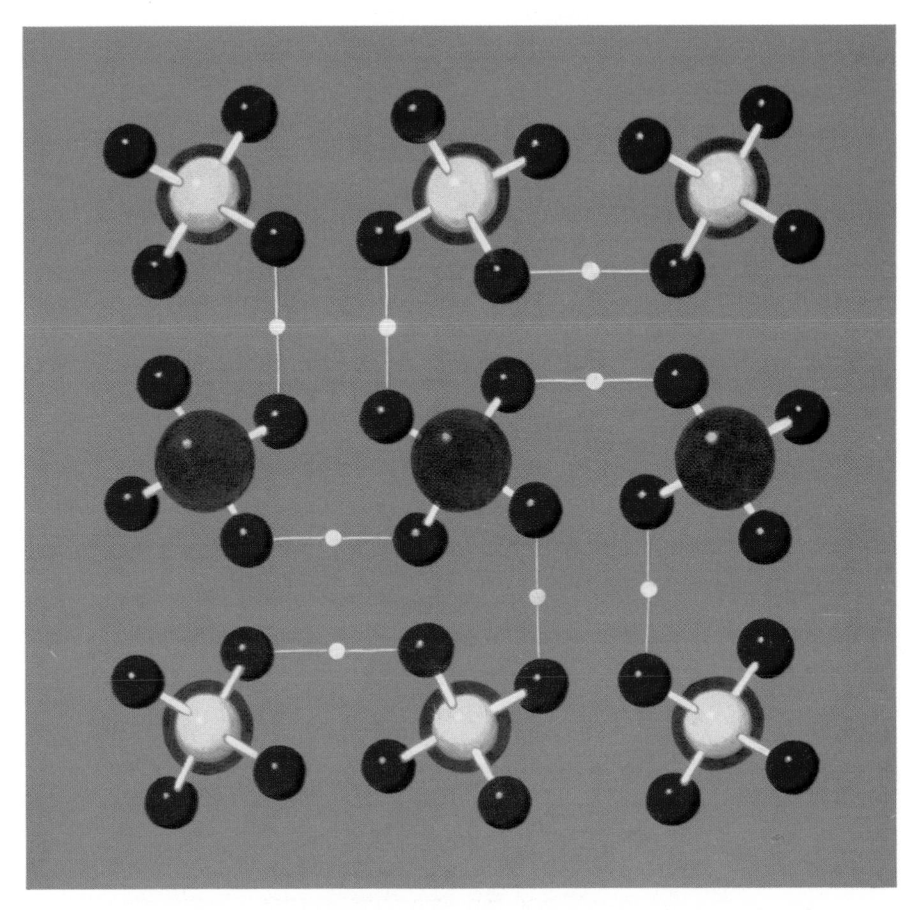

FIG. 1.1.1(b). Crystal structure of potassium dihydrogen phosphate. Phosphorus mauve; potassium blue; oxygen red.

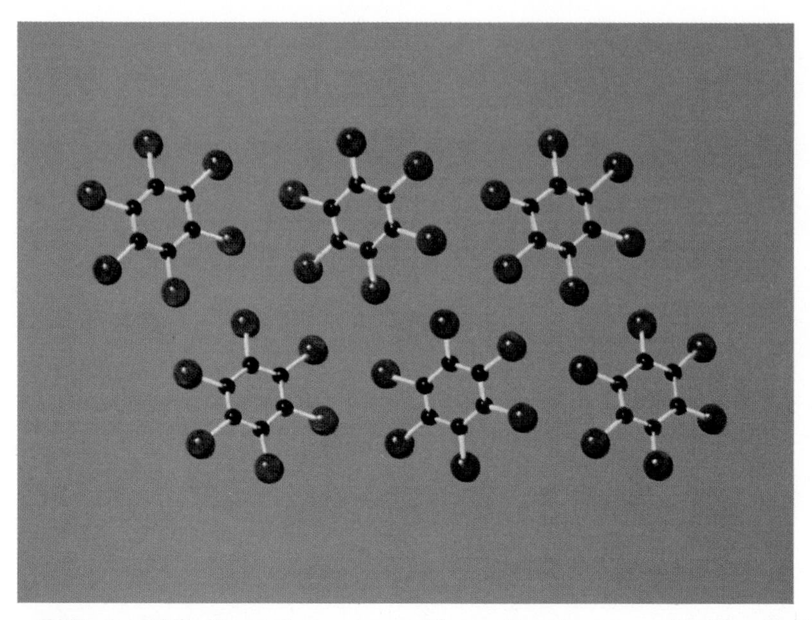

FIG. 1.1.2(b). Crystal structure of hexachlorobenzene. Carbon black; chlorine green.

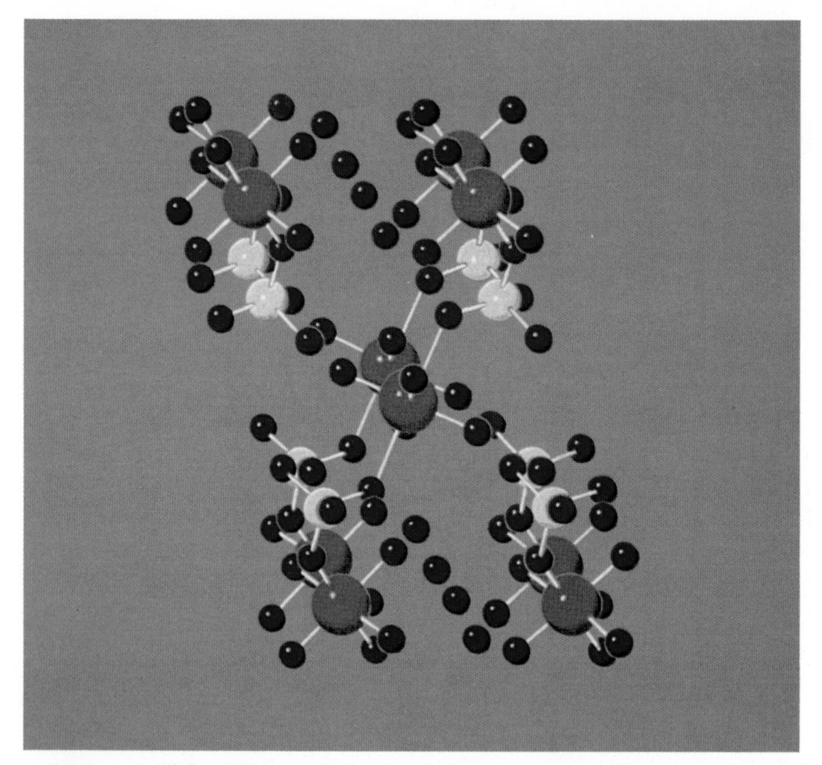

FIG. 1.1.3(a). The crystal structure of copper sulfate pentahydrate. Copper blue-green; sulfur yellow; oxygen red.

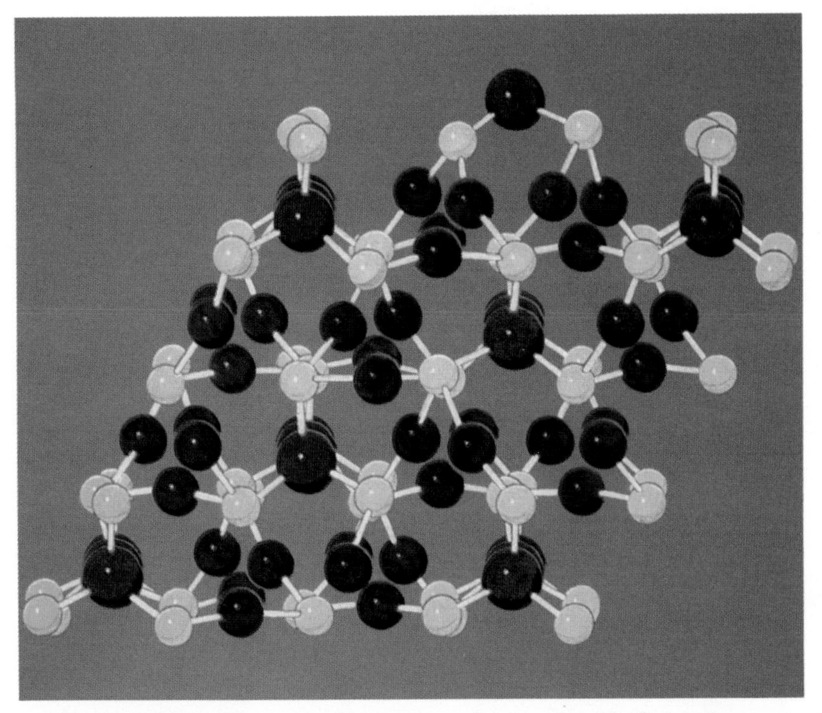

FIG. 2.1.1(b). The crystal structure of proustite. Sulfur yellow; arsenic purple; silver green.

FIG. 2.1.1(d). David and Deborah Harker at the Fox Chase Symposium viewing some proustite crystals kindly loaned by the Smithsonian Institute via Dr Dan Appleman.

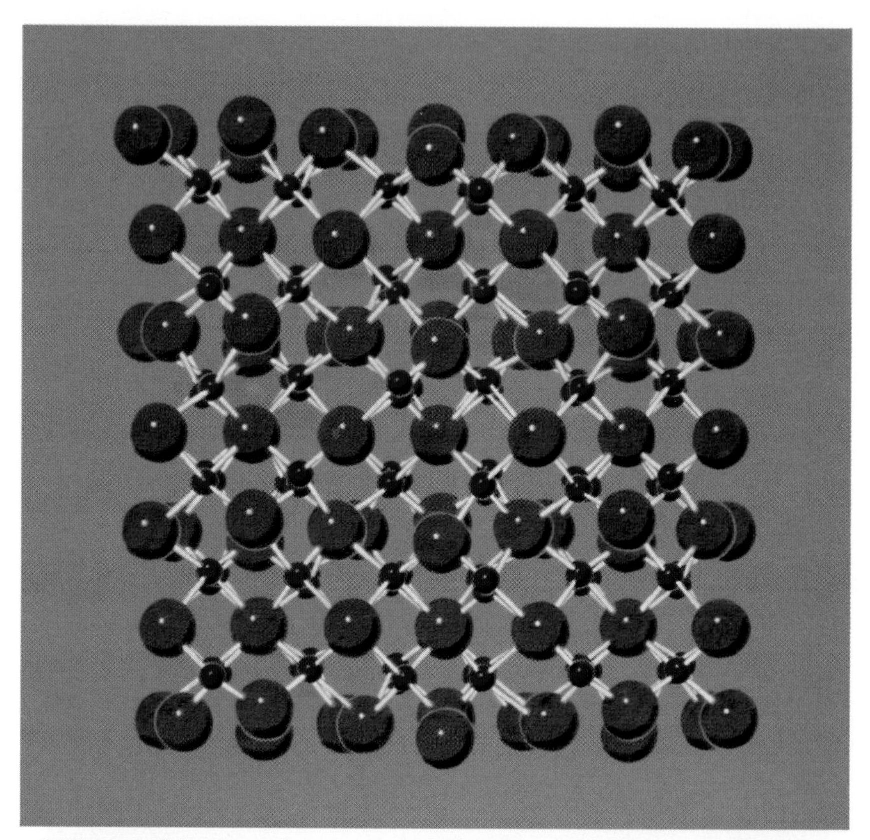

FIG. 3.1.1(b). Crystal structure of bixbyite. Fe,Mn blue; oxygen red.

FIG. 3.1.1(c). Two views of Linus Pauling at the Fox Chase Symposium, lecturing and examining a crystal.

The first and second sets of Patterson map coordinates (and also the third and fourth sets) differ from each other only by $\frac{1}{2}$c. The same is true for the more general peaks listed below.

$x - x_1$	$y - y_1$	$z - z_1$
$x - x_1$	$y + y_1 + \frac{1}{2}$	$z - z_1 + \frac{1}{2}$
$x + x_1$	$y + y_1$	$z + z_1$
$x + x_1$	$y - y_1 + \frac{1}{2}$	$z + z_1 + \frac{1}{2}$

Thus, in the $(h0l)$ Patterson map from $w = 0$ to $w = \frac{1}{2}$ there are six high peaks as predicted and shown in the Fig. 1.1.2(c). These represent Cl ··· Cl vectors.

Vector	u	w	Peak
Cl(1)–Cl(2)	-0.13	0.13	A
	0.69	0.27	B
Cl(1)–Cl(3)	0.54	0.43	C
	0.28	0.21	D
Cl(2)–Cl(3)	0.15	0.34	E
	0.41	0.06	F

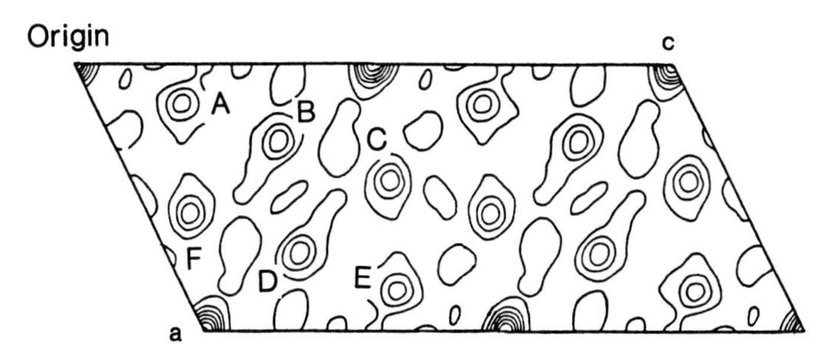

FIG. 1.1.2(c). Patterson map of hexachlorobenzene showing vectors illustrated in (a).

Examples: (c) Copper sulfate pentahydrate

Copper sulfate pentahydrate has great historical interest for crystallographers. A crystal of this was used by Friedrich, Knipping and Laue[35] in 1912 to establish that X-rays may be diffracted. The crystal structure, shown in Figs. 1.1.3(a and b), was determined by Beevers and Lipson[36] in 1934. They used the fact that copper sulfate is isomorphous with copper selenate to establish the location of the sulfur (or selenium) in the unit cell. Thus, it was the first direct use of isomorphism to determine three-dimensional structure. In addition, this was the third structure on which Patterson tested his method[11,12] and finally was convinced that it worked. It was also the first time he sharpened a Patterson map by use of $F^2/\overline{f}^2$ as coefficients; the resulting map was much easier to interpret. The agreement was excellent with the structure as solved by C. A. Beevers and H. Lipson, who sent him their intensity data before publication.[36]

Copper sulfate pentahydrate crystallizes in the triclinic system, space group $P\overline{1}$. Beevers and Lipson[2] measured the unit cell dimensions as a = 6.12, b = 10.7, c = 5.97 Å, α = 82°16′, β = 107°26′, γ = 102°40′. The cell edges were remeasured later as: a = 6.141, b = 10.736, c = 5.986. The copper atoms were found to lie on the centers of inversion at (0 0 0) and $(\frac{1}{2}\,\frac{1}{2}\,\frac{1}{2})$.

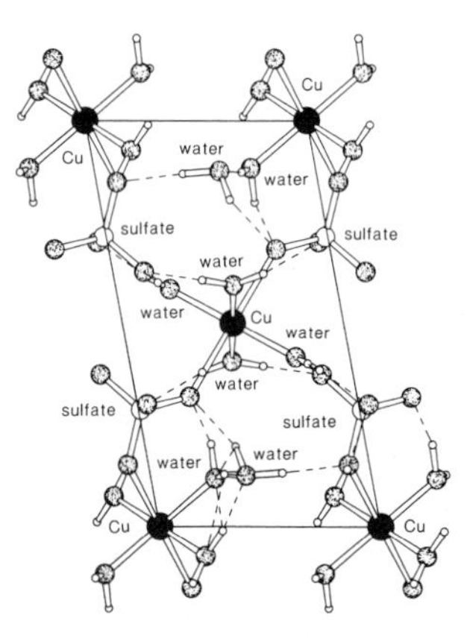

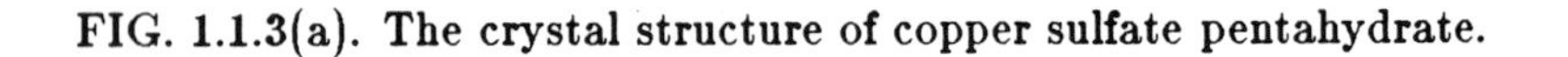

FIG. 1.1.3(a). The crystal structure of copper sulfate pentahydrate.

atom		x	y	z
copper		0	0	0
copper		$\frac{1}{2}$	$\frac{1}{2}$	$\frac{1}{2}$
sulfur		0.011	0.287	0.624
oxygen	1	0.907	0.152	0.675
	2	0.243	0.318	0.797
	3	0.859	0.373	0.635
	4	0.045	0.300	0.384
	5	0.817	0.074	0.154
	6	0.290	0.118	0.149
	7	0.465	0.406	0.299
	8	0.756	0.416	0.019
	9	0.434	0.125	0.630

In the space group $P\bar{1}$, with atomic positions $x\,y\,z$ and $\bar{x}\,\bar{y}\,\bar{z}$, there are peaks in the Patterson map at 0, 0, 0; $2x$, $2y$, $2z$; $x_1 - x_2, y_1 - y_2, z_1 - z_2$; and $x_1 + x_2, y_1 + y_2, z_1 + z_2$. With this information it was possible to interpret the Patterson map in terms of the crystal structure.

The Patterson map, shown in Fig. 1.1.3(c), was 'sharpened' by dividing each F^2 value by the mean value of f^2 at that scattering angle. The result is shown in Fig. 1.1.3(d); it is clear from this that the vectors are accentuated and an analysis became much easier.

In the crystal structure each copper ion is surrounded by six oxygen atoms to give an octahedral arrangement; four of these oxygen atoms are water molecules and two are sulfate oxygen atoms. The sulfate groups are tetrahedral. The locations of the water molecules have been determined[36] and confirmed by a later neutron diffraction study.[37] Four of the water molecules (numbers 5–8) are coordinated to a copper ion and hydrogen bonded to two oxygen atoms. The fifth water molecule has a tetrahedral arrangement of oxygen atoms around it but is not coordinated to copper. This explains the different behavior of one of the five water molecules; it is less readily lost on heating and is not easily replaced by ammonia.

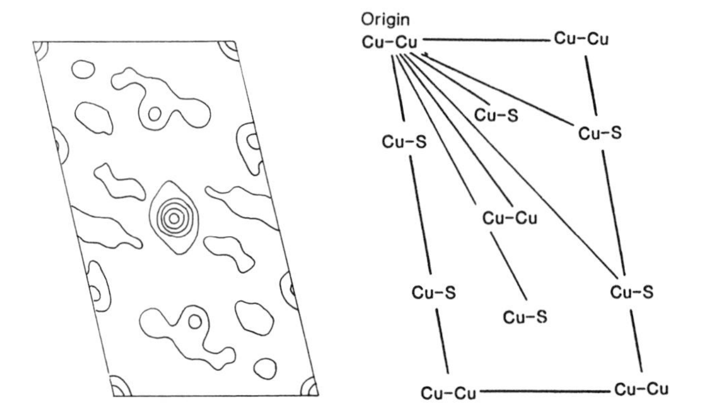

FIG. 1.1.3(c). The unsharpened Patterson map of copper sulfate pentahydrate plus an explanation of the major vectors.

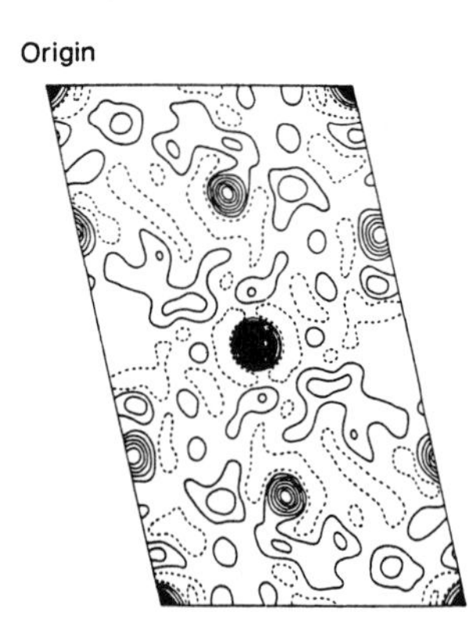

FIG. 1.1.3(d). Patterson map in (c) 'sharpened' by an exponential factor applied to the $| F |^2$ values.

References

1. Bragg, W. L. The structure of some crystals as indicated by their diffraction of X-rays. *Proc. Roy. Soc. (London)* **A89**, 248–277 (1913).
2. Bragg, W. H. and Bragg, W. L. The structure of diamond. *Nature (London)* **91**, 557 (1913).

3. Bernal, J. D. The structure of graphite. *Proc. Roy. Soc. (London)* **A106**, 749–773 (1924).
4. Lonsdale, K. The structure of the benzene ring in $C_6(CH_3)_6$. *Proc. Roy. Soc. (London)* **A123**, 494–515 (1929).
5. Lonsdale, K. An X-ray analysis of the structure of hexachlorobenzene using the Fourier method. *Proc. Roy. Soc. (London)* **A133**, 536–552 (1931).
6. Bernal, J. D. and Crowfoot, D. X-ray photographs of crystalline pepsin. *Nature (London)* **133**, 794–795 (1934).
7. Bragg, W. H. IX Bakerian lecture: X-rays and crystals. *Trans. Roy. Soc. (London)* **A215**, 253–274 (1915).
8. Warren, B. and Bragg, W.L. The structure of diopside, $CaMg(SiO_3)_2$. *Z. Krist.* **69**, 168–193 (1928).
9. Zernike, F. and Prins, J. A. The bending of X-rays in liquid as an effect of molecular arrangement. *Z. Physik* **41**, 184–194 (1927).
10. Warren, B. E. and Gingrich, N. S. Fourier integral analysis of X-ray powder patterns. *Phys. Rev.* **46**, 368–372 (1934).
11. Patterson, A. L. A Fourier series method for the determination of the components of interatomic distances in crystals. *Phys. Rev.* **46**, 372–376 (1934).
12. Patterson, A. L. A direct method for the determination of the components of interatomic distances in crystals. *Z. Krist.* **A90**, 517–542 (1935).
13. Patterson, A. L. An alternative interpretation for vector maps. *Acta Cryst.* **2**, 339–340 (1949).
14. Lipson, H. and Cochran, W. *The Determination of Crystal Structures. The Crystalline State.* Vol. **III.** (L. Bragg, ed.) G. Bell and Sons: London (1953).
15. Glusker, J. P. and Trueblood, K. N. *Crystal Structure Analysis: A Primer.* Oxford University Press: New York. Second edition (1985).
16. Judson, H. F. *The Eighth Day of Creation: Makers of the Revolution in Biology.* Simon and Schuster: New York (1979).
17. Harker, D. The application of the three-dimensional Patterson method and the crystal structures of proustite, Ag_3AsS_3, and pyrargyrite, Ag_3SbS_3. *J. Chem. Phys.* **4**, 381–390 (1936).
18. Buerger, M. J. A new approach to crystal-structure analysis. *Acta Cryst.* **4**, 531–544 (1951).
19. Beevers, C. A. and Robertson, J. H. Interpretation of the Patterson synthesis. *Acta Cryst.* **3**, 164 (1950).

20. Rossmann, M. G. and Blow, D. M. The detection of sub-units within the crystallographic asymmetric unit. *Acta Cryst.* **15**, 24–31 (1962).
21. Nordman, C. E. and Nakatsu, K. Interpretation of the Patterson function of crystals containing a known molecular fragment. The structure of an *Alstonia* alkaloid. *J. Am. Chem. Soc.* **85**, 353–354 (1963).
22. Patterson, A. L. Homometric structures. *Nature (London)* **143**, 939 (1939).
23. Buerger, M. J. Proofs and generalizations of Patterson's theorems on homometric complementary sets. *Z. Krist.* **143**, 79–98 (1976).
24. Franklin, J. N. Ambiguities in the X-ray analysis of crystal structures. *Acta Cryst.* **A30**, 698–702 (1974).
25. Chieh, C. Analysis of cyclotomic sets. *Z. Krist.* **150**, 261-277 (1979).
26. Pauling, L. and Shappell, M. D. The crystal structure of bixbyite and the *C*-modification of the sesquioxides. *Z. Krist.* **75**, 128–142 (1930).
27. West, J. A quantitative X-ray analysis of the structure of potassium dihydrogen phosphate. *Z. Krist.* **74**, 306–332 (1930).
28. Hendricks, S. B. The crystal structure of potassium dihydrogen phosphate. *Am. J. Sci.* **14**, 269–287 (1927).
29. Patterson, A. L. and Tunell, G. A method for the summation of the Fourier series used in the X-ray analysis of crystal structures. *Am. Mineral.* **27**, 655–679 (1942).
30. Lipson, H. and Beevers, C. A. An improved numerical method of two-dimensional Fourier synthesis for crystals. *Proc. Phys. Soc.* **48**, 772–780 (1936).
31. Kekulé, A. Sur la constitution des substances aromatiques. *Bull. soc. chim. Fr.* **3**, 98–110 (1865).
32. Bacon, G. E., Curry, N. A. and Wilson, S. A. A crystallographic study of solid benzene by neutron diffraction. *Proc. Roy. Soc. (London)* **A279**, 98–110 (1964).
33. Stoicheff, B. P. High resolution Raman spectroscopy of gases. II. Rotational spectra of C_6H_6 and C_6D_6 and internuclear distances in the benzene molecule. *Can. J. Phys.* **32**, 339–346 (1954).
34. Tulinsky, A. and White, J. G. Rigid-body torsional vibrations in three typical members of a class of benzene derivatives. *Acta Cryst.* **11**, 7–14 (1958).
35. Friedrich, W., Knipping, P. and Laue, M. Interferenz-Erscheinungen bei Röntgenstrahlen. *Sitzungsberichte der (Kgl.) Bayerische Akademie der Wissenschaften* 303–322 (1912). Translation by J. J. Stezowski in *Structural Crystallography in Chemistry and Biology* (J. P. Glusker,

ed.) pp. 23–39. Hutchinson Ross Publishing Company: Stroudsburg, PA, Woods Hole, MA (1981).

36. Beevers, C. A. and Lipson, H. The crystal structure of copper sulphate pentahydrate, $CuSO_4{\cdot}5H_2O$. *Proc. Roy. Soc. (London)* **A146**, 570–582 (1934).
37. Bacon, G. E. and Curry, N. A. The water molecules in $CuSO_4{\cdot}5H_2O$. *Proc. Roy. Soc. (London)* **A266**, 95–108 (1962).

1.2. Letters from Lindo, 1933–1934

Betty K. Patterson

Introduction: In 1934 when Lindo wrote the paper on the Patterson function he was engaged to Betty Knight and she was, one summer, able to help him do the calculations of the $| F^2 |$ map of copper sulfate. However, he was working at MIT and she was working at the Rockefeller Institute, and, as a result, many letters were written from one to the other.

I would like to read to you some excerpts from letters that Lindo wrote to me when I was in New York and he was at MIT. I was about to give up my job to study for a Ph.D., and he'd given up his job to go up to MIT to work, and we didn't have enough money even to telephone, so all we did was write. I think most of you knew that Lindo (or Pat as some of you called him) in 1929-33 had been working in his spare time on what he called his 'obsession that something was to be learned about structural analysis from Fourier theory.' He needed help from an expert in this mathematical field and decided to resign from his job at the Johnson Foundation at the University of Pennsylvania, and asked his friend Bert Warren, who is right here, if he could have a small place in his lab for a year, coming as an unpaid guest. He had saved enough money for one year. This was, after all, in depression times — he had no idea that he wouldn't get a job for three years — but that's the way it was. He knew that Norbert Wiener, well known for cybernetics, was at MIT and could help him, if anyone could. So he went up in the fall of 1933 and Bert very kindly let him be in his lab, and his first letter I want to read was dated December 17, 1933.

The only excitement was an interview with Wiener, the Fourier series man. I don't know yet what is coming out of that, but I think that I shall get a definite answer as to whether or not my problem will go. So far I have a much more precise statement of the problem which is always an advance. But I got that myself before I spoke to him. He made one or two suggestions, but I only spoke to him for a few minutes, and I hope to see him again next week. He is a queer bird.

Then a little later, he said:

I had another talk with Wiener and he seems to be convinced that there isn't a reasonable solution to my problem from the mathematician's point of view but perhaps from one of the leads he has given me, I may be able to find one to give him. Bert Warren has recently obtained a result which advances the same problem slightly from another direction. I don't think, however, that it will be of much help to me.

On February 23, 1934 he wrote:

I'm afraid I've not been particularly deserving of all your solicitude for my depressed state. [He was like a yo-yo. He went from depressed to the top — up and down again.] Because I started fast for a crest. Bert has been worrying about a theoretical problem which started to break on Tuesday. We had a marvelous evening Tuesday, which resulted in my going without supper until about 11.30. We were both making bright suggestions as fast as possible and it seems that he has made a real advance. It is really a partial answer to the problem I've been talking to Wiener about. It does not answer my problem, but it defines it rather more closely. The whole thing was quite thrilling and still is for that matter. It has cleared up several of my troubles and may open up a lot of possibilities. So that is the cause of my crest.

Then a month later, March 23, 1984:

Things have broken rapidly. There is a paper for Washington and a celebration tonight. Such is theoretical physics.

Well, actually there was a meeting of the American Physical Society in Washington, and he had three days to write an abstract and solve his problem and get it off with a couple of other papers that Bert was sending in, and then he had about a month to do the hand calculations on potassium dihydrogen phosphate which was the first example he used, and I guess he did hexachlorobenzene after that. I can't think that he could have done them both in that short length of time. The letters don't say anything about hexachlorobenzene, so I think it was mostly potassium dihydrogen phosphate. Then on April 17,

1934, which is just about a little less than a month after things had broken rapidly, he said:

> I'm starting a second double series today. I should like to listen to Toscanini, but I think work is indicated. The start of the series is most important as mistakes then are fatal and the lab is peaceful on Sundays. The other series is very good. Just after I'd sealed your letter, I showed the figures to Bert and he was quite enthused, which for him is marvelous, he being a New Englander.

Then we get to August, and he says:

> I have stumbled onto a method of modifying the F^2 series [which is what he always called this] so that the peaks stand out much more clearly than before. However, the theoretical background is just as obscure as it was originally for the F^2 series. It works and that's all I know. Why, I haven't only the vaguest of ideas. However, I think it will improve matters quite considerably, and should increase the practical importance of the method and its general usefulness.

Then we get to November:

> You are not engaged to a bum [that's his designation]. In fact, in my humble opinion at the moment, you are engaged to one of the world's best crystallographers, grudgingly second only to a few who have won the Nobel Prize and possibly one or two others. Last night I was going great guns. I did a dogs amount of work on copper sulfate. At 11 about, I had a very bright idea. Worked on it at Tech until about 1.30 which produced a vague misgiving that the bright idea might not be so good, *i.e.,* good but not practical. Came home, read *Post* [*Saturday Evening Post* in those days] went to sleep, cursing like a trooper. At 4 a.m. world black. No chance of hearing from you. Sure to sleep all morning and waste all day and continue bum. Woke at 8, went to sleep again knowing I was a bum but not doing anything about it. Wakened at 8.30 by alarm, and got up, and therewith ceased to be a bum. Went to lab, completed iron-bound argument for copper sulfate which clicked like no one's business. I couldn't believe it myself. The scale of my contour map was much too large. When I reduced my results to the same degree of accuracy as those I was comparing with, the agreement was much too good to be true. In fact it's grand. I wandered about in a daze for a little. Told several people about it. I did feel pleased.

Then later that day:

I wrote the abstract of my paper for the Pittsburgh meeting, which find enclosed. Bert sends that off tommorrow with his and Burwell's sulfur structure. After that went to be typed, I again wandered about like a lost soul and suddenly started out on a search for Wiener to ask him questions. Before that a talk with G.S. [that, I think, is George Shortley] who was grand about my stuff. Found Wiener and asked him two questions which had been growing for about three or four months about [what he, Lindo called] the 'bugger factor'; [I've learned that that's the sharpening of the F^2 peaks]. He told me just why theoretically it behaved just as I knew, though to my surprise, practically. By that time, I had tea and talked to a lot of people. I then came home and went out to supper. Since I've been home I've planned the micromere [that was a second paper that he was going to write — the second *Zeitschrift* paper] and written nearly all the bull to be attached to it. It is a paper which will be separate from, but I hope will follow, the main paper in the *Zeit. für Krist.* The main work is tables and complicated ones but I think more than 50% of that is done in my notes. During the process of writing it and this, I have drunk a lot of beer. I am at the end of an unbelievable day, one of the kind that one does research for. I haven't done anything new but very many of the things that my 'hopeless', 'fruitless' 'waste' of time for the last six months have come to a successful end today. Today is a step.

1.3. Patterson and copper sulfate pentahydrate

C. Arnold Beevers

Between 1934 and 1938 I was working in the Physics Department of the University of Manchester, in the very room used previously by Rutherford. The room contained two of the historical ionization spectrometers devised by W. H. Bragg, and brought to Manchester by his son, Professor W. L. Bragg. Bragg had been in contact with Patterson, and offered him the data on copper sulfate pentahydrate,[1] which H. Lipson and I had recently measured carefully on one of the spectrometers. In return Patterson sent over the proofs of his paper on his new method.[2] I remember very well this being pored over by several of us together, including R. W. James, the author of the second volume of *The Crystalline State*[3] and later to go to Cape Town as Professor of Physics. It took us some time to make anything of it; it was completely revolutionary in its concepts. In fact it required a year or two before its full implications were realised. The Harker paper[4] coming later was of great value in bringing the uses of the Patterson synthesis to full prominence.

Later, of course, I was able to make considerable use of the Patterson function. It was calculable very readily by the use of the Beevers–Lipson strips, which were already a familiar tool in X-ray crystallography. It was indeed a real challenge to the structure analyst. Many of the straight Fourier syntheses were extremely tentative and uncertain. The Patterson, however, was independent of any assumption, and gave us a certainty and a confidence that was of enormous value. We could recognise the orientation of a molecule in a structure, by looking for its probable Patterson vectors. Once a molecule had been measured up in one structure, its Patterson peaks could be worked out readily, and the pattern of peaks searched for in other structures. We used to make models of the Patterson peaks of important molecules, such as the benzene ring, or a sugar ring. I still have some of these models; they are beautifully symmetrical, and they were of great use in the totally non-computer days of the thirties, forties and fifties.

With the advent of computers the speed and accuracy of X-ray analysis has been revolutionised, but the value of the Patterson has in no way diminished, and it plays its part in hundreds of structure determinations per year — perhaps even per month.

In addition to his scientific work I also much appreciate his personal kindness, and his cheerfulness. Of these things I have the most happy recollections.

References

1. Beevers, C. A. and Lipson, H. *Proc. Roy. Soc. (London)*, **A146**, 570 (1934).
2. Patterson, A.L. *Phys. Rev.* **46**, 372-376 (1934).
3. James, R. W. *The Optical Principles of the Diffraction of X-rays. The Crystalline State*, Vol. **II** (L. Bragg, ed.) G. Bell and Sons: London, UK (1954).
4. Harker, D. *J. Chem. Phys.* **4**, 381 (1936).

1.4. Patterson and copper sulfate pentahydrate

Henry Lipson

My first contact with Lindo was an indirect one, but it was nevertheless of outstanding importance. In 1934 Arnold Beevers and I had just worked out the structure of $CuSO_4.5H_2O$[1] by Fourier methods; as guests in W. L. Bragg's Manchester laboratories — then the world centre of X-ray crystallography — we had come from our own laboratories in Liverpool to spend several laborious weeks making the necessary measurements on the ionization spectrometer.

Lindo had just had an idea. He hoped that he had found a method of taking the guess-work out of crystal-structure determination:[2] he had tried his method out on a simple structure — KH_2PO_4 — and he wanted something more complicated. So he wrote to Professor Bragg to ask if he could suggest something; so Lindo asked us if we would allow our data to be used. Would we? Of course! Being a couple of raw new research students we were delighted to feel that we were really in the swim. The data were sent to Lindo and the rest is history.

The method was not quite so direct as Lindo's paper made it appear. There were, in fact, two sets of Cu and S positions that could explain his peaks, and he should have tried both. But we never pointed this out to him.

The method was however not the alchemist's stone that had been expected. It did not, except in rare cases, lead directly to a complete structure determination, but it gave many people a great deal of fun in trying to devise a routine procedure for analysing Patterson syntheses.

Afterwards I met Lindo many times at conferences and thoroughly enjoyed my contacts with him — scientific and otherwise. It was said that those who did not know him could always recognize him at conferences as the man who talked of the F^2 synthesis when others talked of the Patterson synthesis.

He had an almost elfish sense of humour, and I had the good fortune to be involved in one of his more noteworthy escapades. I happened to be staying at Peter Wooster's home in Cambridge when the morning post provided him with a paper from Robert Evans to referee for *Acta Crystallographica.* Now Wooster was normally an imperturbable scientist, but when he saw this his mouth dropped; it was a very short paper (see *Crystallography in North America*, p. 107)[3] but he could not understand the title, nor recognize many of the words in the paper. He passed it on to me, and I was able to say, with great glee, that although I couldn't understand it either, I knew who had written it.

The background was this. Ramachandran and Wooster had recently written a paper in which they had introduced some 'portmanteaux' words — 'relp' for 'reciprocal-lattice point' for example. The paper for refereeing was written in the same style; the author was A. L. Pon,* and this gave me the basic clue. The initials A.L.P. were famous! With this basis, Peter Wooster, one of his sons and I set about the whole paper and interpreted almost all of it. Lindo had achieved just the effect that he wanted!

This story illustrates perfectly Lindo's refined sense of humour. I cannot think of anybody who would have gone to such lengths to elaborate this sort of point.

References

1. Beevers, C. A. and Lipson, H. *Proc. Roy. Soc. (London)* **A146**, 570 (1934).
2. Patterson, A. L. *Phys. Rev.*, **46**, 372-376 (1934).
3. McLachlan, D. and Glusker, J. P. (eds.) *Crystallography in North America.* American Crystallographic Association: New York, NY (1984).

---

* The paper by A. L. Pon is reproduced in this book at the end of Chapter 41.

2. Harker sections

2.1. Harker sections and proustite: introduction

Jenny P. Glusker

The structure of proustite, a mineral containing silver, arsenic and sulfur, was determined by David Harker in 1936.[1] It was one of the first structures determined by use of the Patterson map. Harker wrote: 'It is shown that the three-dimensional Patterson method can be so simplified by the use of the symmetry properties of the crystal under consideration that its use in determining the positions of atoms in crystals is practicable. This method is then used to determine the positions of the heavy atoms in proustite, Ag_3AsS_3'. Harker showed that structural information may be concentrated in certain areas of the Patterson map.

Proustite occurs as 'tiny, brilliant red. hexagonal prisms.'[1] The space group is $R3c$ and the unit cell dimensions are a =

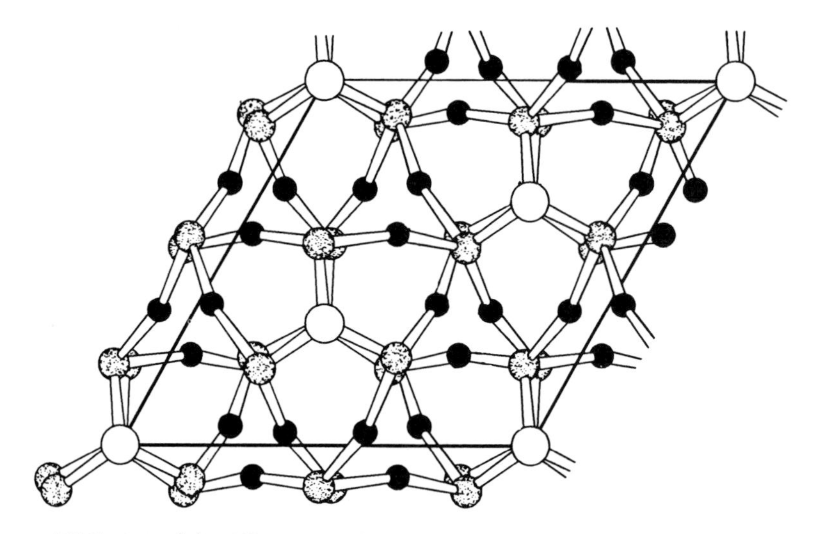

FIG. 2.1.1(a). The crystal structure of proustite. Sulfur atoms are stippled, arsenic atoms are white and silver atoms are black.

10.74, $c = 8.64$Å with hexagonal axes. The structure is shown in Figs. 2.1.1(a and b). Atoms in positions ($2a$) are found at

$$0\ 0\ z;\quad 0\ 0\ \tfrac{1}{2}+z$$

and atoms in ($6b$) appear at

$$\begin{array}{llllllllll} x & y & z & ; & \bar{y} & x-y & z & ; & y-x & \bar{x} & z \\ x & x-y & \tfrac{1}{2}+z & ; & \bar{y} & \bar{x} & \tfrac{1}{2}+z & ; & y-x & y & \tfrac{1}{2}+z \end{array}$$

Thus pairs of atoms appear with the same value of x and values of z differing by $\frac{1}{2}$. Therefore, Harker reasoned, in the Patterson map there should be maxima at

$$(0,\ +(x-2y),\ \tfrac{1}{2});\ (0,\ +(-2x+y),\tfrac{1}{2});\ (0,\ +(x+y),\tfrac{1}{2}).$$

In order to determine x and y (although, of course, no information is given on z) for each atom, it is necessary only to evaluate $P(0, v, \frac{1}{2})$, not the entire three-dimensional Patterson map. This method worked, and Harker solved the structure of proustite.

These areas of the Patterson map were called Harker sections and Harker lines. They were of particular importance in the 1930's to 1960's when computing was an extremely tedious process. Harker's work was extensively used in structure solution by crystallographers and is still used today in protein structure determination. Lindo expressed his chagrin that he had not realized this extension of his work.[2]

The vectors, $P(uvw)$, sampled at only $P(0\,v\,\frac{1}{2})$:

$$\begin{array}{lll} 0 & +(x-2y) & \tfrac{1}{2} \\ 0 & +(-2x+y) & \tfrac{1}{2} \\ 0 & +(x+y) & \tfrac{1}{2} \end{array}$$

gave, for Ag, $x \sim 0.22$ and $y \sim 0.28$. The result is shown in Fig. 2.1.1(c).

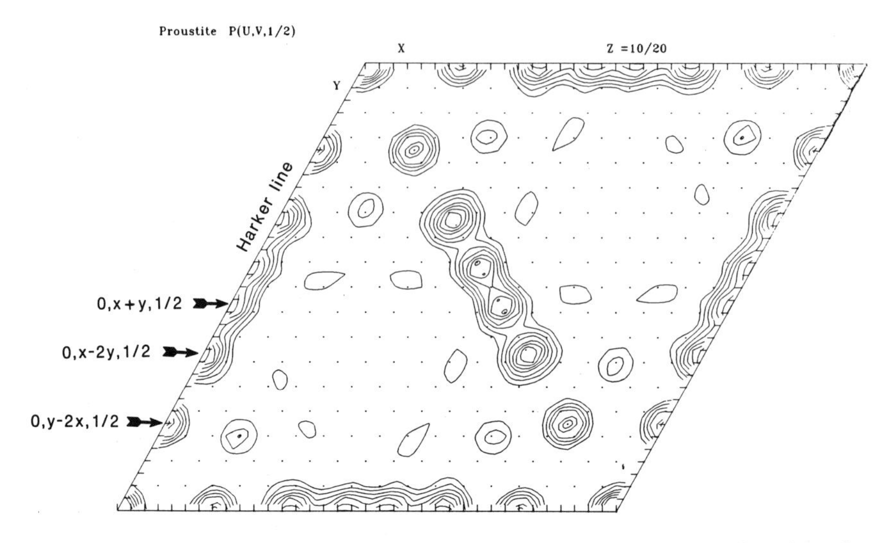

FIG. 2.1.1(c). **Detail of the Patterson map of proustite showing the Harker section at $P(u,\ v\ \frac{1}{2})$. The Ag ··· Ag vectors are designated by arrows (on the Harker line).**

From this information Harker (Fig. 2.1.1(d)) was then able to determine the structure as

2 As in special positions ($2a$), $z = 0.000$
6 Ag in general positions($6b$), $x = 0.246, y = 0.298, z = 0.235$
6 S in general positions($6b$), $x = 0.220, y = 0.095, z = 0.385$

References

1. Harker, D. The application of the three-dimensional Patterson method and the crystal structures of proustite, Ag_3AsS_3, and pyrargyrite, Ag_3SbS_3. *J. Chem. Phys.* **4**, 381–390 (1936).
2. Patterson, A. L.. Experiences in Crystallography —1924 to Date. In *Fifty Years of X-ray Diffraction.* (P. P. Ewald, ed.) pp. 612-622. International Union of Crystallography: Oosthoeks' Uitgeversmaatshapp, Utrecht, The Netherlands (1962).

2.2. My early experiences with the Patterson function

David Harker

David Harker obtained his Ph.D. at Caltech under Linus Pauling in 1936. He had previously worked for the Atmospheric Nitrogen Corporation, and, afterwards, for The Johns Hopkins University, before going to the General Electric Company in 1941, where he eventually headed a Crystallography Division. He then went to the Polytechnic Institute of Brooklyn to head the Protein Structure Project, which was financed by a big endowment dedicated to the determination of the structure of a protein in this country. He moved on to the Roswell Park Memorial Institute in Buffalo with his project in 1959. There the structure of ribonuclease was solved, the first protein structure done by crystallographic methods in this country. Since his retirement he has been working at the Medical Foundation of Buffalo. Dr Harker has had a great influence on crystallography in this country. He extended the use of the Patterson maps to show that certain defined areas of them contain a large amount of information on the structure. He showed how isomorphous replacement could be used to phase proteins if two heavy atom derivatives were available. Together with Kasper he showed how the relative phases of diffracted beams could be obtained directly from the intensities of reflections; and the Harker–Kasper inequalities form the basis for direct methods. The compound they were studying was decaborane. The structure they found by this method was unexpected, but in line with other structures found later by Bill Lipscomb for the boron hydrides. Dr Harker is now studying many colored three-dimensional symmetry and space groups. He was President of the American Society for X-ray and Electron Diffraction, Chairman of the US National Committee for Crystallography, and has received many awards from the the American Crystallographic Association.

In 1933 I started to work under Professor Linus Pauling as a graduate student at Caltech. He started me in X-ray crystallography, and, after an unsuccessful attempt to find the structure of pentlandite, a copper–nickel sulfide, I worked out the structure of tetradymite (Bi_2Te_2S)[1] with much help and instruction from Professor Pauling. The tetradymite structure turned out to require the fixing of two parameters, and was solved by plotting the intensities calculated for each of a number of X-ray diffraction spots against these two parameters as contour maps on Cartesian graphs. My next crystal structure was cupric chloride dihydrate,[2] which was also solved as a set of one- and two-dimensional problems, although three parameters were involved. (We didn't bother with hydrogen atoms in those days.) My next two structures were the minerals pyrargyrite Ag_3SbS_3 and proustite Ag_3AsS_3 — the 'ruby silvers'[3] — an isomorphous pair with the space-group $R3c$. I presume that Professor Pauling expected me to solve their structures using the method of isomorphous replacement, but, I am embarrassed to admit, this notion never occurred to me. These structures were specified by six parameters; a large number in those days. I began to have a very hard time with their solution.

It was now 1934, and the first paper on the F^2 series — the 'Patterson Function' — appeared in print.[4] One of the graduate students (his name was Medlin) gave a seminar on this subject, and we, the structural crystallography group under Professor Pauling's direction, were all thrilled and excited by the idea that the interatomic distances in a crystal could be found directly. I thought so much about this idea that I could talk of little else. And I was getting desperate about my ruby silver structures, because I needed my Ph.D. very badly. Six parameters were a whole lot. You can do a structure with two parameters by plotting out in two dimensions, which I had done, but you couldn't do six, that's in six dimensions.

One night I woke about two o'clock in the morning with the idea in my mind that the vectors between atoms related by a screw axis would all lie in one plane in the Patterson function of that structure. So you did not have to do several weeks of calculations to get a three dimensional Fourier series plotted out. You just plot one plane which might take a couple of days. I sat up in bed as though a spring had been released in me. I was sure I could now solve my ruby silver structures!

By the next day I realized that I needed to compute sections through a three-dimensional Fourier series — a formidable task in those days, before we had any but hand-operated calculating machines available. After a while, I understood that the problem could be reduced to a two-dimensional one for rotational or screw axes, and even to a one-dimensional one for mirror or glide planes. These results followed from the special values of cos $2\pi nx$ and sin $2\pi nx$ for the symmetry-determined values of the variable x (or y, or z) involved. It was accordingly possible to compute the two- or one-dimensional Fourier series with a hand-operated desk calculator in only a couple of days or so! And I solved the ruby silver structures, using these methods, in the ensuing few months! They were published in the *Journal of Chemical Physics* not long afterwards — the title of the paper is 'The use of the three-dimensional Patterson function and the structures of proustite and pyrargyrite.'

The publication of this paper on sections resulted in Patterson writing me a letter saying 'Mr Harker, I'm so glad you pointed this out to me. If I could only have thought a little more, I'd have seen this myself before you did. I could have kicked myself.' And that was my first contact with Patterson. A year or so later Professor M. J. Buerger solved some mineral structures using these methods — he called the special planes and lines in the Patterson function the 'Harker Sections,' and this made me feel greatly honored. Professor Buerger later used these sections in constructing his famous 'Implication Diagrams'[5] which eventually led to the discovery of the so-called 'Direct Methods.'

References

1. Harker, D. *Z. Krist.* **89**, 175–181 (1934).
2. Harker, D. *Z. Krist.* **93**, 136–145 (1936).
3. Harker, D. *J. Chem. Phys.*, **4**, 381–390 (1936).
4. Patterson, A. L. *Phys. Rev.*, **46**, 372–376 (1934).
5. Buerger, M. *Acta. Cryst.* **4**, 531–544 (1951).

3. Homometric structures

3.1. Homometrics and bixbyite: Introduction

Jenny P. Glusker

Homometric structures are different structures with the same vector (Patterson) map. Of course, one expects enantiomers to be homometric, but there are other structures that are not related by handedness or other symmetric operations that are homometric. It follows from this definition that homometric structures must have the same X-ray diffraction intensity pattern. Lindo, in 1939, wrote[1] 'For the sake of brevity, we shall make use of the term "homometric" to describe the relation between two structures which possess the same interatomic distances.' The study of homometric structures then became a major interest that Lindo pursued for the rest of his life.

Lindo's interest in homometric structures sprang from a letter from Linus Pauling about bixbyite. The correspondence on this structure that followed is included here and illustrates how Lindo's ideas on homometrics came about, and the important part that bixbyite played.

Linus Pauling to Lindo Patterson, February 15, 1939

You have without doubt been pleased with the great success of your Fourier series, and you deserve the thanks of the active workers in the crystal structure field. I have the feeling that there is still more to be discovered along the line of your last contribution, and I hope that you are continuing to think about the problem.

Albrecht and Corey have succeeded in making a complete structure determination of glycine, with extensive use of your series. Their paper will appear in the March or April J.A.C.S.

I saw in the program for the New York meeting of the American Physical Society your abstract in which you mention a demonstration that there is only one way in which a given set of atoms can produce a given F^2 series. In *Zeitschrift für Kristallographie* **75**, 128 (1930) Shappell and I stated that Wyckoff's positions **24(*e*)** for the space group T_h^7 gives the same values of F^2 for positive and negative values of the parameter, which however correspond to the different atomic arrangements.

Lindo Patterson to Linus Pauling, February 23, 1939

Many thanks for your letter and also for the prompt mailing of the much needed reprint. The point you raise came as somewhat of a shock to me as I had forgotten the Bixbyite case entirely.

I have been over this carefully and I have convinced myself that the two structures are different as you say. The interatomic distances, however, are identical as shown by the structure factors. I am in the process of trying to find what other space groups show this phenomenon, and shall let you know as soon as I have anything positive to say. It seems that the effect should appear only for the special points of certain space groups and it should be possible to find out quite easily in which groups it may be expected. My contention that there was a unique solution for a given set of X-ray data is therefore subject to certain exceptions. I shall let you know the full details later. At present I am trying to decide whether I still have a paper to be delivered on Friday or not ...

Postscript: I find that the same phenomenon occurs for O_h^{10} (which contains T_h^7 as a sub-group) for Wyckoff's **48(*m*)** (which contains **24(*e*)**). None of the sub-groups of T_h^7 (on the same lattice) have the same property. I have looked at a few other space groups which I thought might show the same thing without much success. I think it should be possible to enumerate the exceptions, but it will be somewhat tedious. It should be done, and I suspect it may be my job since I need it. If you have any suggestions as to appropriate procedures for the search I would welcome them.

Lindo Patterson to Linus Pauling, April 11, 1939

I have not been able to make much progress with the solution of the problem suggested by the Bixbyite structure, *i.e.,* how many structures are there which are different, but have the same interatomic distances. I have therefore decided to send a note to *Nature* stating the problem in the hope that some of the European space group enthusiasts may become interested ...

The structure of bixbyite

Bixbyite is a structure in which two independent atomic arrangements may be derived from the Patterson function; these arrangements are not simply mirror images of each other. Bixbyite was originally studied by Zachariasen[2] who found that it was a cubic crystal, a = 9.35(2)Å, containing $16(Mn,Fe)_2O_3$ per unit cell. The iron and manganese are probably disordered and, because they have nearly the same atomic number, could not be distinguished by X-ray diffraction. This is an example of a series of sesquioxides first studied by Goldschmidt[3,4] and later by Zachariasen.[5]

Zachariasen, neglecting oxygen atoms, assigned them to a space group T^5 with

32 Mn,Fe in	8(*b*)	$t = 0.23$
	12(*c*)	$u = 0.021$
	12(*c*)	$v = 0.312$

with 48 oxygens in two groups of 24 in general positions

$$x_1 = \tfrac{1}{8} \qquad y_1 = \tfrac{1}{8} \qquad z_1 = \tfrac{3}{8}$$

$$x_2 = \tfrac{1}{8} \qquad y_2 = \tfrac{2}{8} \qquad z_2 = \tfrac{3}{8}$$

Pauling and Shappell[6] reinvestigated this structure because they had studied braunite, a tetragonal pseudo-cubic mineral $3Mn_2O_3.MnSiO_3$, which appeared to have a unit cell that resembled two superimposed bixbyite cubes in dimensions. They found, as a result of their reanalysis, that the mineral is, indeed, a solid solution of Mn_2O_3 and Fe_2O_3, as Zachariasen had shown.

Pauling and Shappell measured the unit cell as a = 9.365(20) Å. Systematic absences in the diffraction pattern showed that the lattice is body-centered cubic and the space group is $Ia3$ (T_h^7); this conclusion was originally rejected by Zachariasen, but in a footnote to Pauling and Shappell's paper the authors noted 'Dr Zachariasen had kindly informed us that he now agrees with the choice of the space group (T_h^7)'. In this space group the metal arrangements are:

$$\begin{array}{llll} x\ 0\ \tfrac{1}{4} & -x\ 0\ \tfrac{3}{4} & x\ \tfrac{1}{2}\ \tfrac{3}{4} & -x\ \tfrac{1}{2}\ \tfrac{1}{4} \\ \tfrac{1}{4}\ x\ 0 & \tfrac{3}{4}\ -x\ 0 & \tfrac{3}{4}\ x\ \tfrac{1}{2} & \tfrac{1}{4}\ -x\ \tfrac{1}{2} \\ 0\ \tfrac{1}{4}\ x & 0\ \tfrac{3}{4}\ -x & \tfrac{1}{2}\ \tfrac{3}{4}\ x & \tfrac{1}{2}\ \tfrac{1}{4}\ -x \end{array}$$

with all values at (000 (as given here) and $+\ \tfrac{1}{2}\ \tfrac{1}{2}\ \tfrac{1}{2}$).

Since Mn and Fe have nearly the same structure factors, they may be considered a single entity. A consideration of the intensities led to the result

8(Mn,Fe) in 8(b)
24(Mn,Fe) in 24(d) with $u = \pm 0.030$

There are two physically distinct arrangements of the metal atoms in 24(e), corresponding to $u = 0.030$ and to $u = -0.030$; it was, as the authors pointed out, not possible to distinguish between them with the aid of the intensities of reflections of X-rays which they give. This was illustrated by showing that the structure factor was the same whether u was positive or negative.

These two distinct arrangements are illustrated in Fig. 3.1.1(a and b), together with the results of taking the structure, $u = 0.030$, and looking at the centrosymmetrically-related structure, $(-x, -y, -z)$, which is the same since the space group is centrosymmetric, and at the result of a mirror plane, $(-x, y, z)$. Neither of these correspond to the structure with $u = +0.030$, as shown.

The problem was resolved by determining the probable position of the oxygen atoms. The authors used the assumption that M···O distances would be about 1.80 to 1.90 Å, and O···O distances would not fall below 2.4 Å. The oxygen most likely should be represented by one set of x, y, z, leading, by space group symmetry, to 48 oxygen atoms in the unit cell.

When $u = 0.030$ there was no way in which the oxygen atoms could be introduced without causing interatomic distances smaller than the assumed ones. Therefore they eliminated this arrangement. The second arrangement, with

(i) Correct structure.

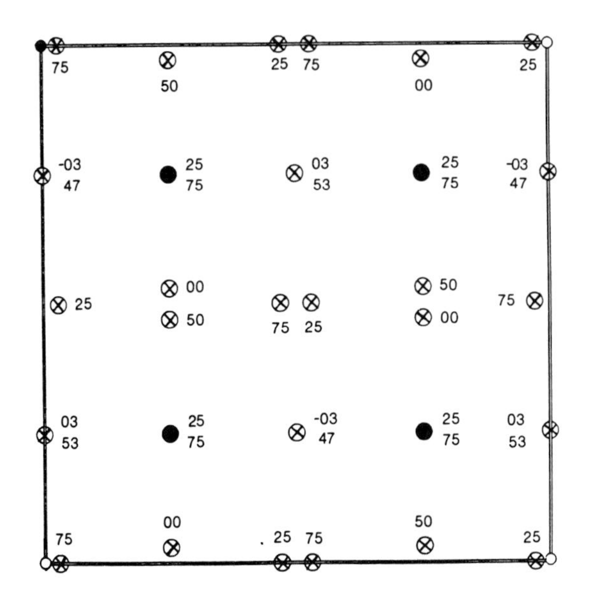

(ii) Wrong structure.

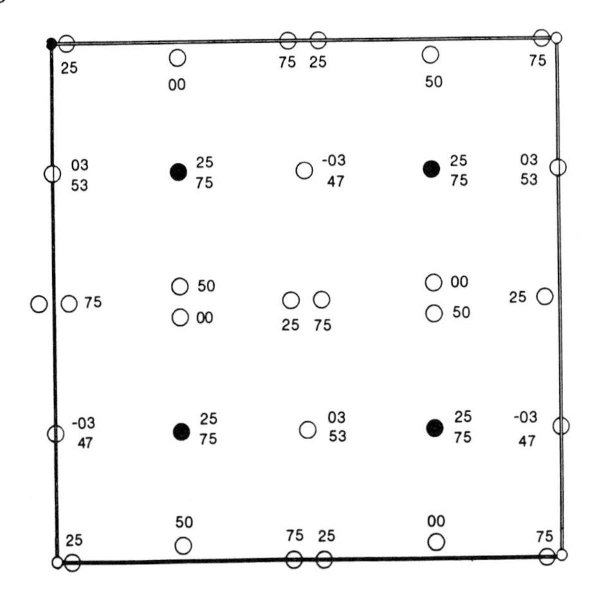

(iii) Vertical mirror plane applied to wrong structure.

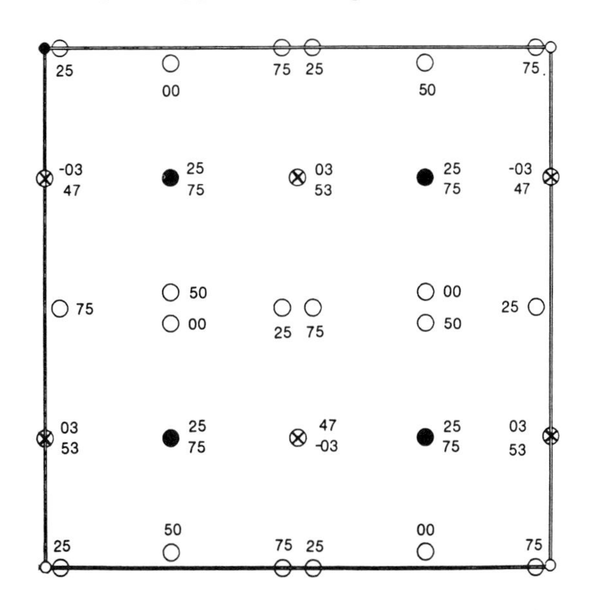

FIG. 3.1.1(a). The arrangements of metal atoms in bixbyite. Shown are (i) the correct structure, (ii) the structure reported by Zachariasen[2] and (iii) the mirror image of the structure in (i). All are different.

$u = -0.030$ could be fit by oxygen atoms at $x = 3/8, y = 1/8, z = 3/8$ ($x = 0.385(5), y = 0.145(5), z = 0.380(5)$). The distances were reasonable. In Fig. 3.1.2 the two possibilities are shown (the metal atoms in 8(e) are stippled). It can be seen that each metal is surrounded by six oxygen atoms (M$\cdots$O = 2.01 Å) with nearly equal distances when $u = -0.030$. When $u = +0.030$ this is not possible.

The structure is, as Zachariasen pointed out, intermediate between fluorite and sphalerite arrangements. The corrected structure for bixbyite probably is valid for other members of the group of *C*-modification of sesquioxides. More recently Dachs[7] has redetermined the structure and finds that $a = 9.40$ Å, and for Fe,Mn atoms $x = -0.0344(30)$, and for the oxygen atoms $x = 0.338(10)$, $y = 0.100(10)$ and $z = 0.125(10)$.

As a result of the correspondence with Linus Pauling, Lindo

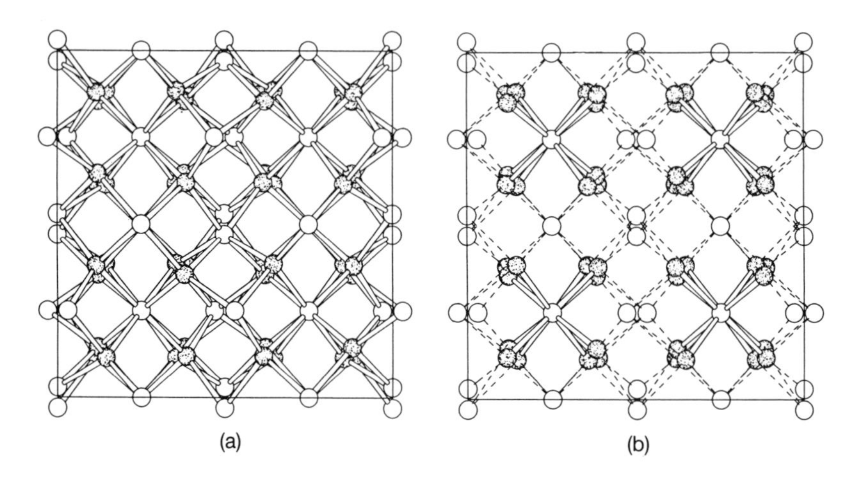

FIG. 3.1.2. The effects of putting oxygen atoms (stippled) into (a) the correct and (b) the incorrect structure of bixbyite. Bonds drawn with solid lines lie between 1.9 and 2.1 Å; broken lines indicate bond lengths outside this range.

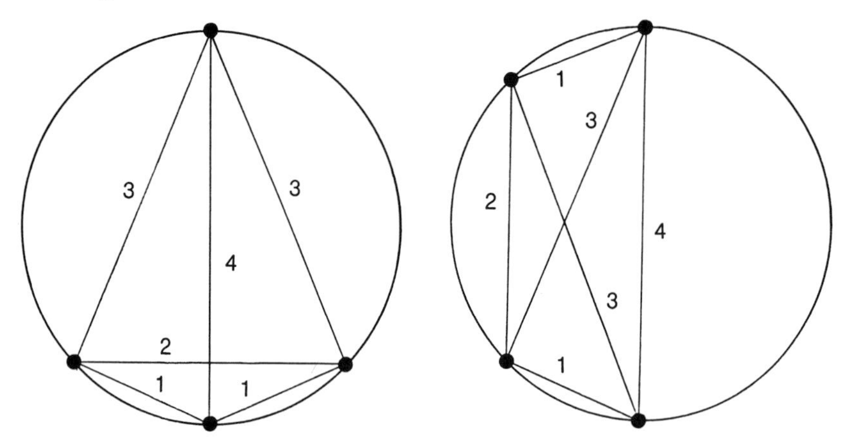

FIG. 3.1.3. A homometric pair. Numbers are shown to emphasize equivalent vectors in the two diagrams.

decided to investigate the systematics of homometric structures. This he did by first considering one-dimensional sets. Since homometric sets are periodic he was able to investigate this problem, as described in Chapters 38, 39 and 40 of this volume, by replacing a set of points on a line by a set of points on the circumference of a circle. This makes it easier to recognize homometric pairs. For example, the points at $x_1 = 0$, $x_2 = a/8$, $x_3 = a/2$, $x_4 = (7/8)a$ form a homometric set with points at $x_1 = 0$, $x_2 = a/2$, $x_3 = (5/8)a$, $x_4 = (7/8)a$ as shown in Fig. 3.1.3. These patterns were called cyclotomic sets. In 2664 sets there were 390 homometric pairs, 7 homometric triplets and 3 homometric quadruplets.[8] Lipson and Cochran wrote at the end of their section on homometric structures: 'Very little has been done to extend this study to two and three dimensions, although it has been shown that every one-dimensional homometric family has its counterpart in two- and three-dimensional sets.'

References

1. Patterson, A. L. Homometric structures. *Nature (London)* **143**, 939 (1939).
2. Zachariasen, W. The crystal structure of bixbyite and artificial manganese oxide. *Z. Krist.* **67**, 455–464 (1928).
3. Goldschmidt, V. M., Ulrich, F. and Barth, T. Geochemical distribution laws of the elements. IV. The crystal structure of the oxides of the rare earth metals. *Skrifter norske Videnskaps. Akad. Oslo. I. Mat.-Nat. Kl.* **5**, 5–24 (1925).
4. Goldschmidt, V. M., Barth, T. and Lunde, G. Geochemical distribution law of the elements. V. Isomorphy and polymorphy of the sesquioxides. The contraction of the 'lanthanums' and its consequences. *Skrifter norske Videnskaps. Akad. Oslo. I. Mat.-Nat. Kl.* **7**, 59 pp. (1925).
5. Zachariasen, W. The crystal structure of the modification C of the sesquioxides of the rare earth metals and of indium and thallium. *Norsk. Geol. Tids.* **9**, 310–316 (1927).
6. Pauling, L. and Shappell, M. D. The crystal structure of bixbyite and the *C*-modification of the sesquioxides. *Z. Krist.* **75**, 128–142 (1930).
7. Dachs, H. Die Kristallstruktur des Bixbyits, $(Fe, Mn)_2O_3$. *Z. Krist.* **107**, 370–395 (1956).
8. Lipson, H. and Cochran, W. *The Determination of Crystal Structures The Crystalline State.* Vol. **III** (L. Bragg, ed.). G. Bell and Sons Ltd: London, UK (1953).

3.2. Patterson and bixbyite

Linus Pauling

I've come to Fox Chase Cancer Center several times, but this time it's a special pleasure for me to be here to celebrate the 50th anniversary of the publication of that extraordinary paper by Lindo Patterson. I remember Lindo and his wife when they came visiting my wife and me in Pasadena, and what a good time we had. But what I remember, especially, is the contribution that he made to science and to our understanding of the world. I don't know if you all realize that it was X-ray crystallography that brought us to the place where we are now in our understanding of the nature of the world and especially of living organisms. The most active field of science in the world right now is molecular biology, DNA biology and the whole field of molecular biology, and that grew out of X-ray crystallography. There's no doubt about it that this is the child of X-ray crystallography, and Lindo Patterson contributed very much to this. I am grateful to Fox Chase Cancer Center for their having recognized his importance as a scientist and given him the opportunity to work here.

I remember very clearly when I first learned about the Patterson diagrams and I had some dim appreciation of what it meant to science to have this straightforward way of interpreting the X-ray diffraction photographs. There was a period of about 20 years when the whole field of X-ray crystallography was dominated by Patterson's contribution[1] and, of course, David Harker's contribution[2] too. This period of about 20 years was very important; in it many contributions were made to our understanding of the nature of the world and especially of the nature of the giant molecules that are present in the human body and in other living organisms.

My personal contact with Lindo was not very great, but it was of some significance, I think, to him. In 1930, I had worked with one of my graduate students, Shappell, in making a crystal structure determination. I knew the crystallographic literature really thoroughly and I was pretty critical in assessing it. Zachariasen had published a paper[3] on the structure of the rare earth sesquioxides, La_2O_3, and all the other rare earth sesquioxides; these were cubic structures. My feeling about crystal structures was such that I thought there was something wrong with these structure determinations. I knew that there was a mineral called bixbyite (the biggest bixbyite crystal in the world is here on display at the Institute for Cancer Research; it is on loan from the Smithsonian Institution). Shappell did his doctoral dissertation with me, but succeeded in getting a Ph.D. in Geology. Others of my students got Ph.D.s in physics and biology and, of course, in chemistry at CIT. California Institute of Technology was a remarkable institution. As David Harker has said, they accepted his dissertation even if it wasn't in the standard form of type written on $8\frac{1}{2} \times 11$ inch paper, and so on. Shappell left the Institute when he got his Ph.D. I don't know what happened to him.

What troubled me about bixbyite was that the interatomic distances weren't right. I didn't have much confidence in these interatomic distances that were reported then. I was sure there was something wrong. I can remember the day in 1930 when I looked over the expressions for the intensities of the X-ray diffraction pattern, and realized that there were two structures which were not physically identical, involving the 32 metal atoms in these sequioxides, that have exactly the same sets of interatomic distances. These were derived from two different structures and, except for enantiomeric pairs, this effect hadn't been recognized before. We found that Zachariasen had picked the wrong one of the two structures, and the right one gave sensible sets of interatomic distances when oxygen atoms were included.[4]

Then, nine years later I opened the issue of *American Physical Society Abstracts* of a coming meeting and saw an abstract[5] by Lindo Patterson saying that he was presenting a proof that no two physically different structures could have the same Patterson diagram. I wrote to him and I said 'Nine years ago we discovered two structures that would give the same Patterson diagram,' and I understand that the rest of his life he worked on the problem of the conditions under which two distinct structures can have the same Patterson diagram!

In 1934 he made his contribution to X-ray crystallography;

the Patterson function initiated the new period of X-ray crystallography so that it was possible to use the intensities of the different diffraction maxima in the X-ray diagram to determine the structure in a straightforward way. Before that it had been a gamble whether you could find the structure or not. After 1934 it was possible by straightforward methods to proceed toward the solution; you could not necessarily find it, that depended on the complexity of the structure, but nevertheless you could proceed and get a long way ahead in attacking this problem. For 20 years this was the only way that we had to attack the problem in a straightforward way. Then Harker and his student Kasper, and the Karles and other people began introducing methods of straightforward analysis of the X-ray diffraction intensities to obtain the solution.

I'm glad that I am able to be here to celebrate this 50th anniversary of the time when Lindo Patterson made this great contribution.

References

1. Patterson, A. L. *Phys. Rev.* **46**, 372–376 (1934).
2. Harker, D. *J. Chem. Phys.* **4**, 381–390 (1936).
3. Zachariasen, W. *Z. Krist.* **67**, 455–464 (1928).
4. Pauling, L. and Shappell, M. D. *Z. Krist.* **75** 128–142 (1930).
5. Patterson, A. L. *Phys. Rev. (abstract)* **55**, 682 (1939).

Part II

Program of the Fox Chase Symposium: A Celebration of the Patterson Function

Symposium Introduction

Molecular structure and function: keys to the ongoing revolution in biology

John Durant

I want to welcome all of you to this Fox Chase Symposium which is a celebration of the Patterson function. This symposium represents, in many ways, what we at the Fox Chase Cancer Center think we are. We are committed to a better understanding of fundamental biomedical processes, so that we can not only understand them for their own sake, but derive information that may be relevant to human disease and apply it for the relief of human suffering.

We are now in the midst of a genuine scientific revolution, in which qualitative and quantitative leaps in our knowledge of life's basic processes are occurring at a rapid pace.Many discoveries in the laboratory are finding clinical applications at a remarkable rate. We can, with confidence, accept our responsibility for attempting to use what we have learned to contribute to improving human existence. The rate of biomedical advances, and the speed with which they affect our lives, have been so overwhelming in the last decade that we have ceased to realize the historical dimensions of this revolution. Yet it has been only in the last 50 years that we have accumulated most of our knowledge of critical biological structures too small to see with microscopes.

A crucial foundation for many discoveries, and the one we are presently honoring, is the seemingly esoteric science of crystallography, which uses the diffraction patterns of X-rays to pinpoint where atoms are and how they are joined together within molecules. These studies of molecular structure help scientists learn how different molecules function and enlarge our understanding of the processes of life. For molecular biologists, crystals are analogous to computer microchips — tiny storage units containing masses of useful data. By unlocking these units, crystallographers provide the necessary first step for advances in a variety of fields. Crystallographic studies made possible the landmark discovery of the structure of DNA in 1953. This, in turn, opened up the field of genetics and resulted in recombinant DNA technology, now being used for medical, agricultural and other commercial applications. Crystallographic studies of drug structures and drug interactions with DNA continue to form the scientific basis for developing better medical treatments. These range from antibiotics to cancer chemotherapy.

Deciphering the structural codes that govern how biological materials interact is the crux of today's revolution in biology. Starting in 1934, a major tool for structural analysis has been a mathematical technique called the Patterson Function, a basic method still used by crystallographers. To celebrate this fiftieth anniversary of the publication of the important paper by Patterson, many of the scientists contributing to the current explosion of knowledge are attending this symposium [November 13-15, 1984] here at the Fox Chase Cancer Center in Philadelphia. Their celebration is taking many forms, from scientific lectures to personal reminiscences. The scientists are gathered here not only to celebrate the achievements of the past but also to share their visions of the future. We trust you will find this symposium rewarding and will enjoy the stimulating atmosphere here at Fox Chase.

4. The impact of the Patterson function

Linus Pauling

Professor Pauling was born in Portland, Oregon, and, as a result of a boyhood love of chemistry, studied chemical engineering at Oregon Agricultural College, Corvallis, Oregon, where he obtained a B.S. degree in 1922. He then went to do graduate work in the Chemistry Department at Cal Tech. This was then a small and lively department. Professor Pauling, who was already interested in the relationship of structure to properties, did a Ph.D. under Roscoe Dickinson and solved some of the earliest crystal structures, including magnesium stannide, the first alloy so studied. He then traveled through Europe visiting and working with scientists who were developing theories on atomic and molecular structure and quantum theory. These included Sommerfeld, in Munich, Schrödinger and Debye who were in Zürich at the time and Bohr in Copenhagen. He returned to Cal Tech where he stayed for 35 years making it a Mecca for many. In 1963 he moved on to the Center for the Study of Democratic Institutions, then to the University of Calfornia, San Diego and to Stanford University. He is now Research Fellow of the Linus Pauling Institute of Science and Medicine and has, in recent years, been particularly interested in vitamins and their role in human health and disease.

In 1954 Professor Pauling won the Nobel Prize for his work on the nature of the chemical bond and its application to the elucidation of the structures of complex substances. This interest began when he was a teaching assistant in Oregon and was extended when he published a benchmark paper on the principles of ionic packing in crystals. In this he deduced the sizes of ions from their interatomic distances and made rules on their mode of packing. He was also interested in the newly proposed quantum mechanics and wrote a book on it with E. Bright Wilson, Jr. He then went on to publish a general text *The Nature of the Chemical Bond* in which he described his theory of resonance. This is the book that we were all raised on and we owe a lot to him for his insight into bonding that it contains. He has since published texts on general chemistry and college chemistry that have been a delight to students.

In his work on peptide conformation, Professor Pauling showed that, from his own ideas on resonance, it could be deduced that the peptide bond is approximately planar. This was later confirmed experimentally. This conformational constraint led to his proposal of α-helices and β-pleated sheets as possible conformations of polypeptides, and again, these were later found experimentally. This work on the α-helix led crystallographers to ask what the diffraction pattern would look like, and a theoretical study of this, which was an outcome of Pauling's work, led to the interpretation of the X-ray diffraction pattern of DNA. Thus Professor Pauling is the father of protein structure and the great-grandfather of DNA structure. His other major contributions to biochemical crystallography include studies of normal and sickle-cell hemoglobin, a general theory of anesthesia and studies of antibodies.

In addition to all of this Professor Pauling has, with the encouragement of his late wife, sought to remind us of our responsibilities as scientists. Because we have some understanding of science it is important to help our fellow citizens understand it as well, and this can be done by explaining our own understanding of the problems that face us in the everyday world. Professor Pauling pursued this idea with great personal sacrifice, and was instrumental in bringing about a treaty that banned atomic bomb testing in the atmosphere because of the hazard that would be caused by radioactive fallout from these tests. He received the Nobel Peace Prize in 1962, and is the first to receive two unshared Nobel Prizes.

The most important thing about Professor Pauling is the great enjoyment of science that he projects to all who listen to him, and his deep scientific curiosity.

It is a real pleasure for me to be back among crystallographers. In recent years I've been forced to associate with physicians, and even with professors of nutrition. This occasion makes me especially glad to be here. Knowledge about crystallography and protein structures, I notice, has been spreading. Just last week I was reading in the journal *Astounding Science Fiction* a story in which one of the characters was a young woman named Alfa Felix.

I have been fortunate in my life. Every once in a while something has happened just through good luck, chance, that as I look back I realize was important. I would never have gone to the California Institute of Technology (CIT) and become a crystallographer if it had not been for the following series of events. In 1922, when I was to get my bachelor's degree from Oregon Agricultural College. I applied to several places for a graduate fellowship. Harvard offered me a half-time instructorship, and pointed out that it would take me six years to get a Ph.D. degree, if I was lucky. I didn't like that prospect; I needed to be making some money before the six years was up, more than this half-time instructorship. I applied at Berkeley, and I was especially interested in Berkeley because I had read G. N. Lewis's 1916 paper on the nature of the chemical bond, and also Irving Langmuir's 1919 papers, and I was really very much taken by them. They influenced my thinking ever after that time. I applied at the California Institute of Technology. Pretty soon I got a letter from the California Institute of Technology, A. A. Noyes, offering me a fellowship of $350, I think, for the year, and with a statement that I should decide and respond immediately so as not to keep someone else from taking such a job if I turned it down. So I wrote accepting it, as I hadn't heard anything from Berkeley. Of course, pretty soon after that it became improper to make such a requirement; when you were made an offer the universities agreed that they should allow you until the first of April to decide something like that. Just two years ago I gave the Hitchcock lectures at Berkeley; I'd been Visiting Professor for the spring semester for five years at Berkeley. Three different people came up to me with a story as to why I hadn't heard from Berkeley. I don't know whether it's apocryphal or not, but each of them said the same thing, that G. N. Lewis had been looking over the applications, perhaps 25 (times were different then), and deciding which applicants should be given fellowships as graduate students. He came to one and read it and said 'Linus Pauling, Oregon Agricultural College. I've never heard of that place,' and down it went into the discard file. Well, I was lucky not to have gone to Berkeley, because if I had I wouldn't have become an X-ray crystallographer, and it is just wonderful, you know, being an X-ray crystallographer.

It was really fine that I went to CIT. The first crystal structures in the United States had been done there at CIT in 1917. A. A. Noyes was responsible. Lalor Burdick had written to Noyes from Germany saying it was necessary to leave Germany because of the First World War; he went to Switzerland and got his Ph.D. in organic chemistry. Noyes wrote to him saying that he ought to go to England and work with the Braggs (about

1916). So he went there, he and Owen published a paper[1] on the structure of carborundum (silicon carbide). Then Burdick came back to the United States and he started building an X-ray spectrometer. I think he built one at MIT but then went with Noyes to CIT. C. Lalor Burdick is still in New Jersey running the Lalor Foundation. So he built a Bragg ionization chamber spectrometer. Now his was not the same as the Bragg apparatus; he had improved it in several ways. One way was this. He put a motor in that would rotate the crystal through the Bragg angle (there were wide slits on the ionization chamber, and it was necessary to rotate through the Bragg angle in order to get an integrated intensity). In 1930 I spent a month in Manchester with W. L. Bragg and I left at the end of that month. I taught, you know at CIT; there were three or four seminars every week. I learned a lot by going to these seminars— chemistry seminars, physics seminars, in 1930 they were even having biological seminars. There were not any seminars at Manchester when I was there with Bragg. I thought, surely they'll want me to talk about my work on the structure of the micas and similar minerals, while I was there. I'd just finished the structure of chlorite and so we had done talc, pyrophyllite and mica, chlorite. Surely they'd be interested to hear me talk about that at a seminar. But they were not. That was the difference between life in the California Institute of Technology (Berkeley was similar with lively seminars there) and life in Manchester.

I put in part of my time running the Bragg Spectrometer. I was working on the structure of epidote. We had our five-year-old son Linus along. One visitor one day said to him, 'What are you going to be when you grow up'? and he said 'A physicist'. 'So what sort of physicist will you be, what will you do?' and he said 'I will determine structures'. and the man said 'What structures will you determine?' He said 'The structure of epidote.' Epidote was what I was working on, on this Bragg spectrometer. I set up the crystal, and began to find a reflection and set it. When you turned the crystal through the Bragg angle, you would start a metronome ticking. There was a little capstan with four arms, and you would go tick, tick, tick, with your two fingers and run the crystal through the Bragg angle in order to get the integrated intensity. That was in 1930. I didn't determine the structure of epidote; it didn't succumb to my efforts.

Crystallography has gone through several periods. The first period, I think, was very interesting. I had the advantage of being taught X-ray crystallography the way it was practiced then by Roscoe Dickinson. He was a wonderful teacher, a very good man, careful, precise. He knew when he knew something. He had a very logical head. I learned a couple of very important things from him. One, that you should think about what you are doing and move ahead and know what you have proved at each step. That was very important.

So I will describe X-ray crystallography and the way that it was carried out then. We used Laue photographs as the basis for the technique. A rotation photograph was taken with an X-ray tube with molybdenum radiation to get one value for the lattice constant for the cubic crystal. You used that from then on as the basis of the analysis of the Laue photograph. This method was developed in part by Nishikawa.[2] Laue, of course, took the first photographs and tried to analyze them.[3] Ewald did analyze a Laue photograph of pyrite[4] to determine a value of the sulfur parameter along the three-fold axis.

Nishikawa introduced space group theory into X-ray crystallography. He was at Cornell with a bright young fellow who wanted to do some work for a doctoral degree — R. W. G. Wyckoff. Wyckoff set to work to determine the structure of cesium dichloroiodide by X-ray diffraction methods. He had been doing some work on the chemistry of cesium, using the Laue technique. Wyckoff[5] wrote *The Analytical Expression of the Theory of Space Groups*, with all of the special positions for all 230 space groups, published around 1922. The book was available when I began work in Pasadena.

The Laue technique was a very interesting one. If you had a cubic crystal and the X-ray beam came in along a cube axis (a cube edge) you got a rather simple pattern which wasn't very valuable. If you tipped the crystal somewhat to get an unsymmetrical photograph, then the photograph was much more valuable. You calculated the interatomic distances (the X-rays were being diffracted by the particular plane for each of the planes) using the gnomonic projection to identify the planes. If the wavelength was in the limit 0.24 to 0.48 Å, as calculated with the edge of the unit cell that you had got from the rotation photograph, then you had a pure first order reflec ion from that plane and you could compare it with another p'ane that happened to be reflecting at the same wavelength. So that you had to make several of these photographs in order to get enough of these somewhat chance comparisons.

Nobody knew what the scattering factors were and their dependence on the interplanar distance angle of reflection $\sin\theta/\lambda$. The form factors hadn't yet been evaluated. The only thing that seemed reasonably certain was that the Lorentz-polarization factors fell off with increasing values of $\sin\theta/\lambda$. So that if you had two planes reflecting at the same wavelength, and the one with the smaller interatomic distance gave a stronger reflection than the one with the larger interplanar distance, you could conclude that the structure factor for the first one was greater than the structure factor for the second one. You could also look for extinctions ($h0l$, for example, being absent when h and l are both odd). In that way you could get a decision as to the space group. You could then look in Wyckoff's book[5] and see what possible positions there were for the atoms on the basis of the assumptions that you had not missed some weak reflections (there are always assumptions involved at each stage). You could then start calculating intensities. In determining the positions of atoms it was easy to evaluate a single parameter, harder to evaluate two parameters, almost impossible to evaluate three parameters.

It was evident that cubic crystals were the ones for which the probability of a successful structure determination were highest. Dickinson had studied ammonium hexachlorostannate and potassium tetrachloroplatinite (the Pt(II) complex) and Wyckoff had studied potassium hexachloroplatinate.[6] They had verified that the hexacoordinate complexes are octahedral and that some of the tetraligated complexes have a square planar structure. Dickinson also determined the structure of potassium tetracyanozincate and found a tetrahedral coordination about the zinc.

All this had been done by 1922, and one more thing had

been done. Dickinson and Raymond[7] had found an isotropic, cubic organic compound called urotropine. It was given to patients with infections of the urinary tract along with ammonium chloride or some acidifying agent which caused it to break down. This cubic compound was not then known by its present name of hexamethylenetetramine. Dickinson showed there was some uncertainty about its chemical formula. It had only two parameters; the nitrogen atom was out in the tetrahedral direction from the origin, and the carbon atoms up along the cube edges. Dickinson had determined those two parameters; this was the first organic compound to have its structure determined.

I was just entranced with the possibility of getting a better understanding of minerals and inorganic compounds. I checked through Groth's *Chemische Kristallographie*,[8] all five volumes, hour after hour, looking for cubic crystals. I started to synthesize them, making them in the laboratory, and carrying out preliminary studies with X-rays. My memory is that I made 14 of them. I'm not sure because my first research book disappeared about 25 or 30 years ago. But I kept getting, even with these cubic crystals, crystals that were too complicated to have their structures determined. I remember I made a calcium mercuric tetrabromide $CaHgBr_4$, and it turned out to have 32 calcium atoms, 32 mercury atoms and 128 bromine atoms in the asymmetric unit. It was too complicated for me to determine this structure. So I built a furnace by getting an Alundum cylinder and wrapping nichrome wire around it. I hooked it up to a source of electric current and put it, with asbestos around it, in a can. With a crucible that I could put down inside this apparatus, I melted up some sodium and added some cadmium to it. The cadmium dissolved and I let it cool slowly over a day. Then I dissolved the solid chunk of metal in absolute ethanol (which attacks sodium producing hydrogen and sodium methanolate) and there were beautiful crystals left of what I thought was sodium dicadmide ($NaCd_2$). I measured the edge of the cubic unit, and took Laue photographs and it turned out that the edge of the cubic unit cell is 30 Å and that it contains about 1200 atoms. For years after that, every two or three years, I would work awhile on sodium dicadmide, and finally 35 or 40 years later, Sten Samson succeeded in determining its structure and locating all of the atoms.[9] It isn't quite $NaCd_2$, but it is a pretty interesting structure.

So I was stuck. I melted up some potassium sulfate, dehydrated some nickel sulfate, and added it to the melt, and let it cool. It contracted as it solidified and crystallized and I could knock off nice crystalline chunks 2 or 3 mm on an edge of a cubic crystal $K_2Ni_2(SO_4)_3$. There are 19 atoms in the formula. As it turned out, when I determined the space group, it was a 19 parameter structure, too; so I gave up on that. The structure is now known. There is a mineral that has the same structure.

So, after nearly three months, I was discouraged. Dickinson went to the stockroom and got a cardboard cylindrical container containing round pebbles of molybdenite, and he cleaved off a chunk of molybdenite. We set it up and took good X-ray photographs and determined the structure, a one-parameter structure.[10] It took about two weeks to determine this molybdenite structure. Two weeks of hard work and it involved something new. The six sulfur atoms around the molybdenum atom were at the corners of a trigonal prism rather than an octahedron. I thought how wonderful it is that in a couple of weeks you can get some additional information about nature. And it's still possible. You can, even now, find some simple substances that I am sure could be handled in the same way and would give some interesting new insight; but it's harder, of course, to find anything new in this field than it was then, when almost every crystal that you examined produced some additional interesting information.

I remember that a few years later Hultgren[11] and I investigated sulvanite Cu_3VS_4. I had just determined the structure of enargite Cu_3AsS_4 and it was like wurtzite but with arsenic replacing a fourth of the metal positions, the zinc positions, and copper replacing three-fourths, so that it had lowered symmetry (orthorhombic symmetry). But each arsenic had four sulfur neighbors and each copper had four sulfur neighbors. It was just a superstructure of the wurtzite tetrahedrally coordinated structure. I thought that since sulvanite, Cu_3VS_4, had a cubic structure it would be a sphalerite analog with vanadium in one of the zinc positions, and copper in the other three, tetrahedrally coordinated. However, that was wrong. The vanadium didn't know it was supposed to be at the origin. It was located at the position $\frac{1}{2}, \frac{1}{2}, \frac{1}{2}$ thereby leaving a hole at the origin. The vanadium atom not only had four sulfur atoms around it, but had six copper atoms bonded to it. These were close enough so that every ten years or so I write a paper about the nature of the chemical bonds in sulvanite. It is still an interesting question as to why vanadium is doing that.

Soon it was getting harder and harder to find crystals. Dickinson succeeded in determining five parameters in a crystal.[12] This Laue method was exceedingly powerful. Herman Mark wrote a paper on tin tetraiodide, saying the unit of structure is a cube containing four SnI_4 molecules. Roscoe Dickinson's Laue photographs showed that that wasn't true. The cube is twice as big on an edge and contains 32 SnI_4 molecules, and five parameters. The SnI_4 molecule is on the three-fold axis, so there are five parameters in all to define tin and iodine parameters. But this was a sort of superstructure on a small unit. So Dickinson calculated the partial derivatives of the structure factors with respect to each of the five parameters, and then set up equations to solve for the five displacements from the positions corresponding to the simple unit. In this way, he was able to determine the structure.

I was getting interested in more and more complicated things, so I think by this time (around 1930) I can say we got into the second stage of crystal structure determination (at least according to my experience). This stage involved the use of what I called the stochastic method, from the Greek *stochastikos*, 'apt to divine the truth by conjecture'. I would say 'if I were a crystal of zunyite or the elements of the crystal of zunyite, $Al_{13}Si_5O_{20}(OH,F)_{18}Cl$, a cubic crystal, what would I do?' Well, by that time I knew that as a silicon atom I would surround myself by a tetrahedron of oxygen atoms; as an aluminum atom I would surround myself with an octahedron of oxygen atoms. In fact, in 1927 or spring of 1928, Holmes Sturdevant[13] and I had used this method for the first time to predict a structure. We noticed that in other forms of titanium dioxide (rutile and anatase) there are TiO_6 octahedra sharing edges, two edges shared in rutile shortened to 2.5 Å and four edges shared in anatase. So we thought that that there is rutile and anatase, and in between them there is another titanium dioxide, which should have three edges shared. We put models of octahedra together sharing three edges in such a way that we could formulate two structures. We shortened the shared edges to 2.5 Å. One of the structures had the dimensions of the orthorhombic crystal. We measured the unit cell to get the values of the parameters and they gave good agreement with the observed intensities, so that was the determination of the structure of that crystal.

Then I determined the structure of topaz[14] by assuming that silicon was surrounded by four oxygen atoms and aluminum, by six (well, four oxygens, plus two fluorines), and this permitted the structure of that mineral to be determined in 1928. I went on having a high old time for a while, predicting structures and checking the structure, predicting complete structure, not just partial structure but the complete structure, and then calculating one set of intensities. If they agreed with the experimental set that was the structure determination. Sometimes this worked, and sometimes it didn't work. It was a great time, and I did not even have to go into the laboratory to do the job. I could just read *Zeitschrift für Kristallographie*, for example. Odd Hassel and Luzanski published a paper[15] on the structure of ammonium hydrogen difluoride, and they published the values of their observed intensities, and the space group (it was an orthorhombic crystal). Then they said that they hadn't succeeded in determining the structure and they weren't going to work anymore on this problem. So I thought 'NH_4HF_2, that ought to be like KHF_2' and Bozorth had determined the structure in Pasadena. We had also determined the structure of the azide KNNN, which is closely similar. But KHF_2 is a tetragonal structure; the difference between KHF_2 and NH_4HF_2 is that the ammonium ion has four hydrogen atoms which can form hydrogen bonds. Potassium in KHF_2 has eight fluorine neighbors, but the ammonium ion in NH_4HF_2 will pull four of them in, and the other four will then move out. This means that they each have to rotate and that makes it an orthorhombic structure. And there you have the structure determination.[16] It was a lot of fun.

Well, time goes on. Pretty soon things began getting too

complicated for me. Then came 1934, when Patterson introduced his Patterson diagrams, and that had a great impact[17]. We moved into the third phase of structure determination when you had a perfectly straightforward direct way of proceeding. You could also use a Fourier with a heavy metal, if there were a heavy metal present. That often permitted a straightforward procedure to be followed, at least for locating the atoms that were near the heavy metal.

The Patterson diagram was a great contribution. In 1937, we made some crystals of ammonium cadmium trichloride, NH_4CdCl_3,[18] and used Patterson diagrams for the first time in my experience to determine the structure. Actually, I had already talked about the Patterson diagrams in my lectures, both in Pasadena and in Berkeley, pointing out what an important step forward it was to have the Patterson diagram, and suggesting that we were moving into this next phase.

In 1937 I tried to determine the structure of α-keratin using what I thought were sound methods of making predictions about ways of folding polypeptide chains. I didn't succeed because I was trying to put an integral number of units into 5.1 Å along the fiber axis, and that was a misunderstanding about what the X-ray diagrams were telling us. The X-ray diffraction photographs were fooling all of us into thinking that there was a sort of repeat in 5.1 Å when that 5.1 Å meridional reflection didn't actually exist. It was off-meridional, a pair of off-meridional reflections, that were so broad and fuzzy that they overlapped on the meridion and so appeared to be meridional.

At the end of 1937 I thought that α-keratin must have a structure,[19] and I had not been able to find it. It must be that I didn't know as much about structural chemistry as I thought I knew, that biological molecules really were different, that the proteins did not have the structure that I thought they had. So we needed to determine the structures of amino acids and simple peptides to find out this elusive bit of knowledge that we didn't have yet. So we set to work, Corey, Gus Albrecht, Verner Schomaker, Dick Marsh and many others, and pretty soon Corey determined the structure of a dipeptide, a cyclic one, diketopiperazine.[20]

So I checked up on what he had done. He didn't use Patterson techniques. This is 1938 by now. He hadn't made use of Patterson diagrams. Instead he had done pretty much as I had been doing. He had said 'We don't know what the bond angles are at the carbonyl carbon atom, or the relation, but I'll put them in as tetrahedral angles the way Debye did in graphite (in his structure of graphite which was wrong — Bernal, of course, settled the question by pushing them up into the plane[21]) and showed that, with reasonable bond lengths in the ring and to the carbonyl oxygen, there was no structure that satisfied the X-ray diffraction pattern. So then Corey made the diketopiperazine ring planar and determined the structure.[20] From then on, Dick Marsh said that every one of the dozen amino acid and peptide structures that were determined in the world (and that means in Pasadena in the next ten years because nobody else had succeeded in determining one) was determined after that first diketopiperazine structure with use of Patterson diagrams. This just gives one indication of how valuable Patterson diagrams were in this period. I asked Peter, my son, in London, what his experience was. He said he had never determined a structure without making use of Patterson diagrams. Then he said, 'and you remember the structure determination that you and I published ten years back of cobalticyanic acid and trisilver cobalticyanide?[22] We used Patterson diagrams in that'. That's a structure where you have either a hydrogen bond or the biligated silver connecting the cobalticyanides along the edges of a cube. Then you have a similar structure a third of the way down the body diagonal and another one a further third of the way down. So you get a rhombohedral structure in which there are three frameworks that are interlaced with one another, like the ice IX or VIII structure that Barclay Kamb[23] determined which is ordinary cubic ice (the ordinary tetraligated hydrogen bonded cubic ice) with a similar cubic ice pushed into the hollows between the atoms; this is the high pressure form. You get two cubic centimeters of ice coalescing into one cubic centimeter to get that structure.

Patterson diagrams are still used. I checked *Acta Crystallographica* and found that 30% of the structural papers published during the last year in *Acta Crystallographica* make use of Patterson diagrams. I checked just by looking at the illustrations to see if there was a Patterson diagram shown in the illustrations. I didn't even have to read the papers to see that these diagrams are still important.

Well, of course we come to around 1935. [*David Harker: I don't know what you're going to say.*] I browbeat Dr Harker into getting busy and getting a doctoral thesis so that he could get his Ph.D. and I could be rid of him. I sent him off so he worked on pyrargyrite and proustite and in the course of that work (I don't remember just how) the idea came up that you could use the symmetry elements of the crystal to get some special information from Patterson diagrams, make some special kinds of Patterson diagrams, Patterson-Harker[24] sections. [*David Harker: I'm supposed to talk about that this evening.*] I'm not going to go on any further with it, but to say that that was an important contribution to this field.

Dick Marsh and Eddie Hughes[25] were involved in the first least squares determination of parameters from observed intensities, and this was a significant step forward. Then other people got involved in developing straightforward methods of crystal analyses from the X-ray diffraction patterns.

That's the fourth phase. That's the present phase that we are in now. One man came in to see me. He said if I had any crystals I wanted the structures determined for, he'd be glad to determine the structures for me. He could do a structure in about a day, had done 50 in the last year — not working full-time. Long before this I had sort of lost interest in it.

It was wonderful back in those days when you didn't know whether you'd be able to determine a structure or not and you had to do a lot of thinking to be successful. But if you had luck, you could get out the structure and learn something more about the nature of the world in a reasonably short period of time. You didn't have to work very hard to get a few X-ray photographs, and even when plotting them out I didn't have to work very hard. My wife made all the gnomonic projections and indexed them and calculated the $n\lambda$ values for me, and in general encouraged me to work hard. Pretty soon you discovered something about the nature of the world, and it was something like solving a crossword puzzle.

By the time I had been at CIT for three months I had material for my first paper with Roscoe Dickinson, and in another couple of months I had material for my second paper by myself,[26] not on sodium dicadmide but on another intermetallic compound which was mentioned earlier, magnesium stannide Mg_2Sn.[15] This had a simple no-parameter structure, the fluorite structure. Every month I could learn something more. I was shocked when I went down as professor for two years to the University of California in San Diego, when I was talking with a young graduate student. I found that her first year she was in the laboratory a sort of handmaiden to an older graduate student, a third-year graduate student, and it would be another year before she could be a sort of handmaiden to the professor in charge of the investigation, and have some autonomy in running the apparatus. I thought this must be awfully discouraging for a young scientist, not to be able to make any discoveries for years. You ought to be able to make a discovery in at the most one year, preferably in a few months, otherwise students are discouraged. Of course sometimes this backfires. I have this student who came to me with an idea about the experiment he wanted to do — the project for his doctor's degree. It involved radioactive tracers in some way. So I said no he shouldn't do that, that he should determine the crystal structure of oxalamide (that has two NH_2 groups in it); and he did, and he got his doctor's degree. But he never has published this structure investigation. During the next year I had some money so I was able to keep him on for a postdoctoral fellowship. During that time he invented the density gradient method of ultracentrifugation, and used it, along with a collaborator in biology, to demonstrate the semi-conservative replication of the gene. You can understand why I don't ask him what it was he was going to do with radioactive tracers. But still I think it probably was worthwhile for him to have got another year or two of experience before tackling that particular problem.

I like X-ray crystallography. The last paper in the field that I wrote was only a year or two ago. It was on the structure of lithiophorite.[27] Lithiophorite ($LiMn_3Al_2O_9 \cdot 3H_2O$) has a complicated formula. It has lithium atoms and some aluminum atoms and some hydroxide groups and some oxygen atoms and some manganese atoms in two different valence states, and nobody was quite sure what its chemical formula was. I thought 'Why does this substance have the complicated formula that it has? I'll try to understand it.' Well, as a first approximation, you can say that lithiophorite contains manganese dioxide layers, and gibbsite or hydrargillite layers ($Al(OH)_3$ layers), and manganese dioxide layers where all of the octahedral positions are occupied by quadripositive manganese atoms. These layers are arranged directly over one another, in the way that layers are in $HCrO_2$ because they form hydrogen bonds. So I thought 'Well, what happens when they form hydrogen bonds?' That transfers some positive electric charge to the manganese dioxide layer, and from the bonds lengths we know that about a seventh of a unit charge is the amount transferred (I think that is in *The Nature of the Chemical Bond* [28] that I refer to whenever I want an authoritative statement that I can quote). So you put a seventh of a positive charge on each of those oxygen atoms, which would mean two sevenths of a positive charge on each manganese atom. How can you make this conform to the principle of electroneutrality? Well, let one out of seven of the manganese atoms be Mn^{2+} and then make a ring of six that are Mn^{4+}; this unit of seven octahedra then takes care of the problem of electronegativity. That might explain why about one seventh of the manganese atoms are bipositive and six sevenths are quadripositive. But what about the hydrargillite layer; we've left it with a double negative charge for every fourteen oxygen atoms. We can take care of that by picking up lithium, which fits into octahedral holes (a third of the octahedral holes are empty in hydrargillite layers) put a couple of lithium ions in those positions. That gives electric neutrality there. So here we have reasonable explanation of this complicated formula; even a prediction as to just what the formula is, which wasn't certain from the chemical analyses. Well, that's fun too. You can see that I'm really at heart a crystallographer, and not just an advocate of vitamin C which, as you know, has been one of my primary interests through the years.

Well, I still like X-ray crystallography, and I'm grateful to Lindo Patterson, Dave Harker and all other X-ray crystallographers who make life worth living for me. I can look forward to getting the next issue of *Acta Crystallographica* and reading the papers. So I'm very pleased to be able to be here to participate in this celebration of the fiftieth anniversary of the publication of Lindo Patterson's great paper.

Discussion

Verner Schomaker: Linus, I think your estimate of the number of papers in *Acta Crystallographica* that use or have used the Patterson function is way way off. It must be more like 90%. Authors don't publish their diagrams, but they say 'We determined the structure by the use of Patterson techniques.' They almost all do.

Linus Pauling: Maybe you are right. I checked two or three papers in *Acta Crystallographica*, and I found the statement 'We made use of Patterson techniques' all right. But I didn't go through to find out how many of them contained this statement because I wanted to do something else. I didn't want to take time to do that. However, Patterson techniques are still important, they are part of the arsenal of the modern X-ray crystallographer. I don't envy him. I think that must be a boring life, compared with the old days. You think I'm wrong.

Verner Schomaker: Absolutely.

David Harker: When it comes to determining the structure of proteins, as far as I know, there is still no way of doing it, except by the heavy atom method; and to get the positions of the heavy atoms, everybody uses the Patterson method. So the method is in use 100 % for proteins.

I have one more remark. When I first came to Caltech, when we were young together, you had published 29 papers on structures that year, all by the stochastic method, if I remember correctly.

Linus Pauling: Thank you, yes. It was a lot of fun. And it is astonishing how much information this technique provided. Most of the information, not all, but most of the information, I would say, about structural chemistry, we got from X-ray crystallography. With organic chemistry electron diffraction, of course, was very important too. But after all, carbon is only one of more than a hundred elements, and so that is a pretty limited part of chemistry.

F. Rubin: This is not a question. It is just a very peculiar insight that I just got hold of when you were talking about crystals. I don't know anything about crystallography. I'm not a chemist. I am an artist, and I'm also studying holistic therapies and *materia esoterica*, which goes back centuries in the millenia. In that system, crystals have been recognized since ancient times to be the basic realities of matter. I think it is fascinating how I am presented, from these discoveries in the twentieth century, with it.

Linus Pauling: Well, that's a good comment. In a sense I agree with it. When I was working in the field of immunology, immunochemistry, and in particular studying the important rule of molecular complementariness, it took me some time before I realized that crystals give the best example. A crystal structure of an organic compound is such that, if a molecule is removed, the surrounding part of the crystal has the structure with the greatest complementariness to the molecule that fits into that position. So crystallization is a phenomenon closely similar to gene replication, enzyme activity, and everything else that we think of as biological and involving biological specificity, and of course its specificity is just great. A crystal will reject everything except the molecules that belong in those positions. In the same way an antibody will refuse to combine with haptenic groups except the appropriate ones that are complementary to the combining region of the antibody. So, I agree that there's a lot that might be said to support the statement that the crystals are a very important part of the whole universe.

James A. Bassuk: Dr Pauling, what was your reaction when Watson and Crick solved the structure of DNA, and how close were you to obtaining that solution?

Linus Pauling: Well, my reaction, if I try to remember, my reaction was 'I wonder if they could have done the job?' It was hard for me to believe. But I think I had thought that sooner or later I would find the structure of DNA, and pretty soon I decided that they had done the job, with of course Jerry Donohue's help (he was the only person whose help they acknowledged). So, then, after a while people began asking me this question, how close was I? I answered by saying 'Well, perhaps I would have discovered it sooner or later.' After all, I had said the gene consists of two mutually complementary strands which can separate and produce duplicates in a semi-conservative manner. Watson and Crick both separately heard me make this statement, too. But I remember 1937, when I should have discovered the α-helix and didn't (not until 1948). Nobody else had discovered it in the meantime. It astonishes me that nobody else discovered it in all these years. I'm also astonished that eleven years went by without my discovering it. So I wouldn't say that my assessment of the probability that I would have discovered the double helix, if Watson and Crick hadn't, is very high. Someone else might have discovered it in this period of eleven years that might well have elapsed before I got around to attacking the problem. Well, of course I'm very pleased with the double helix and with development of DNA biology. I, myself, I think, got it started in Pasadena.

References

1. Burdick, C. L. and Owen, E. A. *J. Am. Chem. Soc.* **41**, 42–50 (1917).
2. Nishikawa, S. *Proc. Math. Phys. Soc. Tokyo* **8**, 199 (1915).
3. Friedrich, W., Knipping, P. and Laue, M. *Sitzungsber. (Kgl.) Bayerische Akad. Wiss. (Munich)*, 303 (1912). Translation by J. J. Stezowski in *Structural Crystallography in Chemistry and Biology* (J. P. Glusker, ed.) pp. 23–39. Hutchinson Ross Publishing Company: Stroudsburg, PA, Woods Hole, MA (1981).
4. Ewald, P. P. *Ann. Physik* **44**, 1183 (1914).

5. Wyckoff, R. W. G. *The Analytical Expression of the Results of the Theory of Space Groups.* Carnegie Institute of Washington: Washington, DC (1922).
6. Dickinson, R. G. *J. Am. Chem. Soc.* **44**, 2404 (1922).
7. Dickinson, R. G. and Raymond, A. L. *J. Am. Chem. Soc.* **45**, 22 (1923).
8. Groth, P. *Chemische Kristallographie.* Engelman: Leipzig, Germany (1906).
9. Samson, S. *Nature (London)* **195**, 259 (1962).
10. Dickinson, R. G. and Pauling, L. *J. Am. Chem. Soc.* **45**, 1466 (1923).
11. Pauling, L. and Hultgren, R. *Z. Krist.* **84**, 204 (1933).
12. Dickinson, R. G. *J. Am. Chem. Soc.* **45**, 958 (1923).
13. Pauling, L. and Sturdevant, H. J. *Z. Krist.* **68**, 239 (1928).
14. Pauling, L. *Proc. Natl. Acad. Sci. USA* **14**, 603 (1928).
15. Hassel, O. and Luzanski, N. *Z. Krist.* **83**, 448 (1932).
16. Pauling, L. *Z. Krist.* **85**, 380 (1933).
17. Patterson, A. L. *Z. Krist.* **A90**, 517 (1935).
18. Brasseur, H. and Pauling, L. *J. Am. Chem. Soc.* **60**, 2886 (1938).
19. Pauling, L. and Corey, R. B. *Nature (London)* **171**, 59 (1953).
20. Corey, R. B. *J. Am. Chem. Soc.* **61**, 1087 (1938).
21. Bernal, J. D. *Proc. Roy. Soc. (London)* **171**, 59 (1953).
22. Pauling, L. and Pauling, P. *Proc. Natl. Acad. Sci. USA* **60**, 362 (1968).
23. Kamb, B. In *Structural Chemistry and Molecular Biology.* (A. Rich and N. Davidson, eds.) p. 507. W. H. Freeman: San Francisco, CA (1968).
24. Harker, D. *J. Chem. Phys.* **4**, 381-390 (1936).
25. Hughes, E. W.*J. Am. Chem. Soc.* **63**, 1737-1752 (1941).
26. Pauling, L. *J. Am. Chem. Soc.* **45**, 2777 (1923).
27. Pauling, L. and Kamb, B. *Am. Mineral.* **67**, 817-821 (1982).
28. Pauling, L. *The Nature of the Chemical Bond and the Structure of Molecules and Crystals.* Cornell University Press: Ithaca, NY (1939).

5. Emerging new methodologies in the imaging of nanostructures

David Sayre

David Sayre was educated at Yale and at Alabama Polytechnic Institute. He then went to Oxford and obtained a D.Phil. degree working with Dorothy Hodgkin. During his graduate work he studied new methods of solving crystal structures and contributed the 'Sayre equation' to the 'direct methods' now used to solve small crystal structures. Dr. Sayre went to IBM after working at MIT, the Office of Naval Research and the Johnson Foundation in Philadelphia (where he worked on the crystal structure of the carcinogen DMBA). At IBM he played an important part in the development of FORTRAN, the language still used by most scientific computer users. He has also had a great interest in the various ways in which the structure of matter can be viewed at high or fairly high resolution. He has pioneered methods for refining structures of large molecules. He was President of the American Crystallographic Association in 1983 and is the present Vice-Chairman of the U.S. National Committee for Crystallography.

It is easy for us to forget how young X-ray crystallography was when A. L. Patterson made his great contribution to the imaging of atom assemblies. Barely twenty years earlier it had still been possible to doubt whether such assemblies had objective reality, and even if it was assumed that they did, little indeed was known of their detailed arrangement. Then in 1911–13 came the *anni mirabili*: the elucidation of the general structure of the atom by Rutherford in 1911, the discovery of X-ray diffraction by Laue in 1912, and the use of X-ray diffraction to solve the first crystal structures by Bragg in 1913. Our story begins with these first crystal structures. Over the next years, as the atomic arrangements were clarified in more and more crystals, all doubt about the reality of structure at the nanometer level disappeared, and the greatest of the 20th century scientific voyages of discovery began.

After 1913, the 1930s brought the first major changes in structure determining technique. Basically, the very difficult step from diffraction pattern to structure was divided, through the work of Bragg, Robertson, and others, into two more manageable steps: from pattern to map, and from map to structure. The change greatly increased the power of the method, and where attention since 1913 had been directed mainly to inorganic structures, it now could be directed also toward organic and ultimately macromolecular structures. With the introduction of the electron-density map, crystallography also became genuinely an imaging technique. This of course was the period of Patterson's own contribution, in which he introduced the Patterson map which, although its interpretation may present difficulty, can always be calculated.

The 1930s was also the period of introduction of the second great technique for imaging the nanoworld, the electron microscope.

With the 1950s came the third major period of new methodology. The direct methods were largely introduced at this time; the arrival of the computer made possible the growth from small- to large-structure and from 2- to 3-dimensional imaging; and the first nucleic acid and protein structures were solved.

It appears possible that the 1980s will be a fourth major innovative decade. From the current work I have picked four examples for brief discussion. The first two have the general character of diminishing the labor and uncertainty associated with the diffraction analysis of structures. The third aims at increasing the information which can be extracted from a diffraction pattern, and the fourth aims at enlarging the class of specimens which can be imaged at the nanostructural level.

Entropy maximization[1,2]

Given the amplitudes and phases of some of the structure factors, the general practice since the 1930s in creating an electron-density map has been to use those structure factors as the terms in a partial Fourier series. Modern theory says that this technique, which is equivalent to declaring that the structure factors not included in the given set are all zero, is not the best practice, and recommends a more complex computation leading to what is called the maximum entropy map.

Without going into detail, such a computation invents amplitudes and phases for additional structure factors and adds this information to the given data before computing the Fourier series. Examples of conventional and maximum entropy maps are shown in Fig. 5.1. The superior interpretability of the maximum entropy maps is evident.

There is more, however. An analysis of the maximum entropy procedure shows that in simple cases it invents in close agreement with the direct methods of crystallography. For example, given a single strong structure factor, it invents the structure factor at double the distance from the origin, and gives it the amplitude and phase called for by the appropriate direct-method joint probability distribution. In more complex cases it invents in a manner which, it may be argued, is better than the normal direct methods. In support of this view, we may note that in the normal direct methods invention of a value for a new structure factor is based on the given values of only a small number of structure factors, namely those lying in what is called the neighborhood of the invariant being employed.[3] In addition, these values are inserted into formulae having a limited region of validity; if the given data lie outside this region, the formulae will be inaccurate. In the maximum entropy method all the given information is used in the prediction of values of new structure factors, and the method automatically recenters for the actual values of the data given.

Another important point is that the maximum entropy technique has recently been extended[1] to allow for a third category of structure factor for which amplitude but not phase is given. For this category, phase invention alone takes place. This form of the technique is particularly suited to the crystallographic situation. It turns out that the maximum entropy map is not unique in this case, and that phase-branching occurs, somewhat similar to that in MULTAN. However, unlike in MULTAN, the branching is not arbitrary, and the only branches generated are ones with a reasonable likelihood of being correct.

In view of the points stated, it is possible that maximum entropy may provide a more powerful direct method than we have yet known, and that before long the method will solve

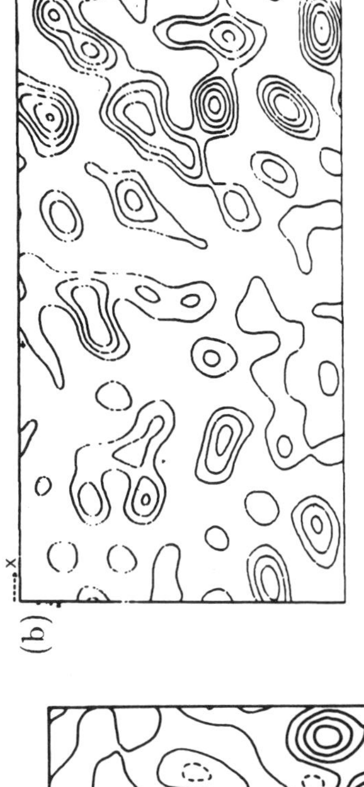

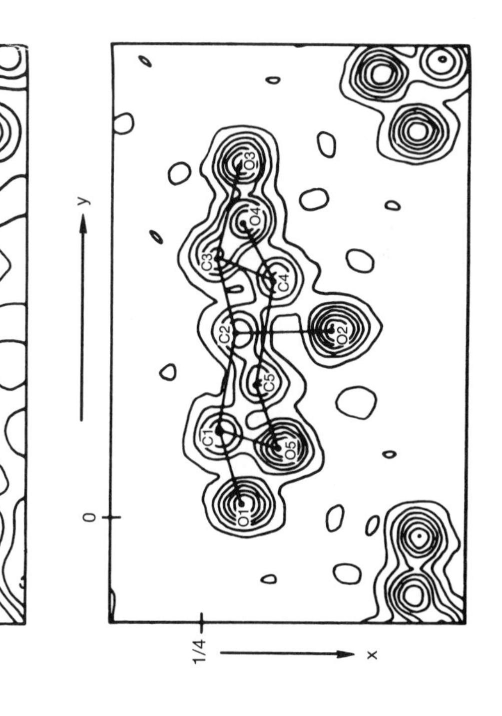

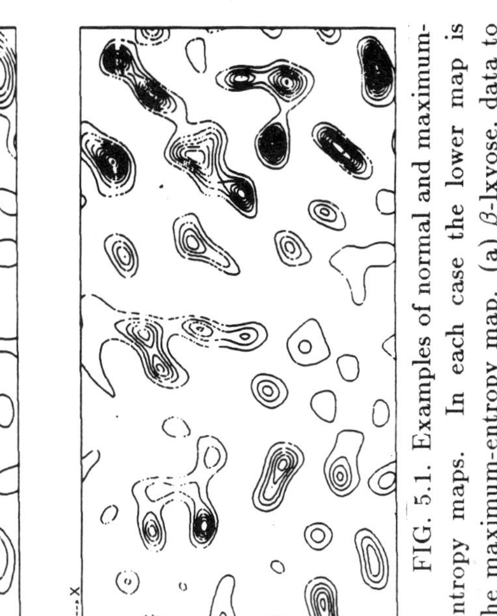

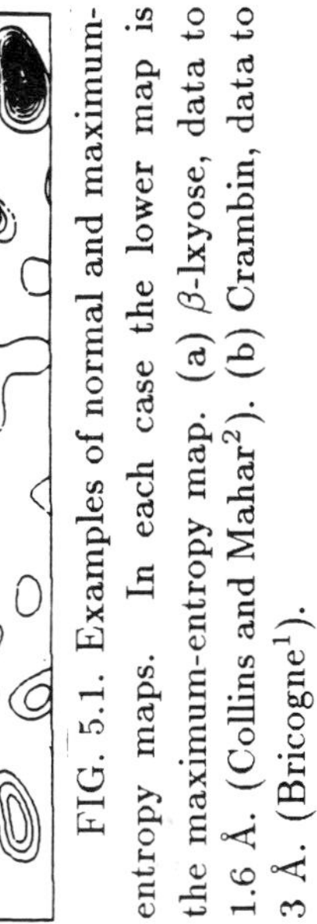

FIG. 5.1. Examples of normal and maximum-entropy maps. In each case the lower map is the maximum-entropy map. (a) β-lxyose, data to 1.6 Å. (Collins and Mahar[2]). (b) Crambin, data to 3 Å. (Bricogne[1]).

proteins from single data sets in much the same way that small structures are solved by direct methods today. In addition, in recent work (unpublished) Bricogne has extended the maximum-entropy formalism to the case of isomorphous-replacement and anomalous-dispersion data.

In summary, maximization of entropy is a form of direct method, but one based on a new and stronger probabilistic technique. Prediction of phases is more globally based, the prediction technique is correctly recentered for the structure factor values actually at hand, and branching of phasing paths is an intrinsic, not an arbitrarily added, part of the technique.

N-beam diffraction

The phasing of the diffraction pattern, a necessary step in the imaging of structures by X-ray diffraction, is carried out today mainly by direct methods in the case of smaller structures, and by isomorphous replacement, anomalous dispersion, or molecular replacement for large structures (with Patterson maps playing a major role in all three latter techniques). But for years there was a feeling among crystallographers that a simpler method than any of these might lie somewhere in the situation which exists when a crystal is so oriented in the X-ray beam that two reciprocal-lattice points lie on the reflecting sphere, *i.e.,* two sets of Bragg planes are in reflecting position simultaneously. In such a circumstance, it was felt, some effect sensitive to the phases of the two reflections should be observable. Attempts to see such effects, however, were without success.

Finally Post[4] correctly predicted and experimentally observed the rather subtle effect. It consists of an asymmetry which develops in the shape of a reflection peak when another reflection is simultaneously being generated. The details of the asymmetry depend on the value of the phase invariant which is based on the two simultaneously scattering reflections, *i.e.,* if they are at h_1 and h_2, on the sum of the phases at h_1, h_2, and $-h_1, -h_2$. Physically, the effect arises from the detailed mode-coupling in the crystal as it undergoes excitation of multiple modes in the X-ray beam. The effect is thus a physical analog of the direct-method triple product relation.

To date the effect has been used to phase reflections in simple centrosymmetric crystals[5] and more recently in simple non-centrosymmetric crystals.[6] Post believes that the effect should be observable in protein and other complex crystals as well, although this does not seem to have been reported to date. A special diffractometer is desirable for detecting the asymmetry,[7] but an ordinary instrument will often suffice for simple crystals.

At the present time there appears to be no definite reason why n-beam phasing should not in time be capable of phasing the diffraction pattern of any crystal, at no expense beyond the diffractometer time necessary for careful exploration of the diffraction intensity in regions near the n-beam crystal settings. The size of structure and the presence or absence of special scatterers would be immaterial, and the quality of phasing could be high and, being directly established by experiment, independent of structural assumptions. Patterns could either be phased entirely by the n-beam method or, because of the logical similarity between n-beam and direct methods, by a mixture of the two techniques. The potential advantages are thus extremely great. At the same time, however, it is too early to rule out major difficulties, especially with large structures where the mode-density is high and the mode-coupling may be correspondingly complex.

Even should the method prove of limited value, however, it is a superb concept that the diffraction intensity, when observed in critical regions in sufficient detail, specifies the diffraction phases. For 70 years the crystal has hidden its phases from direct observation, but now we know that by placing it in particularly testing situations it can be made, in principle at least, to reveal them to us.

Quantum modelling

We turn now to a very interesting recent paper on the structure of beryllium by Massa *et al.*,[8] in which a quantum-mechanically valid electron-density distribution is obtained from the X-ray data. The high-accuracy data were those of Larsen and Hansen.[9] A least-squares procedure was used to search a function space for best fit to the X-ray data, the function space being constructed of atomic orbitals for the beryllium atoms, plus thermal parameters; in this simple case the positions of the beryllium were fixed by symmetry. The result is a description of the state of each atom which is highly informative with respect to the valence orbitals of the atom in its bonded state, together with a quantum-mechanically valid wavefunction for the beryllium crystal as a whole. The squared magnitude of the wavefunction is in close agreement with the electron distribution found by Larsen and Hansen.

Now the principal point is that, given an experimentally measured wavefunction such as the above, one may in principle calculate all physical properties of the crystal. Massa *et al.* illustrate this for the beryllium crystal by calculating the electron-nuclear attraction energy and the electron kinetic energy. The diffraction experiment is a universal experiment, comprehending (at least in principle) the result of every other experiment that can be carried out on a crystal!

By thus placing X-ray analysis in a new light, this paper, it seems to me, breaks important new ground in structure/property relationships, suggesting that with a moderate alteration in technique X-ray analysis can play an even more central role than it now does in the study of the properties of matter.

Soft X-rays: the imaging of biological structures in approximately their natural state

At the present time, thanks to new radiation sources such as electron storage rings, heated plasmas, X-ray lasers, etc., soft X-ray photons (wavelengths 10 – 100 Å) are for the first time becoming available in plentiful numbers to scientists. This section discusses the new imaging opportunities offered by these particles. Briefly, they effect an important extension of the range of specimens which can be imaged at the nanostructural level. In particular among the newly-imagable types of specimen is the fragile, extremely important structure, the intact, even living, biological cell.

We begin by examining the basic properties of the soft X-ray photon which give it its special imaging power. There are

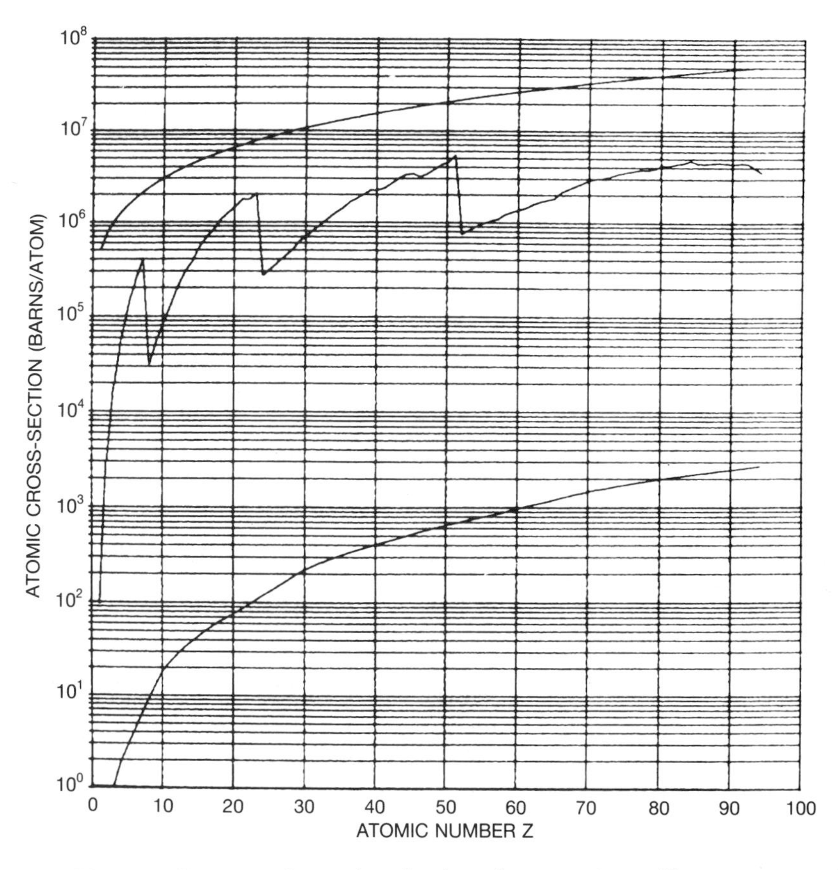

FIG. 5.2. Cross-sections of major imaging reactions. Top curve: total scattering, 100*keV* electrons. Middle curve: photoelectric absorption, 24 Å photons. Bottom curve: coherent scattering, 1.5 Å photons.

three principal properties:

1. The general level of its reaction cross-section with atoms, which at $10^5 - 10^6$ barns/atom is excellently matched to the thickness of the biological cell (Fig. 5.2). The electrons used in electron microscopy have a higher cross-section, which is well matched to the results of cell sectioning or other types of cellular disassembly, but not to the intact cell. The 1.5 Å photon as used in X-ray crystallography is best suited to macroscopic specimens.

2. The strong dependence of its reaction cross-section on Z in Fig. 5.2. With the mild Z-dependence of the electron reaction, the electron microscopist must remove the water from biological structures and add high-Z stain, to produce the contrast needed for imaging. With the soft X-ray photon, such invasive methods of obtaining contrast are not needed. The example shown, the 24 Å photon, images cell structures in the presence of water, due to its low reactivity with hydrogen and oxygen.
3. The fact that photons are bosons. Boson sources can have high brightness, and this in turn permits rapid imaging, which is a necessity with the living cell, which is in rapid motion at the nanometer scale. For example, the heated-plasma soft X-ray source permits imaging exposures in the nanosecond regime, making possible images of living cells with a resolution of a few nanometers (see below and Fig. 5.5).

Formation of the image with soft X-rays has been a problem, but is now being done by both microscopy and diffraction imaging. Early devices for focussing soft X-rays were grazing-incidence mirrors.[10] Today they are more commonly microfabricated Fresnel zone plates.[11] The resolving power of the latter devices stands today at approximately 500 Å, with 100 Å hoped for in the next few years. Normal-incidence mirrors using multilayer coatings are also under development.[12] The focussing elements are used in both imaging and scanning microscopes.[13] There are approximately a half dozen such microscopes in existence today, most of them installed at synchrotron radiation sources. Fig. 5.3 shows an example of the output of the Göttingen imaging microscope installed at the BESSY electron storage ring in Berlin.

At the present time the leading form of soft X-ray imaging is 'contact' imaging. The major reaction of soft X-rays with atoms is photon absorption. Therefore a structural feature in a specimen throws a shadow when illuminated with soft X-rays, which may be recorded by a detector placed close to the exit surface of the specimen. The detector must have adequate spatial resolution to record the shadows of the smal-

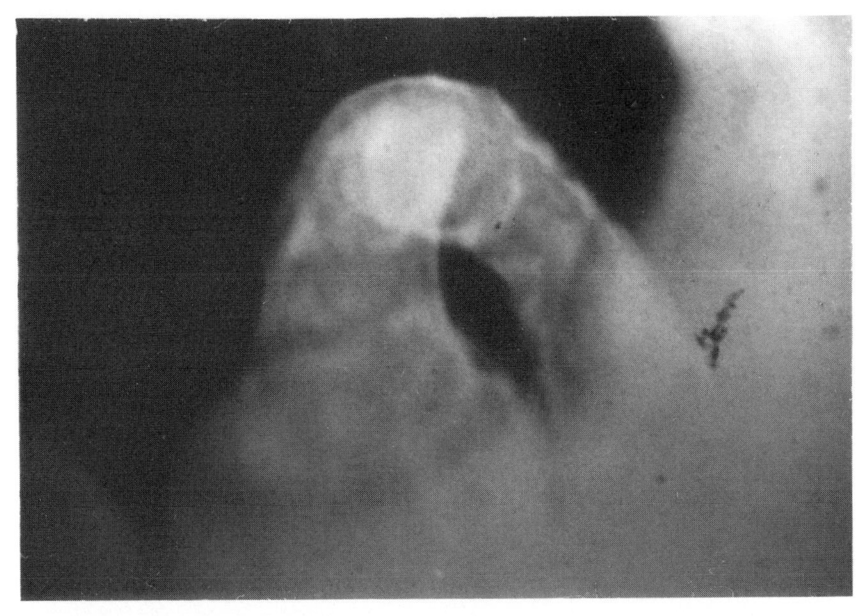

FIG. 5.3. Part of giant salivary gland chromosome from larva of *Chironomus thummi*. Göttingen imaging microscope at BESSY storage ring, Berlin. Wavelength 45 Å. Exposure 12 seconds. Size of field $8.7 \times 5.8\mu$ m.

lest features which it is desired to image. Soft X-ray detectors of the X-ray resist type[14] exist today with spatial resolution of approximately 50 Å. However, due to diffraction effects, it is more realistic to adopt $\sqrt{\lambda t}$ as the resolution of the contact technique, where λ is the wavelength and t is the specimen thickness. For 25 Å X-rays this gives approximate resolutions of 50 Å for $t < 100Å$, 160 Å for $t = 1000$ Å, and 500 Å for $t = 1\mu m$. Figs. 5.4 and 5.5 show images produced by the contact technique. Fig. 5.4 was taken with a fixed-target vanadium L soft X-ray source, the specimen being lightly fixed and critical-point dried to keep it motionless during exposure, but not otherwise processed. Fig. 5.5 was taken with a 100 ns pulse of X-rays from a high-intensity Z-pinch plasma source. The specimen was living as the start of the exposure. Details of approximately 75 Å are visible, suggesting that many structures within the cell had migrated to positions close to the cell surface in contact with the detector prior to exposure.

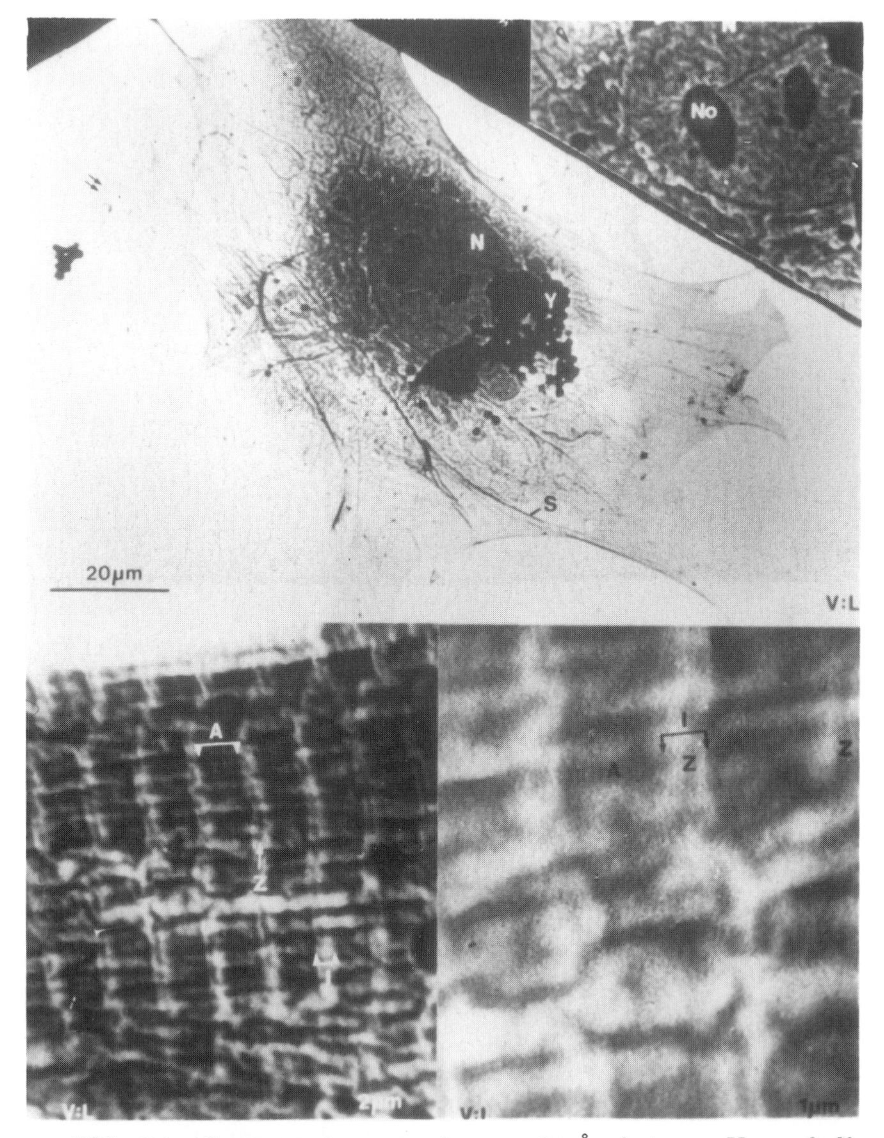

FIG. 5.4. Contact microscopy images, 24 Å photons. Upper half: *Xenopus laevis* fibroblast-like cell. Key: (Y) yolk granule, (S) stress fiber, (double arrow) X-ray dense cell margin, (N) nucleus, (No) nucleolus, (er) endoplasmic reticulum, (m) mitochrondria. Inset: higher magnification view, showing nuclear region. Lower half: *Xenopus laevis* muscle cell. Key: (A) A band of sarcomere, (I) I band, (Z) Z line. Inset: higher magnification view. (Figure courtesy P. C. Cheng.)

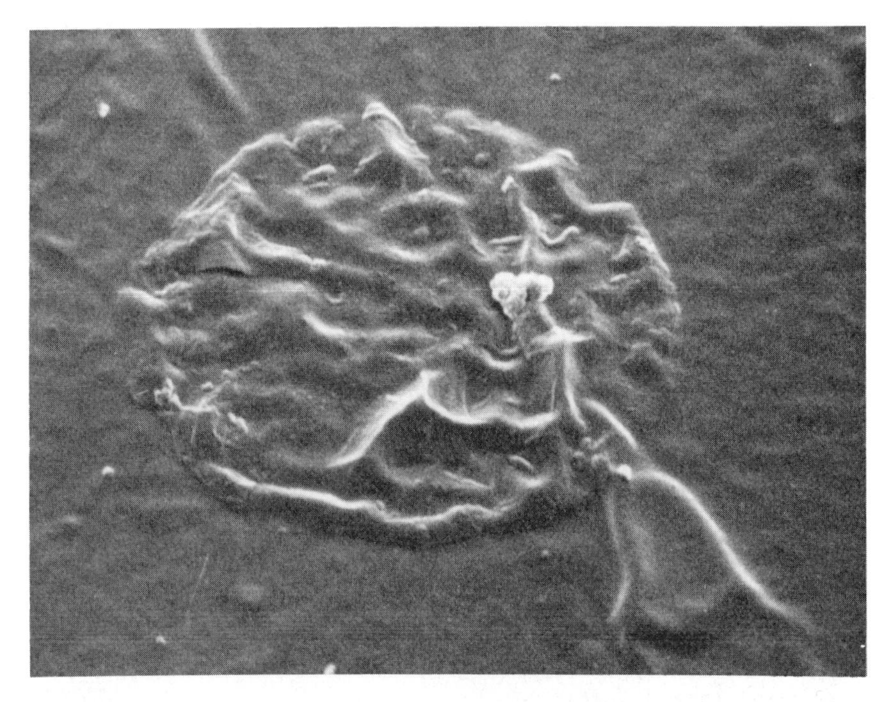

FIG. 5.5. High speed contact microscopy image of a living human blood platelet. The platelet was apparently putting out pseudopods at the moment of imaging. Made with a LEXIS II soft X-ray source, courtesy Maxwell Laboratories, Inc. Size of field $10.4 \times 7.7\mu$ m. (Feder *et al.*[15]).

The contact method of course is diffraction imaging in a simple form. As such it may be thought of as opening up, although as yet only in a crude way, a whole new 'microscopic' form of crystallography, in which the small crystalline specimen is replaced by the microscopically small, biologically more natural non-crystalline specimens of Figs. 5.4 and 5.5. If the implications of this are potentially vast, it must be added that if the contact method has solved the problem of imaging such specimens, it does not yet approach the image quality of X-ray crystallography. One of the principal efforts in soft X-ray imaging at present is to develop better diffraction-imaging methods with the aim of imaging in three dimensions and eliminating the loss of resolution with increasing thickness which occurs in the contact technique. We close our survey of soft X-ray imaging with a brief account of these efforts.

The principal problem in the diffraction imaging of a microscopic non-crystalline specimen is the weakness of the diffraction field set up by such a specimen. Thus one approach to the aim of high-quality imaging would be simply to follow the technique of X-ray crystallography, which involves measuring the intensity of the diffraction field of the specimen under far-field conditions. The experiment should be just feasible with a soft X-ray undulator source.[16] [Note added in proof: A diffraction pattern extending to spacings of approximately 300 Å has now been observed for a single diatom, using a bending-magnet soft X-ray synchrotron radiation source.] Another approach is holography, where low-resolution ($\approx 1\mu$ m) soft X-ray Fresnel holograms of microscopic objects have been recorded and

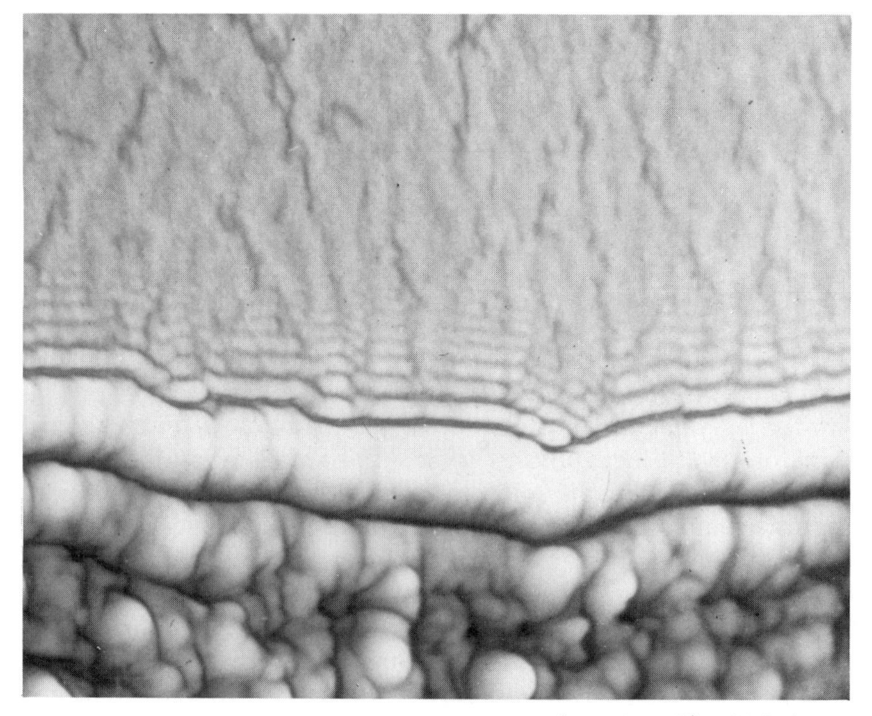

FIG. 5.6. Contact image of a specimen (muscle cell) which is not in good contact with the detector. The effects of diffraction are readily seen. Taken with well-monochromatized 27 Å radiation from the National Synchrotron Light Source, Brookhaven.

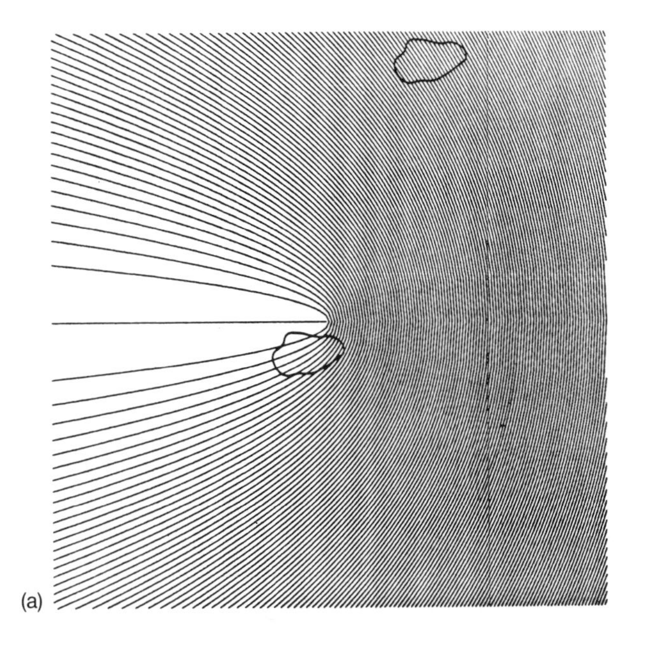

FIG. 5.7. (a) With monochromatic plane-parallel illumination coming from the left, and detection taking place at the focus of the family of paraboloids (parabolas in two dimensions), the paraboloids are the equal-path-length (modulo λ) surfaces for scattering. When the specimen is remote from the detector (upper outline) these surfaces are approximately planar and equidistant, explaining the use of Fourier summation as the basis of image reconstruction. Reconstruction of a specimen close to the detector (lower outline) is similar, but involves summation of paraboloidal fringe-systems. (b) Two cases of reconstruction of a test object from simulated diffraction data. On the left, the object was placed in different orientations in the beam, and in different positions relative to the detector, but the object-detector distance was always 10^3 wavelengths, and reconstruction was virtually Fourier reconstruction. On the right, the distance from the center of the object to the detector was only three wavelengths, and the fringes in the reconstruction were strongly curved. The similarity of the two reconstructions is evident. Each reconstruction has the real part of the object on the left, and the imaginary part on the right. The object was a question mark with a real head and an imaginary dot. The figure shows that near-field diffraction imaging is not fundamentally different from far-field imaging.

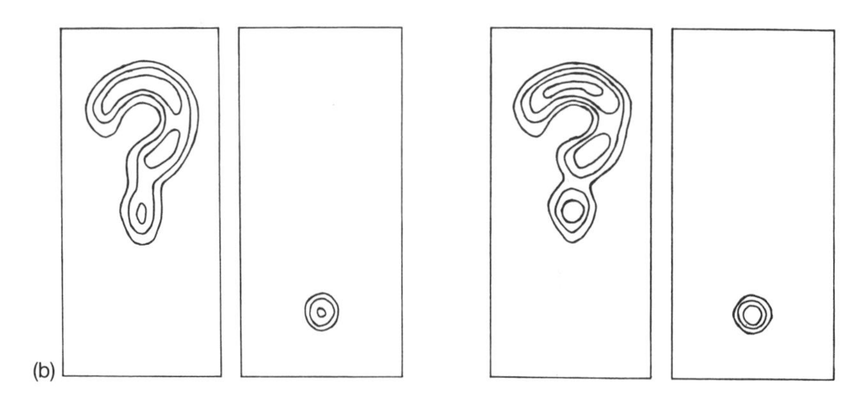

reconstructed.[17] Work is now turning toward higher resolutions. However, the difficulties encountered in high-resolution holography at optical wavelengths suggest that success here is also uncertain.

My own feeling is that the best opportunity lies in developing more sophisticated imaging from the near field. The diffraction nature of the field is clearly illustrated in Fig. 5.6. Basically, what is needed is a technique for phasing the field; assuming that this can be done, Fig. 5.7 illustrates a technique for reconstructing the image.[18] The interesting fact emerges that our familiar Fourier technique of reconstruction from the far field is a limiting case of a more general technique involving the summation of paraboloidal fringe-systems.

Summarizing, the soft X-ray photon offers fascinating opportunities for extending the imaging of nanostructures to new forms of living-system specimens and to new forms of diffraction imaging.

Conclusion

In this article I chose four areas of work for discussion, but might easily have chosen others. The new phasing methods based on anomalous dispersion,[19] the image processing methods introduced into protein crystallography by B. C. Wang,[20] the direct methods being developed by T. N. Bhat,[21] and the application of new spectroscopic methods of imaging[22] could equally well have illustrated the vitality of nanostructural imaging, more than 70 years after its inception. Seventy years is a long time in science, but I think that 70 more years will not exhaust the growth of this subject. The world of nanostructures just has too many fascinating and important things to look at.

Discussion

Ray Pepinsky: What effect does radiation damage here have at these wave lengths? What about secondary electrons?

David Sayre: As noted above, the principal reaction of photons at these wavelengths is absorption in an atom. When this occurs, the entire several hundred eV of photon energy is deposited in the atom, temporarily disrupting the electronic structure, and often altering the valence structure irreversibly. Some of the energy is carried away, as you suggest, by secondary (photo- and Auger) electrons, which spread further damage over a radius of some 50 Å.

Despite these facts. in most situations imaging by soft X-ray photons delivers less energy to the specimen than imaging by electrons. The damaging reaction in the electron case, inelastic scattering, typically transfers only 10 eV or so to the atom involved, but the low contrast of biological materials to electrons makes necessary a very large number of such events. The net result is a mild advantage of soft X-ray imaging over electron imaging in terms of specimen dosage.

David Harker: Why do you say entropy maximization which, in my mind, creates a picture of random blackness?

David Sayre: Entropy measures probability, and the maximum entropy map is the most probable map consistent with the observed data. Without observed data the map would be maximally featureless, as you say. With the observed data, it is far from featureless, as we see in Fig. 5.1. It is, however, as featureless as it can be made without violating the given data and without making the improbable assignment of zero to all unobserved data.

Jack Dunitz: The phrase, I think, comes from a paper by Jaynes.[23]

David Sayre: Jack is right. The methods introduced by Jaynes are gradually being adopted in a number of application areas including crystallography.

References

1. Bricogne, G. *Acta Cryst.* **A40**, 410 (1984).
2. Collins, D. M. and Mahar, M. C. *Acta Cryst.* **A39**, 252 (1983).
3. Hauptman, H. *Acta Cryst.* **A31**, 680 (1975).
4. Post, B., *Phys. Rev. Lett.* **39**, 760 (1977); *Acta Cryst.* **A35**, 17 (1979).
5. Gong, P. P. and Post, B. *Abstracts, American Crystallographic Association meeting (winter 1982)*, paper M6 (1982). Nicolosi, J., *ibid.*, paper M7. Han, F. S. and Chang, S. L., *ibid.*, paper P4.
6. Chang, S. L. and Valladares, J. A. P., *Acta Cryst.* **A40** C346 (1984). Hummer, K. and Billy, H., *ibid.*, C348.
7. Post, B., Nicolosi, J. and Ladell, J. *Acta Cryst.* **A40**, 684 (1984).
8. Massa, L., Goldberg, M., Frishberg, C., Boehme, R. F. and La Placa, S. J. *Phys. Rev. Lett.* **55**, 622 (1985).
9. Larsen, F. K. and Hansen, N. K. *Acta Cryst.* **B40**, 169 (1984).
10. Wolter, H. *Ann. Phys.* **10**, 94 and 286 (1952).
11. Schmahl, G., Rudolph, D., Guttmann, P., and Christ, O. In *X-Ray Microscopy* (G. Schmahl and D. Rudolph, eds.) p. 63. Springer-Verlag (1984).
12. Spiller, E., *ibid.*, p. 226.
13. Rudolph, D., Niemann, B., Schmahl, G., and Christ, O., *ibid.*, p. 192.; Feder, R., Houzego, P. J., Kern, D. P., and Sayre, D., *ibid.*, p. 203; Rarback, H., Kenney, J. M., Kirz, J., Howells, M. R., Chang, P., Coane, P. J., Niemann, R. B., *ibid.*, p. 217.
14. Spiller, E. and Feder, R. In *Topics in Applied Physics* (H. J. Queisser, ed.) p. 35. Springer, New York (1977). There are also electronic detectors which can be used to give a real-time form of contact imaging; see Polack, F. and Lowenthal, S. In *X-Ray Microscopy* (*op. cit.*), p. 251.
15. Feder, R., Banton, V., Sayre, D., Costa, J., Baldini, M., and Kim, B., *Science* **227**, 63 (1985).
16. Sayre, D., Haelbich, R. P., Kirz, J. and Yun, W. B. In *X-Ray Microscopy* (*op. cit.*), p. 314. The recording of diffraction patterns from microscopic crystalline specimens with synchrotron radiation is becoming well established; see Bachmann, R., Kohler, H., Schulz, H. and

Weber, H. P., *Acta Cryst.* **A41**, 35 (1985). For a report of the single-diatom pattern, see Yun, W. B., Kirz, J. and Sayre, D. submitted to *Acta Cryst.*

17. Aoki, S. and Kikuta, S. *Jap. J. App. Phys.* **13**, 1385 (1974). Howells, M. R. In *X-Ray Microscopy* (*op. cit.*), p. 318.
18. Sayre, D. Unpublished. (See *Abstracts , 1985 Meeting of the American Crystallographic Association.* paper M6 (1985)).
19. Hauptman, H. *Acta Cryst.* **A38**, 289 and 632 (1982); Karle, J. *Acta Cryst.* **A40**, 366, 374 and 526 (1984).
20. Wang, B. C., *Acta Cryst.* **A40**, C12 (1984).
21. Bhat, T. N., *Acta Cryst.* **A40**, C15 (1984).
22. Hutchison, C. A., Jr. and Singel, D. J., *Proc. Natl. Acad. Sci. USA* **78**, 6883 (1981).
23. Jaynes, E. T. *Phys. Rev.* **106**, 620 (1957).

6. Chemical information from crystal structures

Peter Murray-Rust

Peter Murray-Rust was a Visiting Scientist at the Institute for Cancer Research in 1982. He is a person who has many of the characteristics that have been described of Lindo—his love of ideas and great interest in his particular science, his chemistry. Peter was educated at Oxford, where he obtained a first-class degree. He then went to the University of Ghana, in Africa, where he was Assistant Lecturer from 1963 to 1965. Then he returned to Oxford and he obtained a D.Phil. with C. K. Prout. He took a position at the University of Stirling, Stirling, Scotland where he helped set up a very active Chemistry department. During that time he spent a sabbatical leave with Jack Dunitz in Zürich, Switzerland and here at the Institute for Cancer Research. He is now working at Glaxo Laboratories outside London. His interests have always lain in the correlation of structure with chemical and other properties, and the use of data bases with their vast store of structural data, such as is to be found in the Cambridge Crystallographic Data File. He is a pioneer in the study of the environment of groups in crystal structures.

I have a very daunting task because what I want to discuss are two of Lindo Patterson's legacies. The first is the one that he has left to chemistry, which is certainly not appreciated by most modern chemists, and the second is the legacy that he has left to Fox Chase, of which I was made vividly aware when I came here, and was communicated by Betty and Jenny in such a marvelous manner; his memory lives on here, both in the memory of him as a person, and in the science which he has initiated here.

I never met Lindo Patterson so I'll talk about the first time I met the Patterson Function. In 1951, when I was 10, Britain was recovering from the war and, as a mark of hope, it set up a festival on the south bank of the river Thames, and this was to show hope for the future. It was to show a combination of Britain's artistic, creative and also scientific achievements. Helen Megaw suggested that one of the things which the festival could do, was to combine art and science, and to produce fabrics which represented crystallographic designs. These very beautiful designs, which were wall coverings, curtains, dress materials, wrapping paper, can be seen in the Science Museum in Britain. So the first Patterson map I ever saw was one which was used by the British people to wrap their parcels in. Also in that Museum is shown the first Fourier synthesis done by Kathleen Lonsdale on hexamethylbenzene.[1] Along with that structure is a letter by the great physical chemist, Christopher Ingold, where he makes a very humble statement: 'The calculations must have been dreadful, but one paper like this makes more certainty into organic chemistry than generations of activity by us professionals.' And that I would like to leave as a memorial to Lindo Patterson's contribution to chemistry.

Let us try and get some idea of what this meant 50 years ago, because modern chemists don't appreciate it. Fig. 6.1 shows a typical chemical formula;[2] the derivation of formulae like this was one of the the great intellectual achievements of all time because chemists who deduced formulae like this had only a thermometer, a Bunsen burner, a beaker, a distillation column, a polarimeter possibly and one or two other pieces of equipment like that. To produce structures like this was a

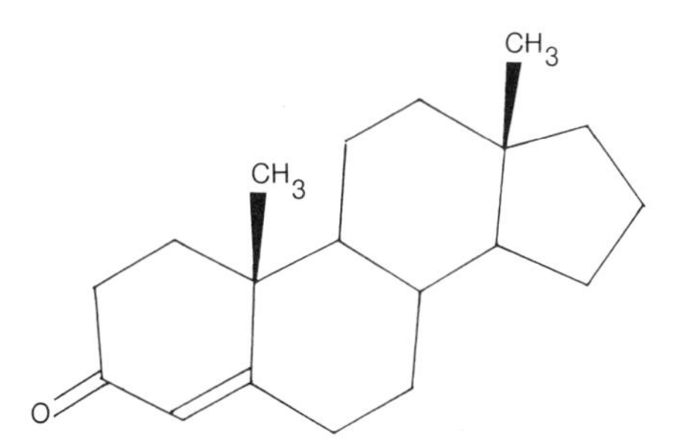

FIG. 6.1. Formula of a steroid (testosterone)

tour de force **that is not really appreciated today. However, this example is probably a little ironic because this is one of a few structures that the organic chemists did not get right. They needed a little help from the crystallographers. But to produce it required a vast amount of intellectual commitment, hard work in the laboratory, and some very careful thought.**

When the crystallographic method came along it became possible to produce pictures such as that shown in Fig. 6.2 of a

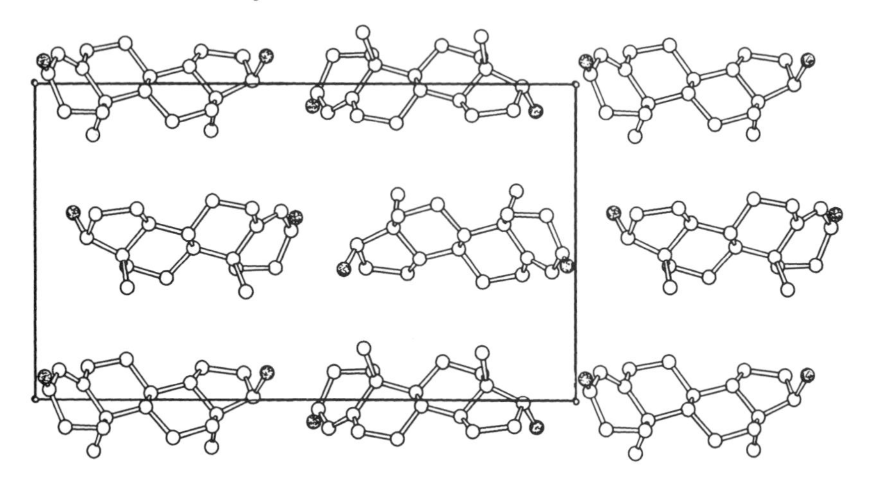

FIG. 6.2. Crystal structure of epitestosterone[2] viewed down an axis.

steroid in a crystal. Here you can see nine individual molecules and you can see, at a glance, the richness of the information that the crystallographic method provides. I want to emphasize that you can observe, as if in the microscope, all the bonds between the atoms, you can work out the formal bond order, you can work out where the hydrogen atoms are, you can work out stereochemistry, you can work out the relationship of one molecule to another. All of that is produced by the X-ray experiment. You get all that information whether or not you're interested in all of it. So an X-ray experiment produces such a richness of information, that I think it must have taken the chemists totally by surprise.

The information in Fig. 6.3 is very familiar to all of us in that it represents the explosive growth that is happening in all scientific subjects today. For many years after the Patterson function in 1934[3] very few crystal structures were available to the average organic chemist. In about 1965 several things came together. The first was the ability to remove the dreadful calculations from the human being to the machine. The second was the development of additional techniques for solving the phase problem, besides the Patterson method. The third was the availability of accurate and reliable instrumentation for collecting X-ray data. All those came together at this stage, with the result that the number of crystal structures determined took

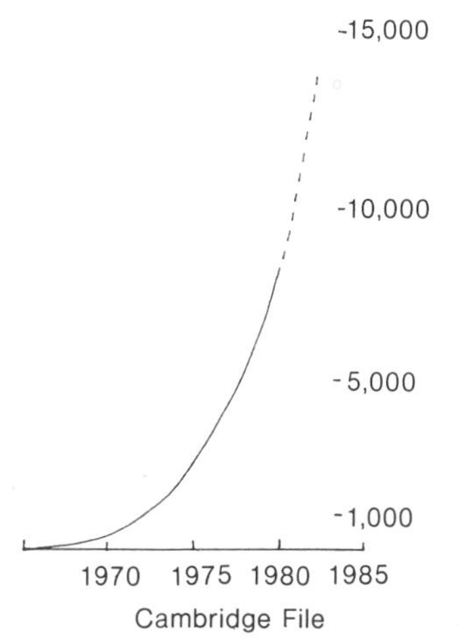

FIG. 6.3. Plot of the number of accurate crystal structures determined, reported, and listed in the Cambridge Crystallographic Data File, as a function of time.

off explosively. Even so, many chemists did not rush to use the X-ray method, because they had invested much time and effort in learning the skills of determination of organic structures by classical means. Professor Cotton had to berate them here, and point out that even though they regarded X-ray crystallography as an unsporting method, they should still use it, and that the other methods of spectroscopy, and degradation, were really superceded by the crystallographic methods. So we have this tremendous growth in the number of structures, and the information obtainable in them.

The second thing that we see with the introduction of new instrumentation is the increase in the accuracy of structure determinations. Structures determined in the 30 years since the Patterson function, had rather large errors because of the difficulty in collecting enough accurate data. There was, of course, pioneering work done by many people, particularly the Glasgow group, but it wasn't until about 1970 that accurate X-ray data became available to everybody, whether or not they appreciated its effects.

We now can get very accurate molecular geometry; at least as accurate as most other physical techniques can provide. We get just as accurate information on intermolecular geometry. We get information on deformation density, which Jack Dunitz will be describing, and we get a lot of information, now, about the atomic and molecular motions in the crystal. I want to concentrate on what chemical information we can get out of these first two fields of information.

A typical, very accurate, modern crystal structure, done by a Dutch group[4] is shown in Fig. 6.4. It was reviewed by two Italians that I've collaborated with — Aldo Domenicano and Sandro Vaciago[5] — who realized that chemical information could be obtained by looking at the accurate structure of molecules containing benzene rings. The structure first of all shows that the rings are not regular hexagons. The angles in the rings may be as much as 3° away from 120° (probably with an error of only 0.1°).

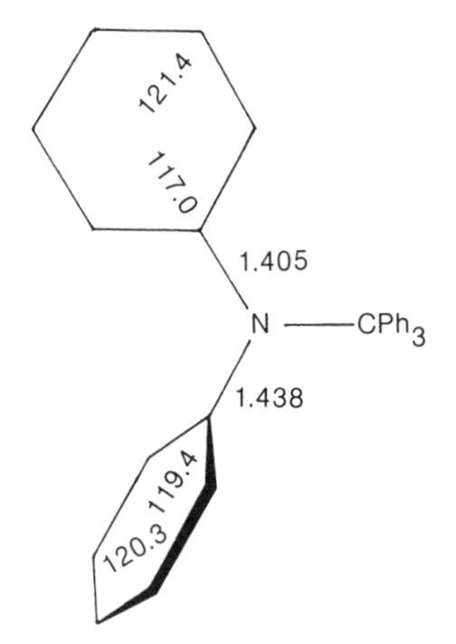

FIG. 6.4. Asymmetry in experimentally determined benzene ring geometry.[4,5]

The second point is that although the rings are chemically equivalent, since they are attached to the same nitrogen atom, they do not have the same geometrical disposition to it. One ring is in the plane of the paper, one lies end on; so the corresponding torsion angles are very different. What Aldo Domenicano and Sandro Vaciago showed was that this information could be used to say something about the electronic properties of the substituent attached to the benzene ring, using the ideas

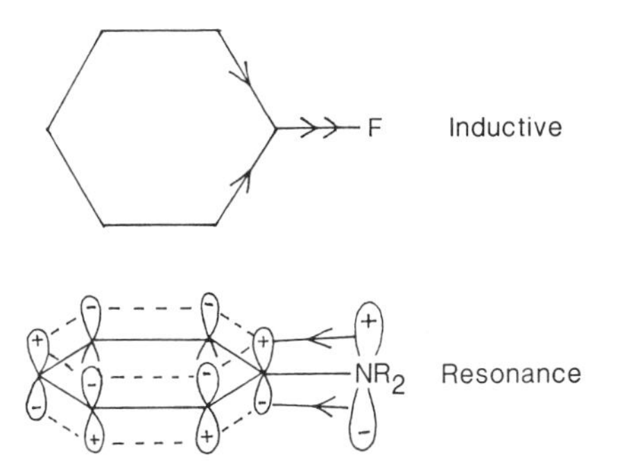

FIG. 6.5. Inductive and resonance effects in benzene derivatives.

of, among others, Christopher Ingold. In Fig. 6.5 we see that the electronegativity of the benzene ring substituent enables it to abstract electrons from a system, thereby distorting the angles around that atom. The 'resonance effect', where electrons are pumped back into the π system of the benzene ring is also shown. By using the geometrical data obtainable from X-ray crystallography, we can show that the geometrical measures of these two effects (which are widely used by organic chemists to talk about the chemical nature of these substituents) are just as good as the measures produced by reaction rates or the pK_a values for the various substituent groups. So a detailed study of the geometry of a molecule leads to an understanding of the chemistry of that molecule.

One of the current problems facing crystallographers is illustrated in Fig. 6.6. This is the amount of discussion per

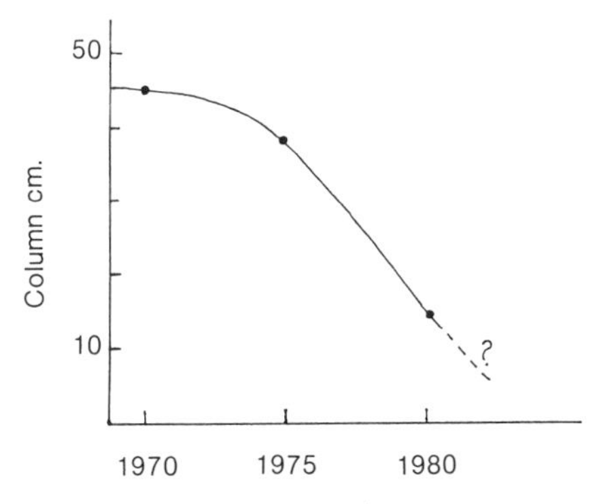

FIG. 6.6. Amount of discussion (in column cm per structure) as a function of year in the *Journal of the American Chemical Society.*

crystal structure that the editor of the *Journal of the American Chemical Society (JACS)* allows, in column centimeters. As you can see, although structures get more interesting, the editor gives us less and less space to write about them, and about now (1984), one does not get anything to say in *JACS*. We have a very real information crisis on our hands, and we need to look to the founding fathers of crystallography for their way out of it.

I would recommend to all research workers who feel that their supervisors are pushing them too hard W. L. Bragg's maxim[6]

> The important thing in science is not so much to obtain new facts, as to discover new ways of thinking about them.

The information crisis was brilliantly recognized, again by one of the fathers of modern crystallography, J. D. Bernal, who twenty years ago made the following remarkable statement,[7]

> However large an array of facts, however rapidly they accumulate, it is possible to keep them in order and to exact from time to time digests containing the most generally significant information while indicating how to find those items of specialized interest. To do so, however, requires the will and the means.

Bernal used his own field of crystallography to put this into practice. With the will and the means of Olga Kennard,[8] they initiated a project to archive crystallographic data in Cambridge, UK. Recall that in 1965 everybody knew every structure in the literature in great detail. To put these into an archive seemed an absurd thing to do. Well, what happened? We have seen how the size of the Cambridge Crystallographic Data File has increased year by year. It now has 45,000 organic structures on it, and, most remarkably, they are all checked. It is, in my view, the best scientific data file in the world, and serves as a model for people in other disciplines to follow. I think that crystallographers as a whole can be extremely proud of this, not just as a crystallographic tool, but as an example to other sciences. The file contains over a million atoms. That is a huge amount of information, and we have to try and find ways of looking at that information, and communicating it not just to crystallographers, but also to chemists.

The Data File was set up in a very farsighted way in that not only crystallographic information, but also chemical information, was put into it. Fig. 6.7 shows the way in which a

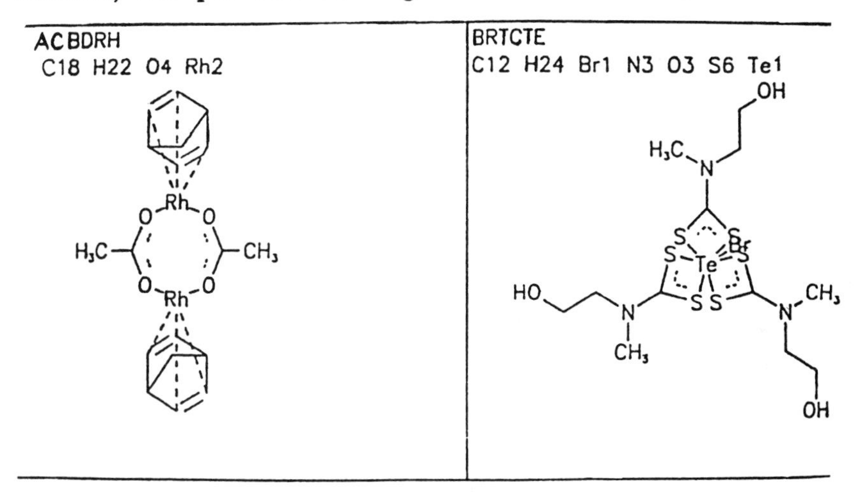

FIG. 6.7. Computer view of two chemical formulæ found from the Cambridge Crystallographic Data File.

chemical formula, as a chemist would understand it, is represented in a computer machine-readable form. As a result, any chemist can, with these diagrams, understand exactly what the chemical nature of the molecule is, and they can use this information to find the molecules that they are interested in. Allied to that is also the crystallographic information (the cell parameters, the space group, and symmetry, the atomic coordinates), those quantities that are required for the reconstruction of the full three-dimensional structure of a crystal; and, very importantly, we also have flags on the accuracy, and comments on the reliability of the structure.

I want to try to give an idea of how this tool can be used. I am going to illustrate it with on-going problems at Fox Chase, which I worked on when here. A very important part of the crystallographic file setup is shown in Fig. 6.8; it illustrates the fact that one does not need to know things like the arcane

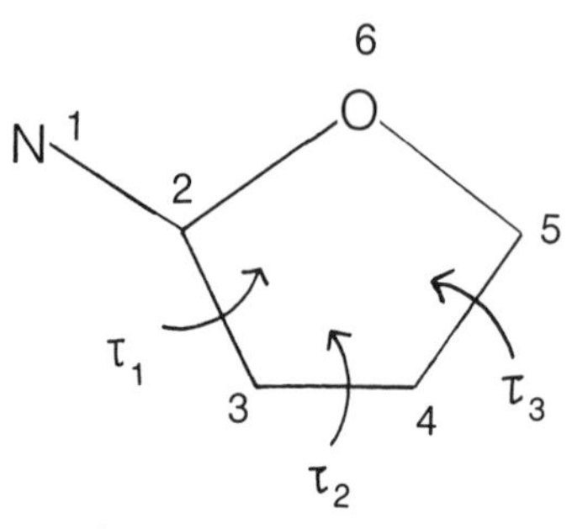

FIG. 6.8. Sugar pucker in nucleosides. Information obtained from the crystallographic file set-up of the Cambridge Crystallographic Data File.

REFCODE	C–N	τ_1	τ_2	τ_3
CDPCHM	1.550	19.6	-35.4	39.5
CTBGLU	1.485	29.9	-38.1	34.3
CYTCYP20	1.479	16.2	5.4	-24.4
CYTIAC	1.475	36.6	-38.8	28.1
CYTIAC01	1.485	34.6	-37.5	27.9

numbering of the steroid molecule or the nucleoside to be able to use the File. The software, written by Sam Motherwell, allows one to input a chemical fragment, and to match that up with all other occurrences of such molecules in the File (without having to know how the authors might or might not have labelled these). I want to take this opportunity to thank the Fox Chase Cancer Center for giving me the computer resources and the encouragement to be able to install the Data File and the software here, and to be able to use it in an efficient way; I would particularly like to thank Bob Stodola and his group in the computing unit, and to congratulate the Center on its marvelous computing setup.

Many of the problems I will discuss stem out of Lindo Patterson's involvement and desire to understand the citric acid cycle. Among the compounds involved is acetyl CoA, which contains a thioester group. David Zacharias and Jenny Glusker had done the structure of two thioesters. It is biologically a rather unusual group (Fig. 6.9(a)) and we wanted to see if we could understand something about it from the geometry to be

(i) alkyl thioesters:

$$R-\underset{\underset{O}{\|}}{C}-S-C(sp^3),$$

(ii) alkyl vinyl/aryl sulfides:

$$C(sp^2)-S-C(sp^3),$$

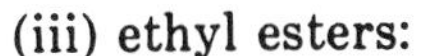

(iii) ethyl esters:

$$R-\underset{\underset{O}{\|}}{C}-O-CH_2-CH_3,$$

(iv) alkyl vinyl/aryl ethers:

$$C(sp^2)-O-C(sp^3),$$

(a)

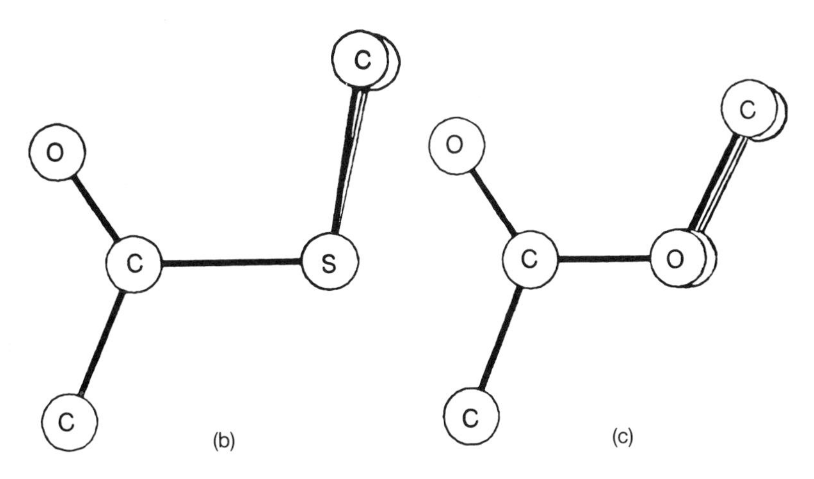

FIG. 6.9. (a) Strategy for determining the amount of resonance in thioesters. (b) Superposition of the averaged alkyl vinyl/aryl sulfide on an averaged thioester. The idealized groups are aligned along the C–S bond and are superposed at C. (c) Superposition in an analogous fashion, of an averaged alkyl vinyl/aryl ether on an averaged ethyl ester. Note the difference in the O (oxygen) position but not the S (sulfur) position. The atomic diameter is 100 e.s.d.

obtained from a crystal structure analysis. It had been claimed from other methods, spectroscopic and chemical, that, whereas in esters there is conjugation in the single C–O bond, resulting in a strengthening of the bond and making it harder to break, in the carbon-sulfur case this conjugation (resonance) is considerably less. One of the ways in which one can measure the degree of resonance is to measure the bond length; the greater the amount of resonance, the greater the shortening that one would expect in the bond length. So what we did was to search for a large number of esters on the Cambridge Crystallographic

Data File, and a large number of ethers in the same way. We then compared the C–O bond lengths in the ester with those in the ether, and we found a shortening of about 0.13 Å. We also did this for the sulfur-containing compounds, and found that the shortening of the C–S bond between thioesters and thioethers was considerably less, only about 0.03 Å, which correlated very nicely with the chemical evidence on the greater reactivity of the acetylCoA, as opposed to normal esters. In Fig. 6.9(b) is shown the superposition of thioesters on thioethers, and Fig. 6.9(c) shows the superposition of the ordinary esters on ordinary oxygen-containing ethers.[9] It is clear that the two types of oxygen-containing compounds do not quite overlap, so that while the effects are very small the accuracy of the X-ray method is so high that the effect that we can observe is extremely significant. Thus, we have the possibility of getting a lot of chemical information from our highly accurate structures.

The second area considered here, one being studied here by Bud Carrell, is the study of steroid molecules. I would also like to thank Bill Duax and his colleagues at Buffalo, who helped me start the study of the variation in the accurate geometry of the steroid nucleus. We were able to find, on the Crystallographic Data File and elsewhere, 15 progesterones whose structures were known with a degree of accuracy greater than 0.01 Å in the bond lengths.[10] These structures can be compared by superimposing them on each other as well as possible by the least squares method. The steroid rings superimpose beautifully when looked at from on top; but when looked at from the side, they do not fit so well. This is a real effect; we must try and find out how to describe that effect, and also try and work out what it is due to. So, although this is a very small effect compared with the overall size of the molecule, we need techniques that will enable us to increase the signal-to-noise ratio of our data. These techniques are well known in the social sciences, and also in the medical sciences, where one is concerned with a lot of observed data and with a number of different independent effects all operating at the same time.

The method which I've used for this is called 'factor analysis' or 'principal component analysis.' Fig. 6.10 shows, in a

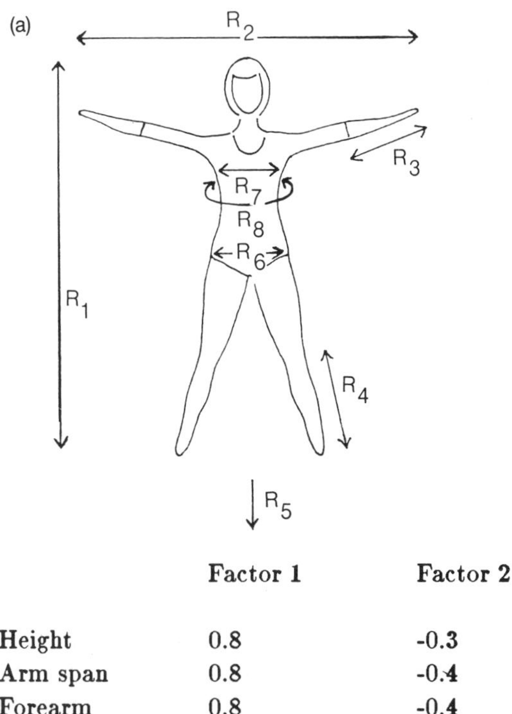

(b)

	Factor 1	**Factor 2**
Height	0.8	-0.3
Arm span	0.8	-0.4
Forearm	0.8	-0.4
Lower leg	0.8	-0.3
Weight	0.7	0.6
Hip	0.6	0.5
Chest	0.6	0.5
Torso	0.6	0.4
Fraction of variance	0.55	0.20
	Body growth	Body type

FIG. 6.10. Factor analysis of body shape. (a) Measurements. (b) Analysis showing two major factors (body growth and body type). Factor analysis for 307 girls (for dress shape).

rather simple way, how such an analysis works. This is a real study done with teen-age girls (not by me)[11] where eight parameters describing size were measured, and a factor analysis was used to analyze these data. One could describe them by only two independent effects, if one hoped to explain three-quarters of the observed variations in these girls. That is the sort of information that is very useful to dress designers, because the first factor shows that all these parameters covary together, so that the height and the arm and forearm all increase at the same rate, which we can call body growth. The second factor shows that we have a negative correlation between such things as height and chest, so that this represents the overall shape.

Fig. 6.11 shows a picture in the Albright-Knox Art Gallery in Buffalo, New York. To quote Maurits Escher, 'this is a static method of illustrating a dynamic fact'. If we apply a factor analysis to the steroid molecules, as shown in Fig. 6.12, we come out with two factors. The first factor represents a distortion of the molecule. This shows that the range of variation of

FIG. 6.11. Movement in a dog illustrating the dynamic effects. Gaicomo Balla. ***Dynamism of a Dog on a Leash*** (1912) Albright-Knox Art Gallery, Buffalo, New York. Bequest of A. Conger Goodyear and Gift of George F. Goodyear, 1964. Reproduced with permission.

one atom, and the range of variation of a second atom, are in phase. So that when one atom goes up, the second atom goes up. Fig. 6.12 shows a second mode of distortion of the steroids that accounts for another 28% of the variance. What we are seeing here then, is a suggestion that the molecule can deform, and that this deformation has got something to say not just about the statics in the molecule, but also its dynamics. We were able to make a film in which we took these distortions, applied a sinusoidal variation to them on a high-performance graphics machine, and used the result to suggest that what we are seeing here, in the crystal, is frozen out molecular vibrations. Also included in the film are studies of deformations in β-loops[12] in peptides, and deformations of the porphyrin ring.[13]

It is now necessary to discuss the relationship between structure and energy. Escher's beautiful represention of the difference between a molecular crystal structure, and the isolated molecules in the gas or liquid phase is shown in his 'Reptiles',

(a)

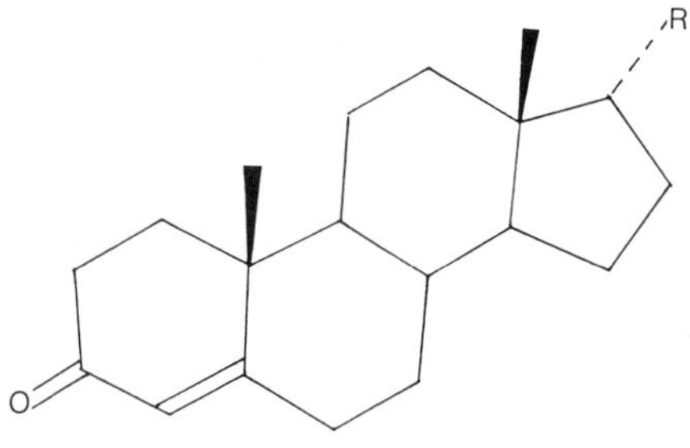

R

O

(b) $R \leq 9\%$. $\sigma(C–C) \leq 0.01$ Å.

20 atoms	r.m.s. per Cartesian	0.05 Å
out-of-plane	r.m.s. variation	0.07 Å
in-plane	r.m.s. variation	0.015 Å

Factor	% variance	(total)	r.m.s./Cartesian
1	47%		0.03 Å
2	28%	(75%)	0.025 Å
3	9%	(83%)	0.015 Å

FIG. 6.12. Results of a factor analysis of 15 Δ-4-en-3-ones.[10] (a) Formula of the Δ-4-en-3-ones studied. (b) Data from the factor analysis.

(c)

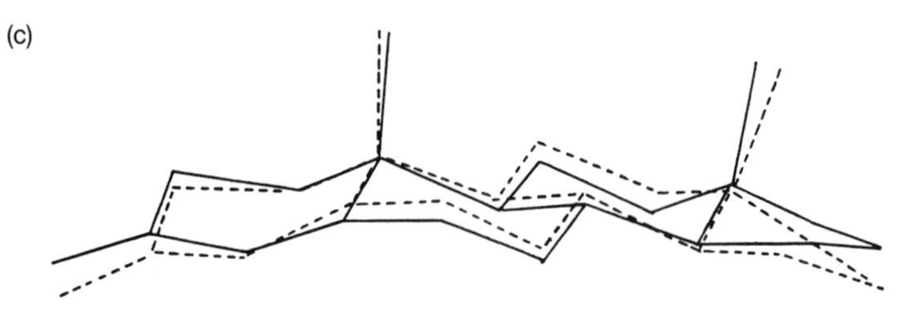

Factor 1 (47%)

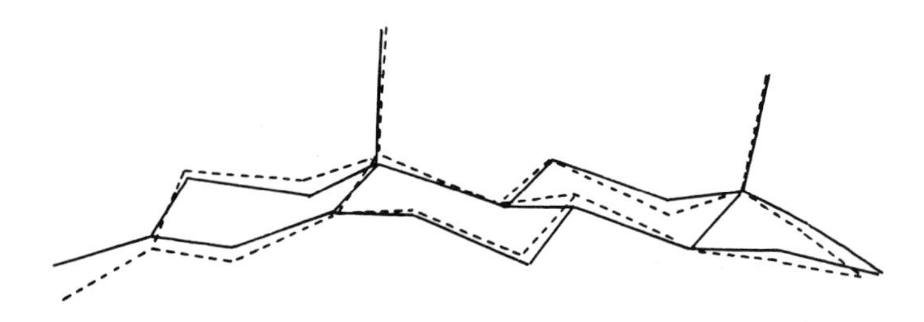

Factor 2 (28%)

(c) Factors 1 (47% of total) and 2 (28% of total). For clarity the two limits of each factor are drawn as solid and broken lines respectively.

shown in Fig. 6.13. He describes it as 'an interplay between the stiff crystallized figures of a regular pattern, and the individual freedom of creatures capable of moving about in space without hindrance.' You therefore pay a price if you crystallize something. You reduce the ability of the molecule to explore the particular conformations and shapes which it would take up in a less ordered phase, which Escher realized. I have done a survey of Escher's work and show in Fig. 6.14 that, in fact, to crystallize these different molecules, we do have to distort them considerably from a symmetrical geometry.

Fig. 6.15 shows a crystal structure I did for a colleague who had no idea what the compound was. The result contains a wealth of other information, and I want to highlight the ethyl ester group,[14] which takes up, as shown in this figure, not the extended conformation that one might expect as lowest energy conformation, but a gauche conformation of about 90° degrees.

I was surprised about this, and when I did the structure I had no idea whether this was an unusual fact or not. How would you go to *Chemical Abstracts* and say 'I want to look up ester groups which are bent in this sort of way'? There's absolutely no way of doing it. The way in which we now can do this is to use the Crystallographic Data File. You can go to the Data File and ask it 'I want to see what the torsion angles are in all ethyl esters.' In Fig. 6.16 is shown the answer that was obtained; it is a histogram showing the frequency of observation of different torsion angles. You can see, in this figure, as you would expect, that the extended form, in which the torsion angle is 180°, has the maximum frequency. But there is also a significant peak at about 85°, which is exactly what was seen in that crystal. So we are first of all able to satisfy ourselves

FIG. 6.13. Maurice Escher's 'Reptiles'. A view of the crystallization process. Reproduced with permission. ©M. C. Escher Heirs c/o Cordon Art – Baarn – Holland.

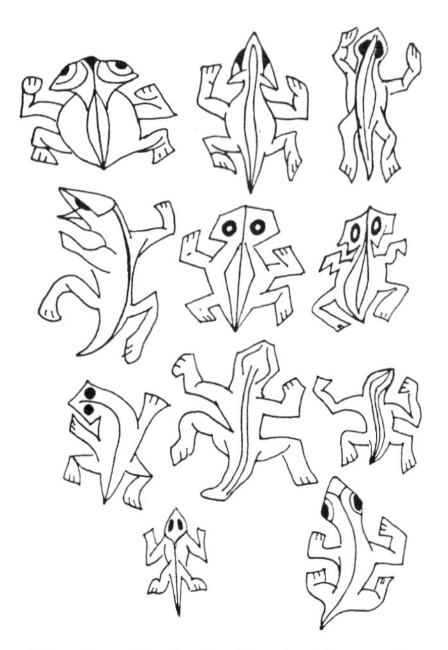

FIG. 6.14. Maurice Escher's illustrations of various reptiles. This shows the conformational variability possible.

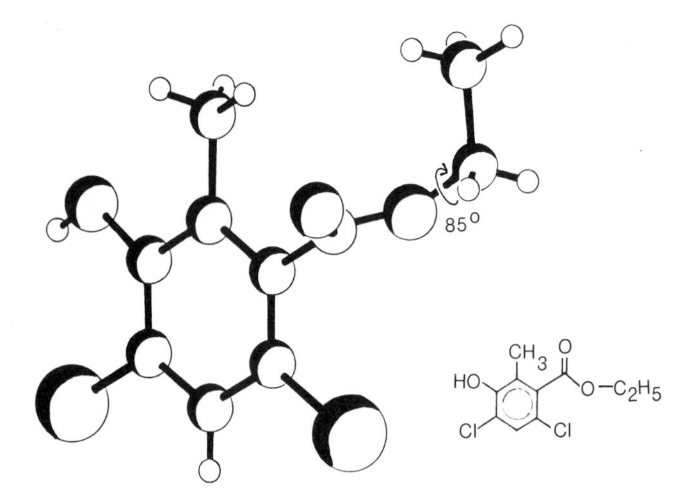

FIG. 6.15. Crystal structure of an ethyl ester.

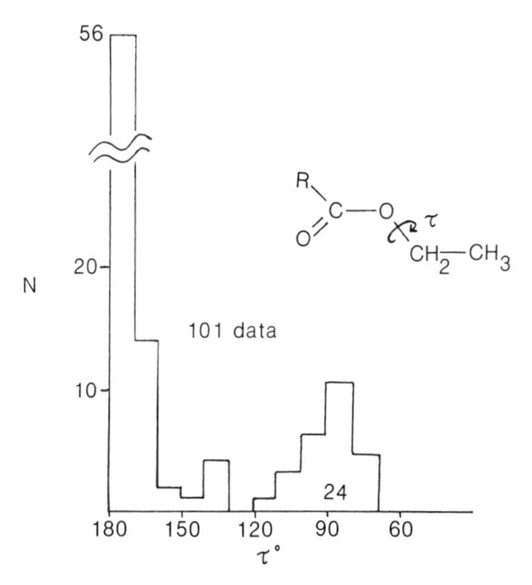

FIG. 6.16. Histogram illustrating the variation in torsion angles in ethyl esters in the Cambridge Crystallographic Data File.

that this is not an unusual phenomenon; but more than that, it is a very suggestive diagram, because what we are seeing here could be described to the chemist as a molecule having two conformations, one of which is in a gauche conformation, and the other of which is in an extended conformation. One might feel, therefore, that the frequency with which one observes a conformation in a molecule is in some way related to the ease of getting that molecule into that conformation, and that ease is measured by the chemist as an energy parameter. Very remarkably, and very beautifully, there is a superb piece of gas phase microwave work on ethyl formate, done by E. Bright Wilson,[15] shown in Fig. 6.17. Wilson has worked out the potential energy curve for the rotation about the C–C bond in the gas phase. We get a very clear correlation between the frequency of an observed structure and the energy corresponding to that particular geometrical configuration.

Now I want to consider not observations of an isolated molecule in a crystal, but the way in which a molecule interacts with its neighbors in the crystal, because that is a most remarkable piece of information that we get from the X-ray experiment. We see molecules interacting, and we are able to

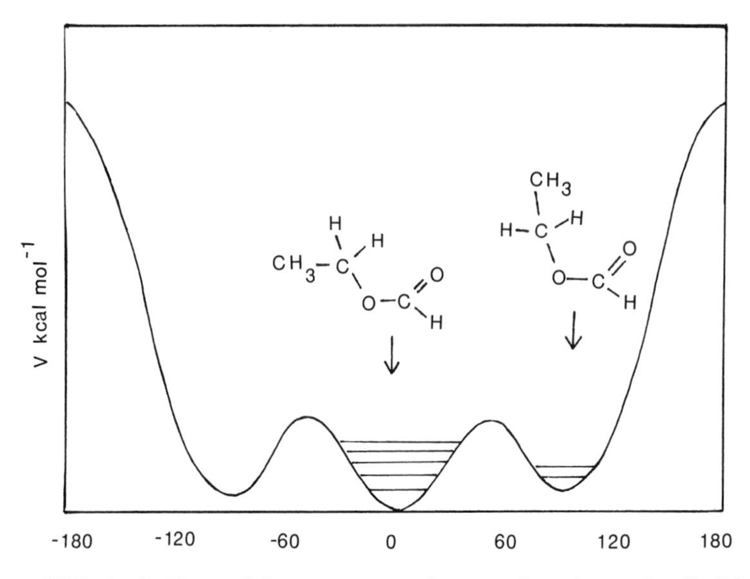

FIG. 6.17. Potential energy curve for rotation about the C–C bond in gaseous ethyl formate (E. Bright Wilson, ref. 15).

see the geometry of it in a way that virtually no other modern technique is able to provide, certainly not at the resolution that we can get from the crystallographic experiment. One of the areas that Lindo Patterson was interested in was the role of citrate. If you put a fluorine atom into citrate, you get one of the most toxic small molecules known. Jenny Glusker and Dave Zacharias have done a number of citrate structures, including two or three fluorocitrate structures, in an attempt to understand what this molecule does. It is known to be an irreversible inhibitor of the enzyme aconitase. Until they get a high resolution structure of this enzyme we can only guess why this fluorine might be important. It seemed to Jenny and her co-workers that it was the interactions that this fluorine made with other groups that made it important. Ethyl fluorocitrate is shown in Fig. 6.18; the structure contains an alkyl ammonium group, RNH_3^+, and this interacts with an oxygen atom in the normal way by forming a hydrogen bond. But it also forms short contacts with the fluorine atom. Now I must admit that I didn't believe that these were hydrogen bonds. The prejudice came about because the way you come across carbon-

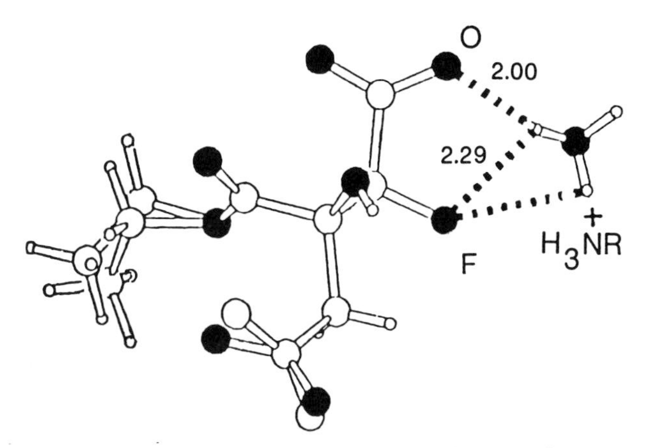

FIG. 6.18. Ethyl fluorocitrate structure showing hydrogen bonding from an RNH_3^+ group to O and F.[16]

fluorine bonds in ordinary life is in non-stick frying pans, and, if you add water to a non-stick frying pan, it runs off. So I had this prejudice that carbon-fluorine bonds would not form hydrogen bonds. However the great thing about the Crystallographic Data File is it allows you to ask questions objectively. So we asked 'Does the carbon-fluoride bond form hydrogen bonds and, in a similar way, does the carbon-fluoride bond coordinate with metal ions?' (because this might be one of the mechanisms of the inhibition of aconitase.[16]) It is not too difficult to abstract all the compounds that contain carbon–fluorine bonds, and then to compute the intermolecular environment in the crystal of those carbon–fluorine bonds. I want to give two examples of what we found. The structure of fluorocortisol[17] has been determined in Buffalo. The molecule contains a carbon-fluorine bond pointing towards a hydroxy group of another molecule. The distance between them is about 3 Å, a bit longer than the normal oxygen–hydrogen–oxygen hydrogen bond which is about 2.75 Å. We had to look at a lot of structures to convince ourselves that this really was a hydrogen bond. As a result I was converted to the idea of carbon-fluorine hydrogen bonds,[16] in what up to then had been a very confused literature. We were also able to show that the carbon-fluorine bond can coordinate to a metal ion. We were able to show that, not only was this fluorine close to the metal ion, but that, if you did a sort of electroneutrality sum (which we heard about from Professor Pauling), that the fluorine was essential in making up the electroneutrality of the metal ion; so that this represents a strong and important interaction.[16]

We also examined hydrogen bonding to oxygen-containing groups. Among current interests at the Fox Chase Cancer Center is the diol epoxide (Fig. 6.19) which is one of the essential intermediates in the carcinogenic mechanism of polycyclic aromatic hydrocarbons. A molecule such as benzo[*a*]pyrene is oxidized to form this diol epoxide and the critical process is the binding of this to DNA and the subsequent reaction so that the diol epoxide acts as an alkylating agent. We wanted to study how the epoxide might bind to other molecules, and what sort of directionality might be involved in hydrogen bonding to the epoxide, because that is going to be critical in how it interacts with molecules such as DNA.

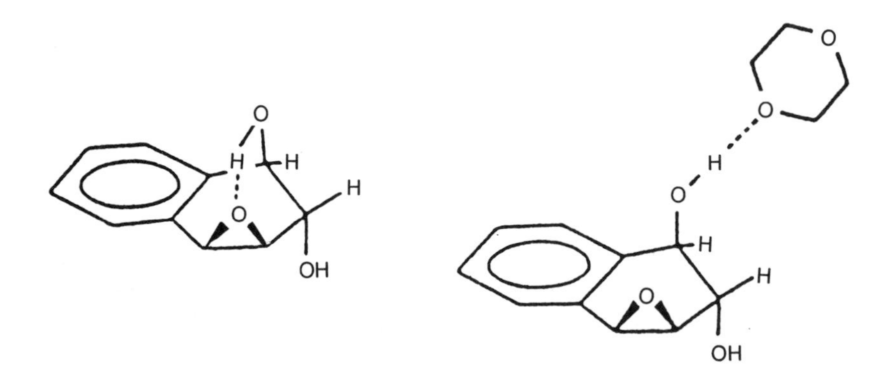

FIG. 6.19. Possible modes of hydrogen bonding of a diol epoxide of a polycyclic aromatic hydrocarbon.

So we studied a number of oxygen-containing proton acceptors.[18] We wanted to look not at the geometry of the hydrogen bond itself, which has been very well studied, but the spatial distribution of the hydrogen bond donor round the acceptor (Fig. 6.20). Can this hydrogen bond come in from any direction, or are specific directions required? This sort of information is going to be vital if we are going to set up quantitative models of these type interactions for both molecular mechanics and molecular dynamics calculations.

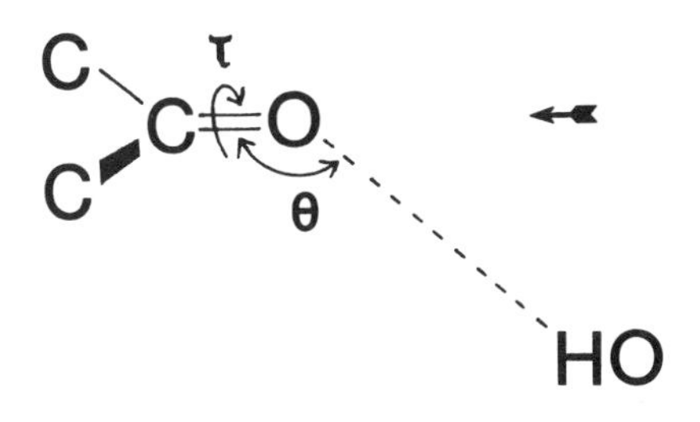

FIG. 6.20. Diagram of the spatial arrangement of hydrogen bonding groups around a ketone group.

We examined three systems: ketones, epoxides and ethers. We looked at a large number of structures on the Data File, for example, several hundred which contain a ketone group; and if they also contain a hydroxy group which is within hydrogen bonding distance of the ketone, we superimposed them all together and plotted out all the different occurrences of hydrogen bonding to the ketone. What we obtained was the probability of a hydroxy group coming in at various specific angles. The result, in Fig. 6.21(a), is by no means isotropic. The maximum concentration of positions of hydroxyl group donors tends to be in the plane of the ketone group, and it also tends to concentrate in two areas. We get similar plots for epoxides, and with the cyclic ethers. It can be a little bit difficult to assess these pictures because your eye gets led astray by outlying points, so I used a piece of software written by Bud Carrell[19] to generate an artificial electron density map from atoms. What we did here is we put a little bit of Gaussian electron density on each of these points (positions of oxygen atoms in hydroxyl hydrogen bond donors), sum up all these densities and then use a contouring program in three dimensions to give us contours of the greatest probability density. This is shown in Fig. 6.21(b). We contoured the greatest density of proton donors, and these positions correspond to where chemists draw the lone pairs in their molecular orbital diagrams. So this has a nice correlation with our simple ideas of electronic properties of molecules. We found that, as shown in Fig. 6.21(c), they come in different positions in space for ketones, epoxides and ethers, and that esters are really quite different from the ethers; epoxides appear to have more directional hydrogen bonding than do ethers.

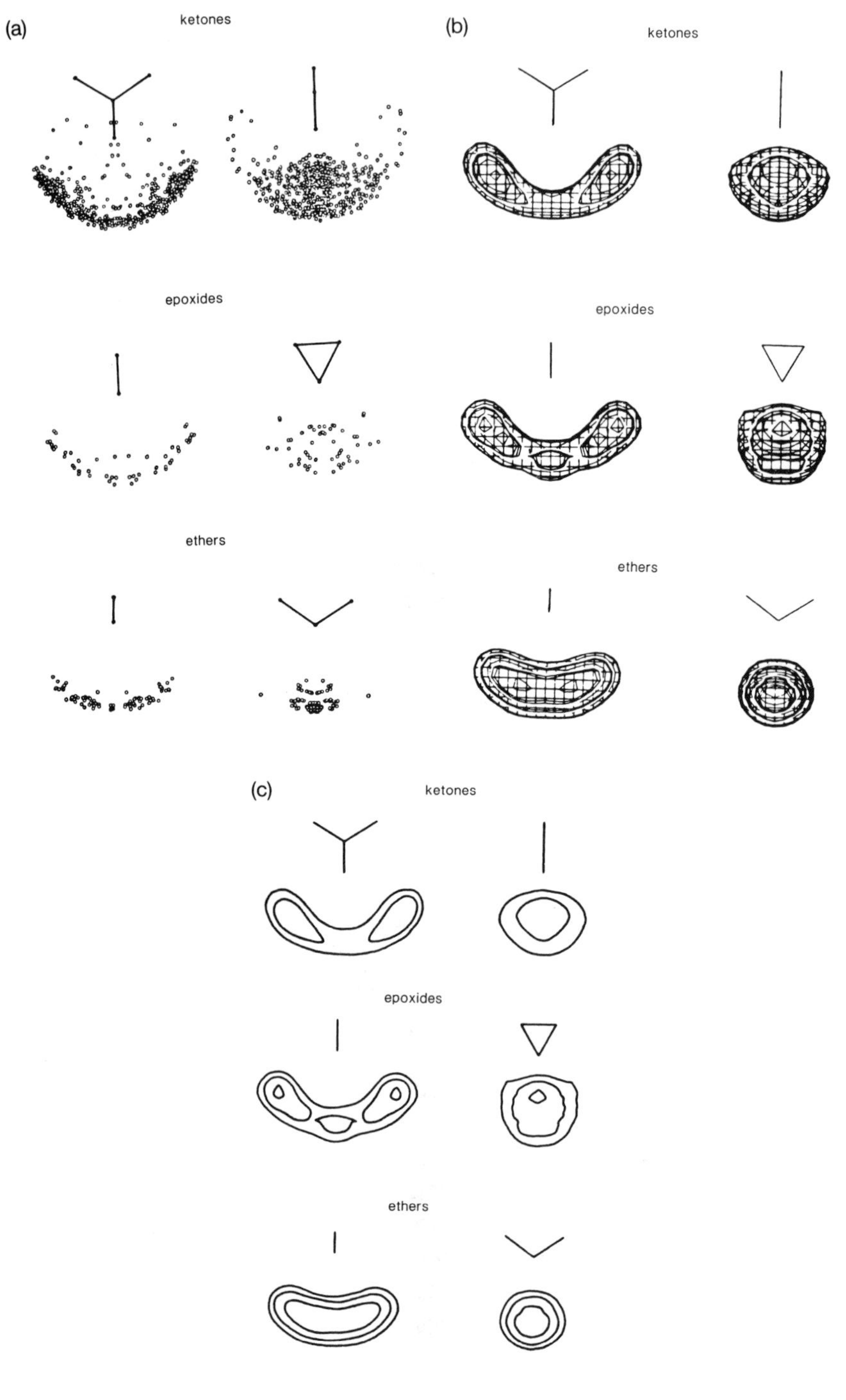
(a)
ketones
epoxides
ethers
(b)
ketones
epoxides
ethers
(c)
ketones
epoxides
ethers

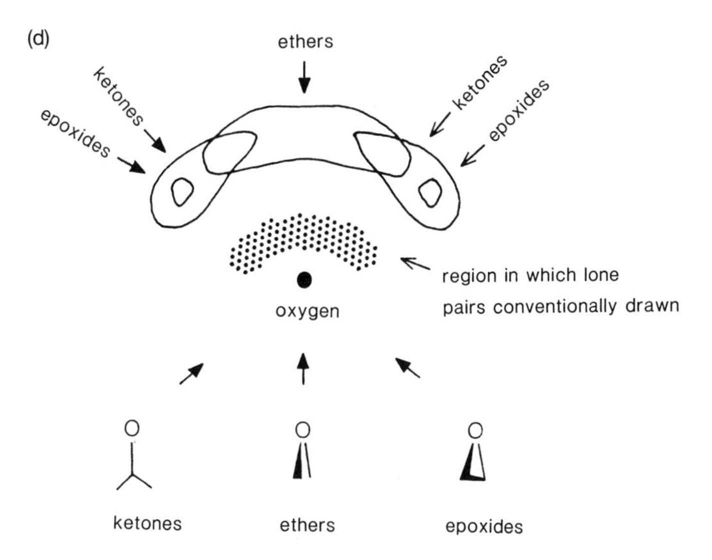

FIG. 6.21. Analyses of functional oxygen environments in crystal structures. (a) Scatterplots of hydrogen bond positions around ketones, epoxides and ethers. (b) Density distribution of hydrogen bond donors around ketones, epoxides and ethers. (c) Simplification of the contour plots in (b). (d) Comparison of distribution of hydrogen bond donors for ethers, ketones and epoxides.

The last thing that I want to discuss is something which relates to macromolecular structures. It is a piece of work I originally started with Sam Motherwell at Cambridge, but whose significance was pointed out to me by Vivian Cody of Buffalo.[20] It relates to the thyroid hormones which are one of a very few groups of naturally occurring molecules that involve the element iodine; we wanted to see in what way the carbon–iodine bond might interact with other molecules. So, we obtained several hundred carbon-iodine containing molecules from the Cambridge Crystallographic Data File, and looked at the distribution of oxygen atoms around this iodine atom, shown in Fig. 6.22. This shows that if you have a carbon-iodine bond, you get a concentration of oxygen nucleophiles along the axis of the bond about 3 Å from iodine. This is not new; it was discovered by Hassel,[21] and was part of the work which led to his award of the Nobel Prize. All we are doing here is putting this on a slightly more quantitative level, and highlighting the

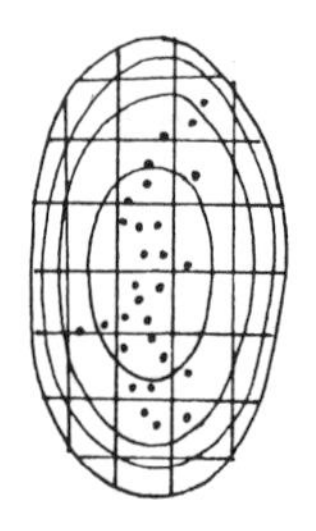

FIG. 6.22. Distribution of oxygen atoms around an iodine atom in a C–I bond.

importance of this effect in the binding of the thyroid hormones to macromolecules. The structure of prealbumin has been determined in Oxford by Colin Blake and Stuart Oatley[22] and others. Down the middle of the molecule is a large channel into which the thyroid hormone binds. The electron density in the region of the thyroid region binding site is shown together with the difference density observed when one soaks thyroid hormone into the crystal. Unfortunately, the analysis is complicated by disorder over a formal two-fold axis, but, for some time, they were troubled by the fact that the iodine atom here made a very short contact of 3 Å with an oxygen, whereas if you look up the radii in *The Nature of the Chemical Bond* [23] one might expect the contact between iodine and oxygen to be 3.7 Å. In this particular instance this corresponds not to a repulsive interaction, but to an attractive interaction; this makes a lot of sense. They were able to feel happy about this interaction, and we have been able to suggest that this might contribute to the binding energy.

In conclusion, I want to come back to the purpose of this celebration. I would be very surprised if in 50 years' time from now, we were not having a 100 year celebration of the Patterson function. I also want to say thank you to everybody at Fox Chase for the great time I had here.

Discussion

Jack Dunitz: In the analysis you made of the directionality of hydrogen bonding to carbonyl groups of esters, you see the main concentrations approximately in the sp^2 lone pair directions with minor concentration along the C=O bond axis. Did you separate the cases where the carbonyl group makes two hydrogen bonds from those where it makes only one? In the former case, the two hydrogen atoms involved are likely to be as far away from one another as possible, and this would tend to put them roughly into the lone pair directions, whereas if there is only one hydrogen bond involved there is no such constrainment. Is this a possible objection to your analysis?

Peter Murray-Rust: It is indeed. And it is one which we were aware of, and we did take care that this was not a serious contaminant of our data. I developed software so that one could, in fact, search for situations where one had just exactly one contact of this sort.

I think it is worth pointing out that, like all new instruments, the Cambridge Data File takes some time before you can drive it safely. Not because of it, itself, but because there are so many possible wrong ways to go if one isn't clear about the nature of the data that you're dealing with. One of the obvious problems is that you want to make sure that you are not looking at the same sort of structure 100 times. That's a very easy trap to fall into where you've got 100 data, and they all relate to basically the same molecule. And there are also problems of accuracy, crystallographic errors which might creep in, but in general, when you've had some practice you learn how to be able to avoid some of these things which would produce artefacts.

Linus Pauling: I just wanted to point out that even before the Cambridge Data File, it was possible to draw conclusions from the crystal structures. The iodine effect that you describe was discovered in 1920 by Ralph Wyckoff who pointed out that if you bring ClI [chloroiodide] up to a chloride ion, the interaction is along the Cl–I axis, and it is so strong, that you get two iodine–chlorine bonds the same length.[24] That was 1920.

Peter Murray-Rust: I quite agree. I mean we have simply been drilling holes where the wood is thinnest, in many of these cases.

Olga Kennard: I'd just like to point out that a more direct use of the data file, which would have delighted Lindo. Yesterday, Isabella [Karle] was alluding to the use of combined Patterson search functions and direct methods, and, with George Sheldrick, we have been working on these, and using the Data File to deduce the geometry of fragments. Peter gave an excellent illustration of how important it is to use the Data File to make the models of the different structural fragments which you can then put into your Patterson search maps, and use it for these rather large molecules, which do not diffract very well, and for which neither the conventional Patterson techniques, nor direct methods, work at the moment. So this is just a use which I think would have delighted Lindo.

George Jeffrey: I saw something in your movie of the β loop that surprised me. It looked to me as if the covalent bonds were vibrating more than the hydrogen bonds. Now, if the hydrogen bond force constants are about fifteen times weaker, then I would expect the hydrogen bond to be moving much more than the covalent bond. Can you explain what happened there?

Peter Murray-Rust: I agree; that is what I expected to see when I did this study. I mean that's what I wanted to prove when I set out to do it, and it didn't happen. I think some of the problem is that there is considerable chemical variation in the structures that we used in the analysis, and some of them certainly come from cyclic peptides, particularly cyclic hexapeptides, where there is a constraint, although a hydrogen bond may or may not occur in some of these structures. So the result is not absolute. But, for example, in the 3_{10} helix, the one with a type 1 turn, where there is no constraint in those structures, the hydrogen bond was remarkably constant. It surprised me. I would have expected it to vary very much more.

Unknown speaker: Is this because of the least squares fit? You tried to superimpose on the model, and it may have happened that these atoms that are hydrogen bonded together do not vary, because of the mathematical method used to superimpose them.

Peter Murray-Rust: Well, that might be a minor factor. But if one analyzes the data by the principal components method, then one has to come up simply with a transformation of the original data. One doesn't destroy information by it (or one shouldn't). So that if there was a real variation in the hydrogen bond angles, and if it was important, then one should see it in the first one or two factors.

Verner Schomaker: There is one danger, I think, in interpreting variations in many structures as a model for the variations in any one structure in the course of its vibration. You have to look at all the different environments of the structure. Bonds can be deformed in different local environments, and the limits to which you can push the structure around in a hundred different situations will not quite be the same as the extent that any one such structure will vibrate in a given setting. So it's a little dangerous to interpret different settings as indicating real vibrational modes.

Peter Murray-Rust: I quite agree. I mean, there are certain types of deformations which the crystal environment is not good at performing. What was remarkable, however, is the result we found when we took the steroid structures to Peter Kollman's laboratory in San Francisco, to do a molecular mechanics calculation. We used the second derivative matrix, from which one can get the predicted frequencies and the predicted normal coordinates. The correlation between the normal coordinates and the deformation that we see in crystals was something like 90%. It was quite remarkable. This shows that the calculations and the crystallography agree. Now both analyses may be wrong, but at the moment our result suggests that the results are both supporting each other. If we look at the situations where one doesn't put on constraints, then the evidence, at the moment, is that the frequency of occurrence of a particular atomic configuration does seem to be related to the energy of producing it. But I agree, there is also the problem that if, for example, one is looking at planar molecules, they will pack in a way that may very well cause them to lie on mirror planes, so one would never see out-of-plane deformations.

William Duax: The thermal parameters in some of the individually well-determined steroid structures also demonstrate a similar kind of motion, so that the free carbonyl is moving primarily perpendicular to the plane of the molecule. This is consistent with the interpretation you make by superimposing a whole series. What you are talking about is an individual molecule's motion in the environment, as Dr. Pauling pointed out yesterday, which it finds for itself, and that's the motion you see in some of the well-determined thermal motion structures.

Peter Murray-Rust: There is a real problem in interpreting thermal displacements as seen in the crystal, which is that they are a combination of at least two effects; one of them is the internal motion of the molecule, one of them is the lattice mode which gives information on how it is moving with respect to its neighbors, and which, at room temperatures, is usually a larger effect than that of the internal modes. This is the work that people like Carol Brock have been doing with Jack Dunitz, and Ken Trueblood and others have shown that, as you cool the structure down, you are able to get a better picture of the internal motion with respect to the high temperature motion. I would love to see some low temperature structures of steroids done, to see if one can see that. I did try and use Ken Trueblood's program on one or two. The result wasn't in favor of, and it wasn't against this idea.

George Jeffrey: There is a circular aspect of molecular mechanics. You derive the parameters from observed structures, so you have to be a little careful, having used those parameters, when you say that you are predicting observed structures. It's this circular nature of molecular mechanics that you have to be careful of.

Peter Murray-Rust: I would agree with that. We did use more than one force field for the steroids. Also, the steroids could be represented by a continuous block of material, without any atoms in it, and one would still get the overall deformation picture, rather like a Chladni plate vibrating. One gets a simple binodal distribution. One does not necessarily need to look at the individual atoms to appreciate what is happening.

Linus Pauling: I'd like to say that what you do is derive the principles from the observed structures and then predict other structures that haven't been observed. That you can do. That's making a contribution.

References

1. Lonsdale, K. *Nature (London)* **122**, 810 (1928).
2. Isaacs, N. W., Motherwell, W. D. S., Coppola, J. C. and Kennard, K. *J. Chem. Soc., Perkin II* 2335–2339 (1972).
3. Patterson, A. L. *Phys. Rev.* **46**, 372–376 (1934).
4. Hoekstra, A. and Vos, A. *Acta Cryst.* **B31**, 1716–1721 (1975).
5. Domenicano, A. and Vaciago, A. *Acta Cryst.* **B35**, 1382–1388 (1979).
6. Bragg, W. L. In *Beyond Reductionism: New Perspectives in the Life Sciences* (A. Koestler and J. R. Smithies, eds.) Hutchinson: London, UK (1968).
7. Bernal, J. D. *Science in History.* Third Edition. p. 943. Hawthorn Books, Inc: New York, NY (1965).
8. Allen, F. H., Bellard, S., Brice, M. D., Cartwright, B., Doubleday, A., Higgs, H., Hummelink, T., Hummelink-Peters, B. G., Kennard, O., Motherwell, W. D. S., Rodgers, J. R. and Watson, D. G. *Acta Cryst.* **B35**, 2331–2339 (1979).
9. Zacharias, D. E., Murray-Rust, P., Preston, R. K. and Glusker, J. P. *Arch. Biochem. Biophys.* **222**, 22–34 (1983).
10. Murray-Rust, P. Unpublished.
11. Harman, H. H. *Modern Factor Analysis* (revised first edition). University of Chicago Press: Chicago, IL (1976).
12. Murray-Rust, P. and Raftery, J. *J. Mol. Graphics* **3**, 50–60 (1985).
13. Murray-Rust, P. and Kratky, C. *Acta Cryst.* **A37**, C-82 (1981).
14. Murray-Rust, P. In *Molecular Structure and Biological Activity* (J. F. Griffin and W. L. Duax, eds.) pp. 117-133. Elsevier Biomedical: New York, NY, Amsterdam, the Netherlands, Oxford, UK (1982).
15. Wilson, E. B. Jr. *Chem. Soc. Rev.* **1**, 293–318 (1972).
16. Murray-Rust, P., Stallings, W. C., Monti, C. T., Preston, R. K. and Glusker, J. P. *J. Am. Chem. Soc.* **105**, 3206–3214 (1983).

Glusker, J. P. *J. Am. Chem. Soc.* **105**, 3206–3214 (1983).

17. Weeks, C. M., Duax, W. L. and Wolff, M. E. *J. Am. Chem. Soc.* **95**, 2865–2868 (1973).
18. Murray-Rust, P. and Glusker, J. P. *J. Am. Chem. Soc.* **106**, 1018–1025 (1984).
19. Carrell, H. L. Program from the Institute for Cancer Research, Philadelphia, PA 19111 (1982).
20. Cody, V. and Murray-Rust, J. *J. Mol. Struct.* **112**, 189–199 (1984).
21. Hassel, O. and Rømming, C. *Acta Chem. Scand.* **21**, 2659 (1967).
22. Blake, C. C. F. and Oatley, S. J. *Nature (London)* **268**, 115–120 (1977).
23. Pauling, L. *The Nature of the Chemical Bond* (3rd edition). Cornell University Press: Ithaca, NY (1960).
24. Wyckoff, R. W G. *J. Am. Chem. Soc.* **42**, 1100–1116 (1920).

7. Protein structure and function

Sir David Phillips

David Phillips was educated at the University of Wales and obtained a Ph.D. from Cardiff University in 1951. He then worked at the National Research Council in Ottawa, Canada, and studied the structures of various forms of acridine. He then returned to England to the Davy–Faraday Laboratories of the Royal Institution in London. In 1966 he went to Oxford University as Professor of Biophysics. Dr. Phillips is best known for his determination of the first structure of an enzyme. This was the enzyme lysozyme which breaks the polysaccharide coating of bacteria. In the *Scientific American* article which is one of the benchmarks of the subject, Dr. Phillips showed the structure of the enzyme, how it might work and how it might fold when first made. David Phillips has continued his work on protein structure and function. Among the proteins he has studied recently are immunoglobulins. He was Biological Secretary of the Royal Society and was knighted by Queen Elizabeth.

The year 1984 was a doubly-important year of jubilee in crystallography for it was the fiftieth anniversary of both the Patterson function and the beginnings of X-ray crystallographic studies of proteins. It was in 1934, thanks to Bernal and Dorothy Crowfoot, that the successful examination of protein crystals by X-rays began.[1] Their paper gave an account of the history behind the critical experiment which illustrates very well the international nature of science and the continuing catalytic role of the itinerant scholar. The important advance was Bernal's realization, following examination of the crystals under the microscope, that it was essential to keep them in contact with the liquid of crystallization, if the three dimensional integrity of the structure was to be preserved and X-ray diffraction observations were to be made.

The fact that these two critical advances, in experimental and theoretical crystallography, came together in one year is symbolic of the lasting relationship between them. For the success of protein crystallography has been based firmly upon the use and development of the Patterson method. Early on, the direct interpretation of Patterson maps of proteins — which could be calculated only with great difficulty — seemed to provide one of the only ways forward and the other approach that seemed promising, the use of the method of isomorphous replacement, also depended on the Patterson method. Not surprisingly, perhaps, in view of work on the alums[2,3] and J. M. Robertson's spectacular success with the phthalocyanines,[4] the possibility of using isomorphous replacement to solve protein structures was considered very soon after the first photographs had been taken: Dorothy Hodgkin[5] has described Bernal's unpublished thoughts on this subject in 1935 and both Bernal and Robertson published suggestions on how to proceed in 1939.[6,7]

On a personal note, I well remember my visits to David Harker's Protein Structure Project in Brooklyn in the very early 1950's when he would talk to me about his hopes of using heavy-atom dyes to prepare the necessary isomorphous derivatives: after all, many dyes were designed to stick to wool and wool is a protein. Clearly all the protein structure groups at that time were trying hard to make the method work, encouraged by A. J. C. Wilson's[8] contribution to understanding what intensity changes might be brought about by a single 'heavy atom' attached to a large protein molecule, but it was not until 1953 that Green, Ingram and Perutz[9] showed that the method really would work in their dramatic solution of a centrosymmetric projection of hæmoglobin.

Bokhoven, Schoone and Bijvoet[10] had by this time already indicated how the method of isomorphous replacement could be extended to determine general phase angles for noncentrosymmetric structures, by the use of two or more different derivatives, and Harker[11] discussed the method in more detail and suggested a neat 'phase circles' construction for deriving the phase angles. This was the method used when John Kendrew and his colleagues in Cambridge and at the Royal Institution[12] produced the first three-dimensional electron-density map of a protein, sperm-whale myoglobin, at low resolution. In parallel with this work, David Blow[13] produced a map of a noncentrosymmetric projection of hæmoglobin, exploring for the first time the complementary use of anomalous scattering data in phase determination, as suggested by Bijvoet;[14] and, together with Crick,[15] he showed how the determination of phases could be formulated to take proper account of experimental errors in the development of a method that has been used in all subsequent work. Analysis of the structure of myoglobin in atomic detail followed in 1959, just twenty-five years ago.[16]

Since then some two-hundred protein structures have been determined, nearly all of them by the method of multiple isomorphous replacement associated with anomalous scattering. Though the procedures that have been used have been continuously improved and refined, often to take advantage of the development of more powerful computers, virtually all of this work has depended upon the Patterson function.

The Patterson function in protein crystallography

The most obvious way in which the Patterson function has been used is, of course, in the determination of the heavy-atom positions in isomorphous derivatives. In this application the coefficients generally used are

$$|\Delta F|^2 = ||F_{derivative}| - |F_{protein}||^2,$$

which give rise to maps, both in centro-symmetric and non-centrosymmetric projections and, especially, in three dimensions, in which the dominant features represent the vectors between heavy-atoms positions — provided, of course, that the degree of substitution and the isomorphism are satisfactory. An early Patterson map of this kind is shown in Fig. 7.1, together with the corresponding map of the heavy-atom electron density — in case the interpretation of the Patterson map is not clear. This illustration of a Patterson projection is taken from Helen Scouloudi's early work on seal myoglobin.[17] Nowadays it is usual to work from the beginning in three dimensions, thanks to the development of both experimental methods for intensity measurement and computers, when the interpretation of Harker sections can be even more straightforward — again provided that the derivative is a good one.

The positions of the heavy atoms can also be determined from the differences in intensities that arise between Bijvoet

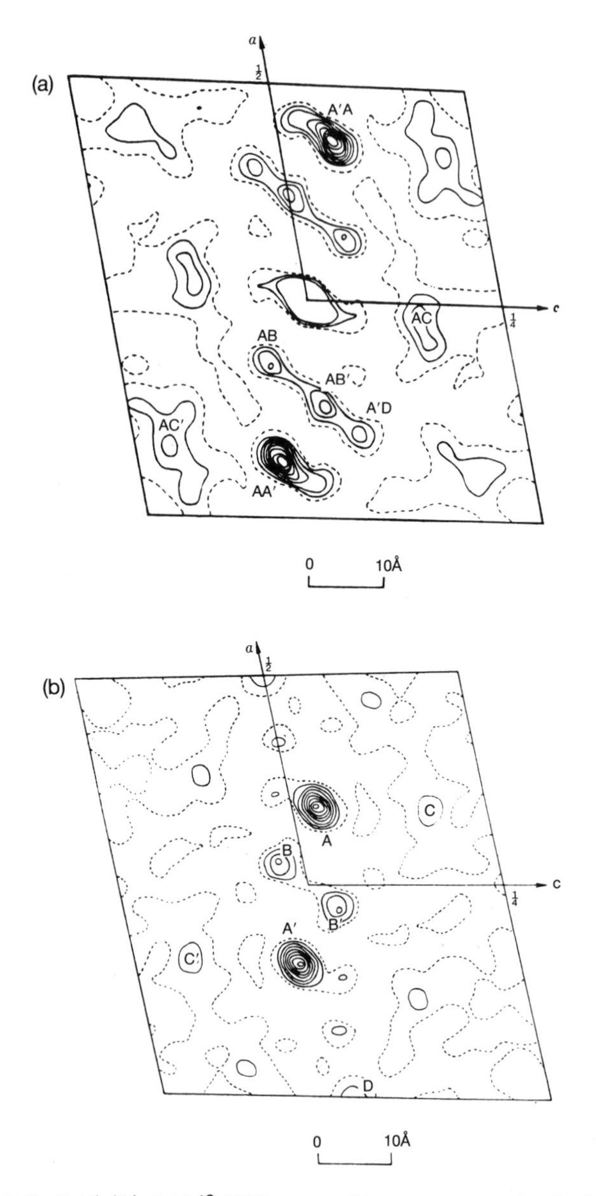

FIG. 7.1(a) | ΔF |^2 Difference–Patterson synthesis for the (010) projection of the mercuri-iodide derivative of seal metmyoglobin.

(b) ΔF-difference synthesis showing the electron density corresponding to (a). In peak A only alternate contours are drawn. (Reproduced by permission of Dr. H. Scouloudi).[17]

pairs of reflexions as a result of anomalous scattering. Here the coefficients that are used in the Patterson function are

$$| \Delta F_{anom} |^2 = || F(hkl) | - | F(\overline{hkl}) ||^2,$$

and the quality of the map no longer depends upon the degree of isomorphism of the derivative since derivative data alone are used. This method has proved useful in the standard method, as a check of derivative and data quality, but it also provides the basis for the determination of protein structures directly from the information provided by intrinsic anomalous scatterers. This method has been used successfully by Teeter and Hendrickson[18] in their study of crambin.

Finally, in this very brief and incomplete survey of the ways in which protein structure analysis depends upon the use of the Patterson function, mention must be made of the ways in which the presence of more than one molecule in the asymmetric unit or the existence of more than one crystal form of the same substance are being used to solve structures. This approach, which was initiated by Rossmann and Blow,[19] depends essentially upon a systematic analysis of the Patterson maps given by protein crystals themselves. The 'rotation function' applied to a simple protein crystal, for example, in effect superimposes the Patterson map upon itself systematically in all orientations so that maxima in some overlap function of features around the origin reveal the relative orientations of the molecules in the asymmetric unit. Similarly a 'translation function'[20] is used to show the relative positions of the molecules in the unit cell. Various methods may then be used to exploit this information in the determination of the structure. This approach has been used most successfully in the study of viruses and a striking example is provided by the work on the crystalline disc derived from tobacco mosaic virus in which 17-fold rotational symmetry was used to improve the observed electron density.[21]

This short introduction will suffice to show that the inclusion of a section on protein crystallography in a celebration of the fiftieth anniversary of the Patterson function is most fitting. The remarkable achievements of protein crystallography over the last twenty-five years have depended in a fundamental way on the use of this function.

Protein structure

Now that a large number of protein structures have been determined it is becoming possible to classify them in various ways, in particular in terms of the famous α helix and β pleated sheet structures described by Pauling and Corey in 1951.[22,23]

Levitt and Chothia[24] have suggested that the first class should comprise the proteins that are made up of α helices, 'the α class' and this, of course, includes myoglobin and hæmoglobin. Myoglobin and hæmoglobin subunits (Fig. 7.2) are made up of eight sections of helix, folded in what looked in 1960 to be a rather complicated tangle, but which seems now to be one of the simpler structures.

The next category is well represented by ribonuclease,

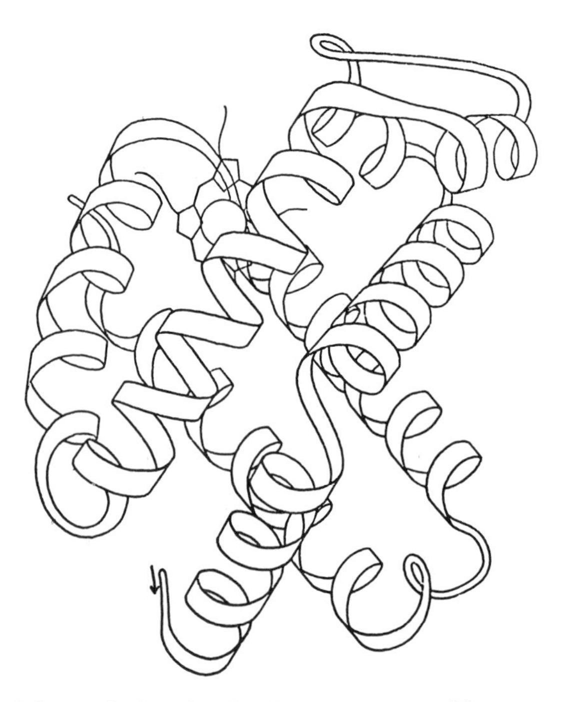

FIG. 7.2. Schematic drawing by Jane Richardson[25] of the main-chain conformation in hæmoglobin β-subunit — an α protein.

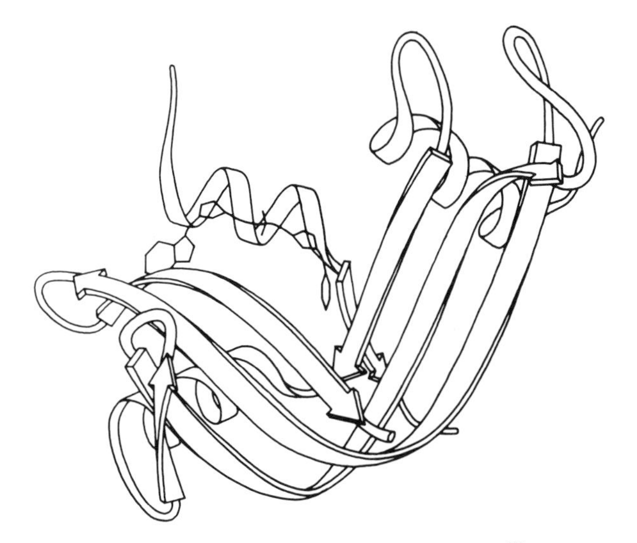

FIG. 7.3. Schematic drawing by Jane Richardson[25] of the main-chain conformation in ribonuclease S — an α + β protein.

shown in Fig. 7.3.[25] This structure includes both β pleated sheet, shown in a rather splendidly twisted form, and some α helices, but with these elements of secondary structure not arranged in any particular relative order. It is referred to as an 'α plus β' structure, a category that includes, among others, both lysozyme and insulin.

Fig. 7.4 shows superoxide dismutase which is made up entirely of β pleated sheet: proteins of this category are known as 'β proteins'. Immunoglobulins are all of this kind with domain structures very similar to that shown in Fig. 7.4.

Finally, in this categorization, there are the α/β proteins which are again made up of both α helices and β pleated sheets, but now with these elements of secondary structure arranged in a regular way. In these molecules a strand of β sheet is followed by a helix, running more or less anti-parallel to it, then a strand parallel to the first one, a helix going back, and so on. In the particular example shown in Fig. 7.5 (triose phosphate isomerase), the strands of β sheet curl around to form a closed barrel-like structure, with a surrounding shell of helices.[26]

Very many intracellular enzymes seem to be of this α/β

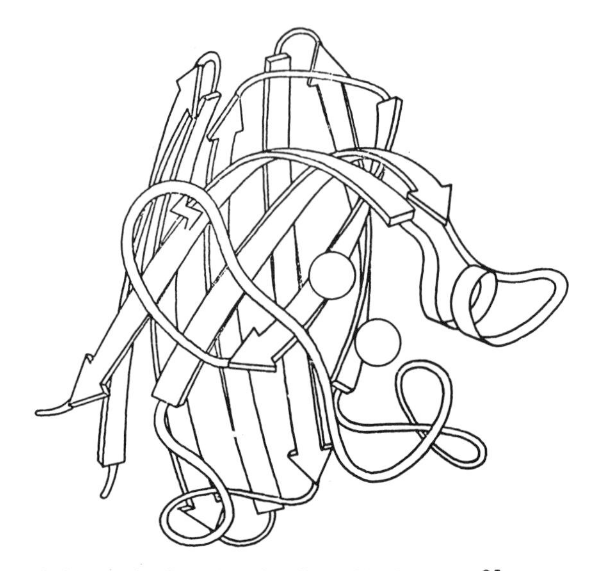

FIG. 7.4. Schematic drawing by Jane Richardson[25] of the main-chain conformation in Cu/Zn superoxide dismutase — a β protein.

kind, with a regular alternation of β strands and α helices making up a variety of different overall structures, and this serves to introduce the next point. Here we are concerned with molecular taxonomy, a topic that may not seem very appealing to physical scientists (though it does call to mind the remark of the physicist who said 'If I could remember the names of all the elementary particles, I should be a botanist'). What is happening is that, as more and more structures are determined, so familiar types keep reappearing. By now, for example, the particular arrangement first seen in triose phosphate isomerase (Fig. 7.5) has been found in at least eight other enzymes, including the xylose isomerase structure determined by H. L. Carrell, Jenny Glusker and their colleagues in this Institute.[27] This kind of observation, and there are many examples, has led to the rather unexpected phenomenon that X-ray crystallography is now making a vital contribution to the study of molecular evolution. At the same time the discovery that proteins with diverse functions and chemical compositions form structural families has provided a possible approach to the pre-

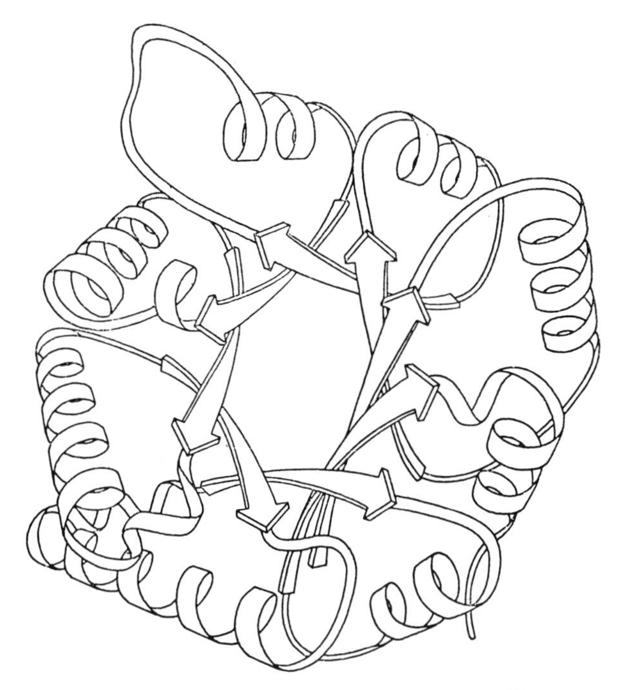

FIG. 7.5. Schematic drawing by Jane Richardson[25] of the main-chain conformation in one subunit of triose phosphate isomerase — an α/β protein.

diction of protein structures in which the first step would be the identification of the family.

Protein structure refinement

Thanks to the development of computers and better ways of making the measurements, including the use of synchrotron radiation, protein crystallography has now been almost completely absorbed into the mainstream of crystallography. There was a time when protein crystallographers tended to regard their subject as distinct but the aim now is to apply to proteins the methods that have been used to study crystals of smaller molecules — modified as appropriate to allow for the differences of scale and complexity. Some indication of how far this trend has developed may be obtained from a comparison of this paper with the one given by Peter Murray-Rust.[28]

Many analyses of protein structures now do approach atomic resolution but the main difference between normal small-molecule crystallography is that, unfortunately, protein crystals do not diffract as well. Nevertheless, protein crystallographers are very much concerned to refine their structures and the results that are being obtained can be illustrated by relatively recent work on lysozyme.

Fig. 7.6 shows part of the electron-density map of human leukæmic lysozyme at 1.5 Å resolution, taken from the work of Peter Artymiuk and Colin Blake[29] in Oxford. It shows clearly that the two sulphur atoms in disulphide bridges are quite well resolved, and that other features are easily recognizable. This is the degree of clarity that can be achieved at the end of a process of refinement at about 1.6 Å resolution, which is becoming

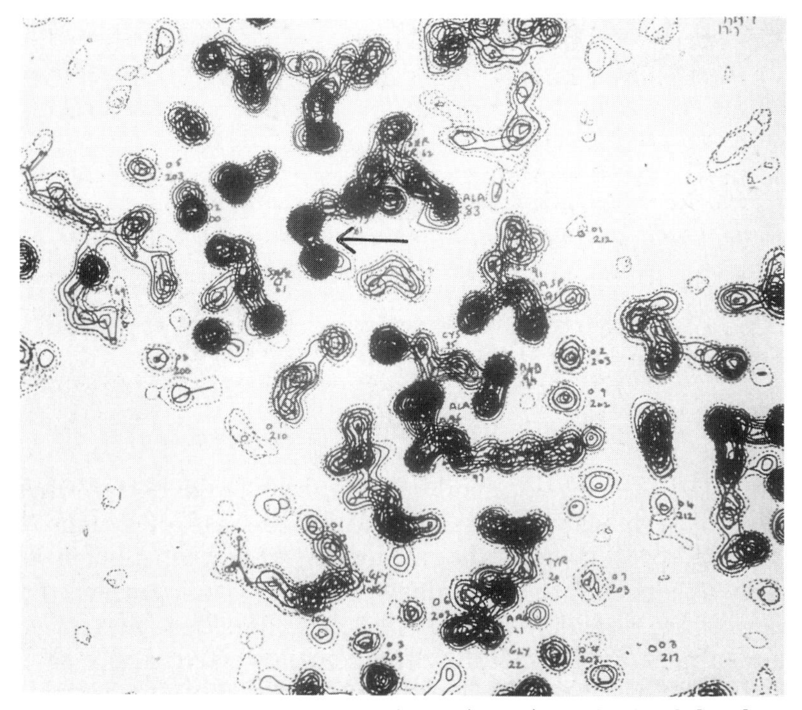

FIG. 7.6. Part of $(2F_{obs} - F_{calc}) \exp (i\alpha_{calc})$ synthesis of the electron density in human leukæmic lysozyme at 1.5 Å resolution. The resolved sulphur atoms of a disulphide bridge are indicated by an arrow (P. J. Artymiuk, personal communication).

quite common in modern analyses. In addition such studies are providing information about the solvent in the crystals, some of which is ordered with water molecules forming a shell around each protein molecule.[30] These observations are, of course, consistent with the finding by Bernal and Crowfoot that water is an essential part of the structure of protein crystals, and they provide an illustration of the growing interest in defining its role and properties.

Protein dynamics

The refinement process clearly involves not only establishing the best mean positions of the atoms, but also defining the fluctuations about these positions. The method used most commonly is restrained least-squares refinement, as programmed by Konnert and Hendrickson[31] following an idea first propounded by Waser[32] and further discussed by Rollett.[33] In this process allowance has to be made for the apparent motion of the atoms and, as everyone knows, this is a dangerous business because the so-called thermal-motion parameters are seriously affected both by experimental errors in the data and by inadequacies in the current structural model. This is one problem, and the second even more fundamental difficulty is that the motion of the atoms in proteins is most unlikely to be harmonic as the usual theory assumes. Yet the description of anharmonic motion requires a very large number of parameters[34] — a much larger number than could possibly be used, given the paucity of experimental data that are generally available. Consequently, with very few exceptions, the work that has been done on the refinement of thermal parameters has rested on the assumption that the motion is not only harmonic but isotropic as well, so that a single B value can be used for each atom (in the usual formulation of the temperature factor). Fig. 7.7 shows that this is a poor approximation: the side chain of tyrosine 52 in human lysozyme[29] is clearly engaged in librational motion. Recently in a very few studies at very high resolution[35] anisotropic motion has been taken into account but so far the problem of describing anharmonic motion has not been solved directly.

Nevertheless, even isotropic B values have proved to be quite illuminating and Fig. 7.8 shows the variation in isotropic

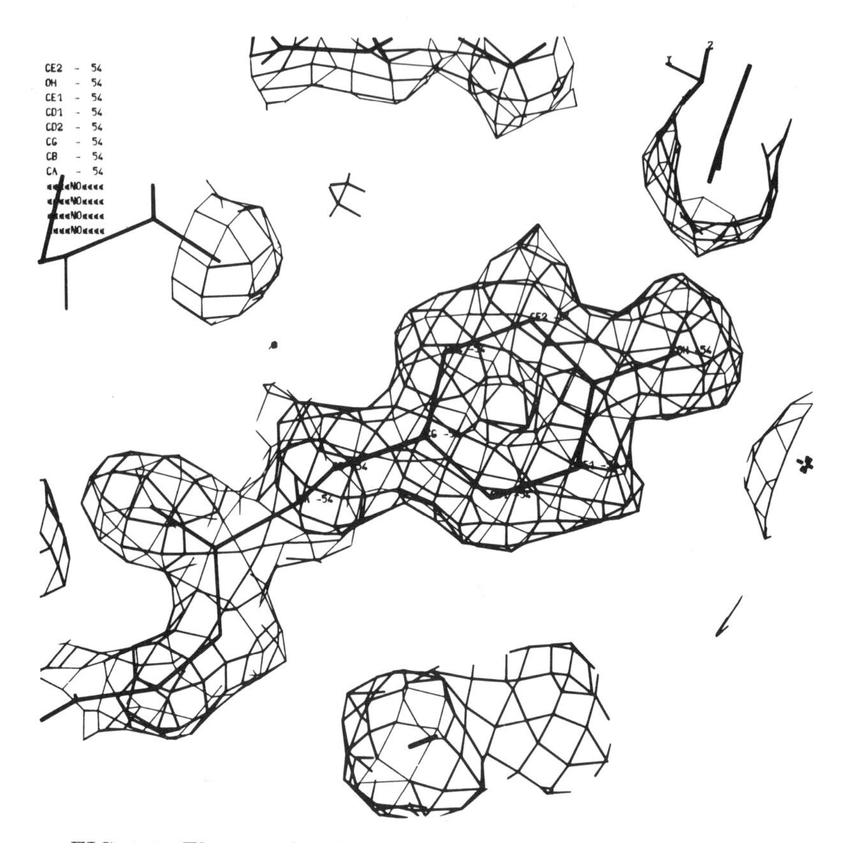

FIG. 7.7. Electron density for Tyr 52 in human leukæmic lysozyme at 1.5 Å resolution showing indications of librational motion. (P. J. Artymiuk, personal communication).

B values that has been observed in studies of lysozyme.[36,37] Clearly there are quite strong variations, particularly near residues 47, 71 and 100.

Given the dangers in the whole process, such a result has to be greeted with scepticism: are the dramatic-looking variations credible, or do they arise from the errors in the analysis? Happily, comparison of the hen-egg white lysozyme[38] results with the corresponding data from human lysozyme (Fig. 7.8) shows that they agree remarkable well. In particular, the prominent spikes appear in the same places and the overall correlation coefficient is about 0.62. Furthermore, there is some correspondence between the variations in B values and the features of the three-dimensional structure. For example, the structure of lysozyme incorporates a region of β-pleated sheet comprising three strands. The strand that is most exposed on the surface of the molecule seems to be moving about the most while the least exposed strand moves about the least. Indeed, even the relatively small variations in B values among the residues that are in α-helices, correlate quite well with the locations of those residues on the surface or in the interior of the molecule.

Hence there is reason to suppose that these slightly improbable-looking B values actually do reveal something about a molecular property and an extension of the analysis tends to confirm this view while, at the same time, bringing out another important issue. The B values taken from independent refinements of different lysozyme structures are generally very similar to one another but they do show some interesting variations. For example, in the triclinic form of the hen egg white lysozyme, the spike near residue 70 is very much reduced.[39] Now the triclinic form is the one which has the least liquid of crystallization and the closest interactions between the molecules in the crystal; these residues in the triclinic form are involved in a molecular interaction. This kind of result has generated renewed interest in the effects of intermolecular contacts on the conformations and properties of protein molecules in crystals.

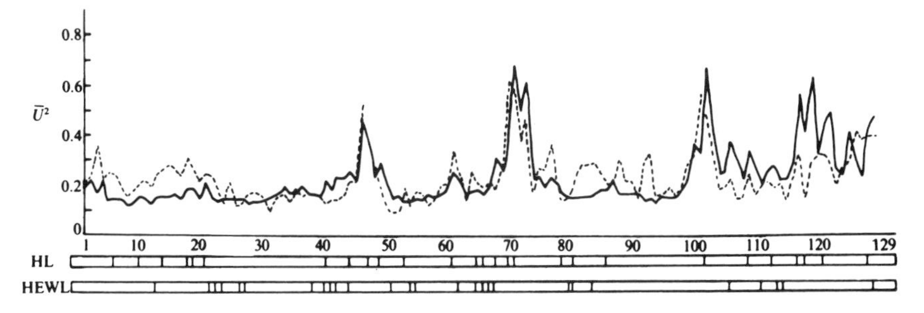

FIG. 7.8. Plot of main chain U^2 values against residue number for human lysozyme (full line) and hen egg lysozyme (broken line). Residues involved in intermolecular contacts in the crystals are indicated in the lower rectangles.

Any protein crystallographer who talks about the function of proteins as derived from the evidence of crystal structures is certain to be asked one question, viz., 'how do you know that what you are looking at in crystals bears any relation to what happens in solution?' That is a question that has to be taken very seriously and it can now be tackled at a deeper level than ever before since the effects of crystallization on dynamic behaviour can be addressed as well as the effects on average conformation. It remains true, of course, that protein crystals are generally well regarded as solutions in the liquid that Bernal and Crowfoot found to be so important, but there is certainly scope for detailed studies of the effects of the crystal environment that may illuminate molecular properties just as they are doing for small molecules.[28] Furthermore, the results are likely to be interesting from a biological point of view, since the behaviour of protein molecules in crystals is likely to be more akin to their behaviour in living cells than is their behaviour in the dilute solutions that are studied most frequently by biochemists.

There is, however, one further difficulty that must be mentioned. As in all crystallographic analyses, the electron-density maps of proteins correspond first of all to time averages of the structures over the lengths of the experiments and, secondly, to space averages over all of the molecules in the crystal specimens. Consequently, it is important to distinguish between genuine motion and static disorder within the crystals, and this is not easily accomplished. Protein maps do occasionally provide seemingly clear evidence that different molecules throughout the crystal have distinguishable conformations and Fig. 7.9 shows a good example. Here the electron-density corresponding to tyrosine 62 in human lysozyme[29] cannot be interpreted in terms of one side chain but can be interpreted, quite convincingly, in terms of a tyrosine side chain that adopts one conformation in some molecules and a different one in others. The alternative conformations appear to occur in roughly equal numbers of molecules. Even in this example, however, the interpretation is not straightforward since the observed electron density would be the same if every side chain spent half of its time in each of the alternative conformations. One way to resolve this difficulty is to study the structure over a range of

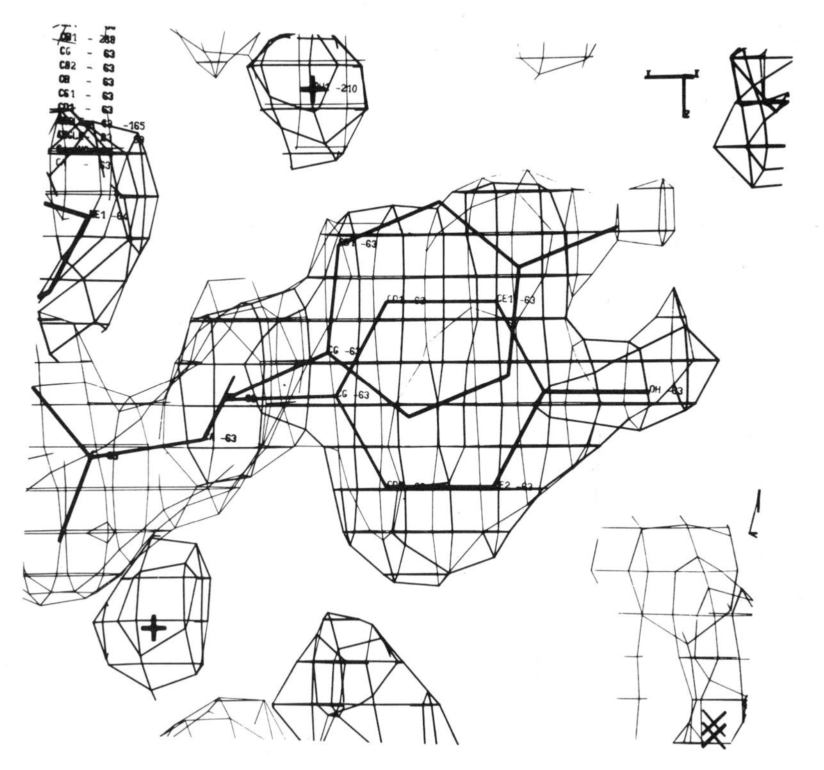

FIG. 7.9. Electron density for Tyr 62 in human leukæmic lysozyme showing two different conformation of the side chain (P. J. Artymiuk, personal communication).

temperatures, as Lonsdale and Milledge[40] suggested for small molecules, and, surprising though this may be, such experiments are practicable. By the use of suitable cryo-solvents, such as methanol-water mixtures, it has proved possible to study protein crystals over a wide range of temperatures and, most remarkably, by the use of a quick-freezing method, Hartmann *et al.*,[41] have determined isotropic B-values for myoglobin at 80 K.

In this way progress is being made towards understanding the apparent thermal motion in protein crystals and it has been established, for example, that static disorder is not the dominant contributor to the observed B-values. The assumptions that the individual atoms in proteins move independently and isotropically in harmonic potential wells are strictly unrealistic, but work over the last few years[41,42] has shown that valuable information on the dynamic properties of proteins can be obtained through crystallographic studies — though it has to be recognized that the time scales of the motions have to be determined by the use of complementary spectroscopic method, particularly NMR.[43] In particular, this work has revealed some details of intramolecular motion that are of biological interest. For example, the activation of serine proteases seems to depend upon the existence of such motion[44] and so does the access of oxygen to the binding sites of myoglobin and hæmoglobin[45,46] Analysis of the lysozyme results in terms of intramolecular motion suggests that some segments of the polypeptide chain are moving relatively vigorously with respect to the rest of the molecule. These apparently mobile regions comprise mainly the two wings of the active site of the molecule, which lies in the groove running down the front face of the structure (as seen in Fig. 7.10). The possible role of this mobility in the enzyme action is considered briefly later.

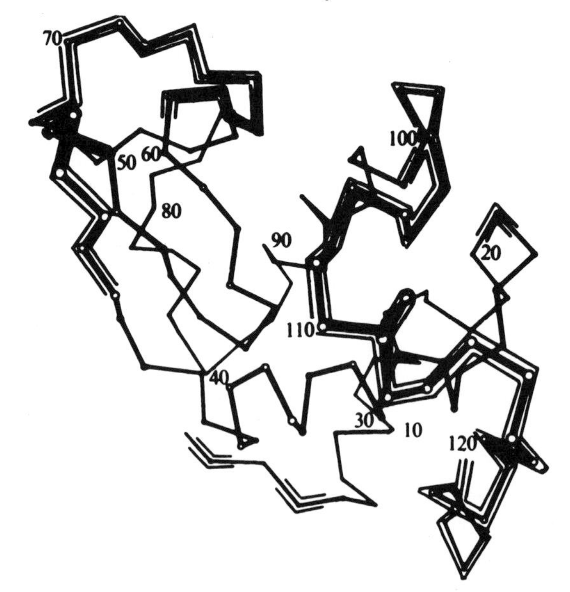

FIG. 7.10. A perspective drawing of the human lysozyme molecule with residues having apparent displacements > 0.2 Å [2] outlined by parallel lines. The active site cleft runs almost vertically down the point of the drawing.

Molecular mechanics

At this stage it is appropriate to mention another development, made possible by increasingly powerful computers, that is contributing to our understanding of protein mobility. In this work, the positions and velocities of the atoms in a protein molecule are calculated as a function of time by numerical integration of the classical Newtonian equations of motion, taking into account detailed interatomic potentials and starting from the observed X-ray coordinates. Such calculations, by Karplus and McCammon,[47] van Gunsteren *et al.*[48] and Levitt,[49] have provided much more detailed indications of the complex anisotropic and anharmonic motion of the atoms in proteins than can be derived directly from the X-ray observations (not least because of the small number of X-ray observations that are generally available in relation to the number of atomic parameters that have to be determined). The descriptions of molecular motion that have been obtained by these calculations agree well, however, with the more limited descriptions derived from X-ray analysis and it seems likely that the use of the two approaches in parallel will lead to a deeper understanding of protein dynamics and, at the same time, provide valuable experimental support for the further development of molecular mechanics.

Recently Levitt *et al.*[50] have developed a method of calculating protein normal-mode dynamics which they have used in studies of trypsin inhibitor, crambin, lysozyme and ribonuclease. Levitt and his colleagues, who have produced a film showing some of the lower frequency modes for these proteins, note that 'when the resulting atomic motion is visualized with computer graphics, it is clear that the motion of each protein is collective with all atoms participating in each mode. the low modes, with frequencies of below 10 cm^{-1} (a period of 3ps), are the most interesting in that the motion in these modes is segmental. The root-mean-square atomic fluctuations, which are dominated by a few slow modes, agree well with the experimental temperature factors (B values). The normal-mode dynamics of these four proteins have many features in common, although in the larger molecules, lysozyme and ribonuclease, there is low frequency domain motion about the active site. It may be added that, in lysozyme, one of these modes involves the opening and closing of the active site cleft that is suggested by Fig. 7.10 — an aspect of lysozyme mobility that has also been studied by McCammon *et al.*[51]

Crystallographic enzymology

These observations on protein mobility lead naturally to crystallographic studies of protein, and especially enzyme function. Happily, the experience of the last twenty years has shown that, once the structure of an enzyme has been determined, information about its activity often can be determined quite easily from the study of isomorphous crystals incorporating complexes with inhibitors or even substrates. The simplest approach is to look at the structure of the complex by the use of the known protein phases in the calculation of Fourier maps with coefficients

$$\Delta F = || F_{(complex)} | - | F_{(protein)} ||.$$

At sufficiently high resolution, such difference maps show the position and orientation of the bound ligand in relation to the protein together with any changes in the protein structure that arise from the ligand binding.

This approach is well illustrated, especially on a nostalgic occasion like this anniversary, by the work on lysozyme in which it proved possible to build up a persuasive model of the enzyme-substrate complex from studies of the complexes between the enzyme and inhibitor molecules closely related to its natural substrate.[52,53] The substrate (Fig. 7.11) is a $\beta(1 \rightarrow 4)$-linked polysaccharide with alternating residues of N-acetylglucosamine (GlcNAc) and N-acetylmuramic acid (MurNAc) and the enzyme promotes the hydrolysis of glycosidic linkages between MurNAc and GlcNAc.

The crystallographic approach was simply to diffuse inhibitor molecules such as N-acetylglucosamine into the crystals, taking advantage of the high liquid content, to measure the diffraction intensities and to calculate a difference map in which the location of the inhibitor with respect to the protein

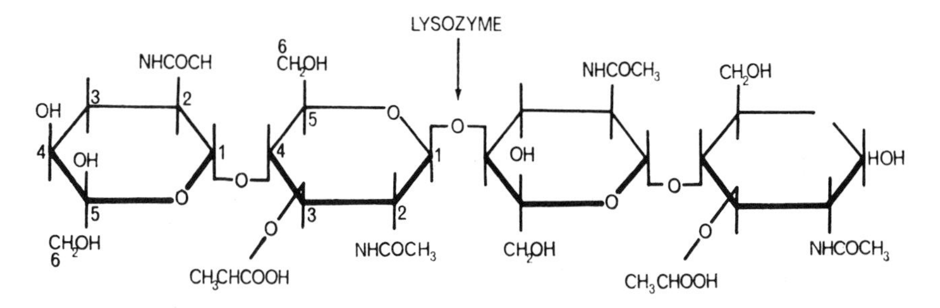

FIG. 7.11. A cell-wall tetrasaccharide with alternate residues of N-acetyl glucosamine and N-acetylmuramic acid (lactyl side chain on C(3)). The $\beta(1 \rightarrow 4)$ glycosidic linkage hydrolysed by lysozyme is shown by an arrow. The formula is drawn in a way that is now unconventional to resemble more closely the actual atomic arrangement.

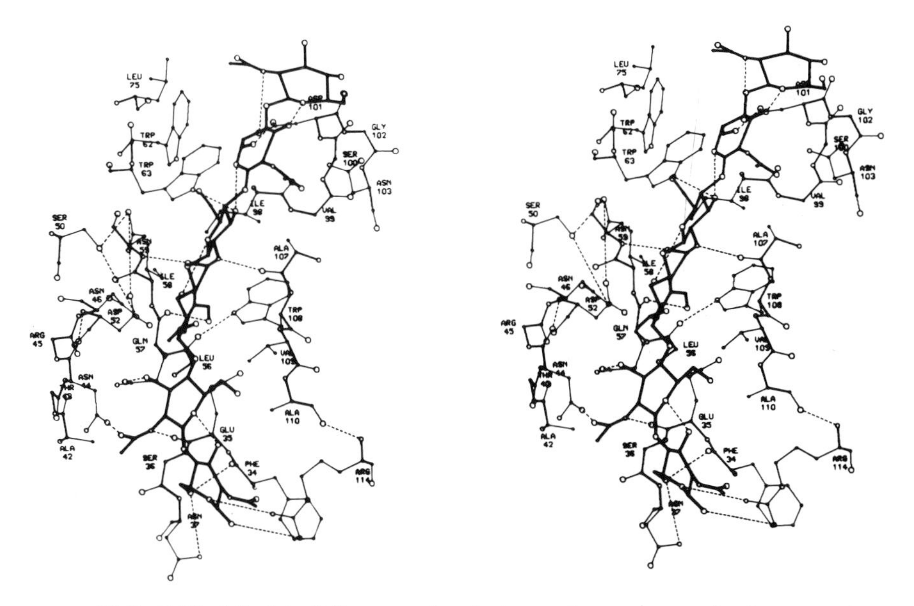

FIG. 7.12. Model of the complex between lysozyme and a hexasaccharide substrate derived from studies of oligosaccharide inhibitor binding.[52]

could be clearly seen. Studies with oligosaccharide inhibitors led to the development of a model of the enzyme-substrate complex (Fig. 7.12) and proposals for the mechanism of action of the enzyme that have been largely substantiated by later experiments.

An essential part of this work was the identification of two carboxylic-acid groups in the enzyme, glutamic acid 35 and aspartic acid 52, that are involved in the catalytic activity and this introduced another field of interest to protein crystallographers that is, to some extent, neglected — though it relates closely to work with small molecules. The carboxylic acid groups of these two amino-acid residues are in noticeably different environments in the molecular structure and these environments affect the reactivities of the groups in ways that are essential to the activity of the enzyme. So much seemed clear at first sight, on the basis of rather general qualitative argument: aspartic acid 52 is in a generally polar environment and seems likely to be ionized at a relatively low pH value whereas glutamic acid 35 is in a non-polar environment and is likely to remain protonated at higher pH. This difference has been confirmed in many subsequent studies of the properties of lysozyme which have shown that the macroscopic pK values of Asp 52 and Glu 35 are respectively 3.4 and 6.1, at 0.1 ionic strength and 25 °C.[54]

Recently, it has proved possible to observe the ionisation states of these residues directly by the use of neutron diffraction, as Graham Bentley and Sax Mason have shown in a study of deuterated lysozyme in the triclinic crystal form at the Institute Laue-Langevin, Grenoble, France. In this respect also, therefore, protein studies are now within the mainstream of crystallographic research — though the resolution achieved so far does leave something to be desired.

The mechanism of action of lysozyme is generally thought to be well understood and there is little point in discussing it here except to introduce a related topic. The enzyme is believed to involve glutamic acid 35 acting as a general acid which donates its proton to the glycosidic oxygen. This brings about the breaking of the C(1)–O glycosidic bond and the development of a carbonium ion on the C(1) carbon atom. The development of this carbonium ion, and its subsequent stabilization, is promoted by the interaction between C(1) and the ionized Asp 52. The positive charge on the carbonium ion is shared to some extent with the neighbouring ring oxygen O(5) — leading to the formation of a carbonium-oxonium ion with a partial double bond between C(1) and O(5).

This calls to mind a proposal by Linus Pauling[55] that enzymes must be expected to work by binding neither substrates nor products as tightly as possible but rather by binding the transition state of the reaction. This idea leads to a possible way of exploring enzyme mechanisms more closely while still avoiding the need to observe very short lived, transient structures: might it be possible to design stable molecules that resemble transition states closely enough to bind to enzymes in similar ways? If so, such analogues should be very good inhibitors of the enzymes and should bind very tightly to them.[56]

In the application of this idea to the study of lysozyme, Secemski and Lienhard[57] prepared the δ-lactone (TACL) derived from tetra-N-acetylchitotetraose, (Glc NAc)$_4$, and found it to be a very effective inhibitor of the enzyme. A crystallographic study by Ford *et al.*[58] showed that it binds in a way that closely resembles the model shown in Fig. 7.12 which may be taken, therefore, to represent the structure taken up in the transition state of the catalysed reaction. (It should be noted that there is still some doubt about the binding of the two sugar molecules at the bottom of the active site cleft which has not yet been observed in detail (cf., Pincus *et al.*).[59]

But this leaves a question. The sugar residue in Fig. 7.12 that is supposed to incorporate the carbonium-oxonium ion during the reaction is in the strained sofa conformation resembling that observed in the lactone: what brings this about? In the original structural work it was supposed that this distortion was brought about on substrate binding as the result of unfavourable contacts between enzyme and subtrate.[52] When the inherent flexibility of the enzyme molecule is taken into account, however, it seems now an oversimple view that such a distortion would be brought about by steric effects alone[60] and electronic effects arising from electric fields in the active site clearly must be taken into account.[61] The study of such electric fields and their effects is another growth area in protein research.[62]

Not all complexes between oligosaccharides and lysozyme are well represented by Fig. 7.12, however, and the differences are interesting and important. Kelly *et al.*[63] have studied the binding of a trisaccharide derived from bacterial cell walls and Fig. 7.13 shows that whereas the transition-state analogue

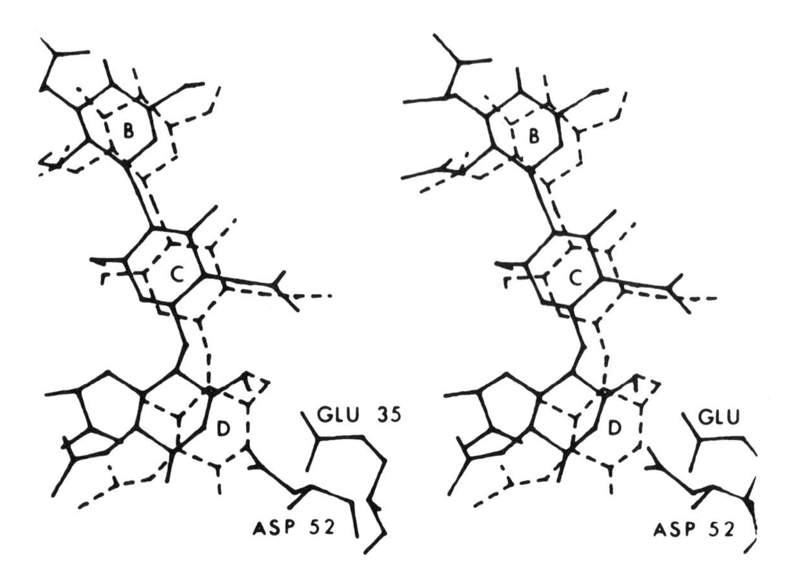

FIG. 7.13. The binding to lysozyme of the cell-wall trisaccharide (solid lines) and tetrasaccharide lactone (broken lines) seen at right angles to the view in Fig. 7.12. The lactone transition-state analogue penetrates more deeply into the active-site cleft.[63]

binds deeply in the active-site cleft, in good approximation to the two carboxylic acid groups that are involved in catalysis, the trisaccharide binds a little way out of the cleft. Arguably the trisaccharide complex is providing a view of the product in the course of being released from the enzyme (or a part of the initial Michaelis complex with the substrate) so that the two studies, taken together, provide snapshots of different stages along the reaction pathway — the initial substrate complex, the transition state of the reaction (or perhaps an intermediate) and the final product complex. These studies, therefore, represent progress towards another goal of protein crystallography, the visualization of all the intermediates and transition states involved in the activity of representative enzymes. So far this has been achieved most completely by Gilbert and Petsko[64] at MIT in their study of ribonuclease, which has involved the use of a transition-state analogue and low temperatures in a most elegant and demanding series of experiments.

Fig. 7.13 also brings us back to the earlier discussion of enzyme flexibility since the importance of such flexibility in permitting or facilitating the interconversion of these structures is easily imagined. With both enzyme and substrate engaged in complex intramolecular motion, it may be supposed that the substrate, tenuously bound in the first place, from time-to-time penetrates more deeply into the cleft and on one of these occasions forms an active complex so that catalysis occurs and products are formed and dissociate. In all of this the role of enzyme mobility is not yet understood in detail but it is the subject of intensive study and speculation.

Future prospects

This brief account of crystallographic enzymology shows that one of the limitations of X-ray crystallography, especially in

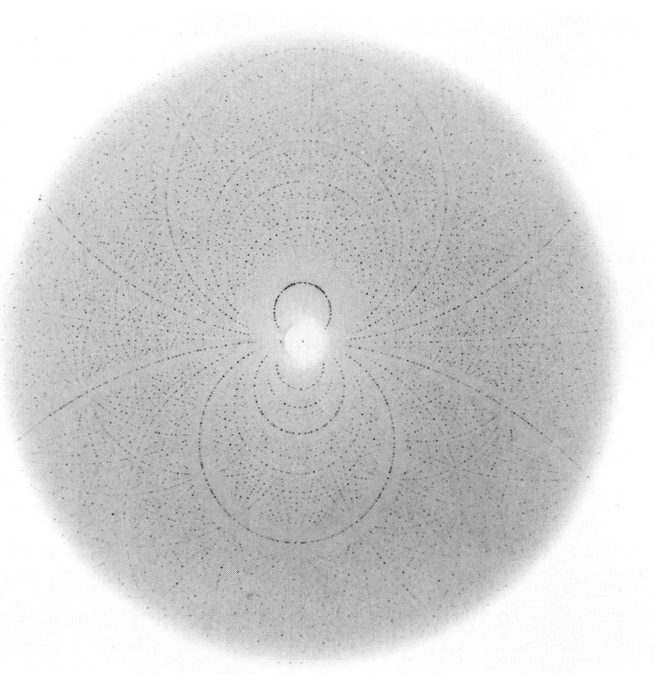

FIG. 7.14. A Laue photograph of a glycogen phosphorylase b crystal recorded on the wiggler beam line at the SERC Daresbury Laboratory Synchrotron Radiation Source.[64]

studies of time-dependent processes and transient complexes, is that the measurements take a very long time and only time (and space) averaged images of what is happening can be obtained. The use of synchrotron radiation as a source of X-rays is now making possible much more rapid data collection and exposure times that approach more nearly the natural periods and lifetimes of critical events and structures. Fig. 7.14 shows a Laue photograph from a phosphorylase b crystal recorded in the Wiggler beam line at the Synchrotron Radiation Source, SERC Daresbury Laboratory.[65] The photograph contains approximately 50% of the three-dimensional data to 3 Å resolution as single-component, spatially-resolved spots. The exposure time was 8 seconds.

More recently even more impressive photographs have been recorded with exposure times of 250 milliseconds. Clearly many processes are now being brought within the scope of time-resolved X-ray crystallography and the prospects for continuing progress are very bright. The centenary of the Patterson function is likely to provide the occasion for another exciting meeting.

Acknowledgements

I am indebted to my colleagues in Oxford and elsewhere for their collaboration and for their permission to quote from their work. The Laboratory is supported by the Medical Research Council and the Science and Engineering Research Council of the UK.

Discussion

Elizabeth Wood: How many spots were there in the Laue photograph?

David Phillips: Ten thousand or so, I think. But I have not counted them.

Jack Dunitz: I see a contradiction. You stress the similarity between what you were saying and what Peter Murray-Rust said this morning, but I see a contradiction in what you showed us at the very end of your lecture. When Peter was showing us what he considered to be the low energy vibrations of the sterol skeleton he emphasized that they were similar to the low energy normal vibrations of flat plate of roughly the same shape as the sterol molecule. By analogy, one might guess that a protein molecule of molecular weight of the order of magnitude of ten thousand would be behaving, to a first approximation, as a little blob of jelly. Now, it was difficult to see in the few picoseconds of the protein motions you showed us, but I had the impression that these were nothing like what one would expect for the low energy vibrations of a blob of jelly. In particular, in what you showed us there seemed to be many instances where neighboring bits of protein were moving with opposite phase, and I do not see how that can really correspond to a very low energy mode. Besides, in your list there was just one mode showing fairly clear low-energy separation at about 4 cm^{-1}, whereas I would have guessed there must be about a thousand or so modes in this low-energy region.

David Phillips: These were computed according to Levitt's calculations for BPTI. This shows that we can, thanks to computers, do this sort of calculation. It all depends very much on what confidence you are prepared to put in the interatomic potentials that go into these calculations. It also shows that one has to make progress in basic physical chemistry, biological chemistry, all at the same time.

David Harker: There are no experimental data?

David Phillips: No one has observed these modes very reliably, so far as I know.

Ray Pepinsky: Would not you see very low frequency Raman resonance?

David Phillips: That has been looked for; there have been some tentative reports, but nothing very substantial, that I know of.

References

1. Bernal, J. D. and Crowfoot, D. M. *Nature (London)* **133**, 794–795 (1934).
2. Cork, J. M. *Phil. Mag.* **4**, 688–698 (1927).
3. Lipson, H. and Beevers, C. A. *Proc. Roy. Soc. (London)* **A148**, 664–680 (1935).
4. Robertson, J. M. *J. Chem. Soc.*, 1195–1209 (1936).
5. Hodgkin, D. M. C. *Biog. Memoirs of Fellows of the Royal Society of London*, **26**, 17–84 (1980).
6. Bernal, J. D. *Nature (London)* **143**, 663–667 (1939).
7. Robertson, J. M. *Nature (London)* **143**, 75–76 (1939).
8. Wilson, A. J. C. *Acta Cryst.* **2**, 318–321 (1949).
9. Green, D. M., Ingram, V. M. and Perutz, M. F. *Proc. Roy. Soc. (London)* **A225**, 287–307 (1954).
10. Bokhoven, C., Schoone, J. C. and Bijvoet, J. M. *Acta Cryst.* **4**, 275–280 (1951).
11. Harker, D. *Acta Cryst.* **9**, 1–9 (1956).
12. Kendrew, J. C., Bodo, G., Dintzis, H. M., Parrish, R.G., Wyckoff, H.W. and Phillips, D. C. *Nature (London)* **181**, 662–666 (1958).
13. Blow, D. M. *Proc. Roy. Soc. (London)* **A247**, 302–336 (1958).
14. Bijvoet, J. M. *Nature (London)* **173**, 888–891 (1954).
15. Blow, D. M. and Crick, F. H. C. *Acta Cryst.* **12**, 794–802 (1959).
16. Kendrew, J. C., Dickerson, R. E., Strandberg, B. E., Hart, R. G., Davies, D. R., Phillips, D. C. and Shore, V. C. *Nature (London)* **185**, 422–427 (1960).
17. Scouloudi, H. *Proc. Roy. Soc. (London)* **A258**, 181–201 (1960).
18. Teeter, M. M. and Hendrickson, W. A. *J. Mol. Biol.* **127**, 219–224 (1979).
19. Rossmann, M. G. and Blow, D. M. *Acta Cryst.* **15**, 24–31 (1962).
20. Rossmann, M. G., Blow, D. M., Harding, M. M. and Coller, E. *Acta Cryst.* **17**, 338–342 (1964).
21. Champness, J. N., Bloomer, A. C., Bricogne, G., Butler, P. J. G. and Klug, A. *Nature (London)* **259**, 20–24 (1976).
22. Pauling, L., Corey, R. B. and Branson, H. R. *Proc. Natl. Acad. Sci. USA* **37**, 205–211 (1951).
23. Pauling, L. and Corey, R. B. *Proc. Natl. Acad. Sci. USA* **37**, 235–240 (1951).
24. Levitt, M. and Chothia, C. *Nature (London)* **261**, 552–558 (1976).

25. Richardson, J. *Adv. Protein Chem.* **34**, 167–339 (1981).
26. Banner, D. W., Bloomer, A. C., Petsko, G. A., Phillips, D. C., Dodson, C. I. and Wilson, I. A. *Nature (London)* **255**, 609–614 (1975).
27. Carrell, H. L., Rubin, B. H., Hurley, T. J. and Glusker, J. *J. Biol. Chem.* **259**, 3230–3236 (1984).
28. Murray-Rust, P. This volume, Chapter 6.
29. Artymiuk, P. J. and Blake, C. C. F. *J. Mol. Biol.* **152**, 737–762 (1981).
30. Blake, C. C. F., Pulford, W. C. A. and Artymiuk, P. J. *J. Mol. Biol.* **167**, 693–723 (1983).
31. Konnert, J.H. and Hendrickson, W. A. *Acta Cryst.* **A36**, 344–349 (1980).
32. Waser, J. *Acta Cryst.* **16**, 1091–1094 (1963).
33. Rollett, J. S. In *Crystallographic Computing* (F. R. Ahmed, S. R. Hall and C. P. Huber, eds.)pp. 167–181. Munksgaard: Copenhagen (1970).
34. Johnson, C. K. *Acta Cryst.* **A25**, 187–194 (1969).
35. Glover, I., Haneef, I., Pitts, J. E., Wood, S., Moss, D. S., Tickle, I. and Blundell, T.L. *Biopolymers* **22**, 293–304 (1983).
36. Sternberg, M. J. E., Grace, D. E. P. and Phillips, D. C. *J. Mol. Biol.* **130**, 231–253 (1979).
37. Artymiuk, P. J., Blake, C. C. F., Grace, D. E. P., Oatley, S. J., Phillips, D. C. and Sternberg, M. J. E. *Nature (London)* **280**, 563–568 (1979).
38. Artymiuk, P. J., Grace, D. E. P., Handoll, H. and Phillips, D. C. Unpublished.
39. Bentley and Mason, personal communication.
40. Lonsdale, K. and Milledge, H. J. *Acta Cryst.* **14**, 59–61 (1961).
41. Hartmann, H., Parak, F., Steigemann, W., Petsko, G. A., Ringe Ponzi, D. and Frauenfelder, H. *Proc. Natl. Acad. Sci. USA* **79**, 4967–4971 (1982).
42. Petsko, G. A. and Ringe, D. *Annu. Rev. Biophys. Bioeng.* **13**, 331–371 (1984).
43. Poulson, F. M., Hoch, J. C. and Dobson, C. M. *Biochemistry* **19**, 2597–2607 (1980).
44. Huber, R. *Trends Biochem. Sci.* **4**, 271–276 (1979).
45. Perutz, M. F. and Matthews, F. S. *J. Mol. Biol.* **21**, 199–202 (1966).
46. Frauenfelder, H., Petsko, G. A. and Tsernoglou, D. *Nature (London)* **280**, 558–568 (1979).
47. Karplus, M. and McCammon, J. A. *CRC Crit. Rev. Biochem.* **9**, 293–349 (1981).

48. van Gunsteren, N. F., Berendsen, H. J. C., Hermans, J., Hol, W. E. J. and Postwa, J. P. M. *Proc. Natl. Acad. Sci. USA* **80**, 4315–4319 (1983).
49. Levitt, M. *J. Mol. Biol.* **168**, 595–657 (1983).
50. Levitt, M., Sander, C. and Stern, P. S. *J. Mol. Biol.* **181**, 423–447 (1985).
51. McCammon, J. A., Gelin, B. R., Karplus, M. and Wolynes, P. G. *Nature (London)* **262**, 325–326 (1976).
52. Phillips, D. C. *Sci. Amer.* **215**, 78–90 (1966).
53. Blake, C. C. F., Johnson, L. N., Mair, G. A., North, A. C. T., Phillips, D. C. and Sarma, V. R. *Proc. Roy. Soc. (London)* **B167**, 378–388 (1967).
54. Kuramitsu, S., Ikeda, K. and Hamaguchi, K. *J. Biochem. (Tokyo)* **82**, 585–597 (1977).
55. Pauling, L. *Chem. Eng. News* **24**, 1375–1377 (1946).
56. Wolfenden, R. *Nature (London)* **223**, 704–705 (1969).
57. Secemski, I. I. and Lienhard, G. E. *J. Am. Chem. Soc.* **93**, 3549–3550 (1971).
58. Ford, L. O., Johnson, L. N., Machin, P. A., Phillips, D. C. and Tjian, R. *J. Mol. Biol.* **88**, 349–371 (1974).
59. Pincus, M. R., Zimmerman, S. S. and Scheraga, H. A. *Proc. Natl. Acad. Sci. USA* **74**, 2629–2633 (1977).
60. Levitt, M. In *Peptides, Polypeptides and Proteins*, (E. R. Blout, F. A. Bovey, M. Goodman and N. Lotan, eds.) pp. 99–113. Wiley: New York (1974).
61. Warshel, A. *Accts. Chem. Res.* **14**, 284–290 (1981).
62. Warwicker, J. and Watson, H. C. *J. Mol. Biol.* **157**, 671–679 (1982).
63. Kelly, J. A., Sielecki, A. R., Sykes, B. D., James, M. N. G. and Phillips, D. C. *Nature (London)* **282**, 875–878 (1979).
64. Gilbert, W. A. and Petsko, G. A. *Biochemistry*, in press (1985).
65. Hajdu, J., Stuart, D. I., Acharya, R., McLaughlin, P., Barford, D., Johnson, L. N. and Helliwell, J. R. Unpublished.

8. Some experiences with the Patterson function in protein crystallography

Warner E. Love, Rufus W. Burlingame, John S. Sack, William E. Royer, Jr and Bradford C. Braden.

Warner Love is a native Philadelphian. He was educated at Swarthmore and obtained a Ph.D. in physiology from the University of Pennsylvania in 1951. He then went to the Johnson Foundation, and then came here to the Institute, and worked with Lindo from 1955 to 1957, before going to The Johns Hopkins University, where he is now Professor of Biophysics. He is particularly interested in the relation of structure to function, and particularly, of course, interested in the structure of hemoglobins (it used to be especially hemoglobin from lampreys, and all the time that he was in the lab here at the Institute he had lamprey hemoglobin stuck away in the refrigerator, and worked on it from time to time). He is also interested in chromosome structure and has recently published an article on the octameric histone core of nucleosomes.

We are pleased to have been given the privilege of contributing to this celebration of the fiftieth anniversary of A. Lindo Patterson's F^2 series.

The Patterson function is indispensable in protein crystallography, and this is to be expected because the Patterson function is the Fourier transform of the diffracted intensities, and they are our fundamental experimental data.

The Patterson function is used in protein crystallography in the methods of multiple isomorphous replacement (MHAIR) and in molecular replacement (MR), and Michael Rossmann[1] has proposed the recovery of 'images' from it (IR) but so far, IR has been but little used.

For MHAIR, an isomorphous series of crystals is prepared consisting of the native protein and at least two different good heavy atom derivatives. In a good derivative, the heavy atoms are bound specifically, and with high occupancy. Three sets of diffraction data, $|F_p|$, $|F_{h1}|$ and $|F_{h2}|$ are obtained and from them 'difference Patterson' maps are calculated, using for example coefficients $(|F_{h1}| - |F_p|)^2$ suggested by Blow[2] and

Rossmann.[3] Such maps de-emphasize the peaks due to vectors between the protein and the heavy atoms, *i.e.*, the protein images in the heavy atoms, so that the vectors between the heavy atoms themselves can be more easily identified. MHAIR has been enormously successful. Anomalous dispersion effects can also be used, although they are smaller and therefore greater care must be taken in measuring them.

The difference Patterson calculated with coefficients $|F_{h1}|^2 - |F_p|^2$ is truly the difference between the Pattersons of the protein with and without heavy atoms attached, and it is from this true difference Patterson that Rossmann [1] has proposed the retrieval of the protein image. Molecular replacement seeks to use the known structure of a molecule to determine the structure of an unknown crystal of that molecule or of an unknown crystal of a close molecular relative. Reciprocal space and Patterson space methods have both been developed.[4] In either case, in principle, an orientation and position are sought for the known structure in the asymmetric unit of the unknown crystal.

Octameric histone core of the nucleosome

An interesting example of heavy atom replacement from our laboratory is the analysis of the tetrakis(acetoxymercuri) methane (TAMM[5]) derivative of the octameric histone core of the nucleosome. X-ray quality crystals of the histone octamer were grown by R.W.B. in the laboratory of Professor Evangelos N. Moudrianakis in the Biology Department.[6] Using diffractometer data at low resolution it was established that TAMM made a good derivative, and then, ably assisted by his staff, we utilized Professor Ng-H. Xuong's multiwire detector facility at San Diego[7] to collect 3.1 Å data. The data sets were collected in 100 hours. Each reflection was observed, on the average, about 3.5 times. The statistical errors of the intensities were ten times lower than our diffractometer data, because the counting times were correspondingly longer.

The space group was either $P3_121$ or $P3_221$. In either case, for a single site derivative, there should be 15 unique non-origin heavy atom self-vector peaks in the unit cell of the difference Patterson function. Fig. 8.1 is the section at $w = \frac{1}{3}$ of the

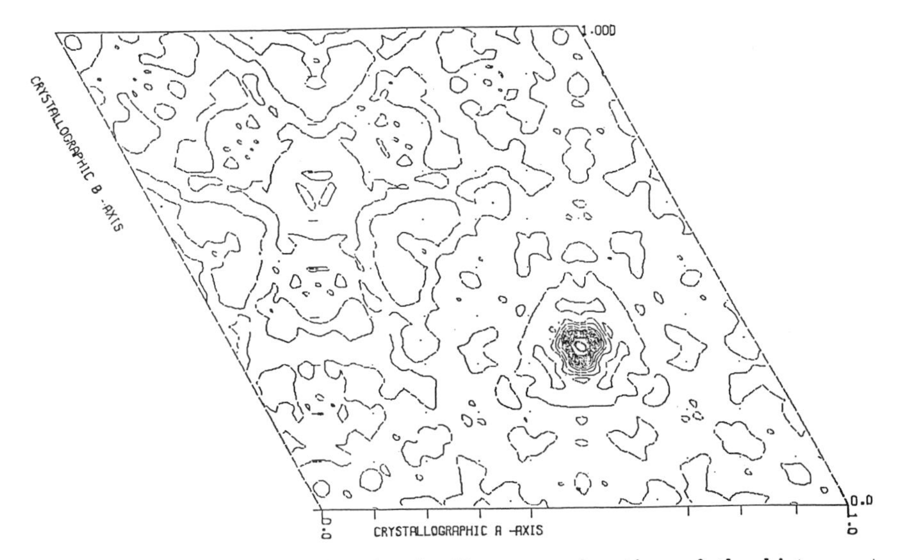

FIG. 8.1. The section in the Patterson function of the histone at $w = \frac{1}{3}$. The contour levels are equally spaced.

map calculated using $(|F_{h1}| - |F_p|)^2$ as coefficients. Except for its centrosymmetric mate, the only non-origin peak in the entire unit cell is shown in this section. The analysis is given in Table 8.1. A single peak at $u, v, w = \frac{2}{3}, \frac{1}{3}, \frac{1}{3}$ will occur, with its centrosymmetric mate at $\frac{1}{3}, \frac{2}{3}, \frac{2}{3}$, if there is a heavy atom at $x, y, z = \frac{2}{3}, \frac{2}{3}, 0$ in space group $P3_121$ or at $x, y, z = \frac{1}{3}, \frac{1}{3}, 0$ in space group $P3_221$ (anomalous dispersion selected $P3_221$ subsequently). The Wyckoff special position in these space groups are of the form $x, x, 0$ and symmetry mates, or $x, x, \frac{1}{2}$ and symmetry mates. The mercury position is thus an additionally specialized Wyckoff position when $x = \frac{1}{3}$. Atoms occupying these 'special' special positions give rise to coalesced multiply weighted Harker peaks in the Patterson function, and conversely contribute only to the sub-set of the general reflections for which $-h + k + l = 3n$. These are the systematic absences for an obverse rhombohedral lattice and indeed, atoms at $\frac{2}{3}, \frac{2}{3}, 0$; $\frac{1}{3}, 0, \frac{1}{3}$; $0, \frac{1}{3}, \frac{2}{3}$ in space group $P3_121$ or at $\frac{1}{3}, \frac{1}{3}, 0$; $\frac{2}{3}, 0, \frac{2}{3}$; $0, \frac{2}{3}, \frac{1}{3}$ in $P3_221$ form an obverse rhombohedral lattice and contribute to reflections only if $h - k + l = 3n$. The details are in Table 8.1. In our case the mercury atom proved to be about 1.5 Å

TABLE 8.1. Analysis of Patterson coordinates

Patterson coordinates*	Space group	Crystal coordinates**	Selection rule
$\frac{1}{3}, \frac{2}{3}, \frac{1}{3}$	$P3_121$	(*i*) $\frac{1}{3}, \frac{1}{3}, 0$	$h - k + l = 3n$
	$P3_221$	(*ii*) $\frac{2}{3}, \frac{2}{3}, 0$	
$\frac{1}{3}, \frac{2}{3}, \frac{2}{3}$	$P3_121$	(*iii*) $\frac{2}{3}, \frac{2}{3}, 0$	$-h + k + l = 3n$
	$P3_221$	(*iv*) $\frac{1}{3}, \frac{1}{3}, 0$	

* Patterson peaks in the unit cell only at $u, v, w = 0, 0, 0$; and $\pm(\frac{1}{3}, \frac{2}{3}, \frac{1}{3})$ or at 0, 0, 0 ; and $\pm(\frac{1}{3}, \frac{2}{3}, \frac{2}{3})$

** Equivalent positions for (i) and (iii) above form right-handed helices that are related by a 2-fold axis parallel to $\vec{c}$. Conversely, positions ii and iv form left-handed helices that are also related by a 2-fold axis parallel to $\vec{c}$.

away from this 'special' special position and therefore it contributes insignificantly to the low order reflections for which $-h + k + l = 3n$, but contributes increasingly significantly to them at higher resolution. In order to obtain phases for the low order reflections and to break the phase ambiguity inherent in a single isomorphous replacement, at Pittsburgh and with his expert help, B.C. Wang's procedures of Iterated Single Isomorphous Replacement and solvent flattening were employed. The excellent quality of the electron density map at 3.3 Å resolution is illustrated in Fig. 8.2 in which two α helices are shown. In the crystals of the histone octamer, $(H2A\ H2B\ H3\ H4)_2$, the molecule is on a crystallographic 2-fold axis. Thus the molecule has 2-fold rotational symmetry. The crystals are about 65% solvent and the molecular boundary is very clear. The molecule is remarkably open with long 4 – 14 Å wide channels running through it. The mass of H3 is defined by its

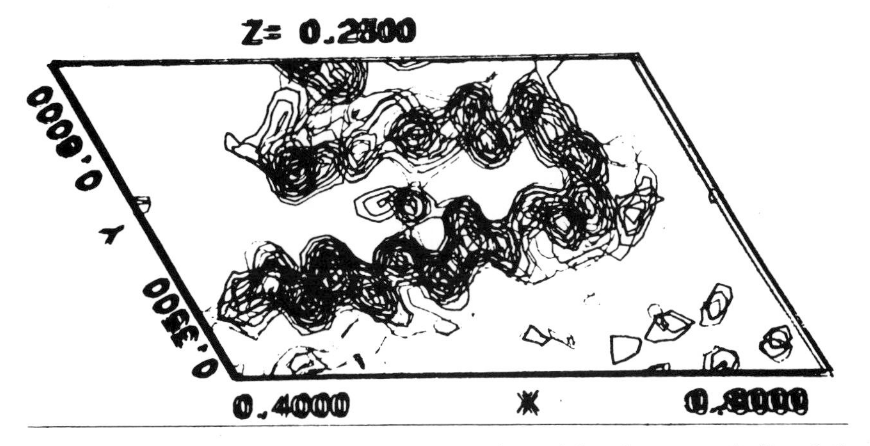

FIG. 8.2. A stack of sections normal to $\vec{c}$ showing an α helix of the histone octamer.

proximity to the mercury binding site because the single cysteine per half-octomer is in H3. Large segments of polypeptide chain are easy to follow, but at this writing we have not yet fitted the amino acid sequences to the electron density map. As expected[8], the octamer is tripartite with H2A H2B dimers flanking an H3 H4 tetramer. Fig. 8.3 is a diagram derived from solution studies,[8] that shows the organization of the octamer. (A more detailed description of this structure is in *Science*.[9,10])

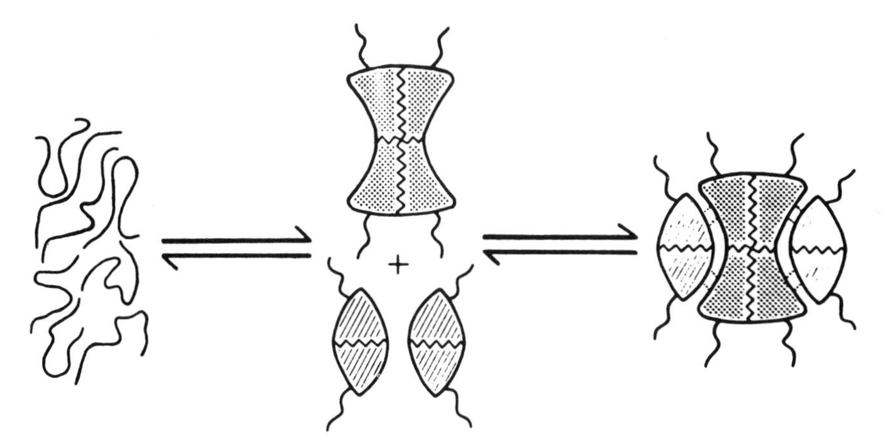

FIG. 8.3. A diagram of the histone octamer showing its tripartite nature, taken from the cover of a reprint by T.H. Eickbush, D.K. Watson and E.N. Moudrianakis.[8]

Molecular replacement

Molecular replacement has been used in this laboratory to determine a number of low resolution crystal structures that in some cases provided starting models for further work at higher resolution; in other cases the low resolution structure itself provided the information that had been sought; and in the remaining cases, the structure analyses are in progress.

The search molecule for the crystal structure of yellow fin tuna myoglobin was sperm whale myoglobin. The tuna myoglobin structure at 6 Å resolution showed that myoglobins from a fish and a mammal are alike.[11]

Normal human deoxy hemoglobin, Hb A, crystallized at high ionic strength,[12] was the test molecule for a variety of human deoxy Hb A structures. It was shown, as a control experiment at low resolution, that the molecular structure of deoxy Hb A is the same in crystals grown by salting-out and in crystals prepared with the then-novel agent, polyethylene glycol.[13] The structure of monoclinic crystals of deoxy Hb S, β6 glu$\rightarrow$val, was initially determined[14] at 5.5 Å and the model has since been refined[15,16] by restrained least squares[17] at 3.0 Å resolution. The intermolecular contacts in these crystals, prepared with polyethylene glycol, are very closely related to those that occur in sickled red cells.[18,19] The structures of two different kinds of crystal of deoxy Hb C, β6 glu$\rightarrow$lys, were determined.[20] These showed that the intermolecular contacts in crystals of deoxy Hb A, S, C and F (F = human fetal) are closely similar even though six different crystal lattices are involved, *i.e.*, 2 of A and C and one each of S and F. In a different orthorhombic crystal form of deoxy Hb S, the intermolecular contacts are rather unrelated to sickling.[21]

Spot (fish) Hb and trout Hb IV have very strong Root effects, *i.e.*, O_2 and CO binding are virtually abolished at low pH, whereas allosteric effectors such as H^+, and organic phosphates are without effect on ligand binding by trout Hb I.[22] Horse metHb was the search molecule for the carbon monoxide derivative of spot Hb and trout Hb's I and IV.

The general molecular replacement problem is six dimensional. There are three angles to be found that define the orientation of the test molecule in the unknown crystal's unit cell, and there are three components of the vector that defines the test molecule's position in the unit cell. The carbon monoxide complex of spot Hb crystallizes with the molecular 2-fold axis coincident with a 2-fold axis of the space group $C2$,[23] and thus the asymmetric unit is half of the hemoglobin tetramer, and thus there is only one unknown angular parameter to be found. There are no positional unknown parameters because there is no unique position on the 2-fold axis in C2. The polarity of the molecular dyad is also unknown. This problem with Spot Hb is close to the case of horse metHb discussed by Rossmann and Blow.[24] Both cases are unusually simple because they are both one-dimensional. However, a self-rotation function in which the Patterson function of the crystal is rotated against itself was used for horse metHb whereas the self-Patterson of horse metHb was rotated against the Patterson of the spot Hb crystal.[21] Thus, for horse Hb, the molecular pseudo-dyad axes were found while for spot Hb, a rotation was found that would bring the horse and spot molecules into coincidence (Fig. 8.4). With the crystallographic dyad antiparallel to the molecular dyad, a somewhat better fit is observed indicating in the parallel case that α chains are being superimposed on β chains, while in the antiparallel case, like is superimposed on like.[21]

A rather straight-forward molecular replacement for trout Hb I is shown in Fig. 8.5. The initial R-value for 10 Å - 6 Å Bragg reflections was 52% when the horse metHb molecule was placed in the trout Hb I unit cell according to the data in Fig. 8.5. After systematic perturbation of the six parameters,[13] and using all the reflections to 6 Å, the R-value dropped to 36%. The changes in Θ_1, Θ_2, Θ_3, X, Y, and Z respectively were $-1.07°$ $+1.05°$, $+0.25°$, $+0.0022$ Å, $+0.0047$ Å, and $+0.0020$ Å.[21]

In contrast to the above successes with molecular replacement we have failed with trout Hb IV. The rotation function for trout Hb IV had a nice strong peak that was used in the translation function which had two virtually equally strong sets of peaks. Packing considerations, difference Fourier maps, and 'hemeless' Fourier maps have all failed to discriminate between these two alternative positions, and we expect resolution of the

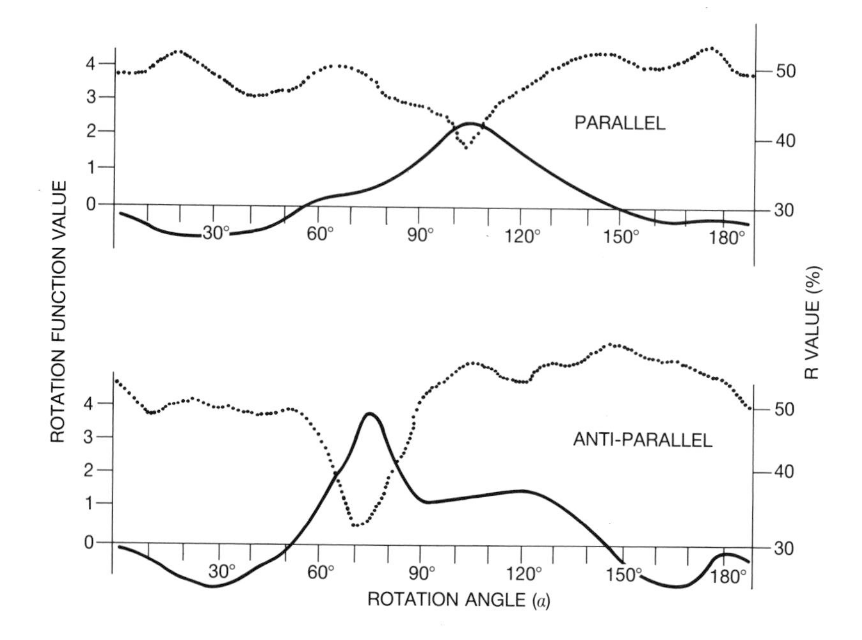

FIG. 8.4. Rotation functions for spot (fish) Hb. For the 'parallel' case, the molecular 2–fold axis was parallel to the $\vec{b}$ axis of the crystal and antiparallel for the 'antiparallel' case. The two solutions are related by a rotation that exchanges the positions of the α and β subunits.

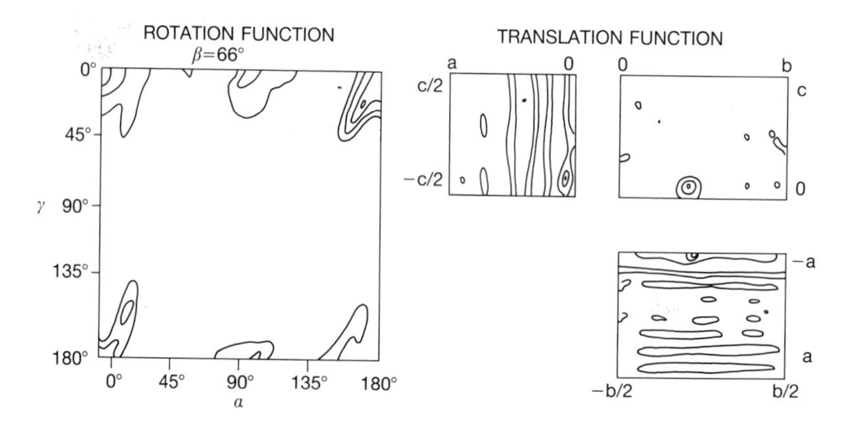

FIG. 8.5. Molecular replacement - trout Hb I. The *R*-value for all reflections to 6 Å resolution, after *R*-value minimization by means of parameter perturbations, was 0.36.

problem will be obtained from an anomalous dispersion Patterson function,[25] or MHAIR.

Tetrameric and dimeric clam hemoglobins

The coelomocytes of blood clams of the *Arcid* family contain cooperative dimeric and tetrameric hemoglobins of about 33,000 and 66,000 molecular weight.[26–32] The dimer is made of two identical polypeptide chains each with a heme group,[30–32] and Hill's n is about 1.5.[29,33] Cooperativity in the dimeric hemoglobin is intrinsic; [34,35] it is not dependent on hemoglobin concentration and therefore is not linked to subunit association, and oxygen binding is unaffected by the allosteric effectors of vertebrate hemoglobin, *e.g.,* H^+, Cl^-, organic phosphate.[30] The same is essentially true for the tetrameric hemoglobin[34] except that the deoxy tetramer can self-associate to make either trimers or tetramers of tetramers, and pH does have a small effect on binding.[36,37]

Crystals of the carbon monoxide complex of both hemoglobins were grown by Faith F. Fenderson from material prepared in Rome and graciously given to us by Dr. Emilia Chiancone. The crystal data are in Table 8.2. The structure of crystals of the tetrameric hemoglobin was determined by MHAIR followed by symmetry averaging, and that of the dimeric hemoglobin by molecular replacement.[38]

Isomorphous heavy atom derivatives of the tetramer crystals were obtained by soaking them in modified mother liquor

TABLE 8.2. Dimeric and tetrameric crystal data

	Ligand	Space group	a	b	c	β	Z^*
dimer	CO	$C2$	93.3	44.1	83.6	122.0	1
tetramer	CO	$C222_1$	102.1	94.4	128.0	-	1

* Z is the number of molecules per asymmetric unit

that was 0.8 to 2.0mM in $KAu(CN)_2$, $UO_2(CH_3COO^-)$ and potassium *cis*-dichloro diamino platinate. Heavy atom binding sites were deduced from difference Patterson maps that were made with the square of the amplitude differences.

The Harker sections could not be interpreted in terms of Harker self-peaks, but the key to the problem proved to be strong peaks due to cross-vectors that fortuitously occurred on the Harker sections. The analysis was aided by space group-specific programs that sampled the difference Patterson map where cross vectors would result in Patterson density. A similar procedure was used in the analysis of heavy atom derivatives of the histone crystals. Thus the likelihood or validity of a postulated pair of atomic positions could be evaluated. The approximate heavy atom positions obtained in this way were refined separately and finally all together.[39]

The Fourier synthesis, with a figure of merit = 0.79 that was made at 5.5 Å resolution was only partially interpretable. The boundaries of the tetramers were identified partly because the solvent channels were clear and partly because of the packing constraints due to the symmetry of the space group. There were rods of density of the size of α helices but it was not possible to identify myoglobin-folded subunits within the tetramers. However, a local molecular two-fold symmetry axis was found by inspection and was used for eleven cycles of symmetry averaging. The map calculated with the improved phases was easy to interpret. Fig. 8.6 is a photograph of a balsa wood model of the tetrameric hemoglobin.

The tetramer is made of four myoglobin-folded subunits, and in each subunit the chain could be traced with no more than three short breaks. Fig. 8.7 is a photograph of balsa wood models of various myoglobin folds and of one of the subunits of the tetramer. Each subunit has an extra helix called 'Pre-A' at the amino end of the chain. In common with lamprey[40] and hagfish[41] hemoglobins and the α chains of shark hemoglobin[42] there are extra residues at the amino end of the clam polypeptide chains,[43–45] and we suppose the Pre-A helix is made of those extra residues and perhaps some of those that usually comprise the beginning of the A helix.

The tetramer is a dimer of dimers. The molecular dyad

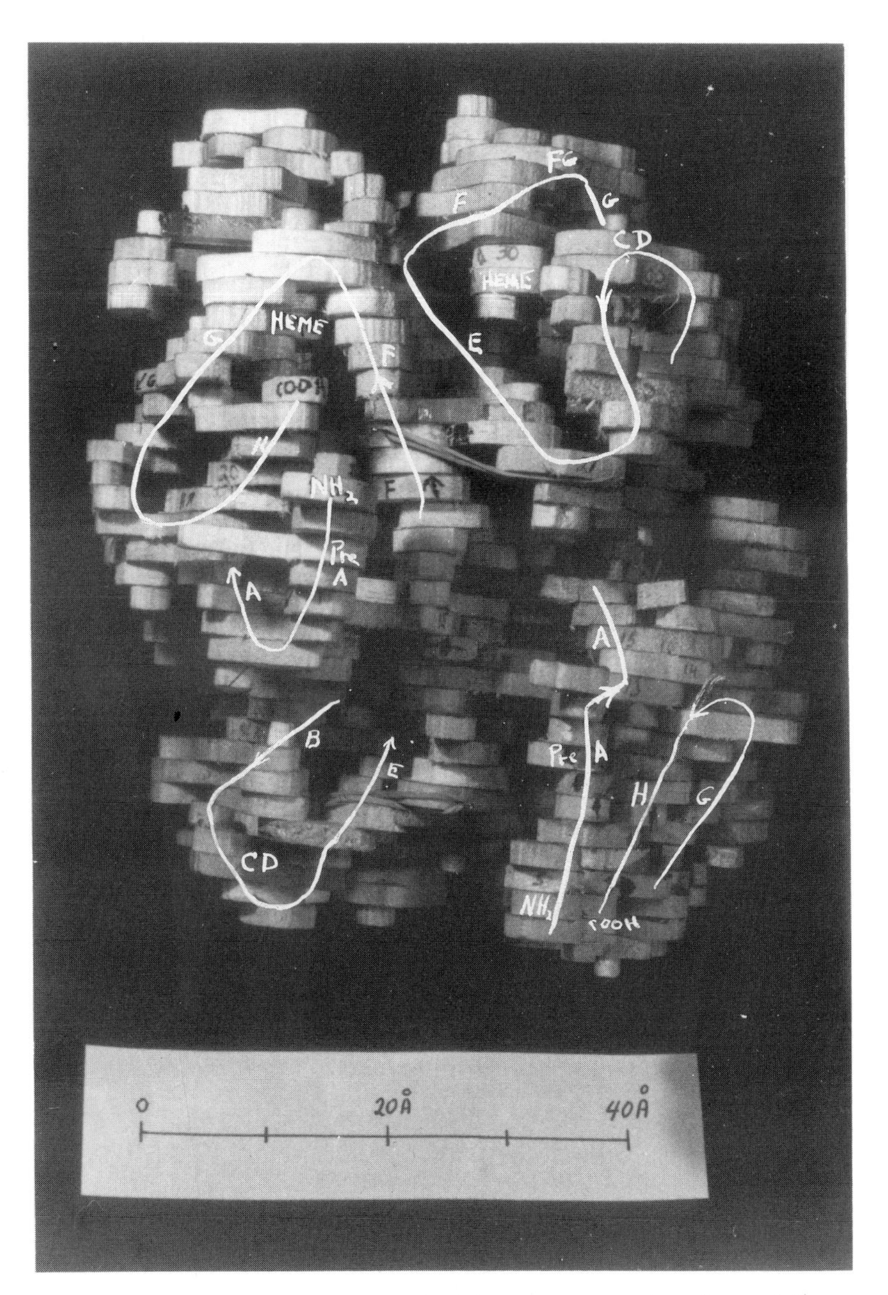

FIG. 8.6. A balsa wood model of the tetrameric clam hemoglobin, at 5.5 Å resolution.

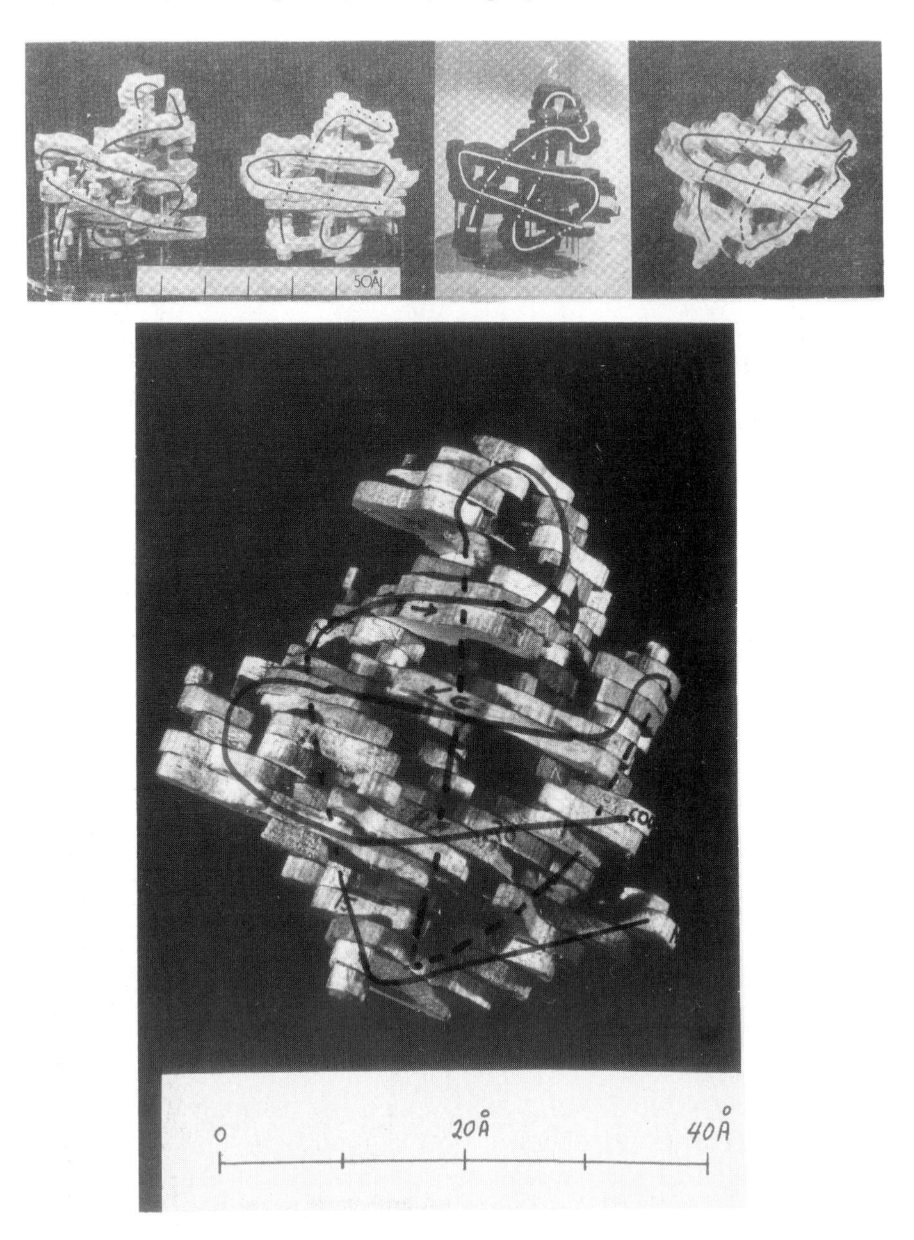

FIG. 8.7. Across the top from the left, sperm whale myoglobin, α and β subunits of horse hemoglobin, and *Glycera* hemoglobin. The larger picture is of a subunit of the tetrameric clam hemoglobin.

axis that relates one half-molecule to the other is the axis that was used in symmetry averaging. The subunits within the dimer are related by a different 2-fold axis that is inclined to the molecular axis by about 75°. In the half-molecule the myoglobin-folds are assembled quite differently from the vertebrate tetramer. Here, the subunits are 'back-to-front' with respect to their relationship in the vertebrate tetramer. The E helix – heme – F helix surfaces face each other in apposition whereas in vertebrate hemoglobin they face outward. The interface between the two half-molecules is much less extensive than that between the subunit within the half- molecule, and here again the contacting surfaces are very different from those in the vertebrate tetramer.

Concomitant experiments with crystals of the dimeric hemoglobin had been disappointing. We had been unable to prepare heavy atom derivatives of the dimer crystals, and we had failed in molecular replacement using various vertebrate α-β dimers as search molecules. However, the compactness of the half-molecule of the clam tetramer suggested that the half-molecule itself might be a good model for a molecular replacement determination of the class dimer crystal structure, and this indeed proved to be true.

With the help of programming obtained from Professor Wayne A. Hendrickson, the electron density of the half-molecule of the tetramer was excised and its 'continuous' Fourier transform was prepared. The 'fast' program of R.A. Crowther was used to compute the rotation function[46] in which the highest peak was 5.3 rms units above the mean. Using this orientation, the translation function of Crowther and Blow[47] was used to find the position of the dimer in its unit cell. The translation function is shown in Fig. 8.8. The highest peak is 5.2 standard deviations above the mean of the map. The solutions of the rotation and translation functions were refined by an R-value minimization procedure of Ward *et al.*[13] in which R changed from 46.2% to 42.8%. The final orientation was 2.2° and 0.2 Å away from the peaks of the rotation and translation functions. Phases from the model were combined with the observed amplitudes to compute an electron intensity map.

The validity of the molecular replacement was checked

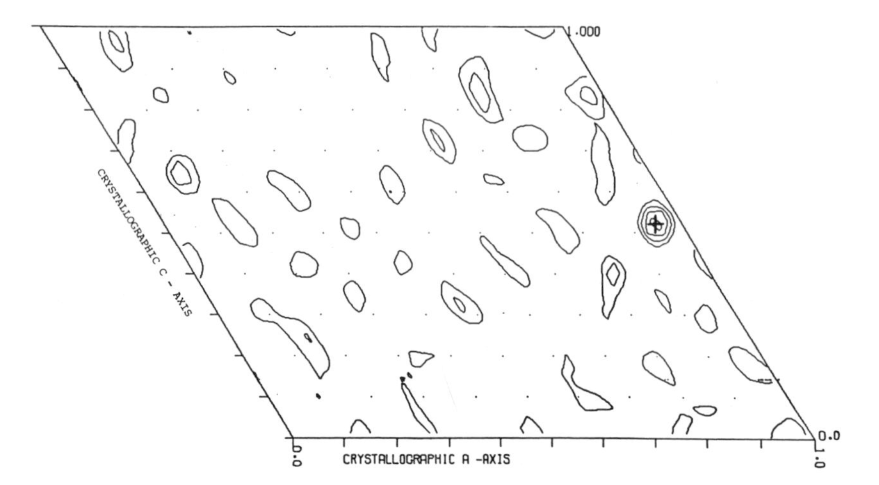

FIG. 8.8. The translation function for the dimeric clam hemoglobin.

against five criteria. First, the R-value, 42.8%, compares well with other successful molecular replacements.[11,13,14,20,21] Second, the molecules did not interpenetrate. Third, a difference map made with coefficients $(|F_o| - |F_c|)\exp(i\alpha c)$ showed easily explained peaks and holes in regions where the tetramer map was less than satisfactory. Fourth, a 'hemeless' Fourier map prepared with the heme contributions deleted from the calculated phases showed good density at the heme positions. Information about the hemes was put into this map only by the observed dimer amplitudes. Finally, an anomalous dispersion Patterson map[25] was calculated from Friedel differences measured from a native dimeric hemoglobin crystal. Good density ocurred in the anomalous Patterson map corresponding to all of the iron-iron vectors.

Within the dimeric Hb and within the half-molecule of the tetrameric Hb, the two subunits are related by a local 2-fold axis that is shown in Fig. 8.9. It is represented by the rod that runs in front of the model diagonally from the upper left. The axis runs across the E helix-heme-F helix face of the monomer. For comparison the α subunit of human carbon monoxide hemoglobin is also shown in Fig. 8.9. The pseudo dyad that relates α_1 to α_2 runs horizontally *behind* the molecule. Fig. 8.10 shows how the 2-fold axes of the vertebrate tetramer are oriented relative to those of the clam tetramer.

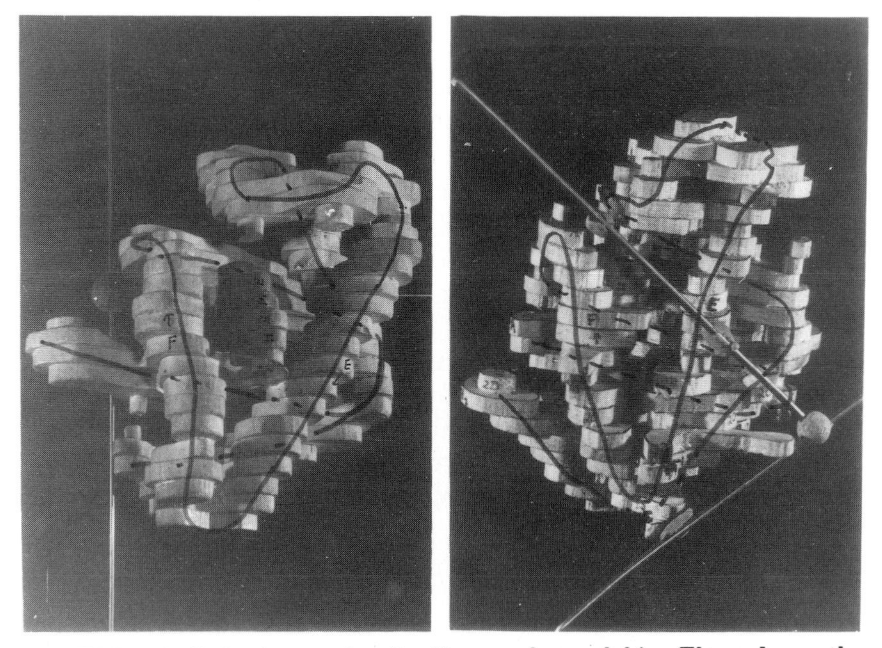

FIG. 8.9. Left: An α subunit of human hemoglobin. The rods are the 2-fold axes that generate the tetramer. Perpendicular to the two axes is the third axis, not shown, of the triple of axes for the point group 222. Right: A subunit of the tetrameric clam hemoglobin. The axis running from upper left to lower right, operating first, produces the dimeric clam hemoglobin and/or the half-molecule of the tetramer. The second axis running more or less horizontally makes the tetramer from the dimer. There is no third axis. The tetrameric clam hemoglobin does ***not*** have 222 pseudo-symmetry.

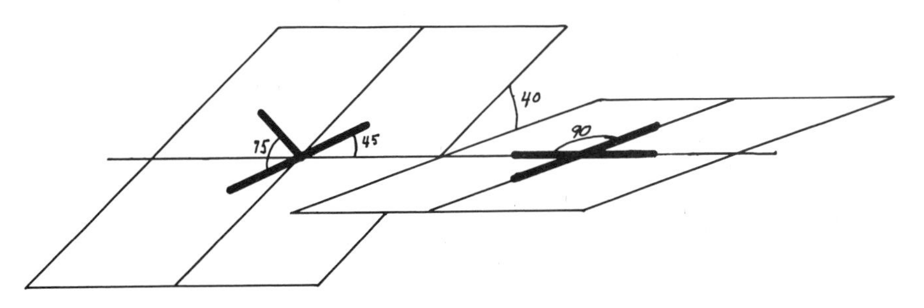

FIG. 8.10. The relation of the axes in vertebrate (right) and tetrameric clam (left) hemoglobins.

Fig. 8.11 is a sketch of the arrangement of the subunits of the clam tetramer that shows that the contacting surfaces of the subunits are very different from those in vertebrate hemoglobin. (A more detailed description of this structure appeared in *Nature*[48]).

According to the divergent view of hemoglobin evolution, the gene for the myoglobin fold appeared in the eukaryote line somewhat prior to the diversification of the phyla and even before the split between plants and animals. The myoglobin-fold has been found in leguminous plants,[49,50] an echinoderm,[51] a gastropod mollusc,[52] an annelid worm,[53] an insect,[54] and in a number of vertebrates.[11,55,56] In the course of evolution at least two modes of monomer assembly have been found that converged to produce a cooperative assemblage. In the vertebrate case, the $\alpha_1\beta_1$ dimer is non-cooperative,[57] whereas in the clam

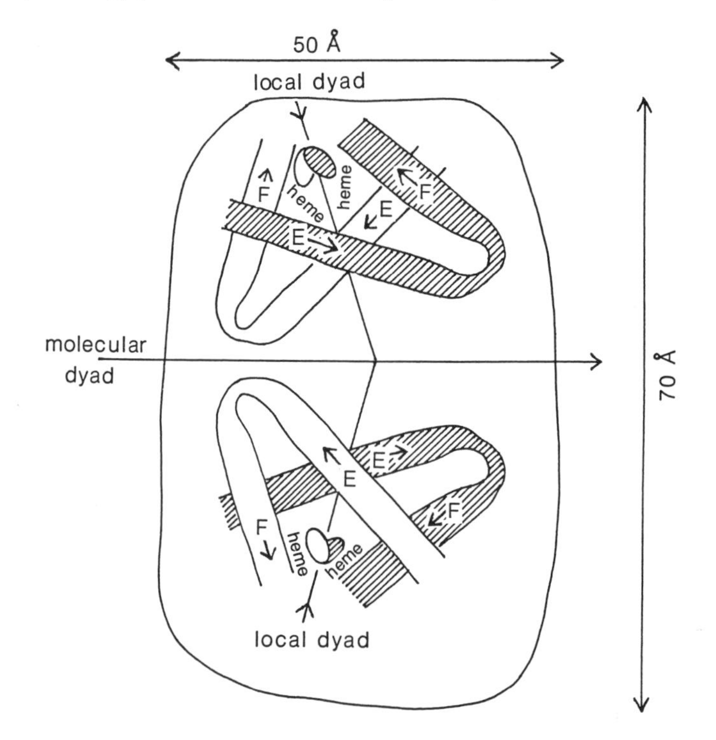

FIG. 8.11. A sketch showing the contacting regions in the tetrameric clam hemoglobins.

the dimer is intrinsically cooperative, with n approximately 1.5.[29,30,33,36] The formation of a clam tetramer from cooperative dimers leads to additional cooperativity with n between 1.8 and 2.1,[29,30,33,36] whereas in the vertebrate tetramer, two $\alpha_1\beta_1$ non-cooperative dimers assemble to form a cooperative tetramer with $n = 2.8$. Other modes of monomer assembly that lead to cooperativity may await discovery. In the 4×10^6 Dalton multiunit cooperative ($n = 4$) hemoglobin of the earthworm[58] one of the chains has a sequence[59] entirely compatible with its being a myoglobin fold. Other cooperative 'hemoglobins' have been found in the holothurian echinoderm, *Molpadia Oolitica* Pourtales,[60] and in the radula muscle of the gastropod mollusc *Nassa mutabilis L.* each of which has two myoglobin-size subunits,[61] and in the clam shrimp *Cyzicus cf hierosolymitanus* (S. Fisher) with 12-13 myoglobin size subunits.[62] Perhaps when the molecular structures of some of these molecules become available further insight will be gained about the various ways cooperativity can be achieved among hemoglobin subunits.

Only after the structures of the deoxy clam dimers and tetramers have been determined will it be possible to learn the structural basis for the cooperativity within the clam Hb molecules. To this end it is planned to crystallize these hemoglobins in the deoxy form and determine their structures.

Acknowledgement

This work has been supported by NIH grants AM-16446 and AM-02528. R.W.B., J.S.S., and W.E.R., Jr. were supported in part by an NIH Training grant, GM-07231. We gratefully acknowledge much helpful guidance from Professor Evangelos N. Moudrianakis. We thank Drs. Emilia Chiancone, Bruno Giardina, and Maurizio Brunori of the University of Rome for gifts of clam and trout hemoglobins. We also thank Professor Wayne A. Hendrickson and Dr. Janet L. Smith for computer programs.

Discussion

Ray Pepinsky: When you do the anomalous dispersion Patterson, are you talking about the square of the difference between the two structure factors, or are you talking about the sine Patterson?

Warner Love: We're not doing the sine Patterson.

Ray Pepinsky: In *that* case the sine-Patterson is not likely to be useful to you!

Warner Love: It's your function!

Ray Pepinsky: If the intensity differences in Bijvoet pairs are clearly observable, it might be well to avoid comments of analysts who disparage the sine-Patterson but who have never used it, and have a look at it yourself. The function is likely to be of some value in structures containing heme groups. Two good examples are type **A** sperm whale myoglobin, which has only two molecules per cell, and the horse hemoglobin with one molecule per cell.

You should know what to look for in the sine-Patterson maps. For hemes, look for a flattish region, with dimensions of a porphyrin, in which there are ***no peaks at all.*** Such a region will be centered at the sine-Patterson origin, but there may be several such superposed. The empty flat region exists because until one reaches the porphyrin side chains, the structure in the porphyrin plane is centrosymmetric. Normal to the disc-like empty region one should see some indication of the imidazole ring of the histidine.

Of course there will be two or more such empty porphyrin regions superposed at some angle to one another; but a modicum of image-seeking, using a sum-function or sum-difference function, should be useful. In horse hemoglobin with one molecule per cell, the heme normals lie within perhaps 30° of one another. Diamagnetic anisotropy observation is a simple physical measurement, and when a hemoglobin species crystal shows such anisotropy, the porphyrin planes ***are*** more or less parallel.

What is the situation in the clam? Are the hemes more or less parallel?

Warner Love: I can't tell you. These are 5.5 Å resolution structures, and you can't tell very well. When you look at the electron density of the heme, it's kind of an ellipsoid, but it's a pretty round ellipsoid.

Ray Pepinsky: A structure analyst has the constitutional right to use any techniques he favors, and to ignore any others. *De gustibus non est disputandum.* But measurements of diamagnetic anisotropy of a crystal are readily accomplished. They have also told us a lot about hemoglobin molecule orientations in the rod-like assemblies which occur in deoxygenated human and oxygenated deer sickle cells. A. J. Adams and I have used these and other examples of magnetic anisotropy extensively. In hemoglobins (or myoglobins) we showed that α-helices are so randomly oriented that they make no significant contribution to that anisotropy; it is essentially all due to the heme orientations. In deoxygenated hemes the iron paramagnetism adds to and strengthens the diamagnetic orientational effect. Of course D. J. E. Ingram's ESR observations are equally useful for deoxygenated heme-containing proteins.

Have any of these methods been of use to you?

Warner Love: No, you just go to higher resolution. These crystals go way out. You really haven't scratched the surface. Obviously, there are two things to be said. One of them, which I failed to say in my talk, there is an obvious explanation for heme-heme interaction in this stuff. In all myoglobins and hemoglobins we know that the E and F helices move with respect to each other, and with respect to the heme when they oxygenate. So if the E F faces of the subunits form the interface between subunits, then that interface must change its configuration when either subunit binds a ligand. The mechanism of communication of one heme and the next, will be, I think simpler to understand than in the vertebrate tetramer. It does point up one thing and that is that there is more than one way, apparently, to put these monomers together to make a cooperative tetramer. There are other non-vertebrate cooperative hemoglobins that may be assembled in yet other manners.

References

1. Rossmann, M. G. Application of the Buerger Minimum Function to Protein Structures. In *Computing Methods and the Phase Problem in X-ray Crystal Analysis* (R. Pepinsky, J.M. Robertson and J.C. Speakman, eds.) pp. 252-265. Pergamon Press: New York (1961).

2. Blow, D. M. *Proc. Roy. Soc. (London)* **A247**, 302–336 (1958).
3. Rossmann, M. G. *Acta Cryst.* **13**, 221–226 (1960).
4. Rossmann, M. G. (ed.) *The Molecular Replacement Method. A Collection of Papers on the Use of Non-Crystallographic Symmetry.* Gordon and Breach: New York (1972).
5. Grdenić, D., Kamenar, G., Korpar-Colig, B., Sikirica, M. and Jovanoski, G. *J. Chem. Soc., Chem. Comm.* 646–647 (1974).
6. Burlingame, R. W., Ph.D. Thesis. The Johns Hopkins University: Baltimore, MD (1984).
7. Hamlin, R., *Trans. Am. Cryst. Assoc.* **18**, 95–123 (1982).
8. Eickbush, T. H., Watson, D. K. and Moudrianakis, E. N. *Cell* **9**, 785–792 (1976).
9. Burlingame, R. W., Love, W. E., Wang, B.-C., Hamlin, R., Xuong, N.-H. and Moudrianakis, E. N. *Science* **228**, 546–553 (1985).
10. Moudrianakis, E. N., Love, W. E., Wang, B.-C., Xuong, N.-H. and Burlingame, R. W. *Science* **229**, 1109–1113 (1985).
11. Lattman, E. E., Nockolds, C. E., Kretsinger, R. H. and Love, W. E., *J. Mol. Biol.* **60**, 271–277 (1971).
12. Fermi, G. *J. Mol. Biol.* **97**, 237–257 (1978).
13. Ward, K. B. Jr., Wishner, B. C., Lattman, E. E. and Love, W. E. *J. Mol. Biol.* **98**, 161–171 (1975).
14. Wishner, B. C., Ward, K. B., Lattman, E. E., and Love, W. E. *J. Mol. Biol.* **98**, 179–194 (1975).
15. Padlan, E. A. and Love, W. E. *J. Biol. Chem.* **260**, 8272–8279 (1985).
16. Padlan, E. A. and Love, W. E. *J. Biol. Chem.* **260**, 8280–8291 (1985).
17. Konnert, J. H. *Acta Cryst.* **A32**, 614–617 (1976).
18. Wellems, T. E., Vassar, R. J. and Josephs, R. *J. Mol. Biol.* **152**, 1011–1026 (1981).
19. Makinen, M. W. and Sigountis, C. *J. Mol. Biol.* **178**, 439–476 (1984).
20. Fitzgerald, P. M. D. and Love, W. E. *J. Mol. Biol.* **132**, 603–619 (1979).
21. Sack, J. S., Ph.D. Thesis. The Johns Hopkins University: Baltimore, MD (1981).
22. Brunori, M. *Curr. Top. Cell. Reg.* **9**, 1–39 (1975).
23. Getzoff, E. D., Tainer, J. A., Sack, J. S., Bickar, D., Richardson, J. S. and Richardson, D. C. *Fed. Proc.* **39**, 2192 (1980).
24. Rossmann, M. G. and Blow, D. M. *Acta Cryst.* **15**, 24–31 (1962).
25. Rossmann, M. G. *Acta Cryst.* **14**, 383–388 (1962).
26. Svedberg, T. and Eriksson-Quensel, I. *J. Am. Chem. Soc.* **56**, 1700–1706 (1934).

27. Yagi, Y., Mishima, T., Tsujimura, T., Sato, K. and Egami, F. *J. Biochem. (Tokyo)* **44**, 1–7 (1957).
28. Komano, T. *J. Japan. Biochem. Soc.* **39**, 405 (1967).
29. Ohnoki, S., Mitomi, Y., Hata, R. and Satake, K. *J. Biochem. (Tokyo)* **73**, 717–725 (1973).
30. Furata, H., Ohe, M. and Kajita, A. *J. Biochem. (Tokyo)* **82**, 1723–1730 (1977).
31. Djangmah, J. S., Gabbot, P. A. and Wood, E. J. *Comp. Biochem. Physiol.* **60B**, 245–250 (1978).
32. Como, P. F. and Thompson, E. O. P. *Aus. J. Biol. Sci.* **33**, 643–652 (1980).
33. Ikeda-Saito, M., Yonetani, T., Chiancone, E., Ascoli, F., Verzili, D., and Antonini, E. *J. Mol. Biol.* **170**, 1009–1018 (1983).
34. Sumita, N., Kajita, A. and Kaziro, K. *J. Biochem. (Tokyo)* **55**, 148–153 (1964).
35. Gattoni, M., Verzili, D., Chiancone, E. and Antonini, E. *Biochim. Biophys. Acta* **743**, 309–316 (1983).
36. Chiancone, E., Vecchini, P., Verzili, D., Ascoli, F. and Antonini, J., *J. Mol. Biol.* **152**, 577–592 (1981).
37. Furata, H., Ohe, M. and Kajita, A. *Biochim. Biophys. Acta* **66B**, 448–455 (1981).
38. Royer, W. E. Jr., Ph.D. Thesis. The Johns Hopkins University: Baltimore, MD (1984).
39. Dickerson, R. E., Weinzerl, J. E. and Palmer, R. A. *Acta Cryst.* **B24**, 997–1003 (1968).
40. Li, S. L. and Riggs, A. *J. Biol. Chem.* **245**, 6149–6169 (1970).
41. Li, S. L. and Riggs, A. *J. Mol. Evol.* **1**, 208–210 (1972).
42. Nash, A. R., Fisher, W. K. and Thompson, E. O. P. *Aus. J. Biol. Sci.* **29**, 73–97 (1976).
43. Como, P. F. and Thompson, E. O. P. *Aus. J. Biol. Sci.* **33**, 653–664 (1980).
44. Furuta, H. and Kajita, A. *Biochemistry* **22**, 917–922 (1983).
45. Fisher, W. K., Gilbert, A. T. and Thompson, E. O. P. *Aus. J. Biol. Sci.* **37**, 191–203 (1984).
46. Crowther, R. A. In *The Molecular Replacement Method* (M. G. Rossmann, ed.) p 174–185. Gordon and Breach: New York (1972).
47. Crowther, R. A. and Blow, D. M. *Acta Cryst.* **23**, 544–548 (1967).
48. Royer, W. E. Jr., Love, W. E. and Fenderson, F. F. *Nature (London)* **316**, 277–280 (1985).

49. Vainstein, B. K., Harutyunyan, E. H., Kuranova, I. P., Borisov, V. V., Sosfenov, N. I., Pavlovsky, A. G., Grebenko A. E. and Konareva, N. V. *Nature (London)* **254**, 163–164 (1975).
50. Ollis, D. L. Ph.D. Thesis. University of Sydney: Sydney, Australia (1980).
51. Kitto, B. G., Hackert, M. L., Mauri, F. and Ernst, S. *Fed. Proc.* **41**, 650 (abs. 2222) (1982).
52. Ungaretti, L., Bolognesi, M, Cannillo, E., Oberti, R. and Rossi, G. *Acta Cryst.* **B34**, 3658–3662 (1978).
53. Padlan, E. A. and Love, W. E. *J. Biol. Chem.* **249**, 4067–4078 (1974).
54. Steigemann, W. and Weber, E. *J. Mol. Biol.* **127**, 309–338 (1979).
55. Takano, T. *J. Mol. Biol.* **110**, 569–584 (1977).
56. Perutz, M. F., Muirhead, H., Cox, J. M., Goaman, L. C. C., Mathews, F. S., McGandy, E. L. and Wells, E. L. *Nature (London)* **219**, 29–32 (1968).
57. Mills, F. C., Johnson, M. L. and Ackers, G. K. *Biochemistry* **15**, 5350–5362 (1976).
58. Giardina, B., Chiancone, E. and Antonini, E. *J. Mol. Biol.* **93**, 1–10 (1975).
59. Garlick, R. L. and Riggs, A. F. *J. Biol. Chem.* **257**, 9005–9015 (1982).
60. Terwilliger, R. C. and Read, K. R. H. *Comp. Biochem. Physiol.* **42B**, 65–72 (1972).
61. Geraci, G., Sada, A. and Cirotto, C. *Eur. J. Biochem.* **77**, 555–560 (1977).
62. Ar, A. and Schejter, A. *Comp. Biochem. Physiol.* **33**, 481–490 (1970).

9. Patterson and Pattersons

Dorothy Crowfoot Hodgkin

Dorothy Hodgkin was born in Cairo, Egypt, since her father, John Winter Crowfoot, an archaeologist and historian, was serving with the Egyptian Ministry of Education, and later became Principal of Gordon College at Khartoum, and Director of Education and Antiquities in the Sudan. Her mother was a leading authority on ancient textiles.

Dr Hodgkin was educated at Somerville College, Oxford, and then went to work with J. D. Bernal in Cambridge in 1932-34. She obtained her Ph.D. in 1936 after she had returned to Somerville College, at first as Research Fellow, then as tutor and fellow. She became Wolfson Research Professor of the Royal Society in 1960. She married Thomas Hodgkin, a tutor in Adult Education, and an African historian, and they had three children.

Dorothy's first X-ray studies were with H. M. Powell in Oxford on thallium dialkyl halides. When she went to work with Bernal, she started an extensive X-ray survey of steroids. Bernal had just shown that the X-ray photographs of steroids were not compatible with the then-accepted Windaus-Wieland formula because the latter could not be fitted into the unit cell of the crystal. The formula was rapidly revised, and then Dorothy studied over 100 steroids, deducing in each case how they packed in the crystal. This was in the days before detailed structure analyses of such molecules became easy. However finally, with Harry Carlisle, she determined the crystal structure of cholesteryl iodide, thus firmly establishing the chemistry and stereochemistry of steroids.

During the second world war she worked on penicillin with Barbara Low, A. Turner-Jones and Charles Bunn. Penicillin was discovered by Fleming at St. Mary's Hospital in London in 1929, and isolated by Florey and Chain and others in Oxford during the second world war. The chemical evidence on the structure was perplexing, and it could not be ascertained whether or not the molecule contained a β-lactam structure. Sir Robert Robinson, the chemistry professor at Oxford was against it, because β-lactam structures are generally stable, unlike penicillin. John Cornforth (then a young research fellow working on the problem) said 'If penicillin turns out to have a β- lactam structure, I shall give up chemistry and grow mushrooms'. Penicillin was shown by Dorothy and her co-workers to have the β-lactam structure, but John Cornforth did not keep his promise. He stayed on in chemistry and won a Nobel Prize eleven years after Dorothy.

the β-lactam structure, but John Cornforth did not keep his promise. He stayed on in chemistry and won a Nobel Prize eleven years after Dorothy.

The power of the X-ray method, aided by Patterson's method, was demonstrated by the determination of the structure of vitamin B_{12}. Crystals of a B_{12} degradation product were obtained in Todd's laboratory by Jack Cannon, who, in frustration, because he'd had such a hard time getting crystals, threw a wide range of solvents into a flask and went on vacation. When he came back there were beautiful crystals of a hexacarboxylic acid, derived from the vitamin. The crystal structure of this, and of the vitamin itself, crystals of which had been given to her by Dr Lester Smith, established the chemical formula of vitamin B_{12}. It contained a hitherto undiscovered ring system, the corrin ring system, and, of course, the vitamin has now been synthesized in the laboratories of R. B. Woodward and A. Eschenmoser.

But Dorothy's main interest throughout her life has been the structure of insulin. She helped Bernal by taking diffraction photographs of crystals of pepsin, of course mounted in its mother liquor which Bernal showed was necessary. The diffraction patterns of pepsin showed that proteins could pack with sufficient regularity in the crystalline state so that an X-ray diffraction analysis would be possible, at least in principle. In August 1969, after years of effort by her and many workers in her laboratory, including Guy Dodson, she announced the structure of crystals of insulin. Dorothy had been invited to give the first A. L. Patterson Memorial Lecture, and she waited until the structure of insulin was solved, and in the spring of 1970 she gave one of the first talks on the structure of the protein here at the Institute.

Dorothy was made a Fellow of the Royal Society in 1947, she was awarded the Nobel Prize in Chemistry in 1964, and the Order of Merit in 1965. She has been Chancellor of Bristol University since 1970, served as President of the International Union of Crystallography, and President of the British Association for the Advancement of Science. Numerous other honors have come her way.

We first heard of the Patterson function[1] in the summer, I think, of 1934, and there was a lot of conversation in Cambridge about it. That peaks in the F^2 series showed interatomic vectors provided a new way to find atomic positions in crystals. As you know, W. L. Bragg was over at the time in the United States to hear the original account, and it was immediately decided that Arnold Beevers should test the method by the solution of the structure of nickel sulfate heptahydrate.[2] The sort of conversation that went on is represented, I think, by the remarks in the *Annual Reports of Chemistry* for 1935,[3] which, while welcoming the new function and saying how useful it was going to be in general for many kinds of structural analysis, went on to say that 'in more complex structures the number of interatomic distances will be so large that only the most prominent between the heaviest atoms will be expected to stand out in a Patterson synthesis. And to the extent that the method has this limitation, it perhaps gives little more information than would be deduced from general considerations by an experienced worker in structural analysis. At the same time the Patterson synthesis does afford the only means of giving an unprejudiced presentation of all the information which may be directly derived from the experimental material'. I think the remarks about experienced structure analysts came from W. L. Bragg, probably somewhat supported at that moment by J. D. Bernal.

I was on my way back to Oxford at the time, and when I got back to Oxford, of course, the full 1935 paper of Patterson in the *Zeitschrift für Kristallographie*[4] came out. I give you from that paper as I saw them then first, two very important Pattersons. Fig. 9.1 shows the electron density projection of copper sulfate as solved by Beevers and Lipson; they gave me a copy of their first paper.[5] As you see, though they were absolutely delighted with it, we wouldn't now draw our electron density maps as simply as they did, and it does imply, if you look at it, that the data that they had collected on this crystal were somewhat limited. However, the map shows the approximate positions of the atoms, including the two non-identical copper atoms. Fig. 9.2 is from Patterson's paper[4] and shows the first Patterson of copper sulfate, and side by side with it, Patterson's calculation of the 'sharpened Patterson', obtained with F^2 values modified to correspond with those from point atoms, as here, or from atoms at rest. The latter brings out the interatomic distances much more clearly even than the electron density map, and certainly than the unsharpened map. Unfortunately we had a tendency to skim this paper, not to notice

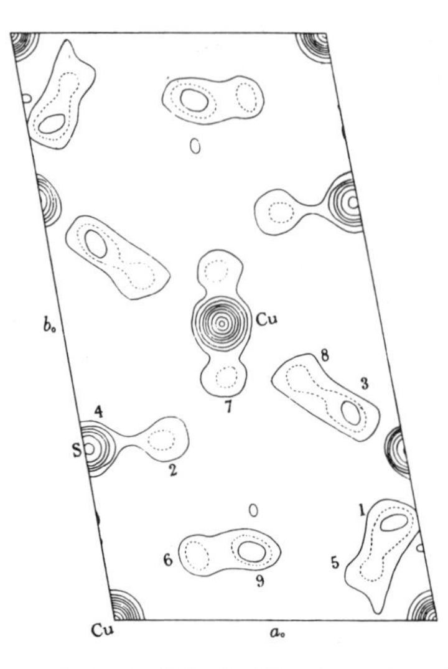

FIG. 9.1. Copper sulphate·$5H_2O$. Electron density projection, $F(hk0)$. (Beevers and Lipson).

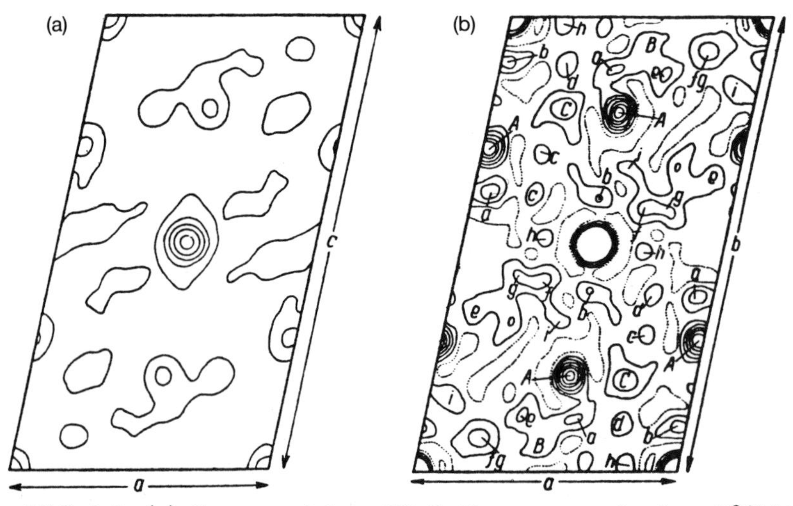

FIG. 9.2. (a) Copper sulphate·$5H_2O$. Patterson projection, $F^2(hk0)$. (b) Copper sulphate $5H_2O$. Sharpened Patterson projection, $F^2(hk0)/f^2$.

all the detail, and people for years went on calculating unsharpened Pattersons, and only gave them up when they found they couldn't solve them, and went slowly over to sharpened Pattersons.

I was no exception to this rule. When I got all the information from the 1935 paper I decided to calculate Pattersons myself, in projection, of two crystals for which I had collected projection data in Cambridge — cholesteryl chloride and cholesteryl bromide[6] — which I thought might be isomorphous, and in which I hoped to see bromine-bromine and chlorine-chlorine vectors. I made a little table, I may say as others round the world were making — cos $2\pi x$ and sin $2\pi x$ at intervals of ten degrees — which, with a slide rule, I used for these calculations. Fig. 9.3 gives these two first projections, and

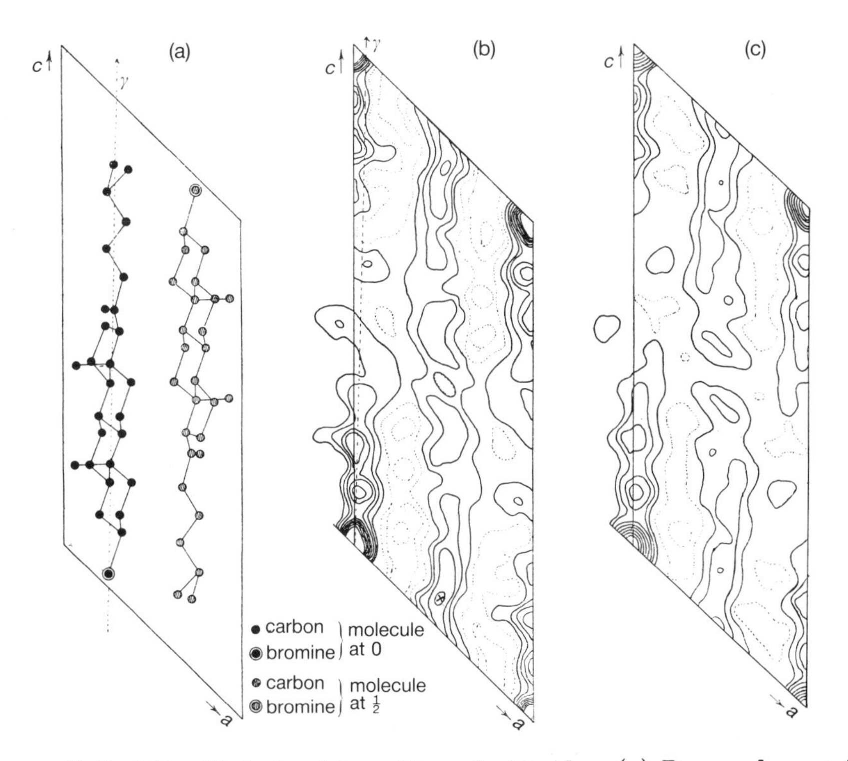

FIG. 9.3. Cholesteryl bromide and chloride. (a) Proposed crystal structure of cholesterol bromide projected on (010). The z-parameter of the two molecules is given by a correlation of the bromine–bromine distance with a peak in the Patterson projection in (b). (b) Patterson projection, $P(uw)$, for cholesteryl bromide. (c) Patterson projection, $P(uw)$, for cholesteryl chloride. Negative contours are dotted in these diagrams, and high contours around the origin are omitted.

what they showed me straightaway, was that Bernal, that experienced X-ray crystallographer, was perfectly correct when he said that the sterols must be long and thin, and have the sort of dimensions that he had given.[7] Next to the electron density, illustrated in Fig. 9.3, I have drawn the idea of the sterol skeleton that had been developed between Bernal's first remarks in 1932 and this calculation in 1935.[8]

I went on, I was fairly sure I could see where the bromine-bromine vector was. I wasn't very sure where the chlorine-chlorine vector was, and it never struck me to sharpen this map, I'm sorry to say. I did some trial calculations to get an electron density map, which was a very poor map which I did publish with the sterol paper, and which did conform roughly to the form of the projection. I thought the two crystals were probably not exactly isomorphous enough to pursue, in which I was wrong. Recent work by Kalyani Vijayan in India, collecting the whole data and calculating the full three dimensional series, has shown they are quite nicely isomorphous,[9] and the sterol structure has come out very well. So my story is an old bit of history — you might call it a child's first efforts of calculating Pattersons.

The next crystal that fell into my hands in the same year, in fact, in October, 1934, was insulin, and again I made mistakes, in spite of being involved with the pepsin experiments in Cambridge six months before with Bernal.[10] I noticed that the microcrystalline preparation, the 10mg that I was given, was brightly birefringent, so I didn't bother to keep the crystals wet. I let the crystals that I grew become dry and took the first X-ray photographs of insulin air-dried, which gave me a very limited diffraction pattern (Fig. 9.4). However, it was easier than the full pattern for someone who hadn't got many computing aids to deal with. Fig. 9.5(a) shows the first Patterson projection of these insulin crystals, in which there appeared to be peaks at about 10 and 20 Å from the origin, immediately suggestive of the kinds of distances Astbury was picking up in fibrous proteins.[11] I thought as this was a projection down 30 Å, I must do a calculation in three dimensions, and I seized on Harker's simplifications[12] to calculate the Patterson function at $z = 0$ and $z = \frac{1}{2}$, because it would give me effectively,

FIG. 9.4. ($hik0$) reflections from air-dried cattle insulin crystals.

owing to the rhombohedral symmetry of the crystals, a three-dimensional distribution, computed at 5 Å intervals along c. Other sections were calculated, perpendicular to these. They show the peaks are not separate individuals, but parts of chains across the unit cell (Figs. 9.5(b and c)).[13]

These particular calculations brought my first letter from A. L. Patterson himself, dated March 13, 1938, asking for reprints, and commenting 'There seems to be so much detail in the F^2 patterns for insulin, that it ought to be possible to get a more detailed description of the shape of the molecule. But I must confess that, for the moment, I do not see how this can be done. I hope very much that you will be able to get further information, and I shall be very interested to hear about it.' I sent him reprints, and he wrote again thanking me for them, and I'm glad to say I was fairly prompt. His reply was on July fifteenth. In this he says 'My great regret at the moment is that our laboratory is now torn into little pieces, and

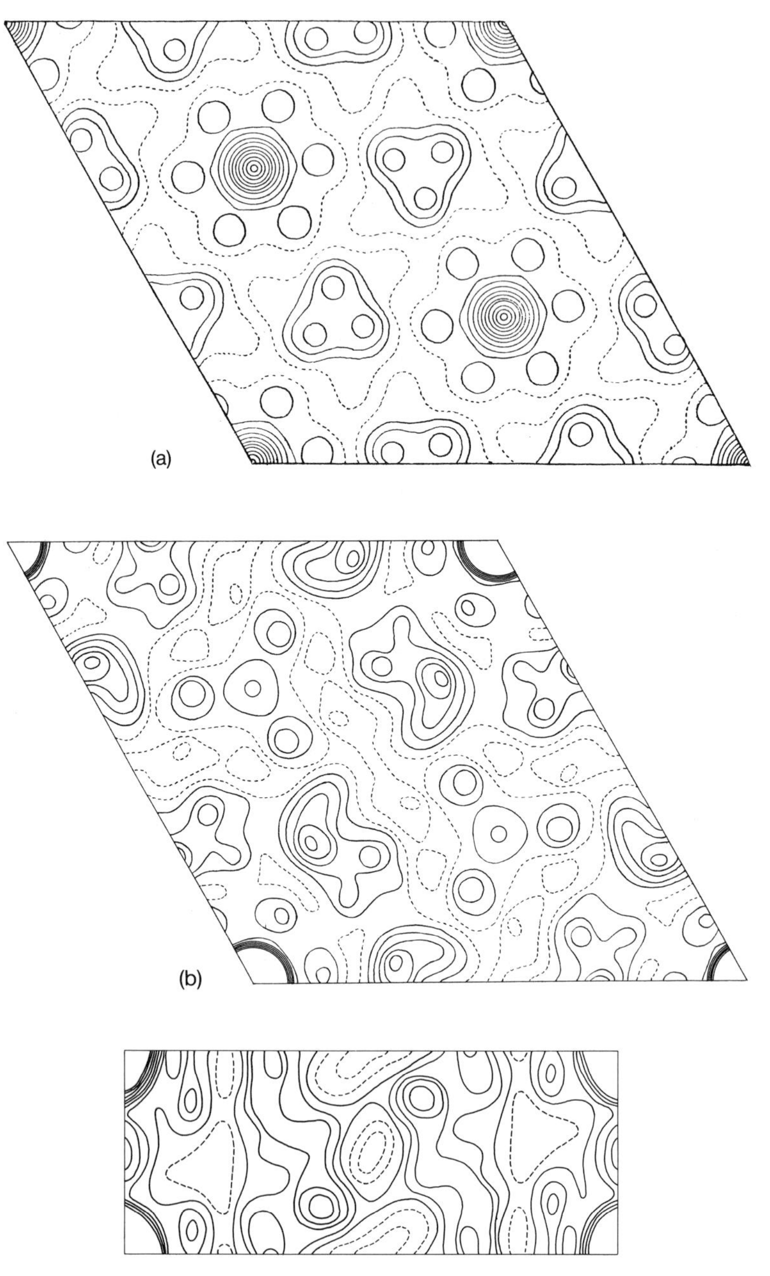

FIG. 9.5. First insulin Pattersons: (a) projection, $P(uv)$, (b) section, $P(uv0)$, (c) section, $P(0vw)$.

it is therefore impossible to start madly calculating, possibly experimenting, on proteins.' I'm always a little sorry, it would have been nice if he had. And what would have happened? Would he have seen some light in these old patterns? These old patterns I showed in the Patterson Memorial Lecture that I gave when I had the actual structure in my hands, do correspond rather well to the pattern you get just from the main chain distribution in insulin, in spite of the complexity of the pattern. Once again I enjoy reliving the structure solution I saw in the Patterson on that occasion.

I am afraid I largely gave up the analysis of insulin for the time being at that point in 1938, though not entirely. I went on to try to grow the crystals again, to get wet crystal data which were much more extensive, but the war gradually overtook me; I did make an effort to measure the wet data by eye estimation, but I had no very great confidence in my results, and I felt sure that, beginner as I was, I should start to work on a simpler structure, and try out the various ideas that we had for structure analysis on a different crystal. The first one, as Jenny mentioned, was chosen by going back to the sterols, giving up cholesterol chloride and bromide owing to my mistake over not sharpening my Patterson, and taking instead cholesteryl iodide and finding, with Harry Carlisle's important assistance, the distribution of the atoms in the sterol molecule.[8]

But before that was finished, work was beginning at Oxford on penicillin, and I was drawn into this. Again another 10 mg of crystals were sent to Oxford, this time from America, of the first sodium benzylpenicillin crystallized by the Squibb Institute. I grew larger crystals with Kathleen Lonsdale watching. The crystals were flown over to Sir Henry Dale, who had asked for them and sent them by hand to Oxford by Kathleen. We followed the directions that we had on a telegram from America, and grew the crystals from 3 mg sodium benzylpenicillin in a little tube. I collected three-dimensional data from these crystals with the help of Barbara Low, and again proceeded to calculate a Patterson. Fig. 9.6 shows the zero section ($v = 0$) in the map of sodium benzylpenicillin. Actually it's a perfectly good one. We should have gone ahead. You see, quite a lot of the peaks are interpretable and are single atom-to-atom

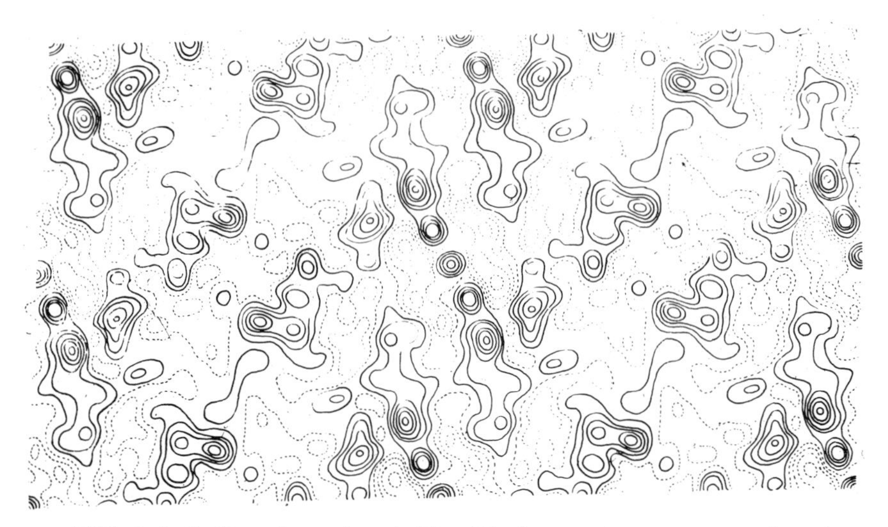

FIG. 9.6. Sodium benzylpenicillin. Harker sections $v = 0$, $P(u0w)$.

vectors. But at the time I was very nervous and I thought it would be better to have an isomorphous series, and asked our local chemists to make me the potassium and rubidium salts. Potassium and rubidium showed up so very much better in the Pattersons that we went ahead on these for the solution of the structure. Even so we got into a rather confused state, and we felt that we should really have a three-dimensional Patterson, and then we would be able to sort everything out. But it was in the middle of the war and our computing aids at the time were Beever and Lipson strips, which we extended ourselves from 6° to 3° whenever we needed three degree intervals. L. J. Comrie came by and offered to put the whole operation onto punched cards. He first of all thought we had to calculate just down lines and sections. He said, now it would be much easier if you wanted full three dimensional series, and we said, of course, it would be very much better to have the full three dimensional series. So he went ahead and got the cards made, but they were not ready for use until the end of the war, and in the meantime the structure came out by rather complicated maneuvers from the projections. We enjoyed calculating the final electron density maps of penicillin in three dimensions,[14] but we never calculated the three-dimensional Patterson. This was why I was congratulating Verner Schomaker, when first I went to Caltech, on having done just this for threonine, because we always felt sorry not to have had it for the penicillin solution.

I am not talking in detail about all of these old stories but I would like to take you on to the next serious structure analysis, the analysis of vitamin B_{12}. It was brought in as small red crystals one day in 1948, by Dr Lester Smith, who wanted his crystals compared with ones that had been reported isolated in America by Folkers in the Merck group. Of course, we set out to collect the full three-dimensional data from these crystals, well not immediately. At first they were a bit small, and the unit cell rather large, and absolutely nothing at all was known about the object within, until in the middle of that summer I was rung up from Glaxo to say this, 'These crystals have cobalt in them, one atom to your molecular weight, your calculated molecular weight. There's a heavy atom for you.'

Everybody knew what was wanted in X-ray analysis at that moment from experiences with penicillin, and Glaxo knew the penicillin story. This encouraged us to collect the full data, and now we had Betty Gittis, Comrie's assistant (who did the penicillin electron density map) and the Comrie program, so we got our first three-dimensional Patterson. Fig. 9.7 shows you a rather beautiful drawing of one of the Harker sections, and in it you can see that as soon as we saw the Patterson we knew that this structure was solvable. The section shows large peaks due to the cobalt-cobalt vectors repeated by Patterson symmetry, and around each four peaks, roughly in a square, one of them extended in a straight line, and when you look backwards and forwards through the Patterson you see a sort of layer of density going backwards and forwards perpendicular to that direction. You are obviously seeing the cobalt-cobalt vectors, and the cobalt to the other atoms vectors. It seems clear that the whole pattern is dominated by the cobalt to other atom vectors, and in so far as it is dominated by them, it should be possible to deduce the structure from a comparison of the distribution around each of the cobalt-cobalt vectors in turn.

I do not remember exactly when Patterson superpositions came in use for structure solving — only that they sprang up in

FIG. 9.7. Vitamin B_{12} (cyanocobalamin) three-dimensional Patterson, Harker section, $P\,(uv0)$.

different laborotories, Martin Buerger leading the exposition. The first day we had the figures of the B_{12} three-dimensional Patterson we began to solve it by drawing to scale on tracing paper the unit cell in the a projection, marking in the cobalt xyz parameters, setting out each cobalt in turn on the origin of the Patterson map and writing around it the figure field extended over the unit repeat. We reduced our numbers to single figures and used different colours for each repeat, hoping to see whether the structure was best derived from the sum of the figures, or the minimum or any other combination. Peaks were clearly visible at atomic intervals, and in great excitement we began to draw the structure as in Fig. 9.8. Unfortunately we got rather tired of this process after a time, and we didn't pursue it very far in the original cyanocobalamin. Instead we calculated a three-dimensional Fourier, which we could do on the punched card machines much more quickly just giving the

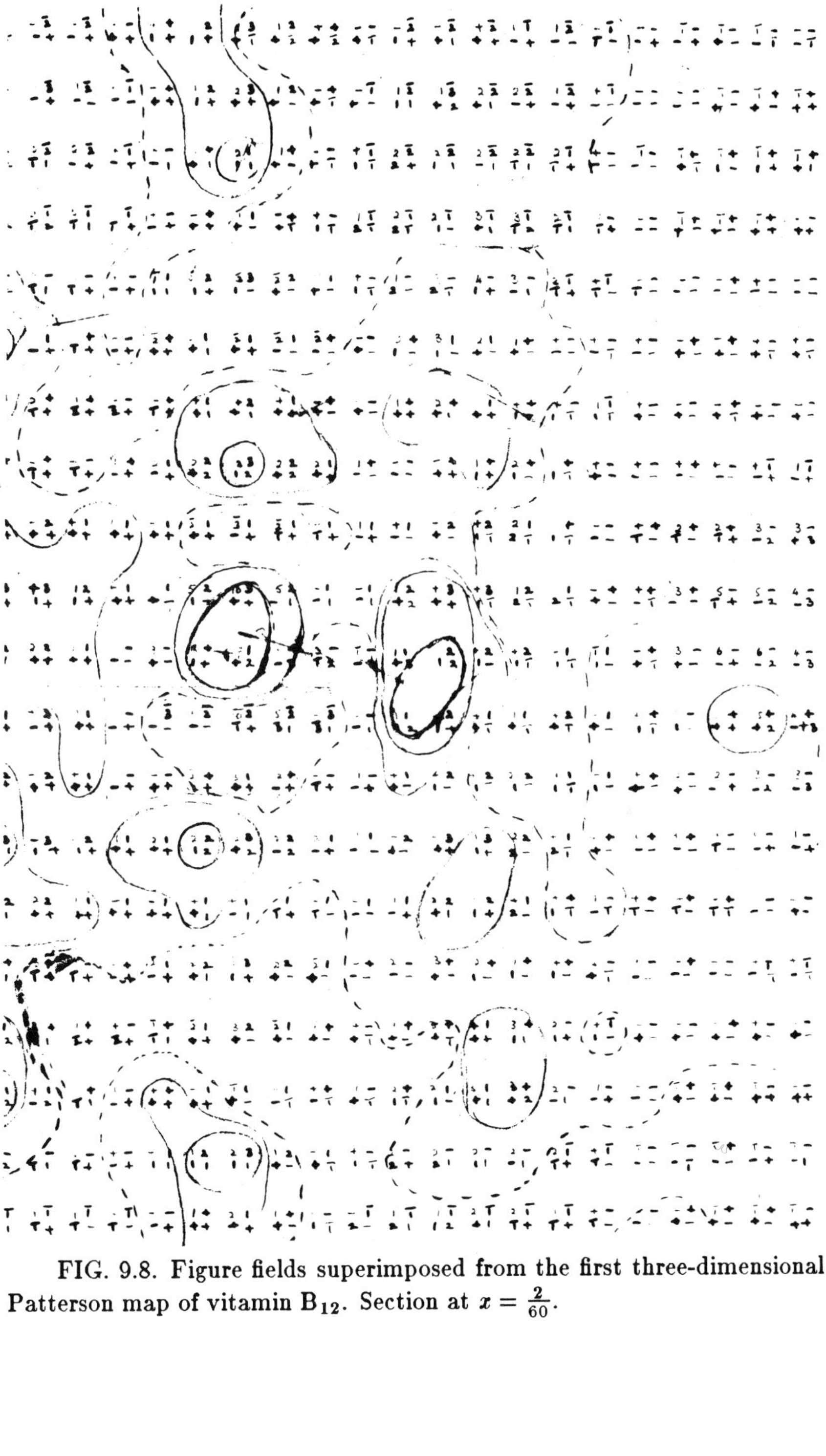

FIG. 9.8. Figure fields superimposed from the first three-dimensional Patterson map of vitamin B_{12}. Section at $x = \frac{2}{60}$.

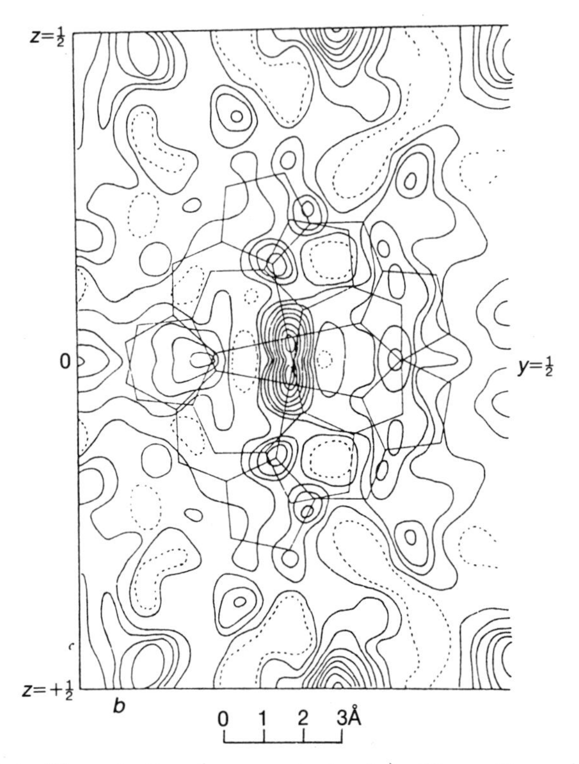

FIG. 9.9. Vitamin B_{12} (cyanocobalamin). Three-dimensional Patterson, section $P\,(\frac{1}{2}\,v\,w)$.

terms cobalt phases, and this was clearly extremely similar to the Patterson version.

Fig. 9.9 shows another view of the structure as seen in the Patterson. It gives the impression that there are five-membered rings attached to the cobalt atoms, and we roughly sketched a porphyrin over the map, which was wrong, of course, but was what you might call a good guess at the time. When John Robertson and Inge Robertson came back from Penn State, we had another crystal which had been made for us by the chemists, a selenocyanide of B_{12}. John and Inge took this one up, and repeated the operations, and together did the whole three-dimensional Patterson superposition map; Fig. 9.10 shows a small section of this.[15,16] From the map they built up a whole three-dimensional structure for the vitamin which had a lot of right atoms in it, but also quite a few wrong ones. We calculated a three-dimensional electron density map on their phases, on phases given by a structure factor calculation which took a most awful long time (about three months), and in the meantime the third crystal came in, the hexacarboxylic acid, of which Jenny described the preparation so dramatically earlier. The hexacarboxylic acid[17] was submitted to the same process, a little shortened. Since the space group was $P2_12_12_1$ and the cell dimensions shorter, the cobalt-to-cobalt atom vectors could be seen, and the cobalt atom positions deduced from the Patterson projections.‡ From the cobalt atom positions Jenny went straight ahead to calculate a three dimensional electron density map in which we could see clearly what we had been suspecting for a year or two, that we had not got

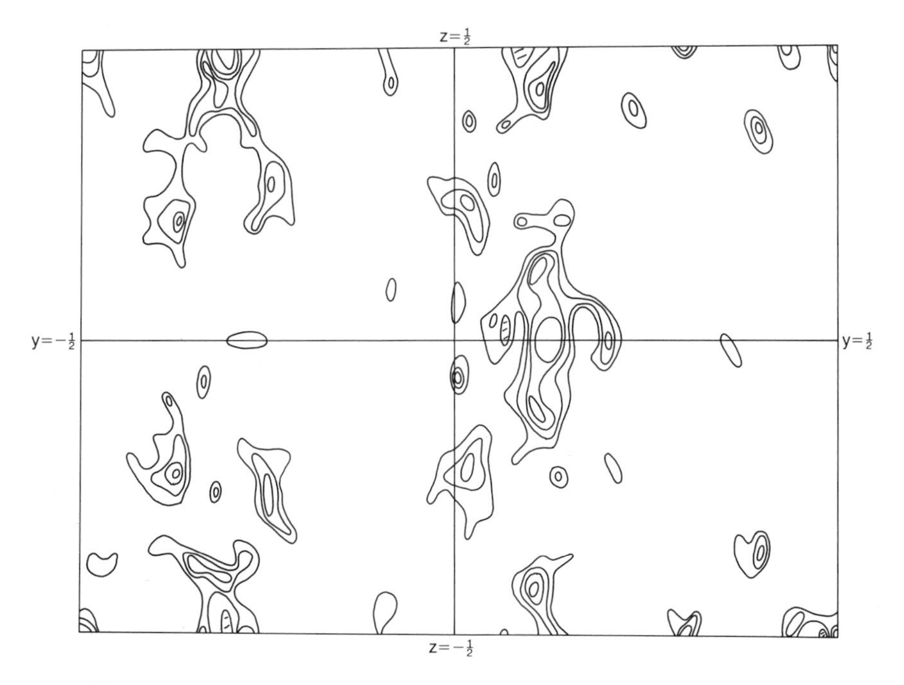

FIG. 9.10. Selenocyanide of cobalamin. Vector convergence diagram based on Co and Se positions. $x = \frac{2}{60}$.

‡ Later, for the record, John Robertson calculated the three-dimensional Patterson.

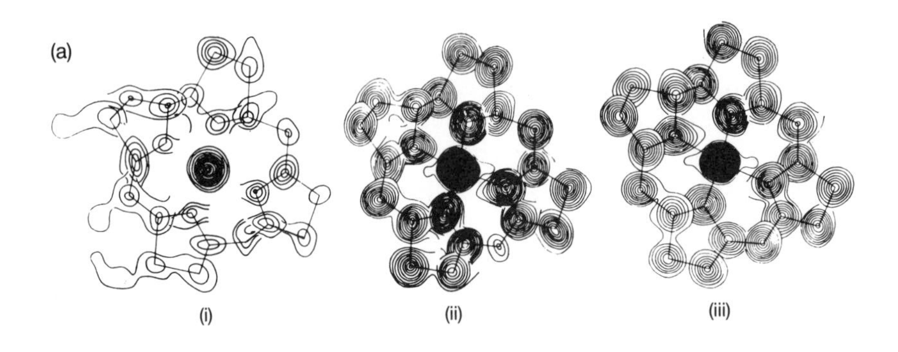

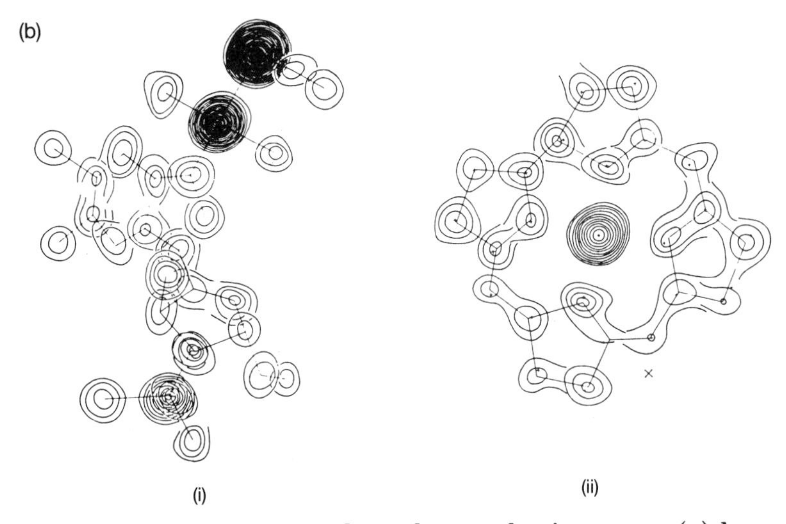

FIG. 9.11. Extracts from three electron density maps. (a) hexacarboxylic acid (i) $\rho(xyz)$ phased on Co. (ii) $\rho(xyz)$ phased on 26 atoms. (iii) final $\rho(xyz)$ (b) selenocyanide of cobalamin (i) $\rho(xyz)$ phased on atoms derived from the Patterson vector convergence diagrams showing Co, Se, CN, 'nucleotide' and the corrin nitrogen atoms. (ii) corrin nucleus.

a porphyrin. We had a quite different ring system, later called by the name corrin.

So in Fig. 9.11 I put together both the corrin ring as observed in Jenny's maps and also part of the structure from the selenocyanide of cyanocobalamin. In both you can see clearly the direct link between two of the five-membered rings in the corrin nucleus.

So I leave this series of structure analyses now, showing in the end in Fig. 9.12 the whole structure, not of cyanocobalamin alone, which was the compound we were working on at first, but of the full vitamin B_{12}, the coenzyme,[18] that was discovered later, whose structure was solved by essentially the same manoeuvers. As I said, the exciting thing about vitamin B_{12} was that we could see straightaway from the Patterson that the structure was soluble, and even on the first day that we looked at the calculation, we knew some of the atoms and their positions in the unit cell. Only it was a large molecule and it took us a long time to work through it, and in the end a long time to be absolutely sure about details of the structure. And in the end we were brought through much more rapidly than we might have hoped by Ken Trueblood and calculations in California that shortened the terrible time the structure factors were taking us in Oxford. Gradually we moved over, therefore, from punch card machines to modern computing.

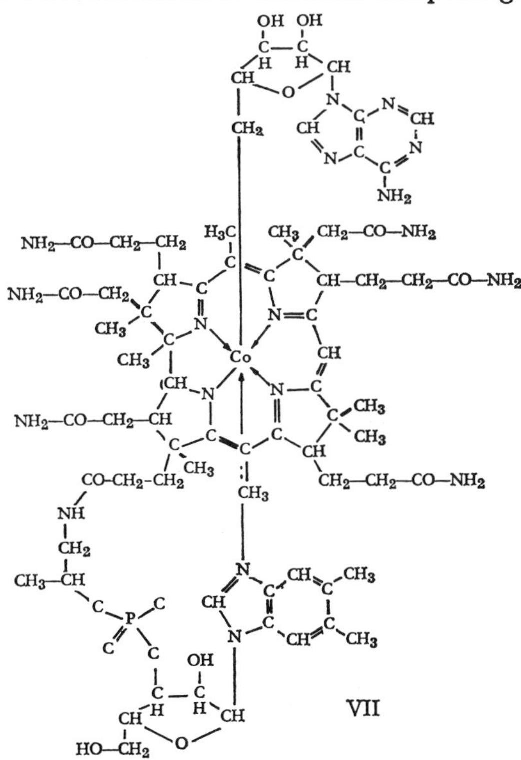

FIG. 9.12. Structure of the B_{12} coenzyme.

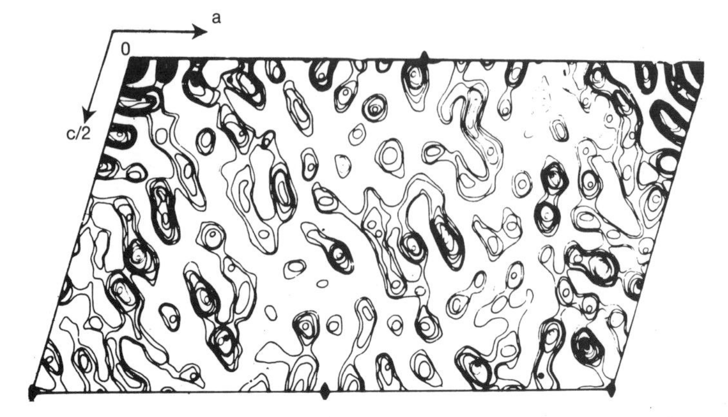

FIG. 9.13. Thiostreptone, three-dimensional Patterson, $P\,(u0w)$.

Fig. 9.13 shows a very different Patterson that was obtained from from a crystal of a large antibiotic called thiostreptone, of a molecular weight of about 1600-1700, of about 120 atoms in the molecule. It was first crystallized in a tetragonal form which, we to this day, have failed to solve. But Viswamitra, an Indian research worker visiting our laboratory, who has a charmed hand as far as growing crystals is concerned, grew another form, a monoclinic form, only two molecules in the unit cell. Fig. 9.13 is the Harker section at $w = 0$ in the three-dimensional Patterson. The molecule was thought to have four or five sulfur atoms in it as the heaviest atoms, and to have a structure which was based essentially on peptide chains, but in some kind of degraded form, and to include some aromatic nuclei, and that the sulfur was probably in thiazoline rings. Brian Anderson and Viswamitra decided they ought to be able to solve this structure by Patterson superposition methods, and they set out with the three-dimensional Patterson drawn on transparent sheets, and did the superpositions from just every peak in the map that you see there in turn, looking for four five membered rings which should be the thiazoline rings. Fig. 9.14 shows some of the bits they thought they picked out, which should identify the sulfur positions. Of these first sulfur positions that they identified, I think three were right, but the others wrong. They kept on going backwards and forwards over the map, trying to find pieces of structure that fitted together, and I never knew how in the end this structure was actually solved. They did at one stage mechanize the superposition process, and put it on to the computer. The first map that Viswamitra calculated was one of the ones that just makes me laugh, if it doesn't make me cry, to think of. They were doing a superposition from five positions, and he obliterated every point at which the density fell below zero (average) on one superpositon. This resulted in a great block of computing output with zeroes all over it, excepting in seven positions. And I said, 'Oh, if you'd only put the figures in, you'd have got something more out. But still in the end, they got the whole structure out.[19] It has been verified by NMR, and I think it is certainly all right; another related molecule has been found and the structure of that solved crystallographically. Fig. 9.15 shows a small part of the electron density map. Fig. 9.16 is the three-dimensional arrangement of

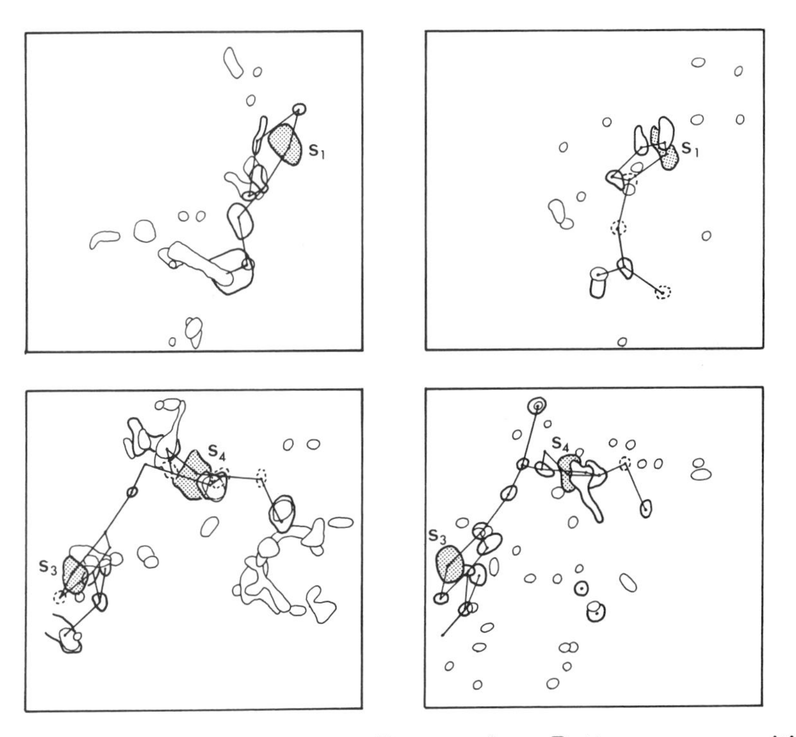

FIG. 9.14. Thiostreptone. Extracts from Patterson superposition maps.

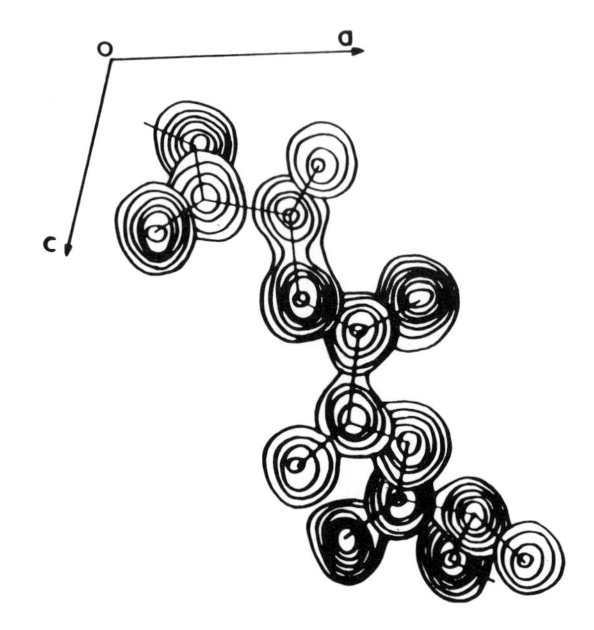

FIG. 9.15. Part of thiostreptone electron density map, showing the alanyl-α-aminoacryl-alanyl moiety.

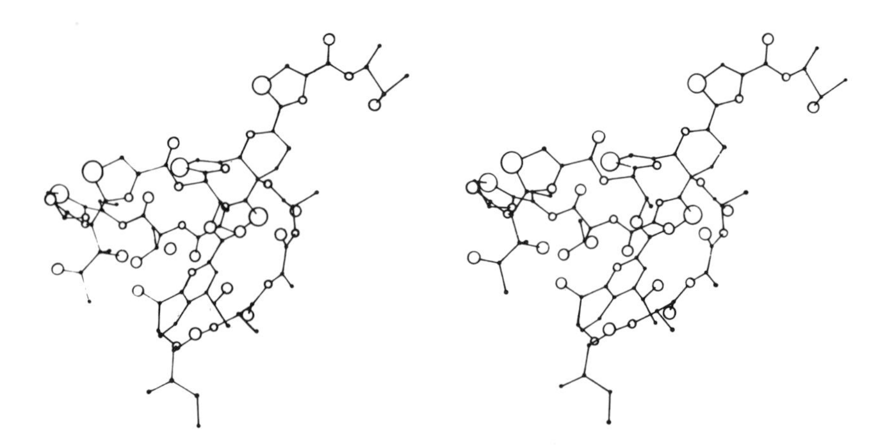

FIG. 9.16. Three-dimensional arrangement of atoms in the molecule of thiostreptone.

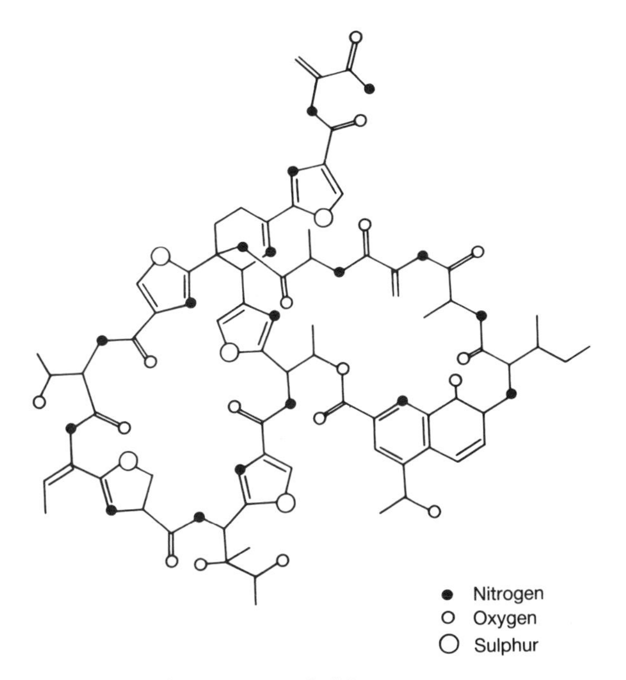

FIG. 9.17. Chemical structure of thiostreptone.

the atoms in the molecule and Fig. 9.17 the formula we wrote. The atoms are linked into a circle in a winding chain, with a branch. In the centre is a water molecule holding everything somewhat steady. It is a complicated object and not one of which you can guess the structure, from the X-ray photograph, nor could any experienced crystallographer, I think, have been able to see, what he was dealing with from the appearance of the X-ray data. But it was all there in the Patterson.

At the end I would like to return to insulin[20,21] and what happened to insulin. Of course, I had the idea that when I returned to insulin after B_{12}, what would be really nice would be to collect data out to a reasonable limit, say 1 Å and calculate a three-dimensional sharpened Patterson from which one could find zinc-zinc vectors and zinc-other atom vectors, and proceed to lead off the structure analysis without necessarily using isomorphous replacement. Though in our case, the isomorphous replacement was obviously nice to do as well, to find the zinc positions, in fact, because we could change the zinc very easily

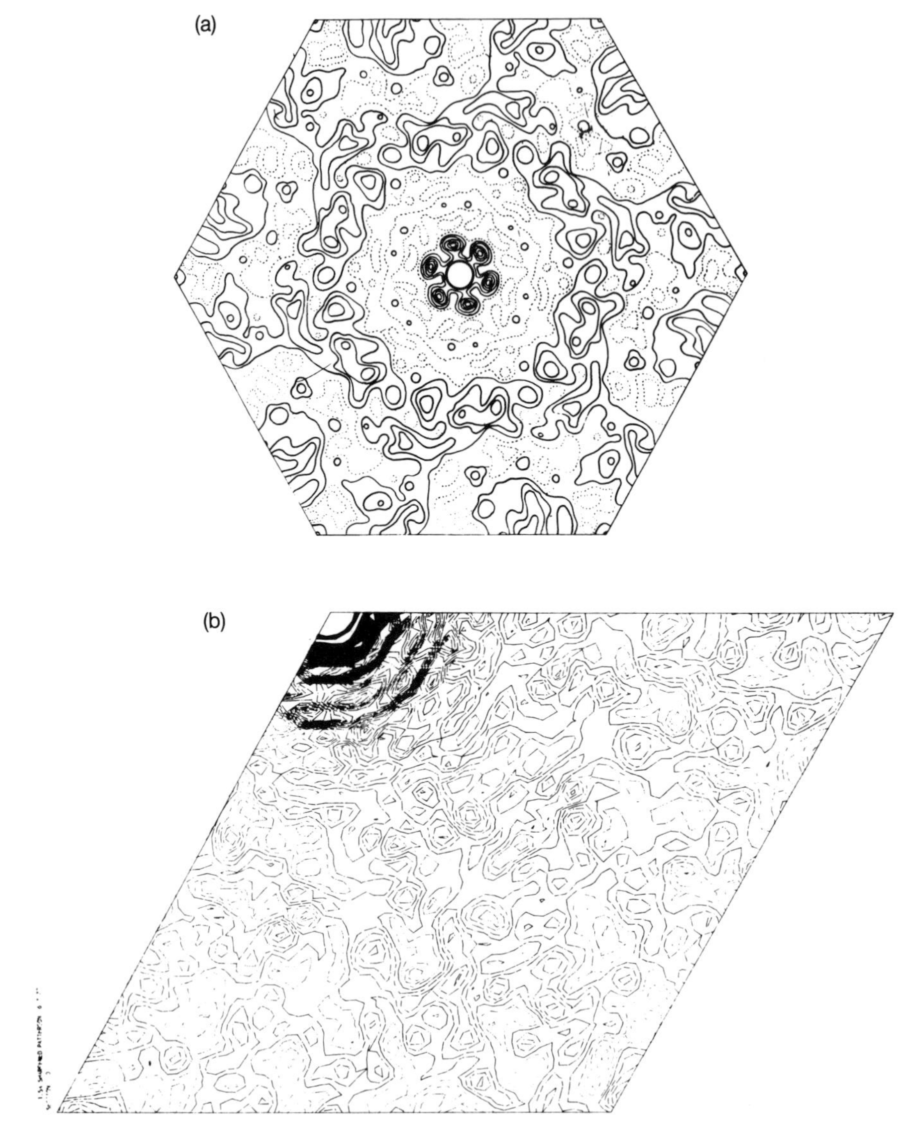

FIG. 9.18. Insulin, three-dimensional Patterson function at $w = 0$. (a) Data limited to 2.6 Å. (b) Data limited to 1.5 Å.

for cadmium, and later also we found for lead, and this did give us directly the zinc positions in the crystals to start off with. But you know it is difficult, it takes a long time to measure all the intensities, and everybody was raring to go, especially after we'd found the zinc-lead replacement. So somehow the data collection stopped short of our hoped-for limit. We found heavy atom derivatives, and solved the structure first by isomorphous replacement, only using Pattersons to find the heavy atom positions.[22]

All the same, it is interesting to see how the insulin Patterson changes as the data are extended. Fig. 9.18 shows three stages. The first, at approximately the 2.5 Å limit, was calculated on the old eye-estimated data during the war. It has some historical interest since the peaks at 5 Å spacing appeared round the origin and the 10 Å vector peaks were clearly seen to be complex, out of the $w = 0$ plane. As the data were extended the field became full of peaks of very similar height. Among these, because we now know where they are, we can recognize in the maps calculated with the data to 1.5 Å the zinc-to-sulfur vectors. Four of them occur together in a small group. The peaks are a little heavier than most of those around but hardly so markedly so that we would have identified them from the Patterson alone. Since Norio and Kiwako Sakabe had measured data to 1.2 Å, I asked them to calculate a sharpened three-dimensional Patterson from their data for this occasion. They calculated a whole series with different 'sharpening' factors. Fig. 9.19 shows two of these — and here the Zn-S vectors are again more visible. But whether one could find the insulin structure directly from this Patterson is very questionable. Perhaps one should try with cadmium insulin.

The methods of solving structures that Patterson made possible still continue to be used and are likely to be used for a long time. There are still different ways in which we can use the function and very very remarkable structures have been solved with their help during this last year, of a kind that one could hardly believe would be possible. If you heard at the International Union of Crystallography Meeting in Hamburg in the summer of 1984 of the work on the photosynthetic centre[23] carried out in Huber's laboratory at Munich you will know that it is quite fantastic. It would be lovely if we had Lindo here today to see the results to which his work has led.

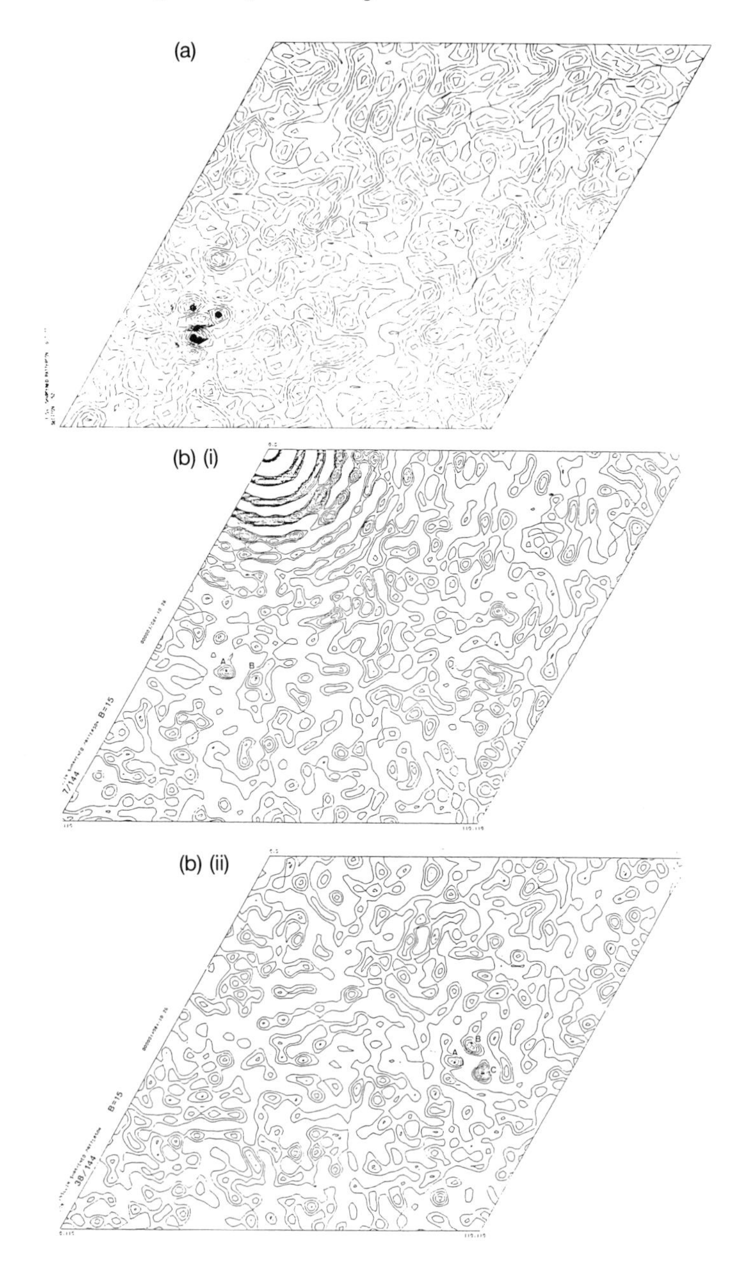

FIG. 9.19. Insulin three-dimensional Patterson function showing Zn–S vector peaks. (a) Data limited to 1.5 Å. (b) Data limited to 1.2 Å, near Zn–S peaks. (i) $w = \frac{7}{144}$ (ii) $w = \frac{38}{144}$.

David Harker: It is such a pleasure to listen to an expert saying what she has done! A work of art!

References.

1. Patterson, A. L. A Fourier series method for the determination of the components of interatomic distances in crystals. *Phys. Rev.* **46**, 372–376 (1934).
2. Beevers, C. A. and Schwartz, C. M. The crystal structure of nickel sulphate heptahydrate, $NiSO_4.7H_2O$. *Z. Krist.* **91**, 157–169 (1935).
3. Bernal, J. D., Crowfoot, D. M., Evans, R. C. and Wells, A. F. Crystallography 1933-35. *Annu. Rep. Progress Chem.*. **32**, 181–242 (p. 186) (1935).
4. Patterson, A. L. A direct method for the determination of the components of interatomic distances in crystals. *Z. Krist.* **A90**, 517–542 (1935).
5. Beevers, C. A. and Lipson, H. *Proc. Roy. Soc. (London)* **A146**, 570–582 (1934).
6. Bernal, J. D., Crowfoot, D. and Fankuchen, I. X-ray crystallography and the chemistry of the steroids. I. *Trans. Roy. Soc.* **A239**, 135-182 (1940).
7. Bernal, J. D. Carbon skeleton of the sterols. *Chem. Ind.* **51**, 466 (1932).
8. Carlisle, C. H. and Crowfoot, D. The crystal structure of cholesteryl iodide. *Proc. Roy. Soc. (London)* **A184**, 64–83 (1945).
9. Vani, G. V. and Vijayan, K. *Mol. Cryst. Liq. Cryst.* **51**, 253–264 (1979).
10. Bernal, J. D. and Crowfoot, D. X-ray photographs of crystalline pepsin. *Nature (London)* **133**, 794–795 (1934).
11. Astbury, W. T. and Woods, H. J. The X-ray interpretation of the structure and elastic properties of hair keratin. *Nature (London)* **126**, 913–914 (1930).
12. Harker, D. The application of the three-dimensional Patterson method and the crystal structure of proustite, Ag_3AsS_3, and pyrargyrite, Ag_3SbS_3. *J. Chem. Phys.* **4**, 381–390 (1936).
13. Crowfoot, D. Crystal structure of insulin. I. Investigation of air-dried insulin crystals. *Proc. Roy. Soc. (London)* **A164**, 580–602 (1938).
14. Crowfoot, D., Bunn, C. W., Rogers-Low, B. W. and Turner-Jones, A. X-ray crystallographic investigation of the structure of penicillin. In *Chemistry of Penicillin* (H. T. Clarke *et al.*, eds.) pp. 310-366. Princeton University Press: Princeton (1949).

15. Pickworth, J., Robertson, J. H., Shoemaker, C. B., White, J. G., Prosen, R. J. and Trueblood, K. N. The structure of vitamin B_{12}. I. An outline of the crystallographic investigation of vitamin B_{12}. *Proc. Roy. Soc. (London)* **A424**, 228–263 (1957).
16. Hodgkin, D. C. X-ray crystallographic study of the structure of vitamin B_{12}. *Bull. Soc. fr. Minér. crystallogr.* **78**, 106-115 (1955).
17. Hodgkin, D. C., Pickworth, J., Robertson, J. H., Prosen, R. J., Sparks, R. A. and Trueblood, K. N. Structure of vitamin B_{12}. II. Crystal structure of a hexacarboxylic acid obtained by the degradation of vitamin B_{12}. *Proc. Roy. Soc. (London)* **A251**, 305–352 (1959).
18. Lenhert, P. G. and Hodgkin, D. Structure of the 5,6-dimethylbenz-imidazoylcobamide coenzyme. *Nature (London)* **192**, 937–938 (1961).
19. Anderson, B., Hodgkin, D. C. and Viswamitra, M. A. Structure of thiostrepton. *Nature (London)* **225**, 223-225 (1970).
20. Adams, M. J., Blundell, T. L., Dodson, E. J., Dodson, G. G., Vijayan, M., Baker, E. N., Harding, M. M., Hodgkin, D., Rimmer, B. and Sheat, S. Structure of rhombohedral 2-zinc insulin crystals. *Nature (London)* **224**, 491–495 (1969).
21. Blundell, T. L., Cutfield, J. F., Cutfield, S. M., Dodson, E. J., Dodson, G. G., Hodgkin, D. C., Mercola, D. A. and Vijayan, M. Atomic positions in 2-zinc insulin crystals. *Nature (London)* **231**, 506–511 (1971).
22. Crowfoot, D. Protein crystals. *Achievements in Chemistry.* (in Russian) Vol. **XVB**, 2 (1946).
23. Deisenhofer, J., Epp, O., Miki, K., Huber, R. and Michel, H. X-ray structure analysis of a membrane protein complex. Electron density map at 3 Å resolution and a model of the chromophore of the photosynthetic reaction center from *Rhodopseudomonas viridis*. *J. Mol. Biol.* **180**, 385–398 (1984).

10. The nanometer world of hydrogen bonds

G. A. Jeffrey

George Jeffrey was born in Cardiff, and educated at Birmingham University where he obtained a Ph.D. degree in 1939, and D.Sc. degree in 1953. During World War II he worked as a research physicist at the British Rubber Producers Research Association and after the war he taught at Leeds. In 1953 he moved to Pittsburgh where he became Professor of Chemistry and Physics, and in Pittsburgh he built up the crystallography laboratory which became the Department of Crystallography, the only one in the United States. Mention has been made of Caltech as a learning ground for crystallographers, but after the war, when more and more crystallogaphers were educated, and it was the rule for European postdoctoral students to spend time in the U.S., the lab in Pittsburgh was, and still is, a well-known location for further education, and numerous now well-known crystallographers spent several useful and pleasant years in Pittsburgh. Not only have there been many postdoctoral students in Pittsburgh, also many graduate students received their education in Pittsburgh and several of them are here in the audience.

Dr Jeffrey has been very active in the American Crystallographic Association and in the IUCr and was President of the ACA in 1963. He also was Chairman of the National Committee for Crystallography.

His intial interest in Pittsburgh was in the structures of clathrate hydrates, which showed a beautiful array of spherical structures based on the dodecahedron. Later on his interests changed to carbohydrates, determined by X-ray diffraction and neutron diffraction. Both subjects are the basis for his interest in the hydrogen bond, the subject of his talk today. Among his many honors are the American Chemical Society Hudson Award in Carbohydrate Chemistry in 1980, and at the present time he is the recipient of the von Humboldt Senior Scientist Award, which he is enjoying very much. Over the years he has also worked hard to build up crystallography groups, for instance in Bangaladesh, Brazil and Portugal.

I was first introduced to the concept of the hydrogen bond by two very distinguished scientists. One was J. D. Bernal who was my external Ph.D. examiner and who, in 1933 with R. H. Fowler, gave a description of the water molecule that is considered to be the starting point of all modern theoretical treatments.[1] The other was Linus Pauling, whose first edition of *The Nature of the Chemical Bond* had just appeared at that time.[2]

A recent inspection of *Chemical Abstracts* suggests that the number of publications to be abstracted under Hydrogen Bonding must now exceed 20,000 and is increasing at the rate of one every twenty minutes of the working day. The reason for this enthusiasm for studying hydrogen bonds, particularly by the life scientists, is certainly connected with the last part of the following statement:

> HYDROGEN BOND. A weak electrostatic chemical bond which forms between covalently bonded hydrogen atoms and a strongly electronegative atom with a lone pair of electrons. ... Life would be impossible without this type of linkage. (Penguin Dictionary of Science, 1977.) [3]

If this is correct, then we need to know as much as possible about the role of the hydrogen bond in the organization, or three-dimensional patterns, of the molecules involved in biological reactions. These patterns are on the nanometer scale, where the nanometer, 10 Å, is equivalent to the meter in the macroscopic world in which we live.

Hydrogen bonds are weak interactions, midway in energy between the weaker covalent bonds and the van der Waals forces. Nevertheless, they are responsible for the recurrent features which are observed in the architecture of biological macromolecules. The best known of these are the α-helix and pleated sheets in proteins and the base-pairing in the nucleic acids. In order to influence the shapes, *i.e.*, conformations, of these giant molecules, these weak forces must function in concert, that is, cooperatively, like the Lilliputians in Swift's *Gulliver's Travels.* Being weak forces, hydrogen-bonds can be rapidly switched on and off with the energies involved in biological processes. In both these respects they resemble the binary operations which can be used to construct a complex program of instructions for the computer. However, because hydrogen bond forces can function in parallel in three-dimensional space, the computer analogy is more with the three-dimensional parallel processors of the future, than with the sequential or one-dimensional parallel processors of the present.

In this talk I will present some relatively new ideas about hydrogen bonding that have come from three sources.
1) Single crystal neutron diffraction analyses
2) Computer readable crystal structure data bases
3) Computer graphics

What is special about these facilities for the study of hydrogen bonds?

Neutron diffraction has the advantage over X-ray diffraction in that it can locate the positions of hydrogen atoms with the same degree of precision as the other major biological elements, carbon, oxygen and nitrogen. It is not a new method, but results come slowly compared with X-ray analyses. It is necessary to visit the neutron sources, and experiments are measured in days and weeks, rather than hours and days as with X-ray analyses. Single crystals suitable for neutron diffraction are more difficult to obtain than those for X-ray diffraction, since they have to be about 1000 times larger.
Crystallographic Data Bases are relatively new. There are two of importance to the biological sciences. One is the Cambridge Crystallographic Data Base, originated by J. D. Bernal and nurtured with great skill and determination by Olga Kennard and her colleagues in Cambridge, England.[4] The other is the Protein Data Base which is maintained by Tom Koetzle and his colleagues at Brookhaven National Laboratory.[5]
Computer Graphics is also a relatively new technique. It permits the display and rotation of molecules or assemblages of molecules or parts of molecules. It has the advantage over the old ball and stick models that one can remove parts of molecules while keeping other parts in their correct spatial relationships. If you have color graphics, different parts of molecules can be colored differently.

Now I want to give examples of the use of these tools to study hydrogen bonding.

Until 1966, crystallographers could not see the hydrogen atoms in organic crystal structures, at least not when they were attached to electronegative atoms, as in the important

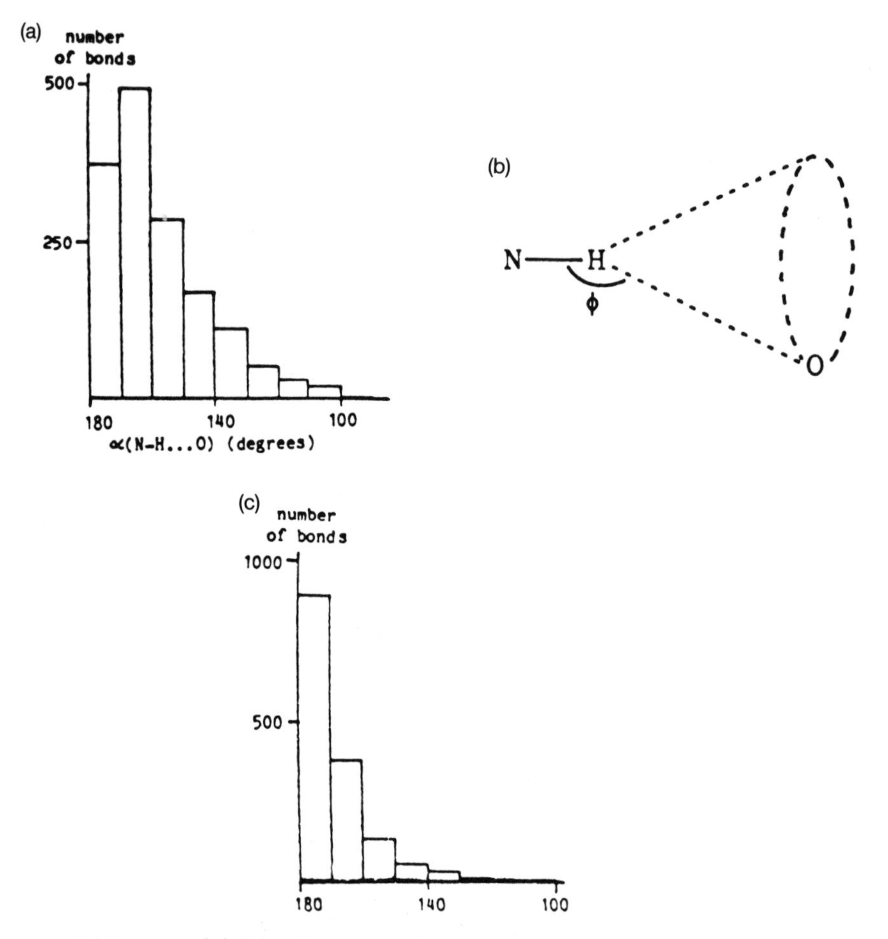

FIG. 10.1. (a) Distribution of N–H···O angles from 1509 N–H···O=C bonds. (b) Geometrical factor affecting the N–Ĥ···O angle.[8] (c) Distribution of N–Ĥ···O angles corrected for the geometrical factor.

hydrogen bonding functional groups, such as O–H or N–H. They therefore ignored them and applied the term 'hydrogen-bond distance' to the separation of the non-hydrogen atoms in **X**–H···**A**, where **X** is the donor and **A** the acceptor atom. An example of such a list for hydrogen-bond lengths is shown in Table 10.1. Since the **X**–H···**A** angle is rarely observed to be 180°,[6,7] see Fig. 10.1,[8] this *old* hydrogen bond distance **X**···**A** is a function of the covalent bond length **X**–H, the *real* hydro-

TABLE 10.1. N–H···X and O–H···X Hydrogen–Bond Lengths and Ranges

(from *X-Ray Structure Determination* by G. H. Stout and L. H. Jensen, Macmillan, NY 1968. 'In memoriam A. L. Patterson')

Bond		Mean, Å	Range, Å
N–H···N		3.10	2.88–3.38
N–H···O			
	ammonia	2.88	2.68–3.24
	amides	2.93	2.55–3.04
	amines	3.04	2.57–3.22
N–H···F		2.78	2.62–3.01
N–H···Cl		3.21	2.91–3.52
N–H···Br		3.37	3.28–3.44
O–H···N		2.80	2.62–2.93
O–H···O			
	oximes, inorganic acids	2.58	2.44–2.84
	carboxylic acids	2.63	2.45–2.75
	H_2O in org.-inorg.	2.71	2.49–3.07
	alcohols	2.74	2.55–2.96
	H_2O in inorg.	2.75	2.49–3.15
	H_2O in org.	2.80	2.65–2.93
	hydroxides	2.82	2.36–3.36
O–H···Cl		3.07	2.86–3.21
O–H···Br		3.30	3.17–3.38

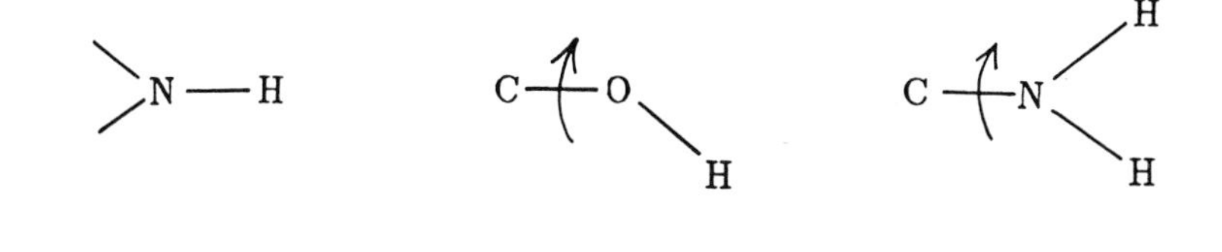

I II III

gen bond length H···A and the hydrogen bond angle X–Ĥ···A. Therefore, to determine the hydrogen bond length, it is necessary to know where the hydrogen is located.

In some circumstances, such as **I**, the position of the hydrogen atom can be guessed from the non-hydrogen atom positions. But in **II** or **III**, it cannot.

Three-center hydrogen bonds

As a result of this quest for hydrogen positions, it was discovered that the most probable X–Ĥ···A angle was not 180°, but ~ 160°. The so-called linear hydrogen bond is not linear. But, more unexpectedly, it was found that there were a number of examples where exploration, by means of computer graphics, in the direction of the bisector of the obtuse X–···A angle, as in **IV**, revealed a third nearest neighbour atom which was also electronegative.[9–12] We refer to this configuration, **V**, as a *three-center bond*, to avoid confusion with **VI**, which was la-

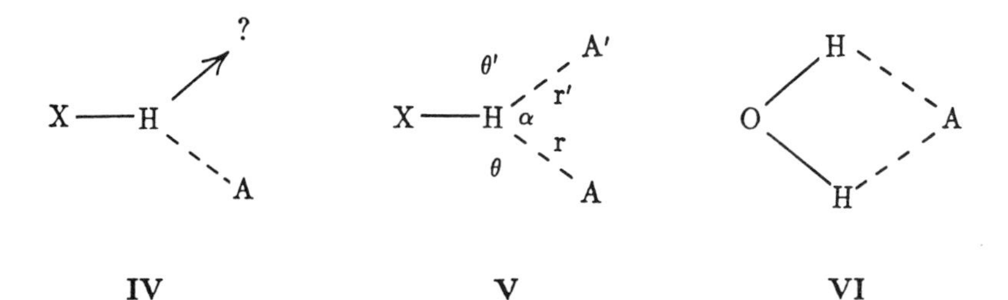

belled '*bifurcated bond*' in the first major text devoted entirely to the *Hydrogen Bond*, by McClelland and Pimental in 1960.[13]

Examples of some typical two- and three-center bonds from the neutron diffraction analysis of methyl α-D-altropyranoside[14] are shown in Fig. 10.2. The criterion for a three-center bond, first used by Parthasarathy in 1969,[15] is that the hydrogen atom be in the plane of the three atoms to which it is bonded, *e.g.*, $\theta + \theta' + \alpha \approx 360°$, or that the hydrogen atom be within 0.25 Å of the non-hydrogen atom plane.[10]

The implication from this planar, or nearly planar, configuration is that the hydrogen atom is being *attracted* by both A and A′. Since electrostatic attractive forces are dependent upon r^{-1}, whereas the attractive component of van der Waals forces is usually represented by r^{-6}, the attractive components between the hydrogen atom and the more electronegative nearest neighbors at long distances will be primarily electrostatic,

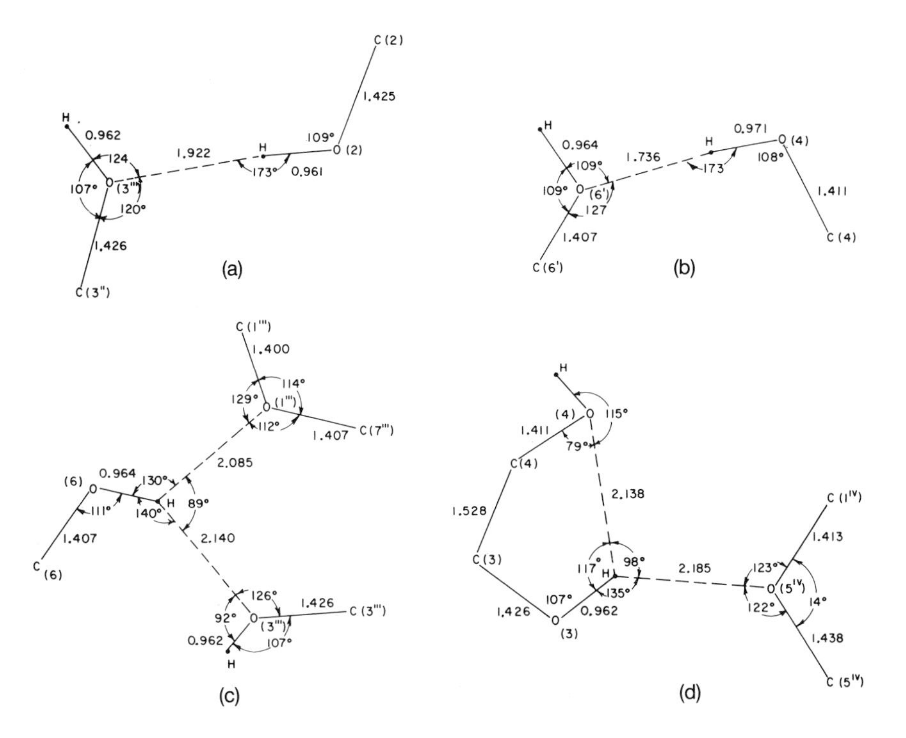

FIG. 10.2. Two- and three-center hydrogen bonds observed by neutron diffraction in the crystal structure of methyl α-D-altropyranoside.[14]

i.e., hydrogen bonds. The concept that a hydrogen bond becomes a van der Waals interaction at longer distances is clearly nonsensical. Therefore it is not appropriate to use sums of van der Waals radii as a criterion for H···A hydrogen bonding. At short H···A distances, non-coulombic interactions will be significant, but for the hydrogen bonds in or between biological molecules, the forces will be primarily electrostatic, becoming predominantly so at the longer H···A distances. Although the H···A distances in three-center bonds are longer than those in two-center bonds, theoretical calculations indicate that the two- and three-center bonds shown in Fig. 10.1 are comparable in energy.[16]

Irrespective of whether there is a distance cut-off applied, there is now evidence that between 20 and 30% of all O–H···O and N–H···O hydrogen bonds are three-centered.[9–12]

Hydrogen-bond lengths are very 'soft' quantities

It is well-known that the C–C single bond length is not an observable constant in organic molecules. These bonds are compressed or stretched by intra- and intermolecular forces, so that C–C single bond lengths are commonly observed between 1.500 and 1.570 Å.[17] Since hydrogen bond stretching force constants are about 15 times weaker than those for C–C bonds, comparable forces will compress or stretch hydrogen bond lengths from their equilibrium or 'most probable' values by 0.2 Å. Fig. 10.3(a) shows just such a distribution of

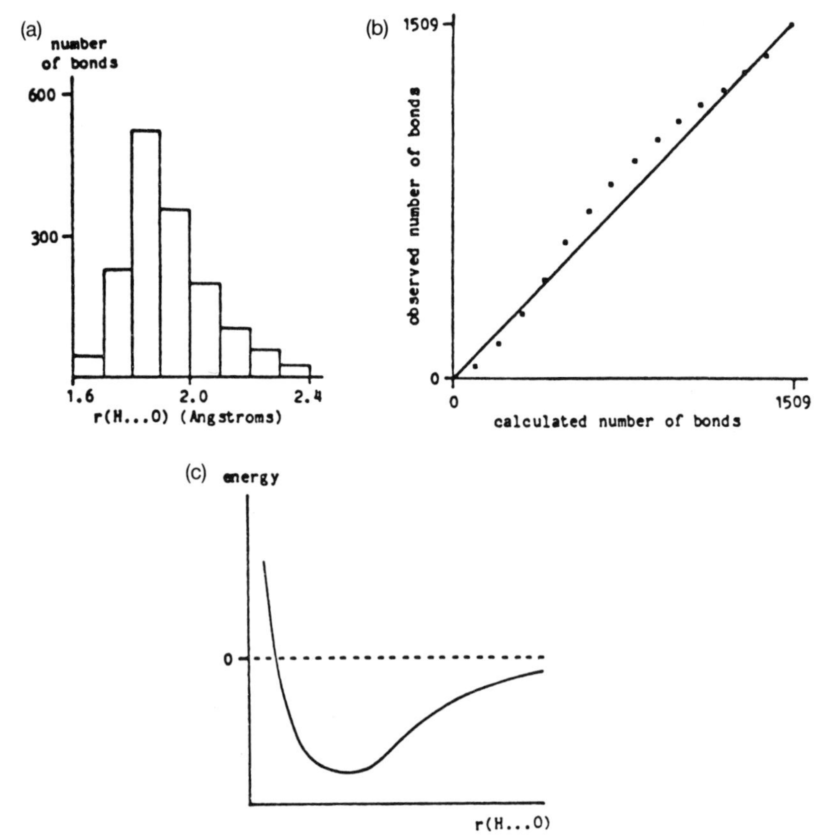

FIG. 10.3. (a) Distribution of H···O bond lengths from 1509 N–H···O=C bonds.[8] (b) Deviation from normal distribution. (c) Corresponding form of potential energy curve.

N–H···O=C bond lengths from a survey of 1509 such bonds from the Cambridge Data File.[8] This distribution is that expected from a potential energy curve such as that shown in Fig. 10.3(c).

Hydrogen bonds have characteristic most probable values

We know that the length of the hydrogen bond is sensitive not only to the donor and acceptor atoms but also to the configurations in which these atoms occur. The 1965 neutron diffraction data on oxalic acid dihydrate in Table 10.2 shows that very clearly.[18]

TABLE 10.2. D···O hydrogen bond lengths in the α and β deutero oxalic acid dihydrates[18]

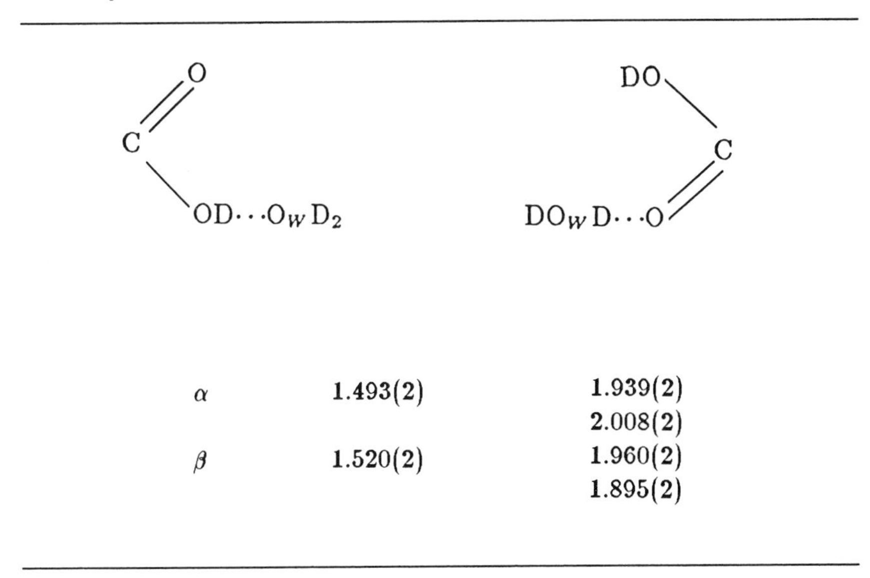

α	1.493(2)	1.939(2)
		2.008(2)
β	1.520(2)	1.960(2)
		1.895(2)

Walter Hamilton had the foresight to realize the importance of studying a whole family of related molecules by single crystal neutron diffraction. An analysis of the hydrogen bond lengths in the amino acids, based mainly on his research, gave the results shown in Fig. 10.4. There were only 168 bonds

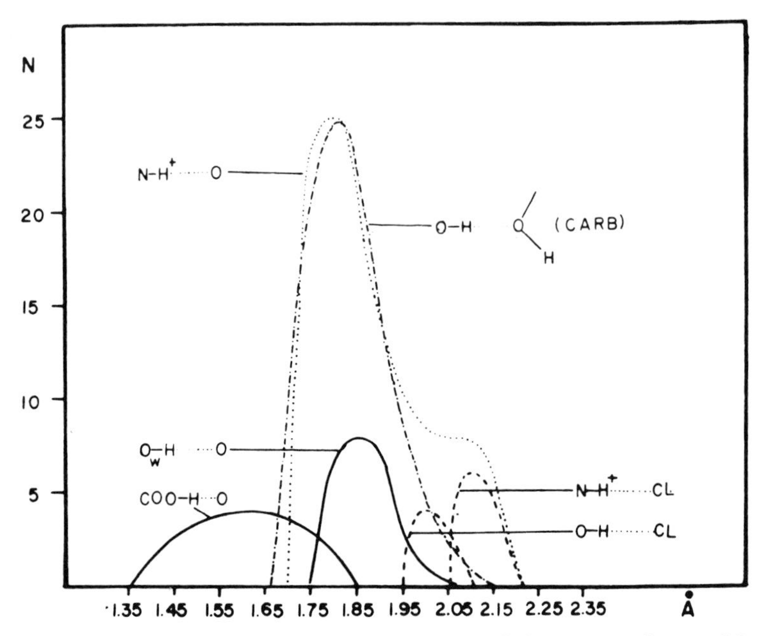

FIG. 10.4. Distribution of hydrogen-bond distances observed by neutron diffraction crystal structure analyses of the amino acids. [Reproduced from Jeffrey, G. A. and Maluszynska, H., *J. Quantum Chem.: Quantum Biol. Symp.* **8**, 231-239 (1981)].

available for this survey, so the statistics were poor, but they show clearly that different bonds have different most probable values.[12]

This question of dependence of bond length on configuration was explored in much greater detail in the paper published earlier this year,[8] from which Fig. 10.3 was obtained. This paper analyzes 1509 N–H···O=C bonds from 889 crystal structures, using the information retrieval capabilities of the Cambridge Crystallographic Data File. The results are shown in Table 10.3. Even with 1509 bonds, the statistics are sometimes poor, but ignoring some minor discrepancies, they reveal exactly what is expected. The more acidic the donor group and the more basic the acceptor, the shorter the hydrogen bond.

Instead of carrying out a survey of a particular hydrogen bond in a variety of crystal structures, as in that analysis, we have surveyed the hydrogen bonds occurring in particular

TABLE 10.3. N–H···O=C Hydrogen Bond Lengths Observed in 889 Crystal Structures[8]

Donor	Acceptor					
	O=C=O	O=C–NH_2	O=C$(CR)_2$	O=C–OH	Row Mean	Wtd. Row Mean
R_3NH^+	11	2	1	0		
	1.722	1.845	1.938		1.835	1.755
$R_2NH_2^+$	47	3	3	6		
	1.796	1.793	1.966	1.887	1.860	1.805
RNH_3^+	226	15	8	68		
	1.841	1.891	1.872	1.936	1.885	1.865
N–H^+	36	12	2	11		
	1.869	1.858	1.844	1.983	1.888	1.887
NH_4^+	56	4	2	13		
	1.886	1.988	1.995	1.916	1.946	1.900
N–H	74	597	38	117		
	1.928	1.934	1.970	2.002	1.959	1.945
Column Mean	1.840	1.885	1.931	1.945		
Weighted Column Mean	1.855	1.931	2.024	1.972		

classes of compounds, *i.e.*, carbohydrates, amino acids, nucleosides and nucleotides.[9,11,12,19] The results from the survey of hydrogen bond lengths in 75 nucleoside and 11 nucleotide structures are shown in Table 10.4.[19] All were X-ray analyses but one, and therefore, as in the N–H···O=C bond survey,[8] all covalent X–H bond lengths were normalized to the neutron diffraction values (O–H =0.97, N–H = 1.03 Å).[20,21] Here the statistics are even worse, but the overall trends make good chemical sense, with the P–OH groups the strongest donors and the P=O groups the strongest acceptors. In this family of structures, the number of three-center bonds was 30%, none of which involved the strong P–OH group donors.

So we are now obtaining data on the bond length distributions and most probable hydrogen bond lengths for all combinations of donor and acceptor configurations in various clases of small molecules which are subunits of the biological macromolecules. Inasmuch as these are different, they will have dif-

TABLE 10.4. Metrical data on two-center hydrogen bonds in the nucleosides and nucleotides[19]

Donor		Acceptor								
		O=P	O_WH_2	O=C	C–O–H	=N–	NH_2	OC_2	Cl^- –	S=C
P–OH	No	4	5		2					
	Min	1.55	1.59		1.65					
	Max	1.57	1.68		1.89					
	Mean	1.56	1.64		1.77					
NH^+	No				2				4	
	Min				1.71				2.11	
	Max				1.81				2.22	
	Mean				1.76				2.14	
C–OH	No	6	19	44	56	27	1	6	12	2
	Min	1.70	1.57	1.74	1.70	1.79		1.95	2.02	2.49
	Max	1.95	2.02	2.65	2.57	2.62		2.70	2.24	2.50
	Mean	1.785	1.760	1.900	1.872	1.910	1.83	2.245	2.132	2.495
HO_WH	No	11	10	6	28	7		1		1
	Min	1.66	1.78	1.81	1.78	1.86				
	Max	2.11	2.18	2.07	2.17	2.04				
	Mean	1.848	1.924	1.910	1.903	1.930		1.92		2.66
N–H	No	6	21	9	19	6		2		
	Min	1.69	1.66	1.72	1.58	1.77		2.02		
	Max	1.85	2.05	2.86	2.46	1.99		2.14		
	Mean	1.768	1.855	1.869	1.955	1.882		2.08		
NH_2	No	11	6		36		14	4	4	3
	Min	1.67	1.71		1.73		1.88	1.95	2.23	2.38
	Max	2.07	2.36		2.73		2.75	2.44	2.30	2.64
	Mean	1.866	1.990		2.047		2.180	2.22	2.27	2.52
C–H	No	2			11					
	Min	2.23			2.08					
	Max	2.55			2.55					
	Mean	2.39			2.32					

ferent equilibrium values and correspond to different potential energy surfaces. They will require different empirical constants in any molecular mechanics program that seeks to calculate the structures of proteins or nucleic acids.[22]

Two examples of pattern recognition

The Carbohydrates. Fig. 10.5 illustrates an infinite chain of ⋯ OH ⋯ OH ⋯ OH ⋯ hydrogen bonds in the crystal structure of methyl α-xylopyranoside.[23] Only the hydrogen bonding functional groups are displayed and the molecules have been oriented so as to give the clearest view on the computer graphics. This is an energetically favorable pattern because of the well-known non-additivity or cooperativity of hydrogen bonds,[20,24]

$$E(\cdots O\text{–}H\cdots O\text{–}H\cdots)_n < nE(O\text{–}H\cdots O).$$

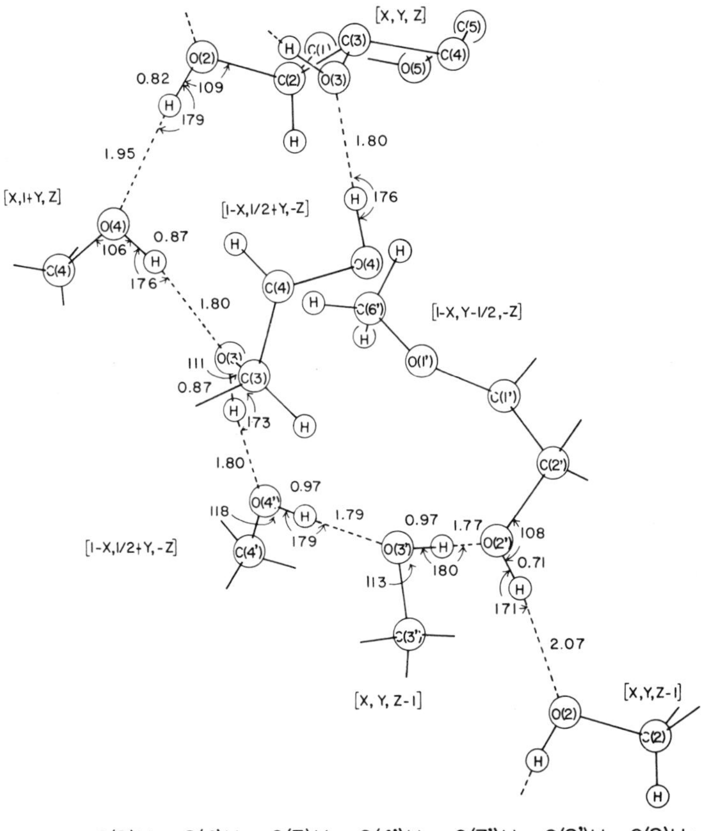

FIG. 10.5. Infinite chain of O–H⋯O hydrogen bonds (Type I) in methyl α-D-xylopyranoside.[23]

where E is the energy of the system. In cyclic carbohydrates such as the pyranosides and furanosides, there is a problem with cooperativity because the ring and glycosyl oxygen atoms (O(r) and O(g) respectively in Table 10.5) are hydrogen bond acceptors but not donors, and therefore cannot participate in infinite chains. From this point of view these cyclic carbohydrates are proton deficient. They have more potential hydrogen bond acceptor oxygens than hydroxyl protons available. The sugar alditols $HOH_2C(CHOH)_nCH_2OH$ do not have this problem and the hydrogen bonding is always infinite chains. The cyclic

TABLE 10.5. Distribution of Hydrogen-Bonding Patterns in the Crystal Structures of Pyranoses and Pyranosides[25]

(I) Omit Acetal Oxygens, O(r), O(g)	(II) Finite Chains Terminating at O(r), O(g)	(III) Infinite Chains with Separate O-H···O(r), O(g)	(IV) Infinite Chains with Three-Center Bonds to Acetal Oxygens
1,6-anhydro mannose	methyl-4-deoxy-4-fluoro-α-glucoside	β-arabinose*	1,6-anhydro-β-galactose
2-deoxy-2-fluoro-β-D-mannose	methyl β-glucoside	β-D,L-arabinose*	1,6-anhydro mannose
α-galactose	methyl α-mannoside*	α-fucose	2-deoxy-β-glucose
β-galactose	methyl β-xyloside*	α-galactosyl rhamnitol	β-fructose*
iso-maltose H_2O	sucrose*	α-glucose*	β-lyxose*
β-maltose H_2O*	α, α-trehalose	β-glucose	α-mannose
methyl β-arabinoside	α-xylose*	α, β-maltose	α, β-melibiose
methyl α-glucoside*	α-galactosyl rhamnitol	α, β-melibiose	methyl α-altroside*
methyl β-maltopyranoside		methyl α-arabinoside	methyl α-galactoside*
methyl α-riboside*		methyl β-glactoside*	methyl α-glucoside*
methyl α-xyloside		α-rhamnose H_2O*	α-rhamnose H_2O
		α-sorbose*	
		α-tagatose	
		α-talose	

* Asterisks indicate neutron diffraction studies. Some structures have more than one type of pattern and appear in two columns.

carbohydrates respond to this dilemma in the four different ways illustrated in Table 10.5. The known carbohydrate crystal structures distribute themselves rather evenly over these four alternatives.[25] Some examples of these alternatives are shown in Fig. 10.6. The mean hydrogen bond lengths, shown in Table 10.6, reveal the differences with the degree of cooperativity, which are small but in the expected direction.

TABLE 10.6. Effect of Pattern on Mean OH···OH Bond Lengths in Carbohydrates[9]

Monosaccharides	No.	H···O	O–H···O
Infinite chains (alditols)	46	1.807 Å	166.5°
Infinite chains excluding C–O–C	29	1.813	165.5
Infinite and finite chains	29	1.823	164.9
Infinite chains with three-center bonds	34	1.830	165.9
Finite chains only	36	1.864	153.8
Disaccharides	86	1.863	161.2

The Amino Acids. The dominant hydrogen bonding functional groups in the amino acid crystal structures are the zwitterion NH_3^+ and COO^- groups. Here is a similar problem, since the surveys of C=O hydrogen bonding acceptor geometries indicate a definite preference for the carboxyl oxygen to accept more than one hydrogen bond.[21,26] However NH_3^+ has only three protons and there are an equal number of NH_3^+ and COO^- groups. This is the dilemma of three husbands and four wives that can be solved in the four ways shown below. The numbers in brackets show how the 49 amino acid crystal structures studied by neutron diffraction are distributed over these four choices.[11] The last one is unpopular due to obvious overcrowding.

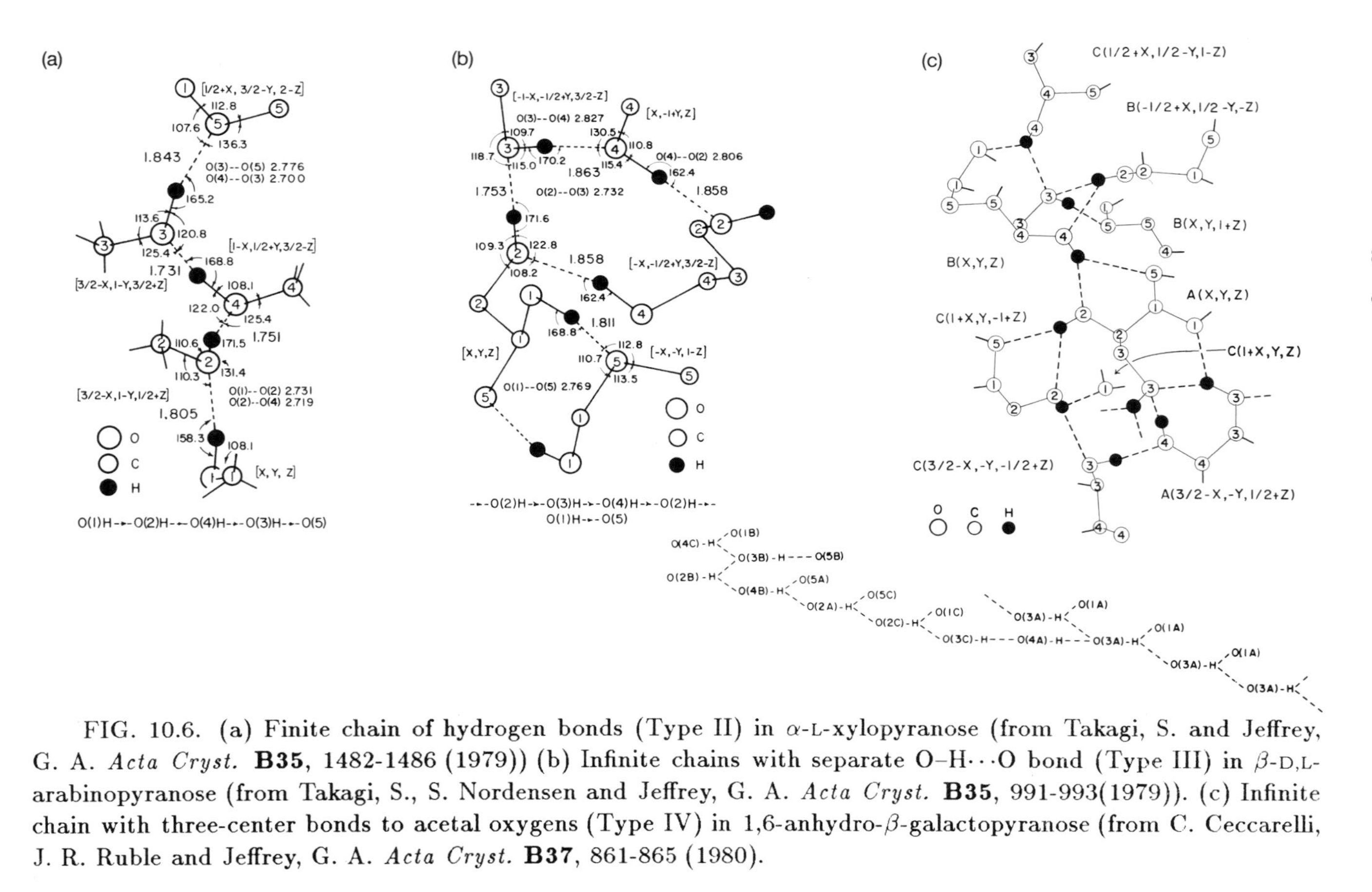

FIG. 10.6. (a) Finite chain of hydrogen bonds (Type II) in α-L-xylopyranose (from Takagi, S. and Jeffrey, G. A. *Acta Cryst.* **B35**, 1482-1486 (1979)) (b) Infinite chains with separate O–H···O bond (Type III) in β-D,L-arabinopyranose (from Takagi, S., S. Nordensen and Jeffrey, G. A. *Acta Cryst.* **B35**, 991-993(1979)). (c) Infinite chain with three-center bonds to acetal oxygens (Type IV) in 1,6-anhydro-β-galactopyranose (from C. Ceccarelli, J. R. Ruble and Jeffrey, G. A. *Acta Cryst.* **B37**, 861-865 (1980).

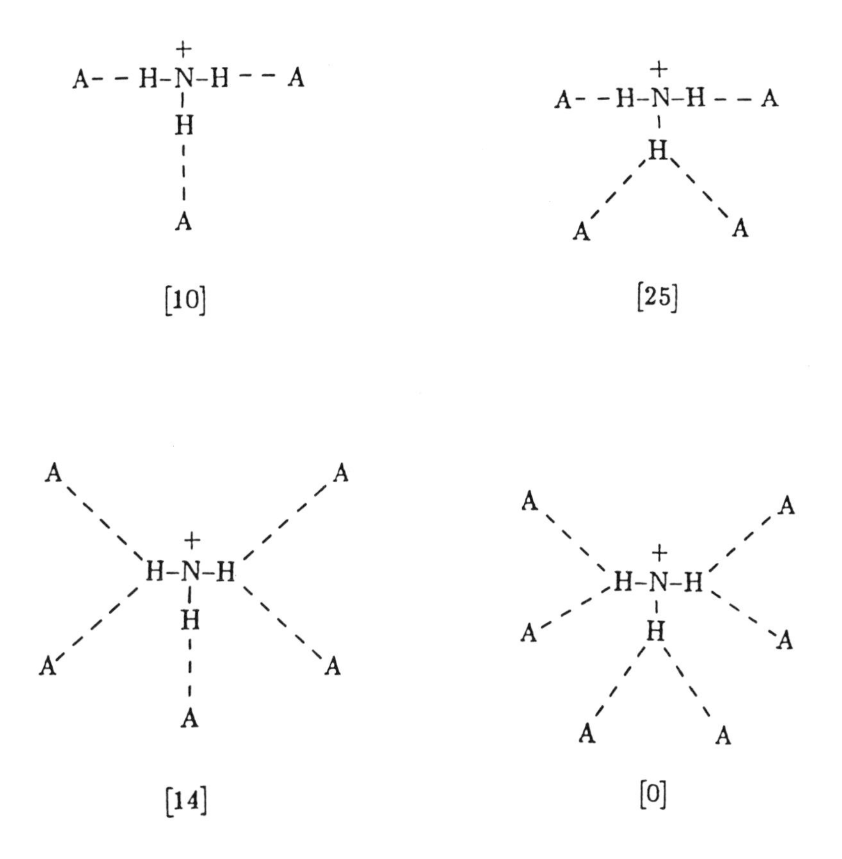

In both the carbohydrates and the amino acids, the three-center hydrogen bonding is a consequence of proton deficiency. Not enough protons are present to satisfy the normal requirements of the hydrogen bond acceptor groups. The term three-center bond is therefore particularly appropriate by analogy with the appearance of the three-center bond in the boron hydrides due to valence electron deficiency.[27]

What about the water molecules?

Water is the most abundant biological molecule, yet its role in the organization of biological structure is poorly understood. Even when it occurs in the crystal structures of small organic molecules, it is frequently ignored, even to the degree of sometimes ignoring its presence in the title of the publication describing the crystal structure.

In the carbohydrates, amino acids, nucleosides and nucleotides, water hydrogen bonds in two ways, **VII** and **VIII**.

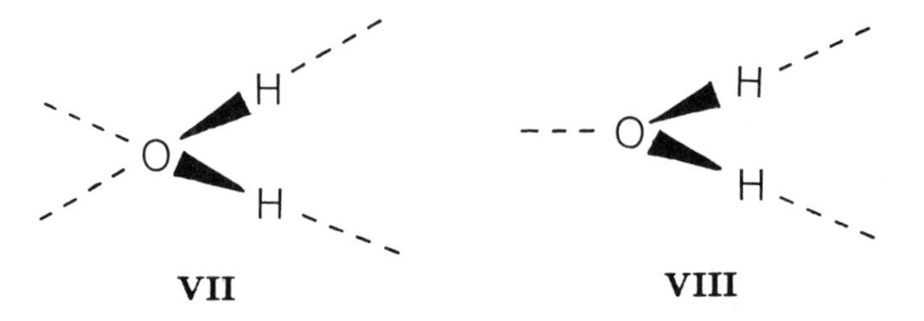

VII **VIII**

In **VII**, the four-coordinated water molecule links finite or infinite chains of hydrogen bonds into three-dimensional nets. This is most common, but not exclusive, in the carbohydrates. An elegant example is that of β-maltose monohydrate,[28] the neutron diffraction results of which are shown in Fig. 10.7.

In **VIII**, the water molecule accepts only one hydrogen bond and lies at the intersection where one chain branches into two. The same type of branching occurs when a hydroxyl group accepts one hydrogen bond and forms a three-center bond.[25] There are no examples where water molecules do not donate two hydrogen bonds, and only in the presence of cations do they never accept at least one hydrogen bond. Similar behavior is observed in the inorganic salt hydrates.[29]

Finally, to illustrate the influence of hydrogen bonding on physicochemical properties, Fig. 10.8 shows the explanation why mucic acid,[30] **IX**, is insoluble in cold water, has a relatively

IX

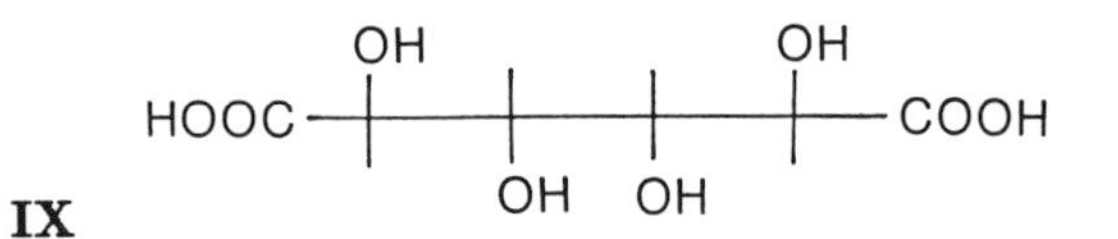

high melting point, 206°, and density 1.831 gm/cc, whereas saccharic acid, **X**, melts at 125° and is deliquescent. The hydrogen

X

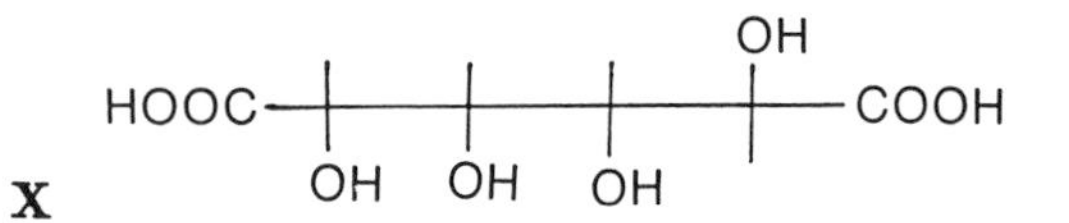

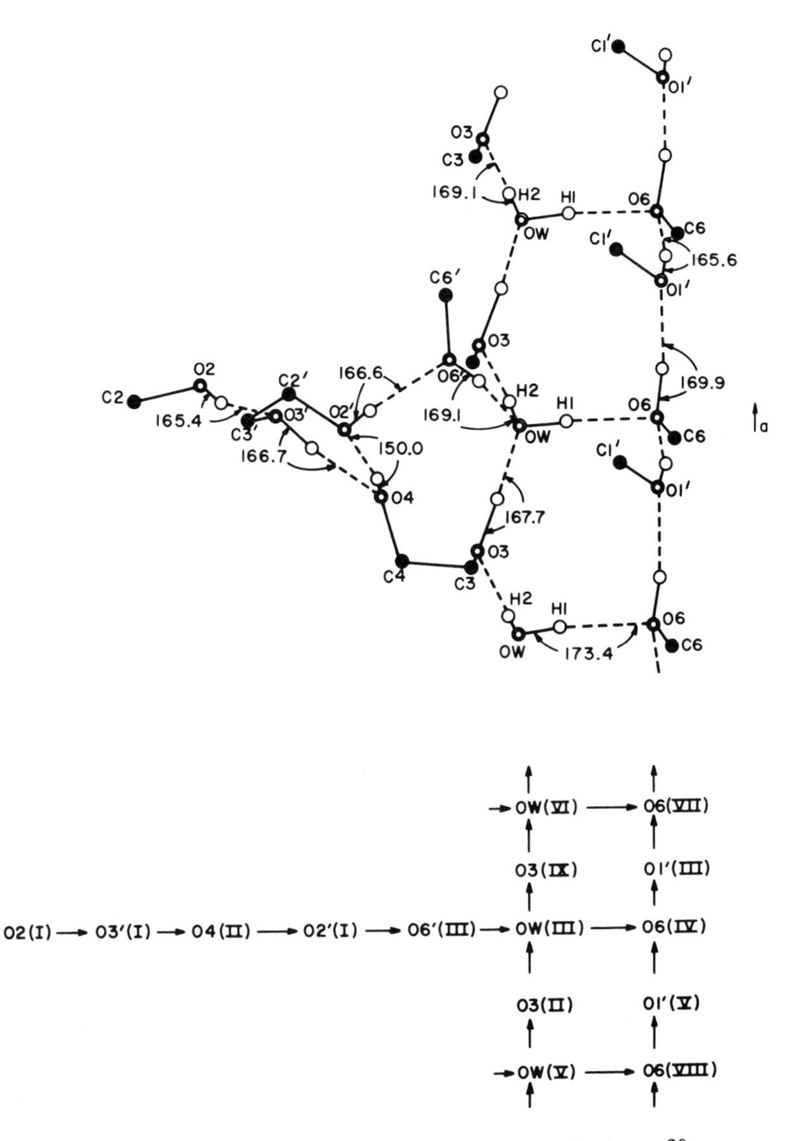

FIG. 10.7. Hydrogen bonding in β-maltose monohydrate.[28]

bonding in mucic acid contains the cooperative cycle of four bonds and the strong carboxylic dimer bonding linked into one system by a three-center bond. This Gulliver is so tied down he cannot escape, even in the presence of water molecules.

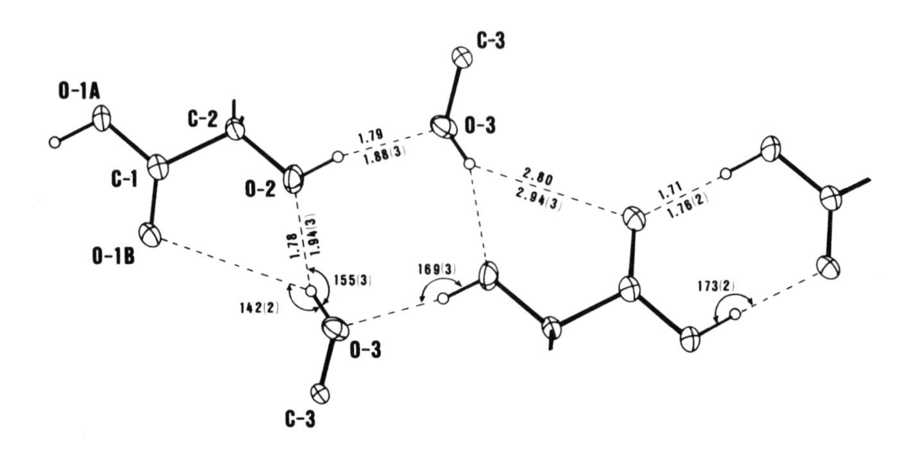

FIG. 10.8. Hydrogen bonding in mucic acid,[30] a dicarboxylic hydroxy acid which is insoluble in cold water.

What conclusions can be drawn which can be extrapolated to the macromolecules, of which these small molecules are the subunits?

1) We can expect that between 20 and 40% of the hydrogen bonds will be three-centered, depending upon the level of proton deficiency. Hydration will alleviate proton deficiency for two reasons. One is that water molecules can more easily orient so as to permit three-center bonding. The second is that each water molecule provides two protons, but quite commonly accepts only one hydrogen bond.
2) It is rare that hydrogen bonding functional groups do not form either two- or three-center hydrogen bonds.
3) There are different equilibrium values and potential energy surfaces for all combinations of groups which are hydrogen bond donors and acceptors. As the crystallographic data base becomes larger, these differences will be better defined and the equilibrium values will become more precise.

It will be interesting to compare these conclusions drawn from the crystal structures of small molecules, where the positions of the hydrogen atoms are known, with those deduced

from the actual structure of protein molecules, where only the positions of the non-hydrogen atoms are known.[31]

Discussion

Dorothy Hodgkin: Members of the same group as did the analysis of hydrogen-bonding in the proteins,[31] that is, the York team, have looked at the hydrogen-bonding in the insulin structure. Guy Dodson, Rod Hubbard and Ted Baker (when he was working there in York) have done the calculation in relation to the insulin crystal structure with calculated hydrogen positions, and occasionally we do see three-center arrangements. For example, we find in the α-helices, the best place for the hydrogens is frequently in a three-center position between nitrogens or oxygens. This makes it uncertain whether it is an α or 3_{10} helix. I have not counted the statistics on two-center versus three-center positions, but it can easily be done from the information available.

George Jeffrey: I am glad you like the terminology three-centered rather than bifurcated. I like it because I believe it arises from proton deficiency and is analogous to Lipscomb's three-center bonds due to the electron deficiency of the boron hydrides.

Dorothy Hodgkin: Yes, I agree, and I feel I can take the word 'bifurcated' out of the paper we are now writing.

I would like to ask a question about history. When Bernal first considered the hydrogen-bond, he stretched the donor and acceptor properties sideways across the periodic table from lithium to boron (B–H) at one end to fluorine (F^-) at the other end. Do you know that paper?

George Jeffrey: Yes, I remember that. It is consistent with present thought. In biological molecules, which I have concentrated on here because of the Fox Chase venue, the strongest hydrogen bond donors will be the P–OH and the strongest acceptors $\bar{P}$=O or F^-. The weakest donors are C–H. The evidence for C–H··· O=$\bar{C}$ hydrogen bonding is good, but not for less electronegative acceptor groups.[32] I doubt if the electronegativity difference between boron and hydrogen is sufficient for B–H to be a donor, even to F^-, but I may be wrong about that.

Richard Dickerson: I'm delighted to see George Jeffrey doing some missionary work for three-center bonds, because the bonds between the drug molecule, netropsin, and the DNA, that I'll talk about this afternoon are quite definitely bifurcated, as you say, three-centered. I've been giving a seminar on this in different places, and physical chemists regularly come up and attack this concept, and say something must be wrong — hydrogen bonds are linear. I've told them that in the α-helix, and 3_{10} helix, that Dorothy just alluded to, this is not so, but I've been put on the defensive time and time again over this drug binding. I'm glad now to have an eminent authority that I can quote.

George Jeffrey: Well, I don't think there is any well-authenticated example of a truly linear hydrogen bond, except for the strong bonds involving HF or H_3O^+.

I would like to mention my objection to the common practice of using sums of van der Waals radii as cut-off criteria for hydrogen bonding. To me, this implies that a hydrogen bond becomes a van der Waals interaction at longer distances. That does not make sense to me. For the short hydrogen bonds, $H\cdots A < 1.5$ Å, the bond character has both exchange and coulombic components, but at $H\cdots A > 2.0$ Å, it must be primarily coulombic. Electrostatic attractions fall off as r^{-1}, whereas van der Waals attractions attenuate as r^{-6}. Therefore if there is structural evidence indicating an attractive force at distances greater than 2 Å, they will be primarily electrostatic, *i.e.*, a hydrogen bond.[3]

Olga Kennard: I'd just like to make one comment on what Jeff said. The work he has described is collaborative work between Robin Taylor and myself. Robin had two scientific fathers. He did his Ph.D. under George Sheldrick, and then went to Jeff in Pittsburgh. I think that all the 'nurturing' that we have done with the Data Base is well worth it if it gives me a chance to work with creative people like Robin and get these types of results. Robin is now working on the statistical tools that you will need if you want to use the Data Base, because the amount of information there, now, is so large that the statistical techniques that we had for normal structural work, we think is not quite adequate. But I just wanted to stress that this should be Robin and Olga or Olga and Robin.

George Jeffrey: Since you are here, Olga, I thought it most polite to refer to you. Had Robin Taylor been here, I might have done it in reverse.

David Sayre: As a non-worker in this field I want to say that I thought this was a wonderful talk, and I was especially fascinated by the existence of a sort of meta-theory of the way in which a shortage or excess of protons expresses itself in recognizable ways in the crystalline state. My question then is whether this sort of accounting can be extended at all into the solution state. From a biologist's point of view the crystal state is certainly an excellent guide to principles, but one would like also to know to what extent vestigial effects of the type you have discussed persist in the solution state.

George Jeffrey: I think that crystallography provides an insight into the most probable situations in the solution state, if one looks at a large number of structures and makes a statistical analysis.

Structure in solution has different meaning from structure in the crystal. One is static and reasonably constant during the long time of the diffraction experiment. The other is dynamic and changing rapidly between a vast number of the more probable structures. Stillinger's paper in *Science* on 'Water Revisited' gives a good picture of how one should view structural concepts in the liquid state. The concept of cooperativity which plays an important role in the hydrogen bonding in the carbohydrate crystal structures is implicit in the flickering cluster concepts of aqueous solutions of Henry Frank.

Betty Patterson: It seems to me that what you are talking about, at least from the point of view of the biochemist, is that in solution you can have stabilization of enzymes by glycols and sugars through hydrogen bonding. This sort of thing you are talking about will finally, for the biochemist, explain what's going on when all the data are in.

George Jeffrey: I would like to ask Dave Sayre when we are going to get the three-dimensional parallel processor computers that are analogous to the three-dimensional hydrogen bond systems in the brain and nerve systems.

David Sayre: I wish I could give you a good answer to that. Three-dimensionality certainly plays a considerable role in integrated chip design today, but I do not know of any overall computer architectures which use three-dimensionality in an essential way. Still, the feeling that the computer of the future must be a structure in which logic and memory are distributed in an interpenetrating fashion in three-dimensional space is very strong in the computer industry. Coexisting with this is the thought that some day such structures will also be miniaturized to the molecular or macromolecular (*i.e.*, molecular electronics) level.

George Jeffrey: It's interesting to me that the artificial intelligence people tell me that they will never build a computer as clever as a man until they have a three-dimensional parallel processor, which is exactly what the hydrogen bonding systems are.

David Sayre: There are proposals to use hydrogen bonded systems as memories.

George Jeffrey: I think Linus Pauling would like that remark, recalling his clathrate hydrate theory of anæsthesia.

Verner Schomaker: I think that if your external examiner, Bernal, were here he might not approve of your getting rid of bifurcated hydrogen bonds in a meeting devoted to history. He was visiting Pasadena one time, I think it was late 1938 or early 1939, when he was shown the crystal structure of glycine. Corey and Albrecht were there, and maybe Pauling, and I was standing in the back row. I think he called it a bifurcated hydrogen bond.

George Jeffrey: Actually it was McClelland and Pimentel who messed it all up, by using the term bifurcated bond to describe

```
   H - - 
O<       >O ,
   H - - 
```

which is quite different from Corey and Albrecht's bond, which was

```
        - -O
N—H <          .
        - -O
```

Verner Schomaker: That's terrible.

George Jeffrey: You are right. It is very important to distinguish between these two, because they are quite different, and it is very confusing when the same descriptor is attached to both.

Verner Schomaker: I liked the old one too. Why not write 'bifurcated-three-centered'?

George Jeffrey: Well, and then the editor will tell you that the description is too long and to take one word out.

Verner Schomaker: I have a real question. How do you get energies for these things from the distribution of distance? You seem to say that that's what you can do.

George Jeffrey: I would like to do it, but I don't know how. You can do that for me, Verner.

Verner Schomaker: I don't know how to do that either. You haven't actually done it, then?

George Jeffrey: No, Olga drew that curve free-hand and printed out the analogy. And I don't know how to do that quantitatively. I asked Murray-Rust if he knew how to do it. I thought he told me at the ACA meeting in Kentucky that he thought he knew how to do it, but maybe he didn't.

Peter Murray-Rust: I think it can be done. The problem is getting the vertical scale.

David Harker: You said that there are no A–H···B linear bonds, but all your dodecahedra in the clathrate hydrates are full of them, *i.e.,* the edges are straight lines.

George Jeffrey: The edges of the polyhedra in the clathrate hydrates are the O··· O separations. We never located the hydrogen atoms which are fully or partially disordered over two sites between the oxygens.

David Harker: Isn't the H–O–H angle in water close to 105°?

George Jeffrey: Yes, in water vapor, it is 104.5°.

David Harker: I know. In ice, of course, the water molecule is surrounded by four others in a tetrahedral coordination and that's approximately 110° (109.5°), and that makes trouble. But as soon as you put something in that can have only a pure van der Waals attraction, you immediately get one of those clathrate hydrates. Then, the angles between hydrogen bonds are close to 105°.

George Jeffrey: Again those are angles between the O··· O separations in the pentagons. All the work on the clathrate hydrates was done with Weissenberg cameras, in cold boxes, at −30°, with eye-estimated intensities. We haven't the slightest idea where the hydrogen atoms are, but guess that they may be closer to the O··· O line of centers when the angles between the O··· O edges are 105° than when they are 110 or 120°.

David Harker: So the O–H··· O bonds are just trying to be linear in the ices and the clathrate hydrates.

George Jeffrey: It could be, but I don't know. You have a hydrophobic guest in the cage. But I don't know what it does to the proton.

David Harker: The guest is not exactly hydrophobic. It's van der Waals attractive. It fills the hollow, so that the hydrogen bonds can do what they like. That was my point.

George Jeffrey: The latest work on the structure of ordinary D_2O hexagonal ice by neutron diffraction at 60 K, 123 K and 223 K by Kuhs and Lehmann using neutron diffraction at Grenoble indicates that this is a very complex problem. Not only are the deuteriums disordered, but the oxygens are also disordered by displacements along the bisectrix of any possible instantaneous D_2O configuration. This shortens the O–D distances and closes up the D–O–D angle from the averaged values obtained by oxygens in ordered positions. This work was reported in an abstract of the Symposium on Neutron Scattering in Berlin that preceded the Hamburg Crystallography Congress. Many of us are eagerly awaiting a full report.

It seems that nothing, not even the ice-cube in your martini, is simple these days.

References

1. Bernal, J. D. and Fowler, R. H. Theory of Water and Ionic Solution with Particular Reference to Hydrogen and Hydroxyl Ions, *J. Chem. Phys.* **1**, 515-548 (1933). In fact, the word hydrogen-bond did not appear in this paper.
2. Pauling, L. The Hydrogen Bond. In *The Nature of the Chemical Bond.* Ch. 12. Cornell University Press: Ithaca, New York (1938).
3. Uranov, E. B., Chapman, D. R. and Isaacs, I. The Hydrogen Bond. *Penguin Dictionary of Science.* Penguin Books: London, UK (1971).
4. Allen, F. H., Kennard, O. and Taylor, R. Systematic Analysis of Structural Data as a Research Technique in Organic Chemistry. *Accts. Chem. Res.* **16**, 147-153 (1983).
5. Bernstein, F. C., Koetzle, T. F., Williams, G. J. B., Meyer, E. F., Brice, M. D., Rodgers, J. D., Kennard, O., Shimanouchi, T. and Tasumi, T. The Protein Data Bank: A Computer-based Archival File for Macromolecular Structures. *J. Mol. Biol.* **112**, 535-542 (1977).

6. Petersen, B. The Geometry of Hydrogen Bonds from Dimer Water Molecules, *Acta Cryst.* **B30**, 289-291 (1973).
7. Kroon, J., Kanters, J. A., van Duijneveldt-van de Rijdt, J. G. C. M., van Duijneveldt, F. B. and Vliegenthart, J. A. O–H···O Hydrogen Bonds in Molecular Crystals. A Statistical and Quantum Chemical Study. *J. Mol. Struct.* **24**, 109-129 (1975).
8. Taylor, R., Kennard, O. and Versichel, W. The Geometry of the N–H···O=C Bond. 3. Hydrogen Bond Distances and Angles. *Acta Cryst.* **B40**, 280-288 (1984).
9. Ceccarelli, C., Jeffrey, G. A. and Taylor, R. A Survey of O–H···O Hydrogen Bond Geometries Determined by Neutron Diffraction. *J. Mol. Struct.* **70**, 255-271 (1981).
10. Taylor, R., Kennard, O. and Versichel, W. Geometry of the N–H···O=C Hydrogen Bond. 2. Three-Center (Bifurcated) and Four-Center (Trifurcated) Bonds. *J. Am. Chem. Soc.* **106**, 244-248 (1984).
11. Jeffrey, G. A. and Mitra, J. Three-Center (Bifurcated) Hydrogen Bonding in the Crystal Structures of Amino Acids. *J. Am. Chem. Soc.* **106**, 5546-5553 (1984).
12. Jeffrey, G. A. and Maluszynska, H. A Survey of Hydrogen Bond Geometries in the Crystal Structures of Amino Acids. *Int. J. Biol. Macromol.* **4**, 173-185 (1982).
13. Pimentel, C. G. and McClelland, A. L. *The Hydrogen Bond.* Freeman: San Francisco (1960).
14. Poppleton, B. J., Jeffrey, G. A. and Williams, G. J. B. A Neutron Diffraction Refinement of the Crystal Structure of Methyl-α-D-altropyranoside. *Acta Cryst.* **B31**, 2400-2404 (1975).
15. Parthasarathy, P. Crystal Structure of Glycylglycine Hydrochloride. *Acta Cryst.* **B25**, 509-518 (1969).
16. Newton, M. D., Jeffrey, G. A. and Takagi, S. Application of *ab initio* Molecular Orbital Calculations to the Structural Moieties of Carbohydrates. 5. The Geometry of the Hydrogen Bond. *J. Am. Chem. Soc.* **101**, 1997-2002 (1979).
17. Dunitz, J. D and Waser, J. Geometric Constraints in Six- or Eight-membered Rings. *J. Am. Chem. Soc.* **94**, 5645-5660 (1972).
18. Coppens, P. and Sabine, T. Neutron Diffraction Study of Hydrogen Bonding and Thermal Motion in Deuterated α and β Oxalic Acid Dihydrate. *Acta Cryst.* **B25**, 2442-2451 (1969).
19. Jeffrey, G. A., Maluszynska, H. and Mitra, J. Hydrogen-Bonding in Nucleosides and Nucleotides, *Int. J. Biol. Macromol.* (1986), in press.

20. Jeffrey, G. A. and Lewis, L.S. Cooperative Aspects of Hydrogen-Bonding in Carbohydrates. *Carbohydr. Res.* **60**, 179-182 (1978).
21. Taylor, R. and Kennard, O. Comparison of X-ray and Neutron Diffraction Results for the N–H···O=C Hydrogen Bond. *Acta Cryst.* **B39**, 133-138 (1983).
22. Weiner, S. J., Kollman, P. A., Case, D. A., Singh, V. C., Ghio, C., Alagona, G., Propata, S. Jr. and Weiner, P. A New Force Field for Molecular Mechanical Simulation of Nucleic Acids and Proteins. *J. Am. Chem. Soc.* **106**, 765-784 (1984).
23. Takagi, S. and Jeffrey, G. A. Methyl-α-D-Xylopyranoside. *Acta Cryst.* **B34**, 3104-3107 (1978).
24. Del Bene, J. and Pople, J. A. Theory of Molecular Interactions. III. A Comparison of Studies of H_2O Polymers Using Different Molecular Orbital Basis Sets. *J. Chem. Phys.* **58**, 3605-3608 (1973).
25. Jeffrey, G. A. and Mitra, J. The Hydrogen Bonding Patterns in Pyranose and Pyranoside Crystal Structures. *Acta Cryst.* **B39**, 469-480 (1983).
26. Murray-Rust, P. and Glusker, J. P. Directional Hydrogen Bonding to sp^2 and sp^3-Hybridized Oxygen Atoms and Its Relevance to Ligand-Macromolecular Interactions. *J. Am. Chem. Soc.* **106**, 1018-1025 (1984).
27. Lipscomb, W. N. *The Boron Hydrides.* W. A. Benjamin: New York (1963).
28. Gress, M. E. and Jeffrey, G. A. A Neutron Diffraction Refinement of the Crystal Structure of β-Maltose Monohydrate. *Acta Cryst.* **B33**, 2490-2495 (1977).
29. Chiari, G. and Ferraris, G. The Water Molecule in Crystalline Hydrates Studied by Neutron Diffraction. *Acta Cryst.* **B38**, 2331-2341 (1982).
30. Jeffrey, G. A. and Wood, R. A. The Crystal Structure of Galactaric Acid (Mucic Acid) at −147°. An Unusually Dense Hydrogen-Bonded Structure. *Carbohydr. Res.* **108**, 205-211 (1982).
31. Baker, E. N. and Hubbard, R. E. Hydrogen Bonding in Globular Proteins. *Prog. Biophys. and Mol. Biol.* **44**, 97-179 (1984).
32. Taylor, R. and Kennard, O. Crystallographic Evidence for the Existence of C–H···O, C–H···N, and C–H···Cl Hydrogen Bonds. *J. Am. Chem. Soc.* **104**, 5063–5070 (1982).
33. Stillinger, F. H. Water Revisited. *Science* **209**, 451–457 (1980).

11. Electron density maps

Jack Dunitz

Jack Dunitz was born in Glasgow, where he also obtained his B.Sc. degree and Ph.D. degree in 1947. He was the first of a long and respected line of Ph.D. students from the laboratory of J. Monteath Robertson. He did postdoctoral work at Oxford, at Caltech, at NIH and at the Royal Institution. In 1957 he went to the ETH, the Swiss Institute of Technology in Zürich where he is Professor of Chemical Crystallography. He is a Fellow of the Royal Society and he received the Centenary Medal of the Chemical Society.

Dr Dunitz has a wide knowledge of crystallography. His scientific interests include the use of crystallogtaphy to study reaction mechanisms or reaction paths, chemical bonding, electron density distributions, physical properties from crystallographic results and also from temperature factors; even recently the possibility of identifying isotopes from X-ray diffraction results. Dr Dunitz is the invited speaker at many important occasions. In 1976 he gave the Baker Lectures at Cornell, and this formed the basis of his invaluable textbook on *X-ray Analysis and the Structure of Organic Compounds.* Every chapter in this book starts with a quotation from *Alice in Wonderland* and Lindo also was very fond of making these quotations. The chapter on electron density distributions in the book by Dr Dunitz starts with the following quotation. 'Let's hear it,' said Humpty Dumpty, 'I can explain all the poems that ever were invented, and a good many that haven't been invented just yet.' Possibly Dr Dunitz will concentrate on the latter part, but otherwise he may prove, to quote again, that 'everything has a moral if only you can find it.'

Thank you for having invited me to this occasion, which so far I have enjoyed enormously; I have not only enjoyed it, but I have also benefitted very much from hearing all these lectures. I first met Lindo on my first visit to the United States, and I remember that I was completely overwhelmed by his friendliness and interest in a young crystallographer whom he'd never heard of until a few days before. And I remember the day that I spent with him and with Betty too, in the evening, and we talked not only about crystallography, but about everything under the sun. When he learned that I came from Glasgow, he showed his intimate knowledge of the Glasgow accent by telling me a story about a distant relation, a small boy who went to school for the first time in that city. The teacher asked him his name; he said ' Pa-erson', the teacher said 'Wha-?', he repeated 'Pa-erson', she said 'There's no such name as Pa-erson', he said 'Sure there is, wi- two T's.'

Yesterday Peter Murray-Rust showed a slide demonstrating the increase in numbers of crystal structures done since 1970 or so, and he marked those which he classified as 'accurate structures.' This number went asymptotically to zero around 1970 and seemed to be approaching infinity around 1990 or so. This recent increase in accuracy is an extremely important aspect of present-day crystallography. The results attainable today are incomparable to what went before. If you want to know anything about the structure of matter, then it's not too great an exaggeration to say that what happened before 1970, although historically very interesting and important, is practically negligible as far as its share of today's information bank is concerned. Today, I should like to take this accuracy aspect a stage further and discuss structures at a level that I might as well call 'superaccuracy.' As for these superaccurate structures, probably even in 1980 there were practically none of them, and there are not too many today, but they are now beginning to increase in number. Partly because of the improved stability of counting systems but mainly because we are learning much more about how to make measurements, data sets are now being produced which are accurate enough to tell us something about the bonding electron density in molecules.

Electron density maps have been in the wind since the early days of crystallography. They were already proposed by Bragg in 1915 and were actually being made in the '20's. In those days they were used mainly for what David Sayre would call 'imaging structures,' for obtaining better images of the structure in terms of the density of scattering power. It has always been realized that these maps, in principle, might also tell us something about the elusive and fascinating problem of the nature of the chemical bond — the title of Pauling's famous book; if only we could make those maps accurate enough, they would illuminate for us the problem of how atoms are held together in molecules and crystals. However, it's my opinion that, until quite recently, most of what had been derived from maps of this kind was wrong, or misleading, or, to put it mildly, not particularly trustworthy. Of course, for some time now, it has also been possible to obtain electron densities in molecules — at least in small ones — by quantum mechanical calculations, but it has been just as difficult for the uninitiated to exercise much judgment about how reliable these calculated densities actually are. It would seem, however, that the two approaches are nicely complimentary since the errors and approximations inherent in the one are virtually independent of those in the other. Thus one can be used to check the other. Nevertheless, comparison should not become an end in itself. It seems unlikely that we are ever going to learn much if all everybody does is to check the activities of the others.

The X-ray experiment gives, in principle, the electron density in the crystal. In practice, the density obtained from X-ray diffraction is tainted by various kinds of systematic error and is anyway difficult to interpret in terms of chemical bonding, because the density close to the nucleus is so enormously larger than the density between the atoms. For many years, it has been recognized that the only hope of doing anything useful with experimental electron densities is to look at so-called 'difference maps.' These can be modeled in various ways, but I will concentrate today on plain, unmodeled maps. The general idea is pretty well known. We define a density $\Delta\rho(\vec{r})$ as the difference between an experimental electron density $\rho_o(\vec{r})$ and an artificial electron density $\rho_c(\vec{r})$ corresponding to the superposition of what we might call '*International Tables* atoms.' These atoms have been accurately described over the years by quantum mechanical calculations, they correspond to isolated atoms, and, to a good approximation, one can regard a crystal or a molecule as a direct superposition of such atoms: the pro-crystal or pro-molecule. We know it's not what the real electron density looks like, but it provides a basis on which to assess the importance of the difference density $\Delta\rho(\vec{r})$, which is then given by the Fourier synthesis

$$\Delta\rho(\vec{r}) = \sum(|F^o{}_H| - |F^c{}_H|)\cos 2\pi(\vec{H}\cdot\vec{r} - \alpha^c{}_H)$$

where the coefficients are the differences between the experimental structure factors and the calculated ones, based on the pro-crystal model, which must include the effect of vibrational motion as well as the atomic positions. The atomic positional and vibrational parameters also have to be determined from an experiment, and there are two approaches to this. One is to determine them from a separate neutron-diffraction experiment, which gives the positions and motions of the atomic nuclei. The other is to determine them from high-order X-ray diffraction data, under the premise that if you go far enough in reciprocal space, then the scattering from the valence electrons has fallen off essentially to zero. What is then left is the scattering from the core, so one hopes to converge on those positions that would be obtained by a neutron-diffraction experiment. Both approaches have their own advantages and disadvantages. Some years ago it was thought that all respectable results had to be based on neutron positions and temperature factors, but I think that's not so unanimous today.

The difference density $\Delta\rho$ also contains the net result of experimental error in the observations and of mistaken assumptions in the model. These may include errors in the atomic form factors, the vibrational (or displacement) parameters, and the atomic positions. If these errors are not too severe, then we may also hope to see something of the redistribution of electron density associated with the formation of the crystal or molecule from separate atoms. Apart from spurious noise, then, the difference map should consist of electron-density features which one would like to interpret in terms of conventional chemical concepts, such as bonding density and lone-pair density. And most of my talk is going to be about the question: is it possible to use this method for obtaining new insights into questions of chemical bonding?

Eight years ago, when I gave the Baker Lectures at Cornell,[1] I wanted an example of an electron-density difference map to illustrate the state of the art in this subject at that time. I chose two maps, made by the Coppens group for 4-nitropyridine-N-oxide[2] and shown in Fig. 11.1. On the left is the X-N map based on the neutron-diffraction model, on the right the corresponding X-X map based on high-order reflec-

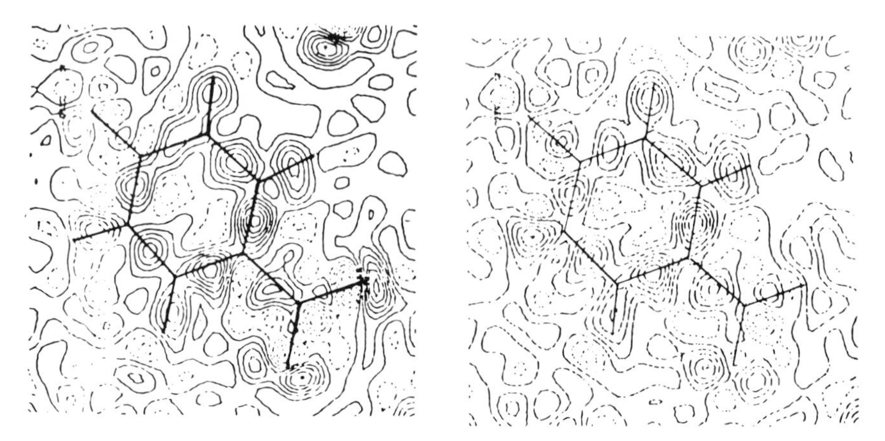

FIG. 11.1. X-N (left) and X-X (right) difference maps for 4-nitropyridine-N-oxide at 30K.

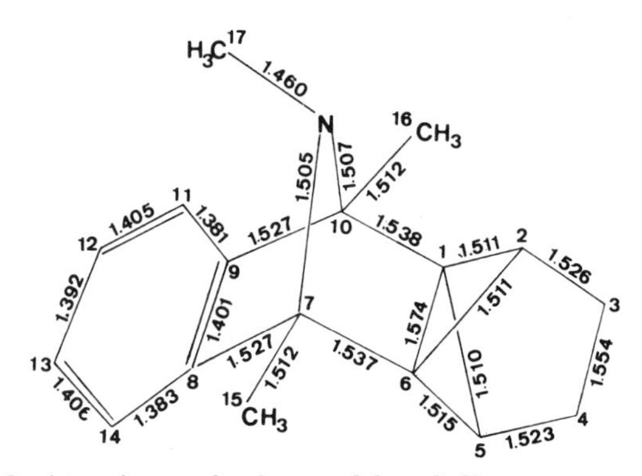

FIG. 11.2. Atomic numbering and bond distances.

tions. You see that the two maps are pretty similar. Both show density between bonded atoms, although it has to be admitted that for the N–O bonds (in the nitro group, and also at the N-oxide end) this 'bonding density' is remarkably weak. One can see features which may be interpreted as lone-pair density of the oxygen atoms. By and large, in spite of considerable noise, the main features make reasonably good chemical sense.

Four or five years ago we began in Zürich to try to learn how to make such maps, only we wanted to make them better. After a while we wanted a test case. The molecule we chose (Fig. 11.2), after some exploratory work, was one of a

series of molecules synthesized for us by Professor Szeimies in München.[3] The interesting feature is the bond between the two inverted carbon atoms 1 and 6. For a normal tetrahedral carbon atom, each bonded neighbor lies on the opposite side of the plane through the other three. An 'inverted' carbon atom is then one where all four bonds lie on the same side of a plane. This is the case for the bond in question, which is common to two three-membered rings and a five-membered ring. Incidentally, the truly astonishing-looking molecule, [1.1.1]-propellane, **I**, was

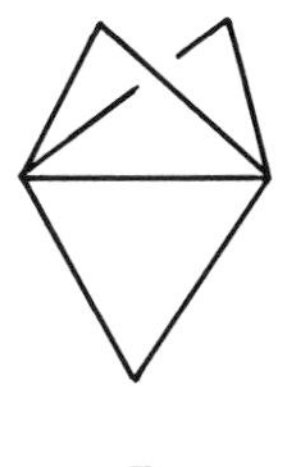

I

in fact, synthesized a couple of years ago by Professor Wiberg at Yale and proves to everyone's astonishment, except Wiberg's, to be quite stable.[4] Everyone thought it could not be obtained or that it would have at most a purely transient existence, except Wiberg, who spent years convinced that if he could make it, it would stay in a bottle. And once he made it, it turned out that he was right.

Everyone knows that there must be bonding density between the atoms in a normal carbon-carbon bond, but what about such a bond between inverted atoms? Do you see density there, or do you not? One disadvantage of the molecule in Fig. 11.2 is that it's rather large for an accurate electron- density study, but, on the other hand, it's got some useful internal checks in it. For example the benzene ring has been looked at in so many other electron-density studies that we know roughly what we ought to see there. Thus the quality of our difference map can be appraised from the result obtained for the benzene ring (Fig. 11.3). The density peaks in the C–H bonds can be made to appear or disappear to some extent, depending on what assumptions are made about the hydrogen atoms: if they are moved inwards, towards their bonded carbon atoms,

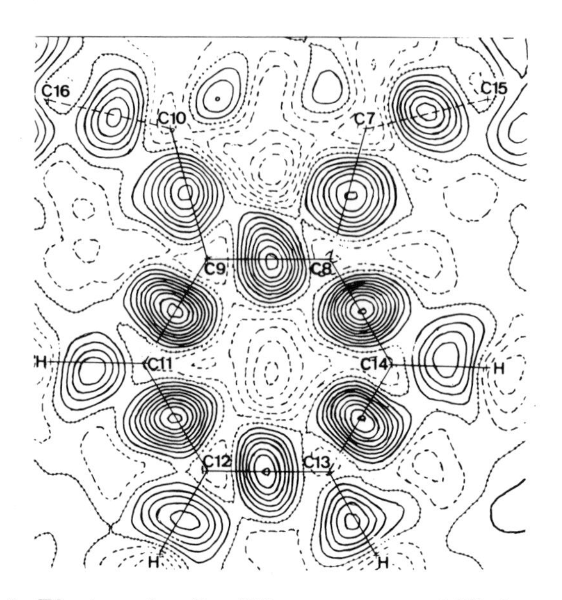

FIG. 11.3. Electron density difference map at 95K through the plane of the benzene ring of the molecule shown in Fig. 11.2. Contour lines at intervals of 0.05eÅ^{-3}, full for positive, dashed for negative, and dotted for zero density.

the difference density peaks become weaker. Our map is based on hydrogen positions obtained by a mixture of experiment and modelling: we keep the C–H bond in the direction in which it was given by the diffraction experiment, but we correct for a systematic error by moving each hydrogen outwards to make the C–H distances equal to 1.08 Å. Otherwise, the atomic positions were determined from high-order X-ray diffraction. Fig. 11.3 shows bonding density in all the C–C and C–H bonds (the C7-C15 and C10-C16 bonds, to the methyl groups, do not lie quite in the plane of the section).

Evidently, the map shown in Fig. 11.3 contains appreciably less noise than those of Fig. 11.1, so we can claim that between 1976 and 1980 we did learn to improve the method somewhat. Our molecule contains a mirror plane, and Fig. 11.4 shows the difference density in that plane. It passes through the midpoints of the bonds C12-C13, C8- C9, C1-C6, and contains the bonds C2-C3, C3-C4, C4-C5, as well as N-C17. With the exception of C1-C6, prominent density peaks are seen for all these

bonds, and also for the lone pair of the nitrogen atom, which is expected to lie in this plane. Although the midpoint of C1-C6 lacks a density peak, Fig. 11.4 does show little accumulations on either side of the midpoint. From Fig. 11.5 it is clear that these accumulations in the mirror plane do not correspond to peaks but are due merely to the overlapping of the bonding densities for C1-C2 and C6-C2 in one three-membered ring and for C1-C5 and C6-C5 in the other. The individual bonding densities appear to lie outside the internuclear lines and so correspond very nicely to 'bent bonds,' similar to the bent bonds in cyclopropane rings seen already by Fred Hirshfeld[5] in the mid-60's in one of the few bonding density studies of that period from which one learned anything.

Figs. 11.4 and 11.5 confirm that there is no bonding density at the midpoint of the C1-C6 bond, the one between the inverted atoms. Because of the many internal checks in this molecule, we can claim that we have indeed an answer to the question: is the bond between the inverted carbon atoms special in any way? The answer is yes. It's special because, unlike

FIG. 11.4. As Fig. 11.3, but in the mirror-plane of the molecule.

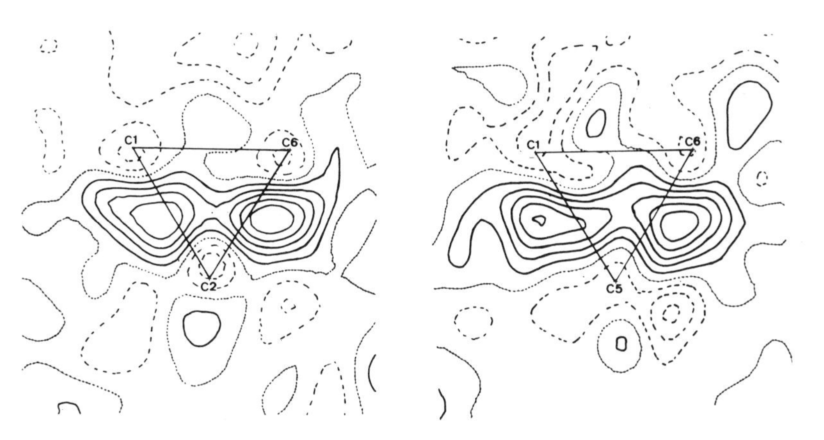

FIG. 11.5. As Fig. 11.3, but in the planes of the three-membered rings.

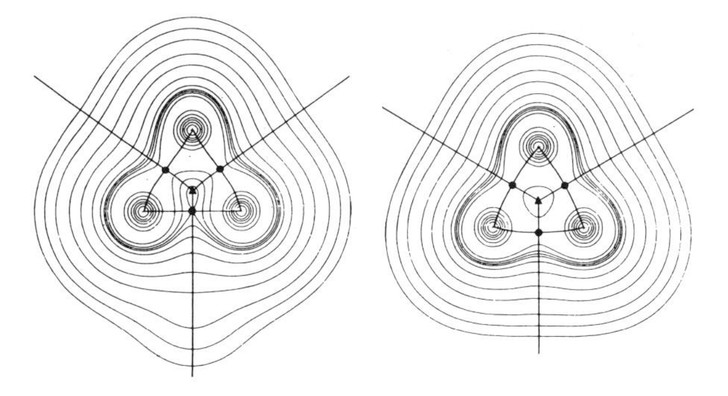

FIG. 11.6. Calculated electron density maps for the three-membered rings of [1.1.1]propellane (left) and cyclopropane (right). The shape of the innermost contour bracketing all three carbons is to be contrasted between the two maps.

all the other bonds in the molecule, it shows no density in the difference maps.

Of course, this result does not mean that there is no electron density between these two atoms. It means merely that the density between them is not distinguishable from the sum of the densities of the isolated atoms. Fig. 11.6 shows calculated electron-density maps for one of the three-membered rings in [1.1.1]propellane and for cyclopropane.[6] In the former, the horizontal bond is the special one, the one between the inverted carbon atoms, and along this direction the electron

density is indeed close to that of the spherical atoms, whereas along the other bonds there is appreciably more density between the atoms.

The next example concerns work that was done a couple of years later, and represents what I think is still further improvement in the experimental techniques. I'm sorry I can't go into these matters in any detail because it would take so long that there would be no time to talk about the results. However, I would like to mention here the name of Paul Seiler in my laboratory, who has devoted a great deal of time to thinking closely about the experimental problems. The quality of our maps is due mainly to his ingenuity and persistence in finding ways to improve the quality of the experimental data. The molecule we chose is tetrafluoroterephthalonitrile (TFT), shown in Fig. 11.7; the room-temperature structure had already been determined in Doyle Britton's laboratory.[7] As there are no hydrogens in this molecule, there is not going to be any very great advantage in doing neutron-diffraction experiments to back up the X-ray ones. One would have to put up with two sets of experimental error instead of one, and this might tend to make the situation worse rather than better. When you do X-ray analysis at about 90 K, the high-order least-squares refinement can lead to

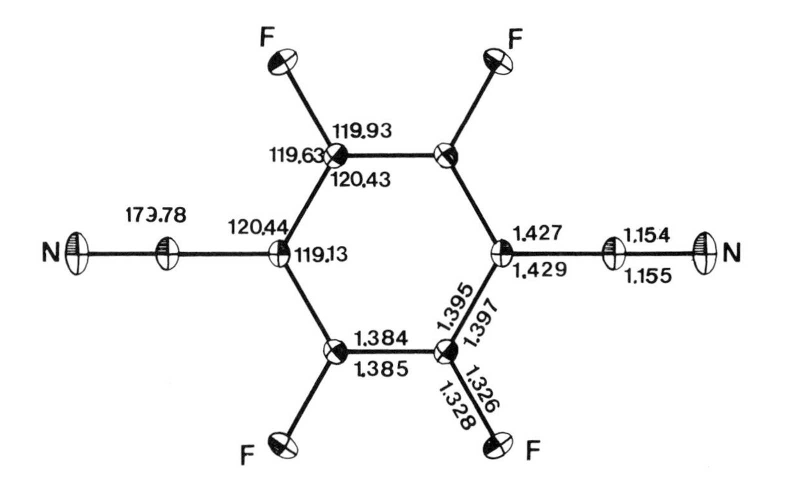

FIG. 11.7. Molecular geometry and vibration ellipsoids for tetrafluoroterephthalonitrile (TFT) at 98K.

bond lengths with an absurdly high number of significant figures, as indicated in Fig. 11.7. Although the nominal standard deviations are extremely small, less than 0.001 Å, the physical significance of these figures is not to be taken too seriously. After all, a bond length determined from a diffraction experiment is an average over a vibrating system. The questions involved in how you define that average lead to an uncertainty that is considerably larger than the official uncertainty as expressed in the standard deviation. You see also that the vibration ellipsoids seem to make sense, as if the whole molecule were wobbling about its long axis.

But our main interest here is in the electron density. For this experiment,[8] we collected the X-ray data out to a $\sin\theta/\lambda$ value about 1.15 $Å^{-1}$ with MoKα radiation. The limiting radius of CuKα radiation is 0.65 $Å^{-1}$, so we can measure nearly six times as many reflections as with that. The atomic positions and displacement parameters were determined from a high-order refinement, including only the reflections higher than

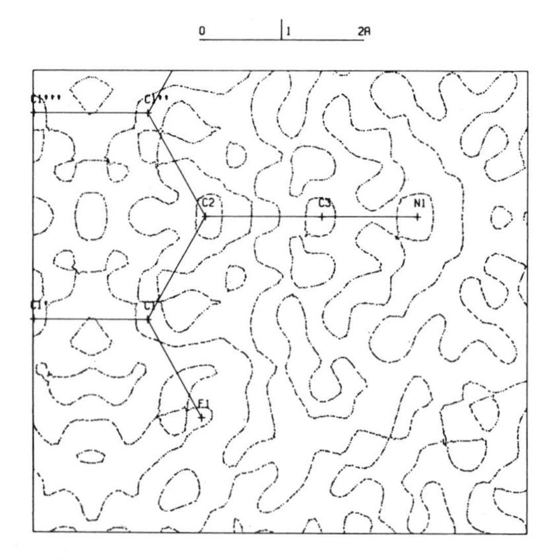

FIG. 11.8. Electron density difference map for TFT calculated with 527 high-order reflections with $F > 20\sigma(F)$ in the range $0.85\,Å^{-1} < \sin\theta/\lambda < 1.15\,Å^{-1}$. Contours at intervals of $0.075\,e\,Å^{-3}$ but the map is so flat that only the zero contour appears.

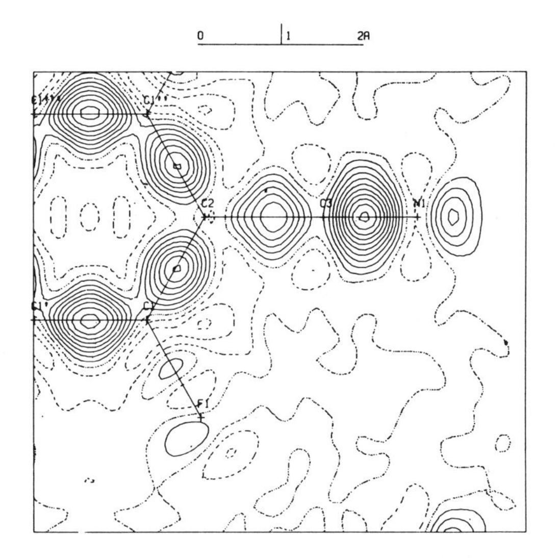

FIG. 11.9. Electron density difference map for TFT calculated with 1269 reflections with $F > 20\sigma(F)$. Contours at intervals of $0.075\,e\,Å^{-3}$, full for positive, dashed for negative, dotted for zero density.

$0.85\ Å^{-1}$. There were 527 such reflections with $I > 5\sigma(I)$, the R factor was 1.3%, and the difference map calculated with these reflections is practically flat (Fig. 11.8). This tells us that the high-order data are compatible with the free-atom formalism. We've also done similar calculations including the still weaker high-order reflections; as we add more and more weak reflections, the map gets noisier, but it still doesn't show anything significant. Compare Fig. 11.8 with the corresponding difference map based on the low-order reflections, those with $\sin\theta/\lambda$ less than $0.85\ Å^{-1}$ (Fig. 11.9). Here you see bonding density in the benzene ring, for the exocyclic C–C bond, for the C≡N triple bond, and for the nitrogen lone-pair density. Notice, however, the remarkably weak bonding density in the C–F bond. We may think of the bond between inverted atoms as a thermochemically weak bond, but the C–F bond is stronger than a carbon- carbon bond. Fig. 11.10 shows the densities in planes perpendicular to the molecular plane and passing through the middle of the bonds. The bonds in the benzene ring are elliptical with the major axis perpendicular to the plane of the benzene ring, corresponding to what chemists

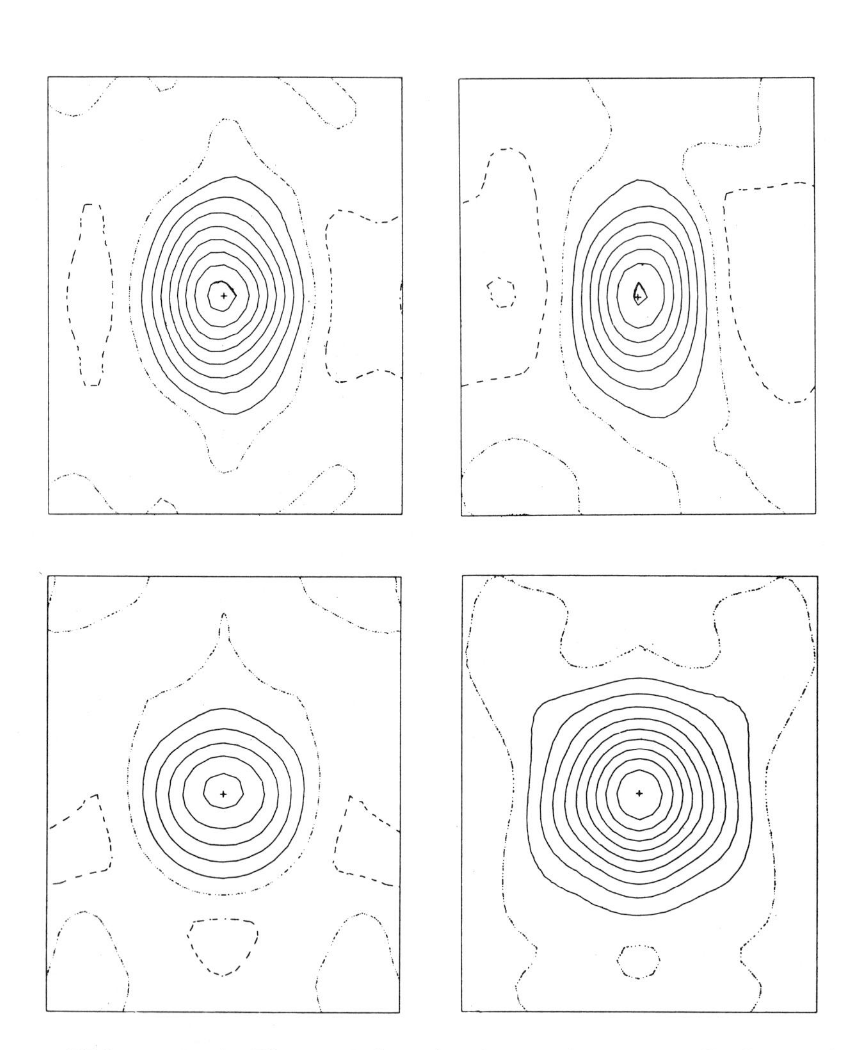

FIG. 11.10. As Fig. 11.9, but showing sections perpendicular to the molecular plane and passing through the midpoints of the bonds (upper left, C1–C1′; upper right, C1–C2; lower left, C2–C3; lower right, C3–N). The direction of the molecular plane is horizontal in each case.

would call the π-electron density, whereas the densities in the exocyclic bond and in the triple bond are much more nearly circular. Thus in this molecule, while most of electron-density features have a more or less obvious chemical interpretation, we have one feature, the low density in the C–F bond, which could come as a surprise.

At this point, we might well ask: what's so special about the C–F bond? Without going into any theory, there are two obvious features which might make it different from the other bonds. For one, it is by far the most polar bond in the molecule. Can the low density in the C–F bond be ascribed to its highly polar nature? Another possible argument could be that the C–F bond has a low difference density because, when you subtract out a fluorine atom with its nine electrons, you are subtracting out something that is more electron rich than a carbon atom; the more you take out, the less you've got left. That seems a plausible argument, except that in the conventional picture of the chemical bond you're supposed to have an excess density between the atoms, regardless of how electron-rich they are. Anyway, here are two possible factors, and a decision between them can be determined experimentally by looking at the difference density in a molecule which contains polar bonds and also non-polar bonds between electron-rich atoms, and asking which have got the larger density.

We decided that molecule **II** would be suitable for this purpose. It contains C–N bonds and C–O bonds, which are of increasing polarity,

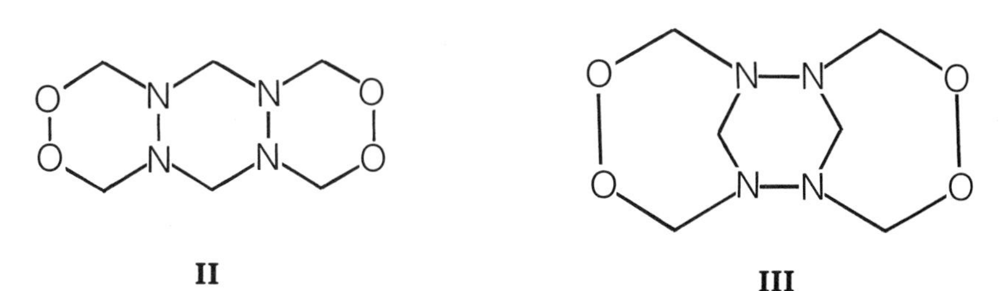

II **III**

and it also has N–N bonds and O–O bonds, which are of increasing electron richness. When we made the compound, as described in the literature, it turned out,[9] however, to have a different structure from what everyone had assumed. Fortunately, the correct structure **III** contains the same kinds of bonds, merely connected in a different way, so it is just as suitable for the purpose we had in mind.

In the crystal, this molecule has a center of symmetry. The two halves of each seven-membered ring, although constitutionally equivalent, are conformationally distinct (Fig. 11.11). For example, the lone pair on the nitrogen atom N1 is pointing roughly parallel to the C1–O1 bond, whereas on the other side of the ring the lone pair on N2 is pointing nearly perpendicular to the C2–O2 bond. These so-called stereo-electronic differences are accompanied by bond-length differences as large as 0.025 Å between constitutionally equivalent bonds, differences that are many times greater than the experimental inaccuracy.[9] In spite of these differences, we expect to see somewhat similar bonding density features on the two sides of the ring.

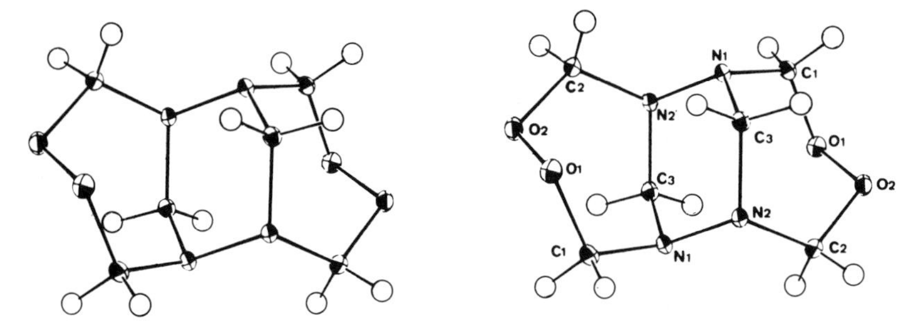

FIG. 11.11. Stereoview of III.

Fig. 11.12 shows several sections through the electron-density difference map.[10] At the top we see the planes passing through the two separate N atoms, and containing the two N–C bonds on either side. They are similar, but not identical. Because of the conformational difference, the C1-H1 bond lies nearly in the C1-N1-C3 plane, while C2-H2 is tipped considerably out of the C2-N2-C3 plane on the other side. All four N–C bonds show difference-density peaks of comparable height. The next pair of maps in Fig. 11.12 show the planes in which we expect to see the lone-pair density of the N atoms, *i.e.*, the planes containing the N–N bond and the mid-points of the other two atoms bonded to each N atom. The difference density peak in the N–N bond is clearly weaker than in the C–N bond — three contours compared to five. Similarly, in the next pair of planes the density peaks in the C–O bonds are again weaker than in the C–N bonds, but the really remarkable

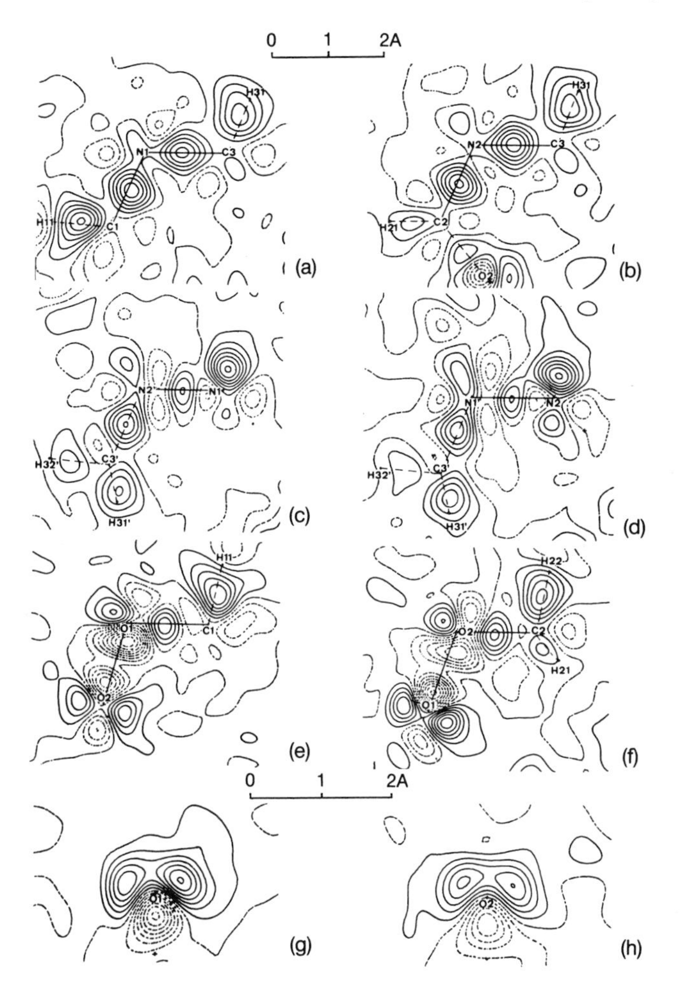

FIG. 11.12. Electron density difference maps for **III**; several sections. Bonds indicated by dashed lines do not lie in the defined planes. Contours at intervals of 0.075 e Å^{-3}, full for positive, dashed for negative, dotted for zero density.

feature is the difference density in the O–O bond, which comes to a maximum near the midpoint but does not even reach base line. There is a *deficit* of density compared with the superposition of the two oxygen atoms. On the other hand, we see positive density peaks corresponding to the lone pairs on the oxygen atoms.

Should one be surprised at these results? Perhaps not. Theoretical difference maps calculated several years ago by Bader, Hennecker and Cade at the Hartree-Fock level show that of the second-row homonuclear diatomic molecules only Li_2 fits the conventional picture of removal of charge from the antibinding regions and a buildup in the binding region.[11] The accumulation between the nuclei in Li_2 is rather diffuse, but then this is not a very good molecule; the dissociation energy is low, and the Li-Li distance is considerably more than 2 Å. For the other homonuclear diatomics, the binding region is increasingly depleted of excess charge density until for F_2 there are two deep negative troughs between the nuclei, separated by a narrow maximum, which is just over base line.

The calculated difference densities differ from our experimental ones in two important respects. First of all, the promolecule density that is subtracted out is based not on spherically averaged atoms but on valence-state atoms. The second point is that the calculated densities are for perfectly stationary nuclei, whereas the experimental densities were made about 90 K and contain the effect of the atomic vibrations.[12] If the theoretical difference density for F_2 were to be convoluted with a Gaussian function to simulate the effect of vibrational motion of the atoms, the result would be qualitatively very similar to the experimental density for the O–O bond seen in Fig. 11.12.

We may also compare some calculated binding energies for F_2 and H_2. As is well known, the classical electrostatic interaction energy for a system of two hydrogen atoms at the correct internuclear distance for H_2 is much too small to account for the binding energy of this molecule. It amounts only to about 2 kcal/mole, compared with a binding energy of more than 100 kcal/mole. Heitler and London[13] were the first to explain the covalent binding between two hydrogen atoms in term of an overlapping of the atomic wave functions, leading to accumulation of charge in the internuclear region for the bonding state. For F_2, however, the situation is very different. Here the classical interaction between two spherically averaged charge clouds is not smaller than the binding energy but considerably larger. According to calculations by Hirshfeld and Rzotkiewicz[14] it amounts to about 133 kcal/mole, compared with the binding energy of about 35 kcal/mole (Hartree-Fock F_2 is actually a metastable species with a negative binding energy). Now, if you need quantum mechanics to account for the stability of the hydrogen molecule compared with the classical model, then one could say that you also need quantum mechanics to explain why real fluorine is less stable than classical fluorine. And if the stability of H_2 is attributed to transfer of electron density from antibinding regions into the binding region (the conventional picture) then the relative instability of F_2 would have to be associated with removal of electron density from the binding region, a conclusion which may make our experimental result for the O–O bond a little more palatable.

The conventional picture of the chemical bond is really a picture of the hydrogen molecule. Most textbooks spend many pages dealing with H_2, and then they say 'Well, that's it, and now we go on to another topic.' Much of the really detailed discussion of the chemical bond has tended to concentrate on the even simpler H_2^+ system, the hydrogen-molecule ion, which has only one electron. Well, whether you've got one electron or even two, as long as they have opposite spin, you don't have to bother too much about the Pauli principle. On the other hand, when you are dealing with the F_2 molecule, with its 18 electrons, you know that you can't have too many electrons in the same region of space, it's not allowed by the Pauli principle. Now, let us think of the approach of the two fluorine atoms in a very elementary way. Each has got seven valence electrons and four valence orbitals: $\frac{7}{4}$ ths of an electron per orbital. For the pro-molecule, the superposition of the two separate charge densities, the region between the nuclei would contain twice $\frac{7}{4}$ ths of an electron. But insofar as this region is limited to only two electrons of opposite spin one can see why the actual molecule has to get rid of electron density from it.

In the last few months we have been engaged in another investigation which I want to mention briefly because it has led so far to one of these perplexing situations where we don't quite know what to make of the results. Very shortly after our results on the charge density of the C–F bond appeared in print, Collins and his co-workers published an account of the difference electron density in Li_2BeF_4, lithium fluoroberyllate.[15] Fig. 11.13 shows a section through this density. Notice that there appears to be appreciable bonding density in the Be-F bonds.

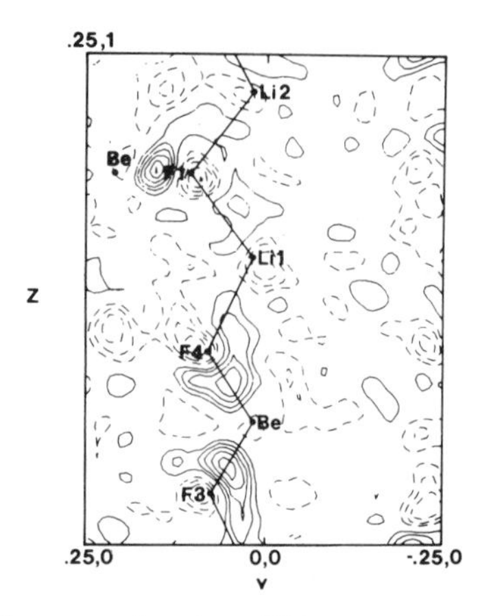

FIG. 11.13. Electron density difference map for lithium tetrafluoroberyllate at room temperature, from Collins *et al.*[15] Contours at intervals of 0.074e $Å^{-3}$, full for positive, dashed for negative, zero not shown.

But if there is no density in the C–F bond of TFT, what on earth is density doing in a Be–F bond? I had the opportunity to discuss this anomaly with Douglas Collins at a Gordon Conference last year, and we agreed that a reinvestigation of the structure was called for. The main difference between his experimental measurements and ours was that his had been made at room temperature. This seemed perfectly in order because the melting point of the crystal is around 1000 °C, much higher than that of the organic compounds which we had been studying. Nevertheless, some of the mean square displacements of the atoms in the Li_2BeF_4 crystal are still quite considerable at room temperature, and we can expect to reduce them by a factor of three or so by cooling to liquid nitrogen temperatures.

We therefore repeated the experiment at 80 K with crystals kindly provided by Collins, and were able to measure many more reflections in the high- order region (out to $\sin\theta/\lambda = 1.35$ $Å^{-1}$). Fig. 11.14(a) shows our version of the difference density, to be compared with that in Fig. 11.13. Like the former, it is based on subtraction of neutral atom densities. As we see, the bonding density in the Be–F bonds has practically disappeared. Quite possibly, the apparent density in the room-temperature analysis is really due to the inadequacy of the Gaussian vibration parameters to describe the atomic motions. Once these

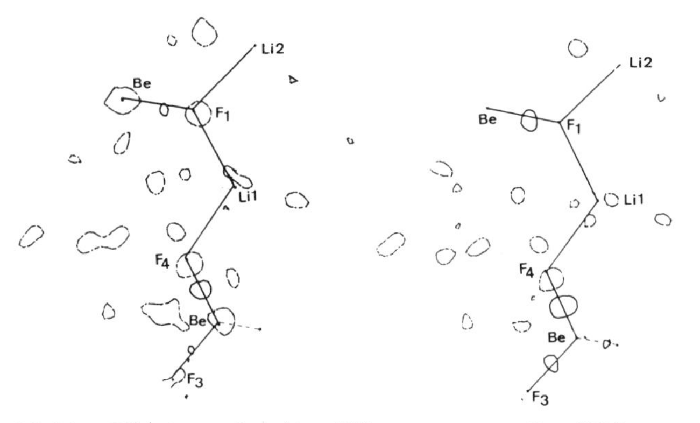

FIG. 11.14. Electron density difference map for lithium tetrafluoroberyllate at 80K. Left: based on neutral atoms; right based on ions as pro-crystal. Contours as in Fig. 11.13.

motions. Once these motions become large enough, anharmonic terms start to get important, and if these are neglected in the least-squares refinement, spurious features can be produced in the difference map. At 80 K°, the atomic displacement parameters are smaller and the anharmonic terms correspondingly less important. As we can see, the difference map based on subtraction of neutral atoms is practically flat. However, it is by no means self-evident that one should consider Li_2BeF_4 as being composed of neutral atoms – rather the contrary. One might ask what would change if the pro-crystal is assumed to consist of ions, Li^+, Be^{2+} and F^-. The answer is that the R factor is the same, the atomic positions and displacement parameters determined from the high-order reflections are practically the same (because for neutral atoms and corresponding ions the scattering curves are practically identical for high values of $\sin\theta/\lambda$), and indeed, as Fig. 11.14(b) shows, the difference map is also practically the same. Does this then mean that with all this fuss about electron density difference maps, we can't even tell the difference between an ionic crystal and one built from neutral atoms?

Fig. 11.15 shows the form factors for F and F^-, for Be and Be^{2+}, and for Li and Li^+. In every case the curves for the neutral and charged atoms are very similar, except in the region of reciprocal space close to the origin. Even the curves

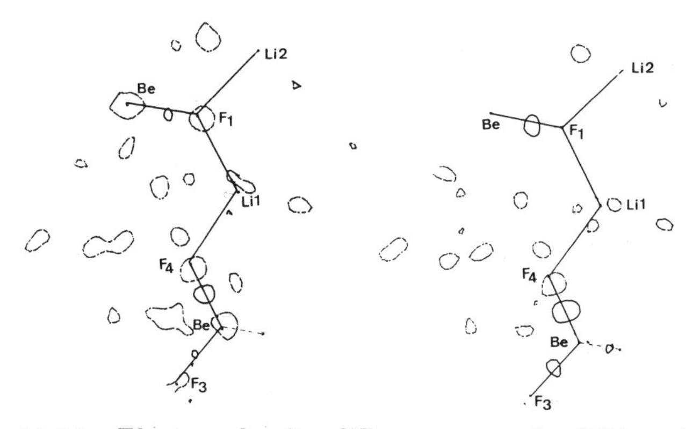

FIG. 11.14. Electron density difference map for lithium tetrafluoroberyllate at 80K. Left: based on neutral atoms; right based on ions as pro-crystal. Contours as in Fig. 11.13.

motions. Once these motions become large enough, anharmonic terms start to get important, and if these are neglected in the least-squares refinement, spurious features can be produced in the difference map. At 80 K°, the atomic displacement parameters are smaller and the anharmonic terms correspondingly less important. As we can see, the difference map based on subtraction of neutral atoms is practically flat. However, it is by no means self-evident that one should consider Li_2BeF_4 as being composed of neutral atoms – rather the contrary. One might ask what would change if the pro-crystal is assumed to consist of ions, Li^+, Be^{2+} and F^-. The answer is that the R factor is the same, the atomic positions and displacement parameters determined from the high-order reflections are practically the same (because for neutral atoms and corresponding ions the scattering curves are practically identical for high values of $\sin\theta/\lambda$), and indeed, as Fig. 11.14(b) shows, the difference map is also practically the same. Does this then mean that with all this fuss about electron density difference maps, we can't even tell the difference between an ionic crystal and one built from neutral atoms?

Fig. 11.15 shows the form factors for F and F^-, for Be and Be^{2+}, and for Li and Li^+. In every case the curves for the neutral and charged atoms are very similar, except in the region of reciprocal space close to the origin. Even the curves

flections are particularly prone to contamination by Renninger reflections and other systematic errors. In his opinion, the reflections in question have not yet been measured with the care and accuracy that would be necessary to make this critical distinction. It will be a very difficult job to do this, but Paul thinks it may just be possible.

In any case, it seems to be extremely difficult to tell the difference between an ionic crystal and one built from neutral atoms. Even if it does turn out that the pro-crystal made of neutral atoms gives a slightly better fit to the total density than a pro-crystal made of ions, we should not forget that the true crystal is built neither from neutral atoms nor from ions, so the question, which model is the better, may not be too meaningful.

In this lecture, I have tried to show you that the use of accurate intensity measurements in crystallography today can not only answer some questions about chemical bonding, but can also raise questions about other aspects of chemical bonding to which we really don't know the answers. I'd like to end by paying tribute to Paul Seiler's ideas and experimental skill. Without them, there wouldn't have been anything for me to talk about.

Discussion

Dorothy Hodgkin: Do you make a correction for extinction?

Jack Dunitz: We do our best. Extinction is certainly one of the major problems in making accurate intensity measurements.

Dorothy Hodgkin: Jack, on a point of history, almost the same experiments carried out with great care and accuracy were done by Bijvoet and Kathleen Lonsdale on lithium hydride, with the same worry about the very few low order reflections, and the almost impossibility of telling covalent and ionic bonding apart.

Jack Dunitz: We had thought about redoing lithium hydride, but decided that the extinction for this crystal is so se-

vere that it is almost pointless to start measuring it at present. What you would see in the difference maps would be mainly the extinction effect.

Chris Nordman: Do you think that, if you are interested in this last question, that you might be better off if you cocrystallized your compound with some innocuous material, whose electron density is not so much in doubt, so as to get it in a much larger unit cell, and thereby increasing your number of low order reflections?

Jack Dunitz: That's certainly a point.

George Jeffrey: On that last slide, you have a crystal that melts at a 1000° so you have got to have extinction there. Extinction affects weak reflections just as much as strong reflections. I believe what you are seeing there are problems of extinction. And it's interesting, because we can never tackle this question of ions versus molecules, because the very ionic crystals we want to study have this extinction problem. We won't solve that until we have a good theory for extinction.

Jack Dunitz: For the weak reflections in question we do not think that extinction is the major problem. In any case, we can at least estimate how serious it is by making our experiments with smaller and smaller crystals, and extrapolating to zero volume.

George Jeffrey: Physicists say the theory that we use for correcting for extinction has no physical reality. It's purely empirical, even though it works.

Jack Dunitz: A measured intensity is compared with the intensity calculated from the kinematic theory of scattering, and the discrepancy between these two quantities is called the extinction error. Extinction isn't really a phenomenon, it's the failure of a theory, the kinematic theory, to take account of multiple scattering events.

Brian Matthews: Would you explain why moving or inter-

changing a few electrons is unlikely to affect the strong reflections and is likely to affect the weak reflections?

Jack Dunitz: O.K. Let's think about lithium fluoride. There are two classes of reflections. With h,k,l even, you add the scattering powers, and with h,k,l odd, you subtract them. Now, suppose we have neutral atoms. Then, for very low-order reflections, the scattering powers are nine and three. Nine plus three is twelve, and nine minus three is ···six? Now suppose we have ions. The scattering powers are ten and two. Ten plus two is still twelve, but ten minus two is eight. For the strong h,k,l even reflections, we get the same intensity, but for the weak h,k,l odd ones the intensities are different for the two models, six squared in one case and eight squared in the other. It's easier to see the difference between 48 and 36···

From the audience: 64

Jack Dunitz: ··· than to see the difference between 144 and 144.

Peter Murray-Rust: I was trying to plot out some sort of relationship between the integrated observed electron density in the bond, and some simple chemical parameter like electronegativity. And, I mean, I remember, from teaching organic chemistry, that the nitrogen-nitrogen and oxygen-oxygen and chlorine-chlorine bonds are exceptionally weak, compared with what one would expect with straightforward smooth function over the Periodic Table. I couldn't get an answer, I mean, can you? Is it a meaningful question to ask?

Jack Dunitz: I don't know. For C–C bonds at least, as Berkovitch-Yellin and Leiserowitz[17] have shown, there does seem to be a correlation between the integral of the bonding density and the C–C bond length, which can also be correlated reasonably well with bond order, and hence with bond energy. Other kinds of bonds may well show similar correlations, but I don't know how to compare one kind of bond with another. The C–F bond is stronger than the C–C single bond, for example, but it shows much less bonding density. I'm not sure I've answered your question.

David Sayre: I'd like to ask a question about extinction. People in electron microscopy and electron diffraction use computer programs based on dynamical theory to handle their extinction-related problems. What is wrong with us, since extinction is an important problem for us too, that we are not willing to face the problem in the same way?

Jack Dunitz: As you know, most crystallographers use a purely phenomenological correction for extinction, as proposed by Zachariasen[18] or by Becker and Coppens.[19] This is just fitting the discrepancies to some sort of model. Whatever the extinction is actually due to, the correction is supposed to take care of it. There are also attempts to make experimental corrections for extinction. As well as varying the crystal size, one can measure the extinction as a function of wavelength and extrapolate to zero wavelength.[20] Another possibility is to use different states of polarization of the incident beam[21] to make the measurements, and still another is to use temperature gradients.[22] In principle, when the extinction is not too severe, all of these can be more or less successful but they are all terribly time-consuming.

David Sayre: That's interesting but you are not answering my question. You're telling me the things that can be done apart from invoking the dynamical theory of X-ray diffraction in crystals. But why not use it, as the electron diffractionists do?

Jerry Karle: I would like to make a remark about a phenomological correction for extinction that I think might be appropriate at this celebration. A little over 20 years ago I wrote a paper with Herb Hauptman, and the name was 'Positivity, Point atoms, and Pattersons'. One of the things that we noticed was that if there were errors in the intensities and a Patterson map were computed and positivized by, for example, wiping out the negative region, then on inversion of the map there would be worthwhile corrections to the intensities that were in error. It was, in effect, in some semi-quantitative sense, a correction for the intensities in error effected by using

the intensities that were measured collectively, plus a positivity criterion. Now, I don't know in these cases, we never made a great study of it, exactly to what extent there would be a correction and whether it would be sufficient for your purposes, but it is a relatively easy thing to do. There are programs for inverting a Fourier series, and so I would suggest that you might try this, and see what effect it had, and perhaps run some tests on it. Thank you, and I'm sorry to take up time when there are other things that seem to be calling us.

David Harker: Just a very short one. The idea that I heard, I think from Nordman, is that most of these simple salts can be made into complexes with sugars, and these give very good crystals. We have wonderful carbohydrate determinations from them.

George Jeffrey: I am not so sure. Because of hydrogen bonding, carbohydrates form crystals with large extinction effects.

Jack Dunitz: And they're non-centrosymmetric.

David Harker: You can put in the D- and L- forms.

George Jeffrey: Since you persuaded Doug Collins with his crystal that melted at 1000°*C* that he should have taken data at liquid nitrogen temperatures, don't you feel you should take your data at liquid helium?

Jack Dunitz: Yes.

Dick van der Helm: You, in a way, take away the difference between a covalent bond and an ionic bond, to some extent. At least it is the way I understand it. On the other hand, I also wonder if we, as diffractionists, don't place too much emphasis on electron density as related to a chemical bond. What is your comment on that?

Jack Dunitz: I have felt myself compelled to make some comments about the possible interpretations, but really, if I've

shown you anything interesting, it's the experimental results themselves. I think that our experimental data are pretty good and that the results are reasonably robust against any slight errors we may have made. But, as far as I'm concerned, anybody's interpretation of what it all means is as good as mine, or better. I haven't got any strong feelings about the interpretation.

Verner Schomaker: Jack, is there more or less than meets the eye in this result about which you are worried? It seems to say that all talk about the electron density deformation densities, is somehow illusory, doesn't really mean anything. It seems to say that the prospect of determining properties by getting the electron density, and deducing everything, the energy and everything else from it, is illusory too. Is there something special about the situation, if you are worried, that makes it a harder problem, aside from extinction?

Jack Dunitz: Well, I think that there are certain concepts in chemistry which are tremendously useful, but which get more and more elusive as one tries to pin them down. Let me give you one example, which I've talked about in Zürich a lot with Dieter Seebach. What is the nature of the lithium-carbon bond? Dieter makes carbanions and combines them with lithium. So we have a bond, carbon at one end and lithium at the other. Whatever the nature of these bonds, the compounds are useful in synthetic chemistry. But we can argue about their nature. Now I only want to make the following point. As I mentioned earlier, if you start off with a neutral lithium atom and integrate the charge density of the $2s$ electron out to a distance of 2.4 Å you only get about 70% of the $2s$ electron inside that shell. The carbon atom is situated about 2 Å away from the lithium. If you integrate out to the middle of the carbon-lithium bond, you are only going to find about 10% of the $2s$ electron inside the shell. Therefore, if you define the nature of the bond in terms of the value of this integral, you come out with the result that it's an overwhelmingly ionic bond, even though you started off with a neutral lithium atom. This seems to me to be a situation where we must be asking

combination of an ion and a neutral atom. I might then be able to say 35% ionic, 65% neutral, for the beryllium, but it could be quite wrong, and possibly dangerously misleading.

Dorothy Hodgkin: Yes.

References

1. Dunitz, J. D. *X-ray Analysis and the Structure of Organic Molecules* Cornell University Press: Ithaca, London (1979).
2. Coppens, P. and Lehmann, M. S. *Acta Cryst.* **B32**, 1777 (1976).
3. Chakrabarti, P., Seiler.P., Dunitz, J. D., Schluter, A.-D. and Szeimies, G. *J. Am. Chem. Soc.* **103**, 7378 (1981).
4. Wiberg K. B. and Walker, F. H. *J. Am. Chem. Soc.* **104**, 5239 (1982).
5. Hartman, A. and Hirshfeld, F. L. *Acta Cryst.* **20**, 80 (1966).
6. Bader, R. F. W., Tang, T.-H. and Biegler-Konig, F. W. *J. Am. Chem. Soc.* **104**, 940 (1982).
7. van Rij, C. and Britton, D. *Cryst. Struct. Comm.* **10**, 175 (1981).
8. Dunitz, J. D., Schweizer, W. B. and Seiler, P. *Helv. Chim. Acta* **66**, 123 (1983).
9. Whittleton, S. N., Seiler, P. and Dunitz, J. D. *Helv. Chim. Acta* **64**, 2614 (1981).
10. Dunitz, J. D. and Seiler, P. *J. Am. Chem. Soc.* **105**, 7056 (1983).
11. Bader, R. F. W., Henneker, W. H. and Cade, P. E. *J. Chem. Phys.* **46**, 3341 (1967).
12. Static deformation density maps may be obtained by expressing the difference density in parametric form and refining the parameters by least-squares analysis along with the usual positional and vibrational parameters. For static deformation density maps of TFT see Hirshfeld, F. L. *Acta Cryst.* **B40**, 484 and 613 (1984).
13. Heitler, W. and London, F. *Z. Phys.* **44**, 455 (1927).
14. Hirshfeld, F. L. and Rzotkiewiez, M. *Mol. Phys.* **27**, 1319 (1974).
15. Collins, D. M., Mahar, M. C. and Whitehurst, F. W. *Acta Cryst.* **B39**, 303 (1983).
16. Amstutz, R., Laube, T., Schweizer, W. B., Seebach, D. and Dunitz, J. D. *Helv. Chim. Acta* **67**, 224 (1984).
17. Berkovitch-Yellin, Z. and Leiserowitz, L. *J. Am. Chem. Soc.* **99**, 6106 (1977).

the wrong question. If you ask me now what is the nature of the carbon-lithium bond, I would probably say 'I don't know,' or 'It's very difficult to find out.' Moreover, if you can't answer this question from a knowledge of the charge density, then how on earth are you going to answer it?

Verner Schomaker: So the answer to my question is that lithium fluoroberyllate is a much more difficult question than benzene.

Jack Dunitz: Yes. Because of the diffuseness of the outer electrons of the neutral atoms.

Verner Schomaker: I am reminded that I once was very proud to have realized that a very nicely directed orbital for covalent bonding, displaced the electrons, it seemed to me, out on the other atom, and I wrote to Pauling and said 'Look, am I not great? I've shown that your beautiful directed bonds are just the same thing as ionic bonds.' And he didn't believe a thing about that. He never did.

Dorothy Hodgkin: I would like one more word about the past. It must have been when I was first calculating electron density maps connected with vitamin B_{12} that W. L. Bragg looked at me deciding from electron counts which atom was carbon, and which atom was oxygen and commented 'We gave up that kind of treatment of maps of ionic crystals when we realized that the existence of extinction made it very difficult to decide even between the ionic and covalent states.' I then find myself thinking, that was a very long time ago. You ought now to be able to measure the extinction correction for 30 inner reflections, Jack, however long it takes.

Jack Dunitz: But, even so, Dorothy, when it comes down to it, the real atoms in the real crystal are neither ions or neutral atoms, they are what they are. Or rather, I should say, the total charge density is what it is. In a sense, I apologize for introducing this problem at all. Let's say I did a least-squares analysis where I assume the actual atom is a mixture, a linear

combination of an ion and a neutral atom. I might then be able to say 35% ionic, 65% neutral, for the beryllium, but it could be quite wrong, and possibly dangerously misleading.

Dorothy Hodgkin: Yes.

References

1. Dunitz, J. D. *X-ray Analysis and the Structure of Organic Molecules* Cornell University Press: Ithaca, London (1979).
2. Coppens, P. and Lehmann, M. S. *Acta Cryst.* **B32**, 1777 (1976).
3. Chakrabarti, P., Seiler.P., Dunitz, J. D., Schluter, A.-D. and Szeimies, G. *J. Am. Chem. Soc.* **103**, 7378 (1981).
4. Wiberg K. B. and Walker, F. H. *J. Am. Chem. Soc.* **104**, 5239 (1982).
5. Hartman, A. and Hirshfeld, F. L. *Acta Cryst.* **20**, 80 (1966).
6. Bader, R. F. W., Tang, T.-H. and Biegler-Konig, F. W. *J. Am. Chem. Soc.* **104**, 940 (1982).
7. van Rij, C. and Britton, D. *Cryst. Struct. Comm.* **10**, 175 (1981).
8. Dunitz, J. D., Schweizer, W. B. and Seiler, P. *Helv. Chim. Acta* **66**, 123 (1983).
9. Whittleton, S. N., Seiler, P. and Dunitz, J. D. *Helv. Chim. Acta* **64**, 2614 (1981).
10. Dunitz, J. D. and Seiler, P. *J. Am. Chem. Soc.* **105**, 7056 (1983).
11. Bader, R. F. W., Henneker, W. H. and Cade, P. E. *J. Chem. Phys.* **46**, 3341 (1967).
12. Static deformation density maps may be obtained by expressing the difference density in parametric form and refining the parameters by least-squares analysis along with the usual positional and vibrational parameters. For static deformation density maps of TFT see Hirshfeld, F. L. *Acta Cryst.* **B40**, 484 and 613 (1984).
13. Heitler, W. and London, F. *Z. Phys.* **44**, 455 (1927).
14. Hirshfeld, F. L. and Rzotkiewiez, M. *Mol. Phys.* **27**, 1319 (1974).
15. Collins, D. M., Mahar, M. C. and Whitehurst, F. W. *Acta Cryst.* **B39**, 303 (1983).
16. Amstutz, R., Laube, T., Schweizer, W. B., Seebach, D. and Dunitz, J. D. *Helv. Chim. Acta* **67**, 224 (1984).
17. Berkovitch-Yellin, Z. and Leiserowitz, L. *J. Am. Chem. Soc.* **99**, 6106 (1977).

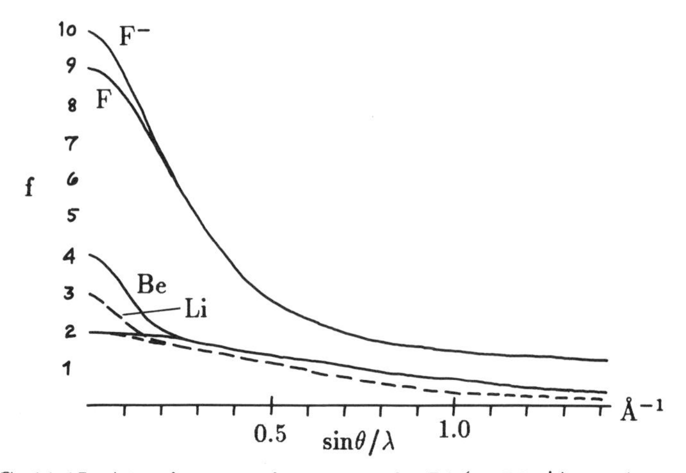

FIG. 11.15. Atomic scattering curves for Li (and Li^+), Be (and Be^{2+}), and F (and F^-) from *International Tables for X-ray Crystallography*, Vol. 4.

for neutral Be, with 4 electrons, and for Be^{2+} with 2, are practically identical for $\sin \theta/\lambda > 0.2$ Å^{-1}. This tells us that the 2*s* orbital of Be is extremely diffuse, and so is the one of Li. Likewise, the charge densities of neutral F and anionic F^- are very similar to each other, except that the anion is more diffuse. Indeed, for Li, the 2*s* density is so spread out that only about 15% of the charge is included within a sphere of radius 1.1 Å (roughly half the distance between a Li atom and its bonded neighbors); some 28% of the charge is still not contained in a sphere of radius 2.4 Å, extending well past the centers of the bonded atoms.[16] However, there are still the considerable differences between the scattering curves in the very low-order region. Can we make use of these? For the Li_2BeF_4 crystal, the unit cell is large enough that there are a few reflections in this region. The most sensitive reflections should be among the weak ones, those where the atoms are scattering out-of-phase with one another. I am tempted to give you the provisional result of a calculation based on the 11 weakest low-order reflections, those with $F < 3.2$ and $\sin \theta/\lambda < 0.25$ Å^{-1}, for both models, the neutral and the ionic pro-crystal. The overall agreement is better for the neutral atom model, with an R factor of 0.043 compared with 0.12 for the ionic one.

I was strongly advised by Paul Seiler against giving you this result, because, as he points out, the weak low-order re-

18. Zachariasen, W. H. *Acta Cryst.* **23**, 558 (1967).
19. Becker, P. J. and Coppens, P. *Acta Cryst.* **A30**, 129 (1974).
20. Lawrence, J. L. *Acta Cryst.* **A33**, 232 (1977).
21. Chandrasekhar, S., Ramaseshan, S. and Singh, A. K. *Acta Cryst.* **A25**, 140 (1969).
22. Seiler, P. and Dunitz, J. D. *Acta Cryst.* **A34**, 329 (1978).

12. Rational design of DNA-binding drugs or How to read a helix

Richard E. Dickerson and Mary L. Kopka

Richard E. Dickerson did his undergraduate work at Carnegie Tech, and obtained his Ph.D. in 1957 at the University of Minnesota. He then went for two years to Leeds and Cambridge in England where he worked with John Kendrew on myoglobin structure. After coming back, he joined the faculty at the University of Illinois for a period, then went on to Cal Tech and a few years ago moved to UCLA to join the Molecular Biology Institute there. He is well known for his studies of the protein, cytochrome *c*, and the evolutionary implication of the structures of different cytochromes *c*. More recently he has been associated with some of the more precise determinations of DNA structure and complexes of DNA and drugs. He has written books on the structure and function of proteins with Irving Geis, articles in the *Scientific American*, a textbook on chemical thermodynamics and a very successful general chemistry textbook.

I did not know Lindo Patterson, so I am of the post-Patterson generation in that regard. I certainly knew of Patterson functions. When I was a graduate student with Bill Lipscomb at Minnesota in the late fifties, I worked on boron hydrides; later I worked on myoglobin with John Kendrew at Cambridge in 1958 and 1959. In those days the Patterson function was the starting point for any crystal structure analysis.

As David Harker mentioned, the simple Patterson function was not directly useful in the early stages of protein structure determination. In fact, the workers on myoglobin and hemoglobin were led astray by a too-simple interpretation of the straight F^2 functions. But, as Harker also said, difference Patterson maps were the key step in finding heavy atom positions for multiple isomorphous replacement phase analysis.

That was a time of great ferment in Cambridge, in that your success in finding heavy atoms depended in part on your ability to come up with better Patterson functions. Michael Rossmann was very skilled in this. One type of function, called a ΔI Patterson, was simply the difference between the Patterson map of the derivative and the protein without the heavy

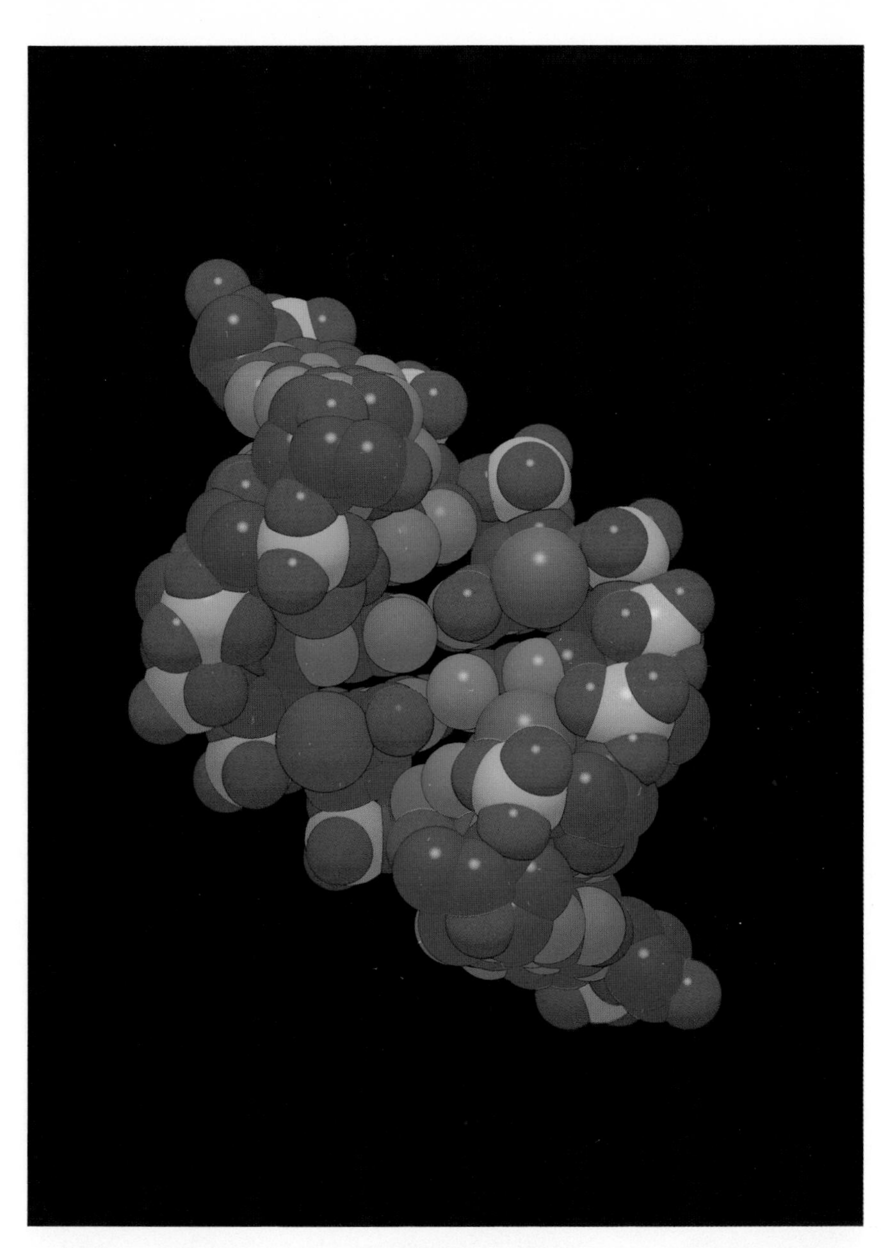

FIG. 12.1. Space filling molecular drawings, by Nelson Max of the Lawrence Livermore Laboratories, of three double-helical DNA oligonucleotides whose structures have been solved by single-crystal X-ray analysis. The view in each case is directly into the major groove.

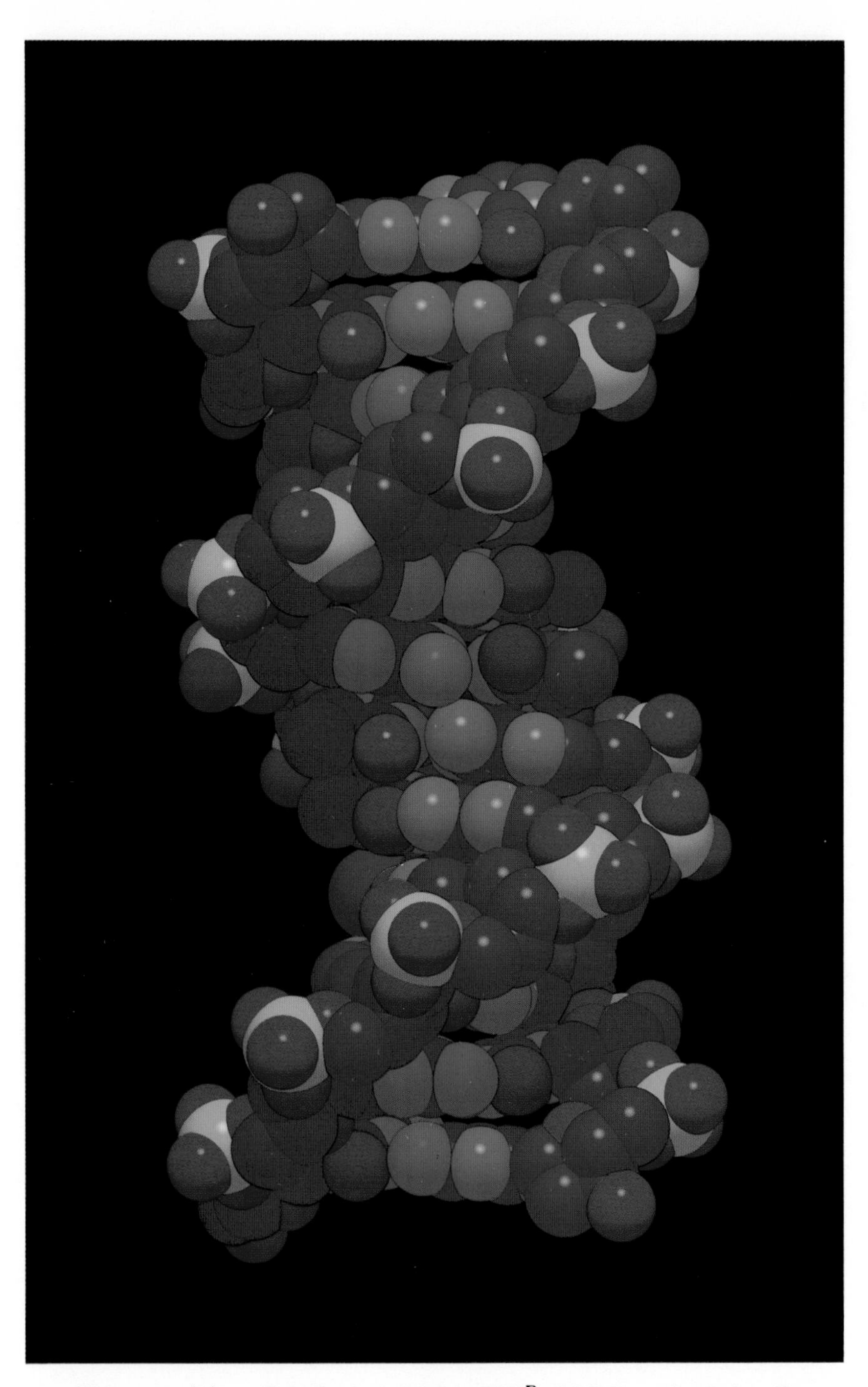

FIG. 12.1(b). B-helical CGCGAATT^{Br}CGCG, solved by Drew, Fratini, Kopka and coworkers in the laboratory of Richard Dickerson at the University of California at Los Angeles. Green bromine atoms label 5-bromocytosines. The view is directly into the major groove. Space filling molecular drawing by Nelson Max of the Lawrence Livermore Laboratories.

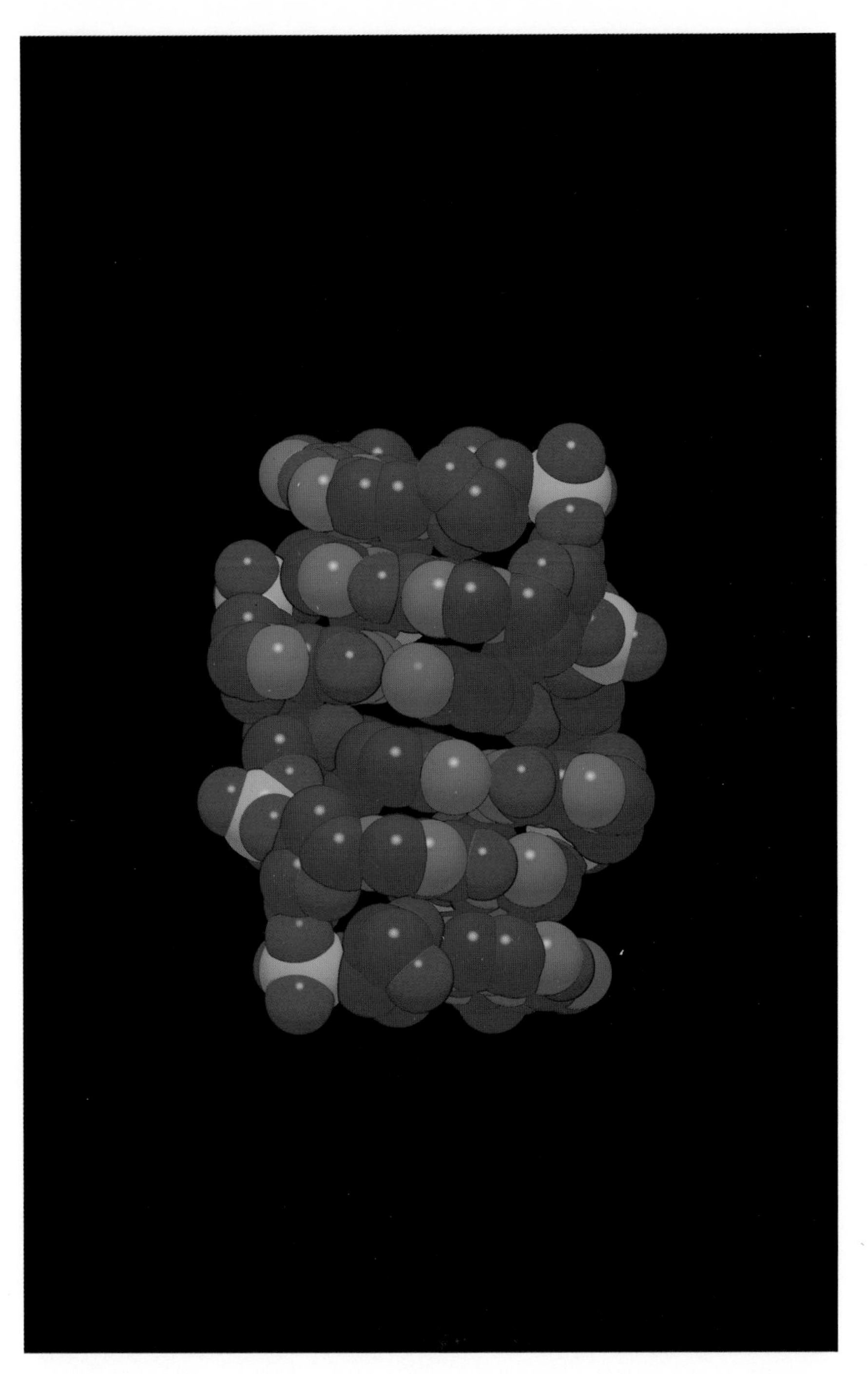

FIG. 12.1(c). Z-helical CGCGCG, solved by Andrew H-J. Wang in the laboratory of Alexander Rich at MIT. The base pairs in Z-DNA are displaced so far from the helix axis that the 'major groove' now is more properly described as a surface. The view is directly into the major groove. Space filling molecular drawing by Nelson Max of the Lawrence Livermore Laboratories.

FIG. 12.7. Painting by David Goodsell of the netropsin molecule (glowing object) nested within the minor groove of double-helical B-DNA of sequence $\mathrm{CGCGAATT}^{Br}\mathrm{CGCG}$. The drug molecule is regarded as the sole source of illumination in this project.

atoms. It was hard to interpret because it was muddied up by protein-heavy atom cross peaks. Another kind called a ΔF^2 Patterson got rid of the cross peaks, but was inexact and in principle should not have worked. I remember that some of the people now present in this room were not at all happy about the use of the ΔF^2 Patterson because it was such an approximate and inexact function. However, several green American postdoctoral fellows did not know enough to know that it would not work, and found that it *did* work fairly well. Michael Rossmann achieved quite a reputation at the time in the area of developing new kinds of Patterson functions and cross correlation functions.

I never expected ever to meet Lindo Patterson. I recall in 1957, when I was first introduced to Sir Lawrence Bragg in London, I suppressed an eerie thought, 'But surely you should be dead'. Anybody who had a law, principle, or a function named after him must have belonged to a by-gone generation. I thought of Lindo Patterson as an historical abstraction. But then one day I discovered something completely unexpected that made Patterson a real person, although as I say, I never did actually meet him.

I had to write a review on protein crystal structure analysis in 1962 for the second edition of Neurath's *The Proteins.* I went back through all the literature of the early days of the myoglobin, hemoglobin, ribonuclease work and fiber diffraction work as well. In about 1949, as I recall, W. T. Astbury of Leeds had written a review on protein structure, and he quoted a limerick by Lindo Patterson at the head of his review. I thought I would share it with you because it brought home to me that Lindo Patterson was real. If my memory is correct, it goes something like the following:

> Amino acids in chains
> Are the cause, so the X-ray explains
> Of the stretching in wool
> And its strength when you pull
> And show why it shrinks when it rains.

I thought: 'Why on earth is a mathematician (obviously Patterson must be a mathematician) writing doggerel on protein structures?' Then I discovered that that was not all he did that was interesting. I would have liked to have met him. I think he would have been an enjoyable person to have known.

DNA and antitumor drugs

My subject today is an aspect of macromolecular structure with dual ties to Lindo Patterson and the Fox Chase Cancer Center: DNA and DNA-Drug complexes. To paraphrase Patterson, I'm going to talk about nucleic acids in chains, and also about a particular antitumor drug, netropsin, which has been studied by Helen Berman here at the Fox Chase Cancer Center. Her X-ray structure analysis of the drug by itself was extremely helpful in our own look at the complex of netropsin with a 12-base-pair piece of DNA.

Our knowledge of DNA structure has changed dramatically in the past ten years because we now can make pieces of DNA, short oligomers of four to twenty-four pairs in any desired sequence. These are of sufficient quality and purity that they can be crystallized and handled in the same way as any protein, that is, it is possible to solve their crystal structures using conventional isomorphous replacement. Sometimes heavy atom complexes can be diffused in or, if not, the oligomer can be synthesized with guanine or cytosine brominated or iodinated at certain positions. Then the oligomer is treated as if it were a small protein.

The DNA structure field today is roughly where the protein structure field was in 1967 or 1968, when theoretical methods of structure determination had been checked out and shown to work on one protein, myoglobin. A flood of new results followed swiftly thereafter: ribonuclease, carboxypeptidase and chymotrypsin. People suddenly realized that this was only the first wave of a great many structure determinations to come. By the same token, many oligonucleotide structures are just now beginning to come out rapidly, mainly at the moment from three laboratories: that of Olga Kennard in Cambridge collaborating with Zippi Shakked and Dov Rabinovich at the Weizmann Institute in Israel, that of Andy Wang and Alex Rich at MIT, and our own lab, formerly at Cal Tech but now at UCLA. The reason why these three labs are doing most of the work right now (and that will certainly change in time) is simply that we have collaborators who are willing to make DNA oligomers. Getting the sequences you need is still the rate limiting step.

From our high resolution X-ray results, we can see things that are not readily found from fiber work. The overall helical structure becomes clear and so does the local sequence-dependent variation in helical structure. This latter must surely be involved in the recognition of DNA by proteins (see the article by Brian Matthews) and by drugs. It is the binding and recognition by drugs that will be discussed here.

Let us look at some of the oligonucleotide structures that have been determined by single-crystal X-ray methods. Three typical structures are shown in Fig. 12.1. The first two, A-DNA and B-DNA, were predicted from fiber work. Z-DNA was a total surprise.

It looks as though the most important aspect in B-DNA is the stacking of bases in each strand as though the other strand did not exist. If you take two stacks of pennies, you can see this. Imagine my arms as being two stacks of pennies. Wrap them around each other in a right-handed helical way and observe the way the wrists cross. You will see that the pennies, or the base pairs in the two stacks then have to have what is defined as a positive propeller twist. That is, the base pairs are not flat, they are rotated so that the nearer base of each pair, sighting down their common axis, is turned clockwise. Every right-handed positive oligonucleotide structure, without exception, has a pronounced right-handed positive propeller twist. It is a natural consequence of the helix rotation. Then, because you do not want the helix to fall apart at $37°C$, you zip it up with sugars and phosphates, but they are really afterthoughts. The most important thing in B-DNA is base stacking.

A portion of right-handed A-DNA of sequence GGTATACC has been studied by Olga Kennard, Zippi Shakked and Dov Rabinovich.[1] It is shown in Fig. 12.1(a). A-DNA has a very wide major groove, and its base pairs are tilted relative to the groove. But the essential feature that distinguishes A-DNA from B- or Z-DNA is none of these facts, nor even the sugar pucker. It is the position of the base pair relative to the helix axis and the stacking of base pairs. In A-DNA, the base pairs are displaced so that the helix axis lies almost entirely on the major groove side of each pair.

An example of B-DNA is shown in Fig. 12.1(b). This view is taken from a study of a 12-mer, done in our laboratory by Horace Drew and others.[2] It has the sequence, CGCGAATTCGCG, selected because it was the EcoRI restriction endonuclease substrate. Again, as you can see, the phosphate groups define a right-handed double helix.

Finally, Z-DNA, shown in Fig. 12.1(c), was studied as a hexamer at MIT by Andy Wang in Alex Rich's laboratory[3] and as a tetramer in our laboratory by Horace Drew.[4] In this case, the major groove is not truly a groove at all. The base pair lies so far away from the helix axis that the minor groove becomes cavernously deep, and the major groove little more than a surface. The phosphate backbone shows it to be a left-handed helical form.

The spine of hydration

Although the subject of this article is drug binding to the B form of DNA, we must preface this with a mention of the spine of hydration down the minor groove of B-DNA.[5,6] This spine plays an important role in stabilizing the B helix, and also is a model for interaction with drugs such as netropsin.

The interconversions that can occur between the three DNA families: A, B and Z, are illustrated in Fig. 12.2. Struther Arnott has popularized the concept that considerable variety or polymorphism exists within each family. Nevertheless, each family has its own distinguishing characteristics, and the concept of a transition from one family to another is a real one. One can go from B to Z in high ethanol or high salt, if the base sequence is permissive, essentially an alternation of purines and pyrimidines without too many exceptions. One can also go from B to A in ethanol, or by drying under low salt conditions. It is this latter transition that we are concerned with here. The A form has a very deep major groove and a very shallow minor groove, while in the B form the major and minor grooves are of equal depth (because the base pairs sit directly on the helix

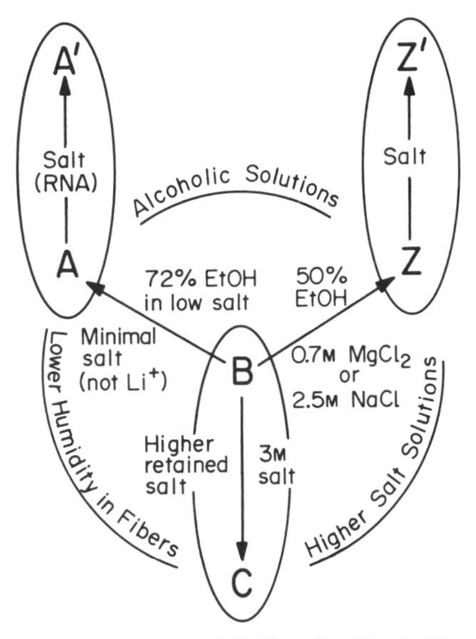

FIG. 12.2. Interconversions of DNA double helices between the three families, indicated by elongated ovals. B-DNA can be converted to Z by ethanol or salt, if the base sequence is a permissive near-alternation of purines and pyrimidines. B can be converted to A by lowering water activity, either by drying or with alcohol, under low-salt conditions. The conversion requires destruction of the chain or spine of hydrating water molecules down the narrow minor groove of the B helix. High salt stabilizes the B form even under dehydrating conditions, since cations can slip into the minor groove and replace the spine of hydration.

axis) but the minor groove is quite narrow. In all of the B structures studied, a spine of water molecules zig-zags its way down the minor groove in A-T-rich regions, as in Fig. 12.3(a). It is stabilized in part because the minor groove is narrower in A-T regions than in G-C regions, a possible consequence of the greater propeller twist of A-T base pairs. The spine connects thymine O2's and adenine N3's on adjacent base pairs, on opposite strands of the helix. The first level of hydration bridges these base pairs, and a second layer of water then links these bridge atoms in such a way as to give them a roughly tetrahedral environment. The result is a one-dimensional 'ice-like' string down the minor groove.

The spine breaks up in G-C-rich regions. The presence of an

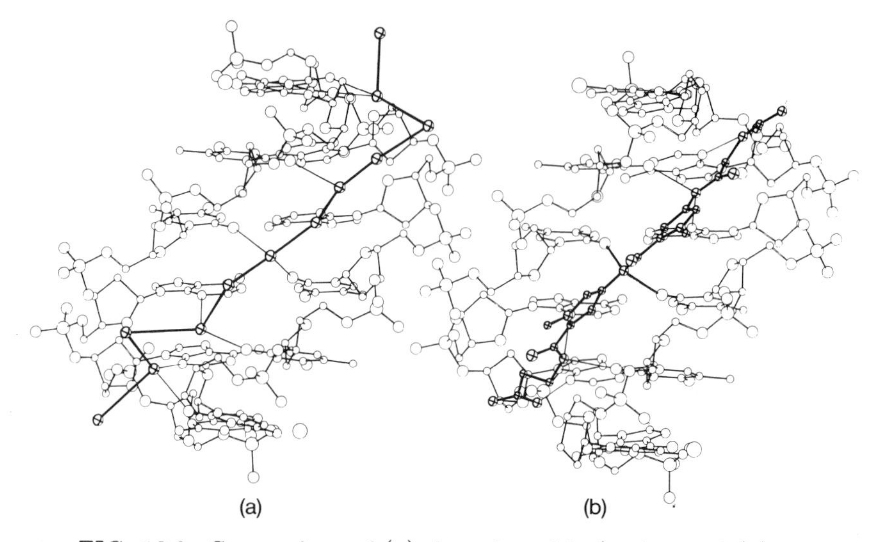

FIG. 12.3. Comparison of (a) the spine of hydration and (b) binding of netropsin, both within the narrow minor groove of B-DNA. Coordinates from the X-ray structure analyses of CGCGAATT^{Br}CGCG[5,6] and its complex with netropsin.[13–15] Only the central six base pairs, GAATTC, are drawn. Open spheres are DNA atoms; crossed spheres are water oxygens or the atoms of netropsin. Note the way in which both water molecules and the netropsin amide NH form hydrogen bonds bridging the O2 of thymines and N3 of adenines on adjacent base pairs and opposite strands of the helix.

$-NH_2$ group on G seems to interfere with the spine of hydration. It is doubly intrusive; it sits where a water molecule of the spine should be, and it provides hydrogen bond donors (Fig. 12.4) in a region where only acceptors are present in A-T base pairs. We believe that the breakup of this spine of hydration is a necessary first step in the B-to-A helix conversion. Hence, one should find that high A-T polymers are more stabilized in the B form than are high G-C polymers, and this is exactly what Struther Arnott finds from fiber diffraction studies.[5–7] Moreover, I-C base pairs, identical to G-C except for removal of the guanine $-NH_2$, behave like A-T base pairs in stabilizing the B-helix.

The binding of netropsin

Netropsin (Fig. 12.5) is an antitumor antibiotic that, in its

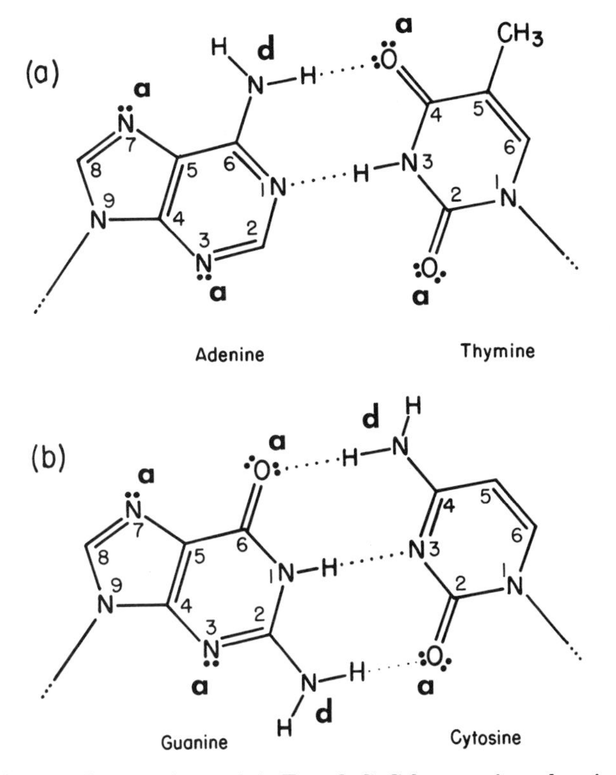

FIG. 12.4. Comparison of A-T and G-C base pairs, showing hydrogen bond acceptors marked 'a' and donors marked 'd'. G-C differs from A-T by having an intrusive -NH_2 amine group in the minor groove (bottom edge as drawn here).

interaction with DNA, appears to displace and mimic the spine of hydration. The drug molecule consists of three hydrogen-bonding amide groups separated by two five-membered pyrrole rings, with cationic charges at either end. Although it has antiviral properties against certain mammalian tumor viruses, it has been found to be too toxic for routine clinical use, or even for sale in this country, and we had to obtain our supply from friends in Italy. It has been extensively studied as the paradigm of a groove binding antitumor drug by people like Christoff Zimmer at Jena and others.[8,9] Remarkably, it has extremely high base specificity. It refuses to bind to G-C pairs, and demands runs of four or more successive A-T pairs. This

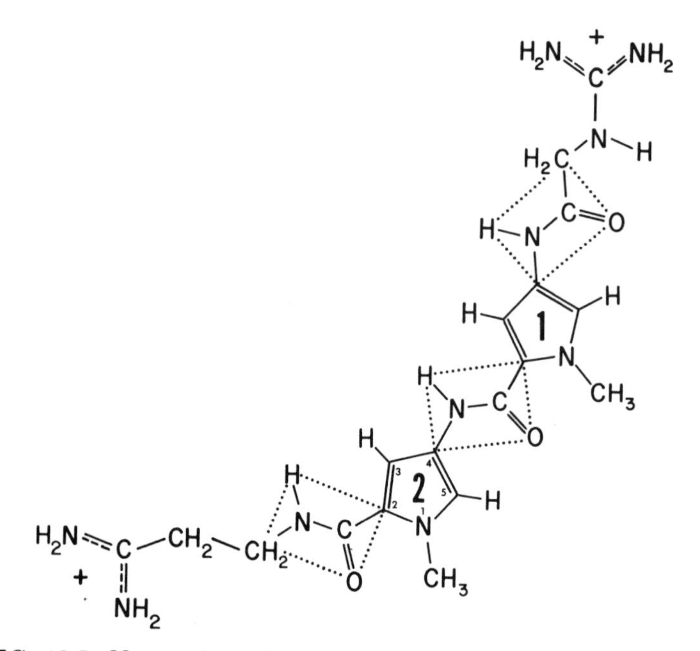

FIG. 12.5. Netropsin molecule, in the crescent shape as found in the single-crystal X-ray analysis of the drug molecule alone.[11] From upper right to lower left the components of the molecule are: guanidinium–amide–pyrrole–amide–pyrrole–amide–propylamidinium.

makes it immediately of interest to a crystallographer because something of an informational nature is to be learned. Since it binds only to double stranded DNA in the B form, not to A or Z DNA, it can in fact, promote the A to B or Z to B transition in solution. It does not bind to double stranded RNA or to single-stranded RNA or DNA.

An analogous class of drug molecules, distamycins, result if you replace the guanidinium tail by an uncharged group. Families of distamycins or netropsins can be synthesized by considering a pyrrole plus an amide as the repeating unit of this polymer. Peter Dervan[10] has observed that if a netropsin or distamycin has x amides, then the optimal binding site on DNA is $x + 1$ successive A-T base pairs. For natural netropsin $x = 3$ and the optimal site is four base pairs. The X-ray analysis makes the reason for this obvious.

The crescent shape of netropsin, illustrated in Fig. 12.6, is

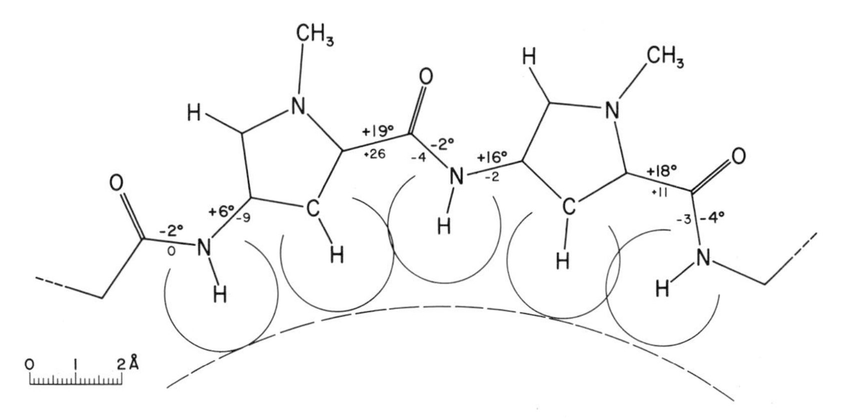

FIG. 12.6. True-scale drawing of the central region of netropsin, using bond lengths and angles as determined by the X-ray analysis of the drug alone. Numbers above and below bonds are torsional deviations from planarity observed in the drug itself (below bonds) and in the DNA-drug complex (above bonds). Van der Waals contact spheres are drawn around lower hydrogens. Dashed crescent below symbolizes floor of minor groove. Note the inability of the central NH to approach within reasonable hydrogen-bonding distance of the base pairs.

the result of a structure analysis of netropsin by Helen Berman and Stephen Neidle.[11] Their structure of the drug by itself was a very great help in solving our own structure of the drug complexed with the 12 base pairs of DNA. As can be seen in Fig. 12.6, all of the amide hydrogen bonding donors point inward toward the floor of the minor groove. This led Berman and Neidle to propose that these amides made hydrogen bonds to acceptors on the floor of the minor groove. Because the difference between the minor groove in A-T and G-C pairs was the absence or presence of the NH_2 group, it was proposed by Wells in the mid-1970's that this group must be important.[12] Somehow or other, the -NH_2 of guanine must be interfering with the binding of netropsin.

The structure of netropsin bound to the dodecamer CGCGAATT^{Br}CGCG is illustrated in Fig. 12.7. This work was done by Mary Kopka with the collaboration of graduate students Dave Goodsell, Chun Yoon and Phil Pjura.[13–15] The drug could not simply be diffused into the DNA; it was necessary to co-crystallize the drug and DNA together. The X-ray pattern from the co-crystallized crystals was very much like that of DNA alone. The space group and unit cell dimensions were similar and the intensity distribution was broadly the same, although different in detail. This was enough to enable Kopka to take the already determined structure of the dodecamer as a starting point, and then use the Jack-Levitt restrained least squares refinement method to refine the structure of the polynucleotide in its new environment. We quickly found that unexplained blobs of density appeared down the minor groove in electron density maps. Water molecules were added in the major groove and around the phosphates, at positions where we were reasonably sure the drug did *not* lie. The crystallographic *R*-factor

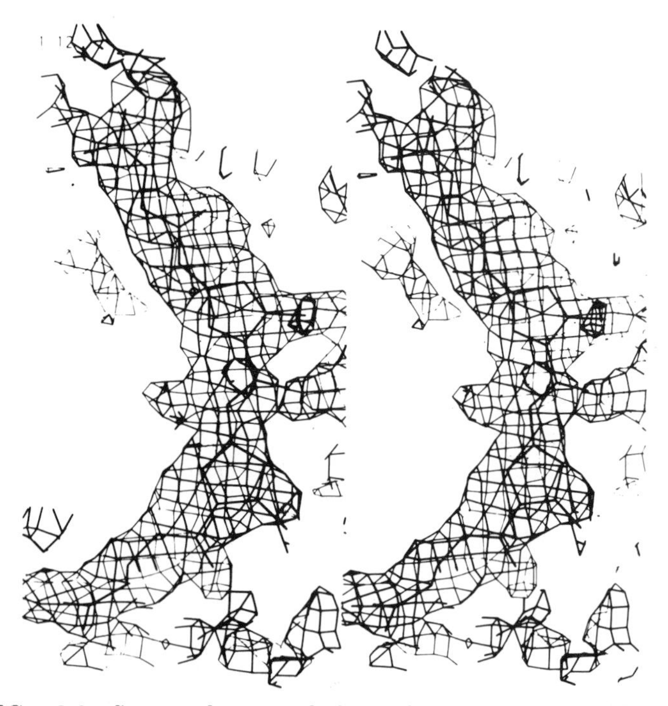

FIG. 12.8. Stereo photograph from the computer graphics screen of the electron density running down the minor groove, that was unexplained by the DNA structure itself. The skeleton inside shows the fitting of a netropsin molecule to this density. The floor of the DNA minor groove would be at left, had the DNA not been subtracted from this map. Note the projections of density where pyrrole methyls and amide carbonyls should be. From reference 15.

started at 46% with just DNA in the unit cell of the complex. It was reduced to 31% by refinement but no further improvement was possible without adding netropsin.

At this point, a difference Fourier map was computed and examined by Kopka and Goodsell, our resident graphics expert, on an Evans and Sutherland picture system. A chicken-wire contour of electron density that was not explained by the partially refined DNA and water molecules is shown in Fig. 12.8. The extra electron density is crescent-shaped and is exactly the right shape to be a netropsin molecule. Coordinates of netropsin from the Berman X-ray analysis were fitted and adjusted into this electron density, and Jack-Levitt refinement was continued until the R-value was then reduced to 26% for all data and 21% for 2σ data.

As a check, Goodsell and Yoon also re-refined the entire structure from the beginning using the Konnert-Hendrickson method. The final R-value was within 0.4% of that from Jack-Levitt refinement with essentially no difference between structures. This is the first time that an oligonucleotide structure has been refined by two different methods. It is very reassuring because it helps to answer the question, 'How do you know that what you are seeing is not dictated by the kind of constraints that you put on the refinement?

How does the netropsin bind? It binds in the minor groove as shown in Figs. 12.3(b) and 12.9. Essentially there are three three-centered hydrogen bonds from netropsin to the floor of this minor groove. They are three-centered because, as Jeffrey has discussed,[16] they involve a proton-poor situation where one amide NH points down in a pocket that has two potential targets for interactions. The central two hydrogen bonds are long for geometric reasons involved in the structure of the drug itself. The two charged ends both interact with the floor of the minor groove, with the propylamidinium end close enough to make a good hydrogen bond to the N3 of the adenine. The guanidinium is farther away from the DNA because of the inherent stiffness of the guanidinium.

Fig. 12.3(a) compares the spine of hydration on the left and netropsin on the right. The drug essentially displaces the hydrogen-bonded spine of hydration and replaces it by a

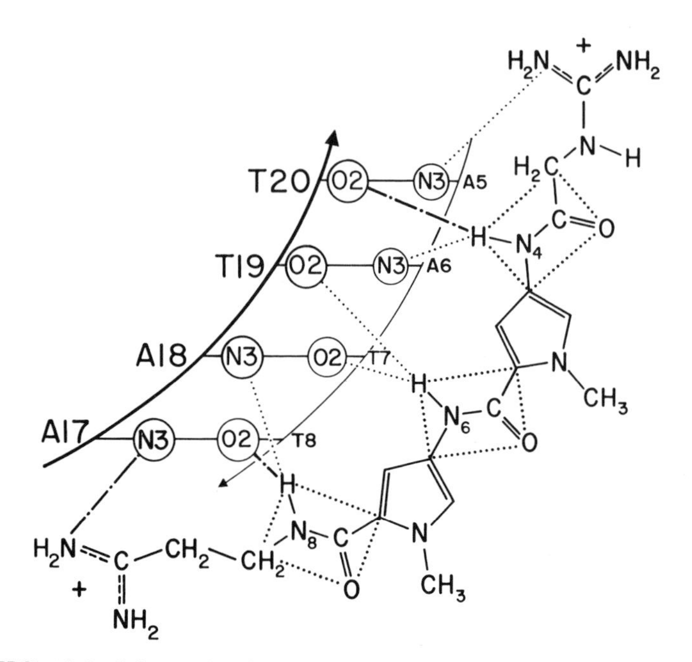

FIG. 12.9. Schematic of the interaction between netropsin and the floor of the minor groove in B-DNA. Bases are labeled by A(adenine)5 through T(thymine)8 in the first strand and A17 through T20 in the second strand. O2 and N3 are atoms of thymine and adenine to which netropsin makes hydrogen bonds (dotted or chain lines). From reference 14.

covalently-bonded 'super-spine.' This is further illustrated in Fig. 12.10 by a computer graphics program from Bob Langridge in San Francisco. It produces a dot sphere representation which we have come to call the 'van der Dot routine'. Fig. 12.10 shows that the propylamidinium makes a close packing contact with the surface of the minor groove, but the guanidinium tail fits less well. Similarly the two outer amides fit snugly, but the central amide buckles away. So the molecule fits imperfectly down in the bottom of the minor groove. It is not quite of the right geometry for a perfect DNA match. Fig. 12.6 shows that without deforming the backbone, one cannot get the central NH close enough to the bottom of the minor groove for a standard hydrogen-bond without pushing two CH bonds into the DNA. It turns out that this steric hindrance is the basis for the sequence specificity of the drug.

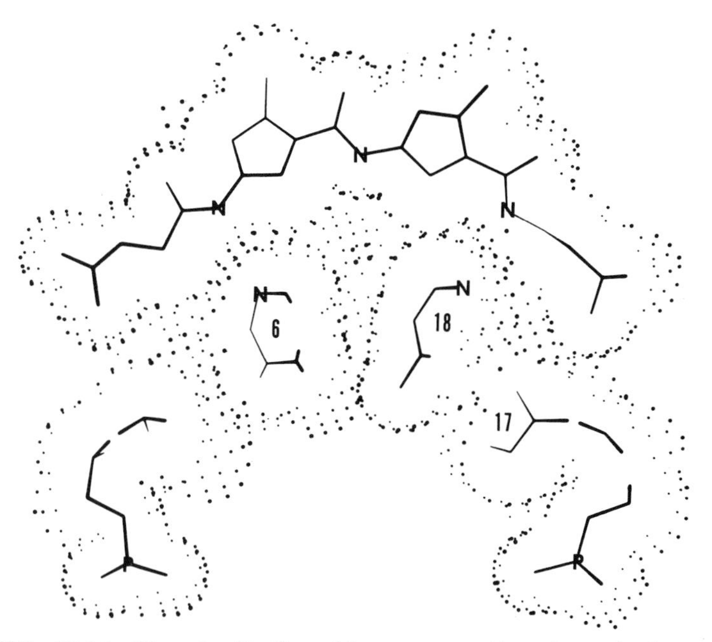

FIG. 12.10. 'Van der Dot' packing cross section through the complex of netropsin with B-DNA. Dots outline van der Waals spheres. The drug molecule is seen in profile, with cross sections below through some of the bases that interact with it. Note that the guanidinium end at left is not as tightly bound to the floor of the minor groove as is the amidinium end at right. Note also that the two flanking amides make tight contacts with edges of adenines 6 and 18, whereas the central amide N is buckled up and away from the bottom of the groove. From reference 15.

Three factors stabilize the drug binding to the DNA: the first is hydrogen bonding, the second is hydrophobic packing, and the third factor is electrostatic. It has been suggested that the cationic ends of this drug might climb up to the surface of the minor groove and interact with one or another of the phosphates, rather in the way that spermine drapes itself from phosphate to phosphate across the major groove in the dodecamer by itself. We found that this is not true at all. The two tails are found down in the bottom of the minor groove, and the more flexible of the two, the propylamidinium, actually makes a good hydrogen bond with an N3 of an adenine.

Why are the tails down in the bottom of the groove?

Bernard Pullman[17] has shown that the negative potential wells actually do not occur near the phosphates. If you move a distance of one typical van der Waals radius away from the surface of the DNA, that is, if you roll a point charge of a radius approximating that of a spherical water molecule over the surface of DNA and calculate the electrostatic potential at its center point by point, and then display this in cylindrical projection, you get the diagram shown in Fig. 12.11. At a van der Waals radius away from the DNA surface, the phosphates cooperate in producing a potential minimum in the bottom of the minor groove, and a less deep minimum in the bottom of the major groove. The cationic ends of netropsin go right to these minima on the floor of the minor groove.

In summary, there are three components to drug binding: hydrogen bonding, electrostatic interaction, and van der Waals

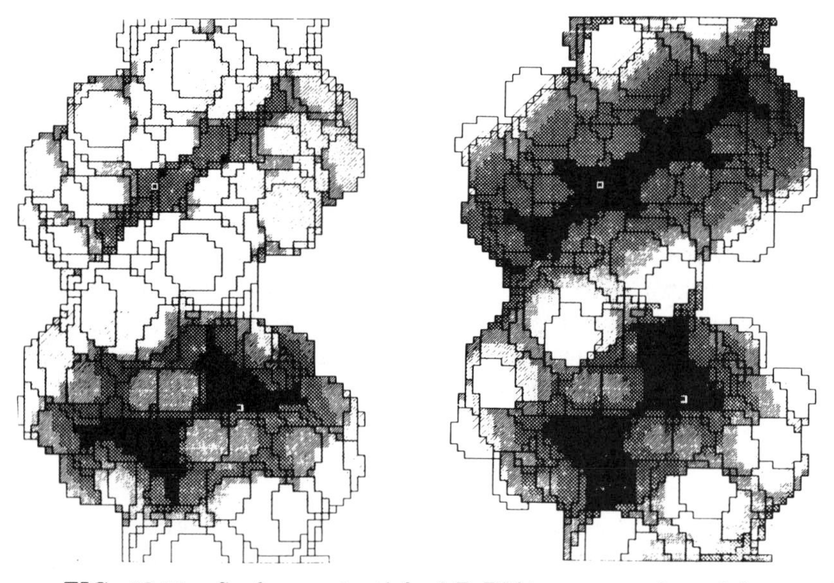

FIG. 12.11. Surface potential of B-DNA, as experienced by a unit positive charge rolling over the surface of the helix with the radius of a water molecule. Darker zones represent deeper potential wells and hence more stable locations for positive charge. Left: unscreened. Right: screened by sodium cations, with one Na^+ localized alongside each phosphate group. The minor groove is in the upper half of each drawing and major groove in the lower half. From reference 17.

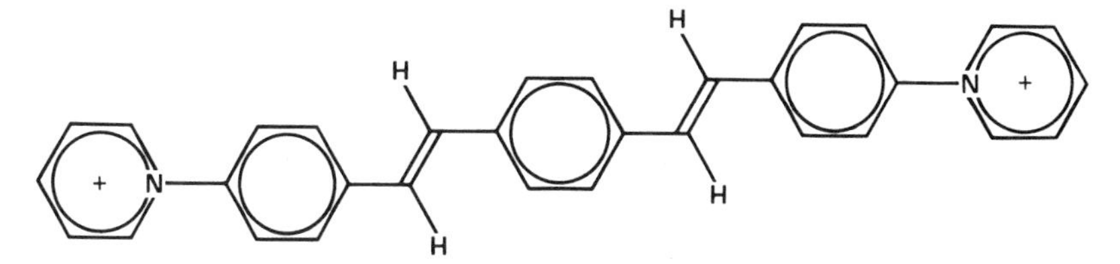

FIG. 12.12. The experimental antitumor agent SN18071, a bisquaternary ammonium heterocycle that has been synthesized at the University of Auckland, N.Z., to test the binding of such drugs to DNA. It has two cationic ends which can interact with the negative charge on DNA, but lacks all potentially hydrogen-bonding possibilities and is a stiff, hyperconjugated molecule that cannot bend easily into a crescent shape.

or hydrophobic packing. However, you can dispense with either one of the first two without harm. Christoff Zimmer has chopped off the two cationic ends of netropsin and has observed that this truncated, chargeless molecule still binds to DNA. It binds less well, of course, but it still has a pronounced A-T preference in specificity. B. C. Baguley has observed that the drug SN 18071 (Fig. 12.12), which has cationic ends but no possibility of hydrogen bonding, also binds preferentially to B-DNA in the minor groove and is A-T base pair specific.[18] Nobody has yet proposed any way of eliminating hydrophobic packing, the third factor.

Breslauer at Rutgers University has measured the thermodynamics of binding of netropsin to DNA, obtaining ΔG and ΔH, and from them calculating ΔS. He used the 10-mer GCGAATTCGC, and found that of the 11.5 kcal/mole of energy driving the binding of netropsin to DNA, about 9 kcal arise from enthalpy, and 2.2 kcal from entropic considerations. The entropy of binding is +7.5 cal/(deg mol).

But how do you bind one macromolecule to another macromolecule to make a single complex and still *create* disorder? The answer is fairly obvious; as is so often the case in macromolecular complex formation, you displace water molecules. From our structure analysis of the DNA alone, we estimate that there must be on the order of 12 water molecules released per netropsin bound. And from Helen Berman's analysis of netropsin, about five water molecules have to come off in order to make hydrogen bonds of the sort that you see to the DNA.

So the real reaction is: hydrated DNA plus hydrated netropsin yield a hydrated complex and about 17 free water molecules. I imagine this is like a Saint Bernard dog running across Trafalgar Square and kicking up the pigeons. That is the source of the 7.5 cal/mole, which contributes a fifth of the free energy of drive toward the complex formation. Thus this entropy comes from the breakup of the spine.

Sequence Specificity

Probably the most important question to a structurally oriented person is 'How is the DNA sequence read?' Seeman, Rosenberg and Rich published a paper ten years ago saying that specificity involving DNA ought to involve hydrogen bonds between side chains of proteins and the edges of base pairs in the minor or the major groove. Most people would say that hydrogen bonding is probably about as good a means of specificity as one could find. This turns out not to be the case here. The reading of the sequence in the netropsin-dodecamer is not being done by hydrogen bonds, but by van der Waals contacts. The role of the hydrogen bonds is to choose the right reading frame for the drug. The van der Waals contacts have a chance to say 'Yes, there is an amine group here' or 'No, there is no amine group here.' So the hydrogen bonds supply the reading frame, and the non-bonded van der Waals interactions do the reading.

The distances between the bumping groups on the drug are shown in Table 12.1. These groups are a methylene carbon #2, a pyrrole carbon #5, another pyrrole carbon #11 and a methylene carbon #16 *vs.* the #2 carbon atom of adenines 5, 6, 17 and 18. All of these are at acceptable van der Waals packing distances. If you were to simulate 2-amino adenine by putting a hypothetical NH_2 group on the adenine, then the distances between the hypothetical NH_2 and these same drug atoms would be 3.1, 3.1, 2.9, and 2.9 A°. They are much too close for van der Waals contacts, and that is why the drug does not bind the G-C pairs.

In summary, the way in which the drug binds to the DNA and reads it is shown in Fig. 12.13. The three heavy lines are the three-center hydrogen bonds between an amide and two hydrogen bonding acceptors, nitrogen or oxygen, on the floor of the minor groove. They provide orientation for the

TABLE 12.1. Close Approach of Netropsin to Adenine Rings

		Distances, netropsin atom to:	
Netropsin atom	DNA adenine	Adenine ring C2	Hypothetical amine N2
C2	A5	3.90	3.05
C5	A6	4.06	3.11
C11	A18	3.87	2.87
C16	A17	3.94	2.92

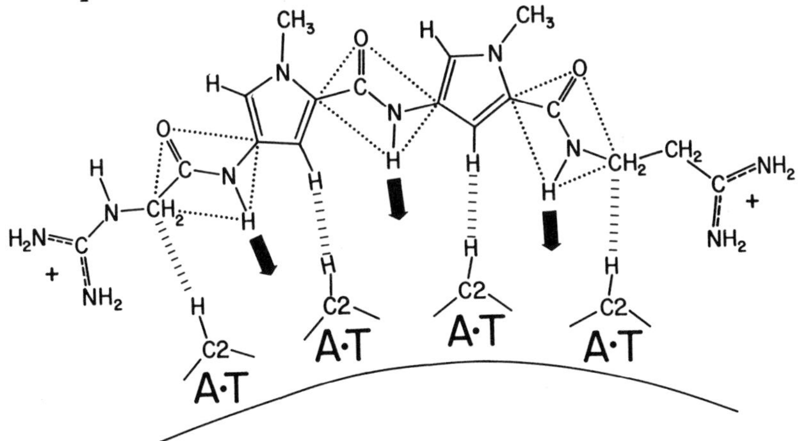

FIG. 12.13. Schematic of binding of netropsin to four successive A-T base pairs. Dark downward-pointing arrows are bifurcated H bonds to adenine N3 and thymine O2 on floor of minor groove. Four 'ladder' interactions are close van der Waals nonbonded contacts between adenine C2 hydrogens and either methylenes or pyrrole CH of the drug molecule. These contacts are so tight that a -NH_2 group of guanine could not be accomodated.

drug along the DNA. The dashed ladders are the four close contacts between C2 of an adenine and the drug. Drug binding is prevented if an NH_2 group is there instead. However, we thought, if this drug is A-T specific, can one synthesize a drug which would be G-C specific, at least in part? Could one design an artificial netropsin which would recognize any desired short DNA sequence? You can make netropsin analogs of longer and shorter length. How could you get them to accept G-C base pairs at certain points?

The answer is to substitute an imidazole or a furan for the pyrrole ring. If, as shown in Fig. 12.14, the first pyrrole is replaced by imidazole, the distances will be just right for a hydrogen bond to the guanine-NH_2. This molecule then would prefer a G-C base pair at that point. In Gdansk, Poland, at a meeting on the chemical basis of chemotherapy last July, I mentioned this idea of substituting furan or imidazole for pyrroles in netropsin. In the question period I was stunned to hear J. William Lown from Alberta get up and say, 'We have

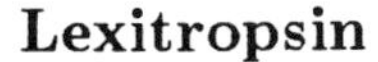

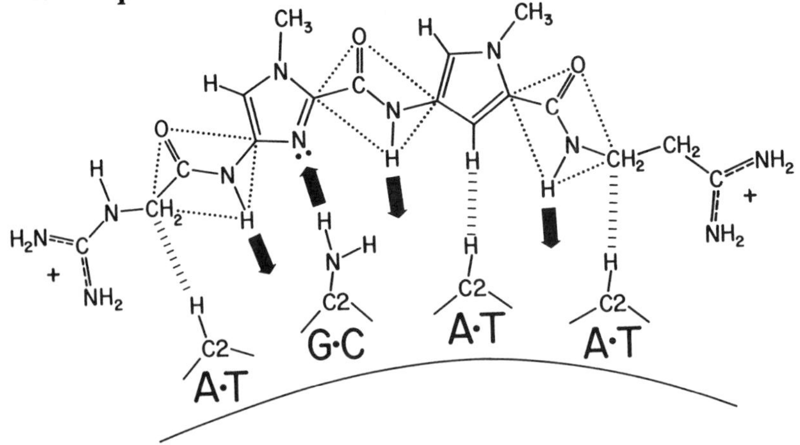

FIG. 12.14. Hypothetical binding of a 'lexitropsin', a sequence-reading netropsin analogue capable of accepting a G-C base pair at the second position of the binding site. A former tight nonbonded contact to a pyrrole ring has now been replaced by a new hydrogen bond between an imidazole ring and the guanine -NH_2 group. The drug molecule should not only accept a G-C base pair; it should prefer it.

had exactly that same idea by looking at CPK models, and we have made 18 of your lexitropsins already'. When the shock was over, we realized we had the basis of a great collaboration. Lown will make the lexitropsins. We will make the DNA. He will look at the DNA. He will look at the complexes with NMR, and we will look at them with X-rays. In the meantime, he has sent some of these lexitropsins out for cell-culture testing. The diimidazole lexitropsin seems to have promising properties against Herpes Simplex I and II. A netropsin in which Lown simply spliced two ends together to make a dimer is now being tested at M.D. Anderson Hospital in Texas against some forms of leukemia.

So it might be possible that by changing the base specificity, we can tame this toxic drug and make it clinically useful. In addition, our wild hope is that if we can target a drug against any short DNA sequence we like, it may be possible to target it to an invader cell rather that those of the host, or a neoplastic cell in reference to a normal cell. These lexitropsins may be reasonable vectors for their own sake, or for the sake of something that can be covalently linked to them. We could do what Peter Dervan has done. We could attach a propyl EDTA-iron to one end or the other of one of our lexitropsins and turn it into a miniature restriction enzyme, cutting any sequence of a DNA that we like. Peter has already done this for distamycins, and he can selectively cleave one side or the other of poly A-T sequences. It would be nice to write down a sequence, go into the laboratory, make a lexitropsin, link EDTA to it, and let it chew up the DNA anywhere you want it.

Other DNA-binding drugs

As a final postscript, we have embarked on a program at UCLA to study a series of groove-binding antibiotics. The first one is Mary Kopka's work on netropsin. Chun Yoon is working on bleomycin, a DNA-binding bisthiazole with a peptide-like tether as shown in Fig. 12.15(c). Unlike many, we do not necessarily believe that bleomycin is an intercalator. All the evidence that suggests that it is an intercalator also can be interpreted otherwise. Yoon has not yet been able to make crystals of the

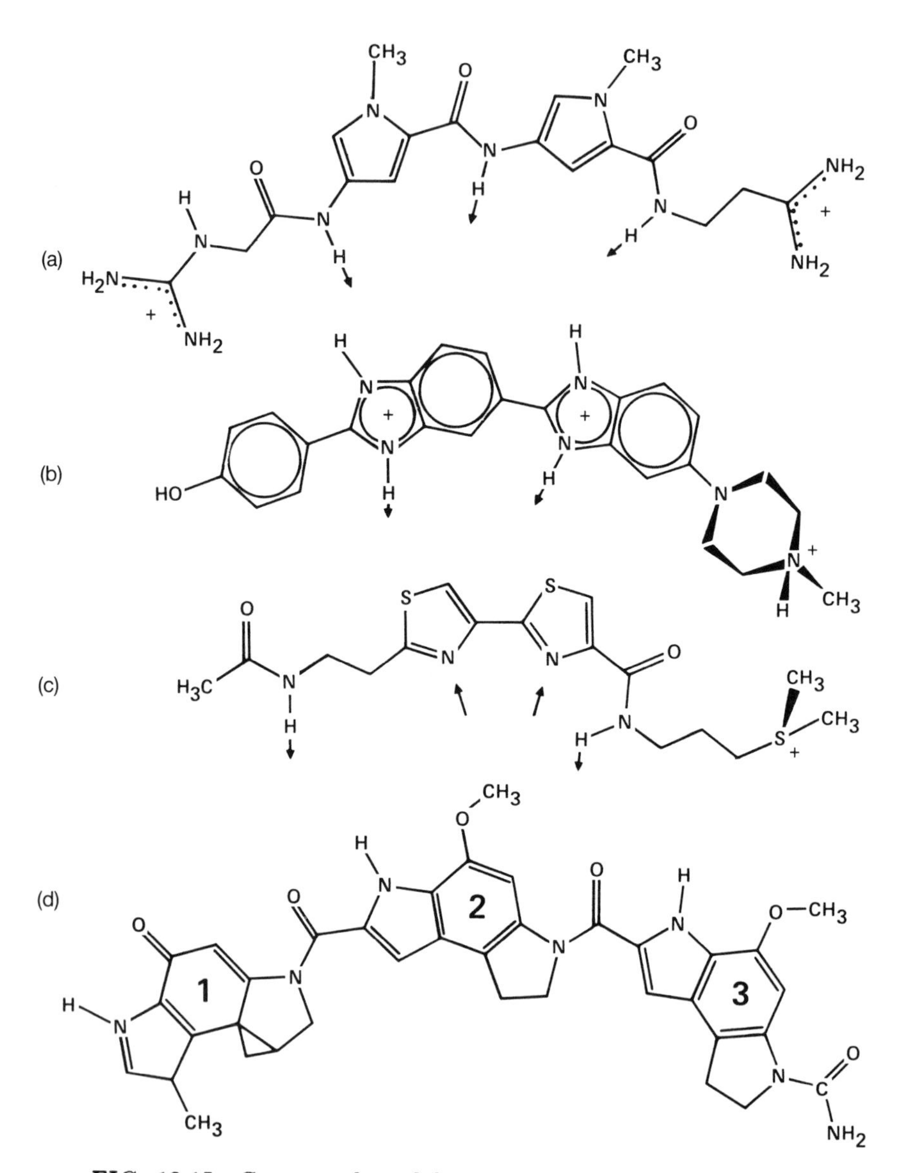

FIG. 12.15. Crescent-shaped heterocyclic molecules, all of which interact strongly with the B-DNA double helix and are thought to do so by displacing the spine of hydration from the minor groove. (a) netropsin, (b) Hoechst 33258, (c) the bisthiazole fragment of bleomycin, (d) Upjohn CC-1065. Arrows up and down indicate donor-to-acceptor directions for possible hydrogen bonds to DNA.

entire molecule of bleomycin; however, the portion which does bind has been crystallized.

In bleomycin there are five groups which provide five of the six coordination sites around the iron to make an EDTA-like cage. The sixth octahedral position is where O2 binds. You do some free radical chemistry, and the drug cleaves the DNA backbone chain. So bleomycin is not simply an inhibitor, it is a cleavage agent. Chun is co-crystallizing the bisthiazole group with the 12-mer used for the netropsin complex. Yoon, Goodsell and Prive have done a theoretical computer energy refinement of the binding of the bisthiazole to the minor groove of the DNA in a way that would explain the observed G – pyrimidine preference of binding of bleomycin. However it has not yet been checked experimentally.

Tammy Smith is an organic chemistry graduate student who is working half with our group and half with Orville Chapman. She intends to synthesize some lexitropsin-like molecules. Chapman suggested that you might get entropy working for you if one were to include a fused ring system, so that the compound will be more rigid than netropsin. So Tammy, with Chapman's tutelage, is developing ideas for making multiple-ring analogs of netropsin. One exists naturally, Hoechst 33258 (Fig. 12.15(b)). Attempts are now in progress to try to co-crystallize the 12-mer with Hoechst 33258 and with CC1065 (Cancer Culture 1065) from Upjohn (Fig. 12.15(d)). Again, this has a multiple fused ring system and is a groove binder which we believe will react more or less like the netropsins.

All organic drugs whose binding to DNA has been worked out have involved the minor groove. This includes our work, daunomycin and adriamycin done by Andy Wang, and triostin A and echinomycin, also from the MIT lab. NMR studies of drug binding have also shown that groove-binding drugs favor the minor groove and tend to be A-T specific. So what we see with netropsin is a good model for almost all groove-binding antibiotics. On the other hand, all of the literature on repressor binding to DNA has suggested that it is the *major* groove that the control protein sees. I have a feeling that this is going to be a general principle, and that the major groove of the B-DNA double helix will be the locus for binding of recogni-

FIG. 12.16. Low resolution diagram of the interaction of the Eco R1 restriction endonuclease with a substrate dodecamer, CGCGAATTCGCGT. Individual amino acid residues in the protein, or individual phosphates, sugars and bases in the DNA, have been approximated by spheres. Dark spheres are DNA; light ones are protein. Note how the DNA lies in a cleft or groove in the surface of the enzyme with its major groove facing the enzyme and how two arms of the enzyme wrap around the DNA from either side, following the major groove. From reference 19.

tion proteins like repressors; whereas the minor groove will be the site of binding and recognition by drug molecules. This is shown nicely in Fig. 12.16 which depicts John Rosenberg's EcoRI enzyme wrapping its arms around our twelve base pair DNA, keeping the arms in the major groove with the minor groove totally untouched.[19]

One of the sequences that we have been considering is CGTGTTAACACG, and we have been looking at it from the standpoint of multiple use. It takes a long time to make this length of DNA, pure and in 50 mg quantities, so you think twice before you start making a particular sequence. However, this sequence is the one that we are considering for the lexitropsins, for distamycin-5 work in collaboration with Dervan at Cal Tech, CC1065, bleomycin and so forth. Our goals over the next five years will be to synthesize several different oligomers, look at the complexes with different groove-binding drugs, and try to look for general principles of the sort that we think we may have found in this first binding of netropsin to the DNA.

Discussion

Olga Kennard: Have you located the water molecules yet in the netropsin complex?

Richard Dickerson: Yes, we have located the water molecules in the minor groove, along the phosphates; the distribution is not seriously different from that in the uncomplexed oligomer. Of course, they are replaced in the spine of the nucleotide.

Olga Kennard: I thought it very interesting that you apparently find very little change in the DNA itself with the groove–binding drug; on the other hand, the intercalating drug, triostin A, which Andy [Wang] and Alex [Rich] have just solved, contains, very surprisingly, a tremendous change in the DNA itself. I wondered if you would like to comment on this?

Richard Dickerson: Yes. The Hoogsteen base pairing in the center of the polynucleotide is the most dramatic change when the triostin A complex is formed. The classical helical parameters, such as torsion angles and local twist angles have changed very little. There are two main changes when netropsin binds. The groove is opened by as much as 1.5 Å, and it is opened most at the propylamidinium end of netropsin. Thus the groove is not quite wide enough to accommodate the drug, and therefore the drug must push its way in. It also bends the helix axis back by 8° as it binds. In other words, netropsin forces its way into the groove a small amount, but without causing any systematic variations in the classically accepted helix parameters. This means that it has not escaped our notice (to paraphrase a more famous paper) that we're going to have a far easier time detemining the structures of a dozen different groove–binding drugs bound to portions of DNA than is Andrew Wang when he studies a dozen different intercalators, each of which will twist and turn and extend the helix and make every DNA packing different.

George Jeffrey: One thing about three-center bonds which you might like to think about. If you are going to get a drug molecule in, you must take the water molecules out. If you slide water molecules across a hydrogen bond donor or acceptor surface using both two- and three-center hydrogen bonds, the potential energy surface is going to be much smoother than if there were only two-center bonds being formed.

Richard Dickerson: Yes, this is a Pangloss model of DNA structure all over again. Yes, that's exactly what you would like.

Louis Marzilli: I have a question about the A-T grooves. What about having runs of As and Ts, and alternating As and Ts and determining the structure. The water structure might stabilize, say, six As binding six Ts each.

Richard Dickerson: We, of course, have only looked at AATT, which is intermediate between polydA, polydT and alternating polydA-polydT. The solution chemists tell us that netropsin binds better to a polydA-polydT than to a strict alternation. I thought this might be because the alternating dAdT does have a slightly different structure and maybe the spine is differently arranged. We have been considering whether one of the sequences we should make would have a strictly alternating ATAT in the center, and look at the binding to that. However, we don't have any data yet.

Unknown speaker: Do you understand the enthalpy in netropsin?

Richard Dickerson: I do and I don't. Rufus Lumry years ago proposed the idea of entropy-enthalpy compensation. The reason why variations in enzymatic reactions tend to have a common ΔG is that if you change the situation so that you have to break more hydrogen bonds, the result is that the enthalpy becomes less favorable. You get more disorder, and that makes the entropy more favorable. So there's compensation. Breslauer has looked at polyA-polyT, and compared it with alternating ATAT and measured the thermodynamics of binding. I had a beautifully intricate theory that would explain that in terms of a poor spine of hydration in the ATAT. However, Breslauer called me up two weeks ago and said 'Don't say anything about that, because we have some more data which I'll send you which says it isn't true.' So the answer is in limbo– I just don't know at the moment.

Unknown speaker: What about the enthalpy?

Richard Dickerson: Well, in this model, if the spine was indeed defective in the alternating AT, because you have a Klug-type alternating B structure, there would not be as much of an enthalpy price to pay in binding netropsin as in the polyA-polyT with a large number of hydrogen bonds to break. But we only have one data point. I can tell you how many water molecules you must pay an enthalpy price for here, but I have nothing with which to compare it.

References

1. Shakked, Z., Rabinovich, D., Cruse, W.B.T., Egert, E., Kennard, O., Sala, G., Salisbury, S.A. and Viswamitra, M.A. *Proc. Roy. Soc. (London)* **B213**, 479–487 (1981).
2. Wing. R.M., Drew, H.R., Takano, T., Broka, C., Tanaka, S., Itakura, K. and Dickerson, R.E. *Nature (London)* **287**, 755-758 (1980).
3. Wang, A.H.J., Quigley, G.C., Kolpak, F.J., Crawford, J.L., van Boom, J.J., van der Marel, G. and Rich, A. *Nature (London)* **282**, 680- 686 (1979).
4. Drew, H.R., Takano, T., Tanaka, S., Itakura, K. and Dickerson, R.E. *Nature (London)* **286**, 567-573 (1980).
5. Drew, H. R. and Dickerson, R. E. *J. Mol. Biol.* **151**, 535-556 (1981).
6. Kopka, M. L., Fratini, A. V., Drew, H. R. and Dickerson, R. E. *J. Mol. Biol.* **163**, 129-146 (1983).
7. Leslie, A. G. W., Arnott, S., Chandrasekaran, R. and Ratliff, R. L. *J. Mol. Biol.* **143**, 49-72 (1980).
8. Zimmer, C., Luck, G., Birch-Hirschfeld, E., Weiss, R., Arcamone, F. and Guschlbauer, W. *Biochim. Biophys. Acta* **741**, 15-22 (1983).
9. Luck, G., Zimmer, C., Reinert, K. E. and Arcamone, F. *Nucl. Acids Res.* **4**, 2655-2670 (1977).
10. Schultz, P. G. and Dervan, P. B. *J. Biomol. Struct. Dyn.* **1**, 1133-1147 (1984).

11. Berman, H. M., Neidle, S., Zimmer, C. and Thrum, H. *Biochim. Biophys. Acta* **561**, 124-131 (1979).
12. Wartell, R. M., Larson, J. E. and Wells, R. D. *J. Biol. Chem.* **249**, 6719-6731 (1974).
13. Kopka, M. L., Pjura, P., Yoon, C., Goodsell, D. and Dickerson, R. E. In *Structure and Motion: Membranes, Nucleic Acids and Proteins* (E. Clementi, and R. Sarma, eds.) pp. 413–432. Adenine Press: New York, NY (1985).
14. Kopka, M. L., Yoon, C., Goodsell, D., Pjura, P. and Dickerson, R.E. *Proc. Natl. Acad. Sci. USA* **82**, 1376-1380 (1985).
15. Kopka, M. L., Yoon, C., Goodsell, D., Pjura, P. and Dickerson, R.E. (1985). *J. Mol. Biol.*, in press.
16. Jeffrey, G.A. and Maluszynyska, H. *Int. J. Biol. Macromol.* **4**, 173-185 (1982).
17. Zakrzewska, K., Lavery, R. and Pullman, B. In *Biomolecular Stereodynamics*. Vol. **1** (R. Sarma, ed.) pp. 163–183. Adenine Press: New York, NY (1981).
18. Baguley, B. C. *Molec. Cell. Bioch.* **43**, 167-181 (1982).
19. Frederick, C.A., Grable, J., Melia, M., Samudzi, C., Jen-Jacobson, L., Wang, B.-C., Greene, P., Boyer, H.W. and Rosenberg, J.M. *Nature (London)* **309**, 327-331 (1984).

13. DNA-protein interactions

B. W. Matthews, D. H. Ohlendorf,
W. F. Anderson and Y. Takeda

Brian Matthews is from the University of Oregon. He was born in a very small town in Australia. He did his undergraduate and his graduate work at the University of Adelaide, where he got his Ph.D. in Physics in 1964. He then went to the MRC lab in Cambridge where he did the structure of α-chymotrypsin with David Blow. Following that he spent two years at the NIH, after which he joined the University of Oregon where he is a Professor of Physics and a member of the Institute of Molecular Biology there. His interests are in the structure and function of proteins. He has solved the structure of thermolysin, a chlorophyll-containing protein, and a form of lysozyme from a bacteriophage. He has also made contributions on molecular evolution, and methods of protein structure determination. More recently he has been interested in DNA-binding proteins.

The structures of three proteins that regulate gene expression have been determined recently and suggest how these proteins may bind to their specific recognition sites on the DNA. One protein (Cro) is a repressor of gene expression, the second (CAP) usually stimulates gene expression and the third (λ repressor) can act as either a repressor or activator. The three proteins contain a substructure consisting of two consecutive α-helices that is virtually identical in each case. Structural and amino acid sequence comparisons suggest that this DNA-binding helix-turn-helix unit occurs in a number of proteins that regulate gene expression.

Although the main theme of this paper is DNA-protein interaction, I would like to begin by acknowledging the central role that the Patterson function has had, and continues to have, in the development of protein crystallography. Indeed, almost every protein structure solved to date depends, in one way or another, on the use of the Patterson function.

As is well known, the critical advance in the history of macromolecular structure determination was the successful application to protein crystals of the isomorphous replacement method by Green, Ingram and Perutz[1] in 1954. However the use of isomorphous replacement required that a way be found to locate the isomorphous heavy atoms and it is in this context that the Patterson method has played such a crucial role.

For a single protein crystal, the usefulness of the 'conventional' Patterson function, as introduced in Patterson's original paper[2] is limited by multiple overlap of the interatomic peaks. However by combining observed structure amplitudes, F_P, from the native protein crystals with structure amplitudes, F_{PH}, from heavy-atom-substituted crystals it was possible to calculate 'difference Patterson' functions that largely eliminated the overlap problem and showed the heavy atom vectors directly.

Although several variations of the difference Patterson function were proposed (*e.g.*, see Perutz[3]), the one that became most popular was actually the simplest, namely one with coefficients $(F_{PH} - F_P)^2$ (*e.g.*, see Blow[4,5] and Rossmann[6]). As can be seen in Fig. 13.1, the amplitude difference $F_{PH} - F_P$ is, in many cases, a poor approximation to the desired amplitude, F_H. Nevertheless, the Patterson function with coefficients $(F_{PH} - F_P)^2$ works remarkably well in practice and has been used in many successful structure determinations. My personal contribution to the subject[7] was to show that the desired magnitude of F_H (Fig. 13.1) can be obtained directly by combining measurements of anomalous scattering with isomorphous replacement measurements. This was also realized independently by Harding[8] and by Kartha and Parthasarathy.[9]

It should also be noted here that two other variations of the

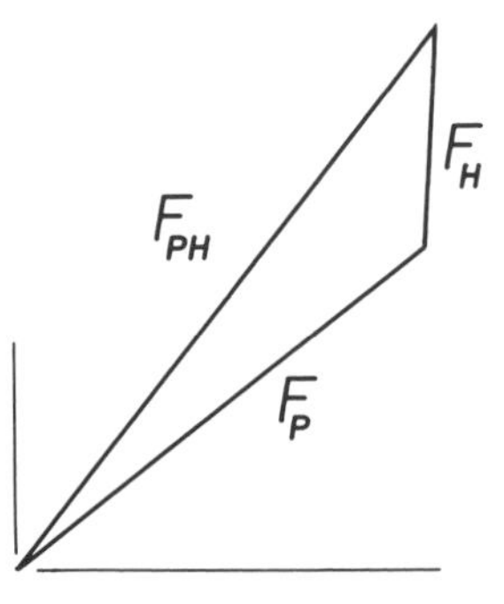

FIG. 13.1. Standard vector diagram showing, for a typical X-ray reflection, the relation between the protein scattering amplitude F_P, the isomorphic heavy atom contribution F_H and the combined scattering amplitude F_{PH} for the protein plus heavy atom.

original Patterson function, namely the 'rotation' and 'translation' functions (*e.g.*, see Rossmann[10]) have become of increasing importance in protein crystallography. These methods have made it possible to use 'known' protein structure to solve other related structures in completely different crystalline modifications (*e.g.*, see Remington *et al.*[11]).

Somewhat ironically (in the context of this meeting) the Cro repressor protein, which I now wish to discuss, is one of the very few protein structures to be solved *without* the use of Patterson methods. Rather, it was necessary to resort to a 'brute force' approach (Anderson *et al.*[12,13]).

DNA-protein interactions.

The control of the genetic information encoded in DNA is of critical importance in all living systems. It has been known for some time that, at least in simple organisms, this control is achieved by proteins ('switch molecules') that recognize and bind to specific sites on the DNA. Until recently, little was known of the structures of these molecules or the way in which they work. However, within the past few years, the three-dimensional structures of the three DNA-binding 'switch molecules' ('Cro,' 'CAP' and 'λ repressor') have been determined and have suggested how these proteins bind to their specific recognition sites on the DNA.

In this paper we will briefly review the structures of these three proteins and discuss the modes of DNA-protein interaction that are suggested. As will become apparent, there are similarities between the three proteins, but there are also striking differences as well. For additional background and more detailed information, reference can be made to the reviews of Ohlendorf and Matthews,[14] Takeda *et al.*,[15] and Pabo and Sauer.[16]

Three DNA-binding proteins

Cro, λ-repressor and CAP are all dimeric DNA-binding proteins but have substantial differences in their overall structures. For both CAP and λ- repressor, the respective polypeptides form two domains. In λ-repressor, the amino-terminal domains bind to the DNA whereas in CAP it is the carboxy-terminal part of the molecule that has this function.

Both Cro and λ-repressor bind to six similar but non-identical 17-base-pair sites on the chromosome of bacteriophage λ. The site at which Cro binds most tightly is the weakest binding site for λ-repressor, and *vice versa.* The two proteins are involved in the adoption by the phage of either the lytic or the lysogenic mode of development. Although this decision-making process is complicated and not yet understood in its entirety, the respective roles of Cro and λ-repressor have been analyzed in detail.[17] Cro is a straight-forward repressor of gene expression. It binds preferentially to the DNA at its highest affinity site O_R3 and, when bound, prevents transcription from the adjacent repressor maintenance promoter. λ-Repressor is more versatile. In common with Cro it can act as a repressor but it can also stimulate the expression of its own gene. CAP ('catabolite gene activator protein,' also known as 'cyclic AMP receptor protein') participates in the regulation of a number of genes in *Escherichia coli.* In the presence of cyclic AMP, CAP promotes transcription of these genes, and, in some circumstances, can act as a negative regulator as well. The DNA sequence that CAP recognizes is approximately 15 nucleotides long and more than ten such sites are known in *E. coli.*[18]

Structures of DNA-binding proteins

A sketch of the structures of Cro,[12] the amino terminal domain of λ-repressor,[19] and the carboxy terminal domain of CAP[20] as determined from the respective crystal structures is shown in Fig. 13.2.

The structure of Cro is quite simple, consisting of three α-helices (α_1, α_2, α_3) and a three-stranded antiparallel β-sheet. In the crystal, four polypeptides associate as a tetramer with approximate 222 symmetry. Ultracentrifugation indicates that Cro is dimeric in solution. This dimer is presumed to be the one shown in Fig. 13.2. Residues 55-61 of each monomer extend and lie against the surface of the other monomer. Phe 58, in particular, makes intimate hydrophobic contact with its partner subunit. The carboxy-terminal residues 62-66 are dis-

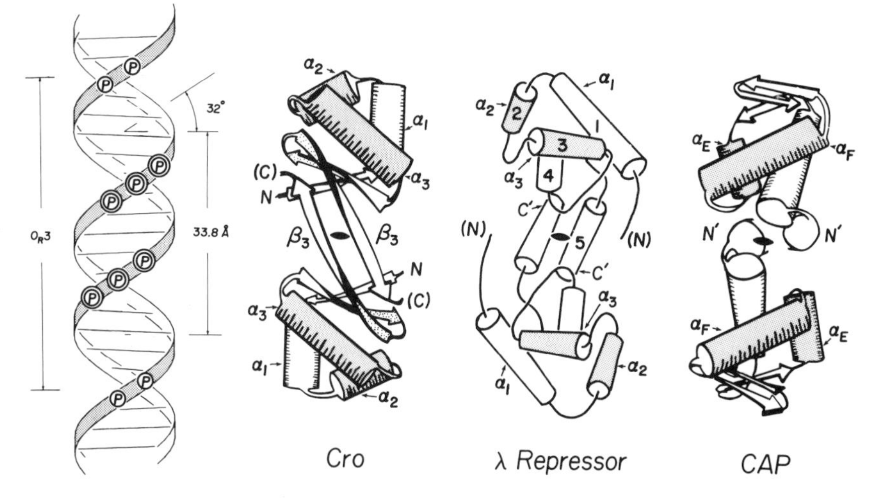

FIG. 13.2. Schematic drawing of a segment of Watson-Crick B-form DNA together with dimers of Cro, λ-repressor amino terminal domains and CAP carboxy terminal domains viewed down their respective two-fold symmetry axes. The corresponding $\alpha_2 - \alpha_3$ (or $\alpha_E - \alpha_F$) helices are shaded. DNA phosphates whose ethylation interferes with binding of both λ-repressor and Cro are indicated by the letter P within a double circle. Phosphates whose ethylation effects λ-repressor (and also P22 repressor) binding, but not Cro, are indicated by a P in a single circle (after refs. 14 and 15).

ordered in the crystals and, presumably, in solution as well.[12,21]

The DNA-binding form of intact λ-repressor is predominantly a dimer. Proteolytic cleavage separates the protein into an amino-terminal domain of 92 residues and a carboxy terminal domain consisting of residues 132-236. The amino-terminal domain is responsible for DNA recognition and can act as both a positive and negative regulator of transcription. Although monomeric in solution, the amino terminal domain dimerizes as it binds to DNA and protects exactly the same bases against chemical modification as intact λ-repressor. It is this amino terminal domain whose structure has been determined[19,22] and is shown in Fig. 13.2.

The structure contains five α-helices with no β-sheet. In the crystal, the two amino-terminal domains make contact via the fifth α-helix in each subunit (Fig. 13.2). Studies of mutants of λ-repressor suggest that a similar helix-helix contact may occur in intact λ-repressor.

The complex of CAP with cyclic-AMP, *i.e.*, the DNA-binding form of CAP, was shown by McKay and Steitz[20] to be a two-domain structure. Fig. 13.2 includes only the carboxy-terminal domains, *i.e.*, the presumed DNA-binding region. Not shown is the larger amino-terminal domain to which the cyclic AMP is bound. The structure in the crystals is dimeric with the amino-terminal domains related by a local two-fold axis. However, the carboxy-terminal domains adopt somewhat different conformations and are not exactly two-fold-related.[23,24]

Models for DNA binding

To date, the only structure that has been determined of a sequence-sequence DNA-protein complex is that of EcoRI endonuclease.[25] In the case of Cro, λ-repressor and CAP, it is not yet possible to visualize the protein-DNA complex directly, and one has to resort to inspection of the un-complexed structures and to model building.

For Cro protein, the 34 Å spacing between the two-fold-related α_3 helices together with their angle of tilt (Fig. 13.2), strongly suggests that these α-helices bind within successive major grooves of right-handed Watson-Crick DNA as illustrated in Fig. 13.3. It is presumed that the flexible carboxy-terminal residues of Cro participate in DNA binding by lying along the minor groove. A characteristic feature of the model is the match between the two-fold symmetry of the protein and the (approximate) two-fold sequence and spatial symmetry of the DNA binding site.[12,13,21,26] Recognition of specific base sequences on the DNA is thought to be due in large part to a multiple network of specific hydrogen bonds between the side chains of the protein and the parts of the DNA base-pairs exposed within the grooves of the DNA.

A similar mode of DNA binding has been proposed for λ-repressor. Here also there is a pair of two-fold-related α_3 helices (Fig. 13.2) that are presumed to bind within successive major grooves of Watson-Crick DNA. Furthermore, the amino-terminal residues of λ-repressor form two long 'arms' with flexible ends that 'wrap around' the DNA when the protein binds.

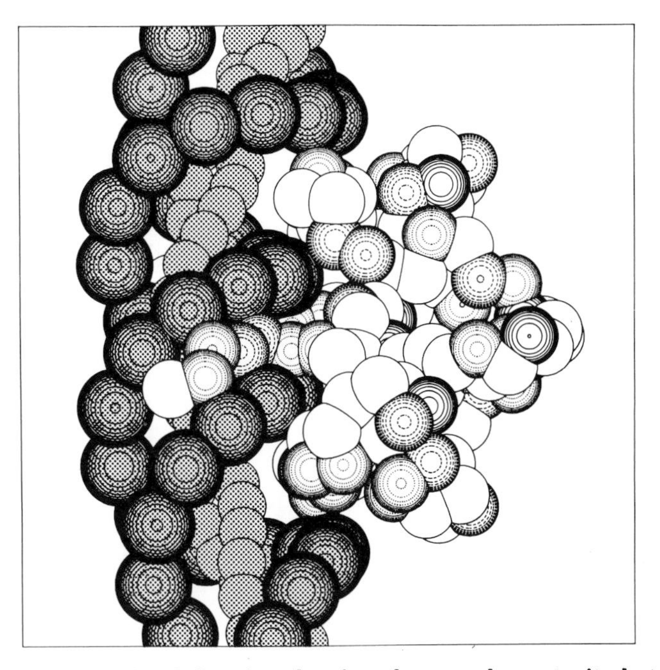

FIG. 13.3. Stylized drawing showing the complementarity between the structure of Cro repressor protein and DNA. In the presumed sequence-specific complex, the protein is assumed to move closer to the DNA, with the α_3 helices penetrating further into the major grooves of the DNA than is shown in the figure. The carboxyl terminal residues of the protein are presumed to bind in the vicinity of the minor groove of the DNA. In order to maximize the contacts between Cro and DNA, the protein may undergo a hinge-bending motion and/or the DNA may bend (as shown), although these are not essential features of the model. The DNA is represented stylistically by large dotted spheres centered at the phosphate positions and small dotted spheres that follow the bottom of the major groove. In the protein, each residue is represented by a single sphere. Acidic residues have solid concentric circle shading, basic residues have broken circle shading, uncharged hydrophilic residues have dotted circle shading and hydrophobic residues have no shading (after refs. 12 and 26).

In this case the 'arms' contact the major groove of the DNA.[27]

Current structural models[28,29] suggest that CAP also binds to DNA in a manner similar to that proposed for Cro and λ-repressor. The amino terminal parts of the α helices (Fig. 13.2) are placed within the major grooves of the DNA and the DNA is presumed to retain its normal right-handed twist. The model of CAP bound to right-handed DNA rather than left-handed[20] is consistent with several lines of evidence.

1) Experiments of Kolb and Buc[30] showing that there is no unwinding of the DNA when CAP binds to its specific binding sites.
2) Studies of mutants of CAP that recognize altered CAP binding sites.[31,32]
3) Electrostatic complementarity between CAP and DNA.[29]
4) Structural correspondences between CAP, Cro and λ-repressor (see below) suggesting that all three proteins might bind to DNA in similar modes.[14,19]

A common helix-turn-helix DNA-binding motif

Following the structure determinations of Cro, CAP and λ repressor, it has become apparent that these three proteins have features in common which extend to a number of other DNA binding proteins.

The suggestion that several DNA-binding proteins might have structural similarities came first from comparisons of their amino acid sequences. In some cases, such as Cro and λ-repressor, the sequence homology is poor, and was not apparent on first inspection. However, with additional sequences available, the overall homology becomes obvious (Fig. 13.4). The sequence homology includes not only repressors and activator proteins from different phages, but also other DNA binding proteins such as the *lac* and *trp* repressors from *E. coli*.[16,33–37]

The region of best sequence homology occurs within the parts of the sequences that align with the α_2 and α_3 helices of Cro and λ-repressor *i.e.,* within the part of the respective proteins that are assumed to interact with the DNA. Thus, it is reasonable to infer that the homologous proteins contain an α-helical DNA-binding supersecondary structure similar to the α_2 - α_3 fold seen in Cro and λ-repressor. Known mutants of *lac* repressor are consistant with such a hypothesis and additional support in this case has subsequently come from NMR studies.[38–40]

α_2 α_3

Protein	Start	Sequence	End
cro λ	14	Phe-Gly-Gln-Thr-Lys-Thr-Ala-Lys-Asp-Leu-Gly-Val-Tyr-Gln-Ser-Ala-Ile-Asn-Lys-Ala-Ile-His	35
cro 434	16	Met-Thr-Gln-Thr-Glu-Leu-Ala-Thr-Lys-Ala-Gly-Val-Lys-Gln-Gln-Ser-Ile-Gln-Leu-Ile-Glu-Ala	37
cro P22	11	Gly-Thr-Gln-Arg-Ala-Val-Ala-Lys-Ala-Leu-Gly-Ile-Ser-Asp-Ala-Ala-Val-Ser-Gln-Trp-Lys-Glu	32
cI λ	31	Leu-Ser-Gln-Glu-Ser-Val-Ala-Asp-Lys-Met-Gly-Met-Gly-Gln-Ser-Gly-Val-Gly-Ala-Leu-Phe-Asn	52
cI 434	16	Leu-Asn-Gln-Ala-Glu-Leu-Ala-Gln-Lys-Val-Gly-Thr-Thr-Gln-Gln-Ser-Ile-Glu-Gln-Leu-Glu-Asn	37
c2 P22	19	Ile-Arg-Gln-Ala-Ala-Leu-Gly-Lys-Met-Val-Gly-Val-Ser-Asn-Val-Ala-Ile-Ser-Gln-Trp-Glu-Arg	40
cII λ	24	Leu-Gly-Thr-Glu-Lys-Thr-Ala-Glu-Ala-Val-Gly-Val-Asp-Lys-Ser-Gln-Ile-Ser-Arg-Trp-Lys-Arg	45
cII 434	24	Leu-Gly-Thr-Glu-Lys-Thr-Ala-Glu-Ala-Val-Gly-Val-Asp-Lys-Ser-Gln-Ile-Ser-Arg-Trp-Lys-Arg	45
cI P22	23	Arg-Gly-Gln-Arg-Lys-Val-Ala-Asp-Ala-Leu-Gly-Ile-Asn-Glu-Ser-Gln-Ile-Ser-Arg-Trp-Lys-Gly	44
LacR	4	Val-Thr-Leu-Tyr-Asp-Val-Ala-Glu-Tyr-Ala-Gly-Val-Ser-Tyr-Gln-Thr-Val-Ser-Arg-Val-Val-Asn	25
CAP	167	Ile-Thr-Arg-Gln-Glu-Ile-Gly-Gln-Ile-Val-Gly-Cys-Ser-Arg-Glu-Thr-Val-Gly-Arg-Ile-Leu-Lys	188
galR	2	Ala-Thr-Ile-Lys-Asp-Val-Ala-Arg-Leu-Ala-Gly-Val-Ser-Val-Ala-Thr-Val-Ser-Arg-Val-Ile-Asn	23
trpR	66	Met-Ser-Gln-Arg-Glu-Leu-Lys-Asn-Glu-Leu-Gly-Ala-Gly-Ile-Ala-Thr-Ile-Thr-Arg-Gly-Ser-Asn	87
Matα1	115	Lys-Glu-Lys-Glu-Glu-Val-Ala-Lys-Lys-Cys-Gly-Ile-Thr-Pro-Leu-Gln-Val-Arg-Val-Trp-Val-Cys	136

FIG. 13.4. Segments of the amino acid sequences of a number of gene-regulatory proteins that appear to be homologous with the helix-turn-helix ($\alpha_2 - \alpha_3$) unit of Cro, λ-repressor and CAP. Amino acids that are identical in two or more sequences are underlined. The symbols at the top of the figure indicate the locations of the residues in the helix-turn-helix unit of Cro; open circles indicate full exposure to solvent, half open indicate part exposure and solid circles indicate buried residues. Stars indicate presumed DNA-contact residues (based on refs. 35 and 36).

As well as the above sequence relationships, it was found that Cro and CAP have a striking structural correspondence in their presumed DNA binding regions.[41] The three α-helices

(α_D, α_E, α_F) in the carboxy-terminal domain of CAP can be approximately superimposed on the α_1, α_2 and α_3 helices of Cro. For the ($\alpha_E - \alpha_F$) helical units, the superposition is striking. There are 24 α-carbons in the respective units that superimpose with an average discrepancy of 1.1 Å. An exhaustive search through all protein structures in the Brookhaven Data Bank (Fig. 13.5) failed to find a similar two-helical unit.

It has also been shown for Cro and λ-repressor that their α_2 and α_3 helices, and parts of of their α_1 helices as well, spatially superimpose.[42] Again, as with Cro and CAP, it is the $\alpha_2 - \alpha_3$ helical units of the two proteins that have virtually identical conformations (Fig. 13.5).

The amino acid sequence comparisons and the structural comparisons both point to a special role for the two-helical '$\alpha_2 - \alpha_3$' unit in DNA recognition and binding. The mode of interaction of this unit with DNA, as inferred from the structure

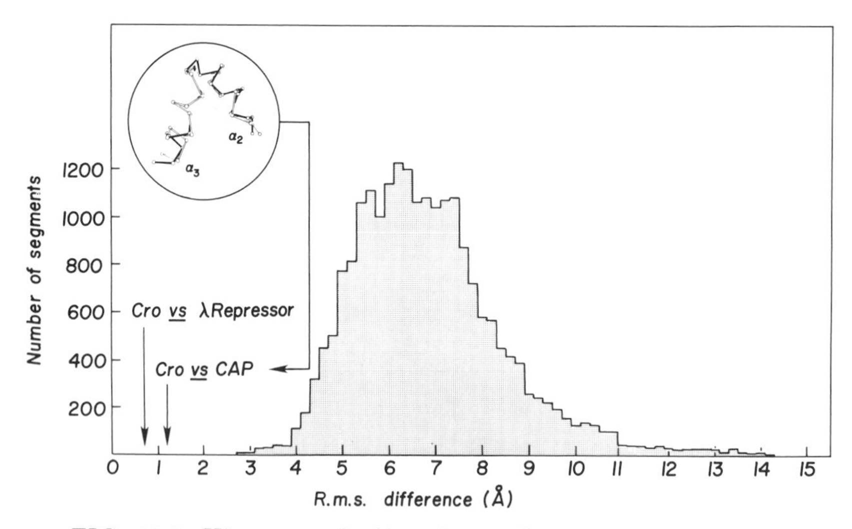

FIG. 13.5. Histogram showing the result of a search through all the protein structures in the Brookhaven Data Bank for a helix-turn-helix as initially seen in Cro and CAP (ref. 41). The root mean square difference is the discrepancy between the 22 α-carbon atoms in the $\alpha_2 - \alpha_3$ unit of Cro and 22 α-carbon segments taken successively from all parts of all known protein structures. The correspondence between the helix-turn-helix units of Cro and CAP is shown in the insert, and the root mean square difference between Cro and λ-repressor is also indicated (after refs. 21,41 and 42).

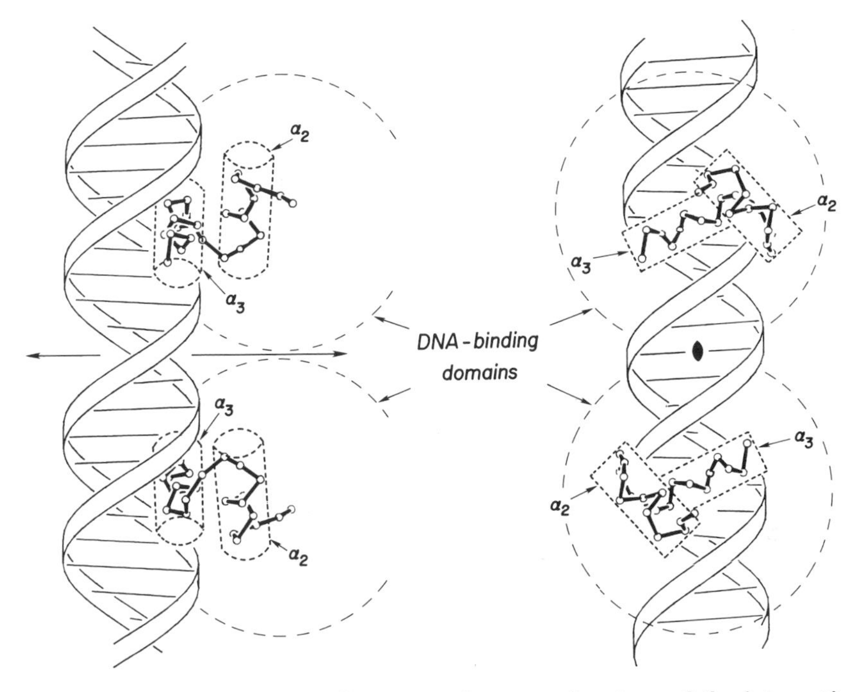

FIG. 13.6. The figure illustrates the general nature of the interaction presumed to occur in many DNA-regulatory proteins between a common $\alpha_2 - \alpha_3$ helical unit and right-handed B-form DNA. At left is a 'side view' with the two-fold axis of symmetry (arrowed) extending from left to right. On the right the view is 'face on' (after ref. 36).

of Cro, is sketched in Fig. 13.6. The α_3 helix occupies the major groove of the DNA with its amino acid side chains positioned so as to make sequence-specific interactions with the exposed parts of the DNA base pairs. Side chains of the α_2 helix are also presumed to contact the DNA, these interactions being primarily to the phosphate backbone.

It is reasonable to anticipate that similar although not necessarily identical modes of DNA binding will be found for a number of other gene regulatory proteins whose sequences have been shown to be homologous with Cro, λ-repressor and CAP.

Evidence in support of the structural models

Considerable evidence, some very recent, provides support for

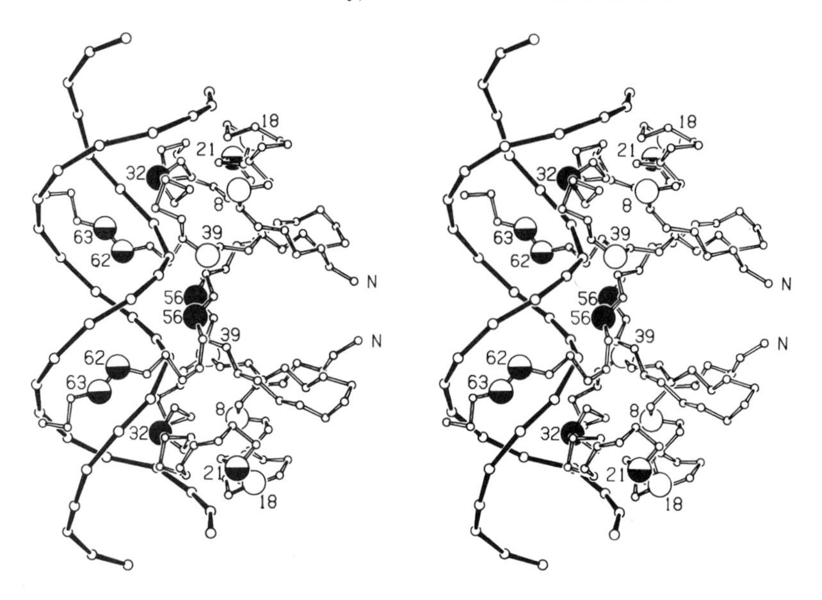

FIG. 13.7. Protection of Cro lysine residues by DNA. The drawing shows the presumed sequence-specific complex of Cro with DNA. All lysine residues are numbered and are drawn as large circles. Solid circles show lysines that are fully protected from ethylation when Cro is bound to DNA; half-solid circles show lysines that are partially protected and open circles indicate that there is no protection of these residues (after ref. 43).

the general features of DNA-protein recognition as summarized above.

a) DNA protection and modification experiments of Johnson *et al.*[17] and Pabo *et al.*[19]

b) Lysine protection and carboxypeptidase digestion experiments for Cro.[43] Fig. 13.7 shows the lysine residues of Cro that are protected from alkylation when Cro is bound to DNA.[43] Consistent with the Cro-DNA complex proposed by Ohlendorf *et al.*,[26] lysines 56 and 32 are fully protected while lysines 21, 62 and 63 are only partially protected. The protection experiments suggest that Cro undergoes a conformational change when it binds to DNA. It also appears that lysines 62 and 63, which are located in the flexible C-terminal 'arms' of Cro, interact somewhat more tightly with the DNA in non-specific Cro-DNA complexes than they do in the complex of Cro with its specific operator site.[43]

c) NMR studies of Cro[44–46] and *lac* repressor.[38–40] In the case of *lac* repressor, it has been shown directly that this protein does indeed contain two α-helices in the position predicted for a DNA-binding helix-turn-helix unit.

d) Mutants of λ-repressor, Cro and CAP that eliminate or modify the ability to bind to DNA.[16,31,32,47–49]

e) Use of site-directed mutagenesis to construct Cro-proteins and phage-repressor proteins that have modified abilities to bind DNA.[50,51]

f) Electrostatics calculations. For CAP and Cro, the known three-dimensional structures have been used to calculate the electrostatic potential surface surrounding the respective proteins. In the case of Cro there is a dominant, elongated positive region that coincides remarkably well, both in position and orientation, with the presumed DNA binding site.[52,53] For CAP there is also strong calculated positive potential in the presumed DNA binding region. The distribution is more consistent with CAP binding to right-handed rather than left-handed DNA.[24,29] For both CAP and Cro the electrostatics calculations suggest that the binding of the protein may cause bending or kinking of the DNA.[29,53]

Presumably the best way to confirm the detailed basis of DNA-protein recognition will be to determine the structures of specific DNA-protein complexes. Suitable crystals of such complexes have been obtained in several laboratories and detailed structural analyses are underway.[25,54,55]

Evolution of DNA-binding proteins

As discussed above, there is accumulating evidence that many sequence-specific DNA-binding proteins contain a common helix-turn-helix unit. However, one of the interesting aspects of this finding is that the helix-turn-helix unit occurs in very different locations within the polypeptide chains of different DNA-binding proteins. Fig. 13.8 shows five examples. We have chosen to include in the figure the three known structures, *i.e.*, Cro, λ-repressor and CAP, together with *lac* and *trp* repressors.

In the case of *lac* repressor there is very strong NMR evidence supporting the existence of the helix-turn-helix motif.[39,40] For *trp* repressor, the helix-turn-helix unit was predicted on the basis of amino acid sequence homology,[36] and is consistent with the location of mutants of *trp* repressor isolated recently by Yanofsky and coworkers,[56] although the existence of the helix-turn-helix unit has not been confirmed directly.

While the structural studies to date suggest that the helix-turn-helix unit has been strictly conserved during evolution, it is to be remembered that it is actually a dimer of helix-turn-helix units that binds to the DNA (Figs. 13.2 and 13.6). If the exact geometry of this dimeric arrangement was of critical importance for DNA binding, then it might be expected that the monomer-monomer interface would have been conserved during evolution. However, inspection of Figs. 13.2 and 13.8 show that this is not the case. In different instances the interface contacts can be either helix-helix or sheet-sheet. Also the contact residues can come from different parts of the

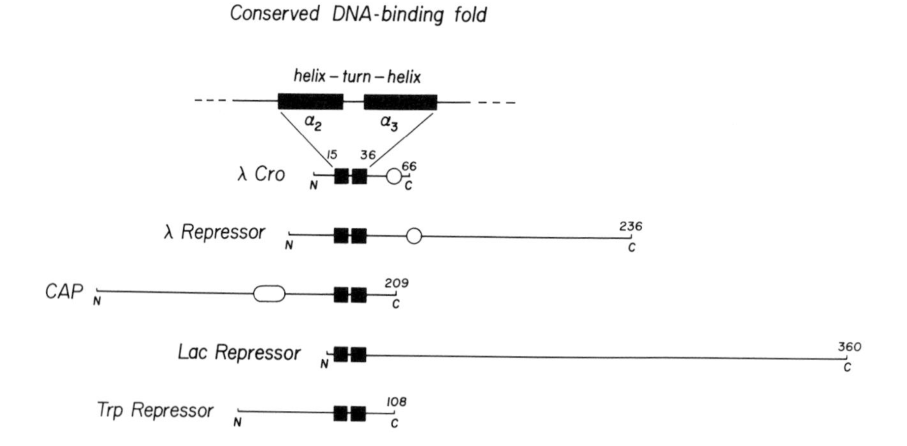

FIG. 13.8. Sketch showing the location (or presumed location) of the helix-turn-helix motif in selected DNA-binding proteins. In the case of Cro, λ-repressor and CAP, the open circles show the locations of the residues that are predominant in forming the dimer interface as seen in the respective crystal structures. (There are thought to be additional dimer contacts within the C-terminal domains of λ-repressor, but the structure of this part of the protein is not known.)

polypeptide backbone. What the DNA-binding domains of Cro, CAP and λ-repressor do have in common is the potential for flexibility.[21] Such flexibility could be of importance in 'sliding' along the DNA. Notwithstanding this flexibility, in the case of Cro and λ-repressor, at least, there still seem to be stereochemical restrictions that prevent the helix-turn-helix units of these two proteins binding to the DNA in exactly the same way.[42]

Another unresolved evolutionary question is posed by the recent structure determination for a DNA-binding protein from *Bacillus stearothermophilus*.[57] This protein does not recognize sequence-specific sites and is presumed to bind to the DNA in a manner very different to that for Cro, λ-repressor and CAP. Strikingly, this protein contains a pair of α-helices arranged in a manner that is very similar to the helix- turn-helix in Cro, CAP and λ-repressor. However, in this new DNA-binding protein structure, the helix-turn-helix unit is far from the apparent DNA-binding region of the protein. This paradox is one that remains to be resolved.

Acknowledgements

This work was supported in part by grants from the NIH (GM-20066 and GM-35114), the M.J. Murdock Charitable Trust and the NSF (PCM-8312151).

Discussion

Unknown speaker: Could the helix-turn-helix unit in the 'DNA-binding II' protein interact with DNA in the same way that is proposed for Cro and cause compaction of the DNA?

Brian Matthews: No, the model suggested for DNA-binding II is that the proteins bind side by side along the DNA and cause the DNA to curve in a type of superhelical coil. The helix-turn-helix units in this structure interact with each other, and face each other, and are not exposed to the DNA in the same way as is the case for Cro, CAP and λ-repressor.

Richard Dickerson: What do you think the function of the α_2- helix is; what is its character, how does it interact with the DNA or is its function just structural?

Brian Matthews: In each of the three protein structures that I've talked about, if you ask which are the parts of the protein that interact with the DNA, in all three it's the α_3-helix (or part of it) that appears to interact with the major groove. The α_2-helix makes interactions with the phosphate backbone or with other parts of the DNA apart from the major groove. It is the side chains of the α_2-helix rather than the backbone that interacts with the DNA.

George Rudkin: Given that DNA in life is associated with many proteins, do you have any insights into how these site-specific ones would find their way without being able to touch the DNA, as it were?

Brian Matthews: Well, there are different levels to your question. In the case of higher organisms, where the DNA is packaged in various sorts of ways, I think in that case the whole recognition process is potentially much more complicated, and I'm not sure at this point to what extent our studies will be directly relevant. But if you take a simpler case of, say of a bacterium, where you have a stretch of DNA and you have many repressors and polymerases and so on, and they're binding in many different places. I think there, it's much simpler. (*Draws on blackboard*). The best example is the case of *lac* repressor, where it has been shown by Peter von Hippel and others that the protein can diffuse along the DNA. This facilitates the location of specific sites on the DNA by reducing a three-dimensional search to a two-dimensional one. Now in part you're asking what happens if there's some other large protein that sits on the DNA and tends to prevent one-dimensional diffusion. One of the interesting observations in the case of *lac* repressor is that the protein is not just a dimer, but it's actually a tetramer. So it has two DNA binding regions, exactly similar, at opposite ends of the molecule. You could ask yourself, 'Why would nature go to the trouble of making a tetramer when it seems that a dimer might do?' One of the intriguing possibilities is that the DNA might bind to a segment of DNA using one end of the molecule, but the other DNA-binding end of the molecule would be free to bind to a second strand of

DNA, promoting 'hopping' or transfer from one strand to another. This would permit the protein to get out of a situation where it was trapped in a relatively short region.

Unknown speaker: Are there special charactersitics in the amino acid sequences of the helix-turn-helix units that might expedite the recognition of such sequences in other bacteria?

Brian Matthews: Yes. In this region there are a number of residues that seem to be important. First of all, there are some hydrophobic residues that are pointing toward the interior of the protein, and may in part tend to make contacts from one helix to the other. These seem to be rather well conserved in that large table that I showed you (Fig. 13.4). Then there is a glycine that has a conformation that is much more likely for a glycine than for residues with a β carbon, which is not impossible for other amino acids, but it's certainly much more likely for a glycine and in the table that I showed you, that glycine is conserved in all of those sequences. We've done searches where we've tried to actually use these requirements like not having prolines in the helices and so on, to improve the selectivity of the search.

References

1. Green, D. W., Ingram, V. M. and Perutz, M. F. *Proc. Roy. Soc. (London)* **A 225**, 287 (1954).
2. Patterson, A. L. *Z. Krist.* **A90**, 517 (1935).
3. Perutz, M. F. *Acta Cryst.* **9**, 867–873 (1956).
4. Blow, D. M., Ph.D. Thesis. University of Cambridge: Cambridge, UK (1957).
5. Blow, D. M. *Proc. Roy. Soc. (London)* **A 247**, 302–336 (1958).
6. Rossmann, M. G. *Acta Cryst.* **13**, 221–226 (1960).
7. Matthews, B. W. *Acta Cryst.* **20**, 230 (1966).
8. Harding, M. M., D. Phil. Thesis. Oxford University: Oxford, UK (1962).
9. Kartha, G. and Parthasarathy, R. *Acta Cryst.* **18**, 745–753 (1965).
10. Rossmann, M. G. (ed.) *The Molecular Replacement Method.* Gordon and Breach: New York, NY (1972).

11. Remington, S. J., Wiegand, G. and Huber, R. *J. Mol. Biol.* **158**, 111–152 (1982).
12. Anderson, W. F., Ohlendorf, D. H., Takeda, Y. and Matthews, B. W. *Nature (London)* **209**, 654 (1981).
13. Anderson, W. F., Takeda, Y., Ohlendorf, D. H. and Matthews, B. W., In *Structural Aspects of Recognition and Assembly in Biological Macromolecules.* (M. Balaban, J. L. Sussman, W. Traub and A. Yonath, eds.) pp. 911–917. Balaban ISS: Rehovot, Israel and Philadelphia, PA (1981).
14. Ohlendorf, D. H. and Matthews, B. W. *Annu. Rev. Biophy. Bioeng.* **12**, 259 (1983).
15. Takeda, Y., Ohlendorf, D. H., Anderson, W. F. and Matthews, B. W. *Science* **221**, 1020 (1983).
16. Pabo, C. O. and Sauer, R. T. *Annu. Rev. Biochemistry* **53**, 293 (1984).
17. Johnson, A. D., Poteete, A. R., Lauer, G., Sauer, R. T., Ackers. G. K. and Ptashne, M. *Nature (London)*, **294**, 217 (1981).
18. de Crombrugghe, B., Busby, S. and Buc, H. In *Biological Regulation and Development,* (B. K. Yamamot, ed.) Vol. **III**, p. 129. Plenum: New York, NY (1984).
19. Pabo, C. O. and Lewis, M. *Nature (London)* **298**, 443 (1982).
20. McKay, D. B. and Steitz, T. A. *Nature (London)* **290**, 744 (1981).
21. Matthews, B. W., Ohlendorf, D. H., Anderson, W. F., Fisher, R. G. and Takeda, Y. *Cold Spring Harbor Symp. Quant. Biol.* **47**, 427 (1983).
22. Lewis, M. Jeffrey, A., Ladner, R., Ptashne, M., Wang, J. and Pabo, C. O. *Cold Spring Harbor Symp. Quant. Biol.* **47**, 435 (1983).
23. McKay, D. B., Weber, I. T. and Steitz, T. A. *J. Biol. Chem.* **257**, 9518 (1982).
24. Steitz, T. A., Wever, I. T. and Matthew, J. B. *Cold Spring Harbor Symp. Quant. Biol.* **47**, 419 (1983).
25. Frederick, C. A., Grable, J., Melia, M., Samudzi, C., Jen-Jacobson, L. Wang, B-C., Greene, P., Boyer. H. W. and Rosenberg, J. M. *Nature (London)* **309**, 327–331 (1984).
26. Ohlendorf, D. H., Anderson, W. F., Fisher, R. G., Takeda, Y. and Matthews, B. W. *Nature (London)* **298**, 718 (1982).
27. Pabo, C. O., Krovatin, W., Jeffrey, A. and Sauer, R. T. *Nature (London)* **298**, 441 (1982).
28. Steitz, T. A., Weber, I. T., Ollis, D. and Brick, P. *J. Biomol. Struct. Dyn.* **1**, 1023–1037 (1983).

29. Weber, I. T. and Steitz, T. A. *Proc. Natl. Acad. Sci. USA* **81**, 3973–3977 (1984).
30. Kolb, A. and Buc, H. *Nucl. Acids Res.* **10**, 473 (1982).
31. Ebright, R. H., Cossart, P., Gicquel-Sanzey, B. and Beckwith, J. *Nature (London)* **311**, 232–235 (1984).
32. Ebright, R. H., Cossart, P., Gicquel-Sanzey, B. and Beckwith, J. *Proc. Natl. Acad. Sci. USA* **81**, 7274–7278 (1984).
33. Anderson, W. F., Takeda, Y., Ohlendorf, D. H. and Matthews, B. W. *J. Mol. Biol.* **159**, 745 (1982).
34. Matthew, B. W., Ohlendorf, D. H., Anderson, W. F. and Takeda, Y. *Proc. Natl. Acad. Sci. USA* **79**, 1428 (1982).
35. Sauer, R. T., Yocum, R. R., Doolittle, R. F., Lewis, M. and Pabo, C. O. *Nature (London)* **298**, 447 (1982).
36. Ohlendorf, D. H., Anderson, W. F. and Matthews, B. W. *J. Molec. Evol.* **19**, 109 (1983).
37. Weber, I. T., McKay, D. B. and Steitz, T. A. *Nucl. Acids Res.* **10**, 5085 (1982).
38. Arndt, K., Nick, H., Boschelli, F., Lu, P. and Sadler, J. *J. Mol. Biol.* **161**, 439 (1982).
39. Zuiderweg, E. R. P., Kaptein, R. and Wuthrich, K. *Proc. Natl. Acad. Sci. USA* **80**, 5837 (1983).
40. Zuiderweg, E. R. P., Billeter, M., Boelens, R., Scheek, R. M., Wuthrich, K. and Kaptein, R. *FEBS Lett.* **174**, 243–247 (1984).
41. Steitz, T. A., Ohlendorf, D. H., McKay, D. B., Anderson, W. F. and Matthews, B. W. *Proc. Natl. Acad. Sci. USA* **79**, 3097 (1982).
42. Ohlendorf, D. H., Anderson, W. F., Lewis, M., Pabo, C. O. and Matthews, B. W. *J. Mol. Biol.* **169**, 757 (1983).
43. Takeda, Y., Kim, J., Caday, C. G., Steers, E. Jr., Ohlendorf, D. H., Anderson, W. F. and Matthews, B. W. *J. Biol. Chem.* in press (1986).
44. Arndt, K. T., Boschelli, F., Cook, J., Takeda, Y., Tecza, E. and Lu, P. *J. Biol. Chem.* **258**, 4177 (1983).
45. Kirpichnikov, M. P., Kurochkin, A. V., Chernov, B. K. and Skryabin, K. G. *FEBS Lett.* **175**, 317–320 (1984).
46. Metzler, W. J., Arndt, K., Teeza, E., Wasilewski, J. and Lu, P. *Biochemistry* **24**, 1418 (1985).
47. Hecht, M. H., Nelson, H. C. M. and Sauer, R. T. *Proc. Natl. Acad. Sci. USA* **80**, 2676 (1983).

48. Nelson, H. C. M., Hecht, M. H. and Sauer, R. T. *Cold Spring Harbor Symp. Quant. Biol.* **47**, 441 (1983).
49. Sauer, R. T., personal communication.
50. Wharton, R. P., Brown, E. L. and Ptashne, M. *Cell* **38**, 361–369 (1984).
51. Eisenbeis, S. J., Nasoff, M. S., Noble, S. A., Bracco, L. P., Dodds, D. R. and Caruthers, M. H. *Proc. Natl. Acad. Sci. USA* **82**, 1084 (1985).
52. Ohlendorf, D. H., Anderson, W. F., Takeda, Y. and Matthews, B. W. *J. Biomol. Struct. Dyn.* **1**, 553–563 (1983).
53. Matthew, J. B. and Ohlendorf, D. H. *J. Biol. Chem.* **260**, 5860 (1985).
54. Anderson, W. F., Cygler, M., Vandonselaar, M., Ohlendorf, D. H., Matthews, B. W., Kim, J. and Takeda, Y. *J. Mol. Biol.* **168**, 903 (1983).
55. Anderson, J., Ptashne, M. and Harrison, S. C. *Proc. Natl. Acad. Sci. USA* **81**, 1307–1311 (1984).
56. Kelly, R. L. and Yanofsky, C. *Proc. Natl. Acad. Sci. USA* **82**, 483 (1985).
57. Tanaka, I., Appelt, K., Dijk, J. White, S. W. and Wilson, K. S. *Nature (London)* **310**, 376 (1984).

Concluding remarks

Christer E. Nordman

This being the end of the program I would like to say a word of appreciation for what has been a remarkable conference. We have seen a combination of the finest there is in up-to-date crystallography-based research and the finest there has been of the role of the value of the human aspect in science as we have experienced it through our associations with Lindo Patterson. I'm sure you will all join me in expressing our thanks to President Durant of the Fox Chase Center, to the Board of Directors of the ICR, to Betty Travaglini for coming up with the idea, and to Betty Patterson and to Jenny Glusker for organizing this beautiful celebration. Thank you very much.

Part III
Contributed Papers

A. METHODS

14. Patterson's 'symmetry map' interpretation of Patterson's function

Edward W. Hughes

As the attendants at this Symposium know full well, A. L. Patterson gave in 1934 his famous interpretation of $|F|^2$ Fourier Syntheses in terms of interatomic vectors. It seems much less well known that in 1949 and 1950 he gave a second interpretation of the same Patterson function. In this he called them 'symmetry maps'. The first paper[1] was with reference only to space group $P\bar{1}$. Unfortunately this paper was left out of the index under the author's name and got into the index only under the heading 'vector maps'. It appears to have been largely overlooked. In 1950, writing for the first time from his new address at this Cancer Research Institute, he produced a full paper discussing all possible space groups. This was read at the Pennsylvania State College meeting at which X-RAC was first demonstrated publically. It was finally published in 1952 in a large paper-back book, along with numerous other papers, put out by the Pennsylvania State College and edited by Professor Ray Pepinsky. This seems to be a rather rare book; I have seen only my own copy and know of no references to Patterson's paper. This paper is republished herewith[2] with the kind permisssion of Professor Pepinsky.

The first paragraph begins: 'The purpose of this paper is to show that the 'vector maps' in common use in X-ray crystallography can also be interpreted as 'symmetry maps',*i.e.,* as maps which indicate the degree to which any point, line, or plane of the crystal cell possesses any one of the symmetry operations of the space-group of the crystal.'

As an example, suppose we have a centro-symmetric molecule, such as anthracene, with its center at vector distance $\vec{R}$ from a crystal center in $P\bar{1}$. There will then be a similar molecule at $-\vec{R}$ and inspection of Patterson's Figure 7 will show that the $|F|^2$ series will show two peaks, at $\pm 2\vec{R}$, which will be as sharp as the origin peak and each with content one-half that of the origin peak. So by simple inspection one obtains the location of the molecule's center in the crystal structure. If however the anthracene has substituents, say 1 methyl, 2 ethyl, 9 propyl anthracene, and is also in $P\bar{1}$, we will still get in the $|F|^2$ series the two sharp peaks at $\pm 2\vec{R}$ but the substituents will not contribute to them, so their content will be less than one-half that of the origin peak. But still by inspection we can locate the center of the anthracene skeleton in the crystal structure. Similar results will obtain for other symmetries, in other space groups, but since the molecular symmetry elements, other than centers, must be oriented the same as those of the space group, it seems likely that centers will prove more commonly useful.

Organic chemists attempting to prepare triphenylfluorocyclobutadiene obtained a dimer, $(C_6H_5)_6F_2C_8$, and we were asked to establish the structure. The work was undertaken by Charles Fritchie, then a graduate student. He found a unit cell containing two of the above dimer molecules in space group $P\bar{1}$. The $|F|^2$ synthesis showed two sharp peaks at $\pm 2\vec{R}$ each with content about one-sixth that of the origin peak. Having only recently discussed these matters with Patterson I recognized at once that the C_8 group must have a pseudo-center governing the 8 C atoms and the 8 other atoms attached to C_8, a total of about one-third of the heavy atoms of the molecule. Thus by inspection alone we found the location of the center of C_8. By superimposing one $|F|^2$ plot with origin at $-2\vec{R}$ on one with origin at $+2\vec{R}$ and taking the smaller of the two densities at each point, one secures a plot due only to those atoms related by the C_8 centers. Fritchie wrote and debugged a computer program to do this, ran it and plotted the result in about two days! At the new origin of the modified $|F|^2$ series there now appeared an excellent approximation to symmetry $2/m$! So we had only to place $\frac{16}{4} = 4$ atoms to explain this new map. This proved to be easy. Subsequent difference maps produced the missing parts of the benzene rings and least squares refinement showed the substance to be *anti*-1,2-difluoro-3,4,5,6,7,8-hexaphenyltricyclo[4,2,0,0(2,5)]octa-3,7-diene. The central C_8 is:

```
8  1  2  3
C—C—C—C
||  | * |  ||
C—C—C—C
7  6  5  4
```

Although the four-member rings are individually nearly planar, they make angles with each other of about 113°, in the *anti* configuration. The asterisk marks the pseudo-center, the pseudo-2-fold axis is vertical in the plane of the paper, through the asterisk, and the pseudo-mirror plane is perpendicular to the 2-fold axis at the asterisk. These pseudo-elements apply not only to the C_8 group but also to the eight atoms directly attached to C_8.

Thus Patterson's 'Symmetry Map' interpretation of the Patterson function led us by straightforward and easy stages to the solution of a structure requiring 138 positional coordinates for the non-hydrogen atoms.

Reference

1. Patterson, A. L. *Acta Cryst.* **2**, 339–340 (1949)
2. Patterson, A. L. In *Computing Methods and the Phase Problem in X-ray Crystal Analysis.* (R. Pepinsky, ed.) pp. 29–42. Pennsylvania State College: State College, PA (1952). Reproduced here (page 701) by permission of Ray Pepinsky.

15. On the paper entitled 'Detection of inversion centers using the Patterson function'

S. C. Abrahams

The three-dimensional Patterson function of a crystal that contains n atoms in a unit cell, space group $P1$, contains $n^2 - n$ vectors. These are vector differences between the coordinates of the atoms. For the space group $P\bar{1}$ there are only $n^2/2$ vectors, of which n have coordinates *double* those of the atoms (*i. e.,* $2x, 2y, 2z$). Expressions, suitable for use on a computer, are developed based on this simple distinction. Routine identification of inversion centers by computer analysis of the Patterson function may hence be made.[1]

Thus, in $P1$ the Patterson function consists of $n^2 - n$ vectors with coordinates $x_i - x_j$, $y_i - y_j$, $z_i - z_j$. By contrast, the space group $P\bar{1}$ gives $n^2/2$ vectors and the presence of inversion centers requires that n vectors have coordinates $\pm 2x_j$, $2y_j$, $2z_j$.

In the method proposed for the detection of an inversion center the following procedure is recommended. The three-dimensional Patterson function is first calculated and all vector peaks in the unit cell greater than a preset level are located, resulting in a list of vector coordinates u, v, w. A second list of coordinates is then produced, in which the measured coordinate values from the Patterson function are replaced by a set with one half values $u/2$, $v/2$, $w/2$. The differences in magnitude between all possible combinations of pairs of coordinates on this second list are now systematically compared with each magnitude on the measured list. When agreement occurs the three sets of coordinates are listed. There should be a list of $4n(n-2)$ u, v, w peaks in space group $P\bar{1}$. In practice this list may not be complete, but it certainly gives an indication of whether the space group is centrosymmetric or not. Incidentally, it also results in a structure solution.

Reference

1. Abrahams, S. C. *Z. Krist.* **125**, 48–51 (1967).

16. A connection between the Patterson function and direct methods

Jerome Karle

The atomic arrangement in the unit cell of a crystal, which is repeated over and over in three-dimensions to form the macroscopic solid, could be found directly if the phases of the X-ray waves scattered by the crystal were preserved in a diffraction experiment. Normally, this is not the case and all that is observed are the intensities of scattering which are proportional to the magnitudes squared of the amplitudes of the scattered waves. Since the phases disappear in the collection of the experimental data, it had been rather generally thought that the values of the phases could not be recovered from the measured intensities and that therefore structures could not be determined directly by means of some analytical manipulation of the intensities. The development of the Patterson function in 1934[1,2] presented an opportunity to question such a view. The literature shows that there was indeed early questioning, even before the advent of the Patterson function, although, at a time when collecting and measuring large numbers of X-ray reflections was very difficult and tedious work and electronic digital computers were almost two decades away, serious thought about the solvability of the structure problem from the intensities alone was not widespread. It should be noted however that Ott[3] derived relationships for determining atomic positions by means of algebraic manipulations of the structure factor equations defined in equation (4) below, Banerjee[4] applied Ott's results in a trial and error self-consistency routine and Avrami[5] made further algebraic investigations in an effort to develop a practical formalism for deriving atomic positions from the measured intensities.

In order to see in what way the Patterson function impacts on the question of solvability, it is helpful to consider the mathematical represention by Fourier series of the electron density distribution, $\rho(\vec{r})$, in the unit cell of a crystal

$$\rho(\vec{r}) = \frac{1}{V} \sum_{\vec{h}=-\infty}^{\infty} F_{\vec{h}} \exp(-2\pi i \vec{h} \cdot \vec{r}), \tag{1}$$

where V is the volume of the unit cell, $F_{\vec{h}}$ is the structure factor which is proportional to the amplitude of the scattered X-ray wave associated with the crystal plane defined by the triple of indices $\vec{h} = (h, k, l)$ and $\vec{r} = (x, y, z)$ is a variable vector that locates the points in the unit cell. The structure factor is, in general, a complex number and may be written

$$F_{\vec{h}} = |F_{\vec{h}}| \exp(i\phi_{\vec{h}}). \tag{2}$$

It is the $|F_{\vec{h}}|$ that are evidently preserved in an experiment since the intensities of scattering are proportional to $|F_{\vec{h}}|^2$. The phases, $\phi_{\vec{h}}$, appear to be lost, giving rise to the so-called phase problem. It is apparent that both the magnitudes and phases of the structure factors are needed in order to calculate equation (1). If the phases as well as the magnitudes are available, the structure would immediately appear from the calculation of equation (1) because the main maxima of the electron density distribution occur at the positions of the atoms in the unit cell.

The Patterson function $P(\vec{r})$ is a Fourier series similar to equation (1) in which the coefficients $F_{\vec{h}}$ are replaced by the $|F_{\vec{h}}|^2$ whose values are directly obtainable from an X-ray diffraction experiment

$$P(\vec{r}) = \sum_{\vec{h}=-\infty}^{\infty} |F_{\vec{h}}|^2 \exp(-2\pi i \vec{h} \cdot \vec{r}). \tag{3}$$

In practice, the infinite range of $\vec{h}$ has to be replaced by the actual range of the measurements of the intensity data. The maxima of equation (3) were shown to be interpretable in terms of the vectors joining atomic positions (interatomic vectors), $\vec{r}_i - \vec{r}_j$, instead of in terms of the atomic positions, $\vec{r}_j$, alone as is the case for equation (1). For our present purposes, we note that for the number of diffraction data normally available, it is straightforward to obtain for the simpler structures a sufficient set of interatomic vectors from the calculation of equation (3) to permit the deduction from them of the values of the atomic positions, $\vec{r}$. This is the desired solution to the structure problem.

It is worthwhile to pursue this matter one step further, however, for conceptual purposes. The phases of the structure factors can be computed from the known structure by use of the definition of the structure factor

$$F_{\vec{h}} = \sum_{j=1}^{N} f_{j\vec{h}} \exp(2\pi i \vec{h} \cdot \vec{r}_j), \tag{4}$$

where $f_{j\vec{h}}$ is the atomic scattering factor for the jth atom in a unit cell containing N atoms. The $f_{j\vec{h}}$ are tabulated and, given values for $\vec{r}_j$, the $F_{\vec{h}}$, as described in equation (2), can be calculated. Equation (4) is seen to be the Fourier transform of equation (1) modified to take into account the discrete distributions of electrons associated with individual atoms.

We have thus shown by use of the Patterson function that, at least in some circumstances, phase information is not really lost in the diffraction experiment in the sense that it can be recovered from the measured intensities. The measured intensities are used in a Patterson function to determine interatomic vectors which in turn can be interpreted to give the desired structure. A further calculation of equation (4) by use of the known structural arrangement of atoms gives values for the phases. This can be diagrammed:

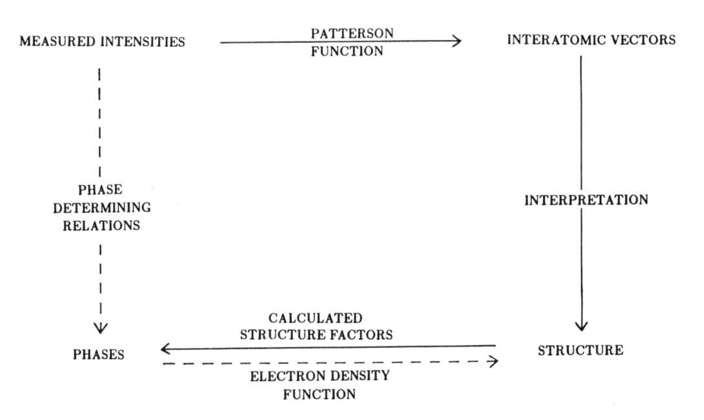

The solid arrows indicate the stepwise path described above. Since phase information is contained in the measured intensities, it is quite legitimate to consider the path indicated by the broken arrows, namely, the determination of phases by use of the intensities and the direct calculation of the structure from equation (1) by use of the measured structure factor magnitudes and the evaluated phases. A strong argument for pursuing the development of a practical procedure for the path described by the broken arrows can also be based on an analysis of the structure factor equations (4).[6]

The development of a practical procedure did in fact occur, starting with the inequality theory of Harker and Kasper[7] and soon followed by the general determinantal inequality theory of Karle and Hauptman.[8,9] The inequalities related phases and magnitudes and it was only necessary to recognize the probabilistic characteristics of the inequality relations to obtain the practical working formulas of the direct method of phase and structure determination. This was done in the early part of the 1950's. It is apparent that the Patterson function in the context of diagram (5) made a worthwhile contribution to the motivational atmosphere in which modern direct methods of phase determination were developed.

References

1. Patterson, A. L. *Phys. Rev.* **46**, 372–376 (1934).
2. Patterson, A. L. *Z. Krist.* **90**, 517–542 (1935).
3. Ott, H. *Z. Krist.* **66**, 136–153 (1928).
4. Banerjee, K. *Proc. Roy. Soc. (London)* **A141**, 188–193 (1933).
5. Avrami, M. *Phys. Rev.* **54**, 300–303 (1938).
6. Hauptmann, H. and Karle, *J. Phys. Rev.* **80**, 244–248 (1950).
7. Harker, D. and Kasper, J. S. *Acta Cryst.* **1**, 70–75 (1948).
8. Karle, J. and Hauptman, H. *Am. Mineral.* **35**, **123** (1950). (Abstracts of presentations at a meeting of the Crystallographic Society of America in Ann Arbor, Michigan, April 1949).
9. Karle, J. and Hauptman, H. *Acta Cryst.* **3**, 181–187 (1950).

17. Computer-aided analysis of multi-solution Patterson superpositions

James W. Richardson, Jr. and Robert A. Jacobson

A method is described in which Patterson superposition maps are computer-analyzed to obtain a 'trial structure' for subsequent refinement. Peaks in the superposition map are first fitted via Gaussian functions to obtain a list of potential atom possibilities. Space group symmetry is then extensively used to validate and subdivide this list into self-consistent subsets. The program is specifically designed to handle multi-solution cases and uses them to good advantage to obtain appropriately averaged results.

The procedure has been successfully applied to over 20 structures. Two examples are given: $(N(CH_3)_3CH_2(C_6H_5))_2Mo_5Cl_{13}$, whose structure proved to be intractable by other techniques, and $C_5H_5Fe(CO)_2(CS)PF_6$, which was solved by this program directly from the Patterson without even resorting to an initial superposition.

Techniques having the most general use for the determination of complex crystal structures have been, and continue to be, based on either direct methods or interpretation of the Patterson function. Direct methods have had considerable popularity in recent years since, when successful, they can be carried out in an automatic fashion using computers, with minimal intellectual effort. The practical reality of fully-automatic analysis of the Patterson function is complicated by the fact that one usually must work with a complex distribution function (the Patterson) which must be digitally represented using a large number of grid points.

We report here a Patterson method, utilizing space group symmetry, which lends itself to computer-aided analysis, while, in similarity to most direct methods, requiring minimal stoichiometric and no stereochemical information from other sources. As such, this approach is complementary to Patterson search techniques such as those used by Nordman[1,2,3] and Simonov,[4] which use stereochemical information to identify fragments of the structure.

It is well known that extensive use of space group symmetry is very important when solving crystal structures with

Patterson-based methods. In the simplest cases, for example, where only one or few heavy atoms are present in the asymmetric unit, a straightforward analysis of the Harker lines and planes will often provide the heavy atoms position(s) which will then be sufficient to initiate the structure determination. For light-atom structures or other nearly equal-atom structures, however, the situation is significantly more complicated.

Some of the earlier Patterson methods having greatest similarity to that proposed here,[5,6] involve the production of a symmetry (or symmetry minimum) map which contains all possible atomic positions consistent with the space group symmetry and with the original Patterson function. These maps were created by combining information from the many symmetry regions (Harker lines and planes, and inversion peaks) of the Patterson function, in such a way as to map Patterson functional values directly onto an electron density space grid. Potential atomic positions are those points which retain significant buildup of intensity from point-wise mappings of the different symmetry regions. Of course peaks on Harker planes, lines, etc. not due to symmetry-caused interactions can produce extraneous peaks on the symmetry map. In addition, since these symmetry maps are produced directly from the Patterson, any origin ambiguity is reproduced; the complete 'unit cell structure' is replicated by half-cell translation and other symmetry elements in the Laue group of the Patterson. This compression of information along with the overall qualitativeness of the symmetry maps can become a significant detriment for complex structures where overlap and/or pseudo-symmetry is present. We have endeavored therefore to develop a method which retains the generalities of Patterson-based methods but avoids these disadvantages.

Details of the Method

This method is distinguished by four important characteristics:

1) A Patterson superposition minimum function technique is used to significantly reduce the number of peaks present in the Patterson (and reduce the number of images of the structure to one or just a few); at the same time the high

symmetry of the Patterson is reduced to the usually lower symmetry of the space group.

2) The remaining peaks are fit using a Gaussian function to obtain the approximate positions of the peak centers effecting a conversion from functional values to a list of vectors between point-atoms.
3) These peak positions are then analyzed using space group (or space sub-group) symmetry to obtain sets of peaks that possess the proper positional inter-relations required by symmetry for symmetry elements at any particular assumed positions in space.
4) A common origin is determined between multiple solution images and these solutions are then averaged to obtain a 'best' solution. Each of these steps in the method is discussed in more detail below.

A Patterson superposition is carried out using a minimum function approach[7,8] usually selecting a peak whose position is relatively unaffected by neighboring peaks and which is significantly removed from the high symmetry planes and lines of the Patterson Laue group. The number of peaks is thereby significantly reduced as is the extraneous symmetry, while at the same time resolution is retained or improved. The superposition map can be viewed as a combination of a relatively few shifted images of the unit cell such that the respective viewing atoms are at the origin of the map.

If the vector selected for the superposition corresponds to a legitimate atom-atom interaction, then the function that is the resultant superposition map should represent a collection of possible atom positions. Therefore the next step is to fit each peak using a three-dimensional Gaussian function to obtain the 'best fit' to peak positions. This has the advantage of incorporating the usually excellent assumption that, to a good approximation, the Patterson density can be represented by a combination of discrete, Gaussian distributions. A list of potential atom positions is thus obtained which is readily amenable to further computer analysis.

The third step is to determine all possible positions for placement of symmetry elements, consistent with the peaks from the various images remaining in the superposition map.

The superposition map coordinates of an atom symmetry related to a second atom, within one of the images represented, can be calculated by

$$\begin{pmatrix} x' \\ y' \\ z' \end{pmatrix} = \left\{ (R) \left[\begin{pmatrix} x \\ y \\ z \end{pmatrix} + \vec{t}_{sup} \right] + \vec{t}_{sg} \right\} - \vec{t}_{sup}$$

where R is the point group rotation operation (3×3 matrix) related to the symmetry element and $\vec{t}_{sg}$ is its associated translation, while $\vec{t}_{sup}$ is the translation needed to shift that image from the superposition map origin to the space group origin. The purpose of this analysis is to determine the components of $\vec{t}_{sup}$. The method used is to systematically test all entries in the 'atom' list to deduce possibilities for translations ($\vec{t}_{sup}$) which could relate the atoms representing the head and tail of the shift vector to their potential symmetry related partners.

To complete step four, for each potential translational choice, the function $Q_s(\vec{t})$ is calculated:

$$Q_s(\vec{t}) = \int_V I \cdot I(s) d\tau$$

where $I(s)$ is the image after applying the symmetry operation and taking into account the origin shift. The calculation can be readily performed in reciprocal space using $G_{\vec{h}}$ values calculated from the 'atom' list obtained as noted above, via:

$$G_{\vec{h}} = \sum_j p_j exp(-2\pi i \vec{h} \cdot \vec{r}_j) / (\sum p_j^2)^{1/2}$$

and p_j is the peak height, $\vec{r}_j$ is the peak position and the sum is over all peaks above a prescribed minimum. Then:

$$Q_s(\vec{t}) = \sum G(-\vec{h}_s) G(\vec{h}) exp(2\pi i \vec{h} \cdot \vec{t}).$$

Changes in the positions of symmetry elements affect only the phase of $G(-\vec{h}_s)$.

The best choices of translation should give the highest Q values. To get the most sensitive discrimination between translational choices, atom positions are averaged using the symmetry, and a figure of merit is computed from the Q values, a conventional residual index value between observed and calculated structure factor magnitudes, and the average deviation between 'symmetry-related' atoms.

In general one would expect more than one solution to be present. Patterson superposition theory predicts that unless a single vector were chosen between two inversion-related atoms, one would expect at least two images of the structure, one inversely related to the other, to be present in the map. In general, multiple vectors are often chosen for shift vectors since the positions of their centers can often be more accurately determined. However this naturally results in multiple images being present. Our procedure recognizes such an occurrence and turns what would be a disadvantage into an advantage. The solutions exhibiting the best figures of merit are averaged together after adjusting to obtain a common origin (Note that origins defined by symmetry elements are often ambiguous as far as translations of a half in unit cell directions are concerned). Averaging tends to produce atomic positions that are more accurate and also helps identify extraneous positions which might not have been eliminated in one of the earlier steps.

Results

To illustrate the effectiveness of this procedure, two examples of its application are presented below.

Researchers over the years have typically assumed that structures containing *many* independent heavy atoms were virtually unsolvable using *ab initio* Patterson analysis. This is, of course, due to the extreme complexity of the Patterson function for such systems. In particular, Harker lines and planes are densely populated by large heavy atom-heavy atom interatomic vectors. We have discovered that performing a single Patterson superposition followed by careful analysis of the resultant map can, in fact, lead directly to a solution of these structures.

The structure of $(N(CH_3)_3CH_2(C_6H_5))_2Mo_5Cl_{13}$ is a case in point. This substance crystallizes in an orthorhombic unit cell, space group *Pcnb*, with dimensions $a = 17.863(2), b = 35.714(4), c = 11.849(1)$ Å and $Z = 8$. The anionic $Mo_5Cl_{13}^{-2}$ group forms in a nearly square pyramidal metallic cluster containing triply bridging, doubly bridging and terminal chlorines. It was readily apparent that more than one image of the structure would, in all likelihood, remain after the superposition. In this case, a multiple Mo-Cl vector was chosen for the superposition. The resultant map therefore contained images as viewed by those molybdenum and chlorine atoms represented by the superposition vector. By retaining only those peaks consistent with the known space group symmetry and combining equivalent images, most if not all false atom positions were eliminated.

There was some uncertainty as to the correct space group initially; the appropriate space group was arrived at by calculating and comparing the sets of $Q_s(\vec{t})$ values for each of the possible space groups. Symmetry analysis of the superposition map revealed two images containing the entire $Mo_5Cl_{13}^{-2}$ unit; one an Mo image and the other a Cl image. These images were averaged together and the composite used in an electron density map calculation to identify the remainder of the structure. From the results of a complete least square refinement, we found that our calculated Mo-Mo bond distances were correct to within 0.02 Å, the Mo-Cl distances to within 0.12 Å, and the Mo-Mo-Mo angles within 0.9°. Some of the nitrogen and carbon atoms in the cation were found to be present in our result but were not used in the initial electron density map calculation. This result demonstrates that this procedure can be used to solve a complex structure without knowing its precise symmetry or its stereochemistry.

This procedure was developed with the view of *always* performing a Patterson superposition as a first step to simplify the interpretation. In the course of this development, we discovered that, in many simpler cases, the superposition process need not be performed. Using this program, interpretation of the Patterson map itself can often lead to the elucidation of the complete structure. As an example, consider the structure of $C_5H_5Fe(CO)_2(CS)PF_6$. The Fe-Fe vectors are immediately identifiable in the Patterson map (The well-known heavy atom technique is, therefore, a viable alternative for this structure solution. The iron atom, however, composes a relatively small fraction of the overall density, and some difficulty could be expected from such an approach.) Application of the procedure described above directly to the Patterson results in the automatic identification of 19 of the 22 non-hydrogen atoms in this structure. The average deviations of the bond distances and angles from refined values are 0.12 Å and 4.5 deg., respectively. The quality of this agreement is strongly affected by the fact that the cyclopentadienyl and PF_6^- groups are partially rotationally disordered and only approximated by the atomic distributions represented.

Conclusion

These results along with comparable results from over 20 other structures provide strong indication that Patterson-based methods are not dead, but in fact just coming of age. There are many structures not readily solved using other methods which are nicely amenable to this approach. Further development along these lines should continue to expand the power, versatility and popularity of these Patterson-based methods.

Acknowledgement

This work was supported by the U.S. Department of Energy, Office of Basic Energy Sciences, Materials Sciences Division, under contract W-7405-ENG-82.

References

1. Nordman, C. E. *Acta Cryst.* **A28**, 134 (1972).
2. Nordman, C. E. In *Computing in Crystallography* (R. Diamond, R. Ramaseshan and K. Venkatesan, eds.) pp. 501-13. Indian Academy of Sciences: Bangalore, India (1980).
3. Nordman, C. E. and Hsu, L. R. In *Computational Crystallography* (D. Sayre, ed.) pp. 141-149. Clarendon Press: Oxford, UK (1982).
4. Simonov, V. I. In *Computational Crystallography*, edited by D. Sayre, pp 150-158. Clarendon Press: Oxford, UK (1982).
5. Mighell, A. D. and Jacobson, R. A. *Acta Cryst.* **16**, 443 (1963).
6. Simpson, P. G., Dobrott, R. D. and Lipscomb, W. N. *Acta Cryst.* **18**, 169 (1965).
7. Buerger, M. J. *Vector Space* John Wiley: New York, NY (1959).
8. Jacobson, R. A. and Guggenberger, L. J. *Acta Cryst.* **20**, 192 (1966).

18. Integration of Patterson search and direct methods— a novel strategy for the solution of difficult crystal structures

Ernst Egert

Although the domain of the Patterson function has always been the location of heavy atoms in crystal structures, its usefulness is by no means restricted to this classical application. If part of the molecular geometry is known, a search fragment (which may consist purely of light atoms) can be correctly positioned in the unit cell by fitting the calculated to the observed vector pattern, even if the expected peaks do not correspond to real maxima in the Patterson map. This so-called Patterson search has been shown by various authors to be a powerful and reliable tool for solving difficult crystal structures, and in a number of cases it succeeded when other methods had failed.[1] Its great advantage over direct methods is that it employs chemical information *directly*, and so can compensate for mediocre precision and resolution of the X-ray data.

On the other hand, the basic ideas and mathematical concepts of the well-known statistical methods have proved outstandingly successful, which is certainly one of the main reasons for the rapid progress in X-ray crystallography during the past two decades. However, there are some inherent weaknesses (*e.g.*, their dependence on a few key reflections during the early stages of phase determination or difficulties with pseudosymmetry, limited resolution and weak data from very small crystals) that make direct methods vulnerable and prevent automatic structure solution in some cases.

In view of our experiences with these two successful but quite different approaches, we wondered if one could combine Patterson search and direct methods into one powerful structure-solving strategy, thereby enhancing the individual strengths and reducing the inherent weaknesses. The rationale behind this attempt was that, in order to solve large problem structures, one should try to exploit *all* the *a priori* available information. Subsequently, we have developed a novel strategy that is efficient, flexible and generally applicable and is designed for a safe and rapid structure solution if the molecular geometry is at least partially known. Since this method is described in detail elsewhere[2] it will be discussed here only briefly, together with some recent results that demonstrate its power and the range of its possible applications.

The search procedure

An almost classical rotation search. The location of a rigid fragment within the unit cell requires the knowledge of up to six parameters. Fortunately, the whole Patterson search can be split into two sub-searches, a rotation and a translation search, because the sphere around the origin of the Patterson function is dominated by intramolecular vectors, which depend on the orientation but not on the position of the fragment. Thus the first stage of the search consists of the determination of the orientation of the model with respect to the cell axes; this is followed by the positioning of the correctly oriented fragment with respect to one of the permissible cell origins. Depending on the space group, the translation search for one fragment may be in fewer than three dimensions or even redundant.

There is no obvious way to improve the efficiency of a Patterson rotation search by the integration of direct methods. Thus the first part of our procedure, although highly automated, does not differ much from other real-space search routines[3,4] and is mainly based on the magnitudes of the Patterson function at positions corresponding to interatomic vectors. The search model, whose geometry may conveniently be obtained from the Cambridge Crystallographic Database or from force-field calculations, is regarded as a rigid ensemble of point atoms with possibly one torsional degree of freedom. The asymmetric unit of a sharpened Patterson function is usually generated by SHELX-86[5] using $E \cdot F$ as coefficients ($51 \cdot 51 \cdot N$ grid points, N arbitrary) and is then densely packed such that each grid value occupies only three bits of computer memory.

For a given orientation, described by a triplet of Eulerian angles, the correlation between the rotated intramolecular vector set and the Patterson function is measured by comparing the weight, w_i, of each vector with the nearest Patterson grid value, P_i. As a figure of merit, *RFOM*, we take the average of the, say, 30% worst-fitting vectors, *i.e.*, those with lowest P_i/w_i:

$$RFOM = \frac{1}{n}\sum_{i=1}^{n}\frac{P_i}{w_i} \quad (n \approx 0.3 \cdot n_{total}).$$

This criterion is related to the 'weighted minimum average',

$$\sum_{i=1}^{n} P_i / \sum_{i=1}^{n} w_i,$$

used by Nordman[6] and, depending on n, resembles either the sum or the minimum function.

The asymmetric unit of angular space that must be investigated is fixed by the Laue group and the model symmetry, but for an optimum use of these relations a specific initial orientation of a symmetric model would be required. Our procedure differs from those mentioned above in that we do not define and systematically scan the respective range of angles, but generate $10,000 - 60,000$ random orientations (corresponding to mean rotation increments of about $7°$) ignoring possible equivalences between angle triplets. Similarity between two orientations is instead inferred from close distances between pairs of equivalent atoms, taking into account all relevant symmetry elements of the Laue group. This procedure not only recognizes all angular symmetry relations (including non-linear ones) automatically, but is also computationally efficient since it has to be applied only to promising solutions, *i.e.*, those with large *RFOM* and physically reasonable packing (judged at this stage from the lattice translations alone).

In order to locate the most prominent maxima in angular space more accurately, a small number of best solutions are refined by testing up to 1,000 additional random points close to each of them, which corresponds to a mean rotation increment of less than $2°$. Thereafter, the correct orientation is usually found among the best two or three, provided the fractional scattering power of the search model is not too small ($p^2 > 0.2$). If the fragment has one torsional degree of freedom, the whole procedure described is repeated for each distinct geometry (conveniently defined by a range of possible torsion angles and an appropriate increment). This provides an opportunity for expanding small fragments and thus for improving the chances of a successful search.

A novel translation search. The location of an oriented fragment is the most time-consuming and least reliable step of a Patterson search. This is because the intermolecular ('cross') vectors, which already tend to be weaker than the intramolecular ones, suffer from errors in both the model geometry and orientation, amplified by the symmetry elements. If the translation search is based on a Patterson sum or minimum function using either a fine grid or interpolation techniques, the computing time rises rapidly with the size of the fragment and the number of symmetry elements. This is exactly the point where, in our opinion, the integration with direct methods could lead to a considerable improvement of the performance of the method.

When an oriented fragment is moved through the unit cell, the calculated phases are a *continuous* function of the shift vector $\Delta\vec{r}$:

$$F(\vec{h}) = F^{\circ}(\vec{h}) \cdot exp(2\pi i\vec{h} \cdot \Delta\vec{r}).$$

Thus, once a structure factor $F^{\circ}(\vec{h})$ for a starting position has been calculated by summing the scattering contributions from all atoms comprising the search model, the structure factor $F(\vec{h})$ for any position is readily obtained by multiplication with a simple phase factor. For the correct position of a complete ($p^2 = 1$) and exact model, the individual phases of the strongest reflections are linked by various statistical phase relations *e.g.*, the extremely important three-phase structure invariants

$$\phi(\vec{h}) + \phi(\vec{k}) + \phi(-\vec{h} - \vec{k}) \cong 0$$

on which direct methods are mainly based. Although the search fragment is usually incomplete and may also be not very accurate, the triple-phase relations are expected to hold approximately, if its scattering power is significant. We measure the triple-phase consistency by

$$TPRSUM = \frac{\sum E(\vec{h}) \cdot E(\vec{k}) \cdot E(-\vec{h} - \vec{k}) \cdot cos\{\phi(\vec{h}) + \phi(\vec{k}) + \phi(-\vec{h} - \vec{k})\}}{\sum E(\vec{h}) \cdot E(\vec{k}) \cdot E(-\vec{h} - \vec{k})},$$

which should be large and positive for the correct solution ($-1 \leq TPRSUM \leq 1$). The two sums are taken over a small number (40 – 60) of strong *and* translation-sensitive three-phase invariants, *i.e.*, those linking the largest E-values to which the oriented search fragment contributes significantly. (The mean E_{calc}/E_{obs} ratio is usually a maximum for the correct orientation.)

Accordingly, a sufficient number (depending on the size of the asymmetric unit of translation search) of random positions are refined by maximizing $TPRSUM$. Because of the limited convergence radius of the refinement procedure (about 0.5 Å for 1 Å data) the starting positions are selected such that they do not give rise to unreasonable packing. Any refined solution with satisfactory triple-phase consistency is tested further for the occurrence of illicitly close intermolecular contacts, which would lead to its rejection. Only then is the fit between the Patterson function and the intermolecular vector set measured, by

$$TFOM = \frac{1}{n}\sum_{i=1}^{n}\frac{P_i}{w_i} \qquad (n \approx 0.2 \cdot n_{total}).$$

For the best (according to both $TPRSUM$ and $TFOM$) non-equivalent solutions, an R-index based on E-magnitudes is calculated as an additional figure of merit and has proved very useful for indicating the correct position. It is defined as:

$$R_E = \frac{\sum\{|E(obs)| - |E(calc)|/p\}}{\sum |E(obs)|}$$

where only positive contributions to the numerator are considered as inconsistencies between observed and calculated E-values. Finally, the solutions are sorted according to a combined figure of merit, $CFOM$:

$$CFOM = \frac{0.2}{R_E} \cdot TFOM \cdot (TPRSUM)^{\frac{1}{2}}.$$

In this expression, $TPRSUM$ is given lower weight because it is the quantity optimized; if a rotation search preceded the translation search, $TFOM$ is replaced by $(RFOM + TFOM)/2$.

This procedure, which operates in direct and reciprocal space simultaneously, is computationally efficient, especially for large structures, because the refinement is based on phase relations derived from a relatively small number of large E-magnitudes independent of the size of the structure. It is important to note that the method relies on the bulk of selected three-phase structure invariants, *not* on single ones. Only when an acceptable solution has been found by this 'direct' search is it necessary to calculate the time-consuming Patterson correlation. Apparent differences from other Patterson translation functions[3,4,7] are the location of the oriented model with respect to *all* symmetry elements of the space group simultaneously and the variety of different criteria employed to judge solutions. One would expect that a position which is in agreement with packing criteria, the Patterson function ($TFOM$), triple-phase relations ($TPRSUM$) and E-values (R_E) is probably correct, and our experience shows that this is indeed the case. Thus this novel combination of Patterson and direct methods should be a powerful structure-solving strategy, if chemical information is available.

Performance of the method

The procedures outlined have been implemented as a fully automatic computer program called PATSEE, which is valid and efficient for all space groups in all settings and is compatible with SHELX-86. The rotation search can find the orientation of a fragment of any size and allows one torsional degree of freedom. The translation search may locate up to two independent search models of any size, taking into account known atoms at fixed positions (*e.g.*, heavy atoms obtained by a 'normal' Patterson interpretation).

Tests with about 30 known structures of different size and complexity (five of them are discussed by Egert and Sheldrick[2]) have confirmed that PATSEE is reliable and widely applicable. If good experimental data are combined with relatively accurate and large ($p^2 > 0.35$) search models, *all* figures of merit often take extreme values for the correct position. With smaller fragments, $TPRSUM$ is usually only a local maximum (which is sufficient for the location of the search model), but $TFOM$

and R_E together are still strongly indicative of the correct solution. Even fragments consisting of just one or two single atoms, which do not necessarily have to be very heavy ($p^2 < 0.1$, *e.g.*, P or S in fairly large organic structures), may be easily located; in such cases, *TFOM* tends to be the most reliable figure of merit, which makes the procedure an almost standard, though automatic, heavy-atom Patterson interpretation. At the other end of the scale, when dealing with very large fragments and low-resolution data, the dominating criteria are the R_E-value and the distance test (although *TPRSUM* and *TFOM* are still useful). With these extremely difficult structures, PATSEE operates similarly to the method of Rabinovich and Shakked,[8] which does not use the Patterson function at all. So, because of the different criteria employed, our procedure seems to be flexible enough to adapt itself to a wide range of possible applications.

In this context, the important question arises: How does the performance of the method described depend on the quality of the experimental data? The Patterson function is certainly not the main problem, because its gross features are less sensitive to, for example, limited resolution than the E-values and thus the number of reliable phase relationships; this is one of the most striking advantages of Patterson methods. In order to find the optimum conditions for our search procedure, we have successively reduced the nominal resolution (d_{min}) for each test structure by applying a 2θ limit to the E-values, without changing the Patterson map. The first result was that the average fragment shift during the triple-phase refinement rises continuously with decreasing resolution and is about $0.4 \cdot d_{min}$. At 1.0 and 1.8 Å, for example, the mean shifts along each cell axis are 0.2 – 0.3 and 0.3 – 0.6 Å and the maximum shifts at least 1 and 2 Å, respectively; however, the latter are sometimes much larger especially when searching for single atoms.

At low resolution, there are, of course, fewer triple-phase maxima, thus requiring less computing time, but they may be very broad. There are also fewer large E-values available, so that the triple-phase relations are less reliable; as a rule of thumb for low-symmetry space groups, the 100 most probable three-phase structure invariants at 1.8 Å resolution are formed by only 50E's> 1.5 compared to 80E's> 2.0 at 1.0 Å. With such highly interlinked phase relations, there is the danger that single E-values may have too much influence on the search (as in direct methods); this should be avoided if possible. At very high resolution, not only are more trials needed for the correct positioning of the fragment (as a result of the increased number of maxima), but the search is often less successful since high-order reflections may suffer considerably from model errors. Therefore it is not surprising that all small-molecule test structures were solved convincingly between 0.8 and 1.2 Å resolution with the best performance at 1.0 Å (although some of them did not cause any problems up to 1.8 Å). It is probably the best strategy to adapt the nominal resolution of the E-values to the accuracy of the search model.

Large structures

Any new method for solving structures should be developed with the narrowing gap between small-molecule and protein crystallography in mind, so that future applications to large structures can be taken into account. Patterson search methods have already been used to solve macromolecular problems;[9] Nordman, for example, has successfully applied vector search procedures to the structure of myoglobin.[10]

In order to demonstrate that PATSEE is able to solve large structures, we have chosen d(G-G-^{Br}U-A-^{Br}U-A-C-C) as a test example.[11] The search model consisted of the whole A-DNA type double helix (altogether 322 atoms, p^2 = 0.7) and was taken from the refined structure; in view of the limited resolution (d_{min} = 1.5 Å) this should not matter too much. The Br atoms were given only half weight to account for their thermal motion at the molecular boundary. The rotation search gave unequivocally the correct orientation; for this application, the short vectors, which are usually ignored, were kept because of their high weights. For the translation search, 60 suitable triple-phase relations were automatically selected from the 100 most probable ones formed by 60$E's$ > 2.0. Since the volume of the search fragment occupied a considerable part of the asymmetric unit, more than 2,000 starting positions had to be generated before 50 promising ones were found. From these, only 20 (of which several were equivalent) gave a reasonable *TPRSUM* value and passed the final distance test. The correct solution was easily recognizable from the combined figure of merit mainly because R_E was substantially lower.

As an attempt to simulate low-resolution structures, the translation search was repeated with successively lower 2θ limits. With a resolution better than 4 Å, the structure was solved convincingly, but even with 9 Å data the correct position was found, although only 15 large E's were available to form 24 suitable phase relations. The mean shifts increased from 0.9 Å (d_{min} = 1.5 Å) to 3.6 Å (d_{min} = 9 Å) and the positional error from 0.4 to 1.8 Å. The method is thus in principle able to solve large structures even if the resolution is rather low.

Acknowledgement

I thank the Fonds der Chemischen Industrie for a Liebig-Stipendium and the Deutsche Forschungsgemeinschaft for financial support.

References

1. Egert, E. *Acta Cryst.* **A39**, 936-940 (1983).
2. Egert, E. and Sheldrick, G. M. *Acta Cryst.* **A41**, 262–268 (1985).
3. Nordman, C. E. *Trans. Am. Cryst. Assoc.* **2**, 29-38 (1966).
4. Hornstra, J. In *Crystallographic Computing.* (F. R. Ahmed, ed.) pp. 103-109. Munksgaard: Copenhagen (1970).
5. Sheldrick, G. M. Unpublished.
6. Nordman, C. E. In *Computing in Crystallography,* (R. Diamond, S. Ramaseshan and K. Venkatesan, eds.) pp. 5.01-5.13. The Indian Academy of Sciences: Bangalore, India (1980).
7. Doesburg, H. M. and Beurskens, P. T. *Acta Cryst.* **A39**, 368-376 (1983).
8. Rabinovich, D. and Shakked, Z. *Acta Cryst.* **A40**, 195-200 (1984).
9. Rossmann, M. G. (ed.) *The Molecular Replacement Method.* Gordon and Breach: New York (1972).
10. Nordman, C. E. *Acta Cryst.* **A28**, 134-143 (1972).
11. Shakked, Z., Rabinovich, D., Cruse, W. B. T., Egert, E., Kennard, O., Sala, G., Salisbury, S. A. and Viswamitra, M. A. *Proc. Roy. Soc. (London)* **B213**, 479-487 (1981).

19. Introduction to rotation and translation functions

D. M. Blow

This is the text of an introductory paper given at a meeting on 'Molecular Replacement' organized by the Collaborative Computing Project on Protein Crystallography of the UK Science and Engineering Research Council at the Daresbury Laboratory, England, February 1985.

Non-crystallographic symmetry

Suppose we have two identical objects in different positions. To superimpose one on the other we must rotate it and translate it. We can specify a rotation about the origin by three variables, which can be two to specify the direction of the rotation axis by its latitude and longitude, and one for the amount of rotation. Then we need another three variables to specify the translation — for example the amount of motion in the x, y and z directions.

There are many other ways the operation can be done, but they will always require six variables to be specified. For example, it is possible to choose the position of the rotation axis in such a way that the only subsequent translation which needs to be done is parallel to the axis. We still need three variables to specify the translation: now there are two to give the position of the axis (in a plane perpendicular to it) and one for the amount of translation parallel to the axis. There are also different ways of specifying the rotation.

This symposium relates to a range of crystallographic problems which arise when such a situation exists in crystals. I shall use the word subunit to represent a diffracting unit which is identical, to sufficient accuracy, to another whether it be a different molecule, a different peptide chain in the same molecule, or two parts of the same peptide chain. The accuracy of the identity will define for the crystallographer the resolution of data for which he can assume the relationship will hold.

If there is more than one subunit in the crystallographic asymmetric unit, the first problem is to find the operation which superimposes one on the other — to determine the six variables which specify the rotation and translation (Case 1). If, on the other hand, the identical subunits are in different crystals, the problem is to find the transformation of one coordinate system which converts the coordinates of the subunit in that crystal to the coordinates of the other subunit in the other crystal (Case 2).

In either case, the problem is intricately entwined with crystallographic symmetry. In Case 1, the rotations and translations which are determined will depend on an arbitrary choice of which particular subunits are superimposed, out of the whole crystal lattice. In Case 2, the same type of arbitrary choice is involved, and in addition the choice of the origin of coordinates is very often (not always) dictated by crystallographic conventions.

There is a lattice symmetry about the posible solutions to the problem. If translation vectors x, y, z give a solution in Case 1, then the addition of any number of lattice translations to x, y, z generates another possible solution. Similarly, if the angles defining the rotation operation are α, β, γ an arbitrary number of 2π can be added to any of them without affecting the solution. So there are an infinite number of equivalent operations

$$\alpha + 2\pi A$$

$$\beta + 2\pi B$$

$$\gamma + 2\pi C$$

$$\vec{x} + U\vec{a}$$

$$\vec{y} + V\vec{b}$$

$$\vec{z} + W\vec{c}$$

where A, B, C, U, V, W can be any integer and $\vec{a}$, $\vec{b}$, $\vec{c}$ are crystal lattice translations parallel to x, y, z.

The possible solutions thus form a six-dimensional lattice, and the existence of crystallographic symmetry means that other solutions exist so that this six-dimensional lattice is in general non-primitive. I know nothing about the six-dimensional space groups, but Rossmann and I considered the three-dimensional lattices of the rotation operations in our original paper.[1] Tollin, Main and Rossmann[2] extended this work, and D. Moss[3] has recently pointed out a further extension to it, which seems previously to have been overlooked.

Eulerian angles

In practice it is convenient to compute the rotation function in terms of Eulerian angles, now referred to as (α, β, γ). I describe here the angular conventions used in the mathematical treatment of Crowther.[4]

We define an orthogonal co-ordinate system $(\vec{X}, \vec{Y}, \vec{Z})$ and arrange the crystal cell so that the axis of highest symmetry is along $\vec{Z}$. In all symmetries higher than monoclinic, the crystallographic **c** direction coincides with $\vec{Z}$, and the crystallographic **a** direction is along $\vec{X}$. In the monoclinic system, the axes are permuted so that the crystallographic **b** is along $\vec{Z}$, and the crystallographic **c** coincides with $\vec{X}$.

The Eulerian rotation operations are defined with respect to the orthogonal system $(\vec{X}, \vec{Y}, \vec{Z})$ (Fig. 19.1).

a) A rotation α of the coordinate system about $\vec{Z}$;

b) A rotation β about the new direction of $\vec{Y}$;

c) A rotation γ about the new direction of $\vec{Z}$.

This system is computationally convenient: in particular it fits well with the symmetry properties of the system. However,

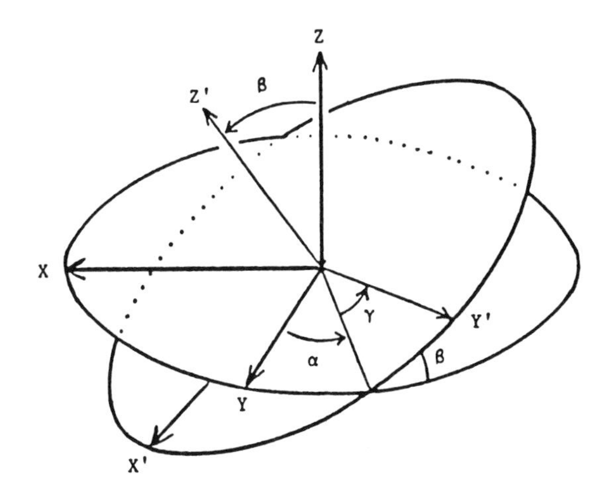

FIG. 19.1. Eulerian angles α, β, γ.

it has some disadvantages, as the three rotations are often far from orthogonal.

1) If β is zero the rotations α and γ have similar effects: all rotations with the same value of $(\alpha+\gamma)$ are identical.
2) If β is π , the rotations α and γ have opposite effects: all rotations with the same value of $(\alpha-\gamma)$ are identical.
3) All possible rotations are included in the range $\alpha = 0,\ 2\pi$; $\beta = 0,\ \pi$; $\gamma = 0,\ 2\pi$.

Another well-known system is spherical polar rotation (χ, ω, ϕ) (Fig. 19.2). Here $\vec{Z}$ is the polar axis and ω and ϕ define the longitude and latitude respectively of the rotation axis. The angle of rotation about this axis is χ. In this system, all possible rotations are included in the range $\chi = 0,\ 2\pi$; $\omega = 0,\ \pi$; $\phi = 0,\ \pi$.

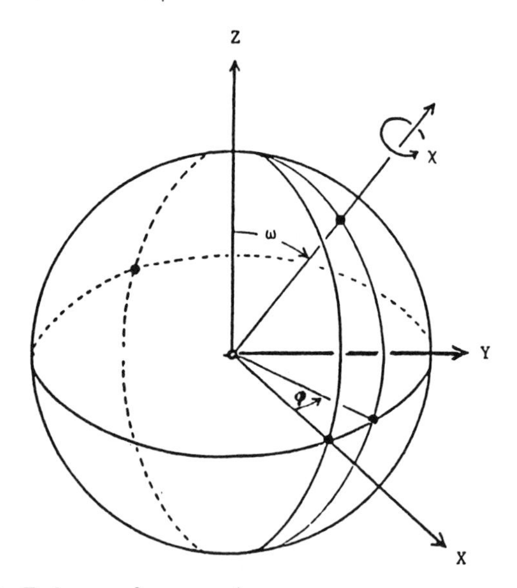

FIG. 19.2. Polar angles χ, ω, ϕ.

Search techniques

The problem as presented so far has been a purely geometric one. If we have two sets of coordinates x_i^A, y_i^A, z_i^A and x_i^B, y_i^B, z_i^B

for the two molecules, the solution is straightforward and can be solved as a standard eigenvalue problem (*e.g.* ref. 5).

The practical problem arises in experimental crystallography when the structure is unknown (or, in Case 2, when one or both of the structures are unknown). The questions are then:

1) How can one detect that non-crystallographic symmetry exists?
2) How can one determine rotational and translational parameters which define the non-crystallographic symmetry operations?
3) Is this knowledge any help in solving structures?

The usual constraints on the practical problem are:

1) The amplitudes of the structure factors are measurable by X-ray diffraction, but their phases are not accessible experimentally ('the phase problem').
2) At the outset, no useful model is available for the structure or any part of it. (The usual situation in protein crystallography, but not in small molecule crystallography.)

In the absence of a direct solution to the practical problem, all methods will involve some kind of search. Searches in six variables easily become very large. It is hard to define how many values of each variable need to be tried in order to locate the solution, but in practice it cannot be less than 10 or so, implying a million search points, increasing to a billion if there are 30 search points in each variable. Such searches are clearly uneconomic, even with the latest computational equipment.

Small molecule crystallographers know well that the angular orientation of a molecule or ring system can often be determined long before the total structure is known. One of very few significant structures solved by a search procedure was the structure of a hydroxyproline.[6] The molecule was treated as having a rigid five-membered ring with two dihedral angles as variables. So the search was eight-dimensional, and it was carried out by a search procedure called a 'method of non-local search.' The search rapidly discovered the angular orientation of the five-membered ring, later the position of the molecule in the unit cell, and finally the proper value of the dihedral angles.

The key to the efficient solution of the problem is to find a way of separating the rotational and translational variables. If this is done, the search over millions of points can be reduced to two consecutive searches, each covering only a few thousand points.

The Patterson function[7] has certain characteristics which are well suited to this problem. Patterson pointed out that it contains *all* the information from the intensities (it has none from the phases): thus it includes all the experimental information, without any interpretive data. As a consequence of this, it is not dependent on any choice of origin: as we all know, it represents the collection of vectors between the elements of scattering density.

The rotation function

The basic idea of the rotation function is that the Patterson function of a molecule, or a subunit, will have a characteristic distribution of densities. In crystals, we are always dealing with arrays of units. If we think of the Patterson function as a collection of vectors, there will be some vectors between scattering density in the same subunit, and some vectors between one subunit and another. We can call these two types of vectors self vectors (S) and cross vectors (X). If we have two identical subunits in different crystals the array S for each will be the same, but the array X will be different. In case 1, with two identical subunits in the same crystal we will have two arrays S_A and S_B which are identical in shape, but in different orientations. A rotated coordinate system in Patterson space can be described, in which S_B is just like S_A in the original system. Multiplying the rotated Patterson function by the original will give a positive contribution from the product of S_A and its identical image S_B. If (following the usual habit) $F^2(000)$ has been omitted from the Patterson function, the other terms in the product will be negative as often as positive.

This positive tendency will exist all over the region containing S_A, so we can enhance the positive effect by looking simultaneously over the whole volume of vector space which it fills. This may be done by integration which defines the rotation function. If we calculate

$$\Re(R) = \int_A P(\vec{u}) \cdot P(\vec{u}_R)\, du$$

as a function of all pure rotation operations R, we expect the function to have significantly positive values when R represents the rotation of subunit B to subunit A. In case 1, there will also be a peak for the inverse rotation which rotates subunit A to subunit B.

What are the limits of summation?

It is important to point out that there are two types of summation (or integration) in the rotation function. The Patterson function itself is a summation over all reflexions, and the rotation function is a summation over a range of Patterson space. As we are in charge of the calculation, we can manipulate the limits of both these sums to our best advantage.

In crystallography there is always a resolution limit, beyond which it is impractical to extend our summation over reflexions. If our two subunits were precisely identical, it would be advantageous to extend our summation to the highest resolution available. In practice, subunits are not precisely identical and this would lead us to terminate the summation at lower resolution: or to be sophisticated, to weigh down high resolution terms. The resolution limit should be at least two or three times the expected mean co-ordinate difference between the two structures. In practice, with existing programs, computational limitations usually enforce a much restricted resolution.

It is also essential to omit very low resolution terms. The reason for this can be understood by considering what a low resolution structure actually looks like. Especially in crystals grown from high salt, the dominant features in the low resolution map are the solvent boundaries. Since the solvent regions are generally smaller in dimension than the molecules, the shapes which you see in maps at 10 Å – 8 Å resolution are the shapes of the solvent regions. Experience suggests that it is best to omit all terms with Bragg spacings greater than 7 or 8 Å.

Turning to the integration in vector space, it is obvious that no self-vector can exceed the largest diameter of the molecule. If the molecule is spherical, the density of its self-vectors forms an almost conical shape falling to zero at the diameter. If the molecule has a very elongated shape, the longest vectors will be equal to the length of the molecule, but there will be many more short vectors, especially shorter than the cross-sectional diameter.

In our summation we obviously want to include regions of vector space where the self-vector density is high, and to omit those where it is zero, but where cross-vector density is high. Just where to make the cut-off is a matter which requires detailed study, but for an approximately spherical molecule I would suggest 75 – 80% of the diameter, which would include about 90% of the integrated Patterson density (Fig. 19.3).

It is not necessary to integrate over a sphere, but simply a matter of convenience. In principle, it should be possible to obtain direct information about molecular shape by finding how the value of the peak in the rotation function varies as the volume of integration changes. In the early days, I tried to discover the shape of the insulin hexamer (represented as a cylinder) by such a method. In those days the computations were too unwieldy and no useful results were obtained.

An important feature of the Patterson function is the origin peak, which is always the maximum value of the function. It expresses the fact that every bit of scattering density has an equal bit of scattering density exactly at a vector zero from it (namely, itself). Because the Patterson function is computed by a Fourier summation, the origin peak has a width characteristic of the resolution of the data, and it is normally spherical in shape. In the rotation function the origin peak is being compared to itself. The two spherical peaks simply add a constant positive term to the rotation function, and cause no particular trouble. They can be subtracted if desired.

There is one situation where origin peaks cause trouble — this is when there is a very flattened or elongated unit cell,

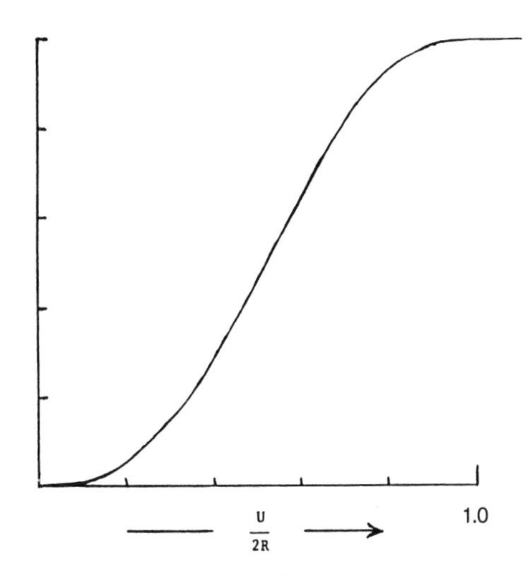

FIG. 19.3. The Patterson function density for a uniform solid sphere of radius R, when integrated over a sphere of radius U about the origin, contains an integrated density given by:

$$\frac{\pi^2 R^6}{36}\left(\frac{U}{2R}\right)^3\left[8-9\left(\frac{U}{2R}\right)+2\left(\frac{U}{2R}\right)^3\right]$$

for $0 \leq U \leq 2R$.

and one of the lattice vectors is comparable to the molecular diameter. An integration in Patterson space over the average diameter of the molecule could possibly include an origin peak one lattice translation from the origin. It is essential to avoid including such an origin peak in the calculation. Either the radius of integration must be less than the smallest lattice translation minus twice the resolution limit or steps must be taken to subtract the origin peaks.[8]

How to detect non-crystallographic symmetry

When calculated, a rotation function usually looks very disappointing at first glance. There are a number of reasons for this. A self-rotation function (Case 1) has its own massive origin peak. Zero rotation obviously superimposes the whole Patterson function, and some other rotation which only superimposes the self-vectors will be miniscule by comparison. The Eulerian angles introduce an inconvenience. As already mentioned, if β is zero, the rotations α and γ have the same effect, so that all rotations with the same $(\alpha + \gamma)$ are identical. On the first page of your rotation function with $\beta = 0$, there is a massive ridge representing $(\alpha, 0, -\alpha)$, which are a whole set of equivalent zero rotations, and a series of other parallel ridges. Suggestions for replotting the rotation function in a different way to remove these problems have been made by Burdina[9] and Lattman.[10]

After we have realised we must ignore the origin peak (in Case 1), it is still often necessary to do some statistics to find whether another peak is significant. I recommend calculating the mean and standard deviation of all the computed values after the origin peak has been removed, then find how many standard deviations from the mean the highest peak represents. In order to know how significant the peak is one other fact is needed, that is, how many independent values the rotation function can have. Let me stress, this number only needs to be known very approximately. It is certainly no greater than the number of reflexions included in the Patterson function, to the appropriate resolution, so will often be a few thousand. In practice, if a peak exceeds 5 standard deviations it means something real; if the value is less than this it must be treated with scepticism.

If one is dealing with a symmetrical molecule, such as a dehydrogenase tetramer with 222 symmetry, there may be a series of peaks which bear a particular relationship to each other. Rossmann *et al.*,[11] devised a 'locked rotation function' which compares the values of the rotation function at different positions. This method can be used to enhance the significance level of a set of doubtful peaks.

Fitting a known structure to an unknown one

An important use of the rotation function is to find the orientation of a molecule of known structure in a new crystal form. The simplest way to deal with this is to compute structure factors for the isolated molecule in a defined orientation.[12] The structure factors are those for a 'model' crystal in which isolated molecules are spaced out on a lattice (usually an orthogonal lattice, but one in which the symmetry is $P1$). In the model crystal, one can arrange that the molecules are so far apart that none of the cross-vectors is as short as the chosen radius of integration. Then there are no cross-vectors in the model at this resolution. Avoid making the model crystal larger than necessary, as the number of structure factors to be computed increases rapidly.

Translation functions

When the rotational relationship of two subunits has been established, the other three variables to determine their relative positions can be found. A number of methods exist. When the two subunits are in the same crystal structure, and are related by a proper rotation (*i.e.*, a rotation of $2\pi/n$), Rossmann *et al.*[12] discovered a relationship between the two sets of cross-vectors which produce peaks in a correlation function based on Patterson functions. This method is, however, difficult to use, and few examples of its use have been published. It remains the only available method for unknown structures.

When one of the structures is known (Case 2), a method due to Tollin[13] can be cast into the form of a Patterson-like convolution function.[14]

Nowadays, the computer power available allows the use of a simpler method, which is to compute the crystallographic R factor as a function of the position of the subunit in the unit cell. This method was used for small molecules in 1964.[15] The molecular transform is computed for a single molecule, and values are added appropriately as the molecule and its symmetry-related copies range through the cell. When coupled with the FFT-based algorithm for structure factor calculation introduced by Agarwal,[16] this method provides a computationally efficient solution to the translation problem for a molecule of known structure.

References

1. Rossmann, M. G. and Blow, D. M. *Acta Cryst.* **15**, 24-31 (1962).
2. Tollin, P., Main, P. and Rossmann, M. G. *Acta Cryst.* **20**, 404-407 (1966).
3. Moss, D. *Acta Cryst.* **A41**, 470–475 (1985).
4. Crowther, R. A. In *The Molecular Replacement Method* (M. G. Rossmann, ed.) pp. 173–178. Gordon and Breach: New York (1972).
5. MacLachlan, A. D. *Acta Cryst.* **A32**, 922-923 (1972).
6. Kayushina, R. L. and Vainshtein, B. K. *Kristallografiya* **10**, 833-844 (1965).
7. Patterson, A. L. *Phys. Rev.* **46**, 372-376 (1934).
8. Lattman, E. E. and Love, W. E. *Acta Cryst.* **B26**, 1854-1857 (1970).
9. Burdina, V. I. *Kristallografiya* **15**, 623-630 (1970).
10. Lattman, E. E. *Acta Cryst.* **B28**, 1065-1068 (1972).
11. Rossmann, M. G., Ford, G. C., Watson, H. C. and Banaszak, L. J. *J. Mol. Biol.* **64**, 237-249 (1972).
12. Rossmann, M. G., Blow, D. M., Harding, M. M. and Coller, E. *Acta Cryst.* **17**, 338-342 (1964).
13. Tollin, P. *Acta Cryst.* **21**, 613-614 (1966).
14. Crowther, R. A. and Blow, D. M. *Acta Cryst.* **23**, 544-548 (1967).
15. Bhuiya, A. K. and Stanley, E. *Acta Cryst.* **17**, 746-748 (1964).
16. Agarwal, R. C. *Acta Cryst.* **A34**, 791 (1978).

20. The relationship between feedback methods and isomorphous difference Pattersons in the solution of biological macromolecules

Michael G. Rossmann

The Patterson function[1,2] has been used in many different ways to effect the solution of crystal structures. While many simple structures have been solved by the direct analysis of Patterson syntheses, nevertheless alternative methods (in particular direct methods) now often supersede Patterson analyses. However, the Patterson function is still used almost exclusively for the determination of the positions of heavy atoms in isomorphous pairs as an essential preliminary to the solution of biological macromolecular structures. In more complex situations, such as crystalline viruses, a systematic approach, aided by a computer, may be necessary. That is especially true when the structure contains non-crystallographic symmetry.[3] Once an initial tentative solution has been obtained it is then possible to check the postulated position or find other heavy atom sites, using 'feedback' procedures.[4] The latter procedure depends on computing phases based on a single isomorphous replacement (SIR) or multiple isomorphous replacement (MIR) data set. However, the one postulated site is omitted. If it is correct, then it should show up in a difference Fourier phased on the remaining sites. If the structure contains non-crystallographic symmetry then all but one of the non-crystallographically related atoms can be used for phasing and the remaining non-crystallographically related site can (in principle) be used to test for the presence of the atom in a difference Fourier. Thus the feedback procedure can be developed into a systematic search procedure akin to a systematic search of a Patterson. It is demonstrated in this paper that the feedback and Patterson search techniques are equivalent but that they each have some advantages and disadvantages. These methods have recently been found useful in the successful solution of the heavy atom positions in the structure determination of the human common cold virus R14.[5]

The Patterson search technique

Patterson search techniques depend on the comparison of the observed Patterson $P_1(\vec{x})$ distribution with one based on a hypothetical point atom multiplied by the crystallographic and non-crystallographic symmetry, $P_2(\vec{x})$. A criterion, C_p, based on the sum of the Patterson densities at all test vectors, within the unit cell of volume V, would be

$$C_p(\vec{X}) = \int_V P_1(\vec{x}) \cdot P_2(\vec{x}) d\vec{x}$$

where X is the test site, within one non-crystallographic asymmetric unit, of a heavy atom. Each different test site corresponds to a different P_2 Patterson. It can then be easily shown that

$$C_p(\vec{X}) = \sum_h \Delta_h^2 E_h^2$$

where the sum is taken over all h reflections in reciprocal space, Δ_h are the observed isomorphous differences, and E_h are the structure factors for the trial point Pattersons.

Let there be n non-crystallographic asymmetric units within the crystallographic asymmetric unit and m crystallographic asymmetric units within the crystal unit cell. Then there are N symmetry-related heavy atom sites where $N = nm$. Let the scattering contribution of the ith heavy atom site $(i = 1, 2, \cdots N)$, have a_i and b_i real and imaginary structure factor components with respect to an arbitrary origin. Hence for reflection h

$$E_h^2 = \Big(\sum_n a_{hi}\Big)^2 + \Big(\sum_N b_{hi}\Big)^2$$

$$= N + \sum_{i \neq j}^{N}\sum (a_{hi}a_{hj} + b_{hi}b_{hj})$$

But $\sum_h \Delta_h^2 N$ is constant and independent of a particular heavy atom test site. Thus the criterion C_p can be re-written as

$$C_p'(\vec{X}) = \sum_h \Delta_h^2 \left[\sum_{i \neq j}^{N}\sum (a_{hi}a_{hj} + b_{hi}b_{hj})\right].$$

More generally, if some sites have already been tentatively determined, and if these sites give rise to the structure factor components A_h and B_h then

$$E_h^2 = (A_h + \sum_N a_{hi})^2 + (B_h + \sum_N b_{hi})^2.$$

Following the same procedure as above it follows that

$$C_p'(\vec{X}) = \sum_h \Delta_h^2 \left[(A_h a_h + B_h b_h) + \sum_{i \neq j}^{N}\sum (a_{hi}a_{hj} + b_{hi}b_{hj})\right] \tag{1}$$

where

$$a_h = \sum_{i=1}^{N} a_{hi} \text{ and } b_h = \sum_{i=1}^{N} b_{hi}.$$

Expression (1) can be used directly, without the prior computation of a difference Patterson (coefficients Δ_h^2) in conducting a systematic search for a heavy atom. The first term corresponds to vectors between the tentatively known atoms and the current test site at $\vec{X}$, while the second site corresponds to the vectors between all the crystallographically and non-crystallographically related sites produced by the test atom at $\vec{X}$.

The feedback method

Inspection of Fig. 20.1 shows that the native phase α will be determined from a single isomorphous pair[6] as

$\alpha = \phi + \pi$ when $| F_N | > | F_{NH} |$

and as

$\alpha = \phi$ when $| F_N | \ll | F_{NH} |$.

Here ϕ is the phase of the structure factor F_H of the heavy atoms alone in the cell, F_N is the structure factor of the native compound and F_{NH} is the structure factor of an isomorphous heavy atom derivative for reflection h. Thus, a SIR difference electron density $\Delta\rho(\vec{x})$ can be computed from a Fourier summation as

$$\Delta\rho(\vec{x}) = \frac{1}{V}\sum_h (| F_{NH} | - | F_N |) m \cos(2\pi\vec{h}\cdot\vec{x} - \phi_h)$$

$$+ \frac{1}{V}\sum_h (| F_{NH} | - | F_N |) m \cos(2\pi\vec{h}\cdot\vec{x} - \phi_h - \pi)$$

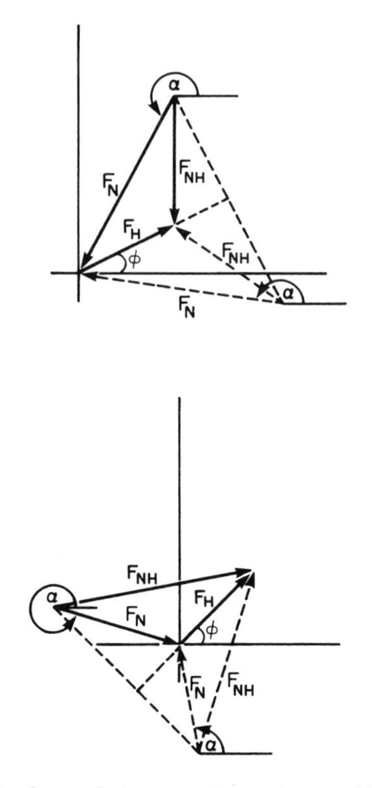

FIG. 20.1. SIR phase determination. The SIR phase α places the native structure factor F_N co-linear with the heavy atom structure factor F_H. Thus when $| F_N | > | F_{NH} |$, then F_N is opposite to F_H and $\alpha = \phi + \pi$. Conversely, when $| F_N | \ll | F_H |$, then F_N is in the same direction as F_H and $\alpha = \phi$.

where the first term relates to positive and the second term to negative differences. All terms are weighted by a figure of merit, m, appropriate for that reflection.[7] Thus quite generally

$$\Delta\rho(\vec{x}) = \frac{1}{V}\sum_h | \Delta_h | \, m \cos(2\pi\vec{h}\cdot\vec{x} - \phi_h).$$

Now if A_h and B_h are the real and imaginary components of the presumed heavy atom sites it follows that

$$\Delta\rho(\vec{x}) = \frac{1}{V}\sum_h \frac{| \Delta_h | \, m}{| F_H |}(A_h \cos 2\pi\vec{h}\cdot\vec{x} + B_h \sin 2\pi\vec{h}\cdot\vec{x}). \qquad (2)$$

If this SIR difference electron density map shows significant peaks at the N symmetry related sites then these sites will be the position of a further heavy atom. Hence a suitable criterion for finding a heavy atom in the presence of a set of tentatively known sites can be

$$C_{SIR}(\vec{X}) = \sum_{j=1}^{N} \Delta\rho(\vec{x}_j) \tag{3}$$

or, by substitution of equation (2) in (3)

$$C_{SIR}(\vec{X}) = \sum_{j=1}^{N} \frac{1}{V} \sum_{h} \frac{|\Delta_h| m}{|F_H|} (A_h \cos 2\pi\vec{h}\cdot\vec{x}_j + B_h \sin 2\pi\vec{h}\cdot\vec{x}_j).$$

But

$$a_h = \sum_{j=1}^{N} \cos 2\pi\vec{h}\cdot\vec{x}_j \quad and \quad b_h = \sum_{j=1}^{N} \sin 2\pi\vec{h}\cdot\vec{x}_j.$$

Therefore,

$$C_{SIR}(\vec{X}) = \frac{1}{V} \sum_{h} \frac{|\Delta_h| m}{|F_H|} (A_h a_h + B_h b_h).$$

Alternatively, if the tentatively known sites are indeed the test sites in all but one of the N symmetry related sites, then equation (2) can be re-written as

$$\Delta\rho(\vec{x}) = \frac{1}{V} \sum_{h} \frac{|\Delta_h| m}{|F_H|} \Big[(\sum_{i\neq j} a_{hi}) a_{hj} + (\sum_{i\neq j} b_{hi}) b_{hj} \Big]$$

from which it follows by using equation (3) that

$$C_{SIR}(\vec{X}) = \frac{1}{V} \sum_{h} \frac{|\Delta_h| m}{|F_H|} \Big[\sum_{i \neq j}^{N}\sum (a_{hi} a_{hj} + b_{hi} b_{hj}) \Big] \tag{4}$$

and then by combining equations (3) and (4),

$$C_{SIR}(\vec{X}) = \frac{1}{V}\sum_h \frac{|\Delta_h| m}{|F_H|}\left[(A_h a_h + B_h b_h) + \sum_{i \neq j}^{N}\sum (a_{h_i} a_{h_j} + b_{h_i} b_{h_j})\right]. \tag{5}$$

This expression thus corresponds to computing an SIR phasing set based on a tentatively known heavy atom set plus a test site at $\vec{X}$. The test site has N equivalent positions of which the jth position is omitted and this operation is done N times, each time omitting another site. The omitted site is sampled at its predicted position to determine the height of the SIR difference density.

Comparison of the Patterson and feedback methods

The Patterson search method corresponds to expression (1) and the feedback search procedure corresponds to expression (5). These expressions are identical except for the coefficients Δ_h^2 and $|\Delta_h| m/|F_H|$, respectively. Since the figure of merit must be small whenever $|F_H|$ is small (the phasing power is small when the heavy atoms do not make a useful contribution to a particular reflection), it is reasonable to consider the term $m/|F_H|$ being roughly proportional to the difference $|\Delta_h|$. Thus, although the coefficients for the Patterson method and the feedback method are not identical, nevertheless they are very similar. Extensive tests on these procedures using different functions for m show little change in the results.[8]

The usual Patterson search procedures in which a Patterson map is stored and then sampled at the vector positions corresponding to a test site is relatively fast,[3] particularly when the programme can be vectorized as in a CYBER 205 computer. Effectively a complete Fourier summation must be performed for every test site in the feedback procedure as expressed by the analytical expressions (1) or (5). Thus, it is a slow method. It has, however, been found useful to explore carefully a series of presumed positions when the effect of data resolution or preparation must be questioned. For instance the 5 mM $Au(CN)_2$

derivative of human rhino (common cold) virus HRV14 lacked isomorphism beyond about 7.5 Å resolution. Inclusion of data to 5.5 Å in the Patterson had a deleterious effect on the search results.

Acknowledgements

The work described here originated from initially disappointing results while attempting to determine heavy atom positions in human rhinovirus R14 (HRV14) using Patterson search procedures. These problems were eventually solved when the data processing was improved and when a better 1mM $Au(CN)_2$ derivative became available. The close equivalence of different crystallographic procedures was of some interest and is reported here as appropriate to a meeting in honour of the Patterson function. Although other methods are sometimes preferred, nevertheless the Patterson concepts are exceedingly graphic and this can provide a depth of understanding which might otherwise be missed. The analyses given here and in particular their application to HRV14 was possible due to the discussions within the rhinovirus team (Eddy Arnold, Gert Vriend, Jim Griffith and Ming Luo) in my laboratory. The observations discussed here, plus some actual results for HRV14, are in press. The work was supported by grants from the NIH (AI), NSF and a Schoewalter grant from Purdue University.

References

1. Patterson, A. L. *Phys. Rev.* **45**, 762 (1934).
2. Patterson, A. L. *Phys. Rev.* **46**, 372-376 (1934).
3. Argos, P. and Rossmann, M. G. *Acta Cryst.* **B32**, 2975-2979 (1976).
4. Dickerson, R. E., Weinzierl, J. E. and Palmer, R. A. *Acta Cryst.* **B24**, 997-1003 (1968).
5. Rossmann, M. G., Arnold, E., Frankenberger, E., Griffith, J. P., Erickson, J. W., Johnson, J. E., Kamer, G., Luo, M., Mosser, A., Rueckert, R. R., Sherry, B. A. and Vriend, G. *Nature (London)* **317**, 145-153 (1985).
6. Blow, D. M. and Rossmann, M. G. *Acta Cryst.* **14**, 1195-1202 (1961).
7. Dickerson, R. E., Kendrew, J. C. and Strandberg, B. E. *Acta Cryst.* **14**, 1188-1195 (1961).
8. Rossmann, M. G., Arnold, E. and Vriend, G. *Acta Cryst.* in press (1986).

21. Searching for non-crystallographic rotational symmetry in Patterson space

Christer E. Nordman

A procedure is described for determining the orientation in the unit cell of a non-crystallographically symmetric molecule by searching the Patterson function with a symmetry grid. The procedure utilizes the entire symmetry of the point group. An improved search algorithm applicable to Patterson functions of higher than $P\bar{1}$ symmetry anticipates the presence of two or more overlapping, symmetric vector sets. Searches of the Patterson function of STNV virus at 12 Å resolution and of hemocyanin are described.

Rotational search of the origin region of the Patterson function is a well established approach to the exploitation of non-crystallographic symmetry in macromolecular crystallography. The standard method is the Rotation Function of Rossmann and Blow[1], based on rotating the Patterson function onto itself, and evaluating the correlation between the two copies. Several improvements[2,3] have further strengthened the method.

In this paper we describe an alternative procedure for retrieving the orientation in the unit cell of the non-crystallographic symmetry elements of any chosen point group. The procedure involves a search, in the three angular dimensions, of a Patterson function stored in the computer memory. The search is done by superposing onto the Patterson function, a *symmetry grid*, that is, an array of points which has the exact symmetry of the point group in question.

A description of symmetry grids and their use in improving electron density functions through phase refinement and phase extension,[4] and a brief account of their application to rotational Patterson search[5] have previously been given.

In this paper we discuss symmetry-grid rotational Patterson searches of STNV virus, at low resolution, and of the hexameric protein hemocyanin. We also consider a modified search algorithm, applicable when the (crystallographic) symmetry of the Patterson function is higher than $P\bar{1}$. In such cases the intramolecular Patterson function consists of two or more overlapping vector distributions, each having the non-crystallographic symmetry. It is reasonable to expect that the search for non-crystallographic symmetry can be made more discriminating if it is carried out in cognizance of the overlapping, crystallographically related vector distribution(s).

Computational aspects

The Patterson function is calculated with its origin peak removed, and with at least moderate sharpening applied to the resulting coefficients. In the origin removal step, reflections which are weak, or unobservably weak, give rise to negative Fourier coefficients, while reflections of unknown structure amplitude are deleted. This is in keeping with the fact that the former contain experimental information, but the latter do not. Unfortunately, macromolecular X-ray data collection procedures, aimed as they are toward structure solution by isomorphous replacement, do not always make a distinction between these two classes of reflections.

In the cases discussed here the Fourier coefficients of the Patterson function were calculated as follows: the raw $F^2(hkl)$ data were first sharpened by a factor $\exp[S(\sin\theta/\lambda)^2]$ where the sharpening parameter S was chosen so as to remove most of the F^2 falloff with sin θ. The data were then subdivided into 'boxes' in *hkl* space, such that each box contained 100-200 reflections. The mean F^2 value within each box was subtracted from the individual values, and a moderate exponential damping applied. Thus, the Patterson coefficients were computed as

$$[F^2_{sh}(hkl) - < F^2_{sh} >_{(hkl)}] \exp(-D \sin^2 \theta\lambda^2)$$

where the $< F^2_{sh} >_{(hkl)}$ are averages of the sharpened F^2-values in local regions of *hkl*.

The subroutines which compute the symmetry grid accept as input the Hermann-Mauguin symbol of the point group, an outer envelope file, or radius, and, optionally, an inner envelope or radius. They also require a step parameter specifying the desired spacing, in Å, of the points in the symmetry grid.

The symmetry grid is subdivided into asymmetric sectors according to the specified symmetry, and into spherical shells of the desired radial spacing. Grid points are placed on each spherical shell so as to give uniform coverage within the asymmetric sector consistent with the specified spacing. A stereo diagram illustrating the design of a symmetry grid is shown in Fig. 21.1.

The symmetry grid is placed in the stored Patterson map,

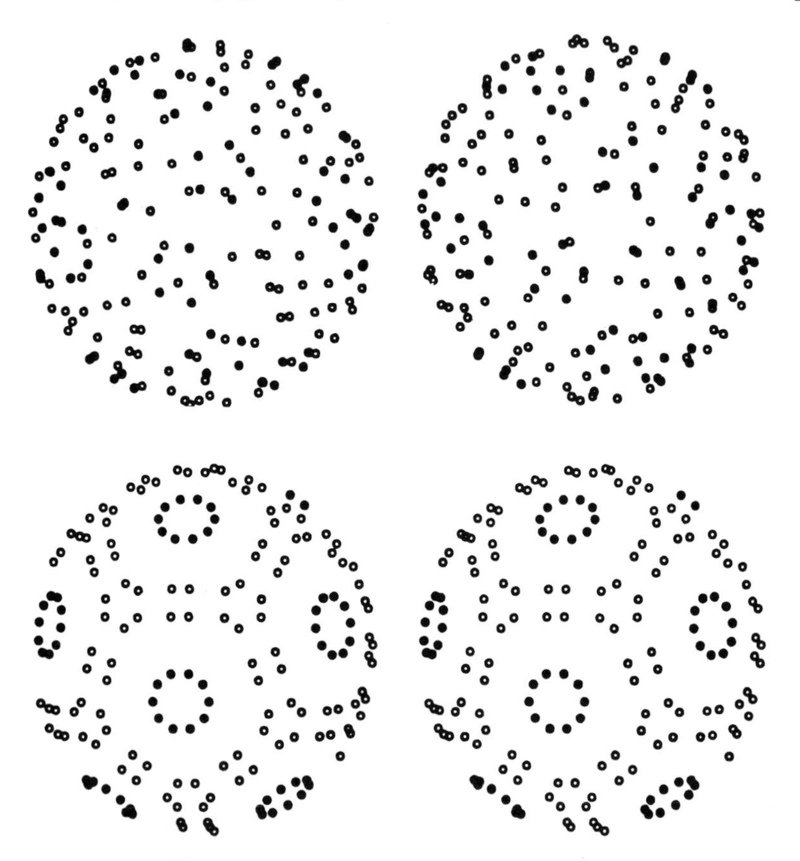

FIG. 21.1. Top: Stereo view of a symmetry grid in the icosahedral point group 532. One set of 60 symmetry-related grid points is accentuated. For clarity, this grid has only 3 points per asymmetric sector, in a single spherical shell. In the symmetry searches of a 12 Å virus Patterson discussed in the text and in Fig. 21.2, a 532 grid was employed with 89 points per asymmetric sector, in 10 shells. Bottom: Front hemisphere of the same grid with the points of the back surface projected onto the front by inversion through the center.

with its center coincident with the Patterson origin. A search is carried out by a stepwise rotation of the symmetry grid through three Euler angles, A, B, and C, which define its orientation relative to the Patterson map. A and B can be thought of as the longitude and latitude (with 0 and 180° at the poles) of a reference direction within the symmetry grid, for example a principal symmetry axis, and C as the azimuth about that axis. It is obviously not necessary to rotate the symmetry grid through the entire ranges of the Eulerian angles. If the order of the grid symmetry group is N_s, there will be N_s rotations which bring the grid into exact coincidence with itself, thus it becomes meaningless to search more than $1/N_s$ of Euler angle space.

At each step in the search the Patterson function is evaluated for every point in the symmetry grid. This is done by interpolation between Patterson map points surrounding the position of the grid point.[4] An eight-point interpolation scheme was used in the calculations reported here.

Let $s = 1, \ldots, N_s$ denote the N_s asymmetric sectors in the symmetry grid, and $i = 1, \ldots, N_i$ the grid points within one asymmetric sector. Let $\vec{x}_{is}$ be the location, in Patterson map space, of the grid point (i, s) at a given step in the search, and $P(\vec{x}_{is})$ the interpolated Patterson value at $\vec{x}_{is}$. To the extent that the symmetry of the grid is present in the Patterson function, the scatter of the Patterson values at $\vec{x}_{is}, s = 1, \ldots, N_s$ will be small, that is, the variance

$$\sigma_i^2 = (1/N_s) \sum_{s=1}^{N_s} [P(\vec{x}_{is}) - \overline{P}_i]^2$$

with $\overline{P}_i = (1/N_s) \sum_{s=1}^{N_s} P(\vec{x}_{is})$, will have a low value. A low mean square, or rms, value of the σ_i's at all points in the asymmetric sector

$$\sigma = \left[(1/N_i) \sum_{i=1}^{N_i} \sigma_i^2 \right]^{1/2} \tag{1}$$

is an indication that the Patterson function symmetry approximates the symmetry of the grid at this point in the search. The

σ-values (equation 1), evaluated at each point in the Euler-angle search, constitute the 'image seeking function' for the search. The significance of 'peaks' (minima, actually) in this function may be assessed in terms of number of standard deviations by which they differ from the mean of the function.

An improved search algorithm

The image-seeking function (equation 1) is a general criterion of symmetry fit, usable in translational and rotational search of (poorly phased) electron density functions as well as in rotational Patterson symmetry search. It may be noted that a Patterson function, being centrosymmetric, should always be searched with non-centrosymmetric symmetry grids, to do otherwise would result in wasteful sampling of the function at pairs of points with identical values.

In any application the objective is to find, embedded in a noisy environment, a significant function component with the searched-for local symmetry. In rotational Patterson searches, the non-crystallographically symmetric component is the intramolecular interatomic vector set of a molecule which possesses the non-crystallographic symmetry.

For a monoclinic crystal with two non-crystallographically symmetric molecules in the unit cell, the origin region of the Patterson, of symmetry $2/m$, contains two overlapping noncrystallographically symmetric intramolecular vector sets. Similarly, an orthorhombic cell with four molecules gives four overlapping vector sets in a Patterson region of symmetry *mmm*. Clearly, in these cases the rotation search will give two, or four, solutions, at orientations related to one another by the crystallographic 2-fold axis or axes. Since the Patterson function is more heavily overlapped with the intramolecular vector sets of more than one molecule, we may expect these searches to be less discriminating than their counterpart in the case of a triclinic cell with one molecule.

However, in the monoclinic or orthorhombic cases the additional vector overlap is not random or unknown noise, but one (or three) vector sets, identical to the one searched for, and in orientation related to the latter by known crystallographic symmetry.

In the following we outline a procedure by which the search for one vector set is carried out in cognizance of the presence of another, identical set, related to the first one by monoclinic two-fold symmetry. The approach is tantamount to a decomposition of the Patterson region of $2/m$ symmetry into two identical overlapping components of known non-crystallographic symmetry, but otherwise of unknown structure.

Consider one spherical shell of a symmetry grid, and let θ_{is} be the symmetry-grid fixed coordinates of grid point i in asymmetric sector s. When the symmetry grid is positioned at Euler angles A B, C, let $\vec{x_{is}}$ be the Patterson map coordinates of grid point is, and $P(\vec{x}_{is})$ the interpolated Patterson function value at this point in the map. If [C] is the transformation matrix for the crystallographic Patterson symmetry, here a 2-fold axis along b, and $\vec{x}'_{is} = [C]\vec{x}_{is}$, it follows that $P(\vec{x}'_{is}) \equiv P(\vec{x}_{is})$. Let $\theta'_{is'}, \phi'_{is}$ be the symmetry-grid shell coordinates of $\vec{x}'_{is}$. The point $\theta'_{is'}\phi'_{is}$ does not in general coincide with a grid point in the symmetry grid.

We represent the decomposition of the Patterson function as

$$P(\vec{x}_{is}) \approx P_1(\vec{x}_{is}) + P_2(\vec{x}_{is}) \tag{2}$$

where the right hand side represents the two overlapping non-crystallographically symmetric components. On account of experimental error and overlaps with intermolecular vectors, this relationship is not exact. The two components are assumed to be identical and related to one another by the crystallographic Patterson symmetry, so that

$$P_2(\vec{x}_{is}) \equiv P_1(\vec{x}'_{is}). \tag{3}$$

Furthermore $P_1(\vec{x}_{is})$, and hence $P_2(\vec{x}_{is})$, is noncrystallographically symmetric, that is

$$P_1(\vec{x}_{is}) = < P_1(\vec{x}_{is}) >_s \tag{4}$$

for all i and s. Substituting equation (3) into equation (2), solving for $P_1(\vec{x}_{is})$, and symmetry-averaging as in equation (4), we get the expression

$$P_1(\vec{x}_{is}) = < P(\vec{x}_{is}) - P_1(\vec{x}'_{is}) >_s \tag{5}$$

which may be used iteratively to obtain an improved approximation to $P_1(\vec{x}_{is})$, from an available, cruder approximation.

The practical implemetation is as follows. For each step in the Euler angle search, the following steps are carried out:

1) For each point *is* in the symmetry grid, identified by θ_{is} and ϕ_{is}, evaluate $\theta'_{is'}, \phi'_{is}$ and $P(\vec{x}_{is})$.
2) Initialize $P_1(\vec{x}_{is})$ to $0.5P(\vec{x}_{is})$.
3) Evaluate an improved approximation to $P_1(\vec{x}_{is})$ by equation (5).
4) Accumulate $\Sigma_{is}[P(\vec{x}_{is}) - P_1(\vec{x}_{is}) - P_1(\vec{x}'_{is})]^2$ to serve as a measure of convergence.
5) Using the improved $P_1(\vec{x}_{is})$ as input, go to step 3.

The calculations reported here were done with five iterations of steps 3 to 5, although three iterations would have given adequate convergence. The square root of the sum accumulated in step 4, at convergence, serves as the image seeking function, or measure of fit, at this point in the Euler angle search. A low value implies that equation (2) is reasonably well satisfied, and indicates a promising fit in Euler angle space.

If this calculation were done for orthorhombic Patterson symmetry, two additional settings, $\vec{x}''_{is}$and $\vec{x}'''_{is}$ is would be used, corresponding to the two added 2-fold axes. Equation (5) would be modified by inclusion of these terms, as would the expression in step 4 and, in step 2, the initialization would suitably be $0.25P(\vec{x}_{is})$. The rate of convergence would have to be assessed by actual computation.

Results and discussion

Satellite Tobacco Necrosis Virus (STNV) crystallizes in a monoclinic cell with one virus particle per asymmetric unit.[6] Thus the entire icosahedral particle symmetry,[7] point group 532, $N_s = 60$, is noncrystallographic. This presents an exceptionally favorable case for a symmetry-grid search. It is to be expected that a search with the entire icosahedral group symmetry will effectively discriminate against false signals arising from the approximately octahedral symmetry of the packing of the virus particles in the cell.[8]

The result of a search with a 532 symmetry grid, performed

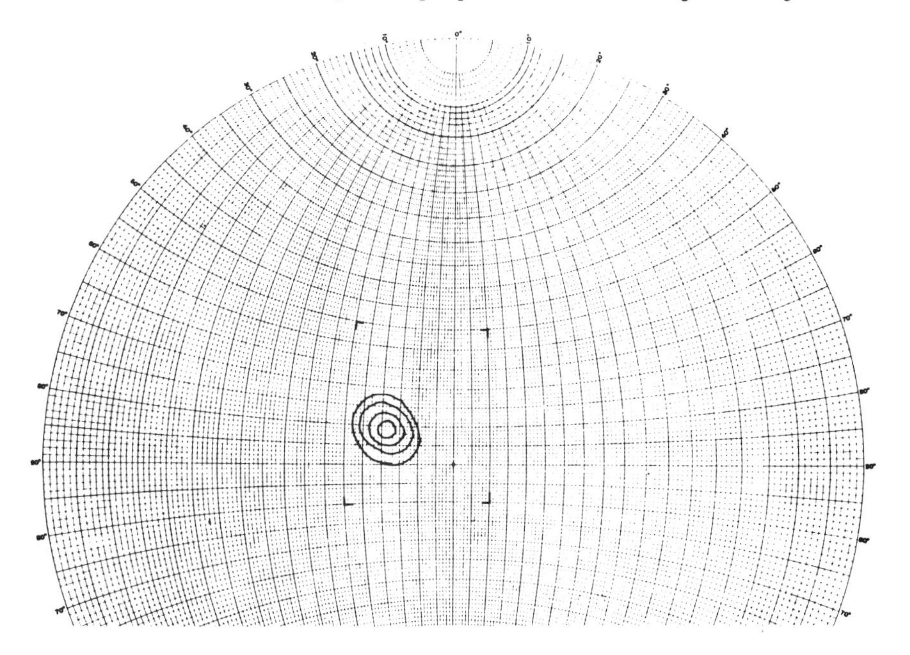

FIG. 21.2. Orientation search of the 12 Å Patterson function of STNV virus with an icosahedral symmetry grid in point group 532. Grid radius 11 – 53 Å. The search covers approximately one icosahedral asymmetric sector in Euler angle space. The (A,B) section near the optimal C is shown, contoured at 1.0, 1.5, 2.0 and 2.5 standard deviations from the mean.

using equation (1) as the image seeking function, is shown in Fig. 21.2. It gives the correct orientation unambiguously. A search of the same Patterson function with an octahedral symmetry grid, in point group 432, with a 55 Å outer grid radius was also carried out. A peak indicative of the interparticle packing symmetry was indeed found, although it was lower (maximum 1.80) and much broader that the peak in Fig. 21.2.

Hemocyanin is a hexameric protein of molecular symmetry 32 (D_3). It crystallizes in a monoclinic cell, space group $P2_1$, with two molecules per cell.[9] Since the point group is of relatively low order, $N_s = 6$, hemocyanin exemplifies a more widely encountered case than does STNV. The conventional symmetry grid search is unambiguous at 5 Å, and probably lower, resolution.[5] Our objective here is to test the improved

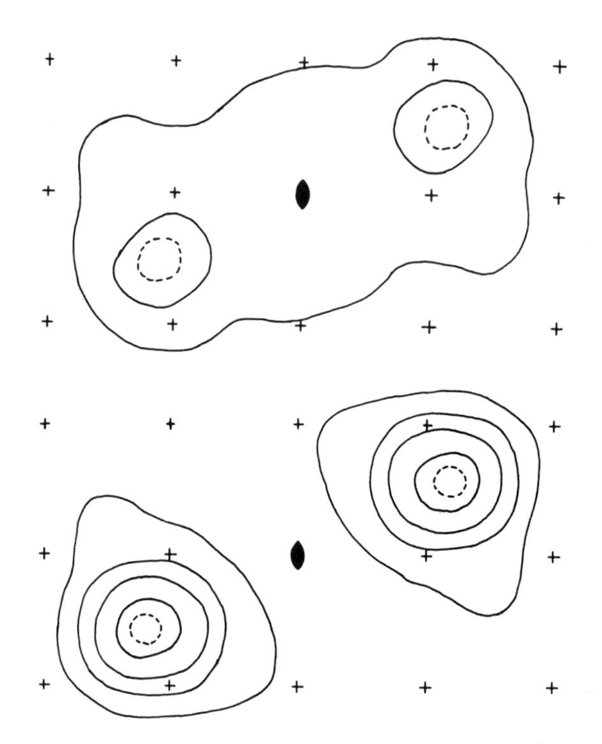

FIG. 21.3. Comparison of two Patterson rotation search algorithms applied to hemocyanin. Top figure: conventional symmetry grid search (equation (1)). Bottom figure: improved algorithm which treats the Patterson function as two overlapping vector distributions, each of 32 symmetry. The maps are (A,B) sections, at the optimal C-value, contoured at 4, 6, 8 and 10 standard deviations from the mean (solid contours). Broken contours are at 7 and 11 standard deviations. Grid markings at 5° intervals.

image seeking function discussed in the preceding section, and to compare it to the conventional one, given in equation (1).

The results are shown in Fig. 21.3, as obtained with a Patterson function at 3.2 Å resolution. The peaks, in **A** (horizontal), B (vertical) space, show the directions of the molecular 3-fold axes of the two molecules. The crystallographic 2-fold axis is also shown. It is clear that the improved version gives the most discriminating peak.

Acknowledgements

The author is indebted to Professor Bror Strandberg, Uppsala, and Professor Wim Hol, Groningen, for the X-ray data for STNV and hemocyanin.

References

1. Rossmann, M. G. and Blow, D. *Acta Cryst.* **15**, 24-31 (1962).
2. Crowther, R. A. In *The Molecular Replacement Method.* (M. G. Rossmann, ed.) pp. 173-178. Gordon and Breach: New York, NY (1972).
3. Lattman, E. E. In *The Molecular Replacement Method.* (M. G. Rossmann, ed.) pp. 179-183. Gordon and Breach: New York, NY (1972).
4. Nordman, C. E. *Acta Cryst.* **A36**, 747-754 (1980).
5. Nordman, C. E. and Hsu, L.-Y. R. (1982). In *Computational Crystallography.* (D. Sayre. ed.) Oxford University Press: New York, NY (1982).
6. Unge, T., Liljas, L., Strandberg, B., Vaara, I., Kannan, K. K., Fridborg, K., Nordman, C. E. and Lentz, P. J. *Nature (London)* **285** 373-377 (1980).
7. Klug, A. *Cold Spring Harbor Symp. Quant. Biol.* **36**, 483-487 (1971).
8. Akervall, K., Strandberg, B., Rossmann, M. G., Bengtsson, U., Fridborg, K., Johannisen, H., Kannan, K. K., Lovgren, S., Petef, G., Oberg, B., Eaker, D., Hjerten, S. and Ryden, L. *Cold Spring Harbor Symp. Quant. Biol.* **36**, 469-483 and 487-488 (1971).
9. Gaykema, W. P. J., Hol, W. G. J., Vereijken, J. M., Soeter, N. M., Bak, H. J. and Beintema, J. J. *Nature (London)* **309**, 23-29 (1984).

22. Automation of rotation functions in vector space

Paul T. Beurskens and Gezina Beurskens
Marianna Strumpel and C. E. Nordman

A fully automated program ORIENT for the determination of the orientation of a known molecular fragment is described. This program is based on the Vector Search programs of Nordman and Schilling.[1]

A slightly sharpened Patterson function is stored. Intramolecular vectors of the known molecular fragment are calculated and selected; vector weights take into account the overlap with neighbouring vectors. The image seeking function is defined as the 'weighted minimum average' function. It is evaluated at each point as the vector set is rotated through the symmetry-independent part of the Eulerian angle space. A coarse scan, using only short vectors, is followed by a number of fine scans to determine the best possible orientations of the fragment.

The program ORIENT is incorporated in the DIRDIF program system. An oriented fragment is positioned by strengthened translation functions[2] and expanded by the DIRDIF phase refinement procedure.

After fifty years of ups and downs in Patterson methods, we now see a renewed interest, especially in the field of 'small'-molecule structure analyses. Patterson methods received relatively little attention during the days when direct methods were computerized and automated. The development of automated direct methods programs induced a significant change in structure-solving-philosophy: routine crystal structure determinations by inexperienced crystallographers became possible, and came into widespread use. Non-automated Patterson programs in vector space were neglected in routine structure analyses.

The present Patterson revival[3,4] has a number of reasons:

1) Patterson methods provide powerful alternatives to direct methods for a large number of different structures (rigid groups, expected geometries, 'medium'-heavy atom structures).
2) Experience in design of user-friendly automatic direct methods programs is transportable to automation of Patterson methods.

3) A vast amount of molecular geometries is readily available in an evergrowing crystallographic database.

In this chapter we describe how *a priori* molecular structure information can be used for an automatic solution of the structure, with computer programs that can compete with direct methods programs in modesty of computer requirements as well as in user-friendliness. No previous experience with Patterson methods or direct methods is needed.

1. Patterson search methods

The molecular replacement method for macromolecular structures is usually based upon finding the correlation between calculated and observed structure amplitudes, as a function of the orientation and the position of a 'known molecule' in the unit cell. Structure amplitudes (or $|E|^2$ values) for low-resolution data, calculated from a large number at atoms, are not very sensitive to errors in the atomic positions, as long as the majority of the atoms are at approximately correct positions. Individual deviations (hopefully of a random nature) are averaged out.

For small molecules the situation is different. As higher-resolution data are used, for a smaller number of atoms, errors in the atomic positions have a much more pronounced effect on the calculated structure factors. Moreover, the molecular fragment is usually very accurately known. For example, a 10 atom, well known rigid molecular fragment in a 100 atom/cell structure in space group $P1$ places restraints on the structure amplitudes which are strong enough to be effectively exploited in the structure solution.

Correlation functions in reciprocal space can be used to find the orientation of the molecule, but there are good reasons to prefer the use of rotation functions in direct space, *i.e.,* to use Patterson vector search methods. Each individual interatomic vector must be 'present' in the Patterson function! Therefore, rotation functions in Patterson space (= vector space) are very reliable. However, they are very sensitive to various run-time parameters, and the use of these methods (and their success!) is dependent on the ability and experience of the crystallographer who is using them.

In addition, and in contrast to a general perception, the calculation of a Patterson rotation function can be very fast. The automatic program ORIENT for the computation of rotation functions will be discussed in Section 3. The program is incorporated in the DIRDIF program system.

The case of translation functions is somewhat different: both direct methods and vector search methods can be used competitively to find the position of the correctly oriented fragment with respect to the symmetry elements. The translation functions used in the DIRDIF system are based on direct methods, and their use is briefly discussed in Section 4.

2. Rotation functions in vector space

Rotation functions in vector space employ *a priori* molecular structure information to get a start at the solution of the phase problem. The approach is briefly described as follows:

1) A rigid molecular group of known geometry (calculated, or retrieved from data in the literature) is defined by a set of atomic positions r_i. The vector set $[\vec{v}_i]$ is obtained from all interatomic vectors $r_i - r_j$, but modified by various weighting schemes and cut-off options.
2) The vector set is rotated stepwise through three Euler angles, A, B and C, in order to scan all possible orientations. For each orientation of the vector set (point (A, B, C) in angular space) its 'presence' is looked up in the Patterson vector map. A measure of fit, denoted by R, is evaluated. Usually R is a minimum and/or a sum function. The rotation function $R(A, B, C)$ thus describes the fit of the model vector set to the Patterson vector map for all possible orientations of model.
3) The highest values (peaks) in the rotation function represent possibly correct orientations of the known fragment. One or more orientations (A, B, C) are selected for further structure elucidation. In all space groups but $P1$, translation functions can be used to find the position of the fragment.

We have used two different vector search programs for solving structures that could not be solved by routine application of direct methods, namely, the program of Braun, Hornstra and Leenhouts[5] and that of Nordman[6] and Schilling.[7]

The program of Braun, Hornstra and Leenhouts uses the sum function

$$R(A,B,C) = \sum P(\vec{v}_i(A,B,C))$$

where P is the Patterson function value for vector $\vec{v}_i$ rotated by angles A, B, C; the summation is over all interatomic vectors.

It has been pointed out by various authors, that the sum function, when all vectors in the structure are used, gives results similar to rotation functions based on $|F|^2$- correlations using all reflections.

The Vector Search program of Nordman and Schilling uses the 'weighted minimum average function'

$$R(A,B,C) = \min(N) = \sum^{N} P(\vec{v}_i(A,B,C)) \,/\sum^{N} W_i$$

where W_i is the weight of vector $\vec{v}_i$ and the summation $\sum^N$ is over the N vectors with the lowest values of P_i/W_i. For $N = 1, R(A,B,C)$ is just the minimum function. For N equals the total number of vectors, $R(A,B,C)$ is the weighted sum function. A good choice for N is about one fifth of the total number of vectors. As the maxima in the $R(A,B,C)$ function depend on a small number of minimum values P/W, the evaluation of the function should be rather accurate. The IBM version uses an 8-bit Patterson function representation and 4- or 8-point interpolation.

For various important features in the program, see the original publications .[8,9]

In the experience of the Nijmegen group, the two programs always gave the correct orientation of the molecular fragment, albeit sometimes in two- or multifold ambiguity.

3. Automation of rotation functions

In this section we describe ORIENT, which is a version of the Nordman and Schilling[1] Vector Search program, modified for the fully automatic determination of the orientation (or possible orientations) of a known molecular fragment.[4] A supervisor routine defines default values for various starting parameters, controls the subsequent branching in the program, and calculates all further control parameters.

Some of the parameters given below are subject to modification, dependent on the peculiarities of the structure at hand. In the following description we use these parameters as if they were constants.

Preliminary Calculations: A slightly sharpened Patterson function will be calculated and stored for fast Patterson function value retrieval. Note that only a relatively small block around the origin is to be stored, as the maximum vector length is fixed to, say, 7 Å. Various sharpening functions are available, and the corresponding function of a single C-C Patterson peak is calculated[1] or approximated from the shape of the origin peak. This peak shape is used for the calculation of vector weights due to overlap of neighbouring vectors. A possible change of weights for different orientations, due to accidental overlap with symmetry related vector sets, is not taken into account (to save computer time).

The important systematic overlap due to the presence of symmetry in the Patterson function is easily accounted for by modifying the Patterson function values on and near to symmetry elements (instead of increasing the weights of vectors approaching a Patterson mirror plane, the stored Patterson values are proportionally decreased; this prevents planar molecules from lining up with the mirror plane).

Definition of the Eulerian angles A, B, C : Several systems of angular variables can describe the orientation of the molecule. Eulerian angles are most commonly used. A second angular system is defined by a spin about an axis given by spherical polar coordinates. It is useful to consider both systems simultaneously.

The conventions for the definition of Eulerian angles used here are as follows (see Fig. 22.1):

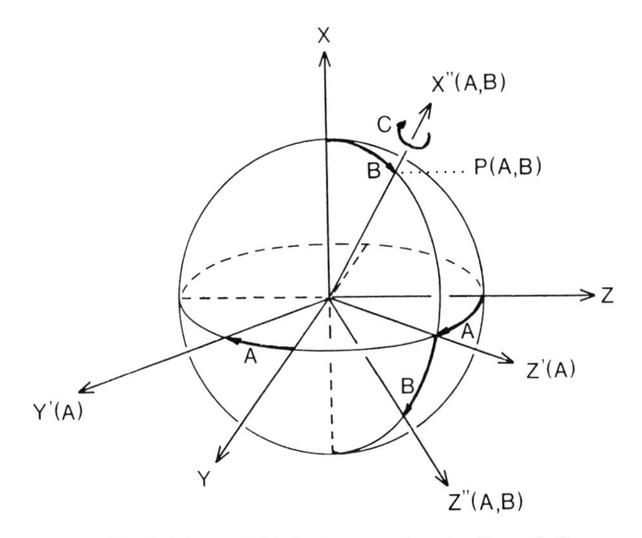

FIG. 22.1. Definition of Eulerian angles A, B and C.

(1)

A is a rotation of the molecule about the Cartesian X axis
B is a rotation of the molecule about the rotated Y axis (Y')
C is a rotation about the rotated X axis (X'')

The Eulerian angles are convenient for computational reasons, but there are two drawbacks: the A and C rotations are not independent, and the optimum sampling step size for A depends on the value of B. Therefore, we also consider an interpretation of A, B and C in terms of polar spherical coordinates. Define:

(2)

A is a rotation of the axial system about the X axis (the molecule remains fixed)

B is a rotation of the molecule and the axial system about the rotated Y axis (Y')

$(C + A)$ is the spin (rotation of the molecule) about the rotated X axis (X'').

The calculations are performed using definitions (1), but the following discussion uses the interpretation (2). Thus $X''(A,B)$ is the direction to which the original molecular X axis has been rotated. This direction is characterized by a point $P(A,B)$ on the surface of a sphere, see Fig. 22.1. The sphere represents all possible angular coordinates A,B.

The angular distance ΔAB between two points $P(A,B)$ and $P(A',B')$ is the smallest angle that rotates $X''(A,B)$ to $X''(A',B')$. For neighbouring points the approximate value of ΔAB (in radians) is calculated from

$$\begin{aligned}(\Delta AB)^2 &= (\sin A \sin B - \sin A' \sin B')^2 \\ &\quad + (\cos A \sin B - \cos A' \sin B')^2 + (\cos B - \cos B')^2.\end{aligned}$$

For the polar regions ($B = 0°, 180°$) the last term may be neglected; near the equator ($B = 90°$) this expression simplifies to

$$(\Delta AB)^2 = (\Delta A)^2 + (\Delta B)^2.$$

Now we consider the spin. In the Eulerian system (1), both rotations A and C contribute to the total spin: $C+A$. The spin difference ΔCA between two points (A,B,C) and (A',B',C') is

$$\Delta CA = C + A - C' - A'$$

The spin $C+A$ is independent of the direction $X''(A,B)$, and the distance ΔABC between two neighboring points in angular space is postulated as

$$(\Delta ABC)^2 = (\Delta AB)^2 + (\Delta CA)^2.$$

This result is useful for the planning and evaluation of rotation functions. Optimum sampling in angular space implies that angular steps $\Delta A, \Delta B$ and ΔC are correlated with the displacements of interatomic vectors in the Patterson map. The step size ΔA certainly should be made dependent on the value of B. The calculation of variable and optimum step sizes ΔA and ΔB leads to a rather cumbersome algorithm. Therefore, it is advisable to cover the sphere (A,B) by points $P(A,B)$ which are only approximately equidistant as measured by ΔAB. The choice of the step size ΔAB is determined by the length of the longest vector, and the resolution of the data. But as the spin is independent of the direction of $X''(A,B)$, the step size ΔC can be chosen independently. When the search model happens to be elongated along X, the longest vector component perpendicular to the molecular X axis ($V\,max_{yz}$) is shorter than the longest vector ($V\,max$). We may then use larger steps for the C rotations. The effective angular distance between two neighboring points is given by

$$(\Delta' ABC)^2 = (\Delta AB)^2 + (q \cdot \Delta CA)^2$$

with $q = V\,max_{yz}/V\,max$.

Strategy of the program ORIENT

The following steps are executed consecutively:

- Initiation
  - read atomic parameters and prepare vectors
  - find symmetry of the vector set
  - read and modify Patterson function
- Cycle 1. Coarse scan
  - generate scan ranges (Note 1)
  - average step size: $\Delta' ABC = 10°$
  - longest vector used: 4.5 Å
  - get $P(\vec{v}_i)$ as largest value from eight stored points around $\vec{v}_i$
  - calculate $R(A,B,C)$ of all points (A,B,C) in the scan ranges
  - collect all points (A,B,C) which have significant values of $R(A,B,C)$; level: 2 sigma.
- Cycle 2. Medium scans
  - define regions covering all significant points (Note 2)
  - step size: $\Delta' ABC = 5°$
  - longest vector used: 6.0Å
  - get $P(\vec{v}_i)$ by 4-point interpolation
  - calculate $R(A,B,C)$ for all points in the regions

collect all points with relatively large values for $R(A,B,C)$; level: 50% of the maximum value of the function $R(A,B,C)$.

- Cycle 3 and 4. Fine scans
  define blocks of $4 \times 4 \times 4$ points around each local maximum, found in cycle 2 (Note 3)
  step size: $\Delta ABC = 2°$, and repeat with step size $= 0.7°$
  longest vector used: 7.5 Å
  get $P(\vec{v}_i)$ by 8-point interpolation
  calculate corresponding $R(A,B,C)$ values
  find peak position in each block; level: 70% of maximum value.
- For each peak (A,B,C): rotate atomic parameters by A,B and C, and output the parameters of the oriented model.
- For the most probable orientation (point of maximum $R(A,B,C)$): prepare input control and data stream for the execution of translation functions.
- Transfer control to the translation function program.

Note 1. Scan ranges, covering the symmetry independent part of angular space[10], are generated. These ranges depend on the symmetry of the Patterson function and on the symmetry, if any, of the search fragment and, in general, on the orientation of the (symmetric) fragment in the Cartesian coordinate system. At present, the program cannot handle all possible combinations of symmetry elements, but, except for the added computing time, failure to find minimal search ranges is harmless, since symmetry-equivalent peaks will be carried forward by the automatic procedure.

The C rotation is executed in the innermost loop, for fixed A and B, and with fixed incremental steps for C. The order of (A,B) values (points on the (A,B) sphere) is irrelevant. It is therefore possible to define a network of points $P(A,B)$ on the sphere minimizing the total number of points to be considered, see Fig. 22.2. Around $B = 90°$ the net is a square grid of 10×10^0 , and the maximum deviation of an arbitrary point in this grid is 7°. The rest of the net, as given in Fig. 22.2, also has a maximum deviation of 7°. This sampling method is very efficient but does not yield undistorted maps as does the

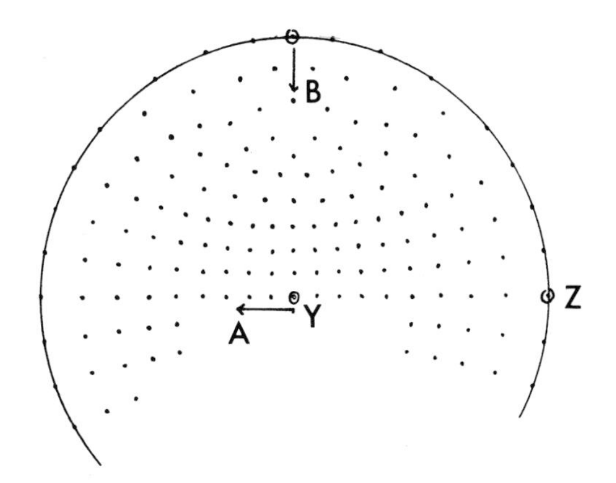

FIG. 22.2. The coarse (A, B) grid plotted on a Wulff net, *i.e.* a stereographic projection of points $P(A, B)$.

orthogonal sampling method of Lattman.[11] However, in our procedure the functions are not visually interpreted.

Note 2. If the scattering power of the search fragment is at least 5% of the total scattering power, the orientation function $R(A, B, C)$ will contain a limited number of peaks which rise significantly above the average value of the function. Some of these points are neighbors in angular space ($\Delta' ABC < 12°$). A block around each neighborhood (one or more points) defines a region.

Note 3. The most crucial step in the recycling procedure is the definition of blocks for the fine scans. There is no problem if well defined peaks result from the medium scan. But we do have a problem when the rotation function shows bananas as a result of orientational Patterson overlap (because of symmetry, or because of the presence of similar fragments, or because the model is not completely correct). We then have many neighbouring points with about equal values for $R(A, B, C)$. For such cases, blocks for fine scans are defined as follows. The point (A, B, C) with the highest R value is accepted and its neighbors ($\Delta' ABC < 8°$) are removed. Thereafter, the next maximum is accepted, and so on.

4. Translation functions

Although the Nordman and Schilling[1] Vector Search programs

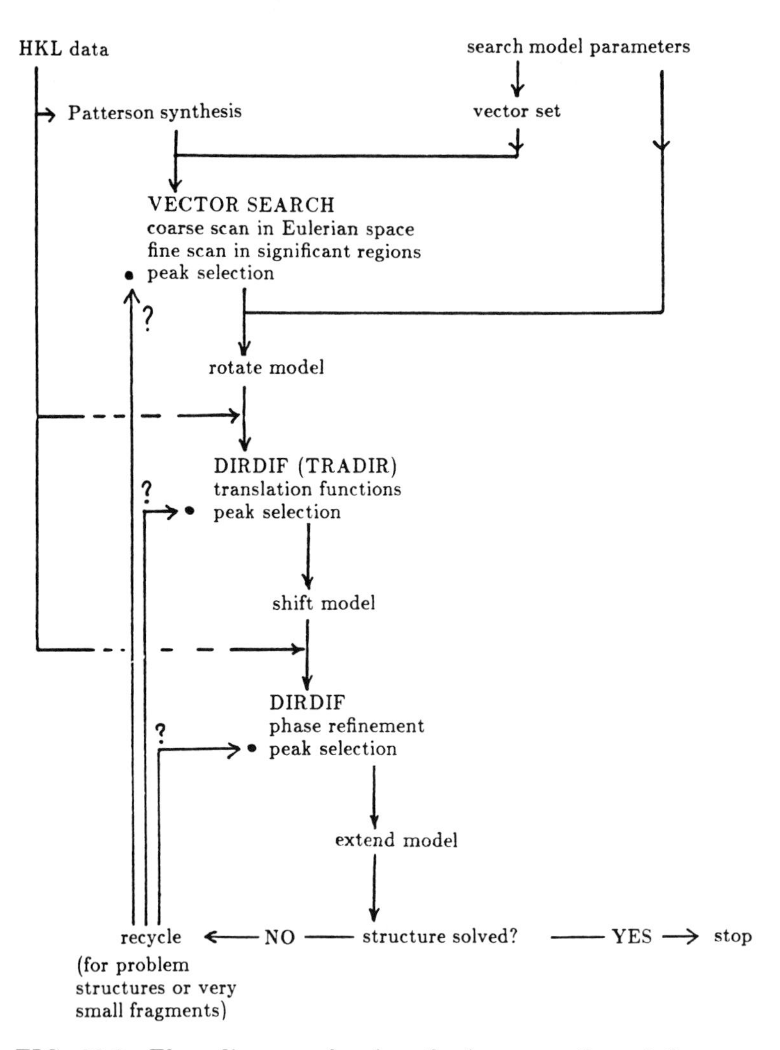

FIG. 22.3. Flow diagram showing the incorporation of the program ORIENT in the DIRDIF program system.

allow the positioning of an oriented fragment with respect to the space group symmetry elements, we have not yet automated this part of the programs. Instead, we can use an automated program, TRADIR, for the calculation of strengthened translation functions.[2] This program applies convolution methods to Fourier coefficients, obtained by application of direct methods to difference structure factors.[12]

The outcome of TRADIR is a shift vector which brings the fragment to its correct position. The shifted fragment is used as input to DIRDIF for further elucidation of the structure. The flow diagram (Fig. 22.3) shows the incorporation of the program ORIENT in the DIRDIF system. The most probable orientation and its most probable shift vector usually lead to the (almost) complete structure in one run. Only for 'problem structures' or with very small fragments (scattering power less than 10% of the entire cell) recycling may be needed.

References

1. Nordman, C. E. and Schilling, J. W. In: *Crystallographic Computing* (F. R. Ahmed, ed.) pp. 110–114. Munksgaard; Copenhagen (1970).
2. Doesburg, H. M. and Beurskens, P. T. *Acta Cryst.* **A39**, 368-376 (1983).
3. Egert, E. *Acta Cryst.* **A39**, 936-940 (1983).
4. Strumpel, M., Beurskens, P. T., Beurskens, G., Haltiwanger, R. C. and Bosman, W. P. *Abstracts, ECM8*, Liege (1983).
5. Braun, P. B., Hornstra, J. and Leenhouts, J. I. *Philips Res. Rept.* **24**, 85-118 (1969).
6. Nordman, C. E. *Trans. Am. Cryst. Assoc.* **2**, 29-38 (1966).
7. Schilling, J. W. In *Crystallographic Computing* (F. R. Ahmed, ed.) pp. 115–123. Munksgaard: Copenhagen, Denmark (1970).
8. Schilling, J. W., Hoge, L. G., Strumpel, M. and Nordman, C. E. Patterson Search Programs — Writeup and User's Guide. The University of Michigan: Ann Arbor, Michigan (1981).
9. Strumpel, M., Thesis. Free University: Berlin, Federal Republic of Germany (1983).
10. Tollin, P., Main, P. and Rossmann, M. G. *Acta Cryst.* **20**, 404-407 (1966).
11. Lattman, E. E. *Acta Cryst.* **B28**, 1065-1068 (1972).
12. Beurskens, P. T., Bosman, W. P., Doesburg, H. M., Van den Hark, Th. E. M., Prick, P. A. J., Noordik, J. H., Beurskens, G., Gould, R. O. and Parthasarathi, V. In *Conformation in Biology* (R. Srinivasan and R. H. Sarma, eds.) pp. 389-406. Adenine Press: New York (1982).

23. Equal atom Patterson methods: practical experience with rotation and translation functions for the solution of the phase problem for crystal structures of complex organic molecules

John J. Stezowski and Emil Eckle

Over the last decade or so, there has been increasing interest in analyzing crystal structures for complex organic molecules. Typically, the asymmetric unit of the crystals in question contains eighty or more C, N and / or O atoms and no particularly 'heavy' atoms. We have determined crystal structures for a number of such systems, even though, in our hands, direct methods had failed. In such cases, we were able to solve the phase problem by the application of Patterson methods (rotation and translation functions), often in combination with direct methods. Since these methods use either a known structure or a model, they are sometimes referred to as 'Molecular Replacement'[1] techniques.

Being convinced of the power and utility of these methods, we hope to demonstrate by a presentation of our experience with them that they are of practical value and hope that we can stimulate increased interest in their use and development. In many ways the level of application of rotation and translation functions to the solution of 'equal atom' problems resembles that of direct methods about ten years ago. We find that these methods are particularly powerful where the data set resolution, crystal quality, molecular symmetry, and/or the size of the asymmetric unit are not necessarily suitable for the straightforward application of direct methods.

The examples discussed are drawn from crystal structure determinations for a number of anthracyclines.

The anthracyclines

As their name implies, the anthracyclines are a family of compounds that contain similar anthraquinone chromophores as part of their chemical structure. We have studied two subgroups of anthracyclines: daunomycin and nogalamycin analogs, and a totally synthetic analog with a modified chromophore.

It is immediately clear to the chemist that the anthraquinone moiety will have a nearly planar conformation. To the crystallographer, this chemical structure is readily recognized as a source of the well-known 'chicken wire' effect frequently encountered in the application of direct methods. Several of the anthracyclines for which we have determined structures crystallized with two molecules per asymmetric unit; in some cases the structure determinations were further complicated by the presence of obvious pseudo-translational symmetry and relatively poor data sets.

General theory and application of Patterson search methods

A review of the techniques and applications of these methods is presented in *The Molecular Replacement Method.*[1] Briefly, when the geometry of all or part of a molecule is known, this information can be used for the solution of the phase problem.

If one defines $\vec{r}_i$ as the crystal coordinates of an atom of a molecule or model and $\vec{r}_{oi}$ as the coordinates of the same atom relative to a molecular coordinate system, then the position of the molecule (or model) can be specified by equation (1):

$$\vec{r}_i = \vec{r}_o + [A]\vec{r}_{oi}, \tag{1}$$

where the matrix $[A]$ serves to orient the molecule in the crystal coordinate system and $\vec{r}_o$ is the origin of the molecular coordinate system. If there are additional symmetry elements present in the unit cell the system is described by equation (2):

$$\vec{r}_{ij} = [S_j](\vec{r}_o + [A]\vec{r}_{oi}) + \vec{t}_j, \tag{2}$$

where $[S_j]$ and $\vec{t}_j$ are the rotation and translation matrices for symmetry operator j. Matrix $[A]$ and vector $\vec{r}_o$ can in principle be determined from an analysis of the Patterson map. Mathematical functions, commonly called rotation and translation functions have been programmed to facilitate this analysis.

Computer programs employed

Numerous computer programs were used in the course of the work described here. We make no claim that any one of them has been fully tested with respect to its potential in the applications we summarize. In alphabetical order, the programs used are: DIRDIF,[2] MULTAN-78,[3] MULTAN-80,[4] PATSEA,[5] PROTEIN,[6] PSR,[7] SHELXS-84,[8] TRADIR,[9] and the XRAY-System.[10]

The procedure used to carry out the rotation and translation function calculations with the PSR programs will be described briefly. These programs implement Nordman's[11,12] vector search method, the application of which is generally similar to those for the Hornstra's[13] vector search technique and for the 'Faltmolekül' method of Hoppe[14] and Huber.[15]

General procedure for the PSR programs

1) Calculation of the Patterson function with an external Fourier program. As fine a grid as is consistent with the quality of the data should be used. The Patterson is then scaled and converted into an internal form by routine DEK.
2) Preparation of a search model in a Cartesian coordinate system.
3) Generation of $N(N-1)/2$ **intra**molecular vectors between the N atoms of the model, weighting them with VSEL.
4) Determination of the orientation with routine ORV by stepwise rotation of the selected vector set through the asymmetric unit of the rotation space group.[16] Estimation of the agreement between the model vector set and the Patterson function for each rotation step (ISF). A fine scan is made around the best peaks to optimize the orientation.
5) Calculation of the $N(N+1)/2$ **inter**molecular vectors (oriented model) for each symmetry element of the space group. Routine VEC is used to weight the vectors and VSEL to select the strongest vectors for subsequent calculations.
6) A stepwise superposition of the intermolecular vector set (for each symmetry element) with the Patterson function is carried out and the agreement established for each step.

Fine scans are run for the best peaks, and the translation determined from the best superpositions.

[7R,9R]-O-Methylnogarol[17]

The compound crystallizes with space group symmetry $C2$ with two independent molecules per asymmetric unit (78 C, N and O atoms). Solution of the phase problem for this structure provided us with our initial experience in the application of molecular replacement methods. The procedure is described with respect to the six steps summarized above.

1) An E^2-1 Patterson function was calculated with the XRAY-System. The resolution of the data set (d_{min} was 0.84Å) readily supported using an $0.2 \times 0.2 \times 0.22$Å grid. Computer central memory limitations prevented using a finer grid and limited the maximum function value in DEK to 511. Under automatic scaling, DEK sets all Patterson values larger than 511 to this value. Because discrimination is reduced, this procedure is not recommended if the model gives rise to numerous coincident vectors resulting in high weights in its vector set.
2) The model and the full chemical structure are presented in Fig. 23.1; 22 atoms (coplanar), of 39 nonhydrogen atoms per molecule, were considered.
3) The 231 **intra**model vectors were weighted with VEC; vectors were considered to overlap if their end points were within a radius of 0.5Å. VSEL selected the 47 strongest vectors (weights greater than 2.0); their end points were required to be at least 0.3Å apart and they were at least 2.0Å long. The longest vector was 8.8Å and the greatest weight was 17.7 (type indicated in Fig. 23.1).
4) The orthogonalization routine in PSR places the Cartesian x axis, x_c, along the crystallographic a axis and y_c in the ab plane; z_c results from the vector product between x_c and y_c. The rotation angles are defined as follows: A is rotation about x_c, B about y_c and C a subsequent rotation about the newly oriented x_c. The rotation matrix is applied to the model with the Patterson held fixed. A clockwise rotation, when viewed in the direction of the origin, is positive. The

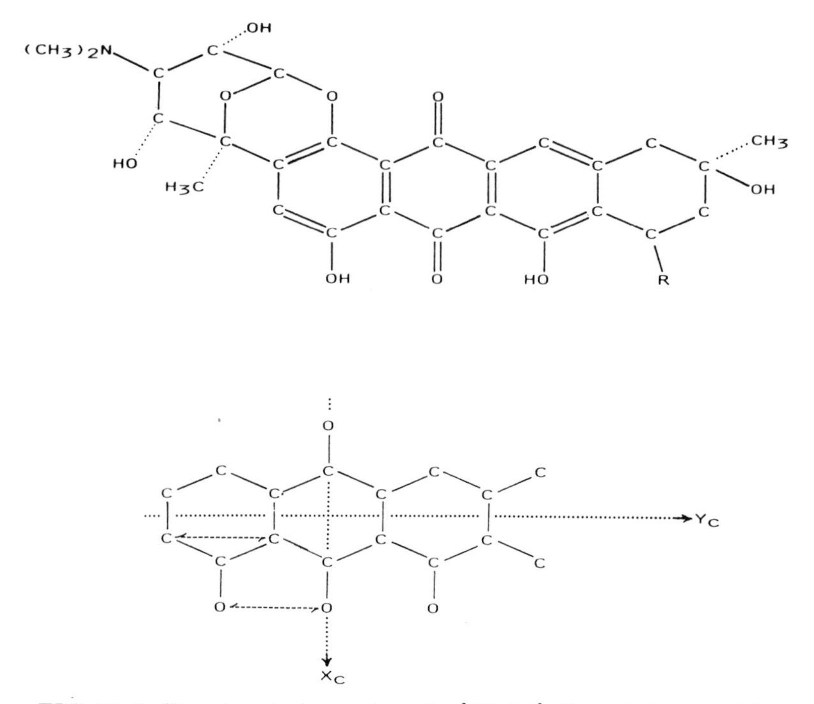

FIG. 23.1. The chemical structure for [7R,9R]-O-methylnogarol (upper image, R = OCH_3 and the model used for the rotation and translation searches (lower image). The dotted lines indicate two of the local coordinate system axes. The dashed lines with arrows are examples of coincident vectors which add to give the vector of maximum weight in the model vector set.

region in space through which the three angles must be varied is determined from the symmetry of the Patterson function and the model vector set.[16] In the structure at hand the following ranges were necessary: $0 \leq A \leq 180°, 0 \leq B \leq 180°$ and $0 \leq C \leq 360°$.

To estimate the goodness of fit, PSR uses equations (3):

$$\min(M)_N = \sum_j (P_j/w_j) \qquad (3)$$

where N is the total number of search vectors and P_j the value of the Patterson function at the end point of vector j with weight w_j. The values are summed over the M smallest

TABLE 23.1. Examples of the min(M) functions from PSR.

Rotation search:

		Orientation angles (in degrees)			
Peak	Function	A	B	C	Height
Coarse scan[a]					
1	min(1)	50	160	280	16
	min(16)	60	160	290	19
2	min(1)	45	55	270	19
	min(16)	45	55	270	19
Fine scan					
1	min(1)	55	159	287	15
	min(16)	60	160	290	19
2	min(1)	47	54	268	18
	min(16)	46	54	267	20

[a] The increment for the coarse scan was 5°, that for the fine scan 1°.

P_j/w_j values. Note: because w_j are large by selection, the smallest values for P_j/w_j will be large only if P_j are large as well. Functions min(1) and min(16) were used in this analysis.

An initial coarse scan (5° increments) revealed two peaks, which were then subjected to fine scans (1° increments, Table 23.1). The results indicated two different orientations for the search fragments. Because the first peak was broader and less well defined than the second, the translation search was pursued only with the latter orientation.

5) The oriented fragment was transformed with the operation $\frac{1}{2} - x, \frac{1}{2} + y, -z$ and all $N(N+1)/2$ **inter**molecular vectors related by the two-fold screw axis were calculated and weighted (with consideration of the mirror plane at $y = \frac{1}{2}$); the 50 highest vectors (at least 0.3 Å apart) were selected.

6) For space group $C2$, the y coordinate of the origin can be chosen arbitrarily, which reduces the translation determination to a two-dimensional problem (determination of the position of the cell contents relative to the 2_1 axis. The necessary translation is determined by establishing the best agreement between the vector set from point 5 and the Patterson function as the former is shifted stepwise (in increments of $\Delta u, 0, \Delta w$) through an asymmetric unit. Steps of 0.01 fractional coordinates were used.

The results from the translation search were more difficult to interpret than those from the rotation search. There were numerous peaks in both $\min(M)$-maps that were more than twice their respective standard deviations. However, there was only one value present in both maps with peak height greater than three sigma (this peak was the strongest in the $\min(16)$-map). The coordinates of the peak were $\Delta u = 0.275$ and $\Delta w = 0.475$. Using the relationships $\Delta u = 1/2 - 2x_o$ and $\Delta w = -2z_o$ gives the translation $x_o = 0.112$ and $z_o = -0.238$.

Because the results for the translation determination were not as convincing as those for the orientation determination, a Karle translation function[18] with $D_1(\delta)$-coefficients, was employed as a check. This function has the same symmetry as the Patterson function and gives peak coordinates that are twice the translation (the sign can be determined by imposing known shifts on the oriented model and examining the resultant peak coordinates). The translation, so determined, was in complete agreement with that from PSR.

The resultant positioned fragment was used for tangent refinement to develop the model. An E-map calculated with 673 reflections, $E > 1.4$ revealed the coordinates of all 78 nonhydrogen atoms of the two independent molecules.

N-Trifluoroacetyl-adriamycin-14-O-valerate (AD-32)[19]

There are two molecules per asymmetric unit (102 C, N, O, and F atoms). This structure determination was especially interesting because the data set was influenced by two features relevant to the solution of the phase problem: a) there was clearly approximate translational symmetry of $\frac{1}{2}$ a and b) the discrimination between space groups $P2_12_12$ and $P2_12_12_1$ was extremely uncertain because of a rather poor data set. The latter problem was subsequently resolved when higher quality crystals were grown, but not before the incorrect space group had been chosen for the initial efforts to determine the structure.

The search model, Fig. 23.2, was identical to that described above except for an additional O-atom bonded to the anthraquinone moiety. The orientation search was conducted with 59 intrafragment vectors with a grid of $0.2 \times 0.3 \times 0.2$Å, the symmetry of the model vector set was $P1$, that of the Patterson function $Pmmm$ (rotation space asymmetric unit: $0 \leq A \leq 90$, $0 \leq B \leq 180$, and $0 \leq C \leq 360°$). The resultant maps contained 4 similar strong maxima consistent with the pseudo-symmetry of the search model. The highest maximum present in both the min(1) and min(20) functions was selected for a fine scan (final angles: $A = 41, B = 52$, and $C = 38°$). The results of the rotation search were fully consistent with the idea that at least the planar moieties of the two independent molecules had the same orientation.

Because the two indicated space groups have the same rotational properties, the uncertainty in selection did not affect the rotation search. It did, however, interfere with the application of the translation function. When the higher quality crystals were grown, and the data set demonstrated that the space group was $P2_12_12_1$, we decided to pursue solution of the phase problem with MULTAN-78. As is often the case with structures built up from planar 6-atom ring systems, initial results provided E-maps with the well known 'chicken wire' pattern.

In order to influence the quality of the E-values, MULTAN-78 was modified to limit the selection of psi zero reflections to a maximum Bragg angle and to limit the selection of reflections with large E-values to those for which $F_o \geq X\sigma(F_o)$ (X became an input variable).

Subsequent attempts to solve the phase problem used all available knowledge concerning the compound under study: a) the results from a crystal structure determination for a close analog (adriamycin-14-O-valerate hydrochloride[20]) provided a

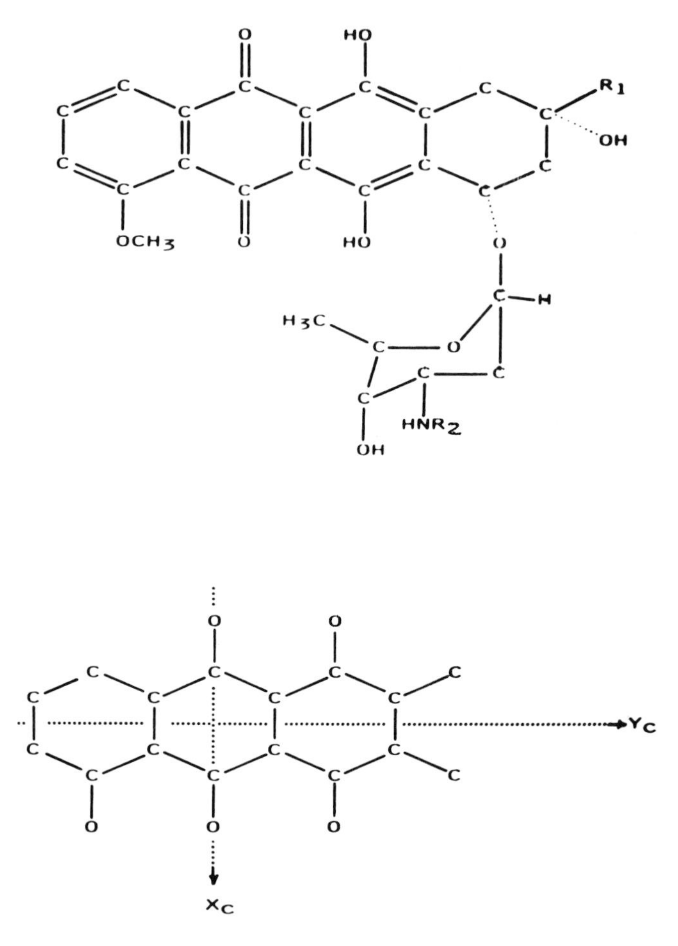

FIG. 23.2. The upper image presents the chemical structure for AD32 (R_1 = $COCH_2OCO(CH_2)_3CH_3$, R_2 = $COCF_3$) and for 9-deacetyldaunomycin hydrochloride (R_1 = R_2 = H). The lower figure presents the model for molecular replacement techniques for both compounds.

molecular conformation which was used as a 'random' fragment and b) the oriented model described above was introduced as an oriented fragment. With $X = 5.0$ in the above equation and limiting E_{min} **to 1.6**, a data set with **242** E**-values** provided a best solution E-map that did not display the 'chicken wire' pattern and allowed the assignment of **54 peaks** to atomic positions. The resultant model was further developed by difference Fourier methods.

9-Deacetyl-daunomycin hydrochloride [19]

There are two molecules per asymmetric unit (84 C, N, , and Cl atoms); space group $P2_1$. An initial attempt to solve the phase problem with MULTAN-80 provided two well-separated molecular fragments (consisting of 19 and 21 atoms), however, the structure could not be expanded in the usual manner. Inspection of a Patterson function indicated that both molecules were probably correctly oriented, which led to a conclusion that they most likely were positioned incorrectly relative to the 2_1 axis.

In this case the program system DIRDIF-TRADIR was used in an effort to determine the translation. The highest peak in TRADIR (1.7 times higher than the next highest) indicated that the fragments were in the correct position. A subsequent Karle translation function calculation also gave the zero translation vector as the highest peak.

The phase problem was eventually solved with SHELXS-84, after which it was found that the fragments were indeed correctly oriented but were **positioned incorrectly with respect to each other**, Fig. 23.3. Subsequently, a translation search with TRADIR was carried out with only one fragment. Because the relative scattering power, P^2, was only 11%, no clear result was expected. Consequently, five possible translation vectors were tested with DIRDIF as potential solutions. One of them led to a complete molecule but the structure could not be further developed. The problem was subsequently traced, once again, to an incorrect position. A TRADIR translation function with the complete molecule gave the zero shift vector as the highest peak but further testing revealed that the fourth highest peak provided the correct vector.

7β-[3′,4′-Di-O-acetyl-2′,6′-dideoxy- L-arabinohexopyranosyl]-6,11- dihydroxyxanthol[2,3-g]tetralin (XANTET)[21]

XANTET is a synthetic anthracycline analog, Fig. 23.4, which

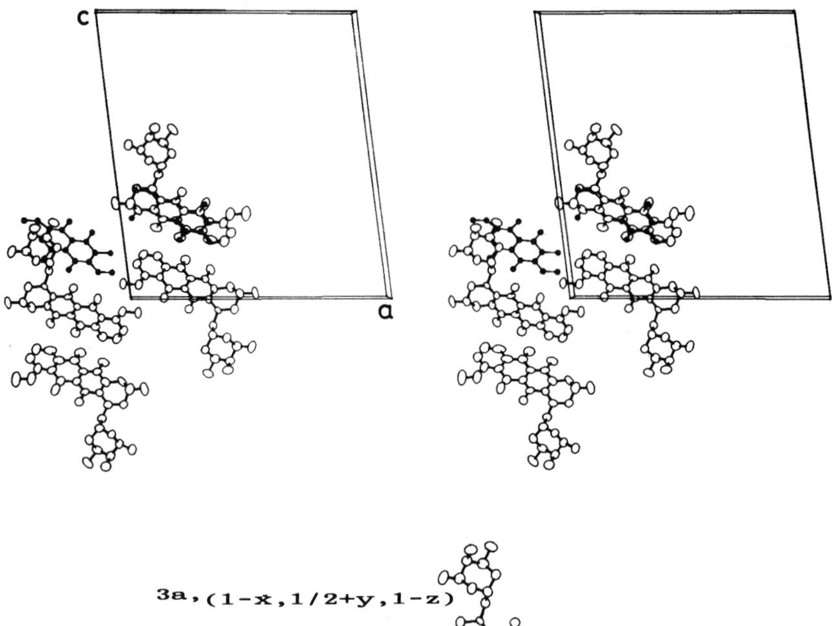

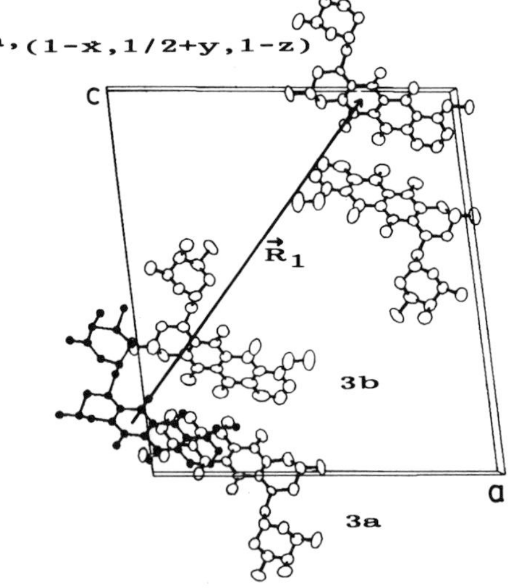

FIG. 23.3. The relationships between the correctly and incorrectly placed fragments encountered during the structure determination for 9-deacetyl-daunomycin hydrochloride. The upper figure, a stereoscopic projection, illustrates the two fragments that were incorrectly placed with respect to each other (first direct methods results). The lower figure displays the incorrectly placed molecule (half the asymmetric unit) from the initial attempts with molecular replacement.

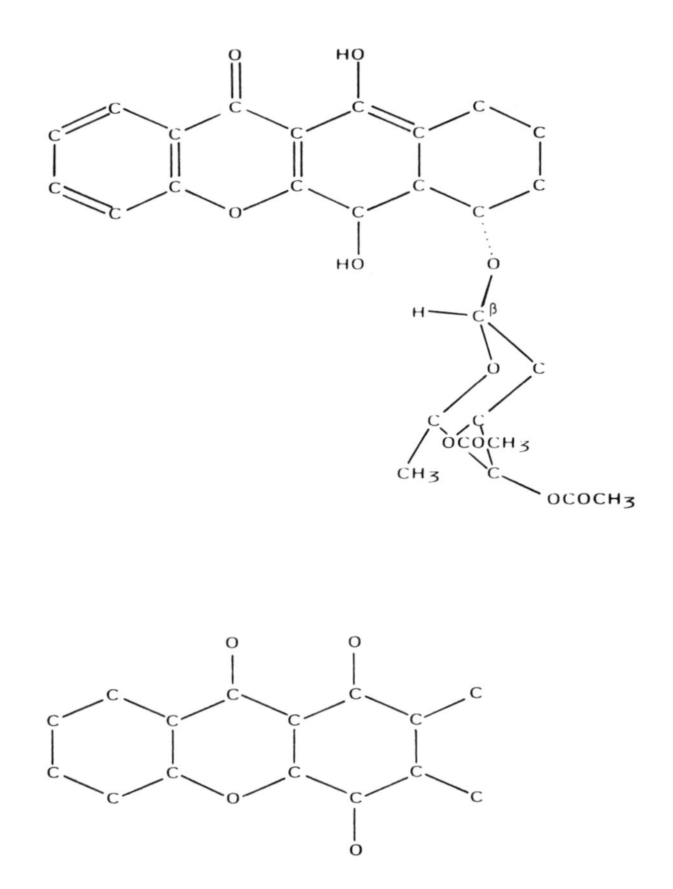

FIG. 23.4. The chemical structure and search fragment for XANTET.

crystallized with space group symmetry $P2_1$ (one molecule per asymmetric unit).

In spite of an unusually good data set for an anthracycline, attempts to solve the phase problem with MULTAN-80 resulted in 'chicken wire' patterns. Attempts were made to use Hornstra's vector search techniques[13] to determine the starting structure. These methods are very similar to those of Nordman; a program version from R. Olthoff-Hazekamp,[5] that had been adapted for the XRAY-System, was available to us.

An E^2-Patterson was calculated. The search model consisted of the fragment shown in Fig. 23.4 (18 of 37 atoms from XANTET). The rotation and translation functions gave solutions which appeared to be clearly indicated based on their goodness of fit, but the structure could not be further developed.

TRADIR, run with the oriented fragment from Hornstra's rotation function, provided two peaks that were significantly stronger that the remainder. Both peaks were tested with DIRDIF and the translation indicated by the second proved to be correct. The weighted Fourier synthesis provided the coordinates of all 37 atoms of XANTET.

Summary

We have found that translation and rotation functions have played a major role in our being able to determine crystal structures for a number of anthracyclines. In total we have determined structures for six members of this family; only two of them were determined without using molecular replacement techniques. In a number of cases, a combination of Patterson and direct methods proved to be effective.

We have used these methods for crystal structure determinations for chemically modified β-cyclodextrins[22] as well. We are convinced that molecular replacement methods will find an ever greater level of applications as the development of user friendly computer programs continues to progress. One of the most serious problems that we have encountered is the lack of similarity (and sometimes clarity) in the definition of rotation matrices in the various computer programs. This severely complicates comparing results from different programs, a technique that we have found useful.

Acknowledgement

This work was supported in part by the Deutsche Forschungsgemeinschaft and the Fonds der Chemischen Industrie e.V. We are most grateful to the numerous colleagues who have provided us with the cited computer programs.

References

1. Rossmann, M. G. (ed.), *The Molecular Replacement Method.* Gordon and Breach: New York (1972).
2. Beurskens, P. T., Bosmann, W. P., Doesburg, H. M., Gould, R. O., van den Hark, Th. E. M., Prick, P. A. J., Noordik, J. H., Beurskens, G. and Parthasarathi, V. DIRDIF: Direct Methods for Difference Structures Technical Report 1981/2, Crystallography Laboratory, Toernooiveld, 6252 ED Nijmegen, Netherlands (1982).
3. Main, P., Hull, S. E., Lessinger, L., Germain, G., Declercq, J.-P. and Woolfson, M. M. MULTAN-78: A System of Computer Programmes for the Automatic Solution of Crystal Structures from X-Ray Diffraction Data, University of York (1978).
4. Main, P., Fiske, S. J., Hull, S. E., Lessinger, L., Germain, G., Declercq, J.-P. and Woolfson, M. M. MULTAN-80: A System of Computer Programmes for the Automatic Solution of Crystal Structures from Diffraction Data, University of York (1980).
5. Hornstra, J. PATSEA: Patterson Search Program, Philips Research Laboratory, Eindhoven, Netherlands; XRAY-System compatible CDC-Version from R. Olthoff-Hazekamp (1976).
6. Steigemann, W. PROTEIN: A Package of Crystallographic Programs for the Analysis of Proteins, Version 79.2, Max-Planck-Institut für Biochemie, Martinsried, Federal Republic of Germany (1982).
7. Schilling, W. J. PSR: Patterson Search Programs, Department of Chemistry, University of Michigan; CDC-Version from B. Hoge (1969).
8. Sheldrick, G. M. SHELXS-84, Universität Göttingen, Göttingen, Federal Republic of Germany (1984).
9. Doesburg, H. M. and Buerskens, P. T. TRADIR: Translation Functions in DIRDIF Fourier Space, Technical Report 1981/1, Crystallography Laboratory, Toernooiveld, 6525 ED Nijmegen, Netherlands (1981).
10. Stewart, J. M., Machin, P. A., Dickinson, C. W., Ammon, H. L., Flack, H. and Heck, H. XRAY-System, Version of 1976, Technical Report TR-446, University of Maryland Computer Center, College Park, MD, USA (1976).
11. Nordman, C. E. and Schilling, J. W. *Crystallographic Computing* (F.R. Ahmed, ed.) pp. 110ff. Munksgaard: Copenhagen, Denmark (1970).
12. Schilling, J. W. *Crystallographic Computing* (F.R. Ahmed, ed.) pp. 115ff. Munksgaard: Copenhagen (1970).

13. Hornstra, J. *Crystallographic Computing* (F.R. Ahmed, ed.) pp. 103ff. Munksgaard: Copenhagen (1970).
14. Hoppe, W. *Z. Elektrochem., Ber. Bunsenges. Physik. Chem.*, **61**, 1076 (1957).
15. Huber, R. *Acta Cryst.* **19**, 353 (1965).
16. Tollin, P., Main, P. and Rossmann, M. G. *Acta Cryst.* **20**, 204 (1966).
17. Eckle, E., Stezowski, J. J. and Wiley, P. F. *Tetrahedron Lett.* **21**, 507 (1980).
18. Karle, J. *Acta Cryst.* **B28**, 820 (1972).
19. Eckle, E., Kiefer, E., Lauser, G. and Stezowski, J. J. *Drugs under Exp. Clin. Res.* **XI**, 207 (1985).
20. Eckle E. and Stezowski, J. J. *Abstracts, ECM6*, 296 (1980).
21. Eckle, E., Stezowski, J. J. and Lown, J. W. *Abstracts, ECM8*, 36 (1983).
22. Pöhlmann, H., Gdaniec, M., Eckle, E., Geiger, G. and Stezowski, J. J. *Acta Cryst.* **A40**, 276 (1984).

24. Computing in crystallography: the early days

Robert Langridge and Jenny P. Glusker

A revolution has occurred in the last thirty years in the determination of crystal structures. Already, in the 1950s, the crystal structure determination of fairly large molecules (of 20 to 30 atoms) was somewhat commonplace. However, each determination comprised three time-consuming tasks: the measurement of the reflection intensities, the induction of a proposed structure to fit the pattern of intensities along with the computation of the relevant structure factors and electron density maps for the proposed structure and, finally, the refinement of the correct structure. So time-consuming were these tasks that it was rare for structures to be studied in three dimensions; most early work involved two-dimensional studies.

With the advent of reliable high-speed computers (resulting from the needs of the new space age) it is now possible to do each of the tasks listed above for a crystal structure determination with relative ease. In other words, the development of crystallographic studies, as we know them today, has paralleled the twentieth century development of computing potential. The developing computer technology has greatly aided the ability of the crystallographer to study molecular structure; likewise the crystallographer, as an extensive user of computers has helped advance the development of computers. Diffraction data collection is now done by a diffractometer under computer control so that only a few initial measurements on the crystal diffraction pattern need be made in order to define the orientation matrix of the crystal and the directions of the diffracted beams; then the computer can be made to control the rest of the data collection. The structure factor calculation is now routine, with many excellent programs available to do this. The structure refinement is usually done by least squares methods (or entropy maximization, see Chapter 5) and, although such refinements may involve several cycles of computation, they are not too taxing to the investigator in terms of time. In fact, the state of the art for computing now is so good that refinement methods may also be attempted for macromolecules if the data have been obtained to a high enough resolution. Finally, with the advent of computer graphics, the display of results has become so efficient that much insight into the stereochemistry of molecular processes has been gained. Thus, much of the drudgery of crystallographic calculations has been removed by using modern-day computers. Liebniz remarked in 1671,[1]

It is unworthy of excellent man to lose hours like slaves in the labour of calculation which could safely be relegated to anyone else if machines were used.

History of computers

The first device recognizably similar in concept and design to the modern programmable computer is the Analytical Engine invented and described by Charles Babbage in 1834. This machine had replaceable loops of punched cards containing information that directed the machine to make specific computations. Babbage had realized the need for both accuracy and speed for calculations such as those of the tables of logarithms that were so much used in those days. The workings of the Analytical Engine were based on the inventions of B. Bouchon (1725), M. Falcon (1728) and Jacques de Vaucanson (1745) that led to the pattern-weaving loom and which were further developed by Joseph Marie Jacquard in 1801 for automatic use. The user of the Jacquard loom could automatically control warp and weft threads in the silk loom by recording the required pattern with punched holes in a collection of cards; the arrangements of holes in these cards were then interpreted (as operations to be done) by an appropriate device that was part of the loom. Such methods are still used today. Although never completed, the drawings of the Analytical Engine are beautifully detailed, and we are fortunate that a contemporary description of the machine was written by L. F. Menabrea, and still more fortunate that this report was translated and extensively, elegantly and eruditely annotated by Ada Augusta Lovelace (neé Byron)[2] who is portrayed in Fig. 24.1. If this steam-driven mechanical device had been completed at that time it would have been the first modern-day computer. It was a great advance over previous calculating machines because it

FIG. 24.1 Ada Augusta Byron, Countess of Lovelace. (1815–1853).

could carry out *any* series of mathematical steps, requiring only the necessary instructions. It could also be made to act on intermediate results and could make defined judgments after comparing numbers. These are the important features that separate a computer from a calculating machine. Menabrea wrote (as translated by Ada Lovelace):[2]

Mr Babbage has devoted some years to the realization of a gigantic idea. He proposed to himself nothing less than the construction of a machine capable of executing not merely arithmetical calculations, but even all those of analysis, if their laws are known... It contains two principal species of cards: first, Operations cards, by means of which the parts of the machine are so disposed as to execute any determinate series of operations, such as additions, subtractions, multiplications, and divisions; secondly, cards of the Variables, which indicate to the machine the columns on which the results are to be represented.

Countess Lovelace added in her notes:

The distinctive characteristic of the Analytical Engine, and that which has rendered it possible to endow mechanism with such extensive faculties as bid fair to make this engine the right-hand of abstract algebra, is the introduction into it of the principle which Jacquard devised for regulating, by means of punched cards, the most complicated patterns in the fabrication of brocaded stuffs.

Ada Lovelace also appreciated, as do crystallographers, the simplification of a system that the use of existing symmetry may effect and wrote:

By the introduction of the system of 'backing' into the Jacquard loom itself, patterns which should possess symmetry, and follow regular laws to any extent, might be woven by means of comparatively few cards.

The report on the Analytical Engine is reproduced in full in the book by Bowden,[3] and, although the high opinion of Lady Lovelace's mathematical understanding which has been formed by most of us who have read her commentary has recently been questioned,[4] her contribution remains a landmark in the bibliography of computer science.

Bernal has described the impact of Babbage's work on modern computing and the reason that such computing can now be done so readily, but was not practicable in Babbage's time:[5]

The mathematical notions behind the modern computer are no more complex than those of the computer ... partly executed by Babbage in the nineteenth century. What brought the idea to life again was the means to carry it out: the components, no longer wooden cogs or even metal ones, as in the earlier machines, were electrical circuits very rapidly switched, first by means of valves [vacuum tubes] and magnetic circuits, and finally by means of semi-conductors.

Bowden wrote in 1953:[3]

> Babbage's ideas have only been properly appreciated in the last ten years, but we now realize that he understood clearly all the fundamental principles which are embodied in modern digital computers. His contemporaries marvelled at him though few of them understood him; and it is only in the course of the last few years that his dreams have come true.

Because of the engineering problems described above by Bernal, digital computers were not developed much further in the nineteenth and early twentieth centuries. Computing was focused on analogue machines, that is computers that use physical quantities, such as lengths of a bar, to represent numbers (in the way that a slide rule uses two wooden bars). They were faster and more practical for the times, but were usually designed for rather specific tasks. The tidal harmonic analyzer of Lord Kelvin is a famous example;[6] by means of this machine Kelvin was able to do a harmonic analysis of tides, *i.e.*, do a Fourier analysis by evaluating the amplitudes of components (fundamentals and overtones). The machine consisted of flexible cord moving over a number of pulleys. Michelson and Stratton improved on this by adding spiral springs and finally built a machine that could handle 80 terms, *i.e.*, sum up to $\cos 80x$.[7] Stratton founded the National Bureau of Standards and later became President of MIT. There he was a supportive colleague of Vannevar Bush[8] who designed a practical way of carrying out Kelvin's ideas on solving differential equations.[9]

Interestingly, the punched card system of the Jacquard loom was modified and introduced by H. Hollerith, acting on an idea of J. S. Billings,[10] for use in the eleventh US Census of 1891. Hollerith also built a series of machines for punching and sorting the ncessary cards and for analyzing them. The Tabulating Machine Company, founded by him in 1896, was the forerunner of the IBM Corporation. It was L. J. Comrie[11] of the National Almanac Office who realized, in 1928, that Hollerith's punched card system could, with advantage, be adapted for scientific work.

Early methods of crystallographic structure calculations

Meanwhile, a new science, X-ray crystallography, was developing. The first X-ray diffraction pattern was recorded in 1912;[12] in retrospect it is amazing how rapidly the methods of analysis of such diffraction patterns were worked out. For example, the equation for structure factors was first given by Sommerfeld[13] in 1913, and W. L. Bragg quickly realized that it should be possible to use a Fourier series in some way in order to obtain a representation of the distribution of the scattering power in a crystal (*i.e.*, the electron density) from the intensities of the diffracted beams.[14] Duane[15] and Havighurst[16] were the first to use one-dimensional Fourier series; they determined the electron density distribution in sodium and chloride ions in sodium chloride. The electron density was computed along a line, even though three-dimensional data were used. Lindo Patterson wrote:[17]

> A very great impetus to our discussions [in Berlin, see Chapter 41 of this volume] was added with the appearance of Duane's paper in 1925 on the crystal as a three-dimensional Fourier series, reviving W. H. Bragg's suggestion from 1915. None of us seems to have known of the latter...

W. L. Bragg, working with B. Warren on the structure of diopside[18] and acting on a suggestion from his father, W. H. Bragg, computed the first electron density projection. The computation of the Fourier projection of this monoclinic structure involved 30–40 reflections summed at 24×12 grid points in the asymmetric unit. W. L. Bragg wrote about the determination of the structure of diopside:[19]

> I had tried without success to see (what seems so obvious now) how such measurements could be used to get a two-dimensional projection as a sort of tartan plaid. My father wrote to me to suggest the solution ought to come from a criss-crossing of Fourier elements in all directions, and a trial with our diopside data showed peaks representing the metal, silicon, and oxygen atoms. I wished to publish the paper with my father, but he insisted I should publish it alone. The first two-dimensional Fouriers appeared therefore as the three principal plane projections of diopside, though credit for its start is due entirely to my father.

Originally scientists like W. L. Bragg could look at an X-ray photograph of a compound containing a heavy atom and immediately be able to tell the coordinates of this heavy atom; he was heard chastising Kathleen Lonsdale at a meeting for not teaching this to her students. Not many crystallographers can now interpret films in this way!

The calculation of electron density maps involves the summation of sine and cosine functions with different amplitudes, frequencies and relative phases. To compute electron density maps it was necessary to find some method of simplifying the complicated equation. For two dimensions and a centrosymmetric structure the electron density at a point $\rho(x, y)$ is

$$\rho(x, y) = \left(\frac{1}{A}\right)\sum\sum F_{hk0}\cos[2\pi(hx + ky) - \alpha_{hk0}]$$

where α_{hk0} is the phase angle of a reflection with indices $hk0$ and F_{hk0} is the structure factor derived directly from the intensity of the reflection. In a centrosymmetric case $F = \pm \mid F \mid$. Each structure factor F_{hk0} contributes a density wave to the electron density map. Analogous expressions, involving $\mid F \mid^2$ instead of $\mid F \mid$ and with $\alpha = 0$ are used for Patterson functions from which interatomic vectors can be derived. The computation of this function is time-consuming, even for two dimensions, but it soon became clear that one only needed to calculate the electron density at selected points (usually, in those days, 1/60 of a cell edge), so that only 3,600 points are studied in two dimensions, 216,000 in three dimensions. But each of these points is obtained by summing many reflections, so that the calculation was still daunting.

A page from Lindo Patterson's notebook, when he was first working on potassium dihydrogen phosphate as an example of the use of the Patterson function, is illustrated in Fig. 24.2, and shows how much had to be calculated by hand in those days. In order to simplify the task of computing structure factors or electron density maps, Beevers and Lipson[20] and Patterson and Tunell[21] (see Chapter 25) introduced 'strips', that is, small pieces of white cardboard about approximately 0.3 inches wide and 7 inches long. They have 15 divisions for values of $F\cos(2\pi hx)$ or $F\sin(2\pi hx)$ for values of F from 1 to 99, of h from 1 to 30 and x at intervals of 1/60 (usually 0 to 15/60 of the

124 264575131 31|568819 59|18349 31

KH_2PO_4 (h00) faltung.

K = 19 H = 1 P = 15 O = 8

Total number of electrons per mol KH_2PO_4 = 19+2+15+32 = 68

" " " " " cell (4 mols) = 272

Table of F' values

sin ϑ	0.0	0.1	0.2	0.3	0.4	0.5	0.6
F(O)	8.0	6.0	3.0	1.65	1.10	0.70	0.50
F(K)	19.0	11.8	8.0	5.3	3.8	2.4	1.2
F(P)	15.0	8.2	5.0	3.7	2.6	1.6	0.8

	0	2	4	6	8	10	12	14
sin ϑ	0.0	0.083	0.165	0.248	0.331	0.414	0.496	0.579
F'(K)	19.0	12.4	9.2	6.7	4.8	3.6	2.5	1.5
F'(P)	15.0	8.8	5.8	4.3	3.3	2.4	1.7	1.0
F'(O)	8.0	6.4	3.9	2.3	1.5	1.0	0.7	0.5
F(K+P) ~~F'(K)~~	34	21.2	15.0	11.0	8.1	6.0	4.2	2.5
F'(8O) ~~F'(P)~~	64.0	51.2	31.2	18.4	12.0	8.0	5.6	4.0
F(4K+P) ~~F'(O)~~	136	84.8	60.0	44.0	32.4	24.0	16.8	10.0
F²(8O)	4096	2621	973	339	144	64	31	16
F²(4K+P)	18496	7191	3600	1936	1050	576	281	100
	73984	8100	576	1600	1024	441	400	196
Σ F²	26688	12433	5546	2614	1338	704	343	132
	47296	4333	4970	1014	314	263	57	64

FIG. 24.2. A page from Lindo's notebook (labelled *A. L. Patterson. 6-414, MIT. Crystal Analysis Data and Fourier Series, December 1933–August, 1934.*)

unit cell, values for the rest of the unit cell being generated by symmetry considerations). The strips were stored, in numerical order, in wooden boxes and it was relatively simple to pull out a strip for each required set of F and h values, for the first summation. Entries were totalled by hand (with all the errors that that can cause), or, more conveniently, by hand-cranked adding machines. Then summations were done for the second dimension to generate a two-dimensional map. Patterson, discussing the use of Beevers–Lipson strips for computing radial distribution functions (which require only sine terms) wrote:[22]

The front of a Beevers–Lipson strip has the appearance

A SE h | 0 Asin $2\pi 2h/120$ Asin $2\pi 4h/120$...... Asin $2\pi 30h/120$ (Sum)

where A is the amplitude, SE stands for sine even, h is the frequency, and the sum in parentheses is the sum of all the values Asin $2\pi h/120$ across the strip. The reverse side of the strip reads

A SO h | 0 Asin $2\pi h/120$ Asin $2\pi 3h/120$...... Asin $2\pi 29h/120$ (Sum)

where SO now stands for sine odd and the other notation is the same. The Beevers–Lipson strips thus facilitate the calculation of sums of the type

$$f(p) = \sum_{h=1}^{30} A_h \sin 2\pi hp/120 \tag{3}$$

Strips are available for the A values $\overline{900}\,(100)\,\overline{100}$; $\overline{99}\,(1)\,99$; $100\,(100)\,900$ and h values $1\,(1)\,30$. For each term in the series (3) a strip (or strips) is selected with the appropriate h, and the correct value A_h. All the strips belonging to the series (3) are arranged parallel to one another and with the front face up. The column totals are then formed to give the values of $f(p)$ for the even values of p. As a check, the column totals are cross-footed to give a sum equal to the total of the sum column.

An analogous explanation can be given for the cosine strips. Such strips are almost never used now, but illustrate rather nicely the principles of the calculation of an electron density map.

Then came the idea of punching the information contained in Beevers–Lipson strips onto punched cards for use with a tabulating machine. In 1940 an application was submitted from Caltech to build a mechanical Fourier machine. One of the referees of this proposal, Dr W. P. Eckert of IBM, suggested that punched cards would be more suitable to do the job. Therefore Pauling's group used an IBM key punch, sorter and tabulator.[23,24] Donnay wrote in 1941,[25]

> When I was in Pasadena, I visited Pauling's laboratory. The first thing that struck me when I entered the office room was an International Business Machine used for Fourier syntheses. It is now in working order and they are working on another modification of it for electron work. As it is they have two hundred thousand grids, eleven such grids stand for one strip of B. and L. [Beevers and Lipson] A quarter of the cell is divided into 125 parts. Any line may be calculated if one wishes a particular line. The summation is carried on as in the B. and L. system. The Miller indices go up to 30. It takes about an hour for an ordinary summation. The trigonometric functions are given with 4 figures instead of 2. The set-up is too accurate for present-day data.

These punched cards could be used for a variety of purposes — reflection data, atomic parameters and Fourier summations. A comparison of a cosine function, displayed graphically, the corresponding Beevers–Lipson strip and one of the eleven IBM cards used at Caltech are shown in Fig. 24.3. In this way with punched cards Donohue and Schomaker were able to compute 600 structure factors for an 8-atom structure in a non-centrosymmetric unit cell in 24 hours; this was a great reduction in time over the 240 hours normally required.[26] In 1955 crystal structures for threonine, serine, hydroxyproline, N-acetylglycine, glycylasparagine and N,N-diglycylcystine were solved (see Chapters 26 and 42.6 by Donohue and Shoemaker respectively). In a similar way, in the mid 50's in Oxford, England, Patterson and electron density maps were computed using punched cards and a Hollerith machine that could only add; for example, a three-dimensional electron density map on the hexacarboxylic acid derived from vitamin B_{12}, phased on the cobalt atom present in the structure, took six weeks to compute in this way.[27]

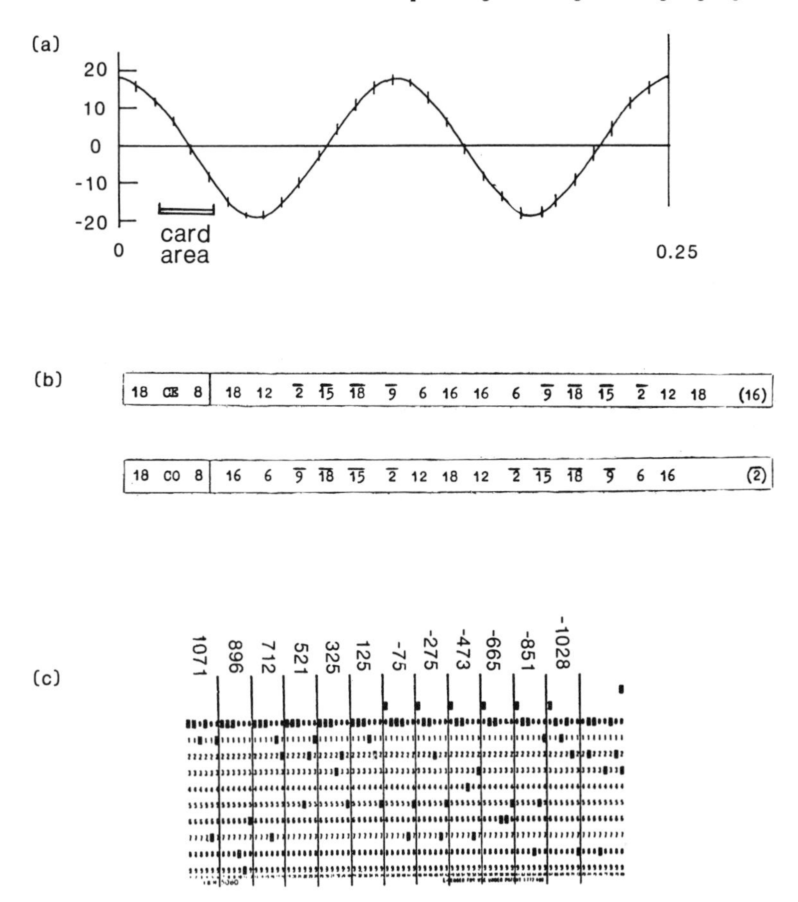

FIG. 24.3. Old style crystallographic computing. (a) A cosine function, (b) its representation on a Beevers–Lipson strip and on (c) one of the eleven cards with the Beevers–Lipson strip information used at Caltech.

Analogue computers in crystallography

Analogue computers, which used physical quantities such as voltage or light intensity to represent numbers, were originally used with considerable success in crystallography to increase the speed and efficiency of calculation. In this way the crystallographers followed the path, described above, of the nineteenth and early twentieth century computer experts. Examples are provided by the optical analogue work of W. L. Bragg and C. A. Taylor and by the analogue computer, X-RAC, built by Ray Pepinsky which was used to display electron density and Patterson maps as contour lines on a cathode ray oscilloscope. W. L. Bragg had suggested the use of diffraction through masks containing holes that represented either the diffraction pattern (with the radii of holes related to the intensities of the reflections) or the arrangement of atoms in the crystal structure. These methods helped solve some crystal structures.[28–30] An example of a Patterson map calculated on X-RAC for Lindo Patterson is shown in Fig. 24.4.

X-RAC was designed by Ray Pepinsky[31,32] (see Chapter 29, this volume). He had been involved in radar oscilloscope studies at MIT during the Second World War. After the war he moved to Alabama and built X-RAC (X-ray Analogue Computer) at Auburn University; it was transported to Penn State College in 1949. Pepinsky first described this machine at the ASXRED meeting at Lake George in 1956. Amplitudes and phase angles were set on dials in front of each of the oscillators. The resulting contour maps were plotted and photographed from the screen. Mina Rees, Director of the Division of Mathematical Sciences of the Office of Naval Research, said in 1950,[33]

> In 1946 when our decision was made about digital machines, I ought to remind you that fully electronic high speed automatic sequence digital machines were just over the horizon. They are still just over the horizon. In the intervening time the X-RAC has come into being and has done important work.

Scientists came from all over the world to Penn State College to compute their Patterson or electron density maps. They changed the signs of the terms in the Fourier summation of centrosymmetric structures and instantly the oscilloscope showed them the effect of such changes on the electron density projection. Any sign change that reduced a deep negative region in the electron density map was considered good. Pepinsky also designed S-FAC, an electronic analogue computer for calculating structure factors.

FIG. 24.4. A section of X-RAC output. The section $v = 0$ of the three-dimensional Patterson map of citric acid is shown. The origin is at the center. This map was computed for A. L. Patterson in June 1954 by Josephine Lombard, X-RAC technician.

Digital computers in crystallography

Thus, computers, or at least calculating machines, have been applied to crystallography for many years[34] beginning with the punched card machines and analogue computers just described. But it was the advent of the *programmable* digital computer that revolutionized crystallography, and in particular protein crystallography. It is hard to realize now that when Bernal, Hodgkin and Perutz pioneered protein crystallography in the 1930's, not only was there no forseeable method for the solution of the phase problem, but, even if there had been, although the mathematics was well understood, the staggering amount of calculation required would have constituted an almost insurmountable barrier.

This barrier was overcome by invention of faster and more efficient computers. Machines based on the binary system of numbers were invented by H. Aiken of Harvard and B. M. Durfee, F. E. Hamilton and C. D. Lake of IBM who built a relay-based device (the Harvard Mark I Calculator)[35] and J. Presper Eckert and John W. Mauchly who built the large electronic computer, ENIAC (Electronic Numerical Integrator and Calculator).[36] The Harvard Mark I Calculator was the first machine to be built that used the principles that Babbage had proposed a hundred years earlier. Aiken had realized that there were several differences between punched card accounting machinery and the required calculating machinery for science; these were the ability to handle positive and negative numbers, to utilize mathematical functions, to be fully automatic and to carry out calculations in mathematical sequence.[37] But it was the all-electronic computer, ENIAC, that changed the way we can calculate. This machine was built at the Moore School of Electrical Engineering of the University of Pennsylvania for the Ballistics Research Laboratory at Aberdeen Proving Ground.[38] It was finished in 1946 and was colossal with 18,000 tubes and 1,500 relays. Von Neumann's group at Princeton further elaborated on these ideas.[39]

The first electronic digital computer of the modern stored program type was MIT's Whirlwind, which contained dozens of huge racks of vacuum tubes. Other early high-speed computers in the USA were the IBM Corporation Selective-sequence Electronic Calculator, SEAC(National Bureau of Standards Eastern Automatic Computer) and SWAC (National Bureau of Standards Western Automatic Computer). SWAC was used for the determination of the structure of vitamin B_{12}.[27] In England the electronic special-purpose computer COLOSSUS was built to help break the German code towards the end of World War II. The Automatic Computing Machine ACE was developed and

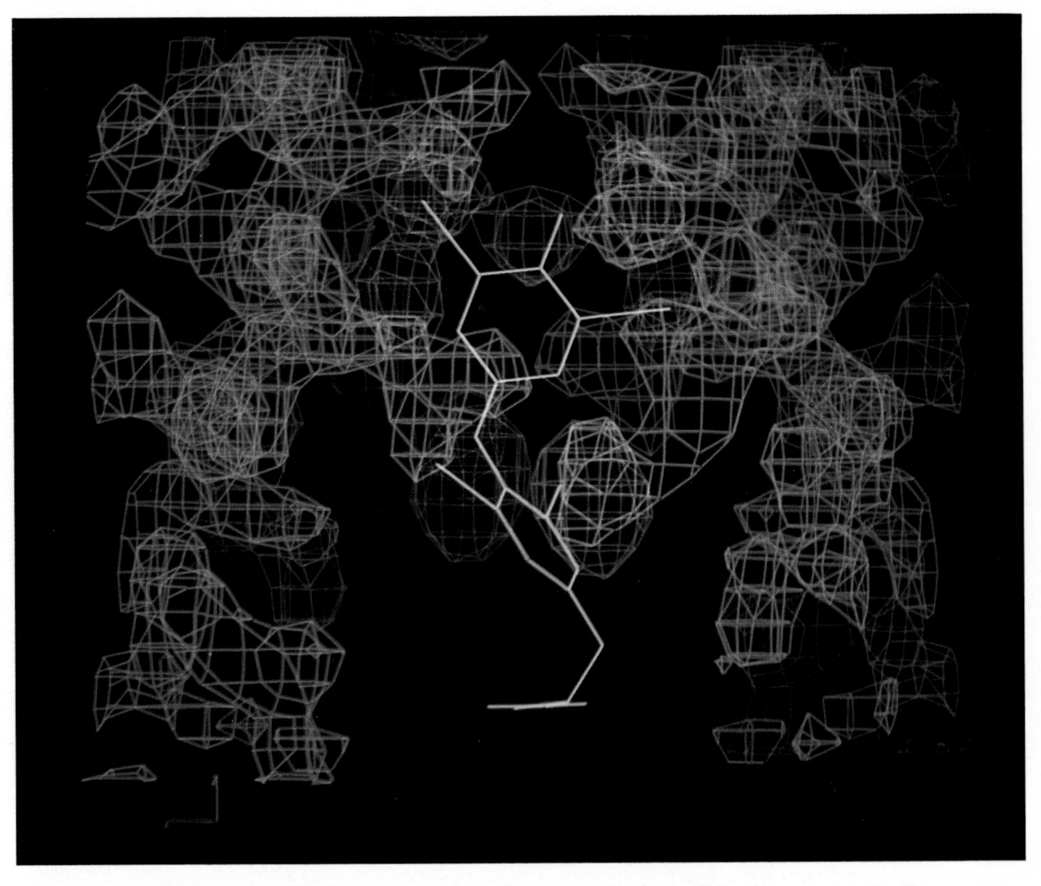

FIG. 24.6. FRODO applied to the thyroxin/prealbumin system. Iodine is red, the rest is blue.[72] (Computer Graphics Laboratory, University of California, San Francico, CA)

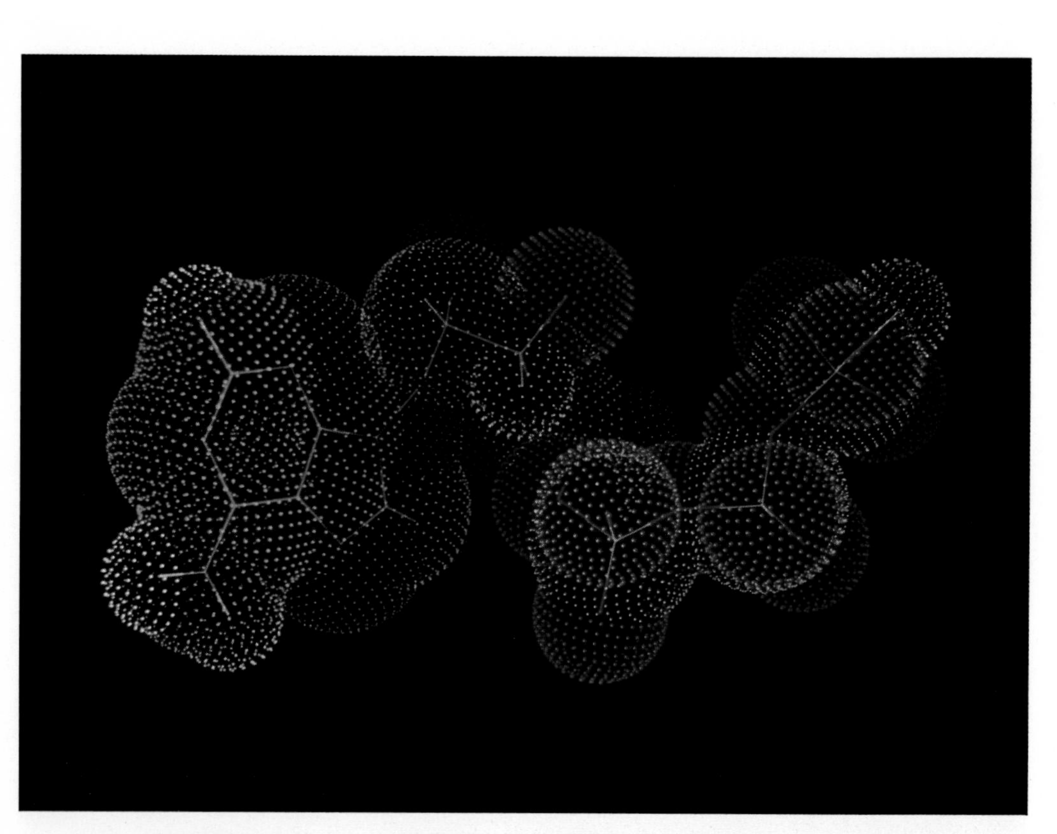

FIG. 24.7. The crystal structure of ATP determined by Olga Kennard and co-workers.[48] (Computer Graphics Laboratory, University of California, San Francisco, CA)

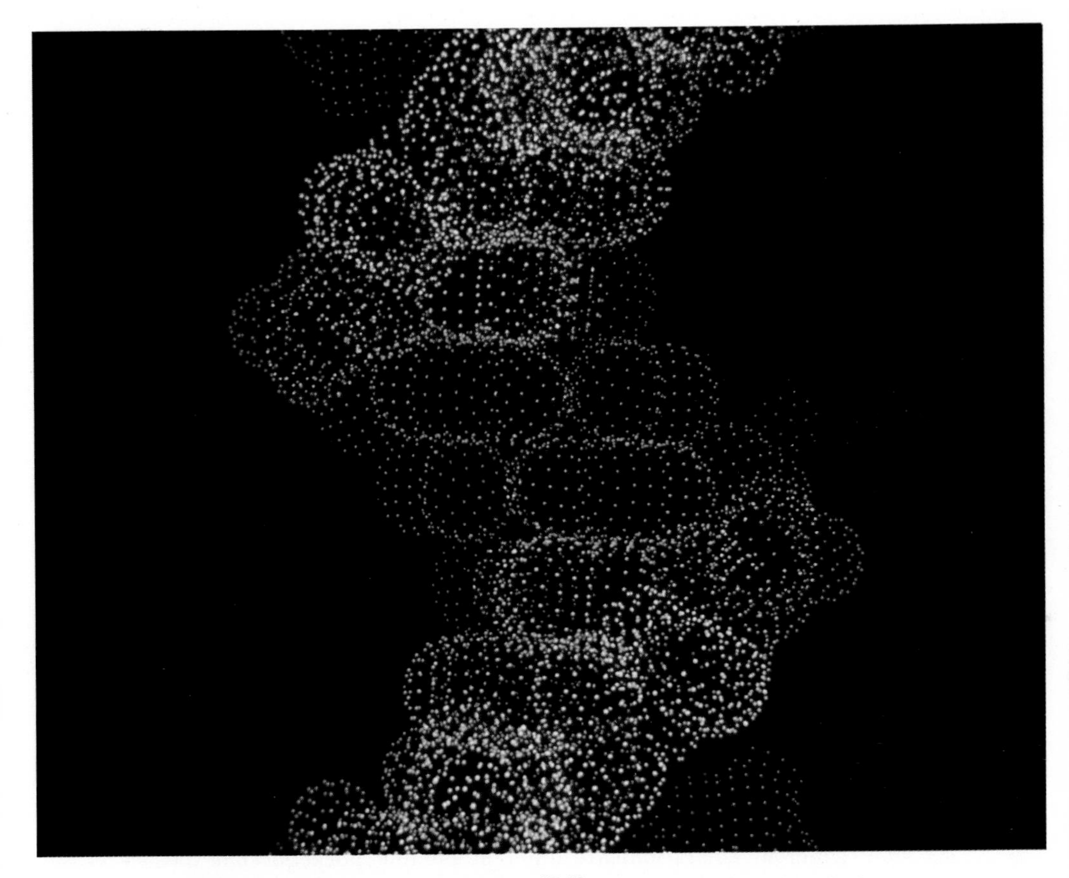

FIG. 24.8. The structure of DNA[49,50] (Computer Graphics Laboratory, University of California, San Francisco, CA)

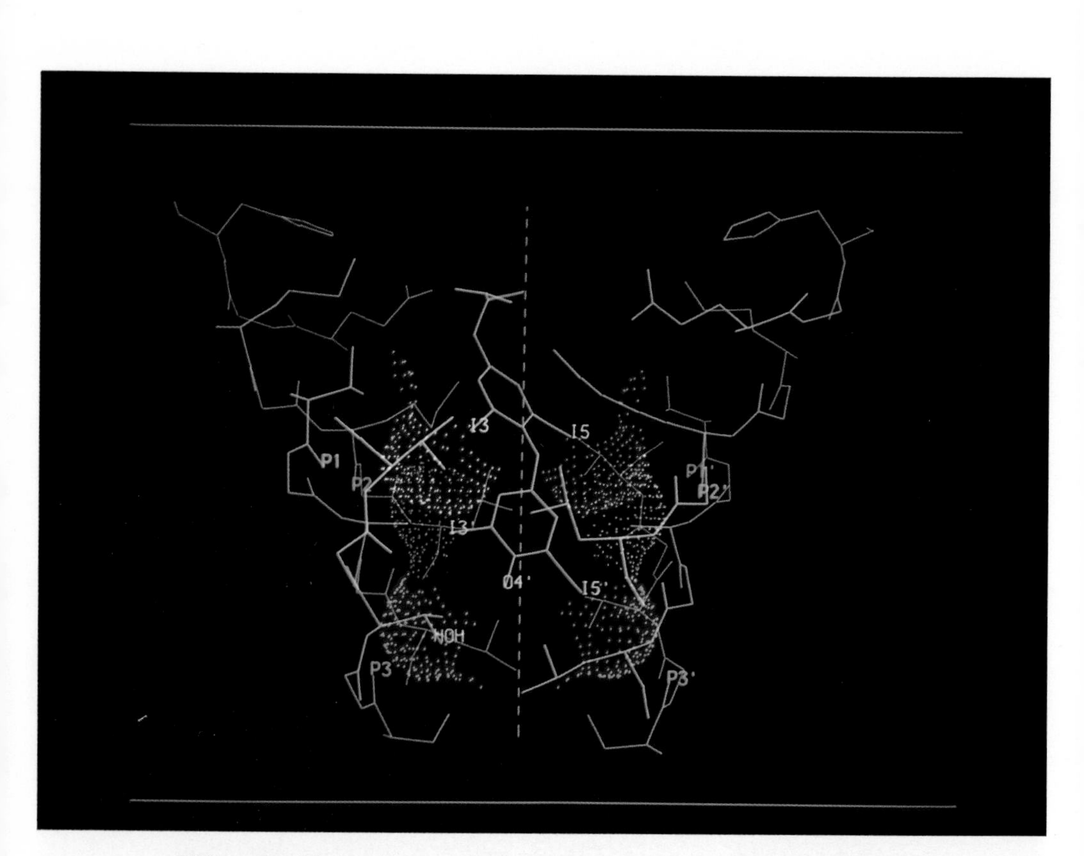

FIG. 24.9. Thyroxine-prealbumin interaction. L-thyroxine in the binding site. Different colors indicate different binding pockets. The dashed line indicates a two-fold symmetry axis.[72] (Computer Graphics Laboratory, University of California, San Francisco, CA)

constructed at the National Physical Laboratory, Teddington, UK, under the direction of Alan Turing. EDSAC, the Electronic Delay Storage Automatic Calculator, was built at Cambridge University, UK by Maurice Wilkes and co-workers and was probably the first stored-program computer. The Manchester Automatic Digital Machine, MADM, was built at Manchester University. By 1960 crystallographers were describing their computer programs[40] for variey of large computers. These included SWAC (mentioned above), the Ferranti Mark I and Pegasus computers, DEUCE (derived from ACE), EDSAC II, the IBM 650 and 704 and many others.

Very early on computers were applied to the study of macromolecular structure. The first digital computer application to protein structure was made by Bennett and Kendrew in England in 1951[41] using the Cambridge computer EDSAC.[42] The output is illustrated in Fig. 24.5. The IBM 650 became a workhorse for many of the small molecule crystallographic calculations of the late 50's and early 60's.[43,44] The first model of DNA was built in 1953[45] but the first application of digital computing to DNA stucture was made in 1956 with the programming of the Fourier transform of a helical molecule for an IBM 650.[46,47]

Of course, a great improvement came about in computers when the vacuum tube was replaced by the transistor and integrated circuits. This made for greater reliability and speed and lower cost. Now a hand-held calculator can do more than many of the early machines, such as the IBM 602A which Lindo Patterson used at Fox Chase in the late 50's. As a result of the improving technology, large computers, especially those in university computing centers, were available in the 60's, manufactured by IBM, UNIVAC and Control Data Corporation. By that time computers had improved so that floating-point hardware, large magnetic core memories, magnetic tape and disk drives and higher speeds were available.

These developments were accompanied by improved software. Of the higher-level programming languages FORTRAN, developed at IBM in 1957 with David Sayre on the original team,[48] has been invaluable to crystallographers. The first FORTRAN compilers were not very efficient and used a large

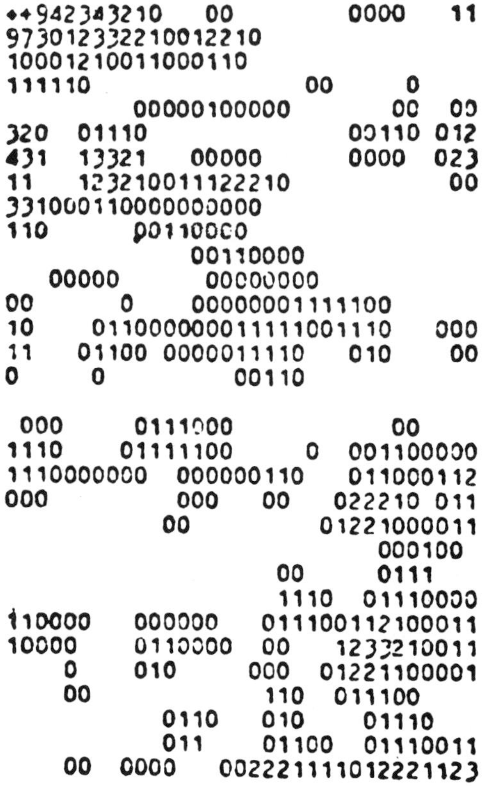

```
++942343210   00        0000    11
973012332210012210
10001210011000110
111110               00     0
         00000100000       00   00
320  01110              00110 012
431  13321   00000      0000  023
11   123210011122210            00
33100011000000000
110      00110000
             00110000
   00000      00000000
00      0    00000001111100
10    011000000011111001110    000
11   01100 0000011110   010     00
0     0         00110

 000     0111000           00
1110    01111100     0   001100000
1110000000  000000110    011000112
000         000   00    022210 011
           00           0122100011
                           000100
                   00      0111
                   1110   01110000
110000   000000    011100112100011
10000    0110000   00   1233210011
    0    010      000  01221100001
    00            110   011100
           0110   010     01110
           011    01100   01110011
    00 0000     002221111012221123
```

FIG. 24.5. Patterson projection of whale myoglobin, printed by ED-SAC. Contour intervals are 32. Every negative value is represented by a space, every positive value 0 to 31 by 0, 32 to 63 by 1, etc. Values greater than 319 are represented by +.

proportion of the available computer memory, leaving very little for data manipulation. In addition, FORTRAN commands were sometimes interpreted by different computers (via their compilers, of course) in different ways. This led Jim Stewart to use 'PIDGIN FORTRAN' which, although somewhat less efficient for a particular computer was interpreted by all computers in the same way. His system of programs, XRAY67,[49] has been widely used. Other well-used programs include ORFLS,[50] UCLASFLS[51] and MULTAN.[52]

Minicomputers were used to control single crystal diffractometers. Then small stand-alone computers began to be used for all aspects of crystallographic computing. So the circle was completed in that crystallographers originally used stand-alone computers in their work, then went on to use the largest computers available; and now that stand-alone computers have such a large capacity for computing, many crystallographers have gone back to a stand-alone computer in their own laboratories or institutes. With the advent of network access to supercomputers, this circle will probably be traversed a second time.

The computations involved in the determination of the strcutures of macromolecules can be massive. J. E. Johnson[53] describes the structure determination of southern bean mosaic virus (SBMV). This required the phasing of 276,000 unique structure factors and Fourier transforms of electron density maps containing 2.7×10^7 grid points in the asymmetric unit. This required the use of a fast Fourier transform algorithm,[54] and finally led to a structure at 2.8 Å resolution.

Computer Graphics and Molecular models

An important part of X-ray crystallographic research is the display of results. The use of the computer to draw pictures began fairly early, if only in the production of figures such as electron density maps by creative use of a printer (usually aided by hand-drawn contours).[41] Illustrations of molecular structures on a cathode ray tube output began in the 1950's, and by 1965 the classic molecule drawing program, ORTEP (Oak Ridge Thermal Ellipsoid Plotter)[55] was available to those with suitable computers and pen plotters.

Originally electron maps could be drawn out on transparent sheets, appropriately space, to give a three-dimensional map.[30] Models could be built to scale in a variety of ways. The first skeletal models of proteins and of DNA were built on a scale of 5 cm to the Å. The protein electron densities were plotted in three dimensions by placing colored clips, coded according to electron density, on flexible vertical rods, with the wire model constructed in the same space.[56,57] Although the first protein models were built in this manner, it is extremely inconvenient. In 1968, Richards[58] proposed a very neat way of keeping the electron density map and the wire model in separate physical spaces, while, to the observer, putting them in the same visual space using a half silvered mirror system. This method was used for a number of years. Computer graphics gradually came into use at about the same time, and have since almost completely taken over the fitting of electron densities to molecular models in protein crystallography.

Almost concurrently, the new age of *interactive* computer graphics began with the seminal work by Sutherland on 'Sketchpad'.[59] Shortly afterward, the first truly *three-dimensional* graphics system was built at Project MAC, MIT in the 1960s.[60] Funded by the Advanced Research Projects Agency of the U.S. Department of Defense, Project MAC had an enormous impact on computing and was the first of the large-scale time-sharing computers (MAC, Multiple Access Computers). The same Project MAC also stood for 'machine-aided cognition,' *i.e.,* attempts to make it easier for humans to 'interface to' computers, particularly by means of interactive graphics rather than by pages of numbers. Thus, this project allowed the user to manipulate, through a 'crystal ball' controlling three rate controllers at right angles, the orientation of a three dimensional object on the screen. Among the first objects to be displayed (somewhat to the surprise of the Defense Department, we suspect) were proteins and DNA[61–63] (it is interesting to note that the paper describing the excellent space-filling Corey-Pauling-Koltun, or C-P-K models,[64] appeared at about the same time).

Early attempts were made to use three dimensional graphics for electron density fitting[65] while working systems such as BILDER[66] and FRODO[67] were first put into general use in the mid 1970's and are now in almost universal use for electron density fitting in protein crystallography laboratories. An example of output from a version of FRODO applied to the thyroxin/prealbumin system by Stuart J. Oatley at the UCSF Computer Graphics Laboratories is shown in Fig. 24.6. Graphics has also come into extensive use for interactive modelling of the these complex structures and their interactions[68,69] particularly in protein engineering and drug design. Examples are given for ATP (Fig. 24.7)[70] and DNA (Fig. 24.8)[69,71] and an early computer-aided drug design paper (Fig. 24.9).[72]

Summary

The expansion of the power of X-ray crystallography since 1950 required a parallel increase in the power of computers. The advances in studies of molecular structure in the last 30 years provide a model example of how scientific developments may be driven by technological advances. As Babbage wrote:[73] 'As soon as an Analytical Engine exists, it will necessarily guide the future course of science.'

Acknowledgements

The authors thank the National Institutes of Health (CA-10925, CA-22780, JPG and RR-1081, RL) for grants.

References

1. Leibniz, G. W. On his calculating machine (translated from the Latin by M. Kormes). Reproduced in *A Source Book in Mathematics.* (D. E Smith, ed.) pp. 173–181. McGraw-Hill: New York, NY (1929). Leibniz designed the 'stepped wheel' still found in some calculating machines; this is a drum containing nine teeth of varying length, one for each digit. This provided an improvement over Pascal's machine.
2. Menabrea, L. F. and Lovelace, A. A. *Sketch of the Analytical Engine Invented by Charles Babbage, Esq., by L. F. Menabrea, of Turin, Officer of the Military Engineers.* Translated and with notes by A. A. L. *Taylor's Scientific Memoirs* **3**, 666–731 (1843). Menabrea attended some lectures given by Babbage in Turin in 1840 and published an account of them in the Bibliothèque Universelle de Genève in 1842. This was translated and annotated by Ada Lovelace.
3. Bowden, B. V. *Faster Than Thought.* Appendix I. Pitman: London, UK (1953). Ref. 2 is also reproduced in *Charles Babbage. On the Principles and Development of the Calculator.* (P. Morrison and E. Morrison, eds.) Dover Publications, Inc: New York, NY (1961).
4. Stein, D. *Ada: A Life and Legacy.* MIT Press: Cambridge, MA (1985). However, we urge the reader to read reference 3 and form their own opinion.
5. Bernal, J. D. After twenty-five years. In *The Science of Science. Society in the Technological Age.* (M. Goldsmith and A. Mackay, eds.) p. 215. Souvenir Press: London, Toronto (1964). Bernal also refers to

Blaise Pascal who invented the first simple digital calculating machine in 1642; however, this did not have the facilities of the modern computer which, as we discuss in this chapter, was first thought of (and partly executed) by Babbage.

6. Thomson, W., Lord Kelvin *Mathematical and Physical Papers.* Vol. VI, pp. 272ff and *Proc. Roy. Soc.* **24**, 266 (1876).
7. Michelson, A. A. and Stratton, S. W. A new harmonic analyzer. *Am. J. Sci.* 4th ser., **5**, 1–13 (1898).
8. Bush, V. The differential analyzer, a new machine for solving differential equations. *J. Franklin Inst.* **212**, 447–488 (1931).
9. Goldstine, H. H. *The Computer from Pascal to von Neumann.* Princeton University Press: Princeton, NJ (1972).
10. Hollerith, H. An electric tabulating system. *School of Mines Quarterly* **10**, 239–256 (1889).
11. Comrie, L. J. The application of the Hollerith Tabulating Machine to Brown's Tables of the Moon. *Royal Astronomical Society, Monthly Notices*, **92**, 694–707 (1932).
12. Friedrich, W., Knipping, P. and Laue, M. Interferenz-Erscheinungen bei Röntgenstrahlen. *Sitzungsberichte der (Kgl.) Bayerische Akademie der Wissenschaften* 303–322 (1912). Translation by J. J. Stezowski in *Structural Crystallography in Chemistry and Biology* (J. P. Glusker, ed.) pp. 23–39. Hutchinson Ross Publishing Company: Stroudsburg, PA, Woods Hole, MA (1981).
13. Sommerfeld, A. Remarks in *La Structure de la Matière. Rapports et discussions du Conseil de Physique tenu a Bruxelles du 27 au 31 Octobre 1913 sous les auspices de l'Institut International de Physique Solvay.* p. 131. (Publication delayed because of the Great War) Gauthier-Villars et Cie, Editures: Paris, France (1921). This reference was kindly supplied by Dr C. A. Taylor.
14. Bragg, W. H. IX Bakerian Lecture: X-rays and crystal structures. *Trans. Roy. Soc. (London)* **A215**, 253–274 (1915).
15. Duane, W. The calculation of the X-ray diffracting power at points in a crystal. *Proc. Natl. Acad. Sci. USA* **11**, 489–493 (1925).
16. Havighurst, R. J. The mercurous halides. *J. Am. Chem. Soc.* **48**, 2113–2125 (1926).
17. Patterson, A. L. Experiences in crystallography — 1924 to date. In *Fifty Years of X-ray Diffraction.* pp. 612–622. (P. P. Ewald, ed.) N. V. A. Oosthoek's Uitgeversmaatschappij: Utrecht, The Netherlands (1962).

18. Warren, B. and Bragg, W. L. The structure of diopside, $CaMg(SiO_3)_2$. *Z. Krist.* **69**, 168–193 (1928).
19. Bragg, W. L. The growing power of X-ray analysis. In *Fifty Years of X-ray Diffraction*. Chapter 8. pp. 120–135. (P. P. Ewald, ed.) N. V. A. Oosthoek's Uitgeversmaatschappij: Utrecht, The Netherlands (1962).
20. Lipson, H. and Beevers, C. A. An improved numerical method of two-dimensional Fourier synthesis for crystals. *Proc. Phys. Soc.* **48**, 772–780 (1936).
21. Patterson, A. L. and Tunell, G. A method for the summation of the Fourier series used in the X-ray analysis of crystal structures. *Am. Mineral.* **27**, 655-679 (1942).
22. Excerpted from an enclosure entitled *Radial Distribution using Beevers–Lipson strips*. Sent by A. L. Patterson together with a letter to Dr Gunnar Ryge, July 7, 1958.
23. Hughes, E. W. Punched card methods in crystal structure calculations. In *Computing Methods and the Phase Problem in X-ray Crystal Analysis*. (R. Pepinsky, ed.) pp. 141–147. X-ray Crystal Analysis Laboratory: The Pennsylvania State College, State College, PA (1952).
24. Shaffer, P. A. Jr., Schomaker, V. and Pauling, L. The use of punched cards in molecular structure determinations. I. Crystal structure calculations. *J. Chem. Phys.* **14**, 648–658 (1946).
25. Letter from J. D. H. Donnay to G. Tunell dated January 24, 1941, quoted in a letter from G. Tunell to A. L. Patterson dated March 17, 1941.
26. Shoemaker, D. P. and Shoemaker, C. B. Chapter 42, this volume.
27. Hodgkin, D. C., Pickworth, J., Robertson, J. H., Trueblood, K. N., Prosen, R. J. and White, J. G. The crystal structure of the hexacarboxylic acid derived from B_{12} and the molecular structure of the vitamin. *Nature (London)* **176**, 325–328 (1955).
28. Bragg, L. Lightning calculations with light. *Nature (London)*, **154**, 69–72 (1944).
29. Taylor, C. A. and Lipson, H. Optical methods of crystal-structure determination. *Nature (London)* **167**, 809–810 (1951).
30. Crowfoot, D., Bunn, C. W., Rogers-Low, B. W. and Turner-Jones, A. X-ray crystallographic investigation of the structure of penicillin. In *Chemistry of Penicillin*. (H. T. Clarke, J. R. Johnson and R. Robinson, eds.) pp. 310–366. Princeton University Press: Princeton, NJ (1949).

31. Pepinsky, R. An electronic computer for X-ray crystal structure analysis. *J. Appl. Phys.* **18**, 601–604 (1947).
32. Pepinsky, R. (ed.) *Computing Methods and the Phase Problem in X-Ray Crystal Structure Analysis.* The X-Ray and Crystal Structure Analysis Laboratory, Department of Physics, Pennsylvania State College: College Park, PA (1952).
33. Reese, M. Introductory Remarks. In *Computing Methods and the Phase Problem in X-ray Crystal Analysis. Report of a Conference held at the Pennsylvania State College. April 6–8, 1950.* (R. Pepinsky, ed.) Pennsylvania State College: State College, PA (1952).
34. Sparks, R. A. and Trueblood, K. N. Digital computers in crystallography. In *Crystallography in North America*, pp. 228–232. (D. McLachlan and J. P. Glusker, eds.) American Crystallographic Association: New York, NY (1983). This is a general description.
35. Aiken, H. H. Proposed automatic calculating machine. *IEEE Spectrum*, pp. 62–69 (August 1964).
36. Hartree, D. R. The ENIAC. *Nature (London)* **158**, 500–506 (1946).
37. Aiken, H. H. Unpublished memorandum, fall 1937. Described by H. H. Goldstine, ref. 9, p. 111.
38. John V. Atanasoff, together with Clifford Berry, built the computer ABC (Atanasoff-Berry Calculator) in 1939. This was possibly the first device to use vacuum tubes for mathematical computations. In 1973 a United States District Court ruled that Atanasoff invented the electronic digital computer (see *Tools for Thought.* H. Rheingold. Simon and Schuster: New York, NY (1985) and *Engines of the Mind.* J. Shurkin. Washington Square Press: New York (1985). However, the impact of ENIAC on computing was enormous.
39. Burks, A. W., Goldstine, H. H. and Neumann, J. von. *Preliminary Discussion of the Logical Design of an Electronic Computing Instrument.* A report prepared for the Ordnance Department, U.S. Army. Institute for Advanced Study: Princeton, NJ. June (1946). Reprinted in *Datamation* **8**, (9 and 10) (1962).
40. Pepinsky, R., Robertson, J. M. and Speakman, J. C. (eds.) *Computing Methods and the Phase Problem in X-ray Crystal Structure Analysis. Report of a Conference held at Glasgow, August 1960.* Pergamon Press: New York, Oxford, London, Paris (1961).
41. Bennett, J. M. and Kendrew, J. C. The computation of Fourier syntheses with a digital electronic calculating machine. *Acta Cryst.* **5**, 109–116 (1952).

42. Wilkes, M. V., Wheeler, D. J. and Gill, S. *The Preparation of Programmes for an Electronic Digital Computer, with Special Reference to the EDSAC and the use of a Library of Sub-routines* Addison-Wesley: Cambridge, MA (1951).
43. Jeffrey, G. A., Shiono, R. and Jensen, L. H. Crystallographic computing on the IBM 650. In *Computing Methods and the Phase Problem in X-ray Crystal Analysis.* (J. C. Speakman, ed.) pp. 25–31. Pergamon Press: London, UK (1961).
44. *Annals of the History of Computing* **8**, 1–88 (Number 1, January 1986). This is a special issue devoted to the IBM 650.
45. Watson, J. D. and Crick, F. H. C. Molecular structure of nucleic acids: a structure for deoxyribonucleic acid. *Nature (London)* **171**, 737–738 (1953).
46. Langridge, R. *Experimental and Theoretical X-ray Diffraction Studies of the Structure of Deoxyribose Nucleic Acid.* Ph. D. Thesis. University of London: London, UK (1957).
47. Langridge, R., Barnett, M. P. and Mann, A. F. Calculation of the Fourier transform of a helical molecule. *J. Mol. Biol.* **2**, 63–64 (1960).
48. Backus, J. W., Beeber, R. J., Best, S., Goldberg, R., Haibt, L. M., Herrick, H. L., Nelson, R. A., Sayre, D., Sheridan, P. B., Stern, H., Ziller, I., Hughes, R. A. and Nutt, R. The FORTRAN automatic coding system. *Proc. Western Joint Computer Conf., Los Angeles.* pp. 188–198 (1957).
49. Stewart, J. M., Kundell, F. A. and Baldwin, J. C. The XRAY 67 system. Computer Science Center: University of Maryland, College Park, MD (1967).
50. Busing, W. R., Martin, K. O. and Levy, H. A. ORFLS. A Fortran Crystallographic Least-squares Program. ORNL-TM 305. Oak Ridge National Laboratory: Oak Ridge, TN (1962).
51. Gantzel, P. K., Sparks, R. A. and Trueblood, K. N. ACA Computer Program No. 317. University of California, Los Angeles: Los Angeles, CA (1962).
52. Main, P., Woolfson, M. M., Lessinger, L., Germain, G. and Declerq, J. P. MULTAN 74: a system of computer programs for the automatic solution of crystal structures from X-ray diffraction data. University of York, University of Louvain: UK and Belgium (1974).
53. Johnson, J. E. Virus structure and crystallography: an historical prespective. In *Crystallography in North America.* (D. McLachlan, Jr. and J. P. Glusker, eds.) Chapter 18. pp. 410–414. American Crystal-

lographic Association: New York, NY (1983).

54. Cooley, J. W. and Tukey, J. W. An algorithm for machine calculation of complex Fourier series. *Math. Comp.* **19**, 297–301 (1965). The first description of the fast Fourier transform is described in a paper by G. C. Danielson and C. Lanczos entitled 'Some improvements in practical Fourier analysis and their application to X-ray scattering from liquids.' *J. Franklin Inst.* **233**, 365–380 and 435–452 (1942) (based on articles by C. Runge *Z. Math. Physik* **48**, 443 (1903) and **53**, 117 (1905)). The Danielson and Lanczos article is described by J. W. Cooley, P. A. W. Lewis and P. D. Welch in an article entitled 'Historical notes on the fast Fourier transform.' *IEEE Trans.* **15**, 76–79 (1967).
55. Johnson, C. K. ORTEP: a Fortran thermal ellipsoid plot program for crystal structure illustrations. ORNL Technical Report 3794. Oak Ridge National Laboratory, TN, USA (1965).
56. Kendrew, J. C., Dickerson, R. E., Strandberg, B. E., Hart, R. G., Davies, D. R., Phillips, D. C. and Shore, V. C. *Nature (London)* **185**, 422–427 (1960).
57. Dickerson, R. E., Kendrew, J. C. and Strandberg, B. E. The phase problem and isomorphous replacement in protein structures. In *Computing Methods and the Phase Problem.* (J. C. Speakman, ed.) pp. 236–251 Pergamon Press: London, UK (1961).
58. Richards, F. M. The matching of physical models to three-dimensional electron-density maps. *J. Mol. Biol.* **37**, 225 (1968)
59. Sutherland, I. E. 'Sketchpad': a Man-machine Graphical Communication System. Ph.D. Thesis. Massachusetts Institute of Technology, Cambridge, MA (1963).
60. Stotz, R. H. and Ward, J. F. *Operating Manual for the ESL Display Console.* Project MAC Internal Memorandum. MAC-M-217. MIT: Cambridge, MA (1965).
61. Langridge, R. and MacEwan, A. W. The refinement of nucleic acid structures. *Proc. IBM Scient. Computing Symp. on Computer Aided Experimentation.* pp. 305–314 (1966).
62. Levinthal, C. Computer construction and display of molecular models. *Proc. IBM Scient. Computing Symp. on Computer Aided Experimentation.* pp. 315–326 (1966).
63. Levinthal, C. Molecular model building by computer. *Sci. Amer.* **214**, 42–52 (1966).
64. Koltun, W. L. Precision space-filling atomic models. *Biopolymers* **3**, 665–679 (1965).

65. Levinthal, C., Barry, C. D., Ward, S. A. and Zwick, M. Computer graphics in macromolecular chemistry. In *Emerging Concepts in Computer Graphics.* (D. Secrest and J. Nievergelt, eds.) p. 231. W. A. Benjamin: New York, NY (1968).
66. Diamond, R. BILDER: an interactive graphics program for biopolymers. In *Computational Crystallography,* pp. 318–325. (D. Sayre, ed.) Oxford University Press: Oxford, UK (1982).
67. Jones, T. A. FRODO: a graphics fitting program for macromolecules. In *Computational Crystallography,* pp. 303–317. (D. Sayre, ed.) Oxford University Press: Oxford, UK (1982)
68. Feldmann, R. J., Bing, D. H., Furie, B. C. and Furie, B. Interactive computer surface graphics approach to study of the active site of bovine trypsin. *Proc. Natl. Acad. Sci. USA* **75**, 5409–5412 (1978).
69. Langridge, R., Ferrin, T. E., Kuntz, I. D. and Connolly, M. L. Real-time color graphics in studies of molecular interactions. *Science* **211**, 661–666 (1981).
70. Kennard, O., Isaacs, N. W., Coppola, J. C., Kirby, A. J., Warren, S., Motherwell, W. D. S., Watson, D. G., Wampler, D. L., Chenery, D. H., Larson, A. C., Kerr, K. A. and Riva di Sanseverino, L. Three dimensional structure of adenosine triphosphate. *Nature (London)* **225**, 333–336 (1970).
71. Langridge, R., Wilson, H. R., Hooper, C. W., Wilkins, M. H. F. and Hamilton, L. D. The molecular configuration of deoxyribonucleic acid. I. X-ray diffraction study of a crystalline form of the lithium salt. *J. Mol. Biol.* **2**, 19–37 (1960).
72. Blaney, J. M., Jorgensen, E. C., Connolly, M. L., Ferrin, T. E., Langridge, R., Oatley, S. J., Burridge, J. M. and Blake, C. C. F. Computer graphics in drug design: molecular modelling of thyroid hormone-prealbumin interactions. *J. Med. Chem.* **25**, 785–790 (1982).
73. Babbage, C. *Passages from the Life of a Philosopher.* Longman: London, UK (1864, reprinted in 1968). Reproduced in *Charles Babbage. On the Principles and Development of the Calculator.* (P. Morrison and E. Morrison, eds.) Dover Publications, Inc: New York, NY (1961).

25. The Patterson–Tunell strip method for the summation of Fourier series

George Tunell

George Tunell is Professor Emeritus of Geochemistry at the University of California, Santa Barbara. This article describes the history of the Patterson–Tunell method for the summation of the Fourier series used in the X-ray analyses of crystal structures by the use of strips carrying numbers, stencils and a counting rack.

In regard to the Patterson–Tunell stencils and strips, I can say that the idea for the use of stencils, strips and counting rack in the calculation of the F^2- and F-series needed in crystals structure determinations was wholly Lindo's. I made one contribution to the improvement of the method as follows. In the first stencils used by Lindo and Dr. Selma Blazer Brody, the odd number and even number patterns were on one stencil, and in consequence some of the stencils had many holes in quite irregular patterns which made the addition of the numbers, visible through the holes, quite difficult. I made the improvement of separating the odd number patterns and even number patterns on different stencils. The holes of the new stencils for only odd number or only even number summations had greatly simplified patterns which made the addition of the numbers, visible through the holes, very much simpler and very much easier.

A comparison of the three methods for the calculation of F^2- and F-series by means of strips carrying numbers is given on pages 678 and 679 of the article by Patterson and Tunell in the *American Mineralogist.*[1] Dr. Booth expressed an unfavorable opinion of the Patterson–Tunell method. He said that it was too difficult to remember that a number on a blue strip seen through a red-ringed hole was minus-minus and therefore plus. Mrs. Tunell had a little trouble with this feature the first day or two that she made the computations, but none thereafter. Several of the graduate students in the Chemistry Department at UCLA who used the improved Patterson–Tunell stencil and strips had similar experiences. Since I have had no experience with the Robertson method or with the Beevers and Lipson method, I am not in a position to judge the relative convenience of the three methods. I can only say that Mrs. Tunell and I were very well satisfied with the use of the improved Patterson–Tunell stencils and strips by means of which we determined the F^2- and F-series maps on projections on a plane perpendicular to the b-axis of calaverite ($(Au,Ag)Te_2$), the F-series map on the projection on a plane perpendicular to the b-axis of sylvanite ($AuAgTe_4$) and the F-series map on the projection on a plane perpendicular to the c-axis of krennerite ($(Au,Ag)Te_2$).

In conclusion I would mention that Professor J. D. H. Donnay and his graduate student Father Tremblay probably made the most careful comparison of the Beevers and Lipson and Patterson–Tunell methods. Their first experience was with the Beevers and Lipson method, so they did not approach the comparison with any prejudice against the Beevers and Lipson method. After experience with both methods, their preference was for the Patterson–Tunell method; in fact, they made up a number of sets of Patterson–Tunell stencils, strips, and counting racks, which they supplied to a number of crystallographers. At present, of course, the use of electronic computers has superseded the three methods in which strips carrying numbers were used for the calculation of F^2- and F-series. However, Professor Gabrielle Donnay wrote in her history of crystallography at the Geophysical Laboratory[2] that the Patterson–Tunell stencils, strips, and a counting rack 'are, to this day, used as a teaching aid to make the student appreciate what really goes on in a Fourier summation.'

References

1. Patterson, A. L. and Tunell, G. A method for the summation of the Fourier series used in the X-ray analysis of crystal structures. *Am. Mineral.* **27**, 655-679 (1942).
2. G. Donnay, with the help of Wyckoff, R. W. G., Barth, T. F. W. and Tunell, G. Fifty years of X-ray crystallography at the Geophysical Laboratory, 1919-1969. *Carnegie Institution of Washington Yearbook* **68**, 278-283 (1970).

26. Early use of the Patterson function at Caltech

Jerry Donohue

Jerry Donohue was Professor of Chemistry at the University of Pennsylvania. He obtained a Ph.D. at Caltech in 1947. His paper on the crystal structure of L_s-threonine with David Shoemaker, Verner Schomaker and Robert Corey was well-studied by subsequent students in the field. He went to Dartmouth (where he had also been an undergraduate) and then to England in 1952–1953 as a Guggenheim Fellow. He was an expert on hydrogen bonding and was the person who set the record straight for Watson and Crick on how the bases in DNA would form hydrogen bonds. Jerry then went to the University of Southern California and then to the University of Pennsylvania. He said that one of the reasons that he came to Philadelphia was because Lindo was there. Bud Carrell, Bill Stallings and Henry Katz all worked for him before they came to the Institute for Cancer Research in Fox Chase.

Jerry was the first person to register for this symposium. He was, however, unable to attend because of the death of his father-in-law. He took an active part in the initial organization of this symposium, and of this volume. On February 13, 1985 he passed away after a brief illness; we miss him greatly.

Because this is an historical meeting commemorating Lindo Patterson's historical formulation of $P(uvw)$ fifty years ago, I would like to recall five structures, the determinations of which were carried out in the 1950's at the California Institute of Technology, and are now part of the history of X-ray crystallography. Trial structures for all of these were obtained by consideration of the complete three-dimensional Patterson functions, not just projections, nor selected sections, then called Harker sections, which are indicated by various symmetry operations. David Harker[1], then at CalTech, shortly after Lindo's original publication on $P(uvw)$, had pointed out that use could be made of particular parts of the complete Patterson function to locate atoms, based on symmetry elements of the space group. These parts were soon called Harker lines and Harker sections. Harker described how he solved the structures of proustite, Ag_3AsS_3, and pyrargyrite, Ag_3SbS_3, by calculating two Patterson lines, $P(0y0)$, which were 'one-dimensional Fourier series and can be completely evaluated in a few hours.' A little later Corey and Albrecht,[2] described how they solved the crystal structure of glycine by considering what they called 'spurious maxima' in the Harker section $P(x\,\frac{1}{2}\,z)$, as opposed to 'real maxima.' The former, which are not 'spurious' at all but arose when two atoms, not related by symmetry, accidentally had the same y-parameter, were twice the magnitude of the latter.

Our approach at the time was that *all* parts of the $P(uvw)$ function were important, not only the Harker sections which were more accessible then because of the primitive computing capabilities available at the time. Somewhat later slightly less primitive computing techniques than the Beevers-Lipson or Patterson-Tunell strips appeared, in the form of files, four feet wide by five feet high, of IBM cards, on which sine and cosine functions were punched; from these files we fished out and fed the appropriate cards into the IBM tabulator, which was really an adding machine with appropriate control cards to tell the machine whether to add, or subtract depending on the parity of h, k, l. Naturally, this took a long time. The calculation of $P(uvw)$, or, later, of $\rho(xyz)$ for L-threonine, four molecules of $C_4H_9NO_2$ in a unit cell $13.6 \times 7.7 \times 5.1$ Å, took about a week, but we were not in such a hurry in those days as people seem to be now. There was no urgency then to get your latest preliminary results published in the *New England Journal of Medicine*, with simultaneous release to the *New York Times*. Some data for five crystal structures are presented in Table 26.1. All of these structures were based on complete sets of visually estimated data from Weissenberg photographs.

L_s-Threonine

Because of the relatively short c-axis, the Harker section $P(u\,v\,\frac{1}{2})$ was first calculated. It turned out to be uninterpretable because the Harker peaks could not be distinguished from the non-Harker peaks. It was therefore decided to calculate the complete Patterson function, with the origin peak removed (as originally suggested by Lindo), and modified by convergence

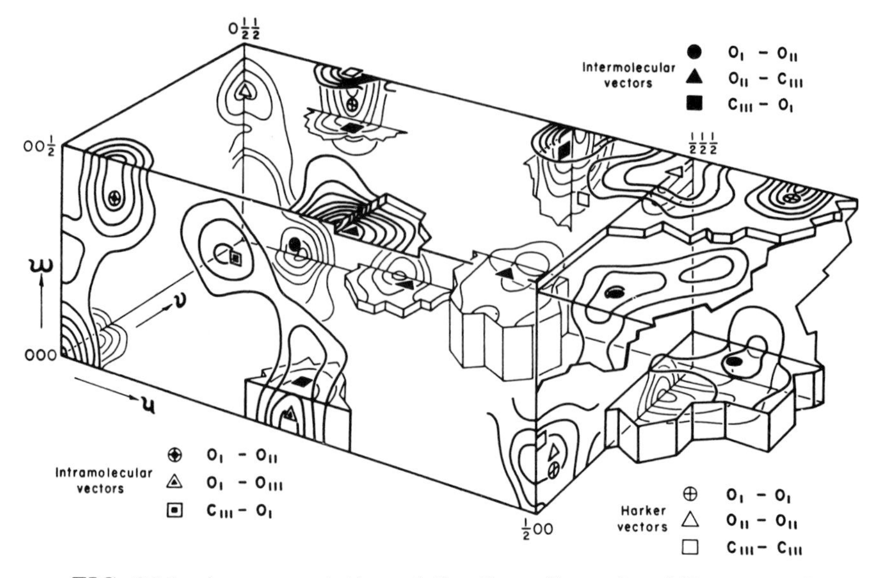

FIG. 26.1. A representation of the three-dimensional Patterson function of L_s-threonine. Only the peaks that correspond to interactions among the atoms O_I, O_{II} and C_{III} are shown.

function. The number of resolved peaks in $P(uvw)$ was only half of that expected, but this turned out to be no hindrance to its interpretation. The unravelling of $P(uvw)$ is shown in Fig. 26.1. The trial structure was refined by successive calculations of $\rho(xyz)$, and by least-squares. The calculation of the 300 coefficients of the normal equations took 50 hours of machine time on an IBM 602 multiplying punch.

The four of us worked on this structure for about three years or more, although we all were doing other things too. After the correct trial structure had been found, Shoemaker and Schomaker took off for a year on Guggenheim fellowships, leaving me to cope with the mountains of punched cards. The paper describing our labors was duly written after they returned to Pasadena, and sent off to the JACS.

As an historical aside, this paper was several times at subsequent meetings of the ACA called by Lindo a 'classic.' It included one and three-fourths pages of $F_{obs}F_{calc}$ tables, set up in type, many pages of discussion of the errors in the $\rho(xyz)$ and least-squares refinements, and what we believe was the first successful interpretation of a complete three-dimensional Patterson function. It was also a milestone in establishing the relative configurations of the two asymmetric carbon atoms in L-threonine. One of the JACS referees had only one comment (I do this from memory, as the original correspondence has disappeared): 'The authors have failed to conform to the convention that in the orthorhombic system $b < a < c$.' (We had $a > b > c$, naughty us).

A month after the appearance of the JACS with our paper we received a post-card, signed merely ALP, showing a photograph of a Philadelphia highway sign: **Penna 309**. That post-card has also vanished.

Hydroxyproline

Following the threonine experience, the complete three dimensional $P(u0w)$, and $P(0vw)$, in which the peaks arise from fortuitous equality, or near equality of z, y, and x parameters, respectively, was computed. The corresponding Harker sections (at w, v, and $u = 12$) will then contain three peaks, a central non-Harker peak separated from two Harker peaks by the vectors observed in the zero sections. The solutions are shown in Fig. 26.2. It is obvious that identification of the Harker peaks as such is impossible, but that looking for linear triples of peaks in the Harker sections based on the peaks observed in the $u, v, w = 0$ section was successful.

There's not enough time to discuss how we did the other structures in Table 26.1, but they're there if you wish to see what happened in the good old days before Karle and Hauptman took the fun out of it by telling us how to do it on a computer. The real guilt, however, lies with Harker and Kasper, who really started this direct methods stuff.

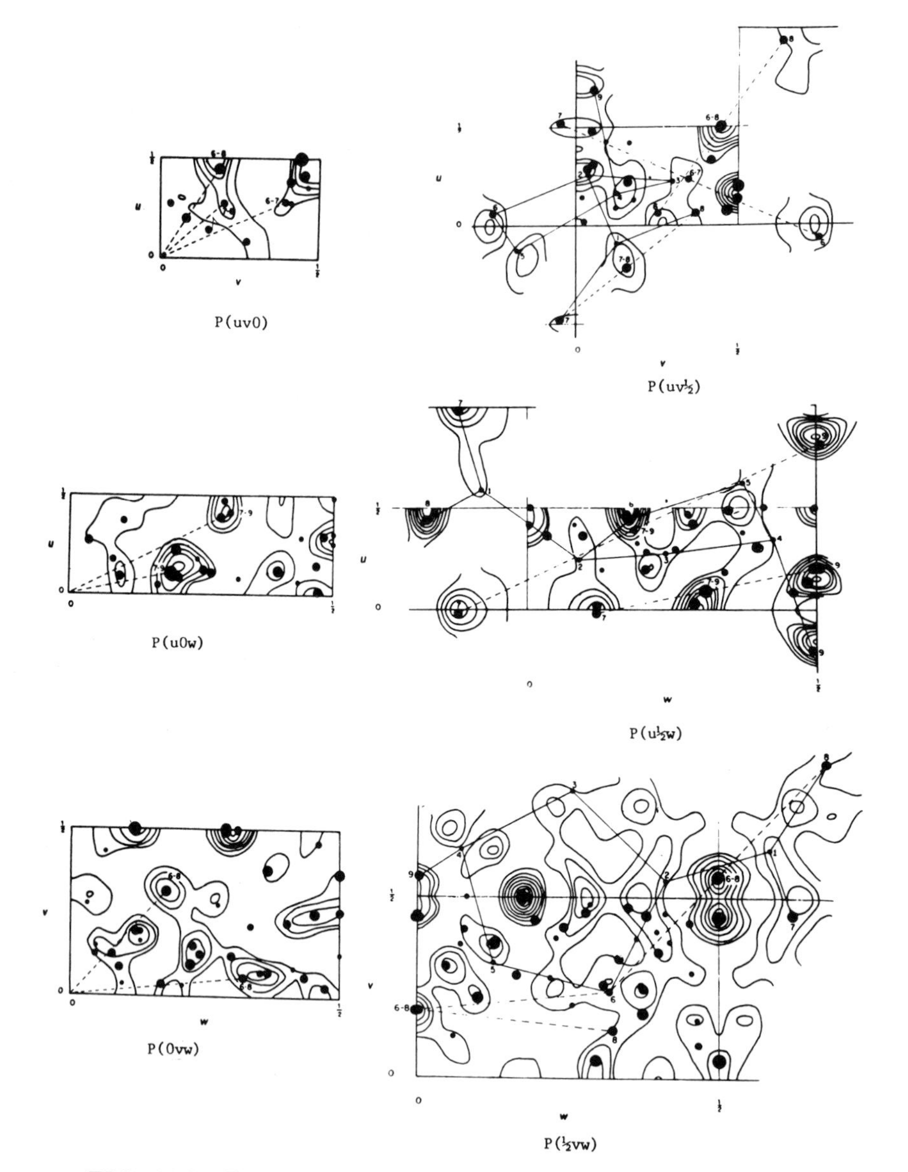

FIG. 26.2. Portions of the three-dimensional Patterson function of hydroxyproline.

TABLE 26.1. Some Crystal Data

Compound	Formula (excluding H)	Space group	Unit cell,Å, β	no. of F_{obs}	R	Ref.
L-threonine	C_4NO_3	$P2_12_12_1$	13.6 × 7.7 × 5.1	555	0.102	3
hydroxyproline	C_5NO_3	$P2_12_12_1$	5.0 × 8.3 × 14.2	646	0.148	4
DL-serine	C_3NO_3	$P2_1a$	10.7 × 9.1 × 4.8, 106°	750	0.145	5
nitroguanidine	CN_4O_2	$Fddd$	17.6 × 24.8 × 3.6	218	0.111	6
histidine·HCl·H_2O	$C_6N_3O_3Cl$	$P2_12_12_1$	15.4 × 8.9 × 6.9	935	0.120	7

Nevertheless, regardless of Harker, Kasper, Karle, Hauptman, Woolfson, Sayre, Cochran, *et al.*, the Patterson function is far from dead: of the 69 structures described in the May 1984 issue of *Acta Cryst.*, volume **C**, 38 of them were solved by the use of Multan (SHELX) and 31 by use of the Patterson function. This means to me that almost half of currently practising crystallographers would rather solve their structures by cerebration, and not with a computer program written by someone else.

Acknowledgement

This work was supported in part by the NSF-MRL Program under Grant DMR-8216718, also NIH Grant GM-20611-08.

References

1. Harker, D. *J. Chem. Phys.* **4**, 381 (1936).
2. Corey, R. B. and Albrecht, G. *J. Am. Chem. Soc.* **61**, 1087 (1939).
3. Shoemaker, D. P., Donohue, J., Schomaker, V. and Corey, R. B. *J. Am. Chem. Soc.* **72**, 2328-2348 (1950).
4. Donohue, J. and Trueblood, K. N. *Acta Cryst.* **5**, 414-418 (1952).
5. Shoemaker, D. P., Barieau, R. E., Donohue, J. and Lu, C.-S. *Acta Cryst.* **6**, 241-256 (1953).
6. Donohue, J. and Bryden, J. H. *Acta Cryst.* **8**, 314-316 (1955).
7. Donohue, J., Lavine, L. R., and Rollett, J. S. *Acta Cryst.* **9**, 655-661 (1956).

27. Patterson methods in the determination of the structure of α-ketoglutarate

Jonah Erlebacher, Henry Katz and H. L. Carrell

Horace L. Carrell obtained a Ph.D. at the University of Southern California in 1966 working with Jerry Donohue. He stayed on as a postdoctoral research associate and came, with Jerry Donohue, to the University of Pennsylvania. He came to the Institute for Cancer Research in 1969 and is at present primarily working on the structure of the enzyme xylose isomerase. Henry Katz also worked in Jerry Donohue's laboratory and came to the Institute in 1982. Jonah Erlebacher is a student from Cheltenham High School, Cheltenham, PA who started working on this project at age 13. He is now adept at interpreting Patterson maps.

α-Ketoglutaric acid is a substrate of isocitrate dehydrogenase, an enzyme of the Krebs cycle. Lindo Patterson's interest in the Krebs cycle, a major metabolic cycle, led to the study of the crystal structures of several citrates[1–8] and an isocitrate.[9–11] The enzyme isocitrate dehydrogenase catalyses the interconversion of isocitrate and α-ketoglutarate.

The crystal structure of potassium α-ketoglutarate was determined in order to find the conformation, mode of ionization, and packing in the crystalline state of the α-ketoglutarate ion. A hypothesis of how isocitrate dehydrogenase catalyses the reaction, converting the isocitrate ion into α-ketoglutarate, is proposed. The crystal structure, solved by determining the location of the potassium ion from a Patterson map, and all other non-hydrogen atoms from a Fourier map, has been reported by Lis and Matuszewski.[12] We report here our determination, done independently and concurrently, in which the locations of *all* non-hydrogen atoms were determined from the Patterson map, and H-atoms from a difference Fourier map.

Experimental

Crystals of potassium α-ketoglutarate were grown by vapor diffusion using acetone and water. Preliminary photographs, taken with a Buerger precession camera, showed that the crystal is monoclinic, space group $P2_1/n$. A least squares fit of 14 centered reflections indicated that the cell dimensions are a = 6.519(1), b = 17.808(2), c = 6.107(1)Å and β = 100.89(1)°, V = 696.2(2). The measured density of the crystal, 1.8 g. cm^{-3}, indicated that only one potasssium ion and one α-ketoglutarate ion form the asymmetric unit.

Diffraction data were collected on a Syntex $P2_1$ four-circle diffractometer using Cu$K\alpha$ radiation (λ = 1.5418Å). The $\theta - 2\theta$ scan technique was used to a 2θ limit of 138°, sin θ/λ = 0.61Å^{-1}, at variable scan speeds from 0.99°min^{-1} to 29.30°min^{-1}, depending upon intensity. For every 50 reflections collected, four check reflections were measured in order to monitor for crystal slippage or crystal decay. Data for 1,273 unique reflections were obtained and used to generate a Patterson map.

Expected vector positions for the space group $P2_1/n$ are listed in Table 27.1. The potassium was found easily because the potassium-potassium vector peaks are high. The potassium coordinates, found from a comparison of vector positions in Table 27.1 with the high peaks positions were x = 0.710, y = 0.482 and z = 0.795.

The rest of the structure was determined by superimposing the Patterson map on itself four times, each time placing the origin at one of the four real space coordinates of the potassium ion. One section of this superposition map is shown in Fig.

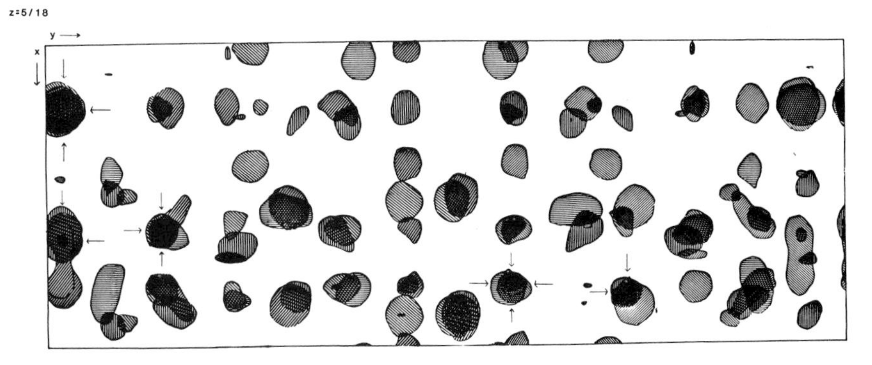

FIG. 27.1. Section of a four-fold superposition map at z = 5/18. Atoms exist where the four maps (indicated by differing angles of shading) overlap; these positions on this section are indicated by arrows.

TABLE 27.1. Vector positions for the structure solution.

(a) Unique vectors in the space group $P2_1/n$. Vectors related to these by a center of symmetry, or by a mirror plane perpendicular to v, also occur.

u	v	w
$2x$	$2y$	$2z$
$\frac{1}{2}$	$\frac{1}{2} - 2y$	$\frac{1}{2}$
$\frac{1}{2} - 2x$	$\frac{1}{2}$	$\frac{1}{2} - 2z$

(b) Highest peaks in the Patterson map.

u	v	w	height
0.500	0.536	0.500	1377*
0.080	0.500	0.910	1206*
0.410	0.000	0.964	761
0.322	0.500	0.038	715
0.420	0.962	0.587	590*

Note: peaks marked with an asterisk are K···K vectors

27.1. This yielded coordinates of a complete α-ketoglutarate ion. The R factor at this point was approximately 26%. The structure was refined using a full matrix least-squares analysis. Hydrogen atoms were found in a difference Fourier map. These were refined isotropically, all other atoms anisotropically. The final R value was 0.032.

Discussion

Refined atomic parameters are shown in Table 27.2, together with values by Lis and Matuszewski.[12] Estimated standard deviations are shown in parentheses. The two structures are iden-

TABLE 27.2. Coordinates of α-ketoglutarate.

Coordinates in fractions of the unit cell. Equivalent isotropic B values are given in Å^2. The upper values are from this work, the lower from the work of Lis and Matuszewski.[12]

Atom	x	y	z	B
K1	0.71038(5)	0.48224(2)	0.79458(5)	2.05(1)
	0.7102 (1)	0.48218(2)	0.7946(1)	1.98(2)
O1	0.6819(2)	0.58512(6)	0.4303(2)	2.53(5)
	0.6816(3)	0.58504(7)	0.4306(3)	2.08(2)
O2	0.8756(2)	0.51584(5)	0.2421(2)	2.29(5)
	0.8762(3)	0.51580(6)	0.2424(3)	1.99(7)
O3	1.0086(2)	0.63944(6)	0.0713(3)	3.74(5)
	1.0085(3)	0.63937(7)	0.0710(3)	2.51(9)
O4	0.9751(2)	0.89280(7)	0.4089(2)	2.64(5)
	0.9753(3)	0.89290(7)	0.4095(3)	2.60(8)
O5	0.7262(2)	0.89323(6)	0.1067(2)	2.50(5)
	0.7258(3)	0.89312(7)	0.1062(3)	2.50(8)
C1	0.8063(2)	0.57693(8)	0.3042(2)	1.68(6)
	0.8067(4)	0.57710(8)	0.3042(3)	1.55(8)
C2	0.8934(2)	0.64745(8)	0.2029(3)	1.90(6)
	0.8933(4)	0.67742(8)	0.2026(3)	1.59(8)
C3	0.8325(3)	0.72328(8)	0.2742(3)	2.08(6)
	0.8307(4)	0.72330(9)	0.2749(3)	1.71(9)
C4	0.9405(3)	0.78562(9)	0.1688(3)	2.49(6)
	0.9410(4)	0.78548(9)	0.1691(4)	1.97(10)
C5	0.8832(2)	0.86219(8)	0.2411(3)	1.88(6)
	0.8833(4)	0.86230(8)	0.2407(3)	1.62(8)
HO5	0.704(4)	0.932(2)	0.166(4)	4.7(6)
H3A	0.871(4)	0.726(1)	0.427(4)	3.4(5)
H3B	0.685(4)	0.728(1)	0.228(4)	3.9(5)
H4A	0.903(3)	0.785(1)	0.002(4)	3.1(4)
H4B	1.097(4)	0.780(1)	0.217(4)	3.7(5)

Note: unit cell dimensions are quoted by Lis and Matuszewski in the space group $P2_1/c$ as $a = 6.517(4), b = 17.816(9), c = 8.050(4)$Å and $\beta =$ 131.79(4)°.

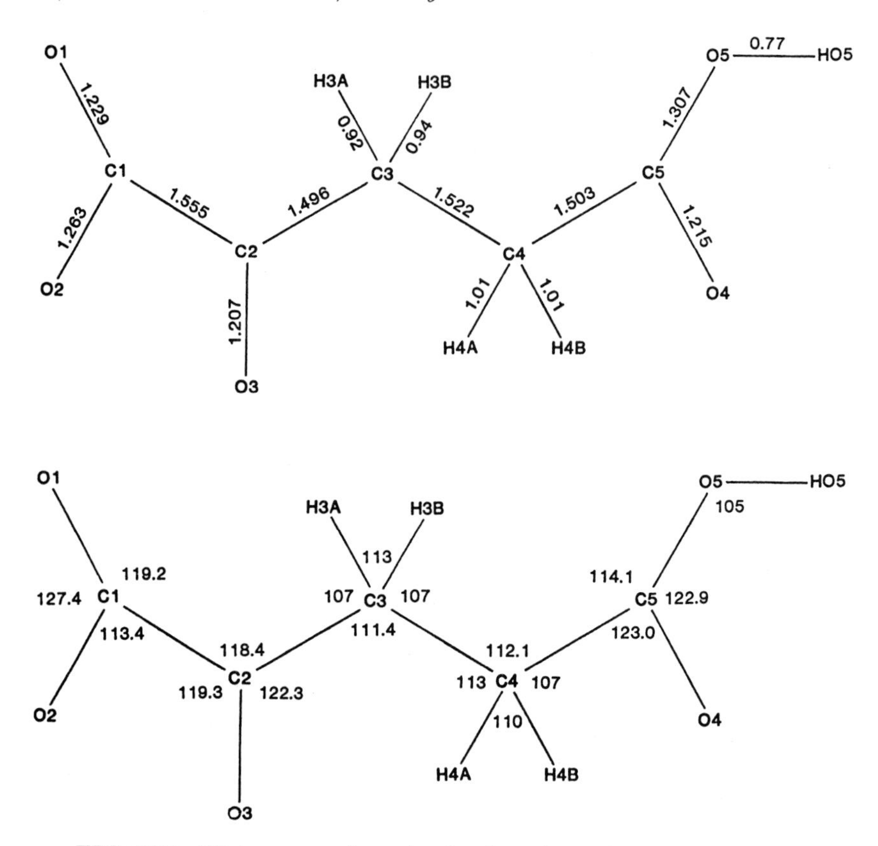

FIG. 27.2. Distances and angles in the α-ketoglutarate ion.

tical within experimental error. The structure of potassium α-ketoglutarate is shown in Fig. 27.2 together with bond distances and angles. The carbon chain of the α-ketoglutarate ion is nearly planar, with the O4-C5-O5 carboxyl group perpendicular to it. This is in line with other crystallographic studies of glutarates.[12–19] There is only one hydrogen atom (HO5) available for hydrogen bonding. The structure of potassium α-ketoglutarate packs with long strands of α-ketoglutarate ions hydrogen bonded between HO5 and O2; these strands are coordinated to the potassium ions that lie between them. This is shown in Fig. 27.3, where it can also be seen that the molecules are bonded head to tail, that is, the head of the α-ketoglutarate ion, O1-C1-O2, is hydrogen bonded to the tail, O4-C5-O5-HO5, of another α-ketoglutarate ion. Unlike many

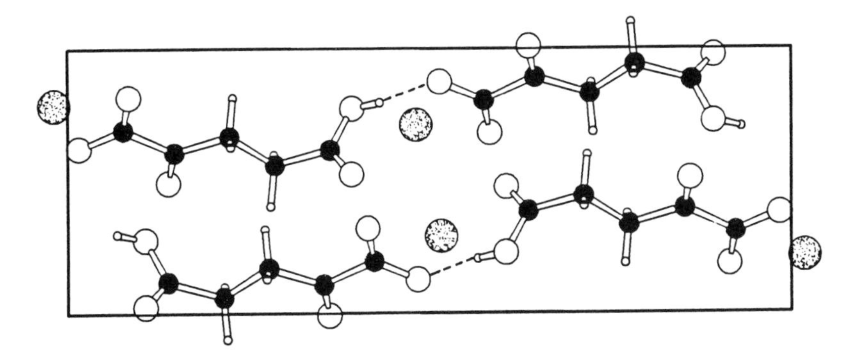

FIG. 27.3. Packing in the unit cell of potasium α-ketoglutarate.

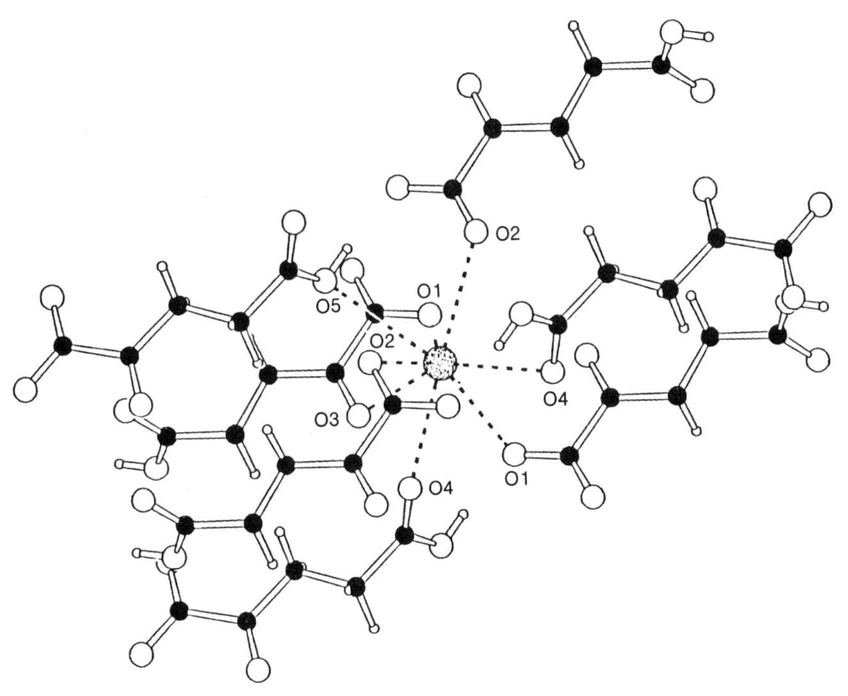

FIG. 27.4. Packing of seven α-ketoglutarate ions around a potassium ion.

similar structures which have two such hydrogen bonds, there is only one hydrogen bond between the carboxyl groups.

The coordination of potassium ions by oxygen atoms of α-ketoglutarate ions is shown in Fig. 27.4. There are 8 oxygen atoms coordinated to one potassium ion and there is a chelate ring (O1-C1-C2-O3-K). The apparent valence on each oxygen can be determined from the information that the valence of the potassium is 1.00 and the net valence of the oxygens is also 1.00. This can be done using the formula

$$s = (R/R_o)^{-N}$$

which gives the valence between two ions;[20] in this formula s is the valence, R is the distance between the atoms, and R_o and N are constants determined by one of the atoms being tested. In the case of a K^+–O bond, $R_o = 2.276$ and $N = 9.1$. The valences between the oxygen atoms and the potassium ion is shown in Table 27.3. The sum of these valences is approximately 0.98 and is well within the uncertainty of the above formula for s.

Isocitrate is decarboxylated to give α-ketoglutarate in the reaction catalyzed by the enzyme isocitrate dehydrogenase as illustrated in Fig. 27.5. While the carbon chain is planar in α-ketoglutarate, it is bent in isocitrate, and in isocitrate there is the second hydrogen mentioned above.

TABLE 27.3. Bond-valence sum for the potassium ion.

Oxygen	Distance from potassium ion	Valence
O2*	2.751 Å	0.172
O2	2.806	0.144
O4	2.847	0.126
O3*	2.856	0.122
O1	2.860	0.121
O4	2.863	0.119
O1	2.922	0.099
O5	3.012	0.075
		0.978

Note: O2 and O3, marked with an asterisk, are in the same α-ketoglutarate ion.

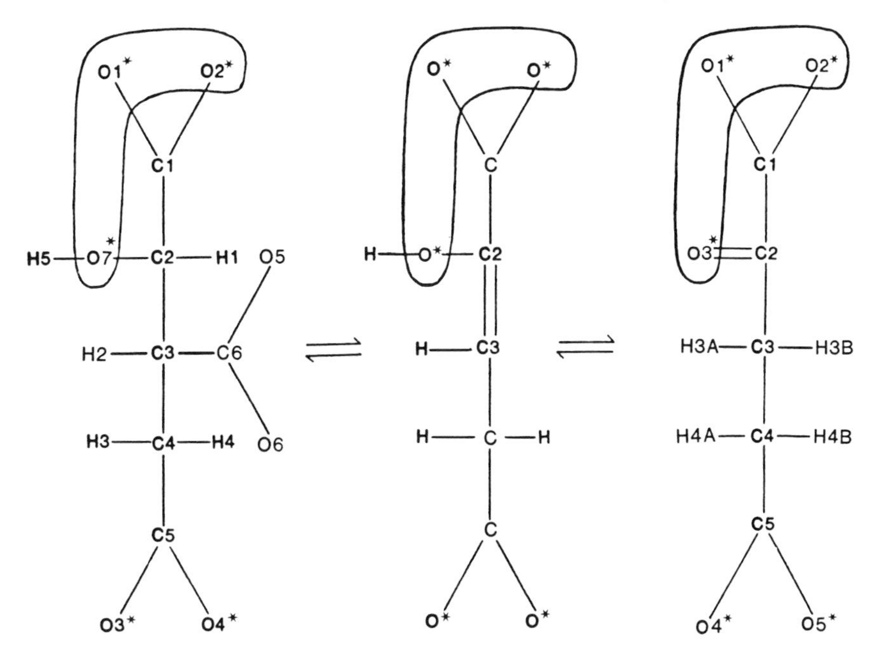

FIG. 27.5. Scheme of interconversion of isocitrate (left) to α-ketoglutarate (right). Asterisked atoms are presumed to bind in the active site of the enzyme.

It is probable that isocitrate dehydrogenase binds the isocitrate and α-ketoglutarate ions at oxygen atoms marked with an asterisk in Fig. 27.5; these atoms have the same spatial arrangement in both molecules. It is hypothesized that this enzyme catalyzes the reaction in the following way:

1) The isocitrate ion binds to the enzyme as described above;
2) the O5-C6-O6 group in the isocitrate ion is then lost as CO_2.
3) In order to keep a stable arrangement of atoms, a double bond forms between C2 and C3; H1 of isocitrate is lost. However, too much energy is expended in keeping this double bond and therefore
4) H5 is transferred to C3, making a more stable double bond between C2 and its adjacent bonded oxygen atom, so forming the α-ketoglutarate ion. This reaction is represented schematically in Fig. 27.6.

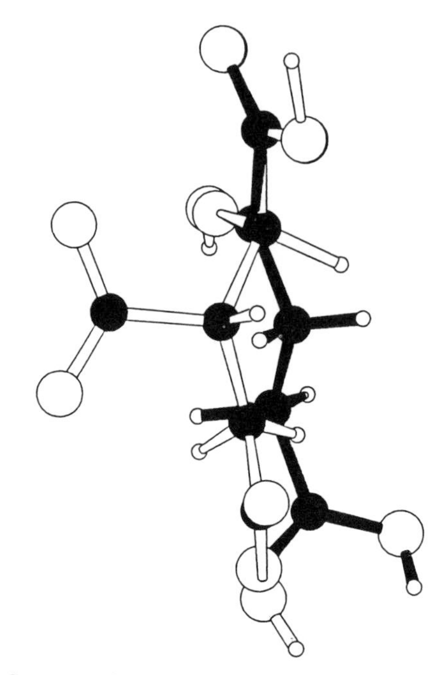

FIG. 27.6. Superposition of α-ketoglutarate (black bonds) on isocitrate (white bonds) via the terminal carboxyl and C–O group; note that the carboxyl group of isocitrate that is lost, as α-ketoglutarate is formed by the action of the enzyme isocitrate dehydrogenase, sticks out to the left in this diagram.

Acknowledgements

This work was supported by grants CA-10925, CA-22780 and CA-06927 from the National Institutes of Health, US Public Health Service.

References

1. Nordman, C. E., Weldon, A. S. and Patterson, A. L. X-ray crystal analysis of the substrates of aconitase. I. Rubidium dihydrogen citrate. *Acta Cryst.* **13**, 414–417 (1960).
2. Nordman, C. E., Weldon, A. S. and Patterson, A. L. X-ray crystal analysis of the substrates of aconitase. II. Anhydrous citric acid. *Acta Cryst.* **13**, 418–426 (1960).
3. Love, W. E. and Patterson, A. L. X-ray crystal analysis of the sub-

strates of aconitase. III. Crystallization, cell constants, and space groups of some alkali citrates. *Acta Cryst.* **13**, 426–428 (1960).

4. Glusker, J. P., van der Helm, D., Love, W. E. Dornberg, M. L. and Patterson, A. L. The state of ionization of crystalline sodium dihydrogen citrate. *J. Am. Chem. Soc.* **82**, 2964–2965 (1960).
5. Glusker, J. P., Patterson, A. L., Love, W. E. and Dornberg, M. L. X-ray analysis of the substrates of aconitase. IV. The configuration of the naturally occurring isocitric acid as determined from potassium and rubidium salts of its lactone. *Acta Cryst.* **16**, 1102–1107 (1963).
6. Glusker, J. P., van der Helm, D., Love, W. E., Dornberg, M. L., Minkin, J. A., Johnson, C. K. and Patterson, A. L. X-ray crystal analysis of the substrates of aconitase. VI. The structures of sodium and lithium dihydrogen citrates. *Acta Cryst.* **19**, 561–572 (1965).
7. Gabe, E. J., Glusker, J. P., Minkin, J. A. and Patterson, A. L. X-ray crystal analysis of the substrates of aconitase. VII. The structure of lithium ammonium hydrogen citrate monohydrate. *Acta Cryst.* **22**, 366–375 (1967).
8. Glusker, J. P., Minkin, J. A. and Patterson, A. L. X-ray crystal analysis of the substrates of aconitase. IX. A refinement of the structure of anhydrous citric acid. *Acta Cryst.* **B25**, 1066–1072 (1969).
9. Glusker, J. P., Patterson, A. L., Love, W. E. and Dornberg, M. L. Crystallographic evidence for the relative configuration of naturally occurring isocitric acid. *J. Am. Chem. Soc.* **80**, 4426–4427 (1958).
10. van der Helm, D., Glusker, J. P., Johnson, C. K., Minkin, J. A., Burow, N. E. and Patterson, A. L. X-ray crystal analysis of the substrates of aconitase. VIII. The structure and absolute configuration of potassium dihydrogen isocitrate isolated from Bryophyllum calycinum. *Acta Cryst.* **B24**, 578–592 (1968).
11. Patterson, A. L., Johnson, C. K., van der Helm, D. and Minkin, J. A. The absolute configuration of naturally occurring isocitric acid. *J. Amer. Chem. Soc.* **84**, 309–310 (1962).
12. Lis, T. and Matuszewski, J. The structure of α-ketoglutaric acid (I), $C_5H_6O_5$, sodium hydrogen α-ketoglutarate (II), $Na^+ \cdot C_5H_5O_5^-$, and potassium hydrogen α-ketoglutarate (III), $K^+ \cdot C_5H_5O_5^-$. *Acta Cryst.* **C40**, 2016–2019 (1984).
13. Macdonald, A. L. and Speakman, J. C. Rubidium hydrogen glutarate and ammonium hydrogen glutarate: X-ray studies of quasi-isostructural crystals involving very short hydrogen bonds. *J. Cryst. Mol. Struct.* **1**, 189–198 (1971).

14. Benedetti, E., Claverini, R. and Pedone, C. Conformation and packing of dicarboxylic acids: the crystal structures of 3,3-dimethylglutaric acid and 1,1-cyclopentanediacetic acid. *Gazzatta Chim. Ital.* **103**, 525–535 (1973).
15. Martuscelli, E., Benedetti, E., Ganis, P. and Pedone, C. Refinement of the crystal and molecular structure of *meso*-α,α'-dimethylglutaric acid. *Acta Cryst.* **23**, 747–753 (1967).
16. Morrison, J. D. and Robertson, J. M. The crystal and molecular structure of certain dicarboxylic acids. Part VII. β-Glutaric acid. *J. Chem. Soc.* 1001–1008 (1949).
17. Martuscelli, E. and Avitabile, G. Crystal and molecular structure of β-ketoglutaric acid. *Ricerca Scient., Scienze Chim.* **37**, 102–110 (1967).
18. Avitabile, G., Ganis, P. and Martuscelli, E. Crystal and molecular structure of $\alpha, \alpha', \alpha, \alpha'$-tetramethyl-$\beta$-ketoglutaric acid (triclinic modification). A model of polydimethylketene with ketonic enchainment. *Acta Cryst.* **B25**, 2378–2385 (1969).
19. Macdonald, A. L. and Speakman, J. C. Crystal structures of the acid salts of some dibasic acids. Part VI. An X-ray study of potassium hydrogen glutarate. *J. Chem. Soc., Perkin II* 942–946 (1972).
20. Brown, I. D. and Wu, K. K. Empirical parameters for calculating cation–oxygen bond valences. *Acta Cryst.* **B32**, 1957–1959 (1976).

28. X-ray imaging of quantum electron structure

L. J. Massa, R. F. Boehme and S. J. La Placa

The same ideas, language and mathematics long used in quantum mechanics are also applicable to interpretation of the X-ray crystallographic experiment.[1–10]

In this connection consider Fig. 28.1. Symbolically the crystallographic experiment may be thought of as the sequence

$$\{\phi\} \Rightarrow \rho(r) \Rightarrow \{F(K)\} \tag{1}$$

where $\{\phi\}$ is a set of orbitals characteristic of the crystal, which give rise to the electron density $\rho(r)$ the source of X-ray scattering which results in measured structure factors $\{F(\vec{k})\}$. Importantly the above sequence may be reversed,

$$\{F(K)\} \Rightarrow \rho(r) \Rightarrow \{\phi\} \tag{2}$$

i.e., given the set of measured structure factors $\{F(\vec{K})\}$, it is

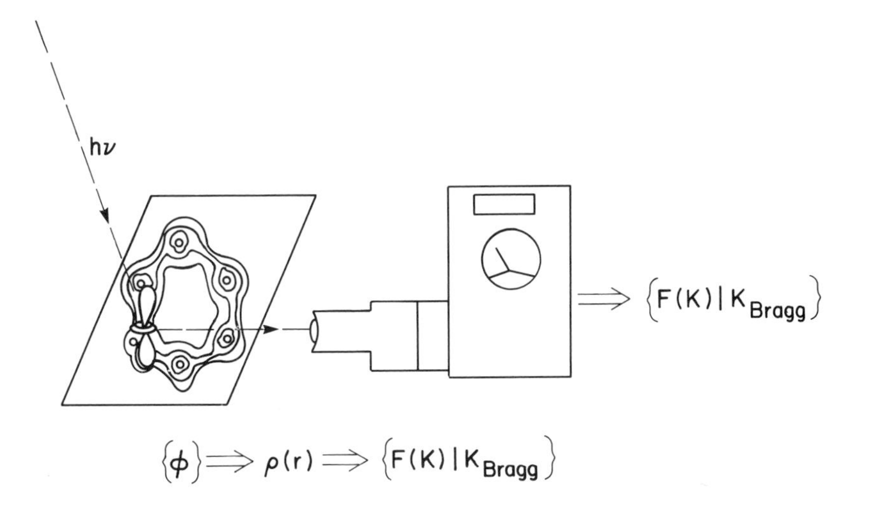

FIG. 28.1. Idealized sketch of X-ray coherent diffraction. X-rays scattered by the electron density are shown counted by a photon detector. Symbols $\phi, \rho(r)$, and $F(K)$ are explained in the text.

possible to recover the electron density $\rho(r)$ and the companion orbitals $\{\phi\}$.

Historically the density model which has dominated crystallography has been that obtained by summing spherical atoms. Some notion of how remarkably good is this approximation and therefore why it has long served crystallography well, may be obtained by consideration of Fig. 28.2. We see there an electron density contour map of a small molecule[11] (N-methyl-acetamide, though our discussion does not depend critically on a particular choice of molecule) obtained in two different ways. In the more accurate case,(a),a theoretical wave

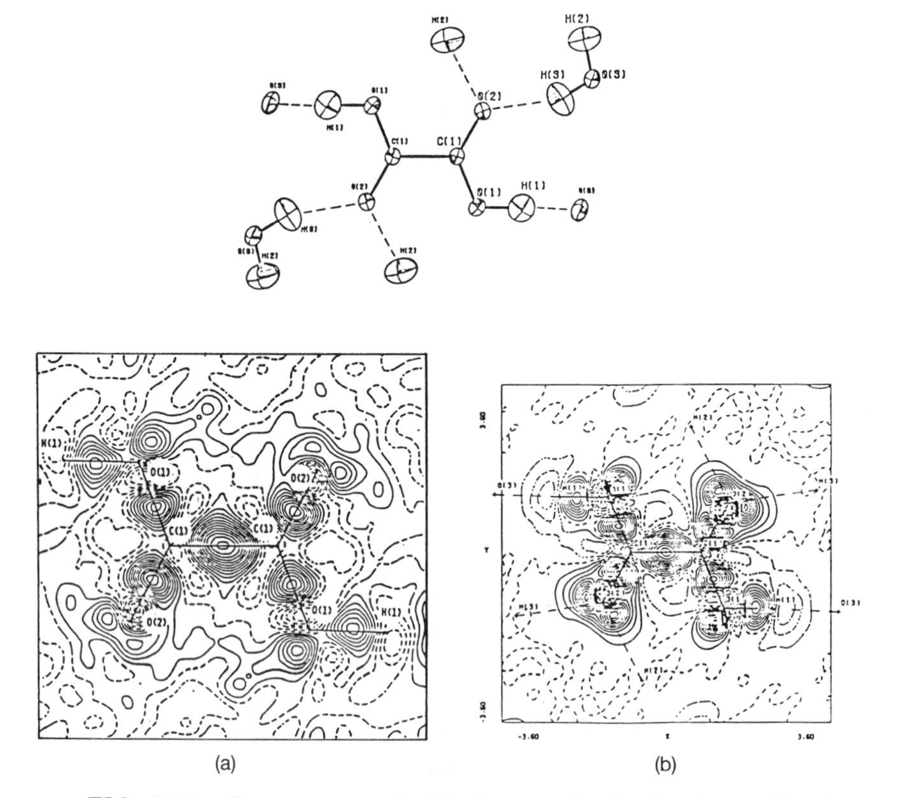

FIG. 28.2. Contour map of the electron density in the amide plane of N-methylacetamide as obtained with the split valence 6-31G basis set.[17] (a) Total density as calculated from molecular wave function. (b) Density obtained from superposition of spherical atoms.

function is obtained for the entire molecule and the molecular density pictured corresponds to that of the molecular wave function. In case (b) the total density is obtained by superposing purely atomic contributions. It is perhaps surprising how closely the sum of spherical atom densities approximates the molecular density. In the context of crystallography, of course, this explains why from that approximation follows one of the principle methods of solving structures. Indeed as long as the principle focus remains that of structure at the atomic level (*i.e.*, finding x, y, z coordinates of atoms) the density model represented by a sum of spherical atoms is adequate. In recent years however great advances made by experimental crystallographers makes possible a new view.

It is now a practical matter to obtain experimental images of electron density containing just those features related to chemical bonding omitted by the sum of spherical atoms approximation. This has been borne out in the case of tens of crystals of many different types. An example is illustrated in Fig. 28.3 where both experimental and theoretical deformation density maps are compared for oxalic acid.[12] The maps indicate a characteristic charge build up in bonding and lone pair regions and the pleasing conformity of theory and experiment are a strong indication that the nonspherical density aspects observed are meaningful. Given the accuracy and reproducibility of such data (which in fact can be expected to continually improve in the near future) it becomes possible to shift attention to structure at the electronic level.[13] Here the natural language, conceptual structure, and mathematics is that of quantum mechanics.

It is a natural evolution for crystallography to image the details of electronic orbital structure. The macroscopic structure and properties of a crystal are understood as due to microscopic structure of atomic positions. In an analogous way the atomic structure itself is based largely upon properties and structure of the electronic system. The electrons conform to restrictions imposed by the Pauli principle and this has structural consequences at the electronic level. That 'structure implies function' is not solely a biological aphorism but applies equally well to the elucidation of chemical-physical electronic

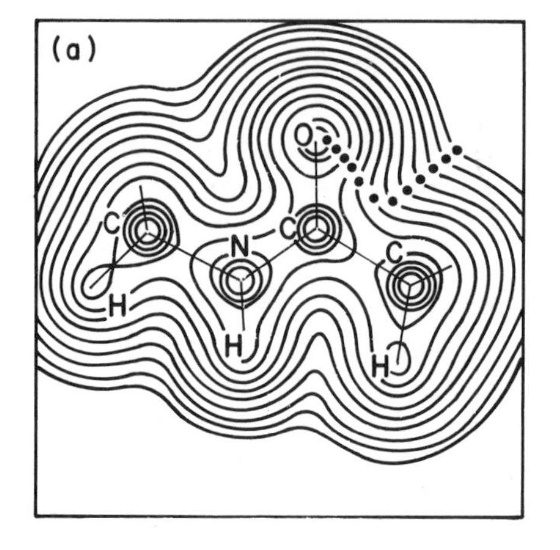

FIG. 28.3(a). Experimental deformation density maps in the plane of the oxalic acid molecule. Contours at 0.05 eÅ^{-3}. Zero and negative contours dashed.

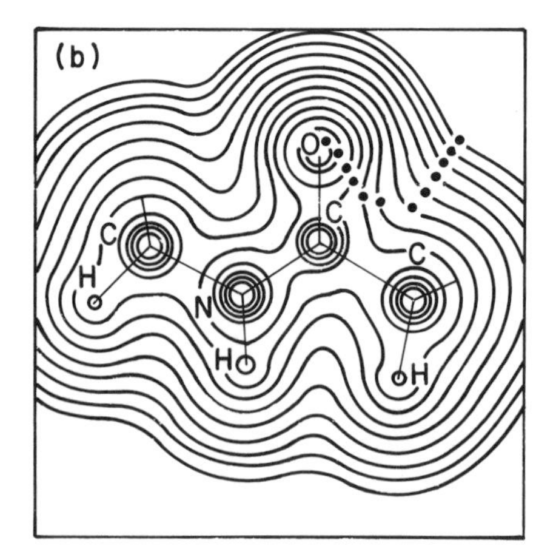

FIG. 28.3(b). Theoretical deformation density maps in the plane of the oxalic acid molecule. Dynamic density. Contours at 0.05 eÅ^{-3} with zero and negative contours dashed.

properties. It is a matter of some interest then to image, consistent with quantum mechanics, the electron density of crystals.

The instinctive notion that electron structure is of paramount consequence to all electronic properties is made rigorous by the fundamental theory of Hohenberg and Kohn (HK) which we now present.

The HK theorem

According to the HK theorem[14] the electron density may be viewed as the basic carrier of all information for a physical system. The proof of Hohenberg and Kohn may be paraphrased as follows. Consider two Hamiltonian operators which differ only in their external potentials, *i.e.*,

$$H_1 - H_2 \equiv v_1 - v_2 \tag{3}$$

Assume that both external potentials lead to exactly the same density, *i.e.*,

$$\begin{matrix} v_1 \rightarrow & \\ & \rho(r) \\ v_2 \rightarrow & \end{matrix} \tag{4}$$

Calculate the expectation value of $H_1 - H_2$ in two ways, first using the exact wave function of H_1 and second using the exact wave function of H_2, obtaining

$$< \psi_1 \mid H_1 - H_2 \mid \psi_1 >= \int \rho(r)(v_1 - v_2) d^3\vec{r} =< \psi_2 \mid H_1 - H_2 \mid \psi_2 > \tag{5}$$

From the extreme right in this last equation we have

$$< \psi_2 \mid H_1 - H_2 \mid \psi_2 >= E_1[\psi_2] - E_2 \tag{6}$$

where E_2 is the exact eigenvalue of H_2 and $E_1[\psi_2]$ is an estimate of the eigenvalue E_1 of H_1 and depends functionally upon ψ_1. The variational theorem asserts that

$$E_1[\psi_2] \geq E_1 \tag{7}$$

so

$$E_1[\psi_2] - E_2 \geq E_1 - E_2 \tag{8}$$

and therefore

$$< \psi_2 \mid H_1 - H_2 \mid \psi_2 > \ \geq E_1 - E_2 \tag{9}$$

From the extreme left, in equation 5, we have, using similar notation

$$< \psi_1 \mid H_1 - H_2 \mid \psi_1 >= E_1 - E_2[\psi_1] \tag{10}$$

The variational theorem asserts

$$E_2[\psi_1] \geq E_2 \tag{11}$$

so

$$E_1 - E_2[\psi_1] \leq E_1 - E_2 \tag{12}$$

and therefore

$$< \psi_1 \mid H_1 - H_2 \mid \psi_1 > \ \leq E_1 - E_2 \tag{13}$$

A contradiction arises since the expected values of the difference Hamiltonian equal to each other, cannot be both less than and greater than the same number $E_1 - E_2$. The initial assumption must be wrong, *i.e.*, two different external potentials cannot lead to the same electron density. Instead, the external potential must be a unique functional of the density. Symbolically

$$\rho \Rightarrow v[\rho] \Rightarrow H[\rho] \Rightarrow \Psi[\rho] \Rightarrow o[\rho], \tag{14}$$

the density implies the external potential which fixes the Hamiltonian, the wave function and every operator property of the system. All quantum information is carried by the electron density. Not all the functional relations symbolized above are known. There does exist an implementation of the HK theorem viz, the Kohn-Sham equations,[15] which generate a single determinant of orbitals, in principle yielding the exact density.

We emphasize the importance of the HK theorem to crystallography. The theorem asserts that the electron density carries all physical information concerning the electronic state and crystallography provides an experimental image of the electron density. The X-ray experiment and its quantum interpretation are now discussed.

X-ray experiment and quantum image of electron density

Consider, in Fig. 28.4, a distribution of charge density $\rho(r)$. The position vector r is measured from an arbitrary point O taken as the origin. An X-ray beam of wavelength, λ, impinges upon the electron distribution from the initial direction indicated by the unit vector $\hat{1}_i$ and is scattered into the final direction indicated by the unit vector $\hat{1}_f$. The path difference between waves scattered from the point r and the origin is clearly $r \cdot (\hat{1}_f - \hat{1}_i)$, which yields a phase difference for the two waves of $(2\pi/\lambda) r \cdot (\hat{1}_f - \hat{1}_i)$. Defining the wave vector $K \equiv (2\pi/\lambda)(\hat{1}_f - \hat{1}_i)$ the phase difference gives for the scattered wave $\rho(r) d^3r \exp(iK \cdot r)$, with the scattering of one electron at the origin taken as the unit magnitude of scattering. Notice $\rho(r) d^3r$ is the amount of scattering charge contained in a volume element at r, and the scattering from r is taken proportional to the amount of charge at the point. Now the amplitude of the total wave scattered is obtained by integrating over the whole electronic distribution to get

$$F(K) = \int \exp(iK \cdot r)\ \rho(r) d^3r. \tag{15}$$

The X-ray experiment measures the intensity, $I(K)$, of the scat-

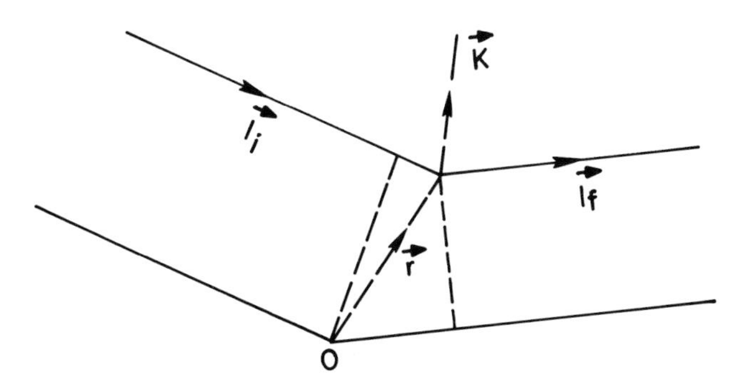

FIG. 28.4. Sketch of the geometry for scattering.

tered wave, which is the square of the wave amplitude, *i.e.*

$$I(K) = | F(K) |^2 \tag{16}$$

This so-called ideal intensity is contaminated in the actual scattering by a host of experimental factors such as absorption, polarization, extinction, etc. In particular the temperature effect associated with the thermal motion of the atomic nuclei will smear the static electron distribution in a time average manner. However, considerable experience indicates that these experimental factors may be modeled within known magnitudes of error and that the experiment yields structure factors, $F(K)$, related to the electron density, $\rho(r)$, characteristic of the crystal.

This Fourier transform relationship connecting the scattering factor, $F(K)$, to the electron density, $\rho(r)$, is the fundamental equation describing the coherent diffraction experiment. The connection between crystallography and quantum mechanics is achieved by ensuring the scattering factor, $F(K)$, comes from an electron density, $\rho(r)$, that is consistent with the restrictions of quantum mechanics. With this in mind obtain a density with an explicit connection to a set of quantum mechanical molecule (or crystal) orbitals.[1-10] For this purpose select a column vector, $\tilde{\psi}(r)$, of orthonormal basis functions and construct molecular orbitals, $\tilde{\phi}(r)$, in the usual 'LCAO' form according to

$$\tilde{\phi}(r) = \tilde{C}\tilde{\psi}(r) \tag{17}$$

The rectangular matrix $\tilde{C}$ contains the linear coefficients which convert the basis into the molecular orbitals. A density matrix associated with the molecular orbitals may be written as

$$\rho(r, r') = tr\tilde{\phi}(r)\tilde{\phi}^+(r') \tag{18}$$

(tr = trace) which is the same as

$$\rho(r, r') = tr\tilde{C}^+\tilde{C}\tilde{\psi}(r)\tilde{\psi}^+(r'). \tag{19}$$

If one identifies the population matrix $\tilde{P}$ as

$$\tilde{P} \equiv \tilde{C}^+\tilde{C} \tag{20}$$

then the density matrix has the simple form

$$\rho(r,r') = tr\tilde{P}\tilde{\psi}(r)\tilde{\psi}^{+}(r'). \tag{21}$$

The electron density required for the X-ray structure factor may be obtained form the diagonal elements of the density matrix, *i.e.*,

$$\rho(r) \equiv \rho(r,r')\,|_{r'\to r} = tr\tilde{P}\tilde{\psi}(r)\tilde{\psi}^{+}(r). \tag{22}$$

This last expression gives for the structure factor

$$F(K) = tr\tilde{P}\tilde{f}(K) \tag{23}$$

where the matrix $\tilde{f}(K)$ contains as elements fourier transforms of basis orbital products. Notice this last equation allows one to think of the elements of P as experimental parameters which can be fit to the experimental X-ray scattering data represented by $F(K)$. However, caution is required since a simple fit of $\tilde{P}$ to the experimental data generally leads to a 'density matrix' $\rho(r,r')$ which is inconsistent with quantum mechanics. This is the problem of N-representability, which has had a long history in the literature of quantum mechanics[16–18] but is less well known in the context of crystallography. In Fig. 28.5 we sketch the essence of the mapping problem involved. Given the set of all antisymmetric N-body wave functions $\Psi(1\cdots N)$, the set of all valid 1-body density matrices is obtained by averaging over all the space and spin coordinates except the spatial coordinates for one particle, *i.e.*,

$$\rho(11') = N\int \Psi(1\cdots N)\Psi^{*}(1'\cdots N')ds_{1\ldots N}d2\cdots N \tag{24}$$

Now the set of all possible one-body functions, $f(11')$, is larger than that of $\rho(11')$. Hence an arbitrary $f(11')$ falling outside the set of $\rho(11')$ cannot be mapped back to any possible wave function $\Psi(1\cdots N)$. Any simple fitting procedure, say a least squares procedure, even with a starting guess within the set $\rho(11')$ will almost always be driven to the invalid wider set of functions, $f(11')$, in the global search for a best possible fit to the experimental data. In brief, simple fits to experimental data do

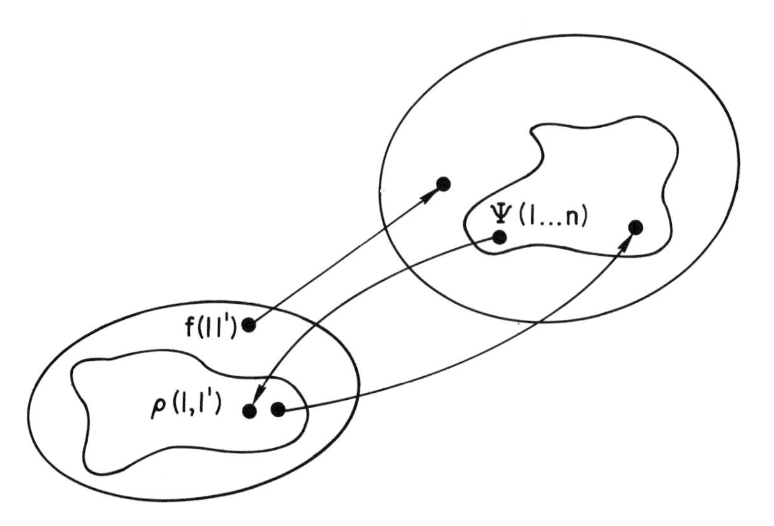

FIG. 28.5. Sketch indicating the mapping problem associated with wave function representability of density matrices.

not ensure the restrictions of quantum mechanics. These must be imposed additionally to fitting procedures. Fortunately for applications to crystallography, the N-representability problem for the one-body density matrix is solved. It is known that the eigenvalues W_K, of $\rho(11')$ must fall within the bounds

$$0 \leq W_K \leq 1 \tag{25}$$

in order for N-representability to be ensured.[16–18] And in particular, the bounds of zero and one are necessary and sufficient for $\rho(rr')$ to be mapped into a single determinant of orbitals, of the form

$$\Psi_{det}(1 \cdots N) = \frac{1}{\sqrt{N!}} \begin{vmatrix} \phi_1(1) & \dots & \phi_N(N) \\ \vdots & & \vdots \\ \phi_N(1) & \dots & \phi_N(N) \end{vmatrix} \tag{26}$$

The eigenvalues mentioned above have the physical interpretation of occupation numbers of the molecular orbitals, so the inequalities above are seen to be a generalization of the Pauli principle. Every quantum density is within reach of a single determinant of orbitals, and therefore we satisfy ourselves with single determinant N-representability of $\rho(11')$. This is tantamount to the requireent that $\rho(11')$ be idempotent, or a projector.

Now we have arrived at the mathematical structure which will ensure an experimental density with an explicit connection to orbitals and wave functions. We seek a population matrix $\tilde{P}$, that is a symmetric, normalized, projector, and which satisfies the experimental scattering factors, *i.e.*,

$$\begin{aligned} \tilde{P} &= \tilde{P}^{+} && (a) \\ tr\tilde{P} &= N && (b) \\ \tilde{P}^{2} &= \tilde{P} && (c) \\ tr\tilde{P}\tilde{f}(K) &= F(K) && (d) \end{aligned} \qquad (27)$$

With these expressions satisfied the density matrix of equation 21 is N-representable as is the experimental density given by equation 22. The advantage of this is perhaps obvious. The crystallography experiment has been described entirely within the formalism of quantum mechanics. The experimental information is obtained in a form that has physical interpretation and can be used for the calcuation of arbitrary physical properties within the usual rules of quantum mechanics, which we take to be the ultimate physical theory relevant to our purpose here.

The methodology[1–10] for solving equation 27 is interesting. There are many numerical procedures that are workable, but these will not be discussed here. We content ourselves with pointing out some of the results obtained to date. The systems discussed are of two types. Firstly, we consider simple theoretical models H, H_2 and Li. In these cases the data used for $F(K)$ are generated mathematically from assumed wave functions and/or densities and then equation 27 is solved to see whether these can be regenerated. The model calculations contain the essential quantum mechanics of the crystallography problem without the contamination of experimental factors not under study. As such, they represent an important test of the mathematical structure represented by equation 27. Secondly, we consider real experimental data, collected with high accuracy, for the beryllium crystal.

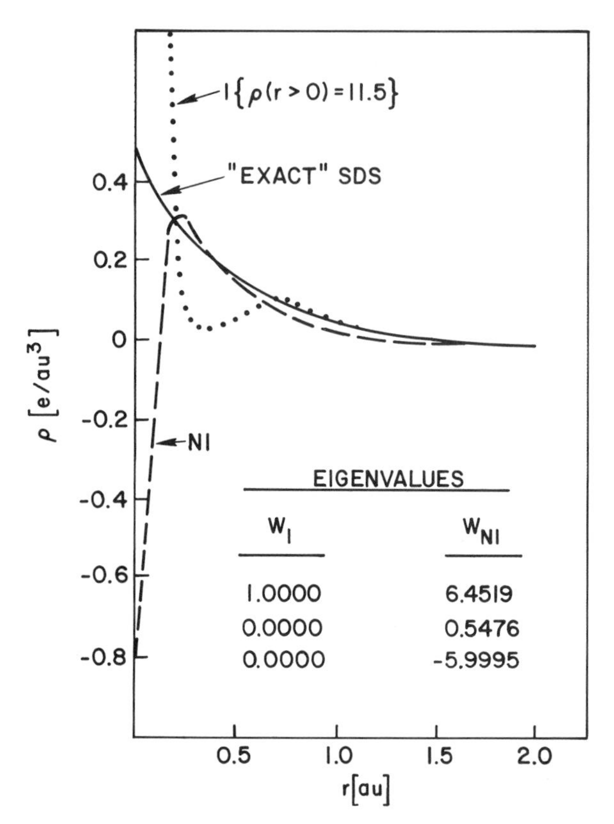

FIG. 28.6. Comparison of an 'exact' density due to SDS with non-idempotent (NI) and idempotent (I) least-squares fitted densities. The eigenvalues of the corresponding 'density matrices' are W_{NI} and W_I.

H atom:[7] In Fig. 28.6 we plot as a solid line an H-atom density due to Stewart, Davidson and Simpson (SDS).[19] Using equation 21 for $\rho(rr')$ we have obtained a fit to this model density with (I) and without (NI) imposition of idempotency. The respective results are plotted as dotted and dashed lines. We also list the eigenvalues (W_I and W_{NI}) for the idempotent and not-idempotent density matrices. The R-factors for the two cases are $R_I = 0.085$ and $R_{NI} = .009$ indicating a good fit has been obtained in both cases. Notice however the eigenvalues W_I are one and zero, whereas W_{NI} fall well outside the allowable values given by the inequalities 25, ranging from $+6.45$ to -5.99. These latter values have no interpretation as possible

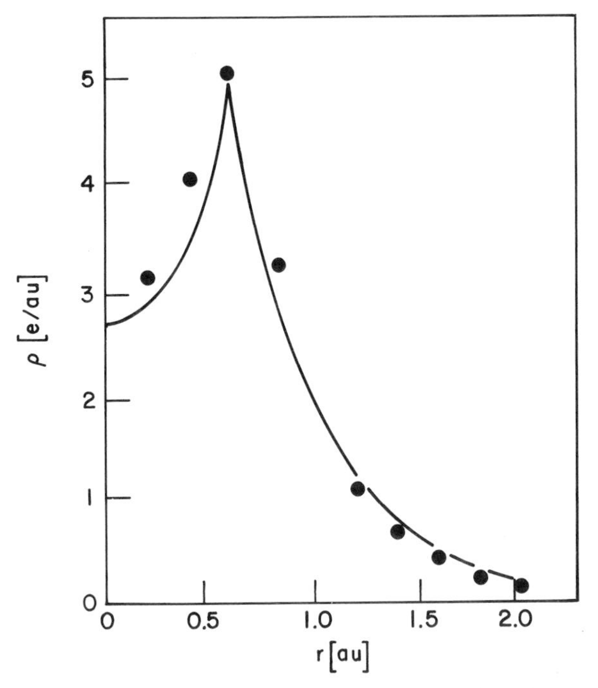

FIG. 28.7. Typical solution of P equations (dots) fit to H_2 molecular density of Stewart, Davidson and Simpson (solid line) (1956). The basis contains two 1s functions with $\xi = 1.1$ and 5.0 and one 2s function with $\xi = 1.1$. The agreement factor is $R = 0.0225$.

occupation numbers of orbitals. This violation of the Pauli principle is simply an example of the fact that density fitting procedures alone do not carry along the requirements of quantum mechanics. In the same vein notice that the density plot associated with the idempotent density matrix is everywhere positive as must be the case for a quantity interpreted as a probability distribution. Contrary to this the non- idempotent density goes negative, a result void of physical interpretation.

H_2 molecule:[7] In order to study the application of our method to bonding effects we have considered an H_2 molecular density (SDS). We have found in numerical experiments with a variety of different basis functions that we can solve equations 27, routinely obtaining a reliable bond distance and reproducing the essential features of the molecular-bonding charge distribution. A typical result for H_2 is displayed in Fig. 28.7.

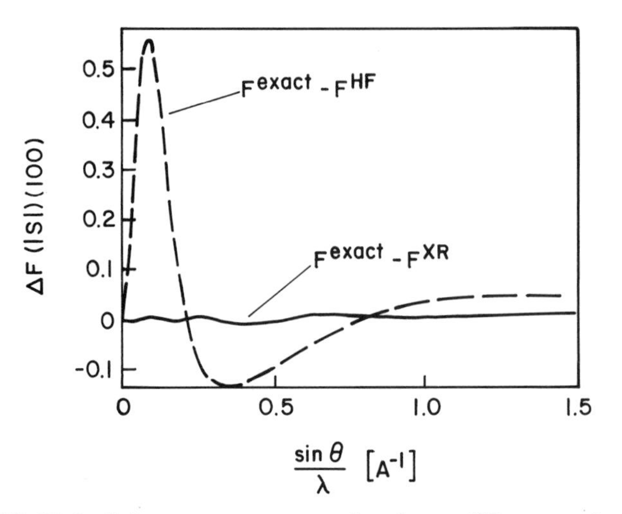

FIG. 28.8. Coherent x-ray scattering factor differences for ground state of the lithium atom.

Li atom:[6] Bennesch and Smith have presented essentially exact $F(K)$ data for the lithium atom, which we have used for equations 27.

In Fig. 28.8 we obtain an indication of how well our method reproduces the scattering factor data. We have plotted $\Delta F \equiv F^{exact} - F^{calc}$, where F^{calc} refers either to our X-ray orbitals (XR) or Hartree-Fock orbitals (HF), the same basis used in either case. Insofar as Hartree-Fock orbitals are generally considered to give a rather good density it is interesting how much better is the fit to F^{exact} using X-ray orbitals.

In Table 28.1 we compare calculated quantum properties of Li using wave functions essentially exact, and Hartree-Fock (HF) plus our X-ray case (XR) on the same basis. For the total energy XR is almost as good as HF, though by definition HF must be lower. The one-electron properties $< -1/2\nabla^2 >$, $< r^{-1} >$, and $< r^2 >$ are all better for XR than HF. The two electron property $< r_{12}^{-1} >$ is slightly worse for XR than XF.

In the discussion of these quantum properties an important point arises. It is, of course, significant that the numerical values of all the XR properties are of reasonable magnitude. A more subtle factor is that because our methodology, encapsulated in equation's 27, is cast in a strictly quantum formal-

TABLE 28.1. Comparison of expectation values from the exact, Hartree–Fock, and X-ray-fitted densities for the lithium atom. Units are in a.u.

	Total energy	$\langle -1/2\nabla^2 \rangle$	$\langle 1/r_{12} \rangle$	$\langle 1/r \rangle$	$\langle r^2 \rangle$
Exact	−7.478025	7.478	2.199	5.71822	18.35034
HF	−7.432749	7.433	2.281	5.71549	18.62610
XR	−7.432594	7.438	2.285	5.71866	18.34609

ism all quantum properties are calculable using the *usual* rules of quantum mechanics. This statement cannot be made for methods of crystallographic analysis which yield the density alone unrelated to wave functions. Along these lines Becker[21] has observed '...the populations that are obtained cannot be interpreted easily, nor can they be transferred to the analysis of Compton experiments.' Compton profiles depend upon the momentum-operator, ∇, and its expectation value cannot within the *usual* rules be evaluated directly from $\rho(r)$; a valid density matrix is required. Hence with our evaluation of $< -1/2\nabla^2 >$ we draw attention to the fact that momentum dependent properties are XR calculable as are all quantum properties.

Be crystal:[10] The image of the X-ray orbital valence density for Be is displayed in Fig. 28.9. The cut shown is through the a-b horizontal plane of the crystal. The $F(K)$ used were from the extremely accurate experimental data of Hansen and Larson. A more complete analysis of our Be results is contained in a recent paper,[10] but we mention a few points of interest. Our weighted R-factor has the small magnitude R_W = .0018. When the principal features of Fig. 28.9 are compared to their analogs obtained from the best available *ab-initio* results for Be there is a pleasing concordance of experimental and theoretical results. There is a general build up of bonding charge in the tetrahedral channels at the expense of charge coming out of the octahedral channels. Our temperature factors, obtained in a simultaneous refinement with all electron parameters, are very close to the best neutron values. The errors in our scat-

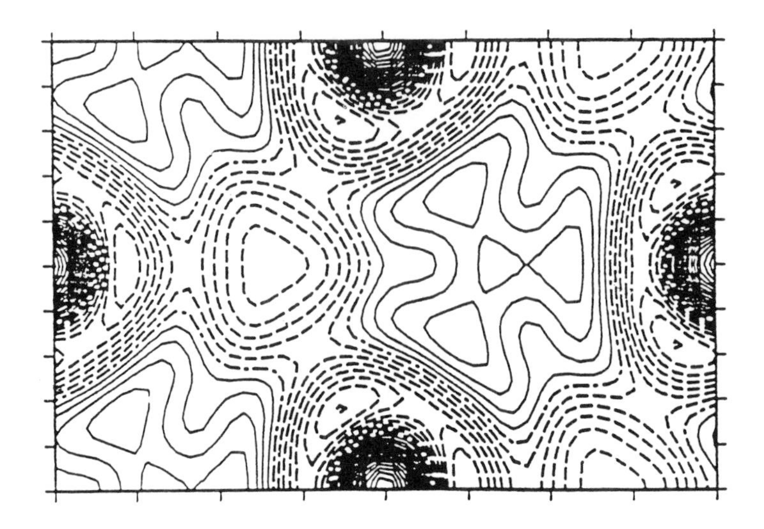

FIG. 28.9. X-ray orbital valence density of Be, calculated from the Fourier summation of 28 thermally smeared structure factors. Contour interval 0.01 (e/Å^3), positive contours solid, zero dot-dash, negative dashed. The zero contour refers to 4 electrons/unit cell volume. Axes in Å.

tering factors follow a random distribution over all values of $\sin\theta/\lambda$. The X-ray orthonormal orbital model of crystallography applied to the case of the best available Be data seems to yield a physically significant image of the Be electron density.

Bloch-Wannier Orbitals

In a recent paper[9] we suggested that the mathematical structure for a quantum description of coherent X-ray scattering, symbolized in equation 27, has relevance for obtaining Bloch and/or Wannier orbitals from crystal data. In this section we extend our suggestion supplying a more detailed and practical formalism. For ease of discussion, we present our development for a one-dimensional crystal lattice but the results are generalizable to three dimensions. Moreover, we assume a case for filled bands that are non overlapping, expecting important aspects of the formalism to carry over in more general cases. Most of the essential ideas are due to Löwdin,[22] it is only their application in the context of the X- ray problem that is new.

In Fig. 28.10 is sketched the essential geometric idea associated with Bloch and Wannier orbitals. First picture the individual 1s functions at each lattice position. Now various linear combinations of these functions can yield orbitals of quite different geometric character. In the Bloch case one has a wave of atomic orbitals; the wave is atomic like at each lattice position but of amplitude that varies as an overall phase factor which is cyclic in some multiple of the latice translation. In general the Bloch orbitals can extend in space over many lattice translations and in effect emphasize the possible 'itinerant character' of the lattice electrons. The number of orthogonal Bloch orbitals is equal to the number of lattice centered atomic orbitals from which they are built. Only one such orbital is pictured here.

In the Wannier case one has a highly localized wave, concentrated in a region near to one lattice position but with small

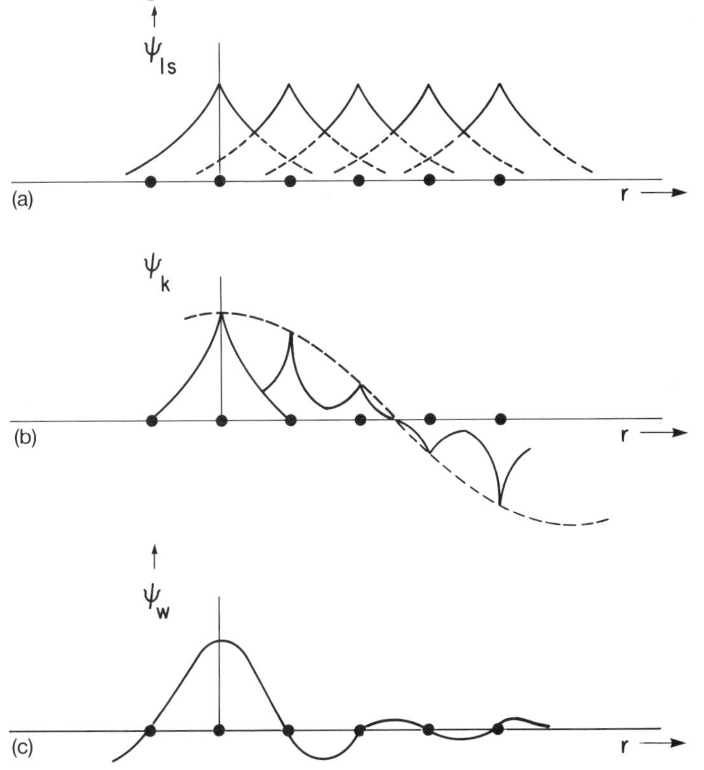

FIG. 28.10. Sketch comparing independent 1s orbitals (a) to their linear combinations in Bloch (b) and Wannier (c) form.

oscillating tails that extend to other lattice positions and ensure orthogonality among neighboring functions. In effect the Wannier orbitals emphasize the possible localized character of the lattice electrons. The number of orthogonal Wannier orbitals is of course also equal to the number of original atomic orbitals. Only one is pictured here.

Bloch and Wannier orbitals, as Fourier transforms of one another, are information equivalents. They are a desirable form for representation of the density since they are useful in describing the properties of a full range of crystals including conductors, semiconductors, and insulators.

Following Löwdin we construct a basis of orthogonalized atomic orbitals. At every lattice postion $R_\ell, \ell = 1, 2, \cdots N_c$ (number of cells) choose a basis set

$$\tilde{\psi}^\ell \equiv \tilde{\psi}(r - R_\ell). \tag{28}$$

Let the $\tilde{\psi}^\ell$ be orthonormal

$$\tilde{\psi}^\ell \cdot \tilde{\psi}^{\ell+} \equiv \tilde{1} \tag{29}$$

where the dot indicates an inner product. If the $\tilde{\psi}^\ell$ are not orthonormal at the outset they can always be made so via

$$\tilde{\psi}^\ell \rightarrow \tilde{S}^{-\frac{1}{2}} \tilde{\psi}^\ell \tag{30}$$

where

$$\tilde{S} \equiv \tilde{\psi}^\ell \cdot \tilde{\psi}^{\ell+} \tag{31}$$

and so we shall assume orthonormalization holds among the functions at a given site R_ℓ for all ℓ. Notice

$$\tilde{\psi}^\ell \cdot \tilde{\psi}^{\ell'+} \neq \tilde{0} \quad \ell \neq \ell' \tag{32}$$

in general, but these overlap integrals can be made to vanish by the following arguments of Löwdin. Define

$$\tilde{\psi} \equiv \begin{pmatrix} \tilde{\psi}^1 \\ \tilde{\psi}^2 \\ \vdots \\ \tilde{\psi}^{N_c} \end{pmatrix} \tag{33}$$

On the right-hand side a superscript is a site index. $\tilde{\psi}$ is a column matrix of site column matrices, and gives rise to the overlap matrix.

$$\tilde{S} \equiv \tilde{\psi} \cdot \tilde{\psi}^{+}. \tag{34}$$

It is convenient nomenclature to write

$$\tilde{S} \equiv \begin{pmatrix} \tilde{1} & \tilde{\epsilon}_{12} & \tilde{\epsilon}_{13} & \cdots \\ \tilde{\epsilon}_{21} & \tilde{1} & \tilde{\epsilon}_{23} & \cdots \\ \tilde{\epsilon}_{31} & \tilde{\epsilon}_{32} & \tilde{1} & \cdots \\ \vdots & \vdots & \vdots & \vdots \end{pmatrix} \tag{35}$$

Each site contributes a unit matrix along the block diagonals of $\tilde{S}$. The off diagonal blocks $\tilde{\epsilon}_{\ell\ell'}$ are the matrices of overlap integrals between sites. We define a matrix the size of $\tilde{S}$ to be

$$\tilde{\epsilon} \equiv \begin{pmatrix} \tilde{0} & \tilde{\epsilon}_{12} & \tilde{\epsilon}_{13} & \cdots \\ \tilde{\epsilon}_{21} & \tilde{0} & \tilde{\epsilon}_{23} & \cdots \\ \tilde{\epsilon}_{31} & \tilde{\epsilon}_{32} & \tilde{0} & \cdots \\ \vdots & \vdots & \vdots & \vdots \end{pmatrix} \tag{36}$$

so that we may write the whole overlap matrix $\tilde{S}$ as a unit matrix plus a matrix $\tilde{\epsilon}$, *i.e.*,

$$\tilde{S} \equiv \tilde{1} + \tilde{\epsilon}. \tag{37}$$

Löwdin obtains $\tilde{\psi}_0$, a completely orthonormalized set of functions as

$$\tilde{\psi}_0 \equiv \tilde{S}^{-\frac{1}{2}}\tilde{\psi}. \tag{38}$$

Notice that

$$\tilde{\psi}_0 \cdot \tilde{\psi}_0^{+} = \tilde{S}^{-\frac{1}{2}}\tilde{\psi} \cdot \tilde{\psi}^{+}\tilde{S}^{-\frac{1}{2}} = \tilde{S}^{-\frac{1}{2}}\tilde{S}\tilde{S}^{-\frac{1}{2}} = \tilde{1} \tag{39}$$

as required. Within the column $\tilde{\psi}_0$ the orbitals at a given site R_ℓ are given by

$$\tilde{\psi}_0^{\ell} = \sum_{\lambda} \tilde{S}_{\ell\lambda}^{-\frac{1}{2}}\tilde{\psi}^{\lambda} \tag{40}$$

Employing a binomial expansion for $\tilde{S}^{-\frac{1}{2}}$ yields

$$\tilde{S}^{-\frac{1}{2}} = \tilde{1} - \frac{1}{2}\tilde{\epsilon} + \frac{3}{8}\tilde{\epsilon}^{2} + \cdots \tag{41}$$

which gives for $\tilde{\psi}_0^\ell$ the expression

$$\tilde{\psi}_0^\ell = \sum_\lambda (\tilde{1} - \frac{1}{2}\tilde{\epsilon} + \frac{3}{8}\tilde{\epsilon}^2 + \cdots)_{\ell\lambda} \tilde{\psi}^\lambda \tag{42}$$

The indices here refer to matrix blocks not individual elements. Simplifying we get

$$\tilde{\psi}_0^\ell = \tilde{\psi}^\ell - \frac{1}{2}\sum_\lambda \tilde{\epsilon}_{\ell\lambda}\tilde{\psi}^\lambda + \frac{3}{8}\sum_{\lambda\mu} \tilde{\epsilon}_{\ell\mu}\tilde{\epsilon}_{\mu\lambda}\tilde{\psi}^\lambda + \cdots \tag{43}$$

We point out Löwdin discusses the conditions under which $\tilde{S}^{-\frac{1}{2}}$ is well defined and for which the binomial expansion can be expected to converge properly. Given these conditions the number of terms that must be retained in the binomial expansion will vary from case to case and is a matter to be decided by numerical trial and error.

It can be expected that the elements of $\tilde{\epsilon}_{\ell\ell'}$ will decrease as $|\ \ell - \ell'\ |$ increases and for some separation of ℓ and ℓ' are neglectable. As an example of the sort of simplification which might occur in favorable cases, at this point let us retain only terms which are of first order in the $\tilde{\epsilon}_{\ell\ell'}$'s, then

$$\tilde{\psi}_0^\ell = \tilde{\psi}^\ell - \frac{1}{2}\sum_\lambda \tilde{\epsilon}_{\ell\lambda}\tilde{\psi}^\lambda \tag{44}$$

Now the blocks of $\tilde{\epsilon}$ can be organized into near-neighbor contributions called 1st, 2nd, 3rd etc.

$$\tilde{\psi}_0^\ell = \tilde{\psi}^\ell - \frac{1}{2}\sum_{\lambda'} \tilde{\epsilon}_{\ell\lambda'}\tilde{\psi}^{\lambda'} - \frac{1}{2}\sum_{\lambda''} \tilde{\epsilon}_{\ell\lambda''}\tilde{\psi}^{\lambda''} \cdots \tag{45}$$

Here the λ' indices refer to 1st near-neighbors of ℓ, λ'' 2nd near-neighbors, etc. Which neighbors can be truncated is again a matter for case by case trial and error, but as an illustration suppose only 1st near-neighbors need be retained, then

$$\tilde{\psi}_0^\ell = \tilde{\psi}^\ell - \frac{1}{2}\sum_{\lambda'} \tilde{\epsilon}_{\ell\lambda'}\tilde{\psi}^{\lambda'} + \cdots. \tag{46}$$

In our one-dimensional case each value of ℓ will have 1st nearest-neighbor values $\ell \pm 1$, so

$$\tilde{\psi}_0^\ell = \tilde{\psi}^\ell - \frac{1}{2}\tilde{\epsilon}_{\ell,\ell\pm 1}(\tilde{\psi}^{\ell+1} + \tilde{\psi}^{\ell-1}) \tag{47}$$

We pause here to notice $\tilde{\psi}_0^\ell$ is a set of orthonormal basis function at site R_ℓ orthogonal to similar functions at neighboring sites (within the approximations noted). We suggest construction of the best (localized) Wannier orbitals within reach of this basis via

$$\tilde{\phi}_w^\ell \equiv \tilde{C}\tilde{\psi}_0^\ell \quad \ell = 1, 2, \cdots N_c \tag{48}$$

The linear weighting coefficients represented by the $\tilde{C}$'s are the same at every site and therefore carry no site label. Requiring

$$\tilde{C}\tilde{C}^+ = \tilde{1} \tag{49}$$

ensures orthonormality of the Wannier orbitals. The density matrix becomes

$$\rho(r, r') = \sum_\ell tr\tilde{\phi}_w^\ell(r)\tilde{\phi}_w^{+\ell}(r') = \sum_\ell tr\tilde{C}^+\tilde{C}\tilde{\psi}_0^\ell(r)\tilde{\psi}_0^{\ell+}(r'). \tag{50}$$

Define

$$\tilde{P} \equiv \tilde{C}^+\tilde{C} \tag{51}$$

and anticipate this matrix is to be hermetian, normalized, and idempotent. The density is obtained from the density matrix in the usual way and is used to write the structure factor

$$F(K) = \sum_\ell tr\tilde{P}\tilde{f}^{\ell\ell}(K) \tag{52}$$

The matrix $f^{\ell\ell}(\vec{k})$ contains elements which are fourier transforms of products of the $\tilde{\psi}_0^\ell$ basis functions, and as this matrix is the same at every site the index ℓ may as well be dropped, writing the complete structure factor as

$$F(K) = N_c tr\tilde{P}\tilde{f}(K) \tag{53}$$

The matrix $\tilde{f}(K)$ is evaluated with functions $\tilde{\psi}_0^\ell$ for arbitrary ℓ. With this last equation one sees we have recovered the entire mathematical structure for the quantum description of crystallography, as in equation 27, but here the resulting orbitals are in Wannier form. The Bloch orbitals corresponding to these Wannier orbitals are obtained in the usual way as

$$\tilde{\phi}_k = \frac{1}{N_c} \sum_{\ell} \exp(ik \cdot R_\ell) \tilde{\phi}_w^\ell = \frac{1}{N_c} \tilde{C} \sum_{\ell} \exp(ik \cdot R_\ell) \tilde{\psi}_0^\ell \tag{54}$$

In this representation the $\tilde{C}$'s are independent of k, as well as independent of ℓ, as stated earlier.

The total wave function can be equally well represented in Wannier or Bloch orbitals as either

$$\Psi_{det}(1 \cdots N) = \frac{1}{\sqrt{N!}} \mid \phi_w^1 \ \phi_w^2 \ \cdots \ \phi_w^N \mid \tag{55}$$

or

$$\Psi_{det}(1 \cdots N) = \frac{1}{\sqrt{N!}} \mid \phi_{(k=1)} \ \phi_{(k=2)} \ \cdots \phi_{(k=N)} \mid \tag{56}$$

Quantum operators which can be expanded in one-body and two-body terms as

$$\sigma = \sigma_1 + \sigma_{12} \tag{57}$$

will have expectation values

$$< \sigma > = < \sigma_1 \rho_1 > + < \sigma_{12} \rho_2 > \tag{58}$$

where (using the Wannier orbitals, say)

$$\rho_1 = \sum_{\ell} tr \tilde{\phi}_w^\ell (1) \tilde{\phi}_w^\ell (1') \tag{59}$$

and

$$\rho_2 = \begin{vmatrix} \rho_1(11') & \rho_1(12') \\ \rho_1(21') & \rho_1(22') \end{vmatrix} \tag{60}$$

Summary and Concluding Remarks

Structure provides one of our fundamental attacks on physical problems. X-ray crystallography is the preeminent method for obtaining structural information. The focus of our attention has long been at the atomic level, and during this period the predominant structural model has been that of the sum of spherical atoms. We have pointed out how remarkably good is this model. The accumulated effects of advances in computer controlled diffractometry, accelerator X-ray sources, accurate modeling of systematic experimental contaminants of measured intensities, and efficient numerical techniques have raised new possibilities. Experimentalists have obtained electron density maps containing detail that is accurate, reproducible, and which demand a theoretical model adequate to their interpretation. The natural evolution of X-ray structural information to the electronic level suggests a quantum model,[23] as electrons are, after all, arch-type quantum objects. Moreover the HK theorem provides powerful motivation in this regard since it asserts that the electron density, *i.e.*, the object imaged by X-ray crystallography, is the fundamental carrier of all quantum information.

From the perspective of the HK theorem a set of orbitals might be considered optimal if they deliver the exact density. The mathematical structure we have discussed here for describing the X-ray experiment has as its purpose[24] obtaining such a set of exact density orbitals. These ideas have been discussed for theoretical models of H, H_2, and Li, and they have been tested with real experimental data in the case of the Be crystal. We have given a suggestion, which we hope will prove of practical value, for obtaining the density in a formalism related to Bloch and Wannier orbitals.

References

1. Clinton, W. L. and Massa, L. J. *Phys. Rev. Lett.*, **29**, 1363–1366 (1972).
2. Massa, L. J. and Clinton, W. L. *Trans. Am. Cryst. Assoc.* 149–153 (1972).
3. Clinton, W. L., Frishberg, C., Massa, L. J. and Oldfield, P. A. *Int. J. Quant. Chem: Quant. Chem. Symp.*, **7**, 505–514 (1973).
4. Frishberg, C. and Massa, L. J. *Int. HJ. Quant. Chem.* **13**, 801–810 (1978).
5. Frishberg, C., Goldberg, M. J. and Massa, L. J. In *Electron Distributions and the Chemical Bond*, pp. 101–110. (P. Coppens and M. B. Hall, eds.) Plenum: New York, NY (1982).
6. Frishberg, C. and Massa, L. J. *Phys. Rev.*, **B 24**, 7018–7024 (1981).

7. Frishberg,C. and Massa, L. J. *Acta. Cryst.*, **A38**, 93–98 (1982).
8. Clinton, W. L., Frishberg, C., Goldberg, M. J., Massa, L. J. and Oldfield, P. A. *Int. J. Quant. Chem.* **17**, 517–525 (1983).
9. Goldberg, M. J. and Massa, L. J. *Int. J. Quant. Chem.*, Vol **24**, 113–126 (1983).
10. Massa, L. J., Goldberg, M. J., Frishberg, C., Boehme, R. F. and LaPlaca, S. J. *Phys. Rev. Lett.* **55**, 622–625 (1985).
11. Hagler, A. Private communication (1977).
12. Commission on charge spin and momentum density. *Project on comparison of Structural Parameters of Electron density Maps of Oxalic Acid Dihydrate.* Project Reporter: Philip Coppens.
13. For a partial list of authors, (theoreticians and experimentalists), contributing over the past thirty years to the idea of using X-ray data to 'see' nonspherical electronic structure consult our ref. 8, footnotes 1–15.
14. Hohenberg, P. and Kohn, W. *Phys. Rev.* **136**, B864 (1964).
15. Kohn, W. and Sham, L. *Phys. Rev.* **140**, A1133 (1965).
16. Lowdin, P. O., *Phys. Rev.* **96**, 1474 (1954).
17. McWeeney, R., *Rev. Mod. Phys.* **32**, 335 (1960).
18. Coleman, A. J., *ibid*, **35**, 3 (1963).
19. Stewart, R. F., Davidson, E. R. and Simpson, W. T., *J. Chem. Phys.* **42**, 3175 (1965).
20. Benesch, R., Smith, V. H. Jr., *Acta Cryst.* **A26**, 586 (1970).
21. Becker, P. In *Electron Distributions and the Chemical Bond*, pp. 153–172. (P. Coppens and M. B. Hall, eds.) Plenum: New York, NY (1982). On this point see also remarks by Löwdin, pp. 8–11, in our ref. 22.
22. Löwdin, P. O., *Advances in Physics* **5** (1956).
23. A point we've not had time to discuss but which we have treated elsewhere (see for example our ref. 8) is the fact that a quantum formalism seems to introduce certain purely numerical advantages in a quite natural way. In particular quantum restrictions reduces the number of experimental parameters and avoids certain correlation of parameters difficulties.
24. We mention in passing, that it seems to us our exact density orbitals have some formal similarity to orbitals of Kohn-Sham theory, and that our method might be viewed as a kind of experimental attack on the HK density functional problem insofar as its view is to begin with experimental density and end in principle with all electronic properties. We intend to discuss this view in another paper.

29. Looking backward: On certain Fourier and Patterson maps

Ray Pepinsky

More than twenty years have passed since I set aside active work in X-ray and neutron crystallography. Among the consequences of that defection, the saddest to me has been my separation from many dear friends: separation both geographic and intellectual. I am deeply grateful to Betty Patterson and Jenny Glusker for the kind invitation to participate in this symposium.

In 1963 I was invited to join in the planning and initial operation of a new university in Florida; after two years, I moved on to a second such institution. There were no adequate library or research facilities at hand, and the only course seemed to be to devise some new research program about which nothing had been written, so that there was nothing to read. I developed a scheme to look at magnetic transitions, comparing effects within the interior of a crystal with those in the very outermost atomic layers as observed by means of exchange scattering of polarized slow electrons. I worked on the theory for a couple of years, and on various implications and related phenomena — in whatever time was available between all the drudgery of building designs, curriculum planning, library acquisitions, fund-raising, faculty assembly, and eventually graduate student instruction on subjects I had long ago studied and largely forgotten. I discovered that polarized electron research was in progress at Mainz and Karlsruhe. I obtained a leave of absence in order to serve for a year as acting director of the crystallographic institute of the late Carl Hermann, my good friend in Marburg, so as to be able to spend time in the Mainz and Karlsruhe institutes for discussions of polarized electron production and scattering. I arrived in Germany in October of 1966, and was travelling between universities, lecturing and exchanging ideas, when Lindo passed away. I did not learn of that sad event until I returned to the United States, almost a year later. It was a great shock — as if an era had ended.

I owed so much to Lindo, for inspiration, direct and most effective help at more than one critical juncture, and unceasing understanding and appreciation of our work.

Time softens sadness. Even without the splendid photographs which line the walls of the symposium auditorium and the adjacent rooms and the corridors – pictures which reveal Lindo so vividly, the noble figure and the handsomely rough-hewn smiling face which radiates such gracious good humor – even without these one remembers him so that he seems still among us. And as long as structural crystallography is practiced, his devising and brilliant first applications of the $|F_H|^2$ Fourier series, which everyone properly knows as the Patterson series, will remain a fundamental step in crystal analysis.

My own work again and again followed Lindo's lead. He was a fine mathematician, and he was always thinking about Fourier transforms and encouraging others to do so. There was no better reward than his interest in what his colleagues, and especially those younger and less experienced than he, were doing in the area of transform theory. He was always the first person to whom one wanted to show some new result. It was especially enjoyable to show Lindo some result particularly pertinent to calculations which were accomplishable on our computer X-RAC, because he was in effect the god-father of that machine; it was his enthusiastic support of my proposal for funds to begin construction of the computer which was most helpful in the success of our request to the Office of Naval Research.

In this paper I shall mention four aspects of transform theory in which Lindo showed his enthusiastic interest. Two are obvious but still useful; the remaining two are a bit more profound. These are: (1) the use of positive kernels for reduction of the numbers of Fourier terms in an electron density or interatomic vector synthesis, without introduction of disturbing ripples or interference with the boundness and especially the non-negativity conditions; (2) the uniqueness or non-uniqueness of density functions derivable from a Patterson synthesis; (3) Patterson maps in neutron analyses, and H-D substitution; and (4) the sine-Patterson or $P_s(u)$ function and its utility in the case of anomalous dispersion of X-rays or neutrons in a structure analysis.

Positive kernels in Fourier syntheses

I introduce the subject of the positive kernels largely because Lindo's response to our telling him about the use to which we put them was so charming. 'Marvelous!' he said. 'Who would have thought that the Cesaro Means would ever be of practical use outside of function theory!'

These kernels are discussed in Appendix V to the paper on X-RAC *and* S-FAC in the 1952 volume on *Computing Methods and the Phase Problem in X-ray Crystal Analysis.*[1] The title of that appendix is *The Use of Positive Kernels in Fourier Synthesis of Crystal Structures.*[2] The problem concerned may be rather of historical than practical significance today, but I suspect it is more than that. Suppose one is studying the diffraction data from a fairly complex molecular structure, and has some idea of the general conformation and perhaps juxtapositions of the molecules. The analyst has some phase information on a limited number of structure factors F_H, and wishes to obtain the best approximation to the structure using these F_H values and others whose phases can perhaps be permuted in a search for a recognizable approximation to the structure. Which of a limited number of F_H values should be utilized, and how should these be artifically modified in order that the *best approximation* to the density function $\rho(r)$ be obtained when all phases of this limited number of coefficients are correctly determined? The problem was particularly pertinent to the use of X-RAC, in those early days, because one could alter phases of coefficients in some systematic way, and immediately observe the effects on a Fourier summation. The chief criterion one used was *non-negativity* of the density function and its approximation. Thus, in selecting the best *approximation* utilizing the limited number of F_H coefficients, one demanded that the approximation also be non-negative.

The analyses in our Appendix V to the 1952 paper were at first restricted for convenience to one dimension, but extension to two or three dimensions was obviously accomplishable, and we indicated how this was done in two important cases.

Consider the Fourier series

$$f(x) = \sum_{k=-\infty}^{\infty} a_k e^{ikx}, \tag{1}$$

where

$$a_k = \frac{1}{2\pi} \int_{-\pi}^{\pi} f(t) e^{-ikt}\, dt; \tag{2}$$

and consider also the partial sum

$$s_n(x) = \sum_{k=-n}^{n} a_k e^{ikx}. \tag{3}$$

Inserting (2) in (3) and expanding, one finds

$$s_n(x) = \frac{1}{\pi} \int_{-\pi}^{\pi} f(x-t) \frac{\sin(n+\frac{1}{2})t}{2\sin\frac{t}{2}} dt. \tag{4}$$

The quantity $\frac{\sin(n+\frac{1}{2})t}{\sin\frac{t}{2}}$ is known as Dirichlet's kernel, which we designate as $D_n(t)$. Then, clearly, $s_n(x)$ is the result of convolution of $f(t)$ with $D_n(t)$. This result is well known in diffraction theory. $D_n(t)$ has a central maximum about $t = 0$ and makes a number of ***ripples***, extending from negative (minima) to positive (maxima) symmetrically about $t = 0$. Peaks in $f(x)$ are thus surrounded by negative-to-positive ripples in $s_n(x)$, and non-negativity of the approximate structure is no longer applicable as a criterion for correct phase assignment for the Fourier coefficients in $s_n(x)$.

Walther Hoppe introduced the use of an isotropic temperature factor as a means of reducing the number of terms appearing in a series approximating $f(x)$.[3] The result is the same as convolution of $f(x)$ with a kernel which is itself expressible as a Fourier series. Calderon has shown that this kernel is the third theta function, as described analytically and graphically in Jahnke and Emde;[4] Calderon designated this as Gauss's kernel, and showed that it is rather similar to Poisson's kernel. Both of these are single-peak functions but with rather broad wings extending in the $\pm t$ direction.

Two other kernels are discussed in that same Appendix V, both of which are superior to those previously utilized. For $f(x)$

and $s_n(x)$ as previously written, one defines the First Cesaro Mean (C,1) of $f(x)$ as

$$\sigma_N^1(x) = \frac{\sum_{n=0}^{N-1} s_n(x)}{N} = \frac{s_0(x) + s_1(x) + s_2(x) + \cdots + s_{N-1}(x)}{N}, \tag{5}$$

which one finds is expressible, upon summation, as

$$= \sum_{n=-N}^{N} (1 - \frac{|n|}{N}) a_n e^{inx}. \tag{6}$$

If we write a_n in the form shown in (2) above, and insert it in (6), we eventually obtain

$$\sigma_N^1(x) = \frac{1}{2\pi} \int_{-\pi}^{\pi} f(x-t) \left[\frac{1}{N} \left(\frac{\sin \frac{Nt}{2}}{\sin \frac{t}{2}} \right)^2 \right] dt. \tag{7}$$

The quantity within the brackets is known as Fejer's kernel, $K_N(t)$, and thus the First Cesaro Mean $\sigma_N^1(x)$ is the convolution of $f(t)$ with $K_N(t)$.

Examination of Fejer's kernel reveals that $K_N(t) \geq 0$, and

$$\frac{1}{\pi} \int_{-\pi}^{\pi} K_N(t) dt = 1. \tag{8}$$

If $f(x)$ is bounded, *i.e.*,

$$m \leq f(x) \leq M, \tag{9}$$

one shows easily that

$$m \leq \sigma_N^1(x) \leq M, \tag{10}$$

so that $\sigma_N^1(x)$ has the same bounds as $f(x)$.*

It is well known in Fourier theory that the First Cesaro Mean $(C, 1)$ converges very rapidly to the function represented by the Fourier series; indeed, no trigonometric polynomial converges faster to $f(x)$ than the First Cesaro Mean.

---

* Actually it can be shown that the bounds of the $(C, 1)$ mean lie *within* the bounds of f(t).

These expressions are easily extended to 2 and 3 dimensions, and that extension is illustrated in the original Appendix V. Indeed, the 3-dimensional expressions become the most interesting and useful, since the boundedness of the 3-dimensional form of $(C,1)$ remains. The Fejer kernel $K_N(t)$ in one dimension is illustrated in that Appendix. It has a maximum of 1 at $t=0$, drops to 0 at $t = \pm n\frac{\pi}{N+1}$, where n is a positive integer, and has small subsidiary positive peaks between successive minima.

It is worthwhile to examine briefly the multi-dimensional applications of the $(C,1)$ mean. For the electron density we can write

$$0 \le \rho(xyz) = \sum_h \sum_{k=-\infty}^{\infty} \sum_l F_{hkl} e^{i(hx+ky+lz)} \le \rho_{max}. \tag{11}$$

Then the $(C,1)$ mean, $\sigma^1_{HKL}(xyz)$, satisfies

$$0 \le \sigma^1_{HKL}(xyz) = \sum_{h=-H}^{H} \sum_{k=-K}^{K} \sum_{l=-L}^{L} \left(1 - \frac{|h|}{H}\right)\left(1 - \frac{|k|}{K}\right)\left(1 - \frac{|l|}{L}\right) \cdot F_{hkl} \cdot e^{i(hx+ky+lz)} \le \rho_{max}, \tag{12}$$

where H, K and L are positive integers.

For $L=1$, the $(C,1)$ mean in three dimensions is equivalent to the $(C,1)$ mean of the *projection* of $\rho(xyz)$ along z onto the x, y plane:

$$\sigma^1_{HKL}(xyz) = \sum_{h=-H}^{H} \sum_{k=-K}^{K} \left(1 - \frac{|h|}{H}\right)\left(1 - \frac{|k|}{K}\right) \cdot F_{hk0} \cdot e^{i(hx+ky)}. \tag{13}$$

We write $d_z(xy)$ for the projection of $\rho(x,y,z)$ along z. Of course

$$d_z(x,y) = \int_0^{2\pi} \rho(xyz)dz = 2\pi \sum_{h}^{} \sum_{K} F_{hk0} e^{i(hx+ky)}, \tag{14}$$

$\qquad\qquad\qquad\qquad\qquad\qquad\qquad -\infty$

where $d_z(x, y)$ is again bounded:

$$0 \le d_z(x, y) \le d_{z(max)}. \tag{15}$$

(For convenience, we ignore dimensional units and normalizing constants. The present x, y, z symbols are in radians. They must be divided by 2π in order to correspond to the fractional cell coordinates usually employed in X-ray crystallography.)

By analogy with (11) and (12), it is then obvious that $\sigma^1_{HK1}(x, y, z)$ is the $(C, 1)$ mean of $d_z(x, y)$, and hence that

$$0 \le \sigma^1_{HK1}(x, y, z) \le d_z(x, y)_{max}. \tag{16}$$

There is nothing unusual about this. But suppose that one has a fairly good *projection* of a structure along some axis. It might well be a projection along a 2 or 2_1 axis, so that d is centrosymmetric. Again, for the axis along z, all the $(C, 1)$ means $\sigma^1_{HK2}(xyz), \sigma^1_{HK3} \cdots, \sigma^1_{HKL}(xyz)$ lie between 0 and $\rho_{max}(xyz)$. If the phases of F_{hk0} coefficients are known over some ranges of $\pm H$, $\pm K$, then one can begin to examine F_{hkl} and $F_{hk\bar{l}}$ phases through the boundedness of $\sigma^1_{HK2}(xyz)$, since, for $L = 2$,

$$\sigma^1_{HK2}(xyz) = \sum_{h=-H}^{H} \sum_{k=-K}^{K} (1 - \frac{|h|}{H})(1 - \frac{|k|}{K}) \cdot (F_{hk0} + \frac{1}{2} F_{hk1} e^{iz} + \frac{1}{2} F_{hk\bar{1}} e^{-iz}) \cdot e^{i(hx+ky)}, \tag{17}$$

which satisfies

$$0 \le \sigma^1_{HK2}(xyz) \le \rho_{max}. \tag{18}$$

In applying the non-negativity condition as one aid in phase determination for the $(C, 1)$ mean of some partial Fourier summation, *all* observable F_{hkl}, within the selected limits of indices $-H \le h \le H$, $-K \le k \le K$, $-L \le l \le L$, should be included in the partial sum; no terms of significant magnitude $|F_{hkl}|$ should be omitted, within the chosen ranges of h, k, and l. If one wishes to omit a non-zero term $F_{h\bar{3}l}$, for example, from a $(C, 1)$ mean $\sigma^1_{HKL}(\vec{r})$, then *no* term with k outside the range $3 \le k \le 3$ should be included in the partial summation.

Following examination of the $(C,1)$ mean $\sigma^1_{HK2}(\vec{r})$ as in equation (18), one can continue to relax the limit on $|L|$ in integral steps, examining in turn $\sigma^1_{HK3}(\vec{r})$, $\sigma^1_{HK4}(\vec{r})$, etc. This adds increasing detail to the approximate Fourier map in the z-direction. Of course the H and/or K limits may be correspondingly relaxed.

The non-negativity criterion had provided direct solutions for several structures with a positive kernel to permit use of partial Fourier summations. One notable case indeed was in the analysis of ferrocene, which was solved in a few minutes, using *all* observed F_{hk0} terms in a centrosymmetric two-dimensional projection, and examining effects on lowest levels of the density pattern by a process of phase changes of Fourier terms in descending order of their amplitudes. No other analytical process was needed, and later refinement processes provided no improvement in the R-factor. The kojic acid structure yielded almost as readily to the same procedure. Caroline MacGillavry and I discussed the use of other simple phase-limiting schemes on X-RAC, based on boundedness and symmetry properties of the density function, and some of those procedures appear in Appendix III to the same chapter in reference (1).

An outstanding advantage of the $(C,1)$ mean is the ease with which the partial sums can be computed in one, two or three dimensions. The kernel does contribute small positive ripples with maxima located between the zeros at $\pm\pi/(N+1)$, $\pm2\pi/(N+1)$, $\pm3\pi/(N+1),\cdots$; but these ripples decrease rapidly in height with increase in x, and crowd in closer to the central peak at $x = 0$ as N increases. The Poisson and third theta function (Gauss's) kernels decrease without ripples from their peak at $x = 0$, but they do not drop to zero at their minima at $x = \pm\pi$, $\pm3\pi$, $\pm5\pi$, etc. Thus any effort to suppress terms beyond a certain index results in a reduction in the heights of low-index terms as well.

My colleague Alberto Calderon sought other positive kernels without ripples and more advantageous in shape than the Poisson or Gauss kernel. The most promising of the kernels he devised was $(\frac{1+\cos x}{2})^m$. This is also discussed in Appendix V to the X-RAC chapter in references (1), and I present a brief account of it here. The kernel is illustrated in Fig. 4 in that appendix; it is very easily plotted. It drops to zero at $\pm\pi$, $\pm 3\pi$, $\pm 5\pi$, etc. When this kernel, which we identify as $g_m(x)$, is convoluted with the density function $f(x)$, all terms in the Fourier series for $f(x)$ with indices greater than m are suppressed. This positive kernel has a maximum of 1 at $x = 0$, obviously; hence peak heights and non-negativity are not altered by the convolution process. Here is a sketch of the proof for the cut-off of terms at index m.

The convolution $f(x) * g_m(x)$ is

$$F(m,x) = \frac{1}{\pi}\int_{-\pi}^{\pi} f(x-t)\cdot\left(\frac{1+\cos t}{2}\right)^m \cdot dt. \tag{19}$$

Introducing the Fourier expansion for $f(x-t)$,

$$F_m(x) = \sum_{-\infty}^{\infty} \frac{a_n}{2\pi}\int_{-\pi}^{\pi} e^{in(x-t)}\cdot\left(\frac{1+\cos t}{2}\right)^m dt, \tag{20}$$

which we write as

$$\sum_{-\infty}^{\infty} A_n e^{inx}, \tag{21}$$

A_n is given by

$$A_n = \frac{a_n}{2\pi}\int_{-\pi}^{\pi} e^{-int}\left(\frac{1+\cos t}{2}\right)^m dt. \tag{22}$$

Expanding $(\frac{1+\cos t}{2})^m$ by the binomial theorem, one obtains

$$A_n = \frac{1}{2}\frac{a_n}{2^m}\sum_{\mu=0}^{m}\binom{m}{\mu}\int_{-\pi}^{\pi} e^{-int}\cos^\mu t\, dt, \tag{23}$$

where $\binom{m}{\mu}$ represents the binomial coefficient $\binom{m}{\mu} = \frac{m}{(m-\mu)!\mu!}$, with m and μ integers and $0 \leq \mu \leq m$.

In order to integrate in equation (23), we write

$$\cos^\mu t = \left(\frac{e^{it}+e^{-it}}{2}\right)^\mu$$

and again expand by the binomial theorem. One obtains ultimately

$$A_n = \frac{a_n}{2^m} \sum_{\nu=0}^{m} \sum_{\mu=\nu}^{m} \binom{m}{\mu} \binom{\mu}{\nu} \frac{1}{2^\mu} \delta_{(n,2\nu-\mu)},$$

with

$$\delta_{(n,2\nu-\mu)} = \begin{cases} 1, & n = 2\nu - \mu; \\ 0, & n \neq 2\nu - \mu. \end{cases} \tag{24}$$

Since A_n depends directly upon m, it is useful to write the coefficient as $A_n^{(m)}$. Values of $A_n^{(m)}$ can be computed from (24) and in this way the convolution

$$F_m(x) = \sum_{-m}^{m} A_n^{(m)} e^{inx}. \tag{25}$$

The computation of $A_n^{(m)}$ is illustrated for $m = 0, 1$, and 2. For $m = 0$:

$$A_n^{(0)} = a_n \cdot \delta_{n,0} = a_0. \tag{26}$$

Hence

$$F_0(x) = f * g_0 = f * 1 = a_0 = f(0). \tag{27}$$

For $m = 1$:

$$A_n^{(1)} = \frac{a_n}{2} [\delta_{n,0} + \frac{1}{2}(\delta_{n,1} + \delta_{n,-1})]. \tag{28}$$

Hence

$$F_1(x) = f * g_1 = \frac{a_0}{2} + \frac{a_1}{4} e^{ix} + \frac{a_{-1}}{4} e^{-ix}. \tag{29}$$

For $m = 2$:

$$A_n^{(2)} = \frac{a_n}{4} [\delta_{n,0} + (\delta_{n,1} + \delta_{n,-1}) + \frac{1}{4}(\delta_{n,-2} + 2\delta_{n,0} + \delta_{n,2})]. \tag{30}$$

Hence

$$F_2 = f * g_2 = \frac{3}{16} a_0 + \frac{a_1}{4} e^{ix} + \frac{a_{-1}}{4} e^{-ix} + \frac{a_2}{16} e^{2ix} + \frac{a_{-2}}{16} e^{-2ix}. \tag{31}$$

As far as computability of terms for the partial summation approximating $f(x)$ is concerned, the $(C,1)$ mean is certainly to be preferred! One can evaluate the $A_n^{(m)}$ terms for this new convolution for various m values once for all, and let a computer do all the work, of course. This is particularly advisable when one needs the convolution for two- and three-dimensional functions.

One can evaluate the convolution also by computing the transform of the kernel $g_m(t)$ directly, sampling this at reciprocal lattice nodes of the density function, and multiplying the sampled value of the kernel by the magnitude and phase of the Fourier coefficient at the node concerned. This procedure is based on the well-known theorem that the transform of the convolution of the density function with the kernel is equivalent to *multiplication* of the *transform* of the density function with the *transform* of the kernel. Although the transform of the kernel $(\frac{1+\cos t}{2})^m$ was calculated and appears in Appendix V of reference (1), and its two- and three-dimensional analogues could likewise be established, the points at which these continuous functions would need to be sampled would differ for each reciprocal lattice, and there would need to be a multitude of transform stored for each set of maximum indices m_H and m_K in two dimensions, or m_H, m_K, and m_L in three dimensions.

One turns back to the First Cesaro Mean, $(C,1)$, with gratitude for its ease of computability!

One matter intrigues me as I write now, with no facilities for crystal analysis available to me. Suppose one would like to obtain a sharpened Patterson map for deconvolution, in order to reduce effects of peak overlapping. If the atoms in the structure are of one kind — *e.g.*, all carbons — one divides each cell structure factor by the atom structure factor at the same $(sin\theta)/\lambda$ value as that for the cell factor, thus essentially obtaining point-atoms. But the Patterson synthesis using these $(F/f)^2$ coefficients contains ripples due to cut-off of the series (and thus convolution with a Dirichlet kernel). The sharpening might have been further enhanced by collecting the X-ray data at low temperature, providing no crystal transition intervened, that is. In order to eliminate the ripples due to cut-off of the series, suppose one computed the convolution with $g_m(x)$ kernel $(\frac{1+\cos t}{2})^m$, in its three-dimensional form. This would provide the best possible approximation to the point-atom Patterson, with terms included to the indices over the range of observable scattering.

A convolution with Fejer's kernel, equivalent to calculation of the First Cesaro Mean $(C,1)$, would be more easily accomplished, and again should provide a better synthesis than that with the unmodified room-temperature structure factors.

Homometry and the continuous transform

Lindo took a rather impish delight in the discovery of homometric structures: those in which more than one density function could be deduced from a single Patterson map. I doubt whether he would have found anything like such enjoyment in exploring puzzles and ambiguities associated with someone else's theory. He was far too much a gentleman, gleefully to question the generality of a colleague's result, particularly if it involved as fundamental and practical a matter as interpretation of an interatomic vector pattern. But no one had a better right to a *caveat* than he. If studying homometric structures had not been so intriguing, I suspect he might have justified his concern as a scholarly duty.

I do not know how far his last studies carried him, but I hope there will be some indication available eventually. He recognized that in centrosymmetric structures homometry could result from the translational symmetry of a crystal. For a positive, real, centrosymmetric non-periodic density function, the Fourier transform has a positive peak at the origin of reciprocal space. Each time the absolute value of the *transform* of the non-periodic centrosymmetric density function *passes through* a zero, it must change its sign. For a two-dimensional centrosymmetric real density function, the zeros of the transform will be seen as closed curves on the plane in transform space, separating positive regions of the transform from negative. The closed curves — we can regard them as zero contour levels of the continuous transform — need not *surround* the origin of the reciprocal plane, although some may do so. In three dimensions the plane curves become closed surfaces, some surrounding the origin, some not; *across* each surface, the three-dimensional transform (of a three-dimensional non-periodic centrosymmetric real density function) changes sign.

If we had full knowledge of the zeros of that continuous transform, there would be no phase problem for a centrosymmetric real density function. At the origin, and until one contacts a zero contour or surface, the transform is positive; whenever a zero contour or zero surface is *crossed*, the transform suffers a change in sign. There can be no centrosymmetric homometric forms of a centrosymmetric real density function for which the continuous transform is known.

The difficulty of phase determination in X-ray crystallography for a centrosymmetric *periodic* structure arises because we only know the magnitude of the Fourier transform at reciprocal lattice nodes, and we generally cannot sample finely enough to know where the zeros of the continuous transform of an individual crystal cell lie. Nice tricks have been used, in a few interesting cases, to explore for information on the transform magnitude in the neighborhood of the nodes; the first practical use of such exploration was achieved in the well-known case of a hæmoglobin crystal in which alteration of the degree of hydration of the crystallized protein altered the monoclinic angle of the crystal without changing the relative orientations of the molecules within the cells.[5] In this way the reciprocal lattice nodes were shifted sufficiently so that one could locate some of the zero contours of the single cell transform. Application of this scheme was later attempted in an effort to locate the zeros of the transform of lipid membrane of neural myelin sheath. Early efforts to attain this information failed because the X-ray analysts failed correctly to determine the zero level of the transform of the centrosymmetric density function.

The above geometrical argument indicates that we can immediately deduce the sign of any region of the transform of a centrosymmetric nonperiodic density function if we know where the zeros of the continuous transform lie. The zeros of the Patterson function for that centrosymmetric nonperiodic density function are of course identical with the zeros of the Fourier transform itself. Thus one can say that if one knows the continuous Patterson function everywhere, there is one and only one centrosymmetric density function which is the solution of the convolution integral equation.

Suppose we write $d(x)$ for a continuous density function in three dimensions; $d(x)$ is of finite mass.† The Fourier transform of $d(x)$ is $F[d(x)]$, which we shall write as $\tilde{d}(y)$:

$$\tilde{d}(y) = \int_{\text{all } x} d(x) e^{2\pi i (x \cdot y)} dv_x. \tag{32}$$

We write $\hat{d}(x)$ for $d(-x)$, and $h(x)$ for the convolution:

$$h(x) = \int_{\text{all } s} d_1(s) d_2(x - s) dv_s. \tag{33}$$

The complex conjugate of a quantity is indicated by a line above the quantity. Since $d(x)$ is real, from equation (32)

$$\tilde{d}(y) = \bar{\tilde{d}}(-y), \tag{34a}$$

and indeed

$$\bar{\hat{d}}(y) = \tilde{d}(-y) = \bar{\tilde{d}}(y). \tag{34b}$$

Also, of course,

$$F(d_1 * d_2) = \widetilde{d_1 * d_2} = \tilde{d}_1 \cdot \tilde{d}_2. \tag{35}$$

If $|\, \tilde{d}(y) \,|$ is known, what can be said about $d(x)$?

$$|\, \tilde{d}(y) \,|^2 = \tilde{d}(y) \cdot \overline{\tilde{d}(y)} = \tilde{d}(y) \cdot \tilde{d}(-y) = F(d * \hat{d}). \tag{36}$$

In words: $|\, \tilde{d}(y) \,|^2$ is the Fourier transform of $d * \hat{d}$; and thus the function $p(x) = d * \hat{d}$ can be obtained by taking the inverse transform of $|\, d(y) \,|^2$. Hence the problem reduces to this: what can be said about the solution of the integral equation

$$p(x) = d * \hat{d} = \int_{\text{all } s} d(s) \hat{d}(x - s) dv_s? \tag{37}$$

Let us assume the following limitation on $d(x)$:

$$\int_{\text{all } x} d(x) e^{\alpha} dv_x < \infty; \tag{38}$$

† x and y are now *vectors*.

that is, the integral is finite for some positive value of α, for example, if $d(x)$ vanishes outside some large circle. Then the Fourier transform $\tilde{d}(y)$ of $d(x)$ is an analytic function of y_1, y_2, y_3 for $|I(y_i)| < \alpha$: that is, for all triples of complex values of y_1, y_2, y_3 whose imaginary parts are less than $\alpha/3$ in absolute value. Thus if $\tilde{d}_1(y)\cdot\tilde{d}_2(y) = 0$, and $\tilde{d}_1(y) \neq 0$ at the point $y = y_0$, then $\tilde{d}_1(y) \neq 0$ in a ***neighborhood*** of y_0 and we must have $d_2(y) = 0$ in that analytic function, this implies that $\tilde{d}_2(y) = 0$ identically.

Thus, for a centrosymmetric (even) function, if $d_1 * d_2 = 0$, either $d_1 = 0$ or $d_2 = 0$. This implies that we cannot have more than one ***even*** solution of the type in equation (37). Suppose that we do have $g = d_1 * \hat{d}_1 = d_2 * \hat{d}_2$. Since d_1 and d_2 are ***even***, $\hat{d}_1 = d_1$ and $\hat{d}_2 = d_2$. Then we would have $d_1 * d_1 = d_2 * d_2$, or $(d_1 + d_2) * (d_1 - d_2) = 0$. This requires either that $d_1 = d_2$ or $d_1 = -d_2$. Thus only one positive solution exists.

This is an analytical proof of our earlier intuitive conclusion.

This result does ***not*** hold, however, for non-centrosymmetric functions. Suppose the density function d is itself a convolution of two other functions d_1 and d_2, ***i.e.***,

$$d = d_1 * d_2. \tag{39a}$$

If d is a solution of the equation $p = d * \tilde{d}$, then

$$d' = d_1 * \hat{d}_2 \tag{39b}$$

is also a solution, since†

$$\begin{aligned}(d' * \hat{d}') &= (d_1 * \hat{d}_2) * (\hat{d}_1 * \hat{\hat{d}}_2) \\ &= d_1 * \hat{d}_2 * \hat{d}_1 * d_2 \\ &= (d_1 * d_2) * (\hat{d}_1 * \hat{d}_2) \\ &= d * \hat{d}.\end{aligned} \tag{40}$$

Hence both d and d' are solutions of p!

† Some aspects of the algebra of transforms are presented in Appendix VIII to reference (1), but these relations are so readily developed that they are not repeated here.

Of course, d_2 should not be even (*i.e.*, centrosymmetric), since if it were, then d and d' would be identical. Furthermore, d_1 also should not be even, because if it were then d would equal $\hat{d}'$ and $\hat{d} = d'$, and d and d' would be related by inversion in the origin.

The homometry is most simply illustrated by various non-centrosymmetric distributions of variously-weighted point masses, one set convoluted with another (unlike) set. A multiply infinite set of homometric pairs of structures can therewith be constructed. I have examined some hundreds of these, choosing arrangements of points pertinent to known structures.

The convolution $d(x) = d_1(x) * d_2(x)$ in density space corresponds to the product

$$\tilde{d}(y) = \tilde{d}_1(y) \cdot \tilde{d}_2(y) \tag{41}$$

in transform (scattering) space. One might explore mathematically the factorization of $\tilde{d}(y)$, which is [in the case of non-centrosymmetry of a real $d(x)$] complex:

$$\tilde{d}(y) = A(y) + iB(y), \tag{42}$$

with

$$A(y) = A(-y), \quad B(y) = -B(-y). \tag{43}$$

In order to demonstrate that both $\tilde{d}(y)$ and $\tilde{d}_2(y)$ are similarly complex, and what limitations are imposed on them, it is best to re-examine the convolution $d(x) = d_1(x) * d_2(x)$, and in three dimensions, in order to avoid the complexities of peak overlapping in a projection of the structure. Each peak in $d_1(x)$ is atom-like, but with fractional density so that when convoluted with a peak in $d_2(x)$ the resulting peak in $d(x)$ is indeed that corresponding to a real atom of the actual structure $d(x)$. One thinks of $d(x)$ as generated when $d_1(x)$ is laid down and then $d_2(x)$ is viewed through the peaks of $d_1(x)$. *Unless* $d(x)$ can be built up entirely by such a process, one cannot obtain the sort of homometry which occurs when $d_1(x)$ is so convoluted with $d_2(x)$, and then a *new* convolution $d'(x)$ generated by convoluting $d_1(x)$ with $\hat{d}_2(x)$ [which latter function is $d_2(x)$ inverted in its origin].

The convolution process by means of which $d(x)$ and $d'(x)$ are formed could have been accomplished just as well by laying down $d_2(x)$ first and then viewing it through $d_1(x)$ to obtain $d(x)$, *or* by laying down $\hat{d}_2(x)$ and viewing it through $d_1(x)$ to obtain $d'(x)$; the results are the same.

The game of designing homometric structures can be carried further, and in fact essentially without limit. Consider a pair of the simplest homomeres, $d(x)$ and $d'(x)$, derived respectively from $d_1(x) * d_2(x)$ and $d_1(x) * \hat{d}_2(x)$. Convolution of $d(x)$ with another non-centrosymmetric density-function $d_3(x)$ will produce a

$$d''(x) = d(x) * d_3(x); \tag{44}$$

and a homomere of $d''(x)$ will be

$$d'''(x) = d(x) * \hat{d}_3(x). \tag{45}$$

Two further homomeres will be

$$d''''(x) = d'(x) * d_3(x) \tag{46}$$

and

$$d'''''(x) = d'(x) * \hat{d}_3(x). \tag{47}$$

Thus we now have a further *quartet* of homomeres

$$d''(x),\ d'''(x),\ d''''(x),\ d'''''(x).$$

These are *new* homomeres, *not* homometric with the pair $d(x)$ and $d'(x)$, however. The process can go on, producing larger and larger sets of homomeres, until one runs out of space for more apostrophes; and with a more sensible notation it can expand indefinitely. The resulting sets of structures will become increasingly unrealistic.

Two final considerations. Firstly, the homometric sets, based on non-periodic non-centrosymmetric density distributions, remain when one convolutes the structures with a periodic lattice in density space, or samples the continuous transform on the reciprocal lattice. The question now surfaces: does any *further* homometry develop merely because of this periodicity? In the centrosymmetric density function case, there is *no* homometry if one knows the zeros of the continuous transform completely. If the reciprocal lattice sampling is not fine enough to permit location of the zeros between the lattice nodes, then two unlike centrosymmetric density functions might, for sufficiently rough sampling, have the same Fourier series for their Patterson functions. This is *pseudo*-homometry, I would say, and yet of prime practical significance if and when it occurs. For the non-centrosymmetric non-periodic case, I have not yet looked for homomeres of the density function which are not of the class described by equations (39) and (40), or (44) through (47), etc.; but I suspect no other kinds exists. If the sampling of a continuous function at the reciprocal lattice nodes is too rough, wide fluctuations might occur in the continuous transform between the nodes and again pseudo-homometry might exist. The case of pseudo-homometry in periodic non-centrosymmetric functions could be of large practical concern.

There is, of course, a significant control over pseudo-homometry due to too rough a periodic sampling of the Fourier transform at the reciprocal lattice nodes. This control is imposed by non-negativity of the density function in the X-ray case; this non-negativity results in the interrelations between Fourier coefficients which are explored and utilized in the 'direct methods' for phase determination. I have not explored what degree of roughness of sampling of the continuous transform does lead to possibility of pseudo-homometry; but undoubtedly the developers and users of the direct methods know all about the matter.

This approach to homometry was first proposed by A. Calderon and myself at the ACA meeting in Chicago (joint meeting with the American Physical Society) in 1951, and discussed in Appendix VIII of reference (1). See references (14) and (15).

On negative scatterers in neutron analyses

This portion of my discussion is introduced chiefly as a bridge to the final section.

When Chalmers Frazer and I began our neutron diffraction studies on the structural mechanisms of the ferroelectric and antiferroelectric transitions in KH_2PO_4 and its relatives, it was immediately apparent that the Patterson maps were particularly interesting. The interaction peaks between the nuclei of monovalent metal ions (or the N of NH_4), P or As, and O atoms, and H to H nuclei, were all positive; but the peaks arising from vectors linking H nuclei and all non-H nuclei were negative. All peaks were positive when all (or almost all) H's were replaced by crystal growth from D_2O ; and when I grew crystals with D's replacing H's until the H to D concentration ratio was the reciprocal of the ratio of the absolute values of their nuclear scattering factors, all coherent scattering from the H's and D's disappeared.

The first copies of photographs of neutron X-RAC maps of the KH_2PO_4, $K(H,D)_2PO_4$ and KD_2PO_4 data were sent to one A. L. Patterson, as a mark of our continued respect and affection. That was in 1951. I presented several discussions of H–D replacement in neutron analyses in 1953 and 1954,[6,7] and we used the technique in every one of the neutron studies of ferroelectric and antiferroelectric structures and transition mechanisms of H-containing crystals for the next dozen years: in Rochelle salt, $LiNH_4$tartrate$\cdot H_2O$, $NH_4H_2PO_4$, $(NH_4)_2SO_4$, $(NH_4)_2BeF_4$, NH_4HSO_4, $LiH_3(SeO_3)_2$, $RbHSO_4$, *tris*-glycine sulfate and fluoroberyllate and the other glycine-containing and sarcosine ferroelectrics, on the ferroelectric alums, on $(NH_4)_2Cd(SO_4)_2$, $NaNH_4SO_4\cdot 2H_2O$, $NaH_3(SeO_3)_2$, $CsH_3(Se)_3)_2$, $CH_2ClCOONH_4$, $Li(N_2H_5)SO_4$ and $NH_4PF_6\cdot NH_4F$. I suppose I have forgotten a good number of others!

Once a ferroelectric or antiferroelectric structure is established, a study of the transition mechanism — if a transition exists — is a matter of structure refinement above and below the transition temperature(s). In H-bonded structures minor changes in H positions are always present and significant. H–D replacement was always a useful tool. It is often stated that one of the advantages of such replacement is that strict isomorphism is therewith maintained; but this is not always the case. In KH_2PO_4 and KD_2PO_4 the transition temperatures differ by 90°K, and refinements reveal significant differences structurally, both between the two crystals in their ferroelectric phase and between their respective paraelectric phases. In numerous cases the deuterated analogue is difficult to crystallize in the same symmetry as the protonated crystal, and it may be metastable.

H–D replacement in Rochelle salt permitted us to demonstrate that the loss of one two-fold screw axis, as the crystal passed into the ferroelectric phase at either the upper or lower Curie point, was chiefly due to a shift in H (or D) atoms; the extinction due to that screw axis was not evident at all in the two non-ferroelectric (orthorhombic) phases when the H and D concentration ratio was such that no coherent scattering was possible from H-bonding protons or deuterons.

I obtained some isotopically pure $Li_2^6CO_3$ and $Li_2^7CO_3$ from Oak Ridge, through the AEC, in order to prepare $Li(NH_4)$ tartrate·H_2O crystals with coherent Li nuclear scattering cancelled out due to presence of the two Li isotopes in proper ratio, so that the participation of the Li ion in the ferroelectric transition could be examined as the translational symmetry changed at the Curie point. There was enough of the separated isotope material left over so that I could play a trick on Willie Zachariasen. I had done the structure of $LiOH{\cdot}H_2O$ under his guidance as my very first analysis, at the University of Chicago. So I prepared some $(Li^6,Li^7)O_2(H,D)_3$ in a good crystal, with the Li isotope ratio such that no coherent Li nuclear scattering was apparent, and grew the material from an H_2O–D_2O mixture such that the (H,D) scattering was *also* not present coherently. Then I presented him both Patterson maps *and* density maps from neutron data and asked him to comment on them. I also gave him some of the crystals and suggested that he might want the Argonne neutron diffractionists to check my data. It was the only time I outwitted him! He could not identify the structure from the neutron data alone, or from the Fourier maps.

The presence of negative scattering nuclei enhances interpretation of Patterson functions, but it mitigates against use of direct methods of phase determination which depend upon non-negativity of the density function. H–D replacement, to the extent that coherent scattering from all protons and deuterons is eliminated, is easily possible in cases where all H's or D's are replaceable in solution, and renders the analysis amenable to the direct methods; and once the phasing of the structure factors for the structure involving non-H or non-D atoms has been accomplished, then the structure for which complete replacement of H's by D's has been accomplished can be easily solved, with interpretation of the Patterson map of the H-containing crystal providing the most useful information. One then computes structure factor phases for the all-D-containing crystal.

To some extent it was our experience with isotopic substitution in neutron analyses which led us to the studies which are discussed in the next (and final) part of this paper.

Anomalous dispersion and the sine-Patterson function

The sine-Patterson is a minor extension of Lindo's original development. I follow somewhat the notation in the homometry section, modified for the periodic structure of a crystal.

$$d(x) = \sum_{H \atop -\infty}^{\infty} F_H e^{-2\pi i H \cdot x}, \tag{48}$$

where x is a vector in direct space,

$$\vec{x} = \vec{a}_1 x_1 + \vec{a}_2 x_2 + \vec{a}_3 x_3,$$

and H is the reciprocal lattice vector

$$\vec{H} = \vec{b}_1 H_1 + \vec{b}_2 H_2 + \vec{b}_3 H_3.$$

F_H is the Fourier coefficient which is now expressed in the form of the discrete atom distribution in the unit cell:

$$F_H = \sum_{j=1}^{N} f_j(H) e^{2\pi i H \cdot x_j}, \tag{49}$$

and $f_j(H)$ is the scattering factor for the jth atom in the cell, and is a function of the reciprocal lattice vector. The Patterson function $p(S)$ is the convolution of $d(x)$ with $d(-x)$ [which later we wrote as $\hat{d}(x)$]:

$$p(S) = \sum |F_H|^2 e^{-2\pi i H \cdot S}. \tag{50}$$

Lindo recognized that $|F_H|^2$ was available directly from the X-ray scattering data, and that the convolution (equation (50)) provided a map of all the interatomic vectors, weighted according to the convolution of the spherical atom electron densities.

The scattered intensity from a plane H is

$$I_H = K F_H F_H^*, \tag{51}$$

where we now use the asterisk to indicate the complex conjugate; and we write $\overline{H}$ for $-H$. One sees immediately from (49) that, if f_j is real so that $f_j \equiv f_j^*$,

$$F_H^* = F_{\overline{H}}, \; F_{\overline{H}}^* = F_H, \tag{52}$$

and

$$I_H = K F_H F_H^* = K F_{\overline{H}} F_{\overline{H}}^* = I_{\overline{H}}. \tag{53}$$

This is 'Friedel's Law' for X-ray scattering from a crystal comprised of real atoms. We cannot discern, merely by examining pairs I_H and $I_{\overline{H}}$, whether or not a crystal has a center of symmetry.

If, however, one (or more) of the f_j terms is (are) *not* real, then $f_j \neq f_j^*$, and

$$F_H^* \neq F_{\overline{H}}, \; F_{\overline{H}}^* \neq F_H \text{ and } I_H \neq I_{\overline{H}}. \tag{54}$$

An atomic scattering factor f_j is *complex* when incident X-rays excite the atom into fluorescence. The atom is then said to scatter 'anomalously' rather than 'normally.' Under this condition

$$f_j = (f_0)_j + \Delta f_j' + i f_j'' \equiv f_j' + i f_j''. \tag{55}$$

Then $f_j^* \neq f_j$.

At the time Lindo developed and interpreted the significance of equation (50), there was no general interest in anomalous dispersion among X-ray crystallographers. What was of pressing importance was to solve crystal structures; and the incoherent X-radiation resulting from fluorescence in anomalous scattering was an inconvenience to be avoided, particularly in those days when one used photographic film for intensity data collection. Besides, why clutter up the Patterson function with complex f_j's ?

If he had written $f_j = f_j' + if_j''$ for the atomic factors, Lindo would have had

$$F_H = \sum_j (f_j' + if_j'')e^{2\pi iH\cdot x_j}, \tag{56a}$$

and

$$F_H^* = \sum_j (f_j' - if_j'')e^{-2\pi iH\cdot x_j}, \tag{56b}$$

so that

$$\begin{aligned} |F_H|^2 &= F_H F_H^* \\ &= \sum_j \sum_k (f_j' f_k' + f_j'' f_k'') \cos 2\pi H \cdot (x_j - x_k) \\ &+ \sum_j \sum_k (f_j' f_k'' - f_j'' f_k') \sin 2\pi H \cdot (x_j - x_k). \end{aligned} \tag{57}$$

The electron density function $d(x)$ in equation (48) is again complex, and the Patterson function is again the convolution of $d(x)$ with $d*(x)$:

$$\begin{aligned} p(S)_{anom.\,scatter} &= \sum_H |F_H|^2 e^{-2\pi iH\cdot S} \\ &= \sum_H |F_H|^2 \cos 2\pi H\cdot S - i\sum_H |F_H|^2 \sin 2\pi H\cdot S \\ &= p_c(S) - ip_s(S). \end{aligned} \tag{58}$$

Since $\cos 2\pi H \cdot S$ is even in H and $\sin 2\pi H \cdot S$ is odd, we have, summing over $+H$ and writing that sum as $\sum'$,

$$p_c(S) = \sum\nolimits_H' (|F_H|^2 + |F_{\overline{H}}|^2) \cos 2\pi H \cdot S, \tag{59a}$$

and

$$p_s(S) = \sum\nolimits_H' (|F_H|^2 - |F_{\overline{H}}|^2) \sin 2\pi H \cdot S, \tag{59b}$$

and from equation (57),

$$|F_H|^2 + |F_{\overline{H}}|^2 = 2 \sum_j \sum_k (f_j' f_k' + f_j'' f_k'') \cos 2\pi H \cdot (x_j - x_k), \tag{60a}$$

and

$$|F_H|^2 - |F_{\overline{H}}|^2 = 2 \sum_j \sum_k (f_j' f_k'' - f_j'' f_k') \sin 2\pi H \cdot (x_j - x_k). \tag{60b}$$

It is only $p_s(S)$ which I shall discuss here. $p_s(S)$ has peaks corresponding to the transform of $f_j' f_k'' - f_j'' f_k'$ at $S = x_j - x_k$, and peaks corresponding to the transform of $-(f_j' f_k'' - f_j'' f_k')$ at $-S = x_k - x_j$. It is antisymmetric, and when wisely used it is a powerful function for analysis of non-centrosymmetric structures.

By suitable chemistry and choice of wavelength, one can generally arrange that there is only one type of anomalous scatterer per cell. Then we have

$$|F_H|^2 - |F_{\overline{H}}|^2 = 2f_m'' \sum_j \sum_m f_j \sin 2\pi H \cdot (x_j - x_k), \tag{61}$$

where f_m'' is the magnitude of the imaginary part of the structure factors for the various anomalous scatterers. The function $p_s(S)$ then has positive peaks corresponding to vectors from the anomalous scatterers to normal scatterers, and negative peaks from normal scatterers to anomalous scatterers. There are no other peaks.

Suppose the cell contains 4 anomalously scattering atoms of the same type, and 4 groups of 20 normal scatterers. The usual Patterson map, and $p_c(S)$, will contain $2N(N-1) = 2 \times 84 \times 83 =$

~ 14,000 peaks, plus one large peak at the origin. The $p_s(S)$ function contains 320 positive peaks and an equal number of negative peaks, and no others; no normal-to-normal atom peaks at all. There is no need to sharpen the f''_m peak, since it is due to electrons close to the nucleus and is essentially independent of H (*i.e.*, independent of scattering angle) [f''_m isotropic means that its transform, which represents the distribution of the electrons concerned, is nearly a delta function.] As equation (61) indicates, all peaks in $p_s(S)$ may be sharpened by dividing each $|F_H|^2$ (*not* each $|F_H|$) by some average f_j for the normal scatterers, with the f_j evaluated at the same H value as the $|F_H|^2$ term; then the Patterson series may be protected from cutoff ripples by use of the First Cesaro Mean, as suggested earlier. In practice we never found cutoff effects to be troublesome, because in three dimensions overlapping was not a serious problem, so that peak-sharpening was unnecesary. Convolution of atom shapes in a normal Patterson results in spread of the interaction peaks; but convolution of the distribution of electrons contributing to anomalous dispersion with normal atom electron densities produces essentially no spread of the normal atom shapes. There is no peak at the origin, in $p_s(S)$.

I denote an anomalously scattering atom by ANS, and a normal scatterer by NS. The positions of positive peaks in $p_s(S)$ correspond to the positions of NS's as viewed from ANS's. The positions of negative peaks correspond to ANS's as viewed from NS's.

Suppose that the distribution of NS's in the neighborhood of a particular ANS is such that some of the NS's are centrosymmetrically disposed about that ANS and some are not. Only the *departure* from centrosymmetry will be indicated by peaks in the neighborhood of the $p_s(S)$ origin. Because the electrons responsible for the imaginary contribution to the ANS structure factor are concentrated close to the nucleus of that atom, the f''_j part of the ANS acts essentially as a point scatterer, and the peak shapes in $p_s(S)$ are essentially those of the density distribution of the NS's, multiplied in scale by the number of electrons responsible for the f''_j part of the ANS structure factor. Any tendency toward centrosymmetry of NS's about the ANS leads to partial or total overlapping of positive and negative peaks in the $p_s(S)$ function; total centrosymmetry reduces $p_s(S)$ to zero, in the neighborhood of the ANS. † That is just what one expects; but such partial or total cancellation of the peaks is accompanied by reduction or obliteration of differences between Bijvoet pairs $|F_H|$ and $|F_{\overline{H}}|$. *If the* $|F_H|^2 - |F_{\overline{H}}|^2$ *values are significant, so is the* $p_s(S)$ *function!*

Let us examine this matter a bit further. One of the earliest applications of the $p_s(S)$ function in my laboratory was to a salt of the $[Co(en)_3]^{+3}$ complex ion: 2D-$[Co(en)_3]Cl_3 \cdot NaCl \cdot 6H_2O$; (en) is ethylenediamine, NH_2-CH_2-CH_2-NH_2. In this ion the Co ion is at the center of a regular octahedron, and each ethylenediamine is coordinated to the Co through its two nitrogens positioned at adjacent vertices of the octahedron. The result is a triple chelate with the (en) groups disposed as if they formed the three blades of a propellor, in either a right-handed or left-handed screw arrangement. We had resolved the two stereomers by a standard procedure, and wished to explore the use of $p_s(S)$ in direct determination both of the crystal structure and the absolute configuration of the complex ion. This was accomplished with ease, and is illustrated in reference (8), p. 288, and in reference (9).

The N atoms at the octahedron vertices are centrosymmetrically disposed about the central Co, and so they do not appear in $p_s(S)$ in the neighborhood of the Co; the -CH_2-CH_2- groups are not centrically distributed, however, and they appear very nicely in the $p_s(S)$ map. We used a generalized project as shown in references (8,9). But the N atoms of each (en) appear in that map nevertheless, very clearly, because they are viewed from *another* Co, in another of the (en) complexes in the same cell. Had there been but one complex ion per cell, this view from another Co would not have been possible; but then the positions of all atoms *not* centrically positioned around the Co would have revealed the structure so easily that the N positions would eventually be readily determined from examination of a normal Patterson.

---

† Contributions in that region *might* come from peaks related to *another* ANS atom in the cell.

The $p_s(S)$ function is an extensively deconvoluted Patterson map, and we have shown that when $p_s(S)$ atom positions are used for image-seeking in the Patterson, complete deconvolution of the latter is achievable.[8,9] The earliest example of this process was the quite straightforward analysis of cobaltous L-aspartate·$3H_2O$ and the analogous zinc salt.[8,9,10] In that analysis a full three-dimensional $p_s(S)$ map was utilized. When a non-centric crystal has one or more centrosymmetric projections, as for example, along a 2- or 2_1-axis parallel to x_3, a generalized projection of the $p_s(S)$ function such as

$$\int_0^1 p_s(S) \cdot \sin(2\pi H_3 S_3) \cdot dS_3$$

is very useful. With $H = 1$, positive peaks represent NS positions whose x_3 coordinates are *higher* than those of the ANS x_3 coordinate, and negative peaks indicate x_3 values for NS's *less* than the ANS x_3 coordinate. A generalized projection with $H_3 = 2$ is shown for 2D-$[Co(en)_3]Cl_3 \cdot NaCl \cdot 6H_2O$ in reference (8) and (9), from which the full structure and absolute configuration could be established.

Reference (8) also illustrates some strikingly useful applications of $p_s(S)$ in studies of the mechanisms of polarization reversal and absolute configurations of polarized ferroelectric crystals. These observations and deductions are especially simple, and the results most accurate, when polarization reversal does not alter the crystal orientation — which is the case for a good fraction of ferroelectric species. In these cases the crystal and X-ray photon counter were set for a particular reflection, and reversal of the direction of the electric field was accomplished periodically, changing H to $\overline{H}$ for the X-ray reflection; scattered photons for the H orientation were stored in one counter-bin and $\overline{H}$ intensities were stored in another. The experimental arrangement is shown in Fig. 6 of reference (8). Generalized projections of the resulting $p_s(S)$ functions are shown in Fig. 4(b) for the KH_2AsO_4 structure, and observed intensities for the (511) and $(5\overline{1}\overline{1})$ plane reflections are shown in Fig. 7 of that reference.

In that experiment MoKα radiation was used to excite the As atoms. The case for structure reversal in the ferroelectric *tris*-glycine sulfate is shown in Figs. (5a) and (5b) (ref. 8); here CuKα radiation excited the S atoms. Use of the $p_s(S)$ function in three dimensions would have been the easiest way of determining the entire structure at the outset; we could also have excited the Se atoms in isomorphous *tris*-glycine·H_2SeO_4, and obtained even more precise results. There is no change in absorption correction required as one inverts H to $\overline{H}$ in these crystals. The absolute configuration change with field reversal was especially significant, physically, in these and other glycine-containing ferroelectrics.

Excitation of K electrons for generation of an f_j'' imaginary component in the ANS structure factor is generally accomplished through the use of Kα radiation from a target whose atomic number is 2 higher than that of the ANS. The resulting f_j'' then has a magnitude somewhere between 3.0 and 3.5. The following data are selected from Dauben and Templeton[11] and Templeton.[12]

Incident radiation		Excited atom		f_j''
CrKα	(24)	Ti	(22)	3.0
MnKα	(25)	V	(23)	3.3
FeKα	(26)	Cr	(24)	3.3
CoKα	(27)	Mn	(25)	3.5
NiKα	(28)	Fe	(26)	3.5
CuKα	(29)	Co	(27)	3.4
ZnKα	(30)	Ni	(28)	3.2
GaKα	(31)	Cu	(29)	3.4
ZrKα	(40)	Rb	(37)	3.2
NbKα	(41)	Sr	(38)	3.2
MoKα	(42)	Sr	(38)	3.0
RuKα	(44)	Zr	(40)	3.0
RuKα	(44)	Nb	(41)	3.3
AgKα	(47)	Ru	(44)	3.6

There are also several useful L emission series target materials: W, Ir, Au, *e.g.*, for Ni, Cu, and Zn, respectively, for example. But values of f_j'' 3 or more times higher than those in the above table can be obtained by L-electron *excitation*.

Dauben and Templeton[11,12] give values for combined Kα and L excitation in elements from Ir(77) to Bi(83) for CrKα , CuKα , and MoKα exciting radiation. The f_j'' magnitudes range from 13 to 17 for CrKα , from 7 to 9 for CuKα , and from 9 to 12 for MoKα , for those 7 elements. But one should study Dauben and Templeton's data carefully before making a decision as to the choice of an ANS element. I suspect there are more modern data on f_j' and f_j'' magnitudes now, but I have not looked for them.

References (13) and (8) discuss the combined use of the normal Pattersons, $p_c(S)$, and $p_s(S)$, and various other matters (*e.g.*, use of two or more different ANS's in a crystal, and alteration of excitation λ's); but I do not consider these now. Anomalous dispersion is accompanied by high absorption and fluorescence, and one must be prepared to deal with these phenomena. Synchrotron radiation permits one to utilize exciting radiation to maximize f_j'' contributions to a structure factor but the limitations in steady access to storage ring facilities do discourage reliance on them somewhat.

Conclusions

Of course, the prime route to crystal analysis is through the 'direct' methods. This has been true for the past twenty years or more. But the foundation stones are still the Patterson functions and chemical reasonability, because the former contain all the available experimental data and provide the starting point for an analysis, and the latter permits escapes from cul-de-sacs and the surmounting of such possible eventualities as homometry. One welcomes any scheme which sets one on the right track, or reduces alternatives at early stages of an analysis. Inclusion of heavy atoms, isomorphous replacement, image-seeking for deconvolution of the Patterson map, and anomalous dispersion techniques, are among the springboards. Bijvoet showed the way to determination of absolute configurations; the $p_s(S)$ function was not essential to that end. I have discussed it here, and related matters, because they are branches on Patterson's tree and he took interest in them, and because they can be used advantageously today.

I think probably the last time Lindo and I were together was at the Villanova meeting of the ACA in June, 1962. We spoke of anomalous dispersion then. He and his associates (including Jenny Glusker) reported their progress on the structure of potassium dihydrogen isocitrate, in which analysis the $p_s(S)$ function was of value. That was over five years before Lindo's death. But the tragic events at Penn State, which within a two-year period led to the departure of almost the entire Physics faculty from that institution, were already irreversibly afoot. I was the last member of that staff to leave, two years later. It was the most terrible of times, and it meant the abandonment of the laboratory I had built there and the dispersion of the splendid investigators whom I had assembled as colleagues. I could have remained only by abandoning my sense of decency and my academic principles. I tried to save what I could from the debacle, but nothing availed. One of the consequences was that I never met with Lindo again.

This symposium, honoring Lindo's memory, has permitted me to meet again with other old friends who shared love and respect for him. It has been a source of great pleasure, tempered with sadness. I again express my gratitude to Betty Patterson and Jenny Glusker for inviting me to participate.

References

1. Pepinsky, R. (ed.) *Computing Methods and the Phase Problem in X-ray Crystal Analysis*. X-ray Crystal Analysis Laboratory, Pennsylvania State College: College Park, PA (1952).
2. Pepinsky, R. and Calderon, A. The Use of Positive Kernels in Fourier Syntheses of Crystal Structures. pp. 319–338 in reference 1.
3. Hoppe, W. *Naturwiss.* **35**, 248 (1948).
4. Jahnke, E. and Emde, F. *Tables of Function*. Dover: New York (1943).
5. Boyes-Watson, J., Davidson, E. and Perutz, M. F. *Proc. Phys. Soc.* **A191**, 83 (1947).

6a. Pepinsky, R. Some Aspects of Neutron Single-Crystal Analysis. *Abstracts, Tenth Annual Pittsburgh Conference on X-Ray and Electron Diffraction, (November 6–7, 1952)* (1952).

6b. Pepinsky, R. Hydrogen-Deuterium Replacement in Neutron Diffraction Analysis. *Abstracts, Eleventh Annual Pittsburgh Conference on X-ray and Electron Diffraction (November, 1953)*. p. 32, (1953).

7. Pepinsky, R. Neutron Diffraction Studies of Hydrogen Bonding. *Acta Cryst.* **7**, 690 (1954).
8. Okaya, Y. and Pepinsky, R. New Developments in the Anomalous Dispersion Method for Structure Analysis. Paper 28. pp. 273-299. In *Computing Methods and the Phase Problem in X-Ray Crystal Analysis.* (R. Pepinsky, J. M. Robertson and J. C. Speakman, eds.) Pergamon Press: New York, Oxford, London, Paris (1961).
9. Pepinsky, R. X-Rays and the Absolute Configuration of Optically Active Molecules. *Record of Chemical Progress (Kresge-Hooker Science Library)* **17**, 145 (1956).
10. Doyne, T., Pepinsky, R. and Watanabe, T. *Acta Cryst.* **10**, 438 (1957).
11. Dauben, C. H. and Templeton, D. H. *Acta Cryst.* **8**, 841 (1955).
12. Templeton, D. H. *Dispersion Corrections for Atomic Scattering Factors.* Report UCRL-9146. University of California, Lawrence Radiation Laboratory, March 1960.
13. Pepinsky, R. and Okaya, Y. Determination of Crystal Structures by Means of Anomalously Scattered X-Rays. *Proc. Natl. Acad. Sci. USA* **42**, 286 (1956).
14. Pepinsky, R. and Calderon, A. On the Uniqueness of Positive Noncentrosymmetric Density Functions for Which the Absolute Value of the Continuous Fourier Transform is Known. *Program and Abstracts, American Crystallographic Association, joint meeting with the American Physical Society, Chicago, Ill., October 25-27, 1951* p. 10. (1951).
15. Calderon, A. and Pepinsky, R. On the Question of the Uniqueness of Solutions of the Phase Problem in Crystal Analysis. Appendix VIII to reference (1) above pp. 356–360.

30. Absolute configuration of natural products using anomalous dispersion of light atoms

Dick van der Helm and M. Bilayet Hossain

Dick van der Helm obtained his doctorate at the University of Amsterdam under Professor Caroline McGillavry. He did postdoctoral work at Indiana University with Lynne Merritt and then came to the Institute to work with Lindo Patterson from 1959–1962. Dr van der Helm then went to the University of Oklahoma where he is George Lynn Cross Research Professor in Chemistry. He is particularly interested in the structure and properties of the naturally-occurring iron-sequestering agents (siderophores) and in peptide conformation. Dr Hossain obtained his Ph.D. from Birkbeck College, University of London. He taught physics and crystallography at the University of Dhaka, Bangladesh. Dr Hossain is at present working as a Research Scientist at the University of Oklahoma.

Most natural product compounds contain asymmetric carbon atoms or other chiral centers. The biological functions of these compounds quite often display certain specificity for a particular enantiomer. Proper enantiomeric recognition of a natural product with a novel skeletal feature can lead to quick discovery of other members in the series and can provide valuable insight into their bio-genetic pathways. Knowledge of the correct absolute configuration is also essential if one desires to synthesize or modify a natural product. It is therefore desirable and at times essential to determine the absolute configuration of these compounds in addition to their gross chemical structure. It is a well established fact that X-ray crystallographic techniques provide the most effective and direct method for determining the absolute configuration of a molecule. The method, first proposed by Bijvoet[1] (1949) and successfully applied in the case of the sodium rubidium salt of (+)-tartaric acid,[2] utilized the anomalous dispersion of X-rays by the rubidium atom in the crystal structure. The anomalous dispersion effect is normally expressed by writing the atomic scattering factor of a dispersive atom as a complex number:

$$f_d = f + f'_d + i f_d''.$$

This consideration results in intensity differences of a Friedel pair, $I(hkl)$ and $I(\overline{hkl})$ for a non-centrosymmetric crystal. These differences, known as Bijvoet differences, can be experimentally measured and compared with the calculated Bijvoet differences obtained on the basis of a given enantiomer. During the early years the application of the Bijvoet method was limited to heavy atom (large f''_d) structures only, for which the Bijvoet differences are large and easily measurable. In fact at that time the solution of a light atom structure itself was a problem. The normal strategy involved heavy atom derivatives of light atom molecules including those of natural product compounds. In our own laboratory, a few early structures of marine natural products and their absolute configurations were determined from their heavy atom derivatives, *i.e.*, eunicin from eunicin iodoacetate,[3] crassin from crassin iodobenzoate[4] and β-gorgonene from its $AgNO_3$ adduct.[5] Although the introduction of a heavy atom facilitates both the structure solution and the determination of absolute configuration, there is always the possibility that in the synthesis of a derivative, the structure of the natural product is altered in a fundamental way. With the advent of direct methods for crystal structure determination, heavy atom derivatives were no longer essential for the structure solution. Simultaneously, groups of researchers embarked on developing procedures to determine the absolute configuration of light atom molecules by using the Bijvoet method. Hope and de la Camp[6] showed that using the very small anomalous scattering power of the oxygen atom one can successfully determine the absolute configuration of a light atom structure. The procedure they followed is related to the 'R method' of Hamilton.[7] The structure of (+)-tartaric acid, $C_4H_6O_6$, was refined to convergence for both the $(2R, 3R)$ and $(2S, 3S)$ enantiomers and final R factors were evaluated using 75 pairs of 'enantiomer sensitive' reflections (ones with the highest $[F_c^2(h) - F_c^2(\bar{h})]/\sigma(F_o)$) giving $R = 0.0328$ for $(2R, 3R)$ and $R = 0.0353$ for $(2S, 3S)$ respectively. A similar modified R method was successfully used by others[8–10] for determining the absolute configuration of light atom structures. Engel[11] determined the absolute configuration of seven compounds containing only C, H and O atoms by a more direct use of the Bijvoet method. Intensities of Bijvoet pairs for each compound were carefully measured using CuKα radiation, and were corrected for absorption. From the successful results for these compounds he concluded that for normal light atom structures the Bijvoet method is more effective than the R method and that the optimum O:C ratio for success is around 0.46, although the absolute configuration of estrone (O:C ratio = 0.16) was also successfully determined. Karlsson[12] reported success for a natural product with a very low O:C ratio (0.067), where he applied a very elaborate and tedious experimental procedure for the measurement of the Bijvoet differences. Rabinovich and Hope[13] have shown that with careful observations, the absolute configuration of a molecule with even a lower O:C ratio (0.059 for 4,4′-dimethylchalcone) can be determined with some confidence. The success in such an extreme case was attributed to excellent crystal quality, perfect diffractometer performance and the data collection procedure (data of a complete set of Friedel pairs rather than the general set of Bijvoet related pairs were collected). On the related subject, Vos[14] explained the possible sources of error for Bijvoet differences and methods for reducing them. A discussion of the methods needed for high-quality data acquisition is given by Hope.[15]

During the structure investigation of natural product compounds in our laboratory we have gradually developed a relatively simple and fast procedure for determining the absolute configuration of light atom structures which is routinely used in relevant cases. The procedure has gone through some modifications over the years but it basically incorporates the following general steps: (i) The structure (one of the two enantiomers) is refined without taking anomalous scattering into consideration, using low-temperature data ($\approx$ 138K). After the refinement has converged, $F_c^2(hkl)$ and $F_c^2(\overline{hkl})$ are evaluated using the anomalous dispersion effect of oxygen (also carbon atoms for Cr-radiation) from an expression similar to one given by

Patterson[16]:

$$\begin{aligned} |F_c(\pm)|^2 &= A^2 + B^2 + 2A\sum_r (H_{dr}f'_{dr} - K_{dr}f''_{dr}) \\ &+ 2B\sum_r (K_{dr}f'_{dr} + H_{dr}f''_{dr}) \\ &- 2\sigma\sum_r\sum_s (H_{dr}f'_{dr}K_{ds}f''_{ds} - K_{dr}f'_{dr}H_{ds}f_{ds}'') \\ &+ \sum_r\sum_s (f'_{dr}f'_{ds} + f''_{dr}f''_{ds})(H_{dr}H_{ds} + K_{dr}K_{ds}) \end{aligned}$$

where, $F_c(+) = F_c(hkl)$ and $F_c(-) = F_c(\overline{hkl})$, A and B are structure factor components for the non-dispersive part of the structure, H_{dr} and K_{dr} are the geometrical components of the rth dispersive atom, $\sigma = 1$ for $F(+)$ and $\sigma = -1$ for $F(-)$ and f'_{dr} and f''_{dr} are the real and imaginary corrections of the scattering factor due to the anomalous dispersion. (ii) The calculated Bijvoet difference, DEL $= (F_c^2(+) - F_c^2(-))/0.5(F_c^2(+) + F_c^2(-))$ is evaluated for all the reflections. As most of these values are small, a selection procedure is employed to find the most enantiomer-sensitive reflections. This is done by evaluating a weighted Bijvoet difference, $(\mathrm{ESF})_{calc}$ (enantiomer sensitive factor) $= \mathrm{DEL}/\sigma(F_o^2)$, where $\sigma(F_o^2) = 2F_o\sigma_F$ and σ_F is obtained from counting statistics. (iii) A number of Friedel pairs (generally 15–20) are selected on the basis of highest $(\mathrm{ESF})_{calc}$ values and their intensities are measured at low temperature (≈ 138K) repeatedly (10–15 times each). For each pair, the intensities of the n-fold related pair (where applicable) are also recorded. These intensities are then averaged and a standard deviation, $\sigma(I)$ is evaluated. $\sigma(I)$ reflects the reproducibility of the measurements and is given by,

$$\sigma(I) = [\frac{1}{(N-1)}\sum_{j=1}^{N}(I_j - I)^2]^{1/2}.$$

N is the number of measurements of a single reflection. (iv) Bijvoet differences (DEL) and weighted differences (observed $(\mathrm{ESF})_{obs} = \mathrm{DEL}/<\sigma(I)>$, $<\sigma(I)> = (\sigma(I)_h + \sigma(I)_{\bar{h}})/2$, are calculated for each of the selected Friedel pairs and these $(\mathrm{ESF})_{obs}$

values are compared with those obtained from the calculated structure factors $(\text{ESF})_{calc}$. (v) The anomalous scattering factors of Cromer and Liberman[17] are normally used in the calculations: $f_0' = 0.047, f''_0 = 0.032$ for Cu-radiation and $f_0' = 0.090, f''_0 = 0.073, f_C' = 0.035$ and $f''_C = 0.021$ for Cr-radiation.

We are, as a part of our research, interested in the structures of natural products from marine sources, primarily because of their potential anticancer activity. The importance of determining the absolute configuration of these compounds is outlined briefly at the beginning. Sometimes there is special interest in the absolute configuration of a certain group of compounds. One such group is a class of diterpenes known as cembranes or cembranoids. In fact the majority of the marine compounds whose structure and absolute configurations were determined in our laboratory are cembranoids or their lactone derivatives, cembranolides. They are all isolated from marine coelenterates and all possess the carbon skeleton of the compound cembrane (I). It is generally accepted that the cembrane skeleton originates from cyclization of geranylgeranylpyrophosphate (II) and that initial cyclization can yield two antipodal

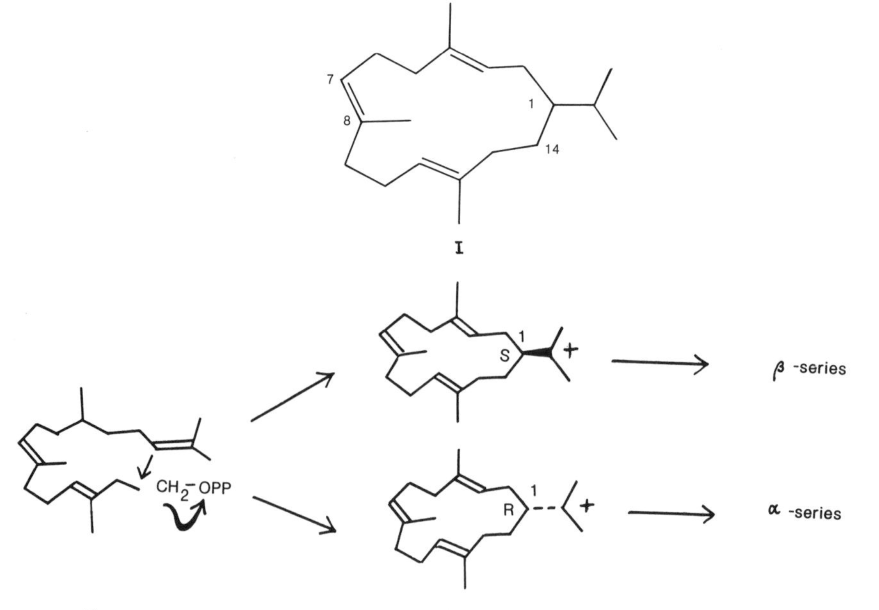

ions.[18] The *S* antipode leads to a series of compounds with a configuration in which the isopropyl group is β (pointing upward), when the (7-8) double bond is located as shown.[19] The *R* antipode leads to a series of compounds in which the isopropyl group is pointing downward and form the α-series. The results of structure and absolute configuration determinations in our laboratory and other places have shown that all cembranes from the order 'alcyonacea' (abundant in the Indo-Pacific region) are members of the α-series. This is also the case for all cembrane derivatives isolated from terrestrial plants. On the other hand, all cembranes isolated from the order 'gorgonacea' (abundant in the Tropical Western Atlantic region) belong to the β-series. Some examples of β-cembranolides [jeunicin (**III**), eupalmerin acetate (**IV**), peunicin (**V**) and cueunicin (**VI**)] are shown in Fig. 30.1. Fig. 30.2 shows some of the α-cembranolides [sinularin (**VII**), dihydrosinularin (**VIII**) and sinulariolide (**IX**)[20]]. The existence of the two series can be explained by the presence of two different enzymatic systems for cyclization of the cembrane skeleton.

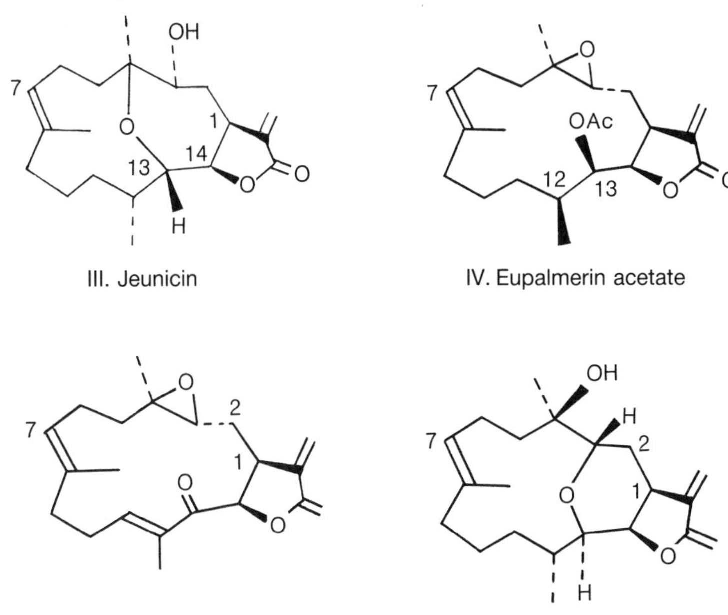

FIG. 30.1. Examples of β-cembranolides.

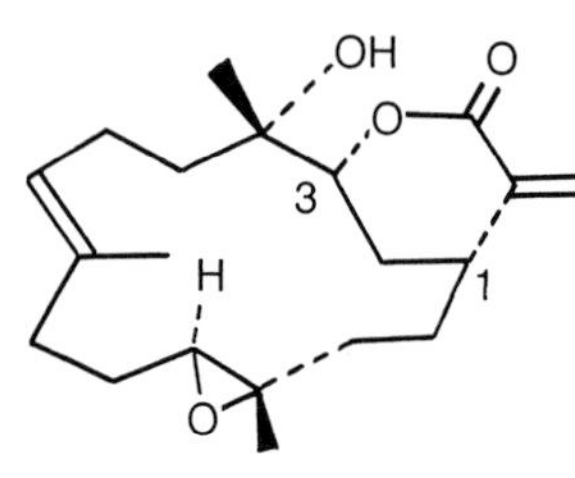

VII. Sinularin

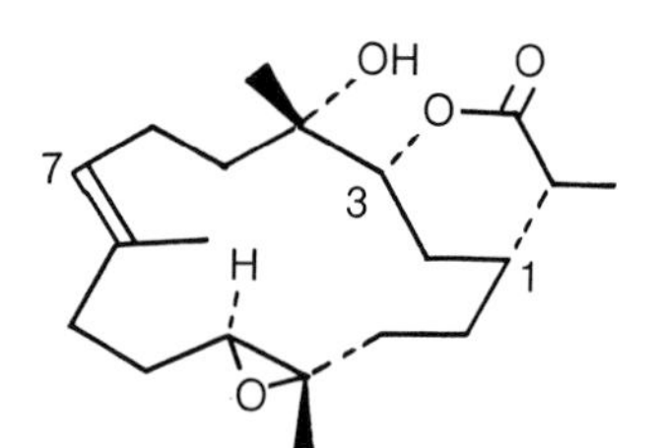

VIII. Dihydrosinularin

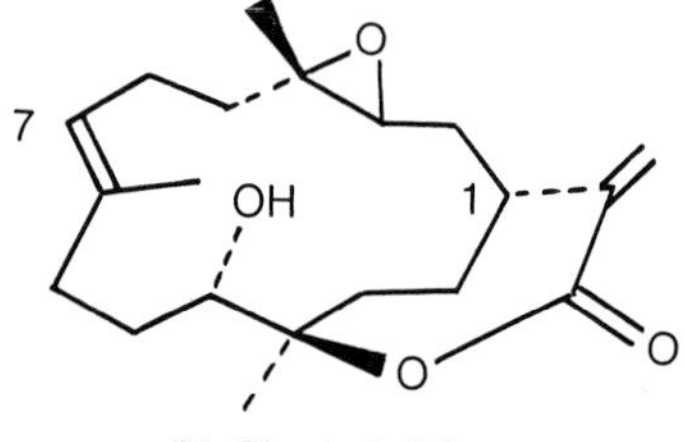

IX. Sinulariolide

FIG. 30.2. Examples of α-cembranolides.

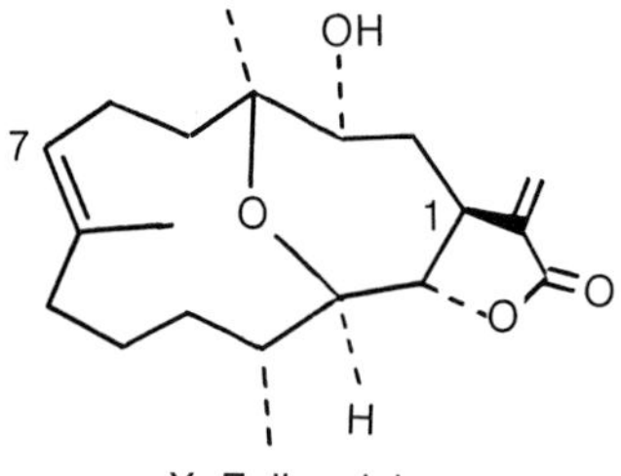

X. Epijeunicin

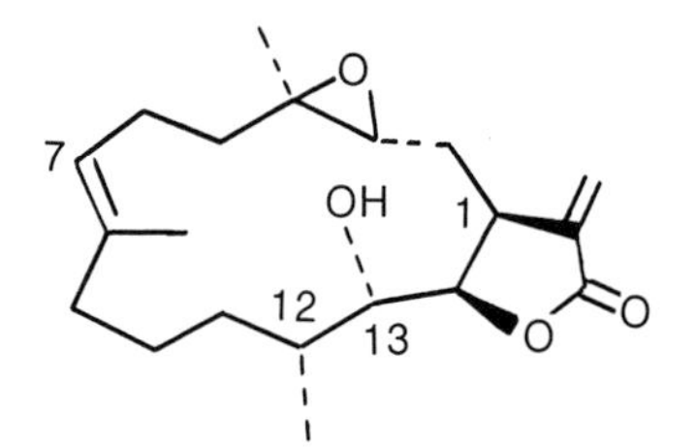

XI. Epieupalmerin

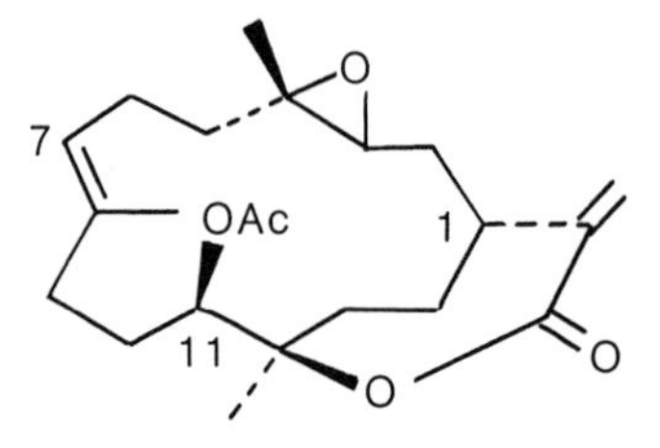

XII. Episinulariolide

FIG. 30.3. Examples of epimeric isomers.

Another common configurational feature of the cembranoids is the tendency of these compounds to form epimeric pairs, isomers with one or two epimeric centers as seen in the case of jeunicin and epijeunicin (**X**), eupalmerin acetate and epieupalmerin (**XI**) and sinulariolide and episinulariolide (**XII**) [Fig. 30.3]. As all of these compounds can be derived from cembrane by various oxygenation reactions, knowing the structure and absolute configuration of one or more members of the family could give valuable information about other possible structures. The existence of many of the isomers mentioned above were actually predicted long before their discovery from the structure and absolute configuration of eupalmerin acetate.[21] The structure and absolute configuration of xenicin (**XIII**),[22] a novel bicyclic diterpene with a cyclononene ring, led to the quick discovery of a large number of compounds with similar backbone.[23,24]

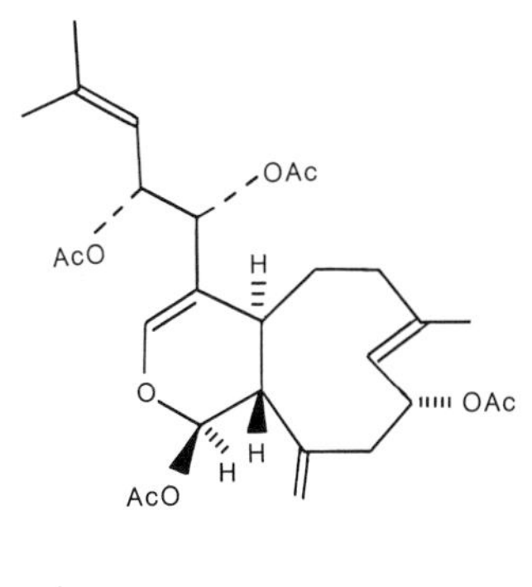

XIII

Table 30.1 gives a list of some of these compounds whose absolute configurations have been determined in our laboratory by using the Bijvoet method and employing only light atom anomalous scattering. Their oxygen contents vary, O:C ratios range between 0.15 in asperdiol A to 0.34 in tedanolide and they are in general much lower than the optimum value (0.46) quoted by Engel.[11] All the structures are fairly well refined with R-factors ranging between 0.032 and 0.063. The agreement factors show that in general 80-100% of the observations indicate one enantiomer over the other. It should be noted that in most cases, where disagreements were observed, the majority of these had lower ESF values. For eupalmerin

TABLE 30.1. List of marine compounds

Molecule	Formula	Space Group	*R*-factor (No. refls)	O:C Ratio	Agreement Factor n(N) *	%	Ref.
1. Xenicin	$C_{28}H_{38}O_9$	$C2_1$	0.046(3015)	0.32	15(20)Cu	75	22
					15(18)Cr	83	
2. Eupalmerin acetate	$C_{22}H_{32}O_5$	$P2_12_12_1$	0.033(2433)	0.23	11(15)	73	21
3. Asperdiol A	$C_{20}H_{32}O_3$	$P2_1$	0.046(2021)	0.15	19(19)	100	26
4. Sinularin	$C_{20}H_{30}O_4$	$P2_12_12_1$	0.036(2221)	0.20	15(15)	100	27
5. Dihydro-sinularin	$C_{20}H_{32}O_4$	$P2_1$	0.043(2087)	0.20	14(15)	93	27
6. Peunicin	$C_{20}H_{26}O_4$	$P2_12_12_1$	0.038(2059)	0.20	14(15)	93	28
7. Plexaurolone acetate	$C_{22}H_{36}O_4$	$P2_1$	0.053(4851)	0.18	17(20)	85	29
8. Epijeunicin	$C_{20}H_{30}O_4$	$P2_12_12_1$	0.063(4191)	0.20	16(18)Cu	89	30
9. Bisepi-eupalmerin	$C_{20}H_{30}O_4$	$P2_1$	0.050(1969)	0.20	22(24)	92	31
10. Tedanolide	$C_{32}H_{50}O_{11}$	$P2_12_12_1$	0.045(2996)	0.34	15(18)	83	32
11. Spongiane	$C_{24}H_{36}O_6$	$P4_1$	0.032(2152)	0.25	16(24)	65**	25

* N: number of Friedel pairs used.
n: number of Friedel pairs agreed with a particular enantiomer.
** no conclusion regarding the enantiomers was made.

(EPA), out of the four pairs that disagreed with the proper enantiomer, three showed no measurable difference in intensities, $(ESF)_{obs} = 0$. The result of EPA was a test case for our method because the absolute configuration of EPA was already determined with greater confidence from a heavy-atom derivative, EPA-dibromide.[21] One of the worst agreements (66%, it was actually concluded to be a failure) was seen in the case of a spongiane derivative (last structure in Table 30.1) even though the structure had the lowest *R* factor and a reasonable O:C ratio. The failure is attributed to two factors, (i) poor crystal quality for the Bijvoet measurements and (ii) symmetry related Bijvoet pairs were not measured. Normally we use fresh crystals for Friedel pair measurements because the crystals of natural product compounds are frequently unstable and there is always a time lag between original data taking and the stage when one is ready to determine the absolute configuration. In the case of the spongiane structure, due to lack of fresh materials, the original data crystal (which deteriorated over a period of six months) was used for the Friedel pair measurements.

A typical set of final results of our procedure is given in Table 30.2. These are for the compound tedanolide, a 18-membered macrolide isolated from the Caribbean sponge,

TABLE 30.2. Tedanolide: absolute configuration

Observed and Calculated Bijvoet Differences			DEL		ESF	
h	*k*	*l*	obs.	calc.	obs.	calc.
2	6	1	0.67	-2.00	0.37	-1.02
3	2	1	-2.59	2.24	-1.57	1.18
5	5	1	2.06	-2.97	0.50	-1.12
7	1	1	-0.90	2.86	-0.26	1.20
8	3	1	-1.83	3.61	-0.62	1.55
11	4	1	12.06	-8.28	0.83	-1.28
11	9	1	-5.04	5.56	-0.56	1.49
2	13	2	-7.95	5.57	-1.50	1.99
5	4	2	-2.27	-2.76	-0.91	-1.23*
8	8	2	0.84	-4.05	0.12	-1.20
2	10	3	1.51	-6.14	0.13	-1.34
5	8	3	-1.73	3.26	-0.50	1.36
2	16	1	1.96	-5.57	0.26	-1.59
12	6	1	-0.61	3.36	-0.12	1.12
2	12	2	4.60	-3.69	0.86	-1.20
3	3	3	3.45	-2.55	1.06	-1.19
5	2	4	5.81	3.24	1.06	1.14*
6	9	4	1.70	2.70	0.42	1.12*

* denotes an alternate indication of absolute configuration.

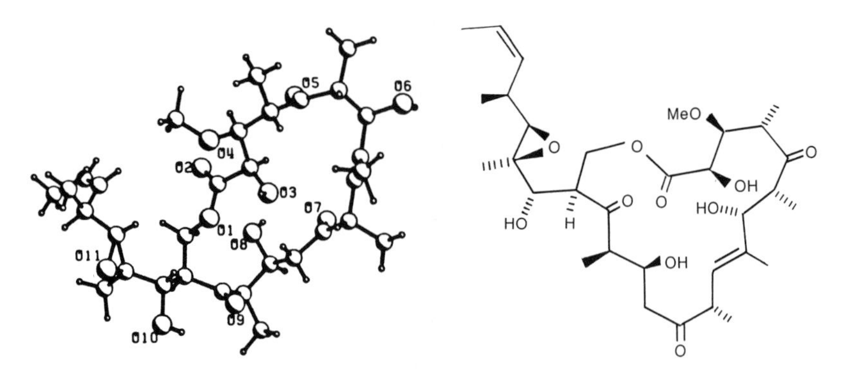

FIG. 30.4. Molecular structure of tedanolide.

Tedania ignis. Tedanolide is highly cytotoxic, exhibiting an ED_{50} of 2.5×10^{-4} μg/ml in KB and 1.6×10^{-5} in PS and is a potential tumor inhibitor. The molecular structure of tedanolide is shown in Fig. 30.4.[25] The large oxygen rich cavity in the molecule appears to make it suitable as an ionophore. It is also a prime target for laboratory synthesis. The relatively high O:C ratio and low R factor make it a good candidate for absolute configuration determination. Eighteen Friedel pairs with highest ESF values were selected, and intensities of hkl, $\bar{h}k\bar{l}$, $\overline{hkl}$ and $h\bar{k}l$ were measured at 138 K repeatedly (15 times for each reflection). Intensities were then averaged, hkl and $\bar{h}k\bar{l}$ on one side, and $h\bar{k}l$ and $\overline{hkl}$ on the other. The final results were obtained following the procedure outlined earlier. Out of the eighteen, 15 pairs indicated the absolute configuration shown in the figure (which was the opposite to the enantiomer refined initially), and 3 pairs indicated the opposite configuration. The results, 83% agreement with one enantiomer, is quite typical.

For two structures, xenicin and epijeunicin, the absolute configuration determination was repeated with Cr-radiation. While for Cu-radiation, only the oxygen anomalous scattering effect was included into the calculation, for Cr-radiation the anomalous scattering of carbon atoms was also used. Although it was expected that the use of Cr-radiation would greatly enhance the results because of larger $\Delta f''$ values (0.073 for Cr compared to 0.032 for Cu) the final results show only marginal improvement. This is probably caused by the greater absorption of Cr-radiation.

The use of light atom anomalous scattering effect in determining absolute configuration shows that the absolute configuration of molecules containing only O, C and H can be obtained with good reliability by using the Bijvoet method properly. The success and the degree of reliability of the results depend on several factors. Very elaborate and at times tedious procedure for data collection and data processing could ensure success even in cases where there are few oxygen atoms in the molecule. We have been using a procedure which is relatively simple which requires very little extra effort or tricks beyond the normal structure determination/refinement routine. The procedure uses a reasonable method to select the most enantiomer-sensitive Friedel pairs by calculating the ESF (enantiomer sensitive factor) for all reflections from a well-refined structure (low R factor). Three factors ensure the accuracy in the intensity measurements of the Friedel pairs (hence the Bijvoet differences): (i) data collection at low temperature; (ii) repeated measurements of a single reflection; and (iii) measurement of symmetry (n-fold) related pairs and averaging of intensities. Although there are a number of areas in which the procedure could be improved, it has proved in its present form to be a practical method which can be routinely used with success for normal light atom structures.

References

1. Bijvoet, J. M. *Proc. Acad. Sci. (Amst.)* **52**, 313 (1949).
2. Bijvoet, J. M., Peerdeman, A. F. and van Bommel, A. J. *Nature (London)* **168**, 271 (1951).
3. Hossain, M. B., Nicholas, A. F. and van der Helm, D. *J. Chem. Soc., Chem. Comm.* 385 (1968).
4. Hossain, M. B. and van der Helm, D. *Recueil. Trav. Chim. Pays-Bas* **88**, 1413 (1969).
5. Hossain, M. B. and van der Helm, D. *J. Am. Chem. Soc.* **90**, 6607 (1968).
6. Hope, H. and de la Camp, U. *Nature (London)* **221**, 54 (1969).
7. Hamilton, W. C. *Acta Cryst.* **18**, 502 (1965).

8. Thiessen, W. E. and Hope, H. *Acta Cryst.* **B26**, 554 (1970).
9. Moncrief, J. W. and Sims, S. P. *J. Chem. Soc., Chem. Comm.* 914 (1969).
10. Neidle, S. and Rogers, D. *Nature (London)* **225**, 376 (1970).
11. Engel, D. W. *Acta Cryst.* **B28**, 1496 (1972).
12. Karlsson, R. *Acta Cryst.* **B32**, 2609 (1976).
13. Rabinovich, D. L. and Hope, H. *Acta Cryst.* **A36**, 670 (1980).
14. Vos, A. In *Anomalous Scattering* (S. Ramaseshan and S. C. Abrahams, eds.) p. 307. Munksgaard: Copenhagen (1975).
15. Hope, H. In *Anomalous Scattering* (S. Ramaseshan and S. C. Abrahams, eds.) pp. 293. Munksgaard: Copenhagen, Denmark (1975).
16. Patterson, A. L. *Acta Cryst.* **16**, 1255 (1963).
17. Cromer, D. T. and Liberman, D. *J. Chem. Phys.* **53**, 1891 (1970).
18. Tursch, B., Braekman, J. C., Daloze, D. and Kaisin, M. *Terpenoids from Coelenterates.* In *Marine Natural Products.* Vol. **II**, (P. J. Scheuer, ed.) pp. 247. Academic Press: New York (1978).
19. Weinheimer, A. J., Matson, J. A., Hossain, M. B. and van der Helm, D. *Tetrahedron Lett.*, 2923 (1977).
20. Karlsson, R. *Acta Cryst.* **B33**, 2027 (1977).
21. Ealick, S. E., van der Helm, D. and Weinheimer, A. J. *Acta Cryst.* **B31**, 1618 (1975).
22. Vanderah, D. J., Steudler, P. A., Ciereszko, L. S., Schmitz, F. J., Ekstrand, J. D. and van der Helm, D. *J. Am. Chem. Soc.* **99**, 5780 (1977).
23. Braekman, J. C., Daloze, D., Tursch, B., Declercq, J. P., Germain, G. and van Meerssche, M. *Bull. Soc. Chim. Belg.* **88**, 71 (1979).
24. Kashman, Y. and Groweiss, A. *J. Org. Chem.* **45**, 3814 (1980).
25. Schmitz, F. J., Chang, J. S., Hossain, M. B. and van der Helm, D. *J. Org. Chem.*, in press.
26. Weinheimer, A. J., Matson, J. A., van der Helm, D. and Poling, M. *Tetrahedron Lett.*, 1295 (1977).
27. Hossain, M. B., van der Helm, D., Matson, J. A. and Weinheimer, A. J. *Acta Cryst.* **B35**, 660 (1979).
28. Chang, C. Y., Ciereszko, L. S., Hossain, M. B. and van der Helm, D. *Acta Cryst.* **B36**, 731 (1980).
29. Ealick, S. E., van der Helm, D., Gross, Jr., R. A., Weinheimer, A. J., Ciereszko, L. S. and Middlebrook, R. E. *Acta Cryst.* **B38**, 580 (1982).
30. Weinheimer, A. J., Matson, J. A., Poling, M. and van der Helm, D. *Acta Cryst.* **B38**, 580 (1982).

31. Gopichand, Y., Ciereszko, L. S., Schmitz, F. J., Switzner, D., Rahman, A., Hossain, M. B. and van der Helm, D. *J. Nat. Products* **47**, 607 (1984).
32. Schmitz, F. J., Gunasekera, S. P., Gopichand, Y., Hossain, M. B. and van der Helm, D. *J. Am. Chem. Soc.* **106**, 7251 (1984).

B. THE PATTERSON FUNCTION APPLIED TO ADVANCES IN THE BIOLOGICAL SCIENCES

31. Model building and structural studies for the DNA-binding protein cII

Mitchell Lewis, Yen-Sen Ho and Martin Rosenberg

A unique model has been developed to explain how DNA binding proteins recognize direct repeat sequences of DNA. In addition, preliminary structural data are presented for the DNA binding protein, *c*II, which recognizes these direct repeats.

The structures of three DNA binding proteins have been determined that recognize DNA sequences with pseudo two-fold axes of symmetry (*i.e.,* inverted repeat sequences). These proteins are oligomeric with dyad axes of symmetry. Model building studies of these proteins complexed to DNA suggest that the two-fold axis of the protein is coincident with the pseudo dyad axis of the DNA.[1–3] A comparison of the three-dimensional structures and the amino acid sequences of several DNA binding proteins suggests that a conserved supersecondary structure, which contains an α helix, a turn, and a second α helix, forms the DNA binding site for this class of proteins .[4–6] The phage λ regulatory protein, *c*II, shares sequence homology with this group of DNA binding proteins and also apppears to have the two-helix motif. The *c*II protein is unique in that it recognizes a tetranucleotide direct repeat sequence rather than an inverted repeat (Fig. 31.1). The four base pair repeat units

```
-40       -30        -20         -10      -1

CTTGCGTGTAATTGCGGAGACTTTGCGATGTACTTGACAC      PI
GTTGCGTTTGTTTGCACGAACCATATGTAAGTATTTCCTT      λ PRE
CTTGCGAGTGCTTGTGAGTTCCATATGTGAACATTCGTGT      P22 PRE
CTTGCTGTAGCTTGCTTGTGCCATTTGTTAATTTACCTAC      21 PRE
```

FIG. 31.1. The activator binding sites for the three lamboid phage activator proteins. The consensus −35 hexamer region recognition sequence is indicated by the solid double lines. The direct repeat tetranucleotide sequences are underlined.

are separated by six base pairs and are thus aligned on the same side of a right handed B-type DNA molecule exactly one turn apart.[7] Assuming that the helix-turn-helix motif that appears characteristic of DNA binding proteins is preserved in *c*II, model building studies were performed to determine how this particular structural unit would recognize a direct repeat DNA sequence.

*c*II is a transcriptional regulatory protein that increases the binding affinity of the RNA polymerase for certain promoters.[8] This transcriptional activator binds to homologous sequences in the −35 region of these promoters.[7] A model of the binding helix of the *c*II protein was constructed by replacing the side chain residues of *c*I with those of *c*II and a three-dimensional representation of the *c*II activator site was built in a right handed B-type conformation. A schematic drawing of two binding helices complexed to the DNA is illustrated in Fig. 31.2. As can be seen, the two related *c*II binding helices interact with the DNA in different ways. The lysine (residue 37) at the N-terminus of the binding helix of one molecule is positioned to interact with the same base sequence as the arginine (residue 42) of the second molecule. Each recognition helix has an approximate two-fold axis of symmetry when the arginine and lysine have similar electrostatic attraction for bases of DNA. Lysine 37 and serine 38 are related to serine 41 and arginine 42 by a 180° rotation about an axis perpendicular to the plane of the page and the helical motifs are related by a second two-fold axis. Thus, two parallel dyad axes result in a single translational repeat.

Mutations in the promoter region which eliminate *c*II binding fall into two small clusters positioned six base pairs apart .[7] Methylation protection experiments have shown that the guanine residues occuring in the two TTGC repeat sequences were in close association with the *c*II protein.[7] The two pairs of protected guanine residues are aligned on one face of the DNA, located in the major groove, and related by a dyad axis in the center of the binding site. The recognition helix of *c*II was positioned so that the two serine residues (38 and 41) could donate their hydroxyl protons to the N(7) position of the protected guanine residues. This model constrains the DNA binding helix to be parallel to the major groove similar to that determined

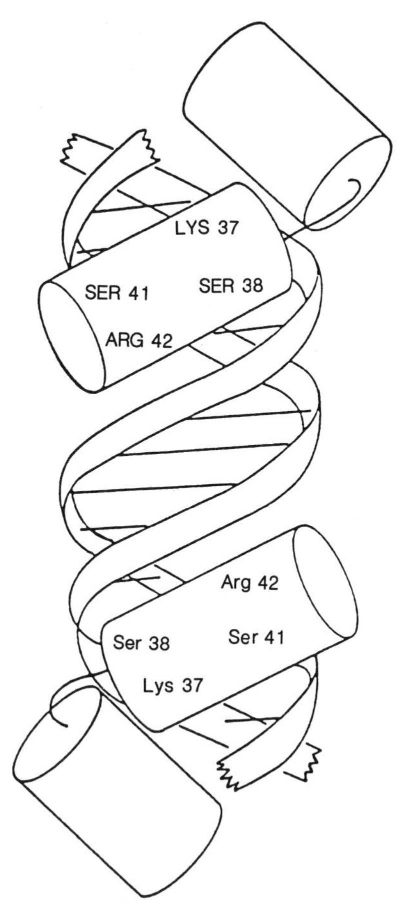

FIG. 31.2. A schematic diagram showing the interaction of the protein helices with DNA. The protein binding helices are related by a two-fold axis of symmetry.

by model building studies of the Cro-operator complex.[9] The side chains of the lysine and arginine were then positioned to optimize favorable electrostatic interactions. If the *c*II protein were to interact with the DNA in the same manner as proposed for *c*I and Cro, then the electrostatic environment that accommodates the lysine residue 37 in one half of the binding site would have to also accommodate the arginine 42 in the second site. The binding of *c*II to DNA is inherently non-symmetric such that the protein should have different affinities for the different half sites which reflects the difference in the binding of the lysine and the arginine. This prediction is fully consistent with the experiments indicating that *c*II binds more tightly to the upstream half of the activator site.[10]

Three related lambdoid phages (λ, P22, and P21) carry homologous *c*II activator genes. Pairwise sequence alignments of these proteins suggest that they are related to each other as well as to the phage repressor and Cro proteins; all sharing common structural features (Fig. 31.3). The three activator proteins are functionally analogous and recognize a common promoter region, P_I (Fig. 31.1), although possibly with dif-

			30					35					40					45
λcII	THR	GLU	LYS	THR	ALA	GLU	ALA	VAL	GLY	VAL	ASP	LYS	SER	GLN	ILE	SER	ARG	TRP	LYS	ARG
p22	GLN	ARG	LYS	VAL	ALA	ASP	ALA	LEU	GLY	ILE	ASN	GLU	SER	GLN	ILE	SER	ARG	TRP	LYS	GLU
21	GLN	ARG	GLY	LEU	ALA	LYS	MET	ILE	GLY	CYS	HIS	GLU	SER	LYS	ILE	SER	ARG	THR	ASP	TRP
				HELIX 2					TURN					HELIX 3						
λ R	GLN	GLU	SER	VAL	ALA	ASP	LYS	MET	GLY	MET	GLY	GLN	SER	GLY	VAL	GLY	ALA	LEU	PHE	ASN
CRO	GLN	THR	LYS	THR	ALA	LYS	ASP	LEU	GLY	VAL	THR	GLN	SER	ALA	ILE	ASN	LYS	ALA	ILE	HIS

FIG. 31.3. Protein sequences of the activator proteins in the region of helices 2 and 3 of repressor and Cro (the DNA binding helices).

ferent affinities. In addition, λ's *c*II will recognize, bind and activate transcription from the P22 activator site and P22 *c*I will recognize λ's recognition site[12] albeit less efficiently than for their homologous sites. The DNA recognition helix of the λ, P22, and P21 activator proteins differ only at their N-terminus. λ has a lysine whereas the other two activators have a glutamic acid side chain at the corresponding position (Fig. 31.3). This amino acid side chain could recognize a dinucleotide sequence. All four activator sites have dinucleotide sequences that are capable of donating two protons to a glutamic acid side chain when the protein forms a complex with DNA. For example, in the P_I activator site the TG dinucleotide (–37,–38) and the CG dinucleotide (–25,–26) (Fig. 31.1) both could interact with a glutamic acid. The TG dinucleotide could donate 2 protons, one from the N(6) of the adenine in the TA base pair and a second hydrogen bond from the N(4) of the first cytosine in the CG base pair and a second hydrogen from the N(4) of cytosine in the GC base pair. In fact, all eight half sites have dinucleotides in the appropriate positions to interact with the glutamic acid side chain, and the dinucleotide hydrogen bond donors are all 4.5 ± 0.2 Å apart. An analogous argument would apply to the interaction of λ's *c*II lysine residue 37 with the negative electrostatic environment of the bases −37, −38 and −24, −25.

Positive activation by *c*II can be modeled by analogy with other activator repressor molecules. The *c*I gene product of λ and the homologue of P22 stimulate transcription by directly interacting with RNA polymerase.[12] If *c*II also recognized DNA using a pair of helices, then a model for positive control can be postulated from the known position of the activator binding site with respect to the promoter site. The operator sites of λ's *c*I are well upstream from the −35 region hexamer recognition sequence for RNA polymerase, whereas the operator site of P22 and the activator sites of *c*II straddle the −35 region recognition sequence. The model for the *c*II tetramer-DNA interaction and the known relationship of the *c*II and RNA polymerase binding sites immediately suggest that λ's *c*II, like the P22 repressor, would stimulate transcription by interacting with the polymerase at the carboxyl end of the DNA binding helix.

Structural studies of the *c*II protein provide an experimental model to test our predictions, further our understanding of protein-DNA interaction, and provide a more detailed model of positive control. The *c*II protein has been obtained in large quantities and in pure form. Crystals have been produced by the standard hanging drop method using DMSO as a precipitation agent. These crystals have been characterized and the space group determined to be $C222_1$ with unit cell dimensions 65.4, 56.5, 127.5 Å. Preliminary data photographs were taken from crystals that diffract to a nominal resolution of 3 Å units. Model building studies of the three analogous lambdoid activator proteins provide an ideal system for studying protein-DNA interactions. The results reported here suggest that the base composition of the DNA binding site is important only with respect to the spatial arangement of its plausible hydrogen bond donors and acceptors. There is naturally a great deal of redundancy in the DNA such that a given protein may recognize very different base sequences, but we anticipate that the pattern of hydrogen bonding interactions will always be conserved.

Acknowledgements

We thank Professor Steven Harrison for useful discussion and access to his X-ray facilities.

References

1. Anderson, W. F., Ohlendorf, D. H., Takeda, Y. and Matthews, B. W. *Nature (London)* **290**, 754–758 (1981).
2. McKay, D. B. and Steitz, T. A. *Nature (London)* **290**, 744–749 (1981).
3. Pabo, C. O. and Lewis, M. *Nature (London)* **298**, 443–447 (1982).
4. Anderson, W. F., Takeda, Y., Ohlendorf, D. H. and Matthews, B. W. *J. Mol. Biol.* **159**, 745–751 (1982).
5. Sauer, R.T., Yocum, R. R., Doolittle, R. F., Lewis, M. and Pabo, C. O. *Nature (London)* **298**, 447–451 (1982).
6. Ohlendorf, D. H., Lewis, W. F., Lewis, M., Pabo, C. O. and Matthews, B. W. *J. Mol. Biol.* **169**, 757–769 (1983).
7. Yen-Sen, H., Wulff, D. and Rosenberg, M. *Nature (London)* **304**, 703–708, (1983).
8. Shimatake, H. and Rosenberg, M. *Nature (London)* **292**, 703–709 (1981).
9. Matthews, B.W., Ohlendorf, D. H., Anderson, W. F., Fisher, R. G. and Takeda, Y., *Cold Spring Harbour Symp. Quant. Biol.* **47**, 427–433 (1983).
10. Ho, Y-S. and Rosenberg, M. Unpublished.
11. Keilty, S., Ho, Y-S. and Rosenberg, M. Unpublished.
12. Hochschild, A., Irwin, N. and Ptashne, M. *Cell* **32**, 319–325, (1983).

32. Active site homology in iron and manganese superoxide dismutases

William C. Stallings, Katherine A. Pattridge, Roland K. Strong, Martha L. Ludwig, Fumiyuki Yamakura, Toshiaki Isobe and Howard M. Steinman

Three metalloproteins observed to catalyze the dismutation of superoxide (Reaction [1]) have been well studied;[1–3]

$$2O_2^- + 2H^+ \rightarrow O_2 + H_2O_2. \qquad [1]$$

These superoxide dismutases have been distinguished, *inter alia*, by the active site metal required for catalysis. Fridovich first recognized that the copper protein, erythrocuprein, possessed superoxide dismutase activity,[4] and he subsequently isolated and characterized dismutases containing iron[5] and manganese.[6] Cu/Zn superoxide dismutases, with Cu^{2+} acting as the reducible catalyst but Zn^{2+} also bound at the active center, are isolated from eucaryotic cells as dimers of identical 16,000 MW subunits. The iron enzymes are usually isolated as dimers of identical 20,000 MW subunits from procaryotes (but see refs. 7–9), and manganese dismutases are also multimers of identical 20,000 MW subunits. In general the mitochondrial manganese enzymes are tetrameric and the bacterial enzymes dimeric, but the thermophile, *T. thermophilus*, which synthesizes a manganese dismutase that assembles as a tetramer, provides one exception to this generalization.

The structures of representatives from these classes of superoxide dismutases have been investigated by X-ray crystallographic methods at varying resolutions. Bovine erythrocyte Cu/Zn superoxide dismutase has been refined at 2.0 Å resolution by the Richardsons and their coworkers;[10] its backbone is folded into a β-barrel with only a short segment of α-helix in one of its three external loops. Iron superoxide dismutases from *E. coli*[11] and *P. ovalis*[12] have been characterized at about

3 Å resolution; the polypeptide chain of the iron enzyme adopts a two-domain, helix-rich structure whose carboxy-terminal domain includes three-strands of antiparallel β-sheet structure with topology: $+1, -2x$. We have also determined the structure of the tetrameric *T. thermophilus* manganese superoxide dismutase.[13] At 4.4 Å resolution our results confirmed predictions from partial sequences[14] which suggest that the subunits of iron and manganese dismutases are folded into homologous tertiary structures (cf., Fig. 32.1) unrelated to the β-barrel fold of the Cu/Zn enzyme. Our results also demonstrated that the interface of the iron dismutase dimer is preserved as one of the interfaces in a manganese dismutase tetramer (cf., Fig. 32.2).

The structure analysis of the *T. thermophilus* manganese enzyme has now been extended to a resolution of 2.4 Å and an atomic model has been constructed using computer graphics.[15] At this resolution, many details of the stereochemistry and non-covalent interactions in the active center can be discerned. In the present work we demonstrate the close structural similarity of the active centers of *T. thermophilus* manganese dismutase and *E. coli* iron dismutase, and combine known chemical sequences with our structural data to identify residues in the metal binding site of iron superoxide dismutase.

Methods

Fourier maps of *E. coli* iron superoxide dismutase at 3.1 Å resolution and *T. thermophilus* manganese superoxide dismutase at 4.4 Å resolution[11,13] were used to determine[‡] a coordinate transformation that would superimpose the two structures.[16] The application of this procedure to the comparison of the backbone folds of iron and manganese dismutase has already been described.[13] The rigid-body transformation determined from the map fitting has now been applied to the model of *T. ther-*

‡ The program for map-to-map fitting, LSQRHO, was written by W.A. Hendrickson.

mophilus manganese dismutase, and the correspondence of the resulting coordinates with the 3.1 Å electron density map of *E. coli* iron dismutase has been examined on an Evans and Sutherland graphics display system using the FRODO software.[17]

The chemical sequence of manganese superoxide dismutase from *T. thermophilus* has not been determined. Construction of the 2.4 Å model started from the sequences of *T. aquaticus* and *B. stearothermophilus* manganese dismutases,[15] but all of the *T. thermophilus* manganese dismutase residue assignments mentioned in the following discussion are conserved in the four known primary structures.[18–21] The conservation of these residues in the *T. thermophilus* sequence is confirmed by their distinctive shape in the electron density function. However, in the absence of sequence we cannot be certain about the exact numbering of the residues.

The active center model of manganese superoxide dismutase

As illustrated in Fig. 32.3, the manganic ion is coordinated by three histidines (residues 28, 83 and 169) and a carboxylate group from an aspartate residue (Asp 165). The histidines bind the metal ion via their N(3) imidazole nitrogens and Asp 165 appears to coordinate the metal ion via a single carboxyl oxygen. The coordination geometry of the oxidized, resting enzyme approximates a distorted trigonal bipyramid[15] with a water molecule occupying a fifth coordination site *trans* to one of the imidazole nitrogens (His 28). The most striking feature of the metal–ligand environment is its strongly apolar nature. The side chains which encircle the metal coordination sphere include: tryptophans 87, 131, and 167; phenylalanine 86; tyrosines 36 and 182; histidines 32 and 33; and glutamine 150. Assuming that the liganding histidines are uncharged and that Asp 165 titrates normally, at pH 7 the Mn(III)-ligand cluster bears a net charge of +2 in this apolar environment.

Several interactions involving side chains in the active center have been described. His 32 is positioned in a manner which allows its imidazole ring to stack against that of liganding His 169. The side chain amide group of Gln 150 serves as a hydrogen bonding bridge between the ring nitrogen of Trp 131 and the hydroxyl group of Tyr 36. The indole ring system of Trp 167, midway in sequence between the third and fourth ligands, may provide a surface against which the water ligand can stack; similar stacking interactions involving nucleic acid bases have been observed.[22] A more complete description of the active center geometry may be found in reference 15.

Comparison of the Mn dismutase model with the electron density of Fe dismutase

Figs. 32.3 and 32.4 show the metal center of manganese superoxide dismutase maneuvered into the electron density of *E. coli* iron dismutase by the rigid-body coordinate transformation determined as described above; no additional model build-

	3rd protein ligand ↓ (D), 4th protein ligand ↓ (H)
P. ovalis FeSOD	P -L -L -T -C -D -V -W-E -H -A -Y -Y -I -D -Y -R -N -L -R -P -K -Y
B. stearothermophilus MnSOD (18)	P -I -L -G -L -D -V -W-E -H -A -Y -Y -L -K -Y -Q -N -R -R -P -E -Y
E. coli MnSOD (19)	P -I -L -G -L -D -V -W-E -H -A -Y -Y -L -K -F -Q -N -R -R -P -D -Y
Human liver MnSOD (20)	P -L -L -G -I -D -V -W-E -H -A -Y -Y -L -Q -Y -K -N -V -R -P -D -Y
Yeast MnSOD (21)	P -L -V -A -I -D -A -W-E -H -A -Y -Y -L -Q -Y -Q -N -K -K -A -D -Y
T. thermophilus MnSOD (15)	P -I -V -G -L -D -V -W-E -H -A -Y -Y -V -K -Y -Q -N -K -R -P -D -Y

FIG. 32.5. Alignment of the central segment of a 37 residue peptide fragment of *P. ovalis* iron superoxide dismutase (FeSOD) with sequences of manganese superoxide dismutases (MnSOD). The full sequence of *T. thermophilus* manganese superoxide dismutase has not been determined by chemical methods; the residues were assigned from the shapes of the side chains in the 2.4 Å electron density map (see reference 15). The 3rd and 4th protein ligands in manganese dismutases were assigned from comparisons of the *T. thermophilus* electron density map (15) and chemically detemined sequences of other manganese dismutases (18–21). Alignment by inspection of 23 residues from a polypeptide fragment of *P. ovalis* iron superoxide dismutase suggests that the 3rd and 4th protein ligands in iron dismutase are also, resepctively, Asp and His residues from a sequence-conserved region of structure.

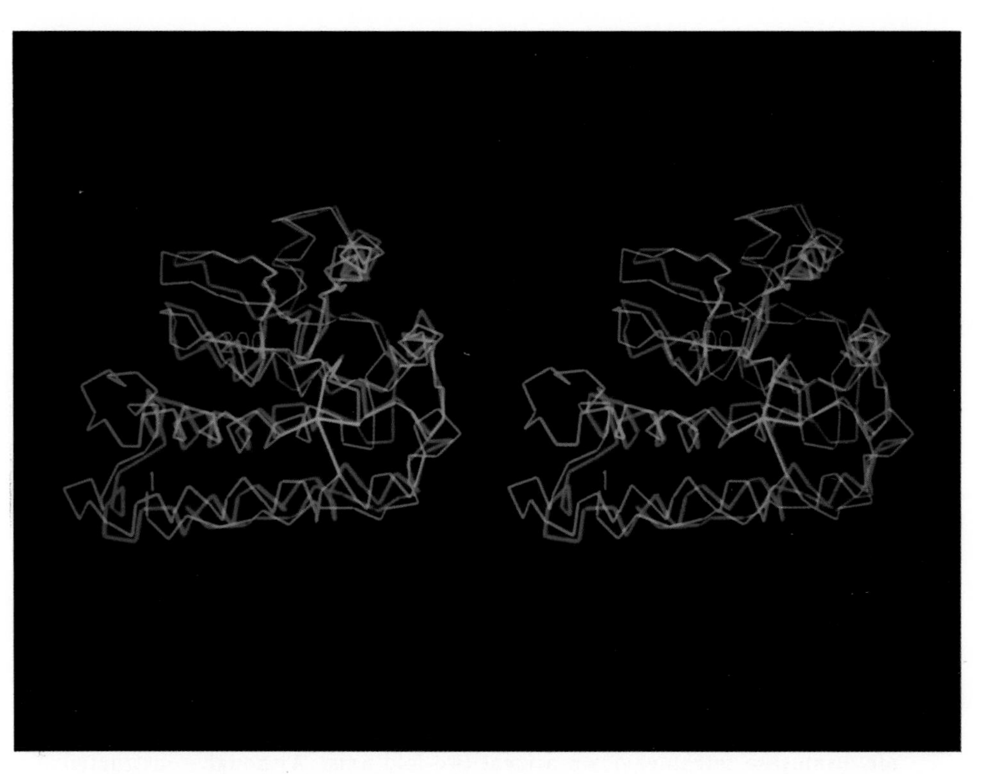

FIG. 32.1. Stereo superposition of the monomer folds of *E. coli* iron superoxide dismutase (red) and *T. thermophilus* manganese superoxide dismutase (green). Virtual bonds (cream) to the metal are also drawn from C_α atoms of the four liganding residues which occupy equivalent positions in the two enzymes. Labels for the first and 200th residues of the manganese enzyme mark the amino- and carboxyl-termini. The proteins are two domain structures with each domain contributing two metal ligands. The principal secondary structural features of the N-terminal domain are two long crossing helices, each of which donates one metal ligand. The C-terminal domains are of mixed $\alpha + \beta$ structure with three strands of antiparallel β sheet; the third ligand is near the C-terminal end of the β structure and is separated from the fourth ligand by only three intervening residues. Relative to iron dismutase, the manganese enzyme contains an insertion of about 10 residues in the region which connects the crossing helices of the N-terminal domain. These residues adopt helical conformations, extend the structure (to the left) in a direction parallel to the first long helix and participate in the formation of the interface not present in iron dismutase dimers.

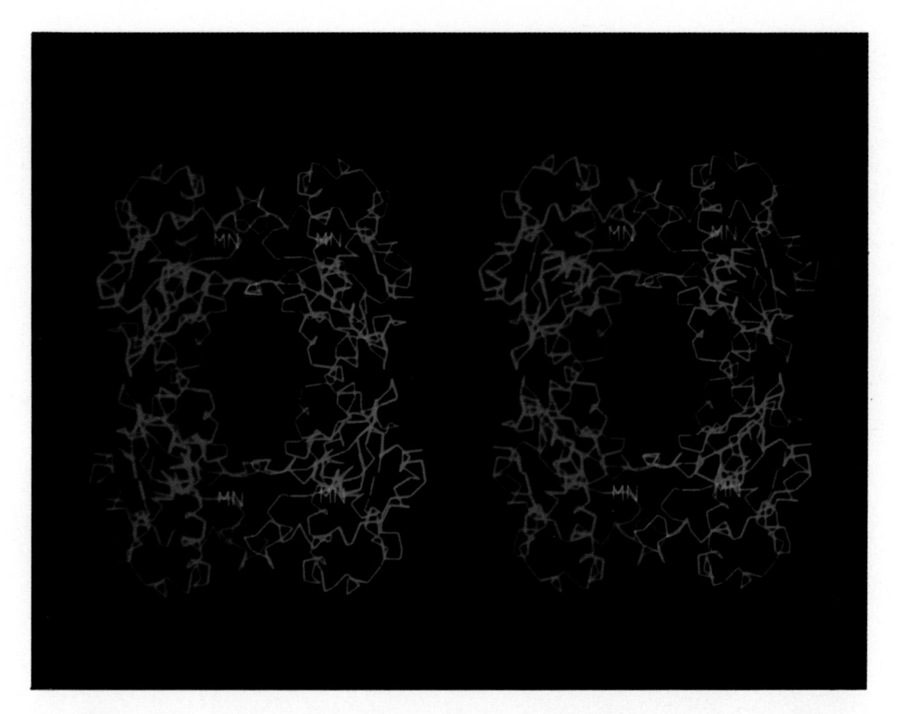

FIG. 32.2. Stereo view of the *T. thermophilus* manganese superoxide dismutase tetramer down a local two-fold axis. Although constructed with 222 molecular symmetry, the tetramer possesses only two unique interfaces with its subunits disposed about the symmetry axes in a flattened arrangement.[13] The center of the molecule is solvent-filled and a channel with an approximate diameter of 17Å passes through the oligomer. The red-green interface, which allows residues from the two metal centers to interact, is preserved in dimeric iron dismutases.[13] The red-red (as well as the equivalent green-green) interface involves an insertion of helical residues not found in the two iron dismutases which have been studied crystallographically.[11,12] See Fig. 32.1.

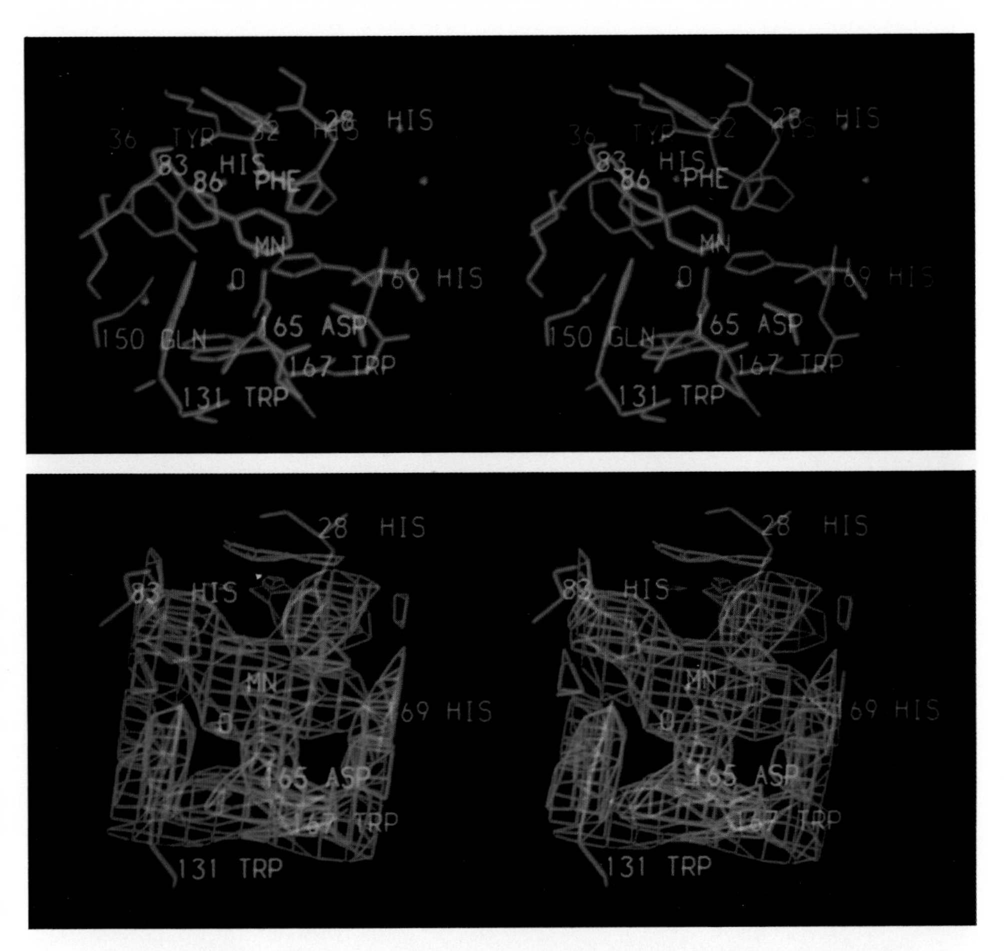

FIG. 32.3. (a) Stereo view of the active center of manganese superoxide dismutase. The manganic ion is coordinated by four protein ligands (His 28, His 83, Asp 165 and His 169). Our current working model of the coordination is trigonal bipyramidal with water (designated 0) as the fifth ligand. His 28 and the water are *trans* ligands but note that the imidazole rings of liganding His 83 and His 169 lie almost parallel to one another, leaving an avenue of approach for an additional ligand. The apolar environment of the ligand cluster includes His 32, Tyr 36, Phe 86, Trp 131 and Trp 167, as well as His 33 and Trp 87 (not shown). The imidazole rings of His 32 and His 169 are stacked; Trp 167 may provide a surface for the stacking of the liganding water; the amide moiety of Gln 150 bridges Tyr 36 and Trp 131 via hydrogen bonding. (b) The active center model of manganese dismutase, illustrated in (a), oriented in the 3.1 Å electron density map of *E. coli* iron dismutase. For clarity only a portion of the model is displayed, but all of the residues included in (a) are accommodated by the iron dismutase electron density. The identities of all of the protein ligands are confirmed from sequences. The calculation of the rigid-body coordinate transformation is described in the text. See also Fig. 32.4.

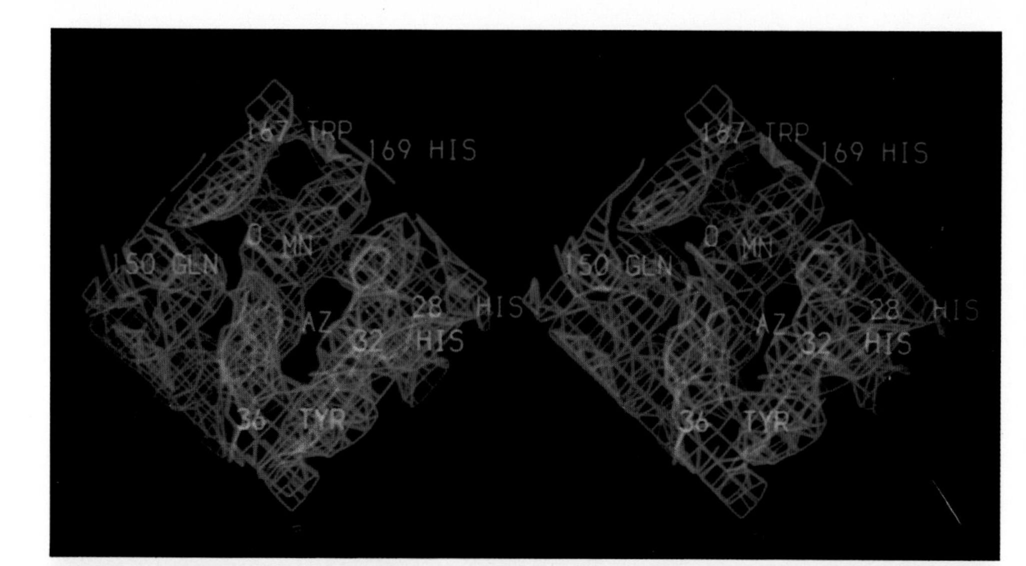

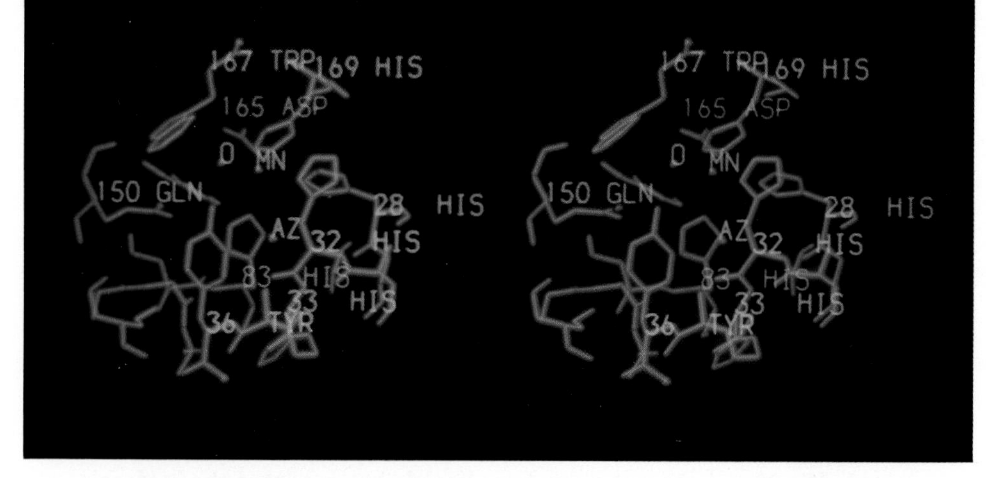

FIG. 32.4. (a) Alternate stereo view of the active center electron density of *E. coli* iron dismutase showing the fit of the manganese dismutase model. In this view the position of the center of mass of a bound azide, detected by difference Fourier techniques in iron dismutase crystals soaked with azide, is also displayed. The manganese dismutase model can be used to interpret the binding of the inhibitor to iron dismutase. Azide binds at an exposed face of the metal ion (see legend, Fig. 32.3) in a cavity formed partially by the van der Waals surfaces of His and Tyr residues. Binding of azide appears to increase the coordination number of the metal ion from 5 to 6 without displacement of the water ligand. (b) Expanded view of (a) without electron density. The strategic positioning of His 32 and Try 36 in *T. thermophilus* manganese dismutase (or homologous His 30 and Tyr 34 in *E. coli* iron dismutase) suggests catalytic roles for these residues.

ing manipulations were employed to achieve this fit. The metal ligation corresponds well in the two structures with the imidazole ring of His 28 occupying a position which appears apical to the three remaining protein ligands. As noted in the manganese dismutase active center, a bulge in the electron density, continuous with the metal and *trans* to the apical imidazole ring, is apparent in the 3 Å iron dismutase map. This feature is consistent with the presence of the water ligand which has been detected in magnetic resonance studies of iron and manganese dismutases.[23,24]

The identity of the first protein ligand in iron superoxide dismutase was established by comparisons of amino-terminal sequences[25] and diffraction studies,[11,12] and shown to be His 26 in the *E. coli* and *P. ovalis* enzymes. Sequences of several polypeptide fragments of the *P. ovalis* iron enzyme have been determined in Japan and can be compared with the four known primary structures of the manganese enzyme. Included among these sequences is a span of 37 residues which show homology with the known manganese dismutase sequences in the region which contains the third (Asp) and fourth (His) protein ligands. The alignment illustrated in Fig. 32.5 indicates that the third and fourth ligands in iron superoxide dismutase are also Asp and His. These assignments are consistent with the shapes in the *E. coli* iron dismutase electron density (cf., Figs. 32.3, 32.4). New sequence information[26] on iron dismutase from the photosynthetic organism, *P. leiognathi*, has been used to align a His residue with the second ligand (also His) in manganese dismutases; the *E. coli* iron dismutase electron density map is consistent with an imidazole ring at this position. Hence all four protein ligands in iron and manganese superoxide dismutases are identical, conserved residues.[26]

Display of the manganese dismutase model in the electron density map of iron dismutase (Figs. 32.3, 32.4) also reveals that the aromatic and imidazole side chains found in the active center of *T. thermophilus* manganese dismutase occur in structurally homologous positions around the Fe(III) coordination sphere of iron dismutase. A residue corresponding to Gln 150 also appears to be oriented, as in manganese dismutase, to provide a hydrogen bonding bridge between residues corresponding to Tyr 36 and Trp 131.† In addition to the ligands, at least nine other residues from the active center of manganese dismutase are located in equivalent electron density in iron superoxide dismutase. Partial sequences of *E. coli* iron superoxide dismutase, determined for several tryptic peptides[27] allow definitive assignment of two of these residues; His 30 and Tyr 34 in the iron dismutase structure correspond to His 32 and Tyr 36 from the manganese dismutase structure. The remaining homologies can be inferred only from characteristic shapes in the *E. coli* iron dismutase map.

The electron density corresponding to Trps 131 and 167 and to several other aromatic residues suggests that in iron dismutase their rings may be oriented somewhat differently than in manganese dismutase, but higher resolution data will be required to establish the exact nature of the changes in orientation. While there may be real but minor conformational differences in the two active center environments, the fit of the unrefined manganese dismutase model to the 3.1 Å map of iron dismutase indicates that the transition metal ion coordination spheres in both proteins are located in essentially equivalent structural environments. We cannot exclude the possibility that a side chain may occupy an equivalent spatial position in both structures, but be attached to different segments of the main chain, as is the case with a Trp residue in cytochrome c551.[28] It will be important to establish the identity of all the residues in both active centers to discern features which may distinguish the two enzymes. For example, substitution of Gln 150 by Glu or Asp in iron dismutase would have important consequences for electrostatic interactions at the active center.

---

† The electron density corresponding to the indole ring of Trp 131 is located close to Mn(III) in manganese dismutase. In previous reports of iron dismutase structures,[11,12] the assignment of this density as a cofactor, not covalently bound to the protein, was considered. We now abandon this interpretation in favor of the Trp assignment as it is in accord with correlations of sequence data[18–21] and the three dimensional manganese dismutase structure.[15]

Possible catalytic residues

Our studies of manganese dismutase have identified residues in the vicinity of the metal ion that could be important in catalysis of the dismutation reaction.[15] Two of these residues, His 32 and Tyr 36, seemed to be particularly good candidates as possible catalytic residues because of their orientations and solvent accessibility. Their similar location in iron superoxide dismutase reinforces the notion that these residues may be essential for enzymatic activity.

In *T. thermophilus* manganese dismutase His 32 and Tyr 36 are positioned in the active center by a distortion in the helix that includes both of these residues as well as the first protein ligand, His 28. The irregularity is clearly seen in the manganese dismutase electron density map as an insertion of an extra residue in the helix structure. The insertion occurs in the turn following His 28; its actual geometry and hydrogen bonding are discussed and illustrated in reference 15, but the result is that between residues 28 and 36, there are 8 residues per two turns of helix instead of the usual 7.2 found in normal α-helical structures. As a consequence of this distortion, the side chains of His 28, His 32 and Tyr 36 all come off the helix in almost parallel directions. Our display of the active center of manganese dismutase in the active center electron density of iron dismutase demonstrates that the residues corresponding to His 32 and Tyr 36 are similarly oriented in the iron dismutase structure.* In addition, examination of the electron density (not shown) of the corresponding helix in iron dismutase reveals that it also accommodates the insertion of an extra residue. Hence, both proteins display a similar structural irregularity in this helix and this observation suggests to us that the distortion is important for strategic positioning of these His and Tyr side chains at the catalytic site.

Additional evidence, which suggests the catalytic importance of these residues, comes from difference Fourier studies of iron dismutase crystals complexed with the competitive inhibitor, azide.[11,12] In accord with observed spectral changes,[29]

* As discussed in the preceding section, these are His 30 and Tyr 34 in *E. coli* Fe dismutase.[27]

the inhibitor appears to bind directly to the transition metal ion with the center of mass of the difference peak located 3 Å from the ferric ion. The rigid-body rotation of the manganese dismutase active center model into the iron dismutase structure has allowed us to interpret this binding in terms of residue interactions. Azide does *not* bind to the metal ion at the water coordination site; instead, the coordination number apparently increases from 5 to 6 with the inhibitor binding at an exposed face of the metal ion between ligands His 83 and His 169, and *trans* to Asp 165. Bound at this location, azide is positioned in a cavity, two of whose walls are formed from the van der Waals surfaces of His 32 and Tyr 36. To the extent that N_3^- is a substrate analog, its position suggests catalytic roles for these residues.

Summary

Display of the active center model of manganese superoxide dismutase in the electron density map of iron superoxide dismutase demonstrates that structural homology between these proteins extends to the level of their active center environments. The metal ions in both resting, oxidized proteins are coordinated by four protein ligands and the maps provide substantial evidence that a water molecule acts as an additional ligand. The chemical environments surrounding the coordination spheres are strongly apolar. Sequence-conserved residues in the active center environment of manganese dismutase appear to be preserved in the metal center of the iron protein. In both proteins these residues are seen to be similarly oriented in the active center and to participate in analogous three-dimensional chemical interactions.

Acknowledgments

We are proud to contribute this work in celebration of the fiftieth anniversary of the Patterson function. The research was supported by grants GM 16429 to M.L.L. and GM 21519 to J.A. Fee.

References

1. Fridovich, I. *Annu. Rev. Biochem.* **44**, 147–159 (1975).
2. Michelson, A. M., McCord, J. M. and Fridovich, I. (eds.) *Superoxide and Superoxide Dismutases.* AcademicPress: New York (1977).
3. Fridovich, I. *Science* **201**, 875–880 (1978).
4. McCord, J. M. and Fridovich, I. *J. Biol. Chem.* **244**, 6049–6055 (1969).
5. Yost, F. J. Jr. and Fridovich, I. *J. Biol. Chem.* **248**, 4905–4908 (1973).
6. Keele, B. B. Jr., McCord, J. M. and Fridovich, I. *J. Biol. Chem.* **245**, 6176–6181 (1970).
7. Kusunose, E., Ichihara, K., Noda, Y., and Kusunose, M. *J. Biochem. (Tokyo)* **80**, 1343–1352 (1976).
8. Kirby, T. W., Lancaster, J. R. Jr. and Fridovich, I. *Arch. Biochem. Biophys.* **210**, 140–148 (1981).
9. Searcy, K. B. and Searcy, D. G. *Biochim. Biophys. Acta* **670**, 39–46 (1981).
10. Tainer, J. A., Getzoff, E. D., Beem, K. M., Richardson, J. S., and Richardson, D. C. *J. Mol. Biol.* **160**, 181–217 (1982).
11. Stallings, W. C., Powers, T. B., Pattridge, K. A., Fee, J. A. and Ludwig, M. L. *Proc. Natl. Acad. Sci. USA* **80**, 3884–3888 (1983).
12. Ringe, D. Petsko, G. A., Yamakura, F., Suzuki, K. and Ohmori, D. *Proc. Natl. Acad. Sci. USA* **80**, 3879–3883 (1983).
13. Stallings, W. C., Pattridge, K. A., Strong, R. K. and Ludwig, M. L. *J. Biol. Chem.* **259**, 10695–10699 (1984).
14. Steinman, H. M. and Hill, R. L. *Proc. Natl. Acad. Sci, USA* **70**, 3725–3729 (1973).
15. Stallings, W. C., Pattridge, K. A., Strong, R. K. and Ludwig, M. L. *J. Biol. Chem.* **260**, 16424–16432 (1985).
16. Cox, J. M. *J. Mol. Biol.* **28**, 151–156 (1967).
17. Jones, T. A. *J. Appl. Cryst.* **11**, 268–272 (1978).
18. Brock, C. J. and Walker, J. E. *Biochemistry* **19**, 2873–2882 (1980).
19. Steinman, H. M. *J. Biol. Chem.* **253**, 8708–8720 (1978).
20. Barra, D. Schinina, M. E., Simmaco, M., Bannister, J. V., Bannister, W. H., Rotilio, G. and Bossa, F. *J. Biol. Chem.* **259**, 12595–12601 (1984).
21. Ditlow, C., Johansen, J. T., Martin, B. M. and Svendsen, I. *Carlsberg Res. Comm.* **47**, 81–91 (1982).
22. Parthasarathy, R., Srikrishnan, T. and Ginell, S. L. *Biomolecular*

Stereodynamics (Sarma, R. H., ed.) Vol. 1. pp. 261-267. Adenine Press: New York (1983).

23. Villafranca, J. J., Yost, F. J. Jr. and Fridovich, I. *J. Biol. Chem.* **249**, 3532–3536 (1974).
24. Villafranca, J. J. *FEBS Lett.* **62**, 230–232 (1976).
25. Walker, J. E., Auffret, A. D., Brock, C. J. and Steinman, H. M. In *Chemical and Biochemical Aspects of Superoxide and Superoxide Dismutase.* (J. V. Bannister and H. A. O. Hill, eds.) pp. 212–222. Elsevier/North-Holland: New York (1980).
26. Barra, D., Schinna, M. E., Bossa, F. and Bannister, J. V. *FEBS Lett.* **179**, 329–331 (1985).
27. Steinman, H. M., Tarr, G. and Fee, J. A. Unpublished.
28. Almassy, R. J. and Dickerson, R. E. *Proc. Natl. Acad. Sci. USA* **75**, 2674–2678 (1978).
29. Slykhouse, T. O. and Fee, J. A. *J. Biol. Chem.* **251**, 5472–5477 (1976).

33. On the structure of a biological crystal determined by its Patterson function

D. B. Litvin

Crystallographers have been using X-rays to investigate the structure of biologically important macromolecules for over forty years.[1] One type of these is the so-called spheroidal or globular macromolecules, such as myoglobin and hemoglobin. Most globular macromolecules can be crystallized. In forming a crystal these macromolecules are not to any large extent distorted, there are in general only a few molecules in the unit cell of the crystal, and the identity of each molecule is preserved. The term 'biological crystal' in the title refers to such a crystal.

The object of investigating the structure of such biological crystals is to determine the structure of the macromolecules. To determine the structure of the crystal one attempts to calculate the electron density $\rho(r)$ of the crystal *via* the Fourier expansion

$$\rho(r) = (1/V)\sum_k F(k)\exp(-2\pi ik \cdot r)$$

where the Fourier coefficients $F(k) = |F(k)|\exp(i\alpha(k))$ are called the complex structure factors, $\alpha(k)$ the phase of the structure factor, and V is the volume of the unit cell of the crystal.

To investigate the structure of these biological macromolecules using only the magnitudes of the structure factors, the so-called 'molecular replacement' method has been developed:[2] This method consists of three steps.

1) The rotation problem: determine the point group of the molecules in the crystal, and their orientation with respect to the translational symmetry vectors of the crystal.

2) The translation problem: determine the vectors between the molecular positions in the crystal. These vectors are called translation vectors.

3) The phase problem: using the information obtained in the first two steps, determine the phases of the structure factors.

The solutions of the rotation and translation problems are based on the analysis of a function which can be calculated from the magnitude of the structure factors $F(k)$. This is the Patterson function introduced in 1934;

$$P(r) = (1/V)\sum_k |F(k)|^2 \exp(2\pi i k \cdot r).$$

The function is the self-convolution (self-correlation) of the electron density of the crystal:

$$P(r) = \int \rho(r')\rho(r'+r)dr'.$$

Because in the biological crystal the macromolecules preserve their identity, the electron density of the crystal can be written as a sum of the electron densities of the molecules, and the Patterson function can be written as

$$P(r) = \sum_{j,k,t} P_{jkt}(r)$$

where j and k index the molecules in the unit cell of the crystal, t the translations of the crystal, and

$$P_{jkt}(r) = \int \rho_{j0}(r')\rho_{kt}(r'+r)dr'$$

the convolution of the electron density of the jth molecule in the unit cell with the kth moleule in the tth unit cell. The convolution function $P_{jkt}(r)$, in general, is called a cross-Patterson function, and in the special case where $j = k$ and $t = 0$, $P_{jj0}(r)$ is called a self-Patterson. There are two problems for consideration:

1. The rotation problem

Here one wants to determine the point group and orientation of the molecules in a biological crystal. The crystals which we consider are assumed to be made up of only one kind of biological macromolecule.

To determine this, one looks for relations between the point group and orientation of the molecules in the crystal and properties of the Patterson function. Consider the electron density $\rho_{j0}(r)$ of the jth molecule in the unit cell. This density is localized because of the finite dimension of the molecule, about the center of mass r_j of the jth molecule. The symmetry point group of the molecule is the set of all proper rotation matrices, P, such that

$$\rho_{j0}(r_j + Pu) = \rho_{j0}(r_j + u)$$

(because the biological macromolecules are made up of 'left-handed' amino acids, the symmetry point group of the molecule consists only of proper rotation matrices). We choose a coordinate system in the crystal; then the group of matrices P, which is the symmetry point group of the jth molecule, is defined with respect to the coordinate system. The orientation of the molecule is the orientation of the rotation axes of the rotations which are represented by these matrices. Another molecule in the crystal has as its symmetry point a group of matrices P' also defined with respect to this coordinate system. The two groups of matrices are in general different but equivalent, *i.e.*, they both belong to the same class of point groups and are denoted by the same symbol in, *e.g.*, international notation.

It follows that if P is the symmetry point group of the jth molecule, then the self-convolution of the electron density of the jth molecule, *i.e.*, the jth self-Patterson function

$$P_{jj0}(r) = \int \rho_{j0}(r')\rho_{j0}(r' + r)dr$$

is invariant under all rotations of P, *i.e.*, $P_{jj0}(r) = P_{jj0}(Pr)$. $P_{jj0}(r)$ is also invariant under inversion, and consequently $P \times \bar{1}$ is an invariance point group of $P_{jj0}(r)$. We shall assume that $P \times \bar{1}$ is the symmetry point group of $P_{jj0}(r)$.

As this self-Patterson is localized in a volume about the origin of the Patterson function, Rossmann and Blow,[3] in order to determine from the Patterson function the point group and orientation of the molecules, introduced (in 1962) the rotation function $R(A)$:

$$R(A) = \int_U P(r)P(Ar)dr$$

where A is a proper rotation, and the integration is over a volume about the origin of the Patterson function. This is an overlap integral of a volume about the origin of the Patterson function with a rotation image of the same volume. Relative maxima of this rotation function, as a function of A, are called peaks of the rotation function. Obviously if $P_{jj0}(r)$ is invariant under a rotation P then there will be a peak in the rotation function at $A = P$. Consequently by determining the peaks of this rotation function, one can obtain information on this point group of the molecules in the crystal. A general method is available to determine systematically the point group symmetry of the molecules from information provided by the rotation function.[4] To drive this method, one needs to apply only very elementary group theoretical arguments.

Consider a biological crystal consisting of identical molecules generated by a space group G from a single molecule at position r_1. Let T denote the translational subgroup of G, and $(R_j \mid \tau_j)\, j = 1, \cdots n$, the coset representatives of T in G. We will consider the case where r_1 is a general position, *i.e.*, the n vectors $r_j = (R_j \mid \tau_j)r_1$ are distinct. We then have n molecules in the unit cell, and the electron density of the molecule at r_j is related to the electron density of the molecule at r_1 by

$$\rho_{j0}(r_j + u) = \rho_{10}(r_1 + R_j u)$$

where $r_j = (R_j \mid \tau_j)r_1$. That is, we have n identical molecules in different orientations in the unit cell, and their mutual orientation is determined by the rotations of the space group of the crystal. It also follows that the self-Pattersons $P_{jj0}(r)$, $j = 1, \cdots, n$, which are all localized about the origin of the Patterson, are identical, in different orientation, and their mutual orientations are also determined by the rotations of the space group of the crystal.

All peaks of the rotation function correspond to rotations which

1) Leave a self-Patterson P_{jj0}, for some j, invariant; or
2) Rotate a self-Patterson P_{jj0} into the orientation of a self-Patterson P_{kk0}, where $j \neq k$.

Let P denote the symmetry point group of the molecule at r_1:

1) The group of rotations $\{R(jj)\} = \{R_j P R_j^{-1}\}$ is the symmetry point group of the self-Patterson P_{jj0}: and
2) The set of rotations $\{R(jk)\} = \{R_k P R_j^{-1}\}$ is the set of all rotations which rotates P_{jj0} into the orientation of $P_{kk0}(r)$.

Therefore, all peaks of the rotation function $R(A)$ correspond to all the rotations contained in the set of rotations:[4]

$$[\{R(jk)\}\ j,k = 1,\cdots,n]. \tag{1}$$

One now has a systematic method to analyze the data obtained from the rotation function.

1) From the rotation function calculate all rotations which correspond to all peaks.
2) Determine the point groups such that the set of distinct rotations in equation (1) is identical with the set of rotations corresponding to peaks of the rotation fuction. If P is such a point group, the conjugate point groups $R_j P R_j^{-1}$, $j = 1,\cdots,n$, where R_j is a rotation of the space group of the crystal, are also such point groups. If there is only one set of such conjugate subgroups $R_j P R_j^{-1}$, $j = 1,\cdots,n$, we say the P is the symmetry point group of the molecules. The orientation of the rotation axes of these conjugate subgroups, with resepct to the translational vectors of the crystal, determine the orientation of the molecules in the crystal. If there is more than one set of such conjugate subgroups, the solution of the rotation function problem is not uniquely determined by this method. However, in practice it is impracticable to search for all peaks of the rotation function, *i.e.*, to calculate $R(A)$ for all possible rotations A. One must then use an alternative method:

1) From the rotation function calculate the rotations corresponding to some of the peaks.
2) Determine the point groups such that the set of distinct rotations of equation (1) includes all those rotations found in step 1 from the rotation function.
3) Determine if there are peaks of the rotation function corresponding to the additional rotations of equation (1).

Example: satellite tobacco necrosis virus (STNV).

A few years ago there was a debate as to whether the protein coat of this 'spherical' virus was of cubic $O(432)$ or icosahedral $J(532)$ point symmetry. A rotation function study was then made of a crystal containing two STNV molecules in the unit cell.[5] The crystal was monoclinic of space group symmetry C_2^3 $(C2)$, and the orientations of the two STNV molecules were related by a rotation of 180°. It was felt that a rotation function study of this crystal would easily determine the point group since there are 4-fold rotations contained in the cubic point group and none in the icosahedral, and 5-fold rotations in the icosahedral and none in the cubic.

A set of strong peaks was found with corresponding rotations which were exactly those proper rotations of a cube, and this was interpreted as meaning that the STNV molecules were of cubic symmetry. There were peaks corresponding to 5-fold rotations, which are characteristic of icosahedral symmetry, but these peaks were much lower than the cubic peaks.

This interpretation was immediately challenged and it was shown that all the peaks of the rotation function corresponding to 5-fold rotations could be interpreted as two molecules of icosahedral symmetry in two different orientations related by the rotation of 180° of the space group of the crystal.[6] There is general agreement that this is the correct interpretation and that STNV molecules do have icosahedral symmetry. However in this reinterpretation the stronger cubic peaks were explained away in an argument that approximated this monoclinic crystal as being cubic!

All these peaks can be explained using the above formalism and taking the point group of the STNV molecules as being icosahedral:[4] if one calculates the set of rotations in equation (1) taking $j, k = 1, 2$, R_2 the rotation of 180° of the space group, and P as the icosahedral symmetry point group of one of the molecules, one finds 240 rotations. These include the icosahedral rotations of both molecules, and a set of cubic rotations, exactly that set of cubic rotations determined from the rotation function! Each of these cubic rotations either leaves both molecules invariant or interchanges the two orientations, explaining the high corresponding peaks, since all other rotations either leave only one molecule invariant, or rotate one molecule into the orientation of the other. One finds also that this set, equation (1), of rotations contains additional rotations which have not yet been determined. However, even without determining peaks of the rotation function corresponding to these rotations, it does seem that the symmetry point group of the STNV molecules is icosahedral.

2. The translation function

Information on the point group and orientation of the molecules in a biological crystal is found, using the rotation function, by considering that volume of the Patterson about the origin of the Patterson function. Information on the translation vectors between molecules is found using a similar method, but considering other parts of the Patterson. One uses a so-called translation function $T(x, A)$ introduced by Rossmann, Blow, Harding, and Coller,[7]

$$T(x, A) = \int_U P(x + r)P(x + Ar)dr.$$

The translation function, like the rotation function, is an overlap integral of a volume U of the Patterson function with a rotated image of the same volume, but unlike the rotation function, the center of the volume is now a variable, and not restricted to be at the origin of the Patterson function.

This translation function is non-zero when:

1) The volume U intersects with a cross-Patterson $P_{jkt}(r)$ and the intersection is left invariant by the rotation A about x; or
2) The volume U intersects two cross-Pattersons $P_{jkt}(r)$ and $P_{j'k't'}(r)$ and one is transformed into the other by the rotation A and X. The relative maxima of a translation function $T(x, A)$ as a function of x, for constant A, are called the peaks of the translation function, and the positions of these peaks are related to the translation vectors between molecules in the crystal. The group theoretical arguments which enter

into determining the relationship between the peaks of the translation function and the translation vectors between molecules are similar to those used in predicting the peaks of the rotation function.

Necessary and sufficient conditions that the translation function $T(x, A)$ has non-zero values associated with the transformation of $P_{jkt}(r)$ into $P_{j'k't'}(r)$ are:[8]

1) the rotation A is such that

$$P_{jkt}(r_k - r_j + t + A^{-1}u) = P_{j'k't'}(r'_k - r'_j + t' + u);$$

2) $A(r_k - r_j + t - x) = r_{k'} - r_{j'} + t' - x$;
3) the vector $\frac{1}{2}y$, where $-y = r_k - r_j + t - x$ is within the volume U.

The first two conditions demand that $P_{jkt}(r)$ is transformed into $P_{j'k't'}(r)$ by a rotation about A about the point x, and the third condition demands that the volume U of the Patterson function centered at x intersects both $P_{jkt}(r))$ and $P_{j'k't'}(r)$. Using group theoretical arguments like those used for the rotation function one can show that all rotations which satisfy the first condition are those denoted by $\{A(jk, j'k')\}$

$$\{A(jk, j'k')\} = [\{R(jj')\} \cap \{R(kk')\}] + \bar{1}[\{R(jk')\} \cap \{R(kj')\}].$$

If this set of rotations is empty, the $P_{jkt}(r)$ and $P_{j'k't'}(r)$ are not congruent. For rotations A contained in $\{A(jk, j'k')\}$, the positions x are calculated from the second condition.

References

1. Bragg, L. and North, A. C. T. In *Progress in Physics, Biophysics, a reprint series.* W. A. Benjamin: New York (1969).
2. Rossmann, M. G. (ed.) *The Molecular Replacement Method.* Gordon and Breach: New York (1972).
3. Rossmann, M. G. and Blow, D. M. *Acta Cryst.* **15**, 24 (1962).
4. Litvin, D. B. *Acta Cryst.* **A31**, 407 (1975).
5. Akervall, K., Strandberg, B., Rossmann, M. G., Bengston, V., Fridborg, K., Johannison, H., Kannan, K., Lovgren, S., Petef, G., Oberg, B., Eaker, D., Hjerten, S., Ryden, L. and Morning, I. *Cold Spring Harbor Symp. Quant. Biol.* **36**, 469 and 487 (1971).

6. Klug, A. *Cold Spring Harbor Symp. Quant. Biol.* **36**, 483 (1971).
7. Rossmann, M. G., Blow, D. M., Harding, M. M. and Coller, E. *Acta Cryst.* **17**, 338 (1964).
8. Litvin, D. B. *Acta Cryst.* **A33**, 62 (1977).

34. Localization of the Cu^{2+} atom in the unit cell of crystals of *Ps. dentrificans* azurin utilizing synchrotron-induced anomalous dispersion

Z. R. Korszun and J. T. Bolin

With the advent of synchrotron radiation which is highly intense and polychromatic, it has been possible to choose the wavelength of radiation used to collect diffraction data in order to enhance the anomalous scattering from metalloprotein prosthetic groups thereby providing phasing information which can be used to solve the protein structure by 'direct methods'. In general, the structure factor F_{hkl} for a reflection, hkl, of a metalloprotein reflection will depend on the energy of the radiation used and may be written as

$$F_{hkl}(\epsilon) = F^{o}_{hkl}(\epsilon) + [\Delta f'_{hkl}(\epsilon) + i\Delta f''_{hkl}(\epsilon)].$$

Here, ϵ corresponds to the energy; $F^{o}_{hkl}(\epsilon)$ is the structure factor for the protein in the absence of any anomalous dispersion; $\Delta f'_{hkl}(\epsilon)$ is the real or dispersive correction to the structure factor; $\Delta f''_{hkl}(\epsilon)$ is the imaginary or absorptive correction to the structure factor. In this equation only one type of anomalous scatterer is assumed so that $\Delta f'(\epsilon)$ and $\Delta f''(\epsilon)$ are out of phase by $\pi/2$; moreover, $\Delta f'$ and $\Delta f''$ are related by the Kramers-Kronig relationship, so that once $\Delta f''(\epsilon)$ is measured as a function of energy, $\Delta f'$ can be calculated. Below the absorption edge of the metal atom, both $\Delta f'$ and $\Delta f''$ are small, on the edge both are large and above the edge only $\Delta f''$ remains large. These differences in anomalous scattering can be exploited in the vicinity of the edge to locate the anomalous scatterers by difference Patterson syntheses. Two different syntheses using coefficients

$$[F^{+}_{hkl}(A) - F^{-}_{hkl}(A)]^2$$

and

$$[F_{hkl}(A) - F_{hkl}(B)]^2$$

can be calculated to reveal the location of the anomalous scatterer. A denotes measurements just above the edge and B

denotes measurements just below the edge. In both of these syntheses $F^{o}_{hkl}(\epsilon)$ varies much more slowly with energy than do $\Delta f'(\epsilon)$ and $\Delta f''(\epsilon)$, and may be taken as constant, once the data sets are scaled. The first Patterson synthesis is only sensitive to $\Delta f''(\epsilon)$, and is calculated using only acentric reflections. The second synthesis is sensitive to both $\Delta f'(\epsilon)$ and $\Delta f''(\epsilon)$ for general reflections, but it is sensitive solely to $\Delta f'(\epsilon)$ if only centric reflections are used. These anomalous dispersion effects can be used to calculate the phases of protein structure factor moduli in a manner analogous to the single isomorphous replacement method where heavy atom anomalous dispersion effects have been measured. In addition, following the method of Sim[1] (suggested by Hendrickson[2,3]), the metal atom partial structure phase contribution can be used to bolster the phases. In the cases where this method can be used, it has advantages over isomorphous replacement because there is no problem due to lack of isomorphism.

Such a strategy has been used in the crystal structure determination of *Ps. denitrificans* azurin, a blue or Type 1 copper protein which crystallizes in space group $P4_122$ (or $P4_322$) with $a = b = 53.6$ Å, $c = 99.3$ Å. Data were collected to 3 Å resolution at CHESS, the Cornell High Energy Synchrotron Source, on stations A1 and A3 which are equipped with cylindrical and sagittal focussing optics, respectively. Precession data were collected below the Cu K-edge (8.00 KeV) for the $h0l, hk0, hhl, h+1h-1l$, and $nhkl$ zones yielding approximately 680 observed reflections most of which are centric reflections. Oscillation data were collected just above the Cu^{2+} K-edge (8.98 KeV). Oscillation photographs were processed by the method of Rossmann.[4,5] Out of 2200 observed reflections, 644 Friedel pairs were fully recorded on the same film. These coefficients were used to calculate the Bijvoet difference Patterson function.

The Harker sections $uv\frac{1}{2}$ and $u0w$ are shown in Fig. 34.1 for the energy difference Patterson function and the Bijvoet difference Patterson function, respectively. The crosses in the maps show the solution to the location of the copper atom. The solution is unique. It is the only consistent solution appearing in both functions, illustrating that useful phase information is available from multiple wavelength measurements and exten-

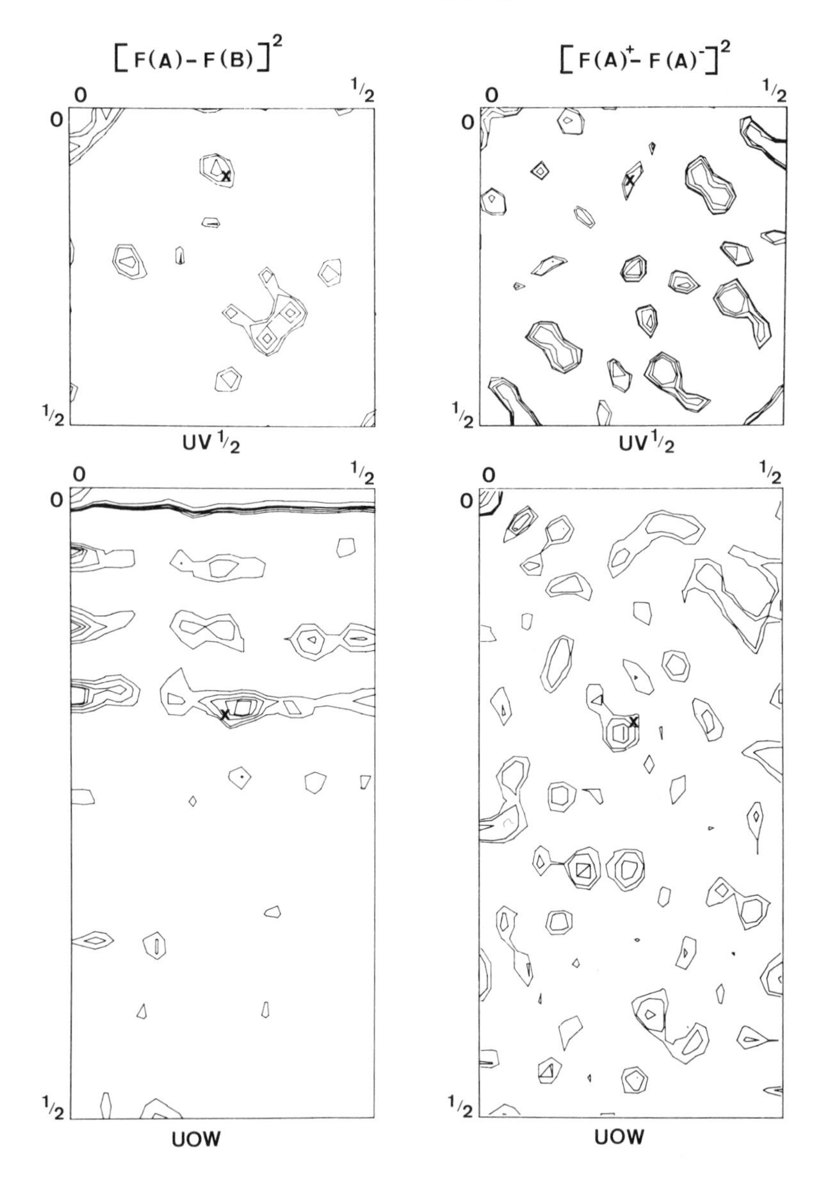

FIG. 34.1. $u\,v\,\frac{1}{2}$ and $u\,0\,w$ Harker sections are illustrated as labeled in the figure. The sections on the left side of the figure are from an energy difference Patterson while those on the right side were calculated for the Bijvoet difference Patterson as described in the text. The solution to the Cu^{2+} location is marked by an X.

ding previous work[6,7] which showed that phasing information can be obtained from energy difference Patterson functions. Currently, anomalous dispersion phases are being calculated and refined. Future work focusses on using the refined phases to build an atomic model of the molecule.

Acknowledgment

This work is supported by NIH Grant GM 32692 to Z. R. Korszun; J. T. Bolin is supported by NIH Grant GM 10704 to M. G. Rossmann.

References

1. Sim, G. A. *Acta Cryst.* **12**, 813-815 (1959).
2. Hendrickson, W. A. and Lattmann, E. E. *Acta Cryst.* **B26**, 136-143 (1970).
3. Hendrickson, W. A. and Teeter, M. M. *Nature (London)* **290**, 107-113 (1981).
4. Rossmann, M. G. *J. Appl. Cryst.* **12**, 225-238 (1979).
5. Rossmann, M. G., Leslie, A. G. W., Abdel-Meguid, S. S. and Tsukihara, T. *J. Appl. Cryst.* **12**, 570-581 (1979).
6. Phillips, J. C., Wlodawer, A., Goodfellow, J. M., Watenpaugh, K. D., Sieker, L. C., Jensen, L. H. and Hodgson, K. O. *Acta Cryst.* **A33**, 445-455 (1977).
7. Phillips, J. C. and Hodgson, K. O. In *Synchrotron Radiation Research.* (H. Winick and S. Doniach, eds.) pp.565-605. Plenum Press: New York, NY (1980).

35. Comparison of X-ray crystallography and solution small-angle scattering: morphological parameters of globular proteins as corroborated by hydrodynamics

Helmut Pessen and Thomas F. Kumosinski

Despite intimations in the literature that protein structures in the crystal and in solution are essentially alike,[1] such an assertion should be regarded with reservation. Because of the different ways in which the relevant techniques probe a system, their results would not be expected to be necessarily identical and, indeed, they are not. It is found that appropriate parameters adapted from small-angle X-ray scattering (SAXS) can be called upon to bridge the information gap between X-ray crystallography (which observes the static molecular shape with its boundary separated from water of hydration) and sedimentation (which observes a dynamic molecule with a surface covering of molecules of bound water.) These two methods give a different view of the molecular structure; the first gives a description of the actual molecular surface, the second gives a description of the surface of the hydrated molecule. The significant differences in results from these two physical chemical methods will be examined here in light of the special characteristics of each and consequently the somewhat different information obtained. The aim is an explanation of the differences in the shape parameters obtained by the two methods.

Analysis of protein structure by X-ray crystallography reveals rugose surfaces comprising clefts, grooves, channels, and protuberances. The contribution of these features to many protein properties is influenced by the fact that globular proteins function in an aqueous environment, where they are fully solvated. Hydrodynamic parameters, sensitive to surface characteristics, have in the past been calculated directly from data-bank coordinates and compared with solution values in attempts at correlation; this has met with only marginal success.[2,3] X-ray crystallographic results contain coordinates for the protein molecule but only some water coordinates are obtained with good precision since solvent is generally somewhat disordered. Therefore such hydrodynamic model calculations did not take into account the bound water, although its presence must moderate the effects of surface irregularities and thus of frictional ratios. Moreover, proteins in solution are in a more dynamic state than that observed by the X-ray diffraction experiment;[4] still other differences may be due to electrostriction in the crystal resulting from charged groups on the protein. There is thus a need for structural information from a solution method which, unlike hydrodynamic ones such as sedimentation analysis, can separate the frictional effects of hydration and structure. SAXS meets that need.[5]

The resulting ability to estimate hydrodynamic parameters from SAXS data affords various insights. Comparison with observed values can remove ambiguities regarding the relative contributions of shape and hydration to the frictional ratio. X-ray data from either diffraction or scattering can provide a measure of molecular surface roughness. Comparison of parameters obtained from the two X-ray methods can illustrate differences between a technique that gives detailed structural information, albeit without regard to hydration, and one that gives less structural detail, but furnishes information on hydration in addition to molecular shape and surface area.

Theory

The criterion for selection of proteins for the present purpose was the availability of two kinds of data in the literature, namely both

1) the sedimentation coefficient, $s^{\circ}_{20,w}$, (or data which allow it to be calculated) and
2) appropriate SAXS data, *i.e.*, reported values of
   a) the radius of gyration, R_G, plus
   b) at least two of three other parameters: the hydrated volume, V; the surface-to-volume ratio, S/V, (required for all proteins in the lower molecular weight range); or the axial ratio, p (or some other shape ratio, depending on the model).

The three available values representing parameters under 2) are the primary parameters for the protein under consideration, as contrasted to secondary parameters, which are any

TABLE 35.1. Structural and hydrodynamic parameters from SAXS

			Auxiliary Parameters		SAXS Parameters		Calculated from SAXS		Observed	Calculated from V and $\bar{v}$
			M[c]		V, Å^3	$(a/b)_1^e$	$(f/f_o)_1^g$	s_1,S[i]	$s^o_{20,w}$,	A_1^k,
	Macromolecule[a]	Model[b]	$\bar{v}$, ml/g[d]	R_G, Å	S/V, Å^{-1}	$(a/b)_2^f$	$(f/f_o)_2^h$	s_2,S[j]	S	g/g
1.	Ribonuclease	PE	13,690[l]	14.8	22,000	1.87	1.036	2.03		
	(bovine pancreas)[7]		0.696[9]		0.29	3.69	1.161	1.81	1.78[10]	0.272
2.	Lysozyme	PE	14,310[m]	14.3	24,200	1.42	1.011	2.07		
	(chicken egg white)[7]		0.702[9]		0.25	2.92	1.107	1.89	1.91[11]	0.317
3.	α-Lactalbumin	PE	14,180[n]	14.5	25,100	1.43	1.012	2.02		
	(bovine milk)[7]		0.704[9]		0.24	2.81	1.099	1.86	1.92[12]	0.362
4.	α-Chymotrypsin	PE	22,000[o]	18.0	37,170[p]	2.0	1.044	2.36		
	(bovine pancreas)[13]		0.736		0.157	2.02	1.045	2.36	2.40[14,15]	0.282
5.	Chymotrypsinogen A	PE	25,000[o]	18.1	37,790[p]	2.0	1.044	2.67		
	(bovine pancreas)[13]		0.736		0.160	2.12	1.051	2.65	2.58[16]	0.175
6.	Pepsin[17,q]	PE	34,160[18]	20.5	54,870[p]	2.0	1.044	3.36		
			0.725[19]		0.26	4.76	1.234	2.84	2.88[20]	0.243
7.	Riboflavin-binding	PE	32,500[o]	20.6	66,500	1.63	1.021	3.12		
	protein, apo (pH 3.0)		0.720		0.203	3.58	1.153	2.76	2.76	0.513
	(chicken egg white)[21]									
8.	Riboflavin-binding	PE	32,500[o]	19.8	55,600	1.76	1.029	3.28		
	protein, holo (pH 7.0)		0.720[r]		0.213	3.62	1.156	2.92	2.92	0.311
	(chicken egg white)[21]									

9.	β-Lactoglobulin A dimer (bovine milk)[22]	PE	36,730[s] 0.751[9]	21.6	60,250[t] 0.166[t]	2.13 2.93	1.052 1.108	3.12 2.99	2.87[23]	0.237
10.	Bovine serum albumin[24]	PE	66,300[u] 0.735[9]	30.6	142,000[v] 0.146	2.90 3.88	1.105 1.174	4.62 4.34	4.30[w]	0.318
10a.	Lactate dehydrogenase, M_4 (dogfish)[x]	OE	138,320 0.741	34.7[x]	253,300[x] 0.0893[p]	0.409 —	1.069 —	7.50 —	7.54[26]	0.362
11.	β-Lactoglobulin A octamer (bovine milk)[22]	OE	146,940 0.751[y]	34.4	215,000[t] 0.125[t]	0.347 0.255	1.097 1.162	7.89 7.45	7.38[23]	0.130
12.	Glyceraldehyde-3-phsosphate dehydrogenase, apo (bakers'yeast)[27]	HOC	142,870[z] 0.737[28]	32.1	264,200 0.0995[p]	0.636 0.389	1.018 1.078	8.15 7.70	7.6[29]	0.377
13.	Glyceraldehyde-phosphate dehydrogenase, holo (bakers' yeast)[27]	HOC	145,520[aa] 0.737[r]	31.7	250,000 0.1016[p]	0.614 0.384	1.024 1.080	8.46 7.97	8.0[29]	0.298
14.	Malate synthase (bakers' yeast)[30]	OE	170,000[o,25] 0.735	39.6	338,000 0.0843[p]	0.363 —	1.089 —	8.40 —	8.25[31]	0.463
15.	Pyruvate kinase, apo[bb] (brewers' yeast)[32]	OEC	190,800[cc] 0.734[27]	43.5	406,000 0.0879[p]	0.321 0.298	1.112 1.127	8.70 8.62	8.70[33]	0.548
16.	Pyruvate kinase, holo[bb] (brewers' yeast)[32]	OEC	192,160[aa] 0.734[r]	42.5	406,000 0.0855[p]	0.349 0.320	1.096 1.113	8.92 8.80	8.81[33]	0.539
17.	Catalase (bovine liver)[34]	PC	248,000[29] 0.730[9]	39.8	420,000 0.0752[p]	1.91 2.24	1.038 1.060	12.20 11.96	11.3[35]	0.290
18.	Glutamate dehydrogenase (bovine liver)[36]	PC	312,000[31] 0.749[31]	47.0	668,000 0.0648[p]	1.98 2.30	1.043 1.064	12.18 11.93	11.4[37]	0.541

(TABLE 35.1, continued)

[a] Superscript numerals following entries indicate references as listed below. Tabulated data were taken from the references thus designated in the first column, unless noted otherwise for a particular parameter.

[b] Geometric model used to describe scattering particle: PE, prolate ellipsoid; OE, oblate ellipsoid; PC, prolate (elongated) cylinder; OEC, oblate (flattened) elliptical cylinder; HOC, hollow oblate cylinder.

[c] Molecular weights, by preference, were based on amino acid compositions and sequences wherever available, except in some cases where the cited authors' values appeared more reliable or consistent with the other parameters under the condition of measurement.

[d] Partial specific volumes were the cited authors' values or, in some cases, more accurate values found in the literature. Corrections for temperature differences between 25° and 20° were not in general made for $\overline{v}$ because resulting differences in $s^{\circ}_{20,w}$ are minimal and do not affect comparisons between the different s values.

[e] From equation (3a). Prolate or oblate cylinders were modeled as equivalent prolate or oblate ellipsoids, respectively.

[f] From equation (3b) or (3c). Cylinder modeled as in Note e.

[g] From equation (2a) or (2b), based on equation (3a).

[h] From equation (2a), based on equation (3b); or equation (2b), based on equation (3c).

[i] From equations (1a) and (1b), based on equation (3a).

[j] From equations (1a) and (1b), based on equation (3b) or (3c).

[k] From equation (9).

[l] From Dayhoff (Ref. 8), p. D-130.

[m] From Dayhoff (Ref. 8), p. D-138.

[n] From Dayhoff (Ref. 8), p. D-136.

[o] Value reported by cited authors (see Note c).

[p] Secondary parameter, calculated with use of indicated model from values of primary parameters of cited authors.

[q] Origin of preparation not stated.

[r] Value for apoenzyme used, since $\overline{v}$ for holoenzyme not available.

[s] From Dayhoff (Ref. 8), Suppl. 1 (1973), p. S-83.

[t] Unpublished data of authors of Ref. (22) (S. N. Timasheff, personal communication).

[u] From Dayhoff (Ref. 8), Suppl. 2 (1976), p. 267.

[v] This value of molecular volume appears to be high, as was the molecular weight of 81,200 reported by the listed authors, pointing to the possible presence of aggregation products. For s_1 and s_2 listed in the table, the amino acid sequence molecular weight was used, together with a proportionally adjusted volume of 115,940. The inconsistent use of $V = 142{,}000$ with $M = 66{,}300$ would result in $s_1 = 4.43$ and $s_2 = 4.06$.

[w] From Miller and Golder (Ref. 25). The value reported in Ref. 24 is 4.1 at 0.75%. Allowing for the concentration dependence according to Ref. 25, the two values are equivalent.

[x] Unpublished data of Pessen, Kumosinski, Fosmire, and Timasheff.

[y] Value for 9 (dimer) used, since $\bar{v}$ for octamer not available.

[z] From Dayhoff (Ref. 8), pp. D-147, D-148.

[aa] Calculated from value for apoenzyme.

[bb] The designations 'apo' and 'holo,' although not strictly correct in this case, are used for brevity. They refer to 'native' and 'fructose diphosphate liganded,' respectively.

[cc] Molecular weight calculated from value for subunits reported by Bischofberger *et al.* (Ref. 33).

that could be derived from the primary ones if not reported independently.

A search of the literature has produced nineteen globular proteins[6] that meet the stated criteria (for references, see Table 35.1, adapted from Table I of reference 6, from which it differs by an additional protein, 10a, and a number of corrections and recalculated values, with the original sequence of numbers retained). The proteins considered here are roughly globular and, in particular, show no significant flexibility. The assumption is made that they can all be approximated by ellipsoidal models (prolate or oblate). This assumption is least exact for protein 12, 13, and 15-18, which are more nearly cylinders; however, it is still a useful approximation and generally considered reasonable.[4]

There are two different expressions available as starting points for the calculation of axial ratios (and thus frictional ratios[38]) of scattering-equivalent ellipsoids from SAXS data. They give different answers:[39] method 1 gives an estimate of the overall molecular shape, without regard to rugosity; method 2 makes use of the surface-to-volume ratio obtained from SAXS and translates it into a hypothetical axial ratio descriptive of the surface area instead of the overall geometry of the molecule. In effect, it provides the molecular model of method 1 with the additional surface required for hydrodynamic equivalence by stretching or flattening it (depending on whether one deals with a prolate or an oblate ellipsoid), thus arriving at a frictional ratio reflecting the extra surface presented by the rugosities. One can use these axial ratios (denoted 1 and 2), by way of the frictional ratios derived from them, to estimate hydrodynamic coefficients, in particular sedimentation coefficients. (Diffusion coefficients and intrinsic viscosities, while showing similar patterns, are less well adapted to this kind of analysis).

For the frictional ratio it is expedient to use Oncley's[40] decomposition of the total ratio, f/f_o, into shape- and hydration-dependent factors, f_e/f_o and f/f_e, respectively. A form of Svedberg's equation suited for calculating theoretical sedimentation coefficients $s^{\circ}_{20,w}$ from SAXS structural parameters is [41]

$$s^{\circ}_{20,w} = \frac{M(1-\bar{v}\rho)}{(f/f_o)6\pi\eta N r_o}, \tag{1a}$$

where the subscript '20, w' denotes reference to water at 20°C, M is the anhydrous molecular weight obtained from amino-acid sequence or composition whenever possible, $\bar{v}$ is the partial specific volume of the protein (calculated or, preferably, experimentally determined), ρ is the density and η the viscosity of water at 20°C, and N is Avogadro's number. For these calculations, r_o, the Stokes radius (in cm), will be related to the scattering volume V of the hydrated macromolecule (in cm^3) instead of the more customary $\bar{v}$, by the relationship

$$r_o = (3V/4\pi)^{\frac{1}{3}}. \tag{1b}$$

Since V, in contrast to $\bar{v}$, already reflects the hydrated molecule, the corresponding frictional ratio is really f_e/f_o, although it was written above (and for simplicity will continue to be written in the following) as f/f_o.

The frictional ratio f/f_o, then, reduces here to the structural factor of the total frictional ratio for the hydrated particle. Since we model all molecules as prolate or oblate ellipsoids of revolution:[38]

$$f/f_o = \frac{(p^2-1)^{\frac{1}{2}}}{p^{\frac{1}{3}} \ln [p + (p^2-1)^{\frac{1}{2}}]}, \quad (p > 1, prolate) \qquad (2a)$$

$$f/f_o = \frac{(1-p^2)^{\frac{1}{2}}}{p^{\frac{1}{3}} \tan^{-1} [(1-p^2)^{\frac{1}{2}}/p]}, \quad (p < 1, oblate) \qquad (2b)$$

where p equals a/b, b is the equatorial radius, and a is the semi-axis of revolution of the ellipsoid (The usage of $p = a/b$ is in agreement with that of Luzzati and co-workers;[39] this p is the reciprocal of the p as defined by Teller *et al.*[42]). The axial ratios p were determined from SAXS parameters by the method of Luzzati,[39] with the use of each of two dimensionless ratios, r_1 and r_2, defined as follows (where V is the volume of the macromolecule, R_G is the radius of gyration, and S is the external surface area):

$$r_1 \equiv \frac{3V}{4\pi R_G^3} = \frac{p}{(\frac{p^2+2}{5})^{\frac{3}{2}}}, \quad (p < 1, p = 1, or\ p > 1), \qquad (3a)$$

and

$$r_2 \equiv R_G \frac{S}{V} = \frac{3}{2p}[1 + \frac{p^2}{(p^2-1)^{\frac{1}{2}}} \sin^{-1} \frac{(p^2-1)^{\frac{1}{2}}}{p}](\frac{p^2+2}{5})^{\frac{1}{2}}, \quad (p > 1), \qquad (3b)$$

or

$$r_2 \equiv R_G \frac{S}{V} = \frac{3}{2p}[1 + \frac{p^2}{(1-p^2)^{\frac{1}{2}}} \tanh^{-1}(1-p^2)^{\frac{1}{2}}](\frac{p^2+2}{5})^{\frac{1}{2}}, \quad (p < 1). \qquad (3c)$$

Equation (3a) incorporates the geometric relationships $V = (\frac{4}{3})\pi ab^2$ and $R_G = [(a^2 + 2b^2)/5]^{\frac{1}{2}}$. Equations (3b,c) express S in terms of a and b for prolate and oblate ellipsoids of revolution, respectively. The ratios r_1 and r_2 may be seen from the limiting case of a sphere to be subject to the approximate constraints $r_1 \leq 2.152$ and $r_2 \geq 2.324$.

It should be emphasized that f/f_o and r_o are derived from solution structural parameters without any assumption regarding the contribution of hydration to the frictional ratio; also, no assumption is necessary concerning the symmetric or asymmetric placement of the water molecule, or concerning electrostriction effects. This contrasts with the use of X-ray crystallographic structures for correlation with sedimentation data of globular proteins, where such assumptions cannot be avoided.[43]

Results

Correlations between sedimentation and X-Ray diffraction data. To ensure that the selection criteria used have resulted in a set of data not very different from those for globular proteins in general, we first test our set of 19 proteins against the sets of Squire and Himmel[44] and Teller *et al.*[42] (selected for a different purpose and according to different criteria), as suggested by the Svedberg relationship[41] for spherical molecules in the form

$$s^{\circ}_{20,w} = [M^{\frac{2}{3}}(1 - \bar{v}\rho)/\bar{v}^{\frac{1}{3}}]\ (3\pi^2 N^2/4)^{-\frac{1}{3}}(6\eta)^{-1}, \tag{4}$$

where all parameters have been previously defined. A plot of $s^{\circ}_{20,w}$ *vs* $M^{\frac{2}{3}}(1 - \bar{v}\rho)/\bar{v}^{\frac{1}{3}}$ is shown in Fig. 35.1 for all 19 proteins. Fitting a least-squares straight line with zero intercept to these points gives a slope of 0.00931 ± 0.00009 S cm g^{-1} mol$^{\frac{2}{3}}$. Also shown in Fig. 35.1 is the theoretical line for molecules considered as smooth spheres, which constitutes an upper limit of slope 0.0120 in the same units, obtained from equation (4) by evaluation of the collection of constants.[42] Squire and Himmel[44] and Teller *et al.*[42] obtained the equivalents of slopes of 0.0108 and 0.010 for their respective sets of proteins. These values are not greatly different from those above. One may take it, therefore, that our set has approximately the same average rugosity as other globular proteins. This statistical correlation is purely empirical and has no structural foundation; frictional coefficients are not explicitly considered.

Frictional ratios may be introduced into this approach by means of the relationships developed by Teller[45] between accessible surface area A_s (in Å^2), packing volume V_p (in Å^3), radius R_p (in Å) from the packing volume, and molecular weight M, which were derived by calculations based on the X-ray crystallographic structures of a set of proteins first used by Chothia:[46]

$$A_s = 11.12 \pm 0.16 M^{\frac{2}{3}} \tag{5a}$$

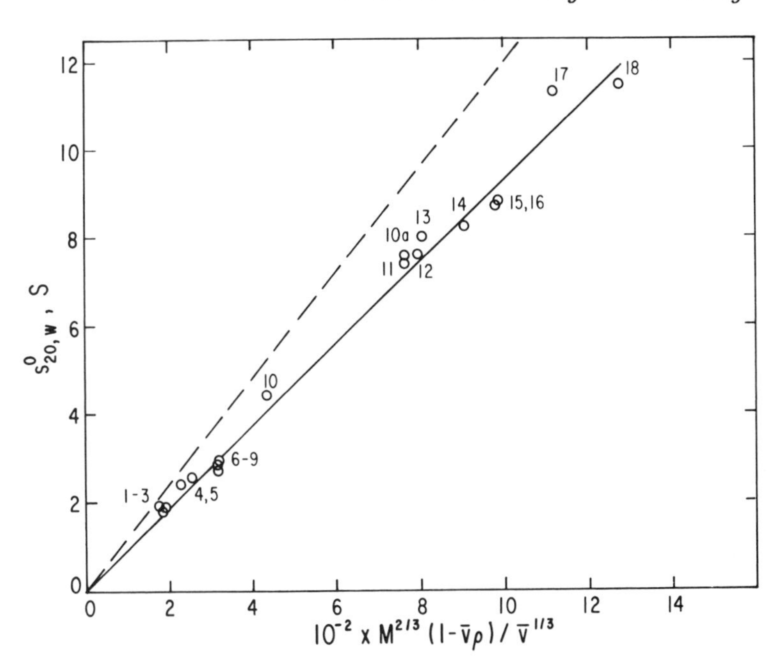

FIG. 35.1. Plot of $s^{\circ}_{20,w}$ *vs* the function $M^{2/3}(1 - \bar{v}\rho)/\bar{v}^{1/3}$ for the 19 proteins, as numbered in Table 35.1. Solid line: linear least-squares fit, with slope 0.00931 ± 0.00009. Dashed line: theoretical upper limit line expected for proteins considered spherical, with slope 0.0120. (Adapted from Fig. 1 of ref. 6).

and

$$V_p \;=\; 1.273 \pm 0.006M. \tag{5b}$$

Equation (5b) is equivalent to

$$R_p \;=\; 0.672 \pm 0.001M^{\frac{1}{3}} \tag{5c}$$

by reason of $V_p = (\frac{4}{3})\pi R_p^3$. R_p is related to the radius of gyration by $R_G = (3/5)^{\frac{1}{2}} R_p$, as may be verified from equation (3a), with $p = 1$.

From these expressions, axial ratios for prolate or oblate ellipsoids of revolution can be calculated by means of equations (3a-c). (S and V here are represented by A_s and V_p, respectively, although it should be realized that these are approximations only, and that V_p, in particular, is not a hydrated volume). The molecular weight cancels out for both $3V/(4\pi R_G^3)$ (the smooth-surface model) and $R_G S/V$ (the rugose-surface model), as it must, these expressions being dimensionless. Since equation (5c) is based on a spherical model,[24] use of $3V/(4\pi R_G^3)$ here will necessarily result in an axial ratio of 1 for the smooth model. The information contained in S, however, is independent of the assumption of such a model and will, therefore, permit calculation of equivalent axial ratios from $R_G S/V$ (prolate: 3.96; oblate: 0.238), and thus frictional ratios from equations (2a) and (2b) (prolate: 1.180; oblate: 1.178). Equations (1a,b), with V again used in place of $\bar{v}$ as a measure of the Stokes radius r_o, then yield

$$s^{\circ}_{20,w} = M^{\frac{2}{3}}(1 - \bar{v}\rho)k \times 10^{-13}, \tag{6}$$

where k, the collection of constants in equation (4), reduces

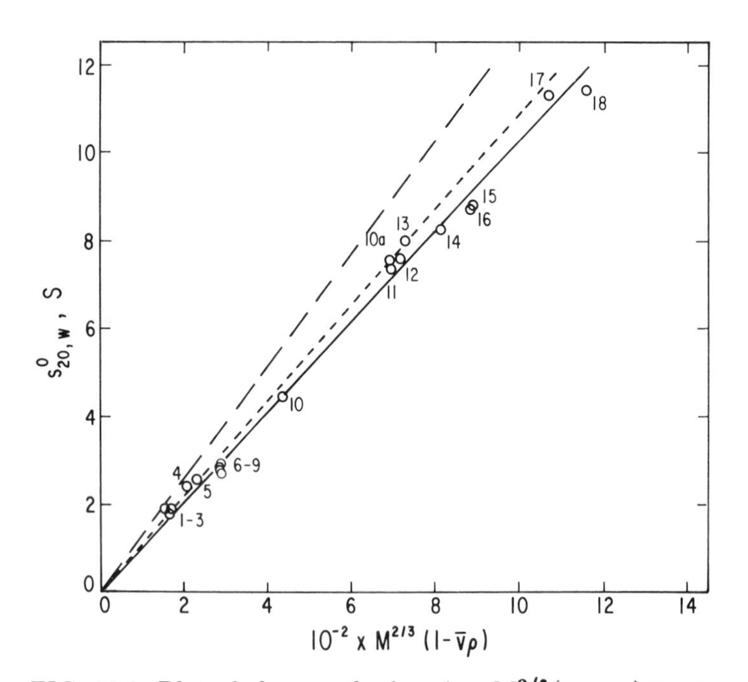

FIG. 35.2. Plot of $s^{\circ}_{20,w}$ *vs* the function $M^{2/3}(1 - \bar{v}\rho)$ for the 19 proteins, as numbered in Table 35.1. Solid line: linear least-squares fit, with slope 0.01030 ± 0.00010; dashed line: theoretical for smooth-surface models; dotted line: theoretical for rugose-surface models. (Adapted from Fig. 2 of Ref. 6).

to 0.01284 for all smooth-surface models and to about 0.0109 (prolate: 0.01088; oblate: 0.01090) for rugose-surface models.

Fig. 35.2 is a plot of $s^{\circ}_{20,w}$ *vs* $M^{\frac{2}{3}}(1-\bar{v}\rho)$ for our 19 proteins. A straight line with zero intercept fitted to the experimental data yields a value of 0.01030 ± 0.00010 for k in Equation (6). Comparison between these experimental values for k and those above, derived from the X-ray crystallographic structures, shows that a prolate or oblate ellipsoid of revolution with an equivalent S/V ratio (rugose-surface model) describes the hydrodynamic behavior of globular proteins to within about 1%, whereas the smooth model is off by nearly 20%. This is in general agreement with the conclusions concerning the rugosity of the surface reached by Teller *et al.*,[42] who used a more elaborate method to calculate the frictional coefficient, with data from a different set of proteins.

Estimation of sedimentation coefficients from SAXS. With these considerations in mind, we turn to using SAXS results in an attempt to predict sedimentation coefficients. In Table 35.1, the radius of gyration, R_G, volume, V, and surface-to-volume ratio, S/V, are listed for our set of 19 proteins. Also tabulated are the partial specific volumes, $\bar{v}$, and the anhydrous molecular weights, M (obtained in most cases from the amino-acid sequence or composition), as well as indications of the geometric model which best describes the scattering particle as determined by SAXS. Experimental values of S/V, however, are available only for proteins 1 through 11. Values for the others had to be calculated from their smooth-surface model. Axial ratios calculated for each protein from the SAXS results of Table 35.1 and equation (3a) for the $3V/(4\pi R_G^3)$ relationship, and equations (3b) or (3c) (as the case may be) for the $(R_G S/V)$ relationship, are listed as $(a/b)_1$ and $(a/b)_2$, respectively.

For each of the proteins for which both values are available, $(a/b)_2$ is larger than $(a/b)_1$ when the model is prolate, the reverse when oblate. The differences become somewhat less as the molecular weight of the protein increases; this would be consistent with the notion that the rugae (which presumably remain of about constant average dimensions) affect the flow lines to a lesser extent as the volume of the particle increases. Frictional ratios for $(a/b)_1$ and $(a/b)_2$, calculated from equations

(2a) and (2b), are listed as $(f/f_o)_1$ and $(f/f_o)_2$, respectively. From the molecular weights, partial specific volumes, and frictional ratios, sedimentation coefficients are obtained for the smooth-surface (s_1) and rugose-surface (s_2) models by means of Svedberg's equation [equation (1a)], with the Stokes radius in this equation calculated from the scattering volume listed in the table. These values as well as the experimentally determined $s^o_{20,w}$ are compiled in Table 35.1.

Whereas s_1 values are consistently larger than $s^o_{20,w}$, s_2 generally is very close to $s^o_{20,w}$, in agreement with Teller's conclusion that the hydrodynamic behavior of proteins is influenced by the rugose accessible surface area.[42] The agreement between s_2 and $s^o_{20,w}$ is particularly close for the holo- and apo-forms of several proteins in this data set, viz., riboflavin-binding protein and glyceraldehyde-3-phosphate dehydrogenase. In these cases, $s^o_{20,w}$ values change owing to some configurational change in the protein, and the calculated s_2 values evidently follow these changes quite faithfully.

It is to be noted that with proteins 12–18 any differences between $(a/b)_1$ and $(a/b)_2$ (and consequently between $(f/f_o)_1$ and $(f/f_o)_2$, and between s_1 and s_2) are not due to rugosity since, in the absence of experimental S/V values, the rugosity could not be taken into account. Instead, in cases 10a, 14, and 18, where S/V was calculated from models of smooth ellipsoids, the information content of $R_G S/V$ must be identical to that of $3V/(4\pi R_G^3)$; therefore only one axial ratio and one s is calculated and listed (designated here as s_1, as the designation s_2 would incorrectly imply that an independent S/V was involved). In cases of other smooth bodies, such as cylinders (Nos. 12, 13, 15-18), there will be a difference between $(a/b)_1$ and $(a/b)_2$, and thus between s_1 and s_2. This is because these bodies have been arbitrarily represented by ellipsoids of equal volume, for the practical reason that frictional ratios for ellipsoids can be readily calculated by means of Perrin's equations. These differences will not, therefore, reflect rugosity but the excess surface due to difference in model (elsewhere[21] termed S_B, the excess surface due to body shape other than ellipsoidal, as distinguished from S_X, the additional contribution to surface area due to rugose surface texture). To the extent that this additional surface affects hydrodynamic properties, s_2 in these cases also should afford the better estimate of $s^{\circ}_{20,w}$.

When it is considered that these SAXS data were compiled from scattered and sometimes fragmentary sources (as illustrated by the references to the table) ranging over a period of nearly three decades, and that these represent observers of varied backgrounds, interests, and familiarity with the technique and using instruments of various types as well as different methods of data evaluation, the agreement shown in the table appears remarkably good.

Discussion

Hydration. Up to this point, no assumptions were made concerning the hydration of the protein. SAXS results, which implicitly contain the hydration term, were simply used to calculate sedimentation coefficients. In order to deal with problems of hydration, it is helpful to use a multicomponent expression for the sedimentation coefficient adapted from Schachman:[47]

$$\mu_{123} = \mu_2 + (km_1 + \alpha)\mu_1 + km_3\mu_3. \qquad (7a)$$

Here μ_{123} is the total chemical potential of the sedimenting unit containing component 1 (water), 2 (macromolecule), and 3 (salt); μ_i and $m_i (i = 1, 2, 3)$ are the chemical potential and molality of the respective component; k is a proportionality constant equal to the ratio of the amount of salt proportionally bound, in moles of salt per mole of protein, to the molality of salt in the bulk of the solution (or, what is equivalent, to the ratio of the fraction of salt bound, to the molality of the protein); and α is the preferential interaction of the protein, *i.e.*, if positive, the hydration beyond that corresponding to the bulk ratio of water to salt, in moles water preferentially bound per mole protein. It is apparent that there is a total of $(km_1 + \alpha)$ moles of water and km_3 moles of salt per mole of protein bound to the macromolecule. The term α is related to the preferential salt binding $(\partial m_3/\partial m_2)_\mu$, used in investigations with other experimental techniques, by the expression

$$\alpha = -(m_1/m_3)(\partial m_3/\partial m_2)_\mu. \qquad (7b)$$

Differentiating equation (7a) with respect to pressure at constant temperature, combining the result with the transport equation (1a) with due regard for the makeup of the sedimenting unit, letting M_i and $\bar{v}_i (i = 1,2,3)$ designate the molecular weight and partial specific volume of the respective component, and making use of the facts that sedimentation coefficients are routinely extrapolated to zero protein concentration (*i.e.*, $\bar{v}_1 \cong 1, \rho \cong 1$) and that the experiments are performed in approximately 0.1 M salt solutions (*i.e.*, $m_3 \ll 1$) so that all terms beyond the first become negligible, yields

$$s^{\circ}_{20,w} \cong \frac{M_2(1 - \bar{v}_2\rho)}{f_{123}N} \tag{8a}$$

and

$$f_{123} = (f/f_o)_{123} 6\pi\eta N (r_o)_{123}. \tag{8b}$$

Here the subscript '123' refers to the sedimenting unit as a whole and $(r_o)_{123}$, the Stokes radius of the sedimenting unit, equals $(3V_{123}/4\pi)^{\frac{1}{3}}$. V_{123}, the hydrated volume from SAXS, is related to $\bar{v}_{123}$, the partial specific volume of the sedimenting unit, by $V_{123} = \bar{v}_{123}M_{123}/N$. Differentiation of equation (7a) as before,[47] followed by replacing the molal units by concentrations in grams per gram of water, g_i (so that k and α are replaced by corresponding quantities in these units, $k' = 1000k/M_2$ and $\xi_1 = \alpha M_1/M_2$, and $(k'g_1 + \xi_1)$ by its equivalent, the total hydration A_1), results in

$$V_{123}N/M_2 \cong \bar{v}_2 + A_1\bar{v}_1. \tag{9}$$

It can be seen that the total hydration lowers $s^{\circ}_{20,w}$ by way of the hydrated volume term in equation (8b). The effect of salt binding in this respect is negligible as long as salt concentrations are of the order indicated above. In solutions of high salt concentration, or even moderate concentration when salt binding is strong (*i.e.*, when the preferential salt binding is positive and the preferential hydration in consequence is negative (*cf.* equation (7b)), salt will contribute to the solvated volume by way of the here neglected third term in equation (7a). Apart from this, salt enters into equations (8a) and (8b) only through its effect on $(f/f_o)_{123}$, with lowered $s^{\circ}_{20,w}$ again the likely result.

In the last column of Table 35.1 are listed the values of A_1 obtained from the SAXS volume (recalculated from emended values of Ref. 6). Here, the first nine proteins have an average value of 0.280 g per g of protein, in fair accord with generally assumed values of about 0.25 g water/g protein.[3,9,40,42,48,49] The last ten, which have higher molecular weights (>100,000) and are actually oligomeric structures, have a value of 0.444 g water/g protein. This higher value might be expected since the phenomenon of trapped solvent (internal solvation) has been observed in such multi-subunit structures as casein micelles, viruses, and aspartate transcarbamylase.[3]

With regard to the solvation effects on $(f/f_o)_{123}$, while it is not possible from the present approach alone to determine exactly where the solvent (water and salt) binding sites on a protein are located, it is possible to compare the (f/f_o) evaluated from X-ray diffraction data with the $(f/f_o)_{123}$ from SAXS.

Structural comparisons of SAXS and X-ray diffraction. A comparison of volumes from SAXS with theoretical volumes derived from the X-ray diffraction structure according to Teller[45] is shown in Fig. 35.3. The SAXS solution volume is seen to be consistently higher than the packing volume from the crystallographic structure. Fitting a least-squares straight line with zero intercept to the SAXS volume *vs* molecular weight plot gives a slope of 1.964 ± 0.045, while the corresponding slope for the diffraction data is 1.27.

Further, the SAXS surface area (Fig. 35.4) can be compared with the accessible surface area according to Teller.[45] Here, the SAXS surface area is slightly lower, and fitting a straight line with zero intercept to the data as a function of $M^{\frac{2}{3}}$ gives a slope of 9.49 ± 0.25, while Teller's value is 11.12. (It may be added that each of the above calculations can also be attempted with a polynominal of degree 2, *i.e.*, with extra terms in M^2 for the volume, and in $M^{\frac{4}{3}}$ for the surface area, but such extra terms are found to result in no statistically significant differences). The SAXS solution volume of a protein, therefore, is found to be larger than the X-ray crystallographic volume, whereas the surface area in solution is slightly lower than the crystallographic accessible surface area. The increase in volume can be expected because of solvation effects; other factors being equal,

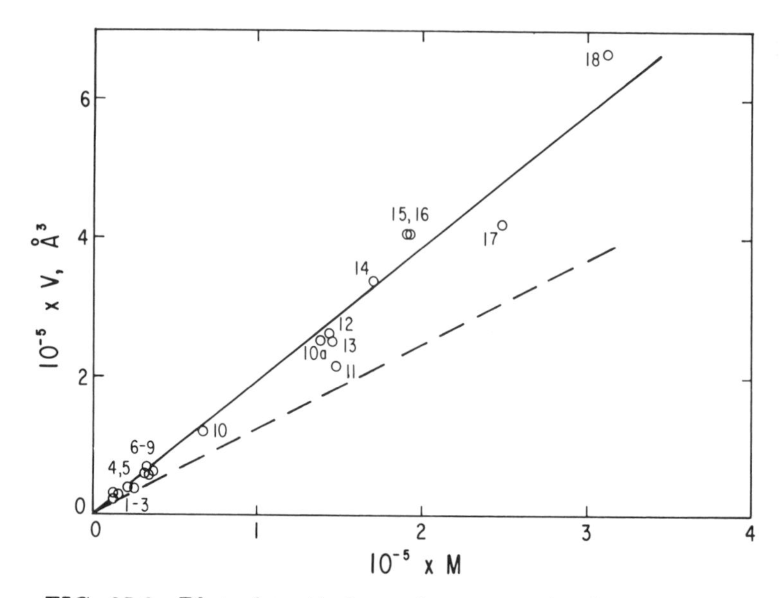

FIG. 35.3. Plot of scattering volume *vs* molecular weight for the 19 proteins in Table 35.1. Solid line (SAXS): linear least-squares fit, with slope 1.964 ± 0.045. Dashed line: from X-ray crystallographic structure data [*cf.* equation (5b)], with slope 1.27. (Adapted from Fig. 3 of Ref. 6).

such an increase would be expected also to yield a correspondingly increased surface area. The contrary area decrease actually observed appears to indicate that the binding of solvent to the macromolecule results in less rugosity, less anisotropy, or a combination of both these effects. In fact, the binding sites may tend to lie within the clefts or grooves of the macromolecule, thus lessening the rugosity. A rough comparison of anisotropies is furnished by $(a/b)_2$, obtained from both the fitted SAXS and crystallographic results, *i.e.*, $A_s = 9.49M^{\frac{2}{3}}$ and $V = 1.964M$, equation (5a), *vs* 11.12 $M^{\frac{2}{3}}$ and 1.27 M, equation (5b) (R_G, needed here only to compare the two types of structure, not to obtain estimates of absolute values for them, has to be approximated by $(3/5)^{\frac{1}{2}}(3V/4\pi)^{\frac{1}{3}}$, equation (5c), though this refers to a sphere; the implication $(a/b)_1 = 1$ would, of course, be contradicted by the data of Table 35.1, where the average $(a/b)_1$ is nearer 1.9). In these circumstances, one finds an average $(a/b)_2$ for a prolate ellipsoid of 2.1 from SAXS, compared to 4.0 from crystallography.

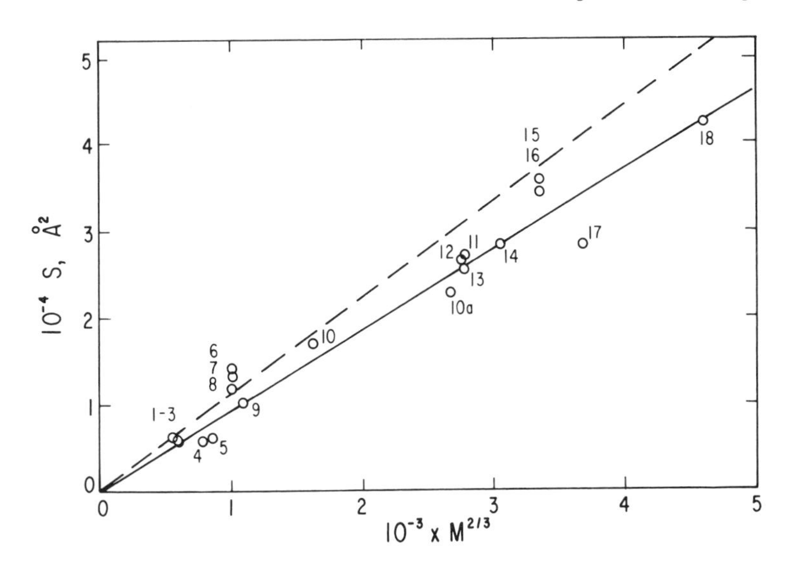

FIG. 35.4. Plot of surface area from small-angle X-ray scattering *vs* $\frac{2}{3}$ power of molecular weight for the 19 proteins in Table 35.1. Solid line (SAXS): linear least-squares fit, with slope 9.49 ± 0.25. Dashed line (X-ray diffraction): accessible surface area computed from three-dimensional X-ray structure [*cf.* equation (5a)], with slope 11.12. (Adapted from Figure 4 of Ref. 6.)

However, the increase in SAXS volume over the X-ray crystallographic volume could be due also to electrostriction of the proteins upon crystallization. The concept of a dynamic alteration of protein conformation in solution ('breathing') has been previously introduced.[5] Whether the observed increase in volume is due to binding of solvent components or to the breathing of the macromolecule cannot be resolved without extensive additional studies. These would include sedimentation in $H_2^{17}O$ and $H_2^{18}O$ for increased solvent density, and small-angle neutron scattering with the use of $H_2^{17}O$ to avoid the increased hydrophobic interactions known to occur in D_2O.

References

1. Kratky, O. *Q. Rev. Biophys.* **11**, 39–78 (1978).
2. Scheraga, H. and Mandelkern, L. *J. Am. Chem. Soc.* **75**, 179–184 (1953).

3. Kuntz, I. D. and Kauzmann, W. *Adv. Protein Chem.* **28**, 239–345 (1974).
4. Creighton, T. E. *Prog. Biophys. Mol. Biol.* **33**, 231–297 (1978).
5. Pessen, H., Kumosinski, T. F. and Timasheff, S. N. *Methods Enzymol.* **27**, 151–209 (1973).
6. Kumosinski, T. F. and Pessen, H. *Arch. Biochem. Biophys.* **219**, 89–100 (1982).
7. Pessen, H., Kumosinski, T. F. and Timasheff, S. N. *J. Agric. Food Chem.* **19**, 698–702 (1971).
8. Dayhoff, M. O. (ed.) *Atlas of Protein Sequence and Structure.* Vol. **5**. National Biomedical Research Foundation: Washington, DC (1972).
9. Lee, J. C. and Timasheff, S. N. *Biochemistry* **13**, 257–265 (1974).
10. Yphantis, D. A. *J. Phys. Chem.* **63**, 1742–1747 (1959).
11. Sophianopoulos, A. J., Rhodes, J. K., Holcomb, D. N. and Van Holde, K. E. *J. Biol. Chem.* **237**, 1107–1112 (1962).
12. Kronman, M. J. and Andreotti, R. E. *Biochemistry* **3**, 1145–1160 (1964).
13. Krigbaum, W. R. and Godwin, R. W. *Biochemistry* **7**, 3126–3131 (1968).
14. Schwert, G. W. *J. Biol. Chem.* **179**, 655–664 (1949).
15. Schwert, G. W. and Kaufman, S. *J. Biol. Chem.* **190**, 807–810 (1951).
16. Wilcox, P. E., Kraut, J., Wade, R. D. and Neurath, H. *Biochim. Biochim. Biophys. Acta* **24**, 72–78 (1957).
17. Vazina, A. A., Lednev, V. V. and Lemazhikin, B. K. *Biokhimiya (Moscow)* **31**, 629–633 (1966).
18. Rajagopalan, T. G., Moore, S. and Stein, W. H. *J. Biol. Chem.* **241**, 4940–4950 (1966).
19. McMeekin, T. L., Wilensky, M. and Groves, M. L. *Biochem. Biophys. Res. Commun.* **7**, 151–156 (1962).
20. Williams, R. C. Jr. and Rajagopalan, T. G. *J. Biol. Chem.* **241**, 4951–4954 (1966).
21. Kumosinski, T. F., Pessen, H. and Farrell, H. M., Jr. *Arch. Biochem. Biophys.* **214**, 714–725 (1982).
22. Witz, J., Timasheff, S. N. and Luzzati, V. *J. Am. Chem. Soc.* **86**, 168–173 (1964).
23. Kumosinski, T. F. and Timasheff, S. N. *J. Am. Chem. Soc.* **88**, 5635–5642 (1966).
24. Luzzati, V., Witz, J. and Nicolaieff, A. *J. Mol. Biol.* **3**, 379–392 (1961).

25. Miller, G. L. and Golder, R. H. *Arch. Biochem. Biophys.* **36**, 249–258 (1952).
26. Pesce, A., Fondy, T. P., Stolzenbach, F., Castillo, F. and Kaplan, N. O. *J. Biol. Chem.* **242**, 2151–2167 (1967).
27. Durchschlag, H., Puchwein, G., Kratky, O., Schuster, I. and Kirschner, K. *Eur. J. Biochem.* **19**, 9–22 (1971).
28. Jaenicke, R., Schmid, D. and Knof, S. *Biochemistry* **7**, 919–926 (1968).
29. Jaenicke, R. and Gratzer, W. B. *Eur. J. Biochem.* **10**, 158–164 (1969).
30. Zipper, D. and Durchschlag, H. *Eur. J. Biochem.* **87**, 85–99 (1978).
31. Schmid, G., Durchschlag, H., Biedermann, G., Eggerer, H. and Jaenicke, R. *Biochem. Biophys. Res. Commun.* **58**, 419–426 (1974).
32. Müller, K., Kratky, O., Röschlau, P. and Hess, B. *Hoppe-Seyler's Z. Physiol. Chem.* **353**, 803–809 (1972).
33. Bischofberger, H., Hess, B. and Röschlau, P. *Hoppe-Seyler's Z. Physiol. Chem.* **352**, 1139–1150 (1971).
34. Malmon, A. G. *Biochim. Biophys. Acta* **26**, 233–240 (1957).
35. Sumner, J. B. and Gralén, N. *J. Biol. Chem.* **125**, 33–36 (1938).
36. Pilz, I. and Sund, H. *Eur. J. Biochem.* **20**, 561–568 (1971).
37. Reisler, E., Pouyet, J. and Eisenberg, H. *Biochemistry* **9**, 3095–3102 (1970).
38. Perrin, F. *J. Phys. Radium* **7**, 1–11 (1936).
39. Luzzati, V., Witz, J. and Nicolaieff, A. *J. Mol. Biol.* **3**, 367–378 (1961).
40. Oncley, J. L. *Annals N.Y. Acad. Sci.* **41**, 121–150 (1941).
41. Svedberg, T. and Pedersen, K. D. *The Ultracentrifuge.* p. 22. Clarendon Press: Oxford, UK (1940).
42. Teller, D. C., Swanson, E. and De Haën, C. *Methods Enzymol.* **61**, 103–124 (1979).
43. Richards, F. M. *Annu. Rev. Biophys. Bioeng.* **6**, 151–176 (1977).
44. Squire, P. G. and Himmel, M. E. *Arch. Biochem. Biophys.* **196**, 165–177 (1979).
45. Teller, D. C. *Nature (London)* **260**, 729–731 (1976).
46. Chothia, C. *Nature (London)* **254**, 304–308 (1975).
47. Schachman, H. K. *Ultracentrifugation in Biochemisty.* p. 229. Academic Press: New York, NY (1959).
48. Cohn, E. J. and Edsall, J. T. *Proteins, Amino Acids and Peptides.* pp. 428 ff. Reinhold Publishing Company: New York, NY (1943).
49. Tanford, C. *Physical Chemistry of Macromolecules.* pp. 359 ff. John Wiley: New York, NY (1961).

36. The Patterson function in fiber diffraction

Gerald Stubbs

The cylindrically averaged Patterson function, the fiber diffraction analog of the crystallographic Patterson function, has not played a major part in the determination of structures by fiber diffraction methods. Part of this is undoubtedly due to its difficulty of interpretation, but I suggest in this paper that by building up a sufficient body of experience, and by carefully choosing the conditions under which it should be used, the function may yet have a significant contribution to offer to fiber diffraction.

Diffracted intensities from a fiber diffraction specimen (such as a rod- shaped virus or a helical nucleic acid structure) are cylindrically averaged. This is because the particles in a fiber or an oriented gel are randomly oriented about their long axes. Solving the phase problem under these circumstances is complicated by the necessity of separating the terms overlapping as a consequence of the cylindrical averaging; it has traditionally been done by model-building and comparing the intensity calculated from the models with the cylindrically averaged diffraction data, but methods such as isomorphous replacement were adapted very early to simple fiber diffraction problems (early examples[1,2]) and in recent years have been extended to determine structures of considerable complexity at moderately high resolutions.[3]

Although a fiber diffraction pattern cannot be used to calculate a Patterson synthesis, MacGillavry and Bruins[4] showed that it can be used to calculate the cylindrically averaged Patterson function. This is

$$Q(r,z) = \sum_{l=-\infty}^{\infty} \int_{0}^{\infty} I_l(R) J_o(2\pi Rr) 2\pi R dr \cos 2\pi \frac{lz}{c}$$

where r and z are real space coordinates, $I_l(R)$ is the diffracted X-ray intensity (suitably corrected for geometric and other factors) on layer line l at reciprocal space radius R; J_o is the zero order Bessel function and c is the repeat distance along the specimen axis. Calculation of the function Q does not require any knowledge of the symmetry of the diffracting system, other than the repeat distance c. The Patterson map can be somewhat simplified by omitting the equator[5] or the centrosymmetric low-radius part of the equator[6] from the synthesis. This has the effect of removing from the map vectors originating in the axially invariant or the cylindrically symmetric parts of the structure. Without some such omission, the map will be dominated by vectors from a cylinder having the same dimensions as the structure under study, and important vectors may be obscured.

Application to DNA

Franklin and Gosling[7,8] calculated the cylindrically averaged Patterson function for the A form of DNA. A-DNA is a crystalline fiber; the molecules are packed in a crystalline array, but microcrystalline regions are randomly oriented about the fiber axis with respect to each other, so that the diffraction pattern is still cylindrically averaged. Such a diffraction pattern contains reflections which are often difficult to index, and Franklin and Gosling[7] were able to use a large peak corresponding to one of the lattice translations to help determine the unit cell parameters and thus index the reflections. The function also showed two sinusoidal distributions of peaks. Such distributions are highly characteristic of helical structures, and in this case clearly indicated the presence of a double-helical DNA structure similar to that which had been found by Watson and Crick[9] in B-DNA.

Tobacco mosaic virus

A cylindrically averaged Patterson map for tobacco mosaic virus (TMV) was published by Rosalind Franklin in 1955.[10] She used data to a nominal resolution of 3 Å. A Patterson map using 4 Å data (the data of Namba and Stubbs[3]) is shown in Fig. 36.1(a). One of the strongest features is a ring of density 11 Å from the origin, and it is possible to discern an array of peaks along the r axis at approximately 11 Å intervals. This array

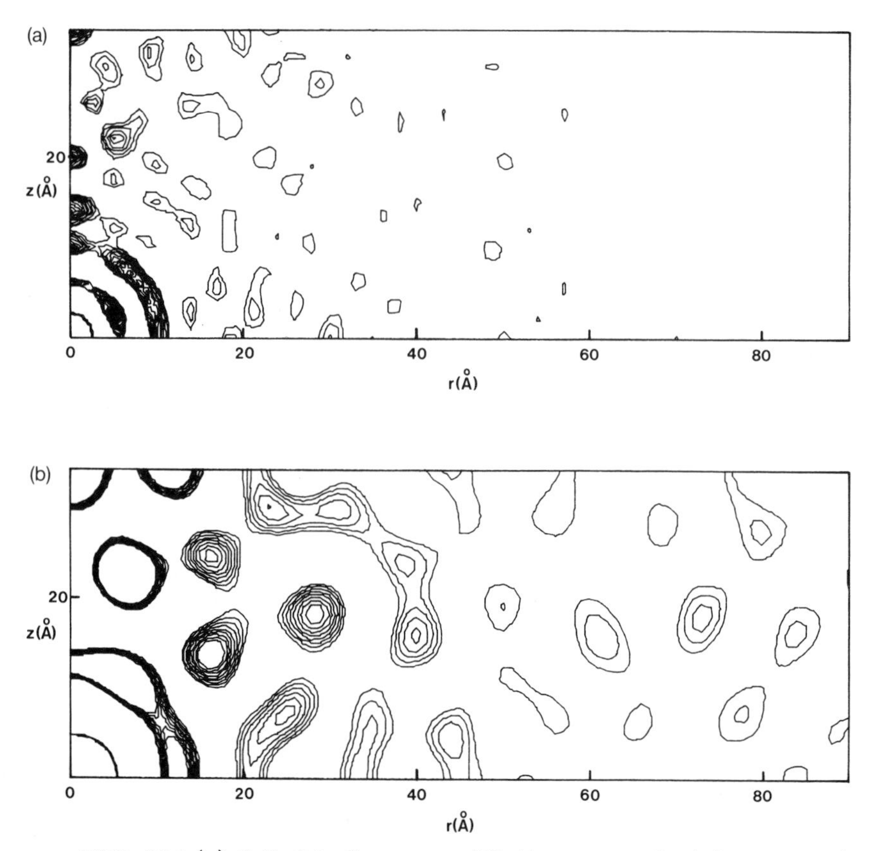

FIG. 36.1 (a) Cylindrically averaged Patterson map for tobacco mosaic virus. Data extending to 4Å resolution were used. Higher contours in the origin peak are not shown. (b) Cylindrically averaged Patterson map for tobacco mosaic virus from the data used for (a), but truncated to 8Å resolution. Contours above the seventh are not shown.

was much clearer in Franklin's map, extending to the seventh peak, and was accompanied by a parallel array 11 Å higher in z. There is a peak at (0, 34.5), which will be discussed below, and a peak at (7, 23) corresponding to the symmetry of the virus ($16\frac{1}{3}$ subunits per turn, with a 23 Å translation in z between turns).

Fraser[11] had suggested on the basis of infra-red measurements that the virus contained α-helices running perpendicular to the virus axis, and Watson[12] surmised that the 23 Å pitch of the virus helix could accommodate a double layer of helices. Franklin correctly interpreted her map as confirmation of these theories, but she assumed from the even spacing of the double layer of peaks that the helices ran tangentially with respect to the axis. The determination of the structure by multidimensional isomorphous replacement methods[13] shows that in fact they run approximately radially.

Fig. 36.1(b) is the Patterson from Fig. 36.1(a), but using data truncated at 8 Å resolution. Franklin's array of peaks is now clearly visible; indeed, it is much clearer in this map than in Franklin's original map. Fiber diffraction data are inherently difficult to measure at high resolution, since disorientation of the sample causes layer lines to overlap in the equatorial region, while data are smeared along the layer lines in the meridional region. While modern techniques permit these problems to be overcome to a considerable degree,[14] it seems likely that the effective resolution of Franklin's data was much lower than the nominal 3 Å, being in fact between the resolution of Fig. 36.1(a) (4 Å) and that of Fig. 36.1(b) (8 Å).

The map in Fig. 36.1(b) can be used to derive the correct packing of the α-helices in TMV. The double row of peaks may be interpreted, as it was by Franklin, as a double row of α-helices. The 11 Å separation of the peaks is typical of the separations expected in α-helices packed in a parallel array. From the very large peak at (0, 34.5), we may conclude that two helices are related by a vertical displacement of 34.5 Å, or $1\frac{1}{2}$ turns of the virus helix. Fig. 36.2 represents a surface lattice of TMV, that is, a surface at constant radius in the TMV cylinder which has been flattened. The parallelogram marks the boundary of an asymmetric unit. With no loss of generality, the two α-helices may be marked as the solid circle in the bottom left corner of the asymmetric unit and the open circle directly above it. The arrow represents the (0, 34.5) vector. The symmetry equivalent of the open circle at the tip of this vector is the open circle in the top right corner of the parallelogram.

If the helices run tangentially, that is, approximately horizontally in the plane of Fig. 36.2, the second layer of helices would appear at low resolution to be directly over the first

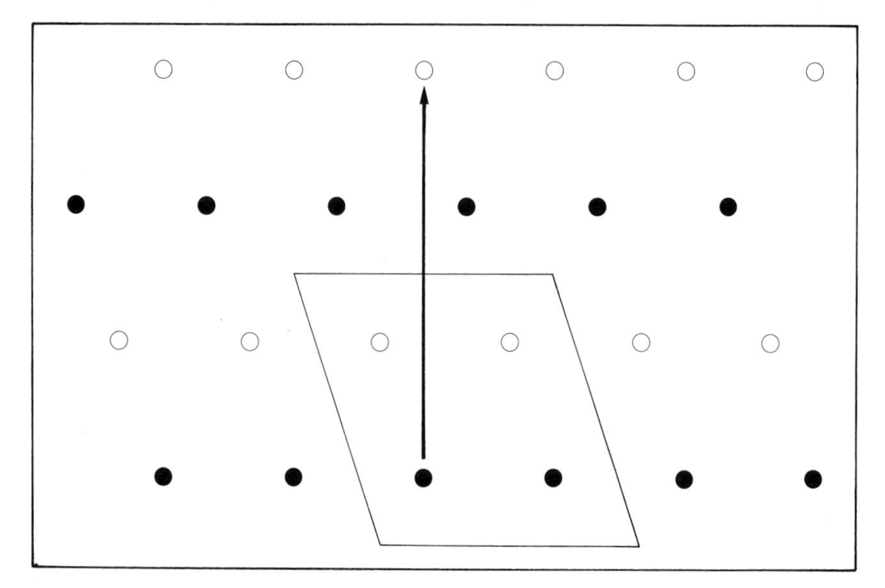

FIG. 36.2 Possible arrangement of features (α-helices) in a surface at constant radius in the TMV cylinder. The parallelogram marks the boundary of an asymmetric unit. The arrow corresponds to the vector at (0, 34.5) in Figure 36.1.

layer of helices, and thus the second layer of Patterson peaks would be directly over the first layer of peaks. This is not the case. It should be noted that this argument was not available to Franklin, since the (0, 34.5) peak in her map is much weaker than the one shown here. There are two possible ways to pack helices to give the required staggered array of Patterson peaks. In one, each helix has as its origin the bottom left solid circle, the top right open circle, or one of their symmetry equivalents. The α-helices run approximately tangentially, but spiralling towards the center of the virus helix. This is a similar packing to that found in the bacteriophage Pf1,[15] except that in Pf1 the α-helices run approximately axially. In this model the α-helices must be very long, in order to provide a sufficiently large array to account for the extended array of Patterson peaks. This seems extremely unlikely, as the Patterson map contains none of the rod-like features which one sees, for example, in the Patterson map of hemoglobin[16] or the cylindrically averaged Patterson map of poly-γ-methyl-L-glutamate.[17]

The remaining possible arrangement has the α-helices running radially. In this case, every circle, filled or open, in Fig. 36.2 represents the origin of an α-helix. The resulting array, four short α-helices running radially, with the second layer displaced laterally about one third of a unit cell from the first, is just what has been found by isomorphous replacement to be the case.

Collagen

In 1955, Yakel and Schatz published a cylindrically averaged Patterson map for collagen.[17] They calculated $rQ(r,z)$ rather than $Q(r,z)$, correctly pointing out that this function reflects the total amount of density being projected onto the Patterson section. However, it produces high levels of noise at large values of r, as well as peaks near the z-axis which were widely misinterpreted at the time, and the method was not used by later workers. The main feature of the collagen Patterson map was a ridge of density running parallel to the z-axis at $r = 5$ Å. Yakel and Schatz were unable to draw any detailed conclusions from this map, although they tentatively favored an open helical structure of average diameter 5 Å.

This work was done before the considerably improved collagen diffraction patterns obtainable by stretching the fiber[18] were widely available. Fig. 36.3 is the cylindrically averaged Patterson map for collagen, calculated using data taken from Rich and Crick.[19] This map reflects the now widely accepted collagen structure: the peaks on the z-axis spaced about 2.9 Å apart correspond to the vectors between residues on the same protein chain, while the row of peaks on the $(5,z)$ line is a clear indication of the parallel-chain structure of collagen. (The super-helical nature of the collagen structure cannot be reflected in this Patterson map, because the map was derived from data which were assumed to lie on evenly spaced layer lines.)

Feather Keratin

Schor and Krimm[20,21] calculated part of a cylindrically averaged Patterson function for feather keratin. This enabled

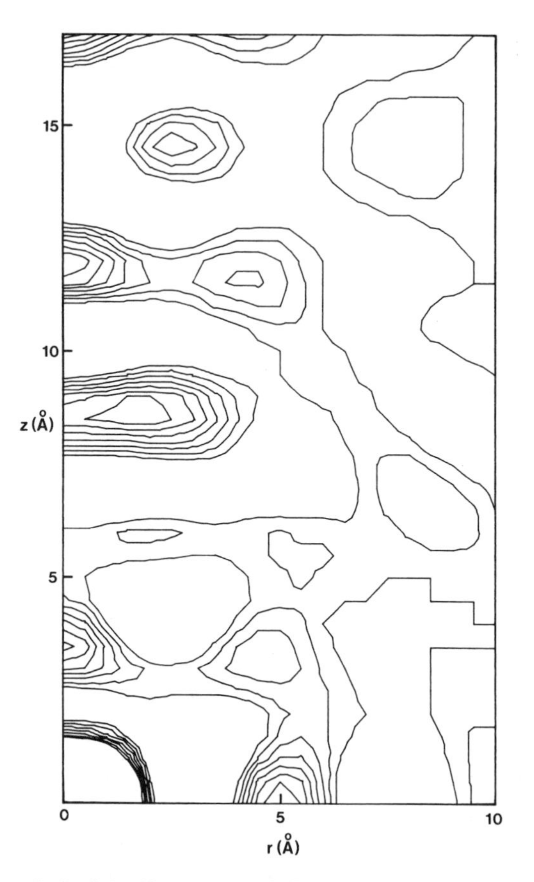

FIG. 36.3 Cylindrically averaged Patterson map for collagen, calculated using data taken from Rich and Crick.[19]

them to show that feather keratin was not α-helical and did not have a collagen-like structure. They recognized that the structure was a helical form of β-structure, and proposed as a model an extended 10-stranded β-barrel. This model accounted for some of the important features of the diffraction pattern, but agreed only moderately well with the Patterson map. In particular, the β-barrel model predicted only one peak on the z-axis, at $z = 19$ Å, whereas there is in fact a rich array of axial peaks.

Fraser *et al.*,[22] proposed an alternative model, a twisted pair of 4-stranded β-sheets. Feather keratin has far too complex a structure to infer from the Patterson function alone, and in fact Fraser *et al.* did not consider the function at all, but it is of interest to note that their model predicts the features of Schor and Krimm's map very well. In particular, it predicts peaks on the z-axis at $z = 14$, 19 and 29 Å and near the z-axis at $z = 7$, 21, and 26 Å. All of these peaks are found.

Recent applications to muscle

Although the cylindrically averaged Patterson function has been little used in the last twenty years, several applications have been made recently, in particular by Japanese groups. Kataoka and Ueki[23] considered the application of the method to diffraction from biomembranes, and two groups have applied it to muscle.

Namba, Wakabayashi and Mitsui[5] calculated cylindrically averaged Patterson maps for the relaxed and rigor states of the thin filaments in crab striated muscle, and used them to determine the sites of attachment of the myosin heads to the thin filaments in the rigor state. They used data to about 50 Å resolution, and obtained fairly simple, typically helical maps. Their analysis was based largely on a significant increase in the intensity of two peaks, corresponding to vectors between the bound myosin heads. Maps calculated for other possible models did not agree with the map based on the observed data.

Tajima, Kamiya and Seto[24] calculated a Patterson map for the thin filament of molluscan smooth muscle, and used it in a derivation of the shape of the actin molecule. The map was very close to that expected from a smooth helix, which supported their claim that actin is elongated along the genetic helix of the filament.

Conclusions

It is evident from these examples that the cylindrically averaged Patterson function can be of considerable value, particularly in guiding the process of model-building in conjunction with other constraints, and in assessing the relative merits of alternative models. Several cautionary considerations should be kept in mind, however. Data should be of as high quality as possible, particularly meridional data. This was of great importance in the case of collagen. Partial Pattersons (that is, maps calculated omitting part or all of the equatorial data) are much more informative than full Pattersons. The figures in the paper by Namba *et al.*[5] illustrate this very well. Finally, it is extremely important to use data at a resolution appropriate to the information expected from the map. Because of cylindrical averaging, these Pattersons contain a great deal of information in a small area, and one can only hope to interpret vectors coming from a small number of molecular features. Thus, high resolution is appropriate for systems like DNA and collagen, which are made up of repeated fairly simple units, but entirely inappropriate for TMV and muscle. The complexity of TMV (the molecular weight of one coat protein subunit and its associated three nucleotides of RNA is about 18,000) is such that anything smaller than a basic unit of secondary structure such as an α-helix would be lost in uninterpretable detail in a high-resolution map. The work described here has demonstrated that 8 Å is of the order of resolution appropriate to such a problem. The molecules that make up muscle filaments are considerably larger again, and the complexity of their arrangements still greater. In the thin filaments studied by the Japanese groups, vectors between whole molecules were all that could reasonably be interpreted, and so 50 Å resolution was appropriate.

With all these considerations allowed, the cylindrically averaged Patterson map is easy to compute and informative, and should always be considered at the start of a fiber diffraction analysis.

Acknowledgement

This work was supported by grant GM33265 from the National Institutes of Health.

References

1. Franklin, R. E. *Nature (London)* **177**, 928-930 (1955).
2. Caspar, D. L. D. *Nature (London)* **177**, 928 (1955).
3. Namba, K. and Stubbs, G. *Acta Cryst.* **A41**, 252–262 (1985).
4. MacGillavry, C. H. and Bruins, E. M. *Acta Cryst.* **1**, 156-158 (1948).

5. Namba, K., Wakabayashi, K. and Mitsui, T. *J. Mol. Biol.* **138**, 1-26 (1980).
6. Holmes, K. C. Ph.D. Thesis. University of London: London, UK (1959).
7. Franklin, R. E. and Gosling, R. G. *Acta Cryst.* **6**, 678-685 (1953).
8. Franklin, R. E. and Gosling, R. G. *Nature (London)* **172**, 156-157 (1953).
9. Watson, J. D. and Crick, F. H. C. *Nature (London)* **171**, 737-738 (1953).
10. Franklin, R. E. *Nature (London)* **175**, 379-381 (1955).
11. Fraser, R. D. B. *Nature (London)* **170**, 491 (1952).
12. Watson, J. D. *Biochim. Biophys. Acta* **13**, 10-19 (1954).
13. Stubbs, G., Warren, S. and Holmes, K. *Nature (London)* **267**, 216-221 (1977).
14. Makowski, L. *J. Appl. Cryst.* **11**, 273-283 (1978).
15. Makowski, L., Caspar, D. L. D. and Marvin, D. A. *J. Mol. Biol.* **140**, 149-181 (1980).
16. Perutz, M. F. *Proc. Roy. Soc. (London)* **A195**, 474-499 (1949).
17. Yakel, H. L. and Schatz, P. N. *Acta Cryst.* **8**, 22-25 (1955).
18. Cowan, P. M., North, A. C. T. and Randall, J. T. In *Nature and Structure of Collagen.* (J. T. Randall, ed.) Butterworths: London, UK (1953).
19. Rich, A. and Crick, F. H. C. *J. Mol. Biol.* **3**, 483-506 (1961).
20. Schor, R. and Krimm, S. *Biophys. J.* **1**, 467-487 (1961).
21. Schor, R. and Krimm, S. *Biophys. J.* **1**, 489-515 (1961).
22. Fraser, R. D. B., MacRae, T. P., Parry, D. A. D. and Suzuki, E. *Polymer* **12**, 35-56 (1971).
23. Kataoka, M. and Ueki, T. *Acta Cryst.* **A36**, 282-287 (1980).
24. Tajima, Y., Kamiya, K. and Seto, T. *Biophys. J.* **43**, 335-343 (1983).

37. The Patterson function and membrane diffraction

C. R. Worthington

The discovery by A. L. Patterson[1] in 1934 of the Patterson function has had a great influence on X-ray crystallography. The Patterson function is the autocorrelation function of the electron density (of the crystal). The properties of the Patterson function and its use in crystal structure analysis have been documented in textbooks.[2,3,4] If the unit cell of the crystal contains n atoms, there are n^2 peaks in the Patterson function. Consequently, when n becomes large, the Patterson function of a crystal of moderate complexity is difficult to interpret in practice[3] due to the multitude of peaks.

In this article the role of the Patterson function in membrane diffraction is examined. Since about 1965 extensive X-ray data (mostly one-dimensional) have been routinely recorded from many membrane systems.[5] Consider the low-angle X-ray diffraction pattern of frog sciatic nerve[6] which shows at low-resolution $h = 5$ orders of diffraction of a unit cell repeat distance of $d = 17$ Å. The Patterson function computed using these data[7,8] shows only three peaks: one at the origin, one at $x = d/2$ and a minor peak at $x \approx d/4$. An interpretation of the Patterson function in terms of an electron density strip model has been presented.[8] Although the interpretation was consistent with all known facts and was very likely to be correct, nevertheless, the question of the uniqueness of the structural solution was not answered. Fortunately, this situation changed completely with the realization of direct methods of structure analysis in membrane diffraction.

Diffraction theory

In order to describe this development it is convenient to review the diffraction theory.[9] As there are important differences in the theory and the methods of X-ray biophysics and X-ray crystallography we do not use the standard notation of X-ray crystallography.[2,3] The unit cell has electron density (e.d.) $t(x)$

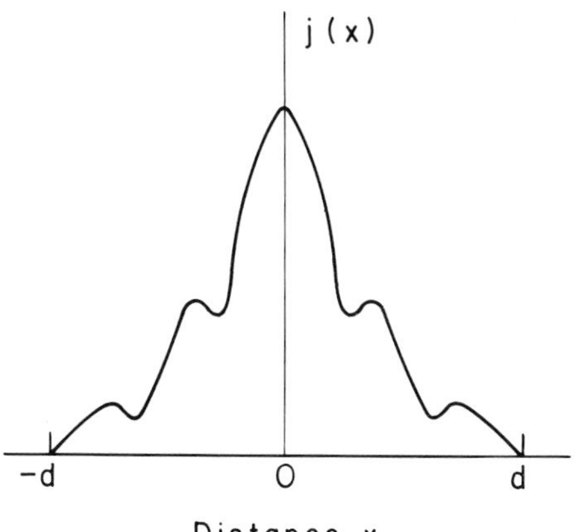

FIG. 37.1. Autocorrelation function $j(x)$ of a single unit cell of electron density $t(x)$ has width $2d$ whereas the unit cell has width d.

and width d. The unit cell is centrosymmetrical, that is, $t(x) = t(-x)$. We use the notation: $t(x) \leftrightarrow T(X)$ where $t(x)$ and $T(X)$ are a Fourier transform pair and x, X are real and reciprocal space coordinates. The autocorrelation function (a.c.f.) of the unit cell $t(x)$ is denoted $j(x)$ and is defined as follows:

$$j(x) = t(x) * t(-x) \tag{1}$$

where $(*)$ is the convolution symbol. It follows that $j(x) \leftrightarrow J(X)$, where $J(X) = |T(X)|^2$. The a.c.f. $j(x)$ has width $2d$ and is shown in Fig. 37.1.

In the general case when more than one unit cell is considered the configuration of the one-dimensional lattice enters into the calculation of the a.c.f. of the assembly. This a.c.f. is denoted $p(x)$: the symbol p is retained to correspond with the Patterson function of X-ray crystallography. The diffracted intensity $I(X)$ of the assembly is the Fourier transform of the a.c.f. and, in our notation, $p(x) \leftrightarrow I(X)$. We treat the case when the lattice points are a constant distance d apart. The lattice generating function is denoted $\phi(x)$ and in the case of N points we can write $\phi(x)$ as

$$\phi(x) = \sum_{j=1}^{j=N} \delta(x - x_j), \tag{2}$$

a finite sum of delta functions where subscript j refers to an arbitrary point in the lattice. The a.c.f. of the lattice is denoted $l(x)$ and

$$l(x) = \phi(x) * \phi(-x). \tag{3}$$

The Fourier transform of the a.c.f. $l(x)$ is $L(X)$, where $l(x) \leftrightarrow L(X)$. In diffraction theory $L(X)$ is called the interference function[2] of the lattice. The a.c.f. of N unit cells of e.d. $t(x)$ is

$$p(x) = j(x) * l(x), \tag{4}$$

and the diffracted intensity $I(X)$ of the assembly of N unit cells is

$$I(X) = J(X)L(X). \tag{5}$$

The relationship between the Patterson function $p(x)$ and the a.c.f. $j(x)$ is important in membrane diffraction.[9] From equation (4) the Patterson function is composed of a succession of a.c.f. $j(x)$ centered at intervals of d. The Patterson function $p(x)$ is shown in Fig. 37.2.

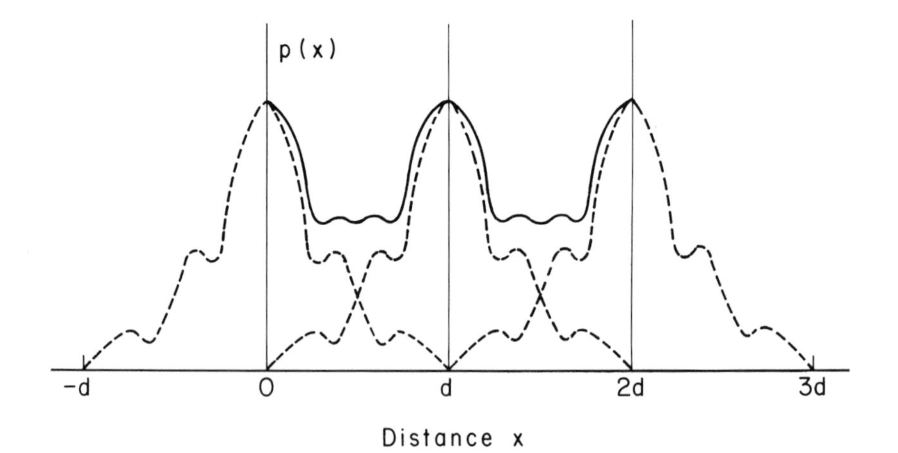

FIG. 37.2. The Patterson function $p(x)$ is a summation of a.c.f. $j(x)$ centered at intervals of d. The $j(x)$ curves are shown as dotted lines and the $p(x)$ curve is shown as a continuous line. The Patterson function $p(x)$ is a periodic function with period d and it has a center of symmetry at half-integral values of d.

Consider the case of N large such that we can ignore the weighting factors of $l(x)$. The Patterson function $p(x)$ in the interval $0 \leq x \leq d$ is

$$p(x) = j(x) + j(d - x). \tag{6}$$

In crystals the width of the a.c.f. $j(x)$ is $2d$ and $j(x)$ is not readily extracted from the Patterson function. In membranes the unit cell may contain a well-defined fluid layer and consequently the two overlapping a.c.f. $j(x)$ tend to separate. Two well known membrane systems that contain relatively wide fluid layers are swollen nerve myelin[10] and retinal photoreceptors.[11] In the case that the unit cell contains a wide fluid layer such that $d \geq 2v$ the Patterson function $p(x)$ is identical to the a.c.f. $j(x)$ in the interval $0 \leq x \leq v$. This is shown in Fig. 37.3.

We recognise that the e.d. $t(x)$ contains a membrane unit of e.d. $m(x)$ of width v and a fluid layer of e.d. F and width $d - v$. It is important in membrane diffraction to consider the minus fluid model $\Delta t(x) = t(x) - F$. In this paper no attempt is made to distinguish between the real structure $t(x)$ and $\Delta t(x)$ the minus fluid model as the diffraction theory of this aspect has been presented elsewhere.[9,12]

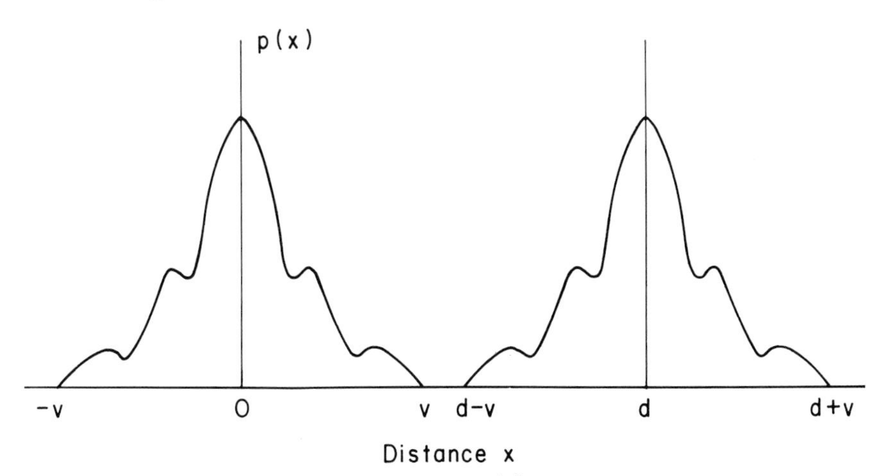

FIG. 37.3. The Patterson function $p(x)$ for a swollen membrane specimen. The Patterson function $p(x)$ is a summation of a.c.f. $j(x)$ centered at intervals of d. As $d > 2v$ the adjacent a.c.f. $j(x)$ curves do not overlap and are resolved.

Deconvolution analysis

Direct methods of structure analysis in membrane diffraction[9] refer to deconvolution analysis of the a.c.f. $j(x)$ or to reconstruction analysis using the sampling theorem or Fourier series expressions. The original theory of the deconvolution analysis of the a.c.f. $j(x)$ was first given by Hosemann and Bagchi.[13] The a.c.f. $j(x)$ can be obtained from the Patterson function $p(x)$ in two different configurations. The first configuration is when there are only a few unit cells in the diffracting assembly.[13] The second configuration is when the membranes are in a swollen state.[12]

Hosemann and Bagchi in a theoretical study[13] first treated the configuration with only a few unit cells in the diffracting assembly. As N is small, the reflections have a broadened line shape. The Patterson function of this assembly has width $2Nd$ such that $j(x)$ can be found from the boundary values of the Patterson function.[9,13] The second configuration was first studied in 1969 when it was recognised that the Patterson function of a swollen membrane specimen was identifiable with the a.c.f. of the single unit cell.[12] In this study certain model parameters were directly determined from the a.c.f. $j(x)$. The structural interpretation centered on finding the model with several parameters which gave the best agreement with the a.c.f. of the single unit cell. Improved methods are now available. The first configuration of a small number of unit cells frequently occurs in the study of oriented lipid bilayers. The feasibility of this method was first demonstrated in 1972 by Lesslauer and Blasie;[14] they studied the X-ray diffraction from a small number of layers of barium stearate.

According to diffraction theory[9,13] deconvolution of $j(x)$ provides an unique e.d. solution in the special case that $t(x)$ is centrosymmetrical. Thus, the a.c.f. $j(x)$ has a one-to-one correspondence with a centrosymmetrical n–strip electron density model. Hosemann and Bagchi[13] suggested two methods for obtaining such a model: the recursion method and the relaxation method. Two solutions $\pm s(x)$ are obtained for the e.d. $t(x)$ but the $\pm$ ambiguity can be removed using density considerations.[9] Other methods[15,16,17] have been proposed as alternatives to the recursion method. The recursion method and these three later methods, however, provide only one solution apart from the $\pm$ ambiguity. Once a solution has been obtained we still want to know whether there are other solutions which might also account for the X-ray data. The recursion method fails in this respect. On the other hand, the relaxation method can be set up to examine each possible solution. The relaxation method is not straightforward to apply in practice and it is seldom used. So far, this method has only been used in membrane diffraction on one occasion[10] although it had been once used in an earlier study of line broadening.[18] Alternative methods to the relaxation method have been proposed. These are the Fourier synthesis method[19] which is a simplified version of the relaxation method and the reconstruction method using the modulus of the continuous Fourier transform.[20]

Data processing

From the above theory, a structural solution can be found using one or more methods of deconvolution analysis. Although, in this article, we are primarily concerned with the role of the Patterson function in membrane diffraction, it is, however, only fair to point out that, even after recording the experimental X-ray data, there is a degree of difficulty involved in obtaining the diffracted intensity $I(X)$. This part of the X-ray method refers to the data processing.

In the configuration when the membranes are in a swollen state the a.c.f. $j(x)$ is obtained from the Patterson function $p(x)$. The Patterson function is the Fourier transform of the diffracted intensity $I(X)$ as defined in equation (5). Experimentally, the integrated intensities $I(h)$ of the reflections are measured. Data processing refers primarily to finding the correction factor.[21,22] Another aspect to consider is the identification and the removal of any diffuse scattering due to disorder within the unit cell.

In membrane diffraction the diffracted intensity $I(h)$ is proportional to $I(h)C(h)$, where $C(h)$ is the correction factor.[21] In our notation the correction factor for crystal diffraction is $(Lp)^{-1}$, where L is the Lorentz factor and p is the polarization factor. Values of L for each crystal method are known.[2,3] In general, the Lorentz factors of crystal diffraction do not apply to membrane assemblies. The evaluation of $C(h)$ requires a knowledge of the transverse size of the specimen, the disorientation within the specimen and the divergence of the X-ray beam.[21] The correction-factor parameters can be estimated from theory[11] using parameters from electron microscopy and from the design of the X-ray camera, or else, they can be determined from an experimental study of the layer-line profiles.[23] Many X-ray workers in this field have simply used the correction factor of crystal powder diffraction without regard to the magnitude of the correction-factor parameters.

Results

As a result of the development of direct methods, the electron density profiles as a function of depth have been established at low resolution (of about 17 Å) for a number of biological membranes. The moderate resolution profiles (of about 7 Å) remain somewhat uncertain, however, until the authenticity of the diffracted intensity $I(X)$ is established. Direct methods of structure analysis can also be applied to lipid bilayers or model membrane systems. All the results to be described refer to X-ray diffraction analysis. Direct methods also apply equally well to neutron diffraction analysis. We note the use of direct methods[24] based upon a difference Patterson deconvolution approach in a neutron diffraction analysis of oriented multilayers of model membranes.

Biological membranes: Some biological membrane systems have naturally occurring multilayered structures and are eminently suited for study by X-ray diffraction. Two such structures are nerve myelin and retinal photoreceptors. The Fourier profile of nerve myelin at low resolution (of 17 Å) was obtained using the relaxation method[25] and, later, in 1978, by Gbordzoe and Kreutz[26] who used a non-linear least squares procedure[17] to interpret the a.c.f. of the unit cell. Two probable moderate resolution electron density profiles of nerve myelin have been determined using the relaxation method.[10] The electron density profile of retinal photoreceptors at 14Å was obtained using the simple recursion method[11] and later, in 1981, the profile was verified using the reconstruction method.[20] Funk *et. al.*,[27] have since obtained a similar profile for retinal photoreceptors using deconvolution analysis.

Biological membranes which do not occur naturally in multilayer form can, however, be artificially prepared in planar multilayer form. Thus, X-ray diffraction suitable for analysis has been recorded from fully hydrated sarcoplasmic reticulum (SR) membrane preparations[28] and wet erythrocyte membranes.[29] Consequently, the electron density profile for the SR membrane at a resolution of 17 Å was first determined in 1973 using direct methods.[28] In later work, the SR membrane profile at a resolution of 10 Å was obtained using the Fourier synthesis method.[30] The electron density profile of the erythrocyte membrane[29] has been determined at low resolution using the non-linear least squares procedure.[17] The Fourier profile of 'membranous cytochrome oxidase' at low and moderate resolution (tentative) has been derived using deconvolution analysis.[31]

Lipid bilayers: The swelling method for obtaining the a.c.f. for lipid bilayers is seldom used mainly because not all lipid bilayers swell appreciably. Another disadvantage with the swelling method is that the lipid layer often changes structure on swelling.[32] We therefore consider the first configuration of a few unit cells in the diffracting assembly. This kind of X-ray specimen is readily obtained by coating multilayers on a curved glass slide in a Langmuir trough after the method of Blodgett.[33] Following this experimental procedure and applying deconvolution analysis to the X-ray data the electron density profiles of barium stearate[14] and magnesium stearate multilayers[34] have been determined. In a similar manner the Fourier profiles of lecithin and of lecithin with incorporated fluorescent probes have been derived.[35]

It is instructive to examine the Patterson functions of three model membrane systems: phosphatidylcholine (DPC), phosphatidylethanolamine (PE) and sphingomyelin (SM). As there are a large number (n) of atoms within the bilayer it might be expected that the large number of overlapping peaks (n^2) would prohibit any direct interpretation. But this is not the case. The Patterson functions computed using high resolution data[32,36] show three peaks.[37] In case of SM the peaks are at

$x = 5$ Å $x = 11$ Å and $x = 16$ Å. These peaks are attributed to the bent conformation of the polar head group of the lipid molecule for in this configuration SM has two high density centers at $x = 2.5$ Å and at $x = 8$ Å. The small differences in the x-coordinates of the peaks of the three Patterson functions can be attributed to differences in the size of the head groups.[37] In previous X-ray studies using different methods it has been shown that the polar head groups of PE[36] and DPC[32] have a bent conformation and lie in the plane of the bilayer.

The above interpretation of the Patterson function essentially ignores the lipid hydrocarbon chains. This suggests a more general approach.[38] If we assume that the lipid chains in the central part of the bilayer have a uniform electron density and if the lipid region is sufficiently wide then it is appropriate to apply deconvolution analysis. This leads to the minus lipid model.[38] The a.c.f. can be directly obtained from the Patterson function provided $d \geq 2v$, where v is the width of the head group region within the bilayer. This direct method[38] has been used to obtain the phases of the first five or six reflections in case of DPC, PE and SM.

Conclusions

In membrane diffraction it is often possible to resolve the a.c.f. of a single unit cell directly from the Patterson function. Membrane structures are commonly centrosymmetrical and in this case deconvolution analysis then provides a unique structural solution. The electron density structures for a number of biological membranes and lipid bilayers have been determined using this approach. The list of structures given here reflects the interest of the author and in no way is it intended to be complete. In future research when the data processing problems have been resolved it is anticipated that the direct methods of analysis based upon the Patterson function will provide electron density profiles of membranes at better than moderate resolution.

References

1. Patterson, A. L. *Phys. Rev.* **46**, 372–376 (1934).
2. James, R. W. *The Crystalline State.* Vol. **2**, G. Bell and Sons: London (1954).
3. Lipson, H. and Cochran, W. *The Crystalline State.* Vol. **3**, G. Bell and Sons: London (1953).
4. McLachlan, D. *X-ray Crystal Structure.* McGraw-Hill: New York (1957).
5. Worthington, C. R. In *The Enzymes of Biological Membranes.* (A. Martonosi, ed.) **1**, 1–29, Plenum Press: New York (1976).
6. Schmitt, F. O. Bear, R. S. and Palmer, K. S. *J. Cell. Comp. Physiol.* **18**, 31–42 (1941).
7. Finean, J. B. *Nature (London)* **173**, 549–550 (1954).
8. Worthington, C. R. *Proc. Natl. Acad. Sci. USA* **63**, 604–611 (1969).
9. Worthington, C. R. King, G. I. and McIntosh, T. J. *Biophys. J.* **13**, 480–494 (1973).
10. Worthington, C. R. and McIntosh, T. J. *Nature (London)* **245**, 97–99 (1973).
11. Worthington, C. R. *Exp. Eye Res.* **17**, 487–501 (1973).
12. Worthington, C. R. *Biophys. J.* **9**, 222–234 (1969).
13. Hosemann, R. and Bagchi, S. N. *Direct Analysis of Diffraction by Matter.* North-Holland: Amsterdam, Netherlands (1962).
14. Lesslauer, W. and Blasie, J. K. *Biophys. J.*, **12**, 175—190 (1972).
15. Pape, E. H. *Biophys. J.* **14**, 284–294 (1974).
16. Moody, M. F. *Biophys. J.* **14**, 697–702 (1974).
17. Pape, E. H. and Kreutz, W. *J. Appl. Cryst.* **11**, 421–429 (1978).
18. Paterson, M. S. *Proc. Phys. Soc.* **63**, 477–482 (1950).
19. Schwartz, S. Cain, J. E. Dratz, E. A. and Blasie, J. K. *Biophys. J.* **15**, 1201–1233 (1975).
20. Worthington, C. R. *J. Appl. Cryst.* **14**, 383–386 (1981).
21. Worthington, C. R. and Wang, S. K. *J. Appl. Cryst.* **12**, 42–48 (1979).
22. Crist, B. and Worthington, C. R. *J. Appl. Cryst.* **13**, 585–590 (1980).
23. Worthington, C. R., Worthington, A. R. and Wang, S. K. *J. Appl. Cryst.* **13**, 273–279 (1980).
24. Blasie, J. K. Schoenborn, B.P. and Zaccai, G. *Brookhaven Symp. Biol.* **27**, III.58–III.67 (1975).
25. McIntosh, T. J. and Worthington, C. R. *Biophys. J.* **14**, 363–386 (1974).
26. Gbordzoe, M. K. and Kreutz, W. *J. Appl. Cryst.* **11**, 489–495 (1978).
27. Funk, J. Welte, W. Hodapp, N. Wutschel, I. and Kreutz, W. *Biochim. Biophys. Acta* **640**, 142–158 (1981).

28. Worthington, C. R. and Liu, S. C. *Arch. Biochem. Biophys.* **157**, 573–579 (1973).
29. Pape, E. H. Klott, K. and Kreutz, W. *Biophys. J.* **19**, 141–161 (1977).
30. Herbette, L. Marquardt, J. Scarpa, A. and Blasie, J. K. *Biophys. J.* **20**, 245–272 (1977).
31. Blasie, J. K. Erecinska, M. Samuels, S. and Leigh, J. S. *Biochim. Biophys. Acta* **501**, 33–52 (1978).
32. Torbet, J. and Wilkins, M. H. F. *J. Theor. Biol.* **62**, 447–458 (1976).
33. Blodgett, K. B. *J. Am. Chem. Soc.* **57**, 1007–1022 (1935).
34. Lesslauer, W., Cain, J. and Blasie, J. K. *Biochim. Biophys. Acta* **241**, 547–566 (1971).
35. Lesslauer, W. Cain, J. and Blasie, J. K. *Proc. Nat. Acad. Sci. USA* **69**, 1499–1503 (1972).
36. Hitchcock, P.B. Mason, R. and Shipley, G.G. *J. Mol. Biol.* **94**, 297–299 (1975).
37. Khare, R. S. and Worthington, C. R. *Biochim. Biophys. Acta* **514**, 239–254 (1978).
38. Worthington, C. R. and Khare, R. S. *Biophys. J.* **23**, 407–425 (1978).

C. HOMOMETRICS

38. Homometrics

Martin J. Buerger

Martin J. Buerger was born in Detroit on April 8, 1903. He grew up in New York and went to MIT to study mining engineering, doing graduate studies in mineralogy and ore deposition, and obtaining a Ph.D. in 1929. He stayed on at MIT and, in 1956, became Institute Professor at MIT and the first Director of its School of Advanced Studies. When he had to retire in 1968 he moved on to the University of Connecticut as University Professor of Geology until 1973. His editorial chores and other writings did not cease until February 25, 1986, six months after participating in the presentation ceremonies of the first *M. J. Buerger Award* of the American Crystallographic Association. He served the crystallographic community with great distinction. When ASXRED was established in 1941, Buerger was an active organizer and served as its third president. He was also the founding president of the CSA. In 1947 he was made prsident of the Mineralogical Society of America. He was a member of the National Academy of Sciences and received numerous awards.

Buerger was inspired by lectures delivered by W. L. Bragg at MIT in 1927. He, with help from Bert Warren, assembled his own laboratory to do X-ray diffraction. In the process he designed and had built many cameras, including the equi-inclination Weissenberg camera and the Buerger precession camera. His studies of crystal structures have involved studies of crystal growth, dislocations and twinning. In the area of crystal structure detemination he developed *implication theory* to replace trial-and-error methods; this foreshadowed the utility of inequalities between measured intensities that proved useful in determining their phases. He also became very interested in the geometrical relations between the vector sets in Patterson space — *image-seeking* methods or the *minimum function*. His insights are recorded in his volumes on *Vector Sets* and *Crystal Structure Analysis*.

He was helpful in providing reprints for use in this volume and we thank Dr Leonid V. Azároff (who wrote this introduction, to be printed as an obituary in *J. Appl. Cryst.*).

When Pauling and Shappell[1] determined the structure of bixbyite they found that there were two arrangements of metal atoms that scattered X-rays with the same intensities and yet were neither congruent (the same) nor enantiomorphic (the mirror images) to each other. This implied that two different arrangements of points could yield the same set of interatomic vectors. Patterson[2] referred to such pairs of structures as *homometric* structures.

At that point Patterson[3] became fascinated by the study of such homometric pairs in periodic sets. He started, as always, from first principles with a study of one-dimensional sets, and,

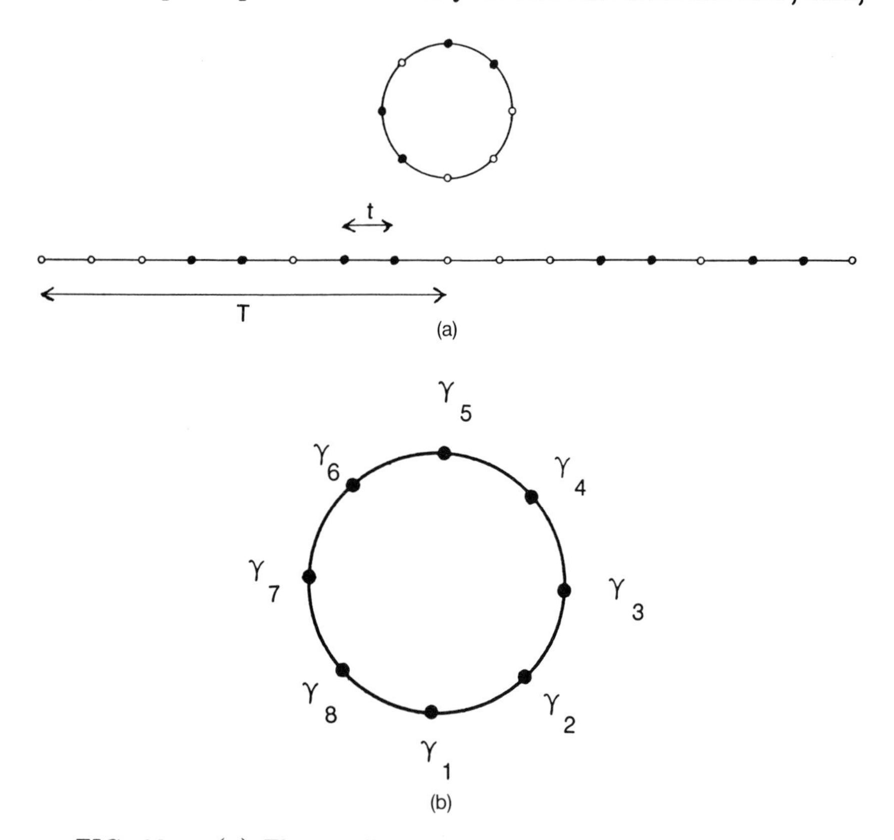

FIG. 38.1. (a) The small circles on the horizontal line make up a periodic one-dimensional set whose translation period is T. In the set shown the points are restricted to some of the points of a lattice, the ***multiple lattice***, with period t. The one-dimensional set is wrapped around a circle with circumference T. This is called the ***circular representation*** of the set. (b) The circular representation of the points of a one-dimensional lattice with $T/t = 8$. The points are then numbered in order as shown.

in order to represent the sets simply within one translation T, he wrapped the set around a circle with circumference T. This translation (or circumference) T was subdivided into N equally spaced points; some of the points were occupied by points of a set Q. Such sets were referred to by Patterson as ***cyclotomic sets***. In such sets, A, only r out of the N possible points are occupied; the other $N - r$ points make up a set A' which is said to be ***complementary*** to A, see Fig. 38.1.

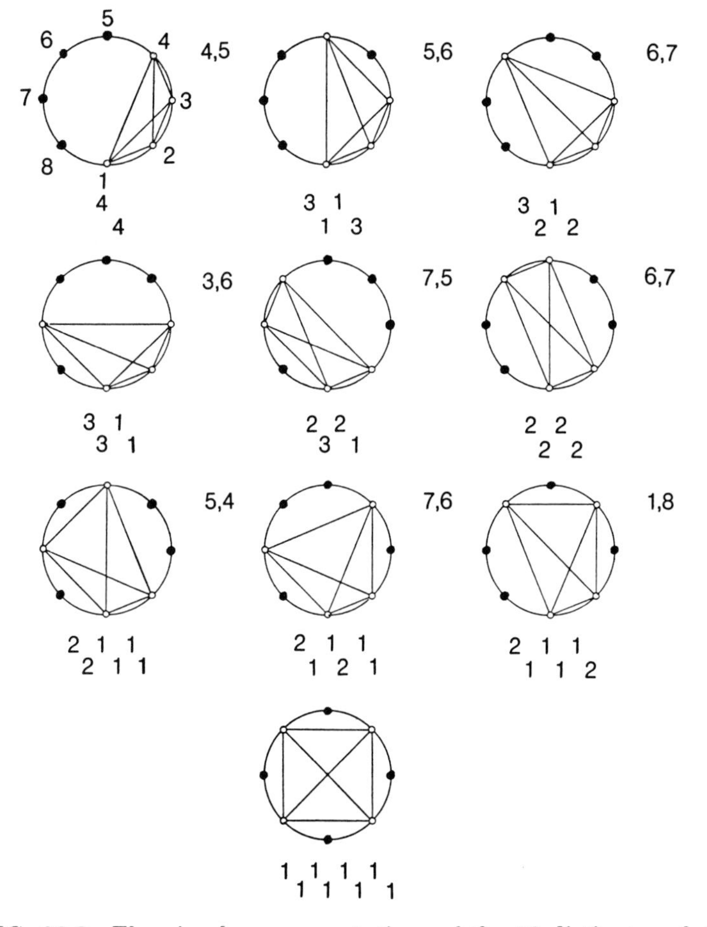

FIG. 38.2. The circular representations of the 10 distinct cyclotomic sets with $N = 8$, $r = 4$. The symbol of the set is given below it. The pairs of numbers between sets indicates the interchange of points to transform a set into the neighboring set.

In this ***circular representation,*** interatomic distances are designated by arcs, but since these would superimpose in drawings, Patterson actually represented each arc by the chord it intersects as shown in Fig. 38.2. He also only dealt with r of the N points obtained by dividing the translation T into N portions of equal lengths. Buerger, who had followed this work with great interest, then extended it. He referred to the set of points which divides T into N segments as a ***multiple lattice***, and introduced the term ***tautoeikonic*** (having the same self-image) to cover two or more sets with the same interpoint distances;[4] this includes congruent (identical), enantiomorphic (mirror image) and homometric sets.

In order to find a simple non-graphical description of cyclotomic sets, Buerger[5] introduced a symbolism that lists the number in a sequence of occupied points (r among N) at an elevated level and those that are unoccupied at a lower level. Thus the sequence $^{5}{}_{2}{}^{1}{}_{3}{}^{2}{}_{3}$ indicates that there are five occupied points followed by a gap of two unoccupied points (see Fig. 38.2), and so forth.

Patterson[3] enunciated, without proof, two theorems on homometric sets:

> It can be shown
>
> (i) that if two sets A and B form a homometric pair then A' and B' [complementary sets] also form a homometric pair, and
> (ii) that every cyclotomic set with N even and $r = n/2$ is either congruent to its complementary set or it forms a homometric pair with it.

It remained for Buerger[5] to prove these theorems with the aid of image algebra and to generalize the theorem to two and three dimensions. He generalized Patterson's theorem (i) as[2]:

> If two sets A and B which occupy the same number of points on duplicate multiple lattices, are tautoeikonic, their complementary sets A' and B' are tautoeikonic.

Similarly Buerger[4] restated Patterson's theorem (ii) as:

> If a periodic set of points A on the nodes of a multiple lattice occupies half the points of the cell of the set, the set and its complementary set are tautoeikonic.

Buerger,[5] systematized the characterization of sets and listed the following definitions:

1) *Tautoeikonic* sets have the same self-image.
2) Tautoeikonic sets whose patterns cannot be transformed into one another by any coincidence operation are *homometric.*
3) Tautoeikonic sets whose patterns can be transformed into one another by a coincidence operation of the second kind are *enantiomorphic.*
4) Tautoeikonic sets whose patterns can be transformed into one another by a coincidence operation of the first kind are *congruent.*

Image algebra was also applied to the study of interpoint distances of cyclotomic sets, and the distance arrays of cyclotomic sets have interesting properties.[6] I was glad that Chieh[7] had been able to program the computer for the analysis of point sets, and that progress has been made by various researchers.[7,8]

For further details the reader is referred to the references below and the chapters by Chieh and Oxtoby that follow this article.

References

1. Pauling, L. and Shappell, M. D. The crystal structure of bixbyite and the *C*-modification of the sesquioxides. *Z. Krist.* **75**, 128–142 (1930).
2. Patterson, A. L. Homometric structures. *Nature (London)* **143**, 939–940 (1939).
3. Patterson, A. L. Ambiguities in the X-ray analysis of crystal structures. *Phys. Rev.* **65**, 195–201 (1944).
4. Buerger, M. J. Proofs and generalization of Patterson's theorems on homometric complemntary sets. *Z. Krist.* **143**, 79–98 (1976).
5. Buerger, M. J. Exploration of cyclotomic point sets for tautoeikonic complementary pairs. *Z. Krist.* **145**, 371–411 (1977).
6. Buerger, M. J. Interpoint distances in cyclotomic sets. *Can. Mineral.* **16**, 301–314 (1978).
7. Chieh, C. Analysis of cyclotomic sets. *Z. Krist.* **150**, 261–277 (1979).
8. Hosemann, R. and Bagchi, S. N. On homometric structures. *Acta Cryst.*, **7**, 237–241 (1954).

39. Homometric structures

Chung Chieh

Homometric crystal structures are different atomic arrangements that have the same Patterson functions. Enantiomorphic pairs are obviously homometric, but there are homometric structures that are not related by handedness. The problem of non-enantiomorphic homometric structures was A. L. Patterson's major concern during the last few years of his life.[1] However, this problem has been neglected and forgotten by almost all but a few crystallographers and mathematicians. Patterson's interest and success in this area is submerged by the great success of using Patterson functions for solving crystal structures. Furthermore, crystal structures derived from the Patterson function have been carefully studied in conjunction with other physical and chemical properties so that ambiguity is eliminated. The great creator of the useful F^2 synthesis that enables one to derive useful results from a set of experimental data in an otherwise then hopeless situation, carefully and seriously studied the problem, which seems to question the value of the Patterson function. This may appear ironic, but it is really an indication of the greatness of Patterson's honesty and integrity as a person, and his sincerity for the truth as a scientist.

The question of uniqueness of an X-ray crystal structure was directed to Patterson's attention by Linus Pauling.[2] In his classical paper Patterson wrote:

> 'Pauling and Shappell[3] have shown ··· while parameter values $+u$ and $-u$ correspond to the structures which are not identical nor are they mirror-image of one another.' ··· 'No direct method seems to offer itself for the enumeration of the number of possible sets of nonidentical homometric structures. ··· The following brief account ··· is presented in the hope that other workers in the field may make further progress.'

After the Fox Chase symposium, Professor Linus Pauling sent me his correspondence with Patterson on the homometric problem. The three letters have some historical significance and inspiring value, and they are included in Chapter 3.2.

Since then, the problem on homometric structures has been studied by crystallographers and mathematicians.[4–6] Professor M. J. Buerger investigated this interesting problem from 1976 to 1978,[7–9] and he inspired this writer to spend a few rewarding and exciting months of sabbatical leave with him working on the problem. Although no method has been found for the enumeration of the number of possible sets of non-enantiomorphic or non-complementary homometric structures, algorithms for computer systematic search of some simple models have been developed. At some point, I remember Professor Buerger saying: 'Had Patterson been able to see your results today, he would be very happy.' Thus, it is appropriate to dedicate this paper to the 50th anniversary of the Patterson function. At the symposium, Dr. E. K. Patterson told this writer that A.L.P. left a lot of computer output on the subject of homometric sets. Apparently, he did employ computers in those days to analyze the homometric sets.

Patterson studied the homometric structures by first investigating the one-dimensional *crystal structures.* He invented the cyclotomic point sets for the study of their homometry. He had discovered some two-dimensional homometric patterns, which were privately communicated to Hosemann and Bagchi.[4] The special topic in the *International Tables for X-ray Crystallography* on close packing[10] employed some concept of cyclotomic point set, but there was no mention of homometric cyclotomic sets. This article summarizes and reviews the development of homometric structures for a wide readership, so that the problem may receive further attention in the crystallographic community; a special effort is made to link the results of cyclotomic sets to those in two- and three-dimensional spaces.

Transformation of one-dimensional crystal structures to cyclotomic sets

Patterson[11] simplified the complicated problem of searching for homometric structures by dealing with one-dimensional structures, which may be represented by line lattices. He further constructed one-dimensional structures by placing points

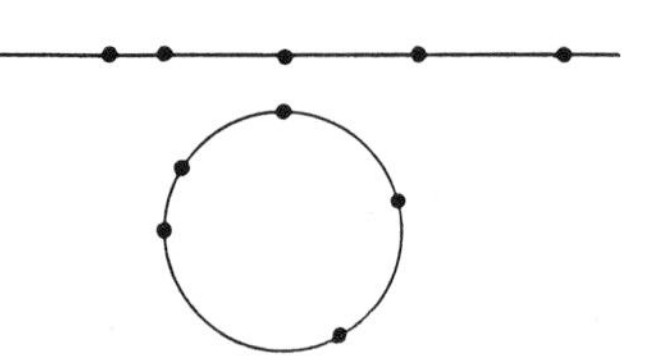

FIG. 39.1. Transformation of a linear structure into a cyclotomic set.

within a period of line lattices. To facilitate the analysis for homometry, *i.e.*, compare the Patterson maps of these structures, he transformed these structures into circular representations as shown in Fig. 39.1. The distribution of atoms is represented by points on the circumference of a circle, the linear coordinate x_i is converted to angular coordinate by $\theta_i = 2\pi x_i/n$, where n is the period. Thus, distances $D_{i,j}$ (mod n) for a periodic set on the line given by

$$D_{i,j} = x_j - x_i \pmod{n}, \quad (i, j = 1, 2, ..., n/2),$$

may be represented by another set $A_{i,j}$ expressed in terms of angles,

$$A_{i,j} = \theta_j - \theta_i \pmod{2\pi} \quad (i < j).$$

There is an one-to-one correspondence between the A's and the D's, and homometric structures will give the same set of A's or D's (Patterson map).

In short, Patterson wrapped one-dimensional periodic structures around circles, whose circumferences equal the translation periods n, and thus transformed them into **cyclotomic (point) sets**, a geometric approach to modular arithmetic.

Cyclotomic sets

In order to search systematically for homometric structures, Patterson arranged r points on the n vertices of an inscribed regular polygon, resulting in an r, $n - r$ cyclotomic set Q and the complementary $n - r$, r cyclotomic set Q'.

Patterson examined the diagrams of cyclotomic sets with

$r < 9$ and $n < 17$, and he found a number of homometric structures. This process was 'so time-consuming, tedious, and subject to possible error' that Buerger[8] developed a numerical representation suitable for a manual synthesis and search. Buerger's representation consists of a series of numbers, each representing a continuous subset of ***occupied*** and ***unoccupied*** positions of a r, $n - r$ cyclototomic set. The r occupied points in a set is divided into several continuous uninterrupted subsets containing a, b, c, d, ... points interspaced by continuous subsets of unoccupied m, n, o, p, ... points. This set Q and its complementary set Q' are given below:

$$Q = \begin{Bmatrix} a\,b\,c\,d\ldots & \\ & m\,n\,o\,p\ldots \end{Bmatrix} \qquad Q' = \begin{Bmatrix} m\,n\,o\,p\ldots & \\ & a\,b\,c\,d\ldots \end{Bmatrix}$$

These notations are related to those of Zhdanov[6] who used underlines to mark the unoccupied or 'negative' sequences in the consideration of layer packing. Over the years, various symbols have been used to represent this type of sets, *e.g.*, Franklin[5] included the number if the position is occupied in a bracket as 0, 1, 4, 7. The evolution in the printing industry plays an important role in terms of notations.

In the course of the examination, Patterson enunciated two important theorems on homometric sets, and the proof was later provided by M. J. Buerger[7] and by J. E. Iglesias[12] using two different methods. These theorems state:

(*i*) that if two sets Q and R form a homometric pair, then their complementary sets Q' and R' also form a homometric pair, and

(*ii*) that every r,r cyclotomic set (*i.e.*, numbers of occupied and unoccupied points are equal), is either congruent to its complementary set or it forms a homometric pair with it.

If a cyclotomic set does not have a center (or mirror plane) of symmetry, the enantiomorphic set Q_e and its complementary set Q'_e are also homometric. Thus, the above theorems indicate that for each r,r (or $n/2,n/2$) cyclotomic set Q, there are three other homometric sets, and **these four homometric sets are**

called a quartet,[13] as shown by the Buerger[8] transformation:

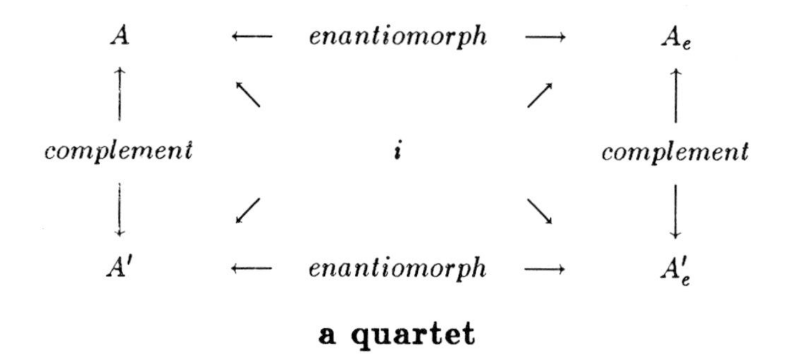

a quartet

where *i* indicates a symmetry operation due to **inversion**. A diagram showing a typical quartet is given in Fig. 39.2.

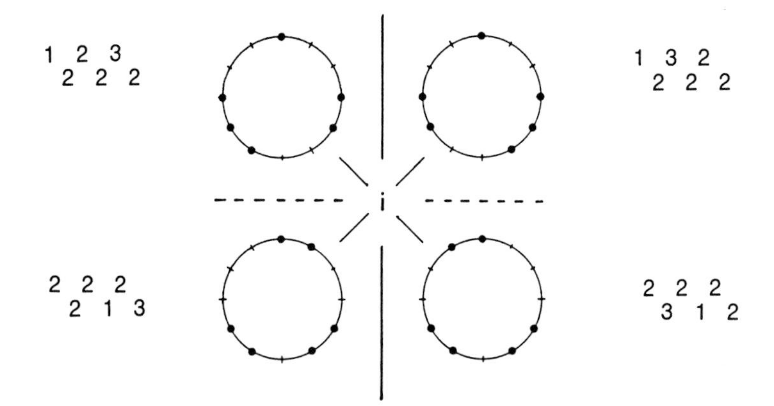

FIG. 39.2. A typical quartet, with Buerger's numerical representations indicated.

A quartet has four sets if it has no (center of inversion) symmetry, and if it is distinct from its enantiomorph, however, a quartet may degenerate due to the presence of symmetry, and thus, there are five categories as classified by the following scheme:

Type	symbol	condition	no. of distinct sets
Asymmetric homometric,	a_H:	Q, Q_e, Q', Q'_e	4
Symmetric homometric,	s_H:	$Q(=Q'_e), Q_e(=Q')$	2
Enantiomorphic,	E :	$Q(=Q_e), Q'(=Q'_e)$	2
Asymmetric congruent,	a_C:	$Q(=Q'), Q_e(=Q'_e)$	2
Symmetric congruent,	s_C:	$Q(=Q'=Q_e=Q'_e)$	1

The short notations in the second column were given by Buerger.[8]

While working on cyclotomic sets, Patterson found another type of homometric sets in addition to those listed above, for example see Fig. 39.3. However he did not give a new classification for this type, but simply called them **homometric.**

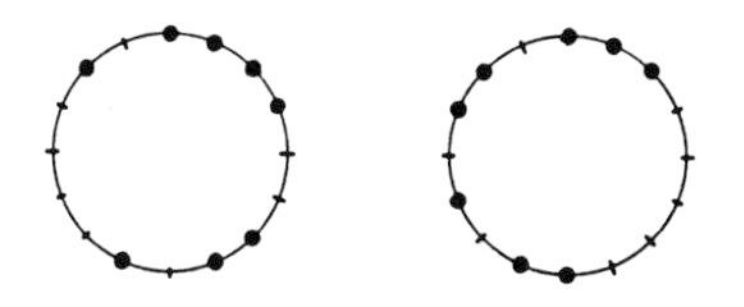

FIG. 39.3. A pair of tautoeikonic cyclotomic sets or two representative sets of homometric quartets.

To distinguish this new type from those related by complementarity or enantiomorphism, Buerger[8] suggested the term **tautoeikonic sets** for all sets that have the same self-image. This term covers all the sets which have the same Patterson function, but they are neither enantiomorphic nor complementary. Thus, any two sets that are tautoeikonic will result in a pair of **tautoeikonic quartets**.

In the crystallographic community, the term 'homometric sets' is more widely used than the term 'tautoeikonic sets'. Since the complementary and enantiomorphic sets are grouped into a quartet, it leads to no ambiguity to use the term **homometric quartets**. However, this term occasionally would be misunderstood as enantiomorphic structures. Patterson[11] gave some tautoeikonic cyclotomic sets, but the existence of quartets was recognized by Chieh.[14] There are cases where three and four quartets have been found to be tautoeikonic,[14] and **tautoeikonic double, triple** or **quadruple quartets** have been used. It should be noted, however, that quartets are present only in r,r cyclotomic sets. In cyclotomic sets, where the number of occupied points (black dots) is not equal to the number of unoccupied points (white dots), *i.e.*, p, q cyclotomic sets, the complementary sets are not homometric. Thus, the quartets are broken up into two **duets**, which may degenerate into a single set if it is indistinguishable from its enantiomorph.

The dots on the cyclotomic sets may bear various physical significances. For example, in the study of layer structures, we may consider the relationship between adjacent layers by a black or a white dot if there are only two types of relationships. This aspect of the cyclotomic sets has been applied to study the close packing of spheres as will be discussed later. The same have been applied to the study of enhanced symmetries of diffraction patterns of layer structures.[15]

Algorithm for the analysis of homometric quartets

Buerger's[8] numerical representation gives a systematic approach to the analysis of cyclotomic sets. These representations may be transformed to yet another numerical representation in order to employ a computer as a tool for the systematic analysis of homometric quartets. This representation is based on the notion that the n locations of the circumference may be labelled consecutively, starting at an arbitrary origin. If the location is occupied, its label is kept in the series, otherwise, it is omitted. For example, the representation **1 3 9 10** is equivalent to $^{1}{}_{1}\,^{1}{}_{5}\,^{2}{}_{2}$ of Buerger's representation in a 4, 8 cyclotomic set. This set is homometric with the (4, 8 cyclotomic) set **1 3 4 9**. The interpoint-distance matrices may then be calculated as shown below. Franklin,[5] as was brought to my attention by Professor Verner Schomaker at the Patterson symposium, also used this type of representation.

Interpoint-distance matrices of two tautoeikonic cyclotomic sets are:

	1	**3**	**9**	**10**
1	0	2	8*	9*
3		0	6	7*
9	4*		0	1
10	3*	5*		0

	1	**3**	**4**	**9**
1	0	2	3	8*
3		0	1	6
4			0	5
9	4*			0

Because the distance matrices are symmetric, the lower triangles of the matrices are left empty, unless converted values are required. Both the converted values in the lower triangles and the corresponding values in the upper triangles are marked by '*' signs (*e.g.*, $4^* = | 8^* - 12 |$). The two sets of interpoint distances are the same (1 2 3 4 5 6 numerically, or in units of $2\pi/12$ as angles), but if the distance d is greater than $n/2$, then $| n - d |$ is taken as the distance, because the Patterson function always has a center of symmetry.

Since Patterson's major concern was crystal structures, the modulation by the lattice, in this case mod$n$, is implicit for the interpoint distances. For two- or three-dimensional structures, the Patterson maps may be obtained through modulation of the appropriate lattice translations.

The algorithms and strategies required for distance calculation, sorting, and comparison are given by Chieh.[13,14] The number of homometric triple, and double quartets for cyclotomic sets with various p and q are given in Table 39.1.

TABLE 39.1. Number of tautoeikonic double, and triple duets (of some p, q cyclotomic sets) or quartets (in some p,p cyclotomic sets)

q/p	4	5	6	7	8	9	10
7	0	3	2	0			
8	1	0	6	10	3		
9	1	5	21	42, 3		63	
10	0	4	29, 1	26			138, 2
11	0	10	16	135, 16			
12	2	58	52, 6	58			

If the number is not available, the space is left blank in Table 39.1. Obviously, the occurrence of tautoeikonic quartets and duets or homometric sets is quite common.

As stated in the introduction, Patterson was concerned about a method to enumerate the number of homometric cyclotomic sets. This problem is solved by Iglesias[12,15] in his study

Two enantiomorphic sets

```
0−+ − 4−6        0−+ − +−5
|  |   |  |        |  |   |  |
1−+ − +−+        +−2 − +−6
|  |   |  |        |  |   |  |
2−+ − 5−+        1−+ − +−+
|  |   |  |        |  |   |  |
+−3 − +−+        +−3 − 4−+
   (a)               (b)
```

Singular afine transformed linear sets

```
6 - 4  - +  -  0 - + - + - + - 1 - + - 5 - + - 2 - + - + - 3 - + (-6)   (al)
       ← • ⟶
       b    a
5 - +  - +  -  6 - + - + - 2 - + - + - + - + - 1 - + - 4 - 3 - + (-5)   (bl)
+ - +  - +  -  5 - + - 2 - + - 6 - 1 - + - + - + - + - 3 - 4 - + (-+)   (bm)
```

Convolution-squares maps

```
1 1  1 1 1 1 2 1 1 1 1 2 1 6 1 2 1 1 1 1 2 1 1 1 1  1 1  (al-cs)
1 1  2 1   2 1 1 1   3 1 1 6 1 1 3   1 1 1 2   1 2  1 1  (bl-cs)
     1 1 1 1 1 2 2 1 1 2 2 6 2 2 1 1 2 2 1 1 1 1 1       (bm-cs)
```

Patterson maps

```
12 2 6 4 2 4 4 6 4 6 4 4 2 4 6 212  (al-P)
12 2 4 8   6 4 2 8 2 4 6   8 4 212  (bl-P)
12 4 4 2 2 6 6 4 4 4 6 6 2 2 4 412  (bm-P)
```

FIG. 39.6. Transformation of two-dimensional patterns into linear structures: Patterson's two enantiomorphic planar patterns (a) and (b), their corresponding linear structures transformed from origin O (al) and (bl), and another linear image based on origin O′ with b axis reversed. The corresponding cyclotomic sets are labelled as (al), (bl), and (bm), whereas the convolution squares and Patterson function maps are labelled with additions of (-cs), and (-P) respectively.

Because the distance matrices are symmetric, the lower triangles of the matrices are left empty, unless converted values are required. Both the converted values in the lower triangles and the corresponding values in the upper triangles are marked by '*' signs (*e.g.*, $4^* = |\,8^* - 12\,|$). The two sets of interpoint distances are the same (1 2 3 4 5 6 numerically, or in units of $2\pi/12$ as angles), but if the distance d is greater than $n/2$, then $|\,n - d\,|$ is taken as the distance, because the Patterson function always has a center of symmetry.

Since Patterson's major concern was crystal structures, the modulation by the lattice, in this case mod$n$, is implicit for the interpoint distances. For two- or three-dimensional structures, the Patterson maps may be obtained through modulation of the appropriate lattice translations.

The algorithms and strategies required for distance calculation, sorting, and comparison are given by Chieh.[13,14] The number of homometric triple, and double quartets for cyclotomic sets with various p and q are given in Table 39.1.

TABLE 39.1. Number of tautoeikonic double, and triple duets (of some p, q cyclotomic sets) or quartets (in some p,p cyclotomic sets)

q/p	4	5	6	7	8	9	10
7	0	3	2	0			
8	1	0	6	10	3		
9	1	5	21	42, 3		63	
10	0	4	29, 1	26			138, 2
11	0	10	16	135, 16			
12	2	58	52, 6	58			

If the number is not available, the space is left blank in Table 39.1. Obviously, the occurrence of tautoeikonic quartets and duets or homometric sets is quite common.

As stated in the introduction, Patterson was concerned about a method to enumerate the number of homometric cyclotomic sets. This problem is solved by Iglesias[12,15] in his study of diffraction enhancement of symmetry. His recursive formula which gives the number of distinguishable arrangements of p 'black' and q 'white' dots equally spaced on a circle is

$$a(p,q) = \frac{(p+q-1)!}{p!q!} - \sum \frac{1}{d_i} a(p/d_i, q/d_i)$$

where the summation runs over all integers $d_i \neq 1$ which exactly divide p and q, and where arrangements having subperiods have not been counted. Sets of this type are calculated in the recursive function $a(i, j)$. Results calculated from this formula agree with those generated by computer systematic analyses.

Two-dimensional homometric structures

Hosemann and Bagchi[4] received some two-dimensional patterns from Patterson through private communications. The patterns they studied were either complementary or enantiomorphic. As a result of Patterson's theorem (*ii*) given above, these patterns have to be homometric. The two patterns given by Patterson were enantiomorphic; thus, they have the same Patterson function as well as the **convolution-squares** map, which is the interpoint vector image map of the finite pattern, whereas the Patterson map is that of a 'crystal structure' using the pattern as a unit cell. The difference between the two is the modulation after the lattice matrices or translations. The convolution squares and the Patterson map of Patterson's enantiomorphic pair in a redefined unit cell are given in Fig. 39.4. These diagrams are output from a computer program by this author for analyzing two-dimensional patterns for this article. The two patterns communicated to Hosemann and Bagchi by Patterson had a smaller cell compared to the one outlined in Fig. 39.4, thus, this may result in a different Patterson map from its original, but Patterson maps of enantiomorphic patterns are always the same.

The following pair (Fig. 39.5) is an example of complementary two-dimensional homometric patterns. The convolution squares of the complementary pairs are different, but the Patterson maps are identical. Complementary pairs are not limited to square or regular lattices. In fact they are still com-

```
+-- - -+        11
|    o o |     11111
| o      |    111111
| o   o  |    11611
|  o     |   111111
+-- - -+     11111
  (a)          11
              (cs)
```

```
+-- - -+       61216
|      o |     11331
|  o   o |     23232
| o      |     13311
|  o o   |     61215
+-- - -+         (P)
  (b)
```

FIG. 39.4. Two enantiomorphic two-dimensional patterns (a) and (b), their convolution square (cs), and Patterson map (P) for the cell as outlined. Numbers indicate the weight of the vector image.

```
+-- - -+        121
|  o o   |     1331
| o o    |     13431
|  o o   |       484
|  o o   |     13431
+-- - -+        1331
  (a)           121
              (a-cs)
```

```
+-- - -+       1 13 12
| o    o |     2114112      84 48
|    o o |     3 181 3      64246
| o    o |     2114112      64246
| o    o |     21 31 1      64246
+-- - -+       1  2  1      84 48
  (ac)         (ac-cs)     (a-ac-P)
```

FIG. 39.5. Two complementary patterns (a) and (ac), their convolution squares (a-cs) and (ac-cs), and their Patterson map (a-ac-P).

Two enantiomorphic sets

```
0 — + — 4 — 6          0 — + — + — 5
|   |   |   |          |   |   |   |
1 — + — + — +          + — 2 — + — 6
|   |   |   |          |   |   |   |
2 — + — 5 — +          1 — + — + — +
|   |   |   |          |   |   |   |
+ — 3 — + — +          + — 3 — 4 — +
     (a)                    (b)
```

Singular afine transformed linear sets

```
6 - 4  - +  -  0 - + - + - + - 1 - + - 5 - + - 2 - + - + - 3 - + (-6)   (al)
       ← • →
       b   a
5 - +  - +  -  6 - + - + - 2 - + - + - + - + - 1 - + - 4 - 3 - + (-5)   (bl)
+ - +  - +  -  5 - + - 2 - + - 6 - 1 - + - + - + - + - 3 - 4 - + (-+)   (bm)
```

Convolution-squares maps

```
1 1  1 1 1 1 2 1 1 1 1 2 1 6 1 2 1 1 1 1 2 1 1 1 1  1 1  (al-cs)
1 1  2 1   2 1 1 1   3 1 1 6 1 1 3   1 1 1 2   1 2  1 1  (bl-cs)
     1 1 1 1 1 2 2 1 1 2 2 6 2 2 1 1 2 2 1 1 1 1 1       (bm-cs)
```

Patterson maps

```
12 2 6 4 2 4 4 6 4 6 4 4 2 4 6 212  (al-P)
12 2 4 8   6 4 2 8 2 4 6   8 4 212  (bl-P)
12 4 4 2 2 6 6 4 4 4 6 6 2 2 4 412  (bm-P)
```

FIG. 39.6. Transformation of two-dimensional patterns into linear structures: Patterson's two enantiomorphic planar patterns (a) and (b), their corresponding linear structures transformed from origin O (al) and (bl), and another linear image based on origin O′ with b axis reversed. The corresponding cyclotomic sets are labelled as (al), (bl), and (bm), whereas the convolution squares and Patterson function maps are labelled with additions of (-cs), and (-P) respectively.

plementary pairs if both undergo the same nonsingular afine transformation such as that of distorting a square lattice to a rectangular or oblique one. Two-dimensional patterns may even be converted into linear structures by a singular afine transformation. Hosemann and Bagchi[4] gave an example of this transformation, but they did not mention the consequence of homometric properties. The present analysis found that two 2-dimensional enantiomorphic patterns undergoing the same singular afine transformation result in two linear structures that correspond to neither enantiomorphic nor tautoeikonic cyclotomic sets. This is demonstrated in Fig. 39.6, where the singular afine transformation is defined by having two linear base vectors in the opposite directions. For convenience of identification, the corresponding points on the grids and those on the linear structures are labelled by the same numbers

Not only is homometry not conserved for enantiomorphic pairs, depending on the choice of origin and directions of axes, the transformed linear structures from a single planar homometric pattern may give rise to cyclotomic sets, which are not necessarily homometric. The three transformed images have their unique convolution squares and Patterson maps as seen from Fig. 39.6.

The singular afine transformation between planar patterns and linear structures does not result in one-to-one correspondent images. Thus, cyclotomic sets and planar patterns have to be studied separately for their homometric properties.

Cyclotomic sets and three-dimensional crystal structures

Patterson not only employed cyclotomic sets for the study of possible homometric linear structures, he also applied the concept to the investigation of close packings of equal spheres. Patterson and Kasper[10] considered the spheres packing as made up of layers. The relative positions of layers are labelled as A, B, or C, and they assigned a '+' sign (or black dot) to it if the transition in an arbitrary but fixed direction is one of A→B→C→A type, a '−' sign (or white dot) if the change is of the A→C→B→A type. For example, a packing of spheres with a sequence of letters as given below may be translated to the corresponding cyclotomic representation:

A B A B A C B A C A B A (C...(B...(A	Sequence of layers
+ − + − − − − − + + − − (+...(+...(+	dot representation
0 1 2 3 4 5 6 7 8 9 10 11 ...	z in 36ths
$1_1 1_5 2_2$	Buerger representation

In reality, the above translation was made in reverse order. We start with the Buerger representation of a set from a tautoeikonic duet, translate it into '+' and '−' signs, then begin the sequence with an A layer and follow that with an appropriate sequence. This tautoeikonic duet has the shortest repeating sequence, $n = 12$,[14] the other representative in the duet is ${}^{1}{}_{1}{}^{2}{}_{4}{}^{1}{}_{3}$. We may further follow the method of Patterson and Kasper[10] to assign the space groups for these two close packings of spheres. Since the difference (mod 3) between the positive and negative signs in the set is -1 (4 - 8 = -4 mod 3 = -1), the beginning of the next cyclotomic sequence of '+' signs corresponds to the change from an A to a C layer as indicated by (C ... (B ... (A above. Thus, the crystallographic periodicity in the z direction is three times the periodicity of that of the cyclotomic set. This close packing pattern belongs to a rhombohedral lattice. Since there is no center of inversion, nor mirror line symmetry in either cyclotomic set, both belong to space group $R3m$ (No: 160). For convenience, we may use a threefold centered hexagonal cell of period $3n$, and specify coordinates for atoms of a cyclotomic period. The coordinates for atoms in the A layers in the first set are 0, 0, z, with $z = 0$, 2, 4, 7, 9, 11, (12, 14, 16, ⋯, 24, 26, 28, ⋯) expressed in 36ths, those for atoms in the B layers 2/3, 1/3, z, with $z = 1$, 3, 6, 10, (13, 15, 18, ⋯, 25, 27, 30, ⋯), and those in the C layers 1/3, 2/3, z, with $z = 5$, 8, (17, 20, 29, 32).

There is a symmetry center in the set ${}^{(1)}{}_{1}{}^{2}{}_{(5)}{}^{2}{}_{1}$, which is tautoeikonic with ${}^{1}{}_{2}{}^{1}{}_{1}{}^{3}{}_{4}$, that does not have any symmetry. The space groups of the close packings of spheres belonging to these cyclotomic sets are $R\bar{3}m$ (No: 166) and $R3m$ (No: 160) respectively. The atomic coordinates may be assigned in a manner in accordance with the example given above.

Although tautoeikonic sets have the same Patterson map in a one-dimensional space, it remains to be investigated if the three-dimensional structures resulting from close packing of spheres are indeed tautoeikonic or homometric.

Conclusion

The study of planar and spatial configuration with certain interpoint distance distribution is not limited to crystallography. Results derived from these studies may be useful for the design of spatial configurations where interpoint travel, communication, and location of stations are important. Patterson initiated the study of a crystallographic field which has a yet-to-come far reaching impact as does his many contributions to crystallography.

The homometric problem has attracted the attention of many mathematicians; its implication in structural science has yet to be fully studied. It is a great pleasure to be given this opportunity to write an article on this interesting problem to honor the 50th anniversary of the Patterson function, and writing this article has rekindled my interest in it.

The terminology regarding homometric structures may have left a reader confused. We have used the terms **congruent, enantiomorphic, complementary, homometric**, and **tautoeikonic** to describe the relationships among crystal structures based on their interatomic vectors. The enantiomorphic pairs and the seemingly different finite patterns resulting from the various choice of origins or axes of congruent crystal structures are well known to crystallographers; but there are no real complementary crystal structures, except that the stacking sequences of the layers in crystal structures may be complementary to each other. Thus, complementary homometric structures are not of major concern to practicing crystallographers. It should be pointed out, however, that complementary relationships among real layer-packed crystal structures have not been investigated in detail.

In a paper to prove Patterson's theorems on homometry, Buerger[7] used the term 'tautoeikonic' to describe point sets that have the same self image; they may be congruent, enantiomorphic, or homometric. To most crystallographers, the term 'homometric' may mean the same; congruent and enantiomorphic point sets (or structures) are special homometric sets. Thus, the two terms become synonomous, perhaps the latter being more commonly used. Thus, **homometric crystal structures** have the same Patterson function and, if they are further related, can be called **enantiomorphic**, or **congruent**. While enantiomorphic pairs of any point sets are always homometric, complementary pairs are homometric only if they have the same number of points.

Two finite, non-periodic patterns may be represented by two point sets r and s. Their vector images, (limited) Patterson functions, or convolution-squares[5] may be represented by $D(r)$ and $D(s)$ respectively. **Homometric sets** have the same vector image, *i.e.*, $D(r) = D(s)$, and may be made into homometric crystals by translations of a lattice, but non-homometric sets may also become homometric crystal structures by translations, *i.e.*, $D(r) = D(s)$ *mod lattice matrices. Thus, a distinction must be made between* **homometric sets** *and* **homometric crystal structures**.

References

1. Glusker, J. P. In *Crystallography in North America* (D. McLachlan and J. P. Glusker, eds.) pp. 103–107. American Crystallographic Association: New York (1983).
2. Patterson, A. L. Homometric structures. *Nature (London)* **143**, 939–940 (1939).
3. Pauling, L. and Shappell, M. D. The crystal structure of bixbyite and *C*-modification of the sesquioxides *Z. Krist.* **75**, 128–142 (1930).
4. Hosemann, R. and Bagchi, S. N. On homometric structures *Acta Cryst.* **7**, 237–241 (1954).
5. Franklin, J. N. Ambiguities in the X-ray analysis of crystal structures. *Acta Cryst.* **A30**, 698–702 (1974).
6. Zhdanov, G. S. *Compt. Rend. Acad. Sci. USSR.* **48**, 39–42 (1945).
7. Buerger, M. J. Proofs and generalizations of Patterson's theorems on homometric complementary sets. *Z. Krist.* **143**, 79–98 (1976).
8. Buerger, M. J. Exploration of cyclotomic point sets for tautoeikonic complementary pairs. *Z. Krist.* **145**, 371–477 (1977).

9. Buerger, M. J. Interpoint distances in cyclotomic point sets. *Canad. Mineral.* **16**, 301–314 (1978).
10. Patterson, A. L. and Kasper, J. S. Close packing. In *International Tables for X-ray Crystallography*. Vol.**II**, p. 342. Kynoch Press: Birmingham, UK (1959) (or 1972 Third Edition).
11. Patterson, A. L. Ambiguities in the X-ray analysis of crystal structures *Physic. Rev.* **65**, 195–121 (1944).
12. Iglesias, J. E. On Patterson's cyclotomic sets and how to count them. *Z. Krist.* **156**, 187–196 (1981).
13. Chieh, C. Analysis of cyclotomic sets. *Z. Krist.* **150**, 261–277 (1979).
14. Chieh, C. Tautoeikonic sets in some *p*,*q* cyclotomies. *Z. Krist.* **160**, 205–217 (1982).
15. Iglesias, J. E. Diffraction enhancement of symmetry enhancement cases for crystals formed by stacking of two kinds of layers *Z. Krist.* **150**, 279–285 (1979).

40. Some recollections about Patterson and homometric sets

John C. Oxtoby

Lindo Patterson had been aware of possible ambiguities in the X-ray analysis of crystal structures for some time before I became a colleague of his in the fall of 1939. But it was in the early 1940's that he began to study cyclometric sets and to find the remarkable examples which he described in his 1944 paper.[1] These presented interesting mathematical questions, and we had many discussions about them over a period of years. We never found any practicable way to identify all homometric sets, but sometimes a family would emerge as the solution of a system of modular equations representing some correspondence between distance arrays.

Fairly early we came upon the monograph by Sophie Piccard *'Sur les ensembles de distances des ensembles de points d'un espace euclidien'*.[2] This contained many interesting results, as Patterson noted in his 1944 paper. One of these asserted that two finite subsets of the line are isometric if they share the same distance set and all of these distances are distinct. We tried long and hard to find a proof that we could understand. Then one day Lindo came up with a counterexample! I can't remember just when this was, perhaps around 1946 or 1947. He never published this result; I think he intended to return to the problem at some future time, but he never got around to doing so.

Many years later when Joel Franklin became interested in the problem of finite sets he and Eddie Hughes asked Betty Patterson to see what she could find among Lindo's papers. At that time (February 1974) I suggested that she look for a sheet that might be labelled 'finite homometric pair' or might mention Piccard's theorem. The most I could recall was that it was a pair consisting of, I thought, six points, with relatively small integer coordinates. She was unable to find any record of such an example, and I didn't know any way to reconstruct it, but a few years later Gary S. Bloom published a note entitled

'A counterexample to a theorem of S. Piccard'[3] in which he gave the example

$$X = \{0, 1, 4, 10, 12, 17\},$$

$$Y = \{0, 1, 8, 11, 13, 17\}.$$

I cannot say for sure whether this is the same as Patterson's, but it agrees perfectly with my recollection. In any case, Bloom deserves full credit for the discovery and for making it known. It is not the only such example. Solving the equations that represent the corresponding permutation of the distance set leads to the family

$$X = \{0, a, a+b, 4a+2b, 6a+2b, 8a+3b\}$$
$$Y = \{0, a, 5a+b, 5a+2b, 7a+2b, 8a+3b\}.$$

For arbitrary positive real numbers a and b these have the same distance set and are not congruent; the distances are distinct if and only if $b \neq a$ and $b \neq 2a$. This and another such family

$$X = \{0, a, a+b, 4a+2b, 6a+4b, 8a+5b\}$$
$$Y = \{0, a, 3a+b, 3a+2b, 7a+4b, 8a+5b\}$$

$a > 0, b > 0, a \neq b$, were found by Bloom and S. W. Golomb.[4] No other Piccard counterexamples have as yet been found, but there are many finite homometric pairs with some distances repeated, for example those found by Franklin.[5] On the other hand, F. A. Grünbaum[6] has shown that a set of integers $x_1 < x_2 < \cdots < x_N$ is uniquely determined up to congruence if, in addition to the N^2 differences $x_i - x_j$, one is given the $2N^3$ numbers of the form $2x_i - x_j - x_k$ or $-2x_i + x_j + x_k$, each set being presented merely as a collection of integers with corresponding multiplicities.

References

1. Patterson, A. L. *Phys. Rev.* **65**, 195–201 (1944).
2. Piccard, S. *Mem. Univ. Neuchâtel* **13**, (1939).
3. Bloom, G. S. *J. Combinatorial Theory (A)*, **22**, 378–379 (1977).
4. Bloom, G. S. and Golomb, S. W. *Proc. IEEE.*, **65**, 562–570 (1977).
5. Franklin, J. N. *Acta Cryst.* **A30**, 698–702 (1974).
6. Grünbaum, F. A. *Adv. Math.* **26**, 1–7 (1977).

Part IV
Patterson—The Man and the Scientist

41. Biography of A. L. Patterson

1902–1966

Arthur Lindo Patterson (Lindo) was born July 23, 1902 in Nelson, near Auckland, New Zealand.[1] His parents, Nellie Tweedale Slack and Arthur Henry Patterson were English. Lindo's grandparents on his father's side died at an early age in the yellow fever epidemic in Bermuda where his grandfather was an engineer in charge of building the fortification. Their two sons were sent to live with uncles; Lindo's father went to Ireland where at age 14 he was apprenticed to the dry goods business. Lindo's mother was the eldest daughter of a family of 13 and, being anxious to leave her home in the North of England, had the courage 'to marry a man she hardly knew'[2] who took her immediately to New Zealand where he started a business of importing commodities from England. Mrs Patterson's diary[3] of their long voyage around Cape Horn is fascinating; as a pretty nineteen year old she apparently spent her time flirting with the passengers while her new husband played whist.

The story of the name 'Lindo' is characteristic of Mrs Patterson's enterprise. At the time that Lindo, her third child, was born she had made friends with a 12-year-old neighbor, Lindo Levien, a Jewish boy from Jamaica. She converted him to Christianity, and he was baptized at the same time as her baby, so she honored him by giving the baby the middle name, Lindo.[1] Lindo recounted, with a chuckle, that *his* only memory of New Zealand was of lying in his crib and seeing a fly on the ceiling![4] When the baby was $1\frac{1}{2}$ years old, Mr Patterson accepted the job of starting an agency for the English Jaeger Wool Company in Montreal, Canada. So Lindo spent his early years in Canada. However, his mother found that she could not stand the way the local Canadian boys pronounced her son's name Ar-thur, so she decided that he should be called Lindo. Then she had to put up with the constant call 'Lindo, come to the window'.[1]

When Lindo was 14, Mr Patterson was promoted to director of the Jaeger Company and the family moved to a suburb of London, England. Lindo was sent to Tonbridge, a well known

English public school (to Americans, a private school) from 1916–1920. After an easy time in the Montreal public schools, life at Tonbridge was a real shock. His horrendous experiences as a tall, red-headed and somewhat spoiled Canadian-speaking boy in such a school taught him to curb his temper. When he graduated at 18, he essentially had completed the equivalent of the first two years of college and was especially advanced in mathematics.[1]

In 1920 Lindo was admitted as a junior at McGill University in Montreal (1920–1924), where his sister, Dorothy, and brother-in-law, Frank Pratt, lived. His older brother, Nolan Tweedale Patterson, had died from gangrene from a leg injury in Belgium in World War I;[4] Nolan was a Lieutenant in the Canadian Forces, and one of his tent mates was Raymond Massey, the film star, a fraternity (KA) brother from the University of Toronto.[5] Lindo was never old enough to serve in the armed forces. However, later, he always hated peanut butter, a reminder of the poor diet at that time.[4]

After a year at McGill, Lindo shifted his major from mathematics to physics, dropped back a year and received a B.Sc. in 1923, but with Second Class Honours. In *Fifty Years of X-ray Diffraction*[6] Lindo describes 'the depth of ignominy' attached to this in British and Colonial Universities. 'This disgrace was correctly ascribed by my professors to too many friends in Montreal and an addiction to skiing, bridge, dancing and other related activities. Some quite fantastic suggestions were made as to how I might *live this down* but knowing myself these suggestions were a little difficult to take.'[6] At this time he was, briefly, engaged to the sister of his brother-in-law Frank Pratt (as his brother Nolan had been engaged to another sister) and the parties were continuous. Also he was house treasurer of his fraternity. But he did receive a good training in physics with emphasis on classical theoretical physics. He was required to study a lot of mathematics but 'a great deal of 20th century continental mathematics had not penetrated'[6] to McGill. He went on further at McGill and received an M.Sc. in 1924 with a thesis 'On the production of hard X-rays by the I ^{95}-rays from a radium active deposit',[1] done under the supervision of A. S. Eve. Thus started an interest in X-radiation and its uses.

In 1924 McGill announced the endowment of two Moyse Travelling Fellowships and, since Lindo's parents were in England, he wrote to W. H. Bragg and was able to work with him at the Royal Institution in London from September 1924 until September 1926.[1] Characteristic of Bragg was his short letter which Lindo found in his ship cabin as he was leaving for England: 'Dear Mr Patterson, I will expect you in London between the 23rd and the 26th, when I shall be glad to see you. Yours sincerely, W. H. Bragg'.[6] Since Bragg was interested in naphthalene and anthracene, while Muller and Shearer were studying long chain compounds, Patterson worked on determinations of the unit-cell dimensions and space groups of various phenylaliphatic acids. To do this, he had to build his own X-ray equipment. At that time no structure determinations for the phenylaliphatic acids were contemplated. Bernal described how, at the Royal Institution 'research workers came and went. They were accommodated wherever an inch could be found for them in the old building. I remember Patterson settling down as best he could between the glass cases that enshrined the historic apparatus.'[7] But he did work hard building his own X-ray equipment, a gas tube designed by Shearer, and finished it by November. He first had to make a vacuum tube of soft glass, then was given Pyrex (the new hard glass) to blow the two he needed. He described the apparatus in detail.[6] He then studied the structure of phenylpropionic acid using Bernal's new rotating crystal method. At this time, little structural information could be obtained — only space groups, unless the molecule had symmetry.[6]

Meanwhile many exciting scientific studies were being reported. Duane's paper in 1925, wrote Lindo, had described a crystal 'as a three-dimensional Fourier series, reviving W. H. Bragg's suggestion from 1915.'[6] At the Royal Institution, George Shearer used one-dimensional Fourier series to explain 'intensity distributions in long-chain paraffins, acids and histones.'[6] So Lindo was familiar with Fourier series but it was not until his quantum mechanical friends in Germany suggested he study the book by Courant and Hilbert[8] that he discovered that 'something called the Fourier Transform existed and at the same time learnt what a potent mathematical entity an orthogonal set of functions really was ... I had been taught about a Fourier Integral Theorem but always in its double integral form. The fact that it [the Fourier integral] could be split in two, one the spectrum of the other, was something very startling and illuminating.'[6]

Lindo's stories of this happy time in London show his delight not only in the city but in his friends at the Royal Institution who included Astbury, Bernal, Burgers, George, Gibbs, Jackson, Knaggs, Mathieu, Müller, Orelkin, Plummer, Ponte, Shearer and Yardley (Lonsdale). The chemists, Saville, Smith and Lawrence, provided 'wonderful chemical backing.'[6] He was practically adopted as a son by W. H. Bragg who delegated to Lindo the task of entertaining and fending off the many visitors.[1] In addition, Lindo spent quite some time at the Bragg's home. He mentions taking Gwendy Bragg to dances and says he almost wore out his white tie and tails and morning coat (cutaway) at that time.

Instead of returning to McGill in 1926 as originally planned, Lindo took up the suggestion from the Pulp and Paper Industry Research Laboratory at McGill which 'had become interested in the work of Herzog and Jancke on the X-ray diffraction from cellulose'[6] to go to Dahlem (Berlin), Germany, and join the group headed by Hermann Mark in Herzog's Institute at the Kaiser Wilhelm Institut für Fasserstoffchemie. He received a National Research Council of Canada Fellowship for this purpose. Mark was away for part of the year so Lindo was nominally working under Herzog on the particle size of cellulose.[6] Herzog and Mark suggested that von Laue's paper on the diffraction of X-rays by small particles might be useful in the accurate determination of the 'particle size' of cellulose. Lindo was 'fascinated by the notion which Laue had introduced of the space surrounding a reciprocal lattice point'[6] and struggled to see how methods other than those of approximation could be used. This resulted in Lindo writing a brief note for *Zeitschrift für Physik* which was the basis for his later work on particle-shape functions. He did not publish any interpretation of the experimental work he had done on line-broadening in Dahlem because he doubted the 'meaning of a particle size determination.'[6]

A diary[9] that Lindo kept during his stay in Germany starts with the train trip to Berlin during which his compartment companion advised him that the best way to learn German (Lindo had only a high school course in this language) was to get a girl and she would teach him fast. No doubt he took this advice. He roomed in a house with two old maid landladies who were very kind to him. His triumph was when he could put together German verbs 'I am very proud to say that yesterday [September 19] I separated my first prefix successfully. I think that the sensation must be similar to that of a doctor removing his first appendix.'[9] In a September 29th entry he says: 'Today I think for the first time I succeeded in cracking a joke in German without having to explain three times. I also succeeded in making a successful pun, so I am feeling pretty good.'

Lindo also recorded in his diary,[9] 'Tonight my landladies have given me some light reading — 'Fürst Bismarcks Briefe an seine Braut und Gattin' [Count Bismarck's Letters to his Bride and Wife]. I really cannot face it so I think that bed is indicated'.[9] However he also wrote of his enthrallment at the gorgeous music he heard at every opportunity and his excitement at attending the weekly Physics Colloquia with von Laue as chairman and Einstein, Planck and Nernst in the front row with Bethe, Hahn, Meitner, Pringsheim and Wigner there. 'Wednesday was a great day. In the afternoon, I went to my first Physics Colloquium in the University. Einstein, Laue, Planck, Nernst, form the nucleus, with goodness knows how many other stars. It really was wonderful. I understood most of everybody but Laue, and the Germans themselves find him difficult to follow. Einstein is quite a cheery soul. I met Ewald who is a first rate sort, who speaks, as do most people, about 3 languages fluently. I don't think that even in the R.I. [Royal Institution] or at the B.A. [British Association] I have ever been in the same room with so many scientific bloods. They all seem quite ordinary people, but Einstein, and he seems extremely human in character, but he has one of the most marvellous faces of anyone I have ever seen. A perfectly marvellous head as well. There is a photograph of him in a shop in the Uhlandstrasse that I would like to buy, but I don't know if such is allowed. When you see the man you realize how it is possible that he has done such wonderful work.'[9]

Lindo gave his first seminar in German at Eitel's Institut für Silikatforschung down the street where he had made many friends as he had at Haber's Institut für Physikalische Chemie. His subject was W. L. Bragg's paper 'The Structure of Certain Silicates.' Herzog said after his talk: 'Vorlesen können sie vielleicht, aber Deutsch leider nicht.'[6] [Translation: You can, perhaps, lecture but, unfortunately, not in German.]

Lindo's love of music was given a terrific boost by his stay in Germany. From his diary[9] it seemed that he went to a concert almost every other day. And music formed a background to his thinking.

He wrote: 'This leads me rather to a discussion which I had with Mark and a comparison, I think due to him in the discussion, between music and Crystallography [sic]. He is not so particularly keen on Bach, from the music point of view, and although he acknowledges him as the founder of modern music, only from the mechanical point of view. The comparison is between the mathematical theory of Crystallography as developed by Schoenflies and von Fedorow, and Bach. The mathematical theory of music was provided by Bach, and just as Fedorow and Schoenflies would not have been able to make the beautiful applications of their theories to the interpretations of crystal structure, so was Bach unable to write the Ninth Symphonie [sic]. The Goldberg Variations fall very nicely into the analogy, as being the space groups of a certain symmetry class. The Well Tempered Clavichord [sic], is more or less the set of examples appended to show what can be done with the beautiful new mathematics. The St. Matthew Passion, and the Mass, and the Brandenburg Concertos are more difficult to fit in. The two, three and four piano concertos are more or less the examples of the cubic system in extension of W.T.C. Carrying the analogy further, von Laue, Friedrich and Knipping compare with Mozart and Haydn, who found out what could be done with the mathematics in applying it to the expression of these ideas. W. H. and W. L. Bragg acquire the cloak of Beethoven. W. L.'s *Can. Phil. Soc.* paper and the structure determinations of salt and diamond compare with the first two symphonies of Beethoven, the first, which also might have been written by Mozart, and the second showing the hand of the master. The Eroica, when the artistic sense of the master begins to take hold, one can compare with the application of the work to the organic substances. Fedorow, Schoenflies or even Laue would have wanted a little more to go on, but the application of the touch of the artist brought about this beautiful work, which may lack the preciseness of geometrical definiteness, but which is truth for all that. The work of W. L. Bragg on beryl, I should say, ranks very high in the string quartets of Beethoven when we carry the comparison on, but I am afraid, that the Vth and IXth Symphonies have yet to be written. Weissenberg comes in, in my opinion, as the Mahler of the IIIrd Symphony, the young man who has found this beautiful machinery, as well as the more beautiful methods of applying it, and who tries to see how far he can possibly carry it. Wyckoff I had thought of as the Wagner of the piece, but it won't go. If Wagner had accentuated Bach instead of Beethoven, it would have been all right, but Wyckoff as an exponent of the Braggs is, to say the least, a little strained as an analogy. Wyckoff's love for the one method of attack viz., Laue diagrams, can be slightly compared with Wagner's love of a single theme for each character. I am afraid Jancke's work is jazz in its worst form. Some good has come out of jazz just as the structure of cellulose came out of Jancke's photography. I wonder if the analogy can be made to bring in any others, perhaps if my knowledge of music were a little greater, but this is certainly not the place to do it.'[9]

It took until Easter time in 1927 before Lindo got to know von Laue. On Good Friday Lindo received a copy of Davisson and Germer's *Nature* paper on the diffraction of electrons by nickel. He had all weekend to study it in detail. Von Laue reported the paper at the next colloquium and Lindo brashly pointed out where he disagreed with von Laue's interpretation. Von Laue, unlike most German professors, did not stand on his dignity and asked Lindo to get up and review the paper. 'This I did, so scared that the first line I drew on the blackboard came out dotted.'[6] Von Laue later invited Lindo to his home where they had a long discussion about electron diffraction, work on particle size and on Lindo's efforts on the Fourier interpretation of the reciprocal lattice. After that they were very friendly and had frequent talks.

Lindo returned to McGill in the Fall of 1927 and received his Ph.D. in Physics in 1928. Before receiving his degree he had a graduate student, Thomas N. White, Jr. Tom relates in his thesis,[10] 'On the resumption of X-ray work at McGill University, the question arose as to what substances might best be chosen as a field for research. Both Dr. Patterson and the writer had been interested in the problem of organic structure and a survey was made of the state of knowledge in this field at that time.' Other workers appeared to be successfully studying aliphatic compounds, complex substances such as rubber and cellulose, and unsaturated aromatic compounds. 'Little work had been done on the saturated compounds, and none on those saturated with the lighter atomic groups, such as hydrogen and the hydroxyl group, such as the hydroxy substitution products of cyclohexane. It seemed, then, that this might be an excellent field for X-ray investigation.' They collaborated on a study of cyclohexane hexols. Lindo says that his reading for this work convinced him of the 'important future' of X-ray diffraction in biochemistry.[6] Lindo was appointed Lecturer in Physics at McGill and taught for a year before leaving for New York to work with Ralph Wyckoff as an Associate at the Rockefeller Institute. Tom White went along on a National Research Council postdoctoral fellowship. They continued their work on the cyclohexane compounds and 'looked at a number of other substances.'[6]

Lindo, wrote,[6] '*my obsession with the notion that something was to be learned about structural analysis from Fourier theory continued.*' He went to the New York Public Library and asked for a stack permit. Request denied. So he told them he wanted to see all their mathematical journals, starting with A and going on through Z. After a day of bringing him journals when he looked through the table of contents and only looked at promising papers, reading only a few carefully, he received his permit and continued his search. He admits that he really didn't understand a lot of the papers he tried to read and that he missed the point of many papers. But he did pick up some ideas and wrote two papers for *Zeitschrift für Kristallographie* about location of maxima of Fourier series and 'the enhancement principle.'[6]

In 1931, Lindo accepted a job at the Johnson Foundation for Medical Physics in Philadelphia in the hope of doing something with X-ray diffraction on biological materials. He took some powder photographs of horse hemoglobin and 'had a good time collaborating with Ray Zirkle in some experiments on the effect of X-rays on the growth of fern spores, but my other excursions into biology were not too happy and I decided in 1933 after two years in Philadelphia that somehow I had to get back into crystallography and Fourier series.'[6]

He had saved enough money to support himself for a year, but he had hardly anticipated the effect of the depression which kept him out of a job for three years. In the midst of this period, in the fall of 1933, he became engaged to Elizabeth Lincoln Knight (Betty), a research scientist he had met in the Rockefeller Institute. They were not married until September 1935.[1]

Knowing that Bert Warren had an X-ray lab at the Massachusetts Institute of Technology (MIT) and that Norbert Wiener at MIT 'probably knew as much about Fourier integrals as anyone in the world,'[6] Lindo asked Warren to take him on as an unpaid guest with free use of the lab. Warren accepted and Lindo moved to Boston in September 1933. His letters to Betty (see Chapter 1.2, this volume) showed how happy he was to again be with physicists and crystallographers such as K. T. Compton, John Slater, J. Stratton, George Kimball and in Bert's group, Gingrich, Hultgren, Serduke and G. G. Harvey, in addition to bright seniors and good Ph.D. candidates.[6] Although they worked hard, there were many parties; they went to concerts and even movies frequently.[11]

Lindo started on particle-size work because he had been asked to give a seminar on this subject.[6] He worked on this problem on and off during his three years at MIT. 'The particle-size problem is what is technically known as a bastard. It isn't really a series of hurdles but just one large barbed wire entanglement and when you think you've arrived somewhere as various people have done in the past fifteen years, you find that there are many places still to go. One has to make so many compromises with integrals that no answer that one ever gets is in any way final. All I can hope for out of my present efforts is to get rid of the compromises and possibly develop some of my trouble into a critical review of the field which someone may publish.'[12]

He 'had many opportunities to talk with Wiener. This latter was a laborious process but a very intriguing one. There was then and is now no subject which can be brought up on which Wiener does not have something interesting to say and with him the subject is always changing. I estimate I got about one question on Fourier theory per two or three hours conversation, but the answers were usually pay dirt.'[6]

At a party while singing Gilbert and Sullivan with Wiener, Lindo threw in some questions in the songs on what function could be used to solve the phase problem in crystallography. Wiener replied 'The Faltung.'[1] Lindo later said it took him a very long time to realize the full potential of the Faltung, which came 'from the work on liquids and their radial distributions.'[6] Warren and Gingrich have perfected techniques of Debye and Menke based on earlier suggestions of Zernike and Prins for the X-ray scattering from liquids. Warren and Gingrich thought that 'these methods, applied to powders, would give the radial distribution in a crystal.'[6] While studying their work Lindo 'noticed that the mathematical form of the theory given by Debye and Menke [for X-ray scattering from liquids] would be identical with that of the Faltung if the integrations over random orientation were left out and the randomness of choice of origin was left in'. This, combined with the (singing) conversations with Wiener, enabled Lindo to put the pieces together so that Fourier theory could, indeed, be used in elucidating crystal structures. 'What was immediately apparent was that the crystal contained atoms and that the Faltung of a set of atoms was very special in that it would consist of a set of atom-like peaks whose centers were specified by the distances between the atoms in the crystal. It was fortunate that this was clear from the beginning and it was in this form that the interpretation of $|F|^2$-series was proposed. It is unfortunate that the notion arose later that the maxima of the Faltung were determined by the distances between the maxima of the Fourier series. This is clearly untrue in general and was never suggested by me.'[6]

Lindo wrote on March 23, 1934: 'Things have broken rapidly. There is a paper for Washington and a celebration tonight. Such is theoretical Physics.' He told Betty[1] that it was on a Tuesday that he finally saw how the Faltung could be used to put together two portions of his equations that had been, so far, disconnected. Unfortunately, that Friday was the deadline for abstracts for the spring meeting of the American Physical Society in Washington.[6] He rapidly prepared an abstract to be submitted in with that of Gingrich and Warren on the radial distribution in powders, basic to Lindo's work, and a paper of Warren's on radial distribution in carbon black. 'The only $|F(h)|^2$-series which I was able to compute in the month between the deadline and the meeting was the $(hk0)$ of KH_2PO_4 and a one-dimensional series for a simple layer structure. All three papers were very well received and had very full discussion with A. H. Compton in the chair and W. L. Bragg in the audience to ask the right questions.'[6]

Bragg visited MIT after the meeting and told Lindo about the copper sulfate structure of Lipson and Beevers. They had good intensity data, which was a requirement for the use of what Lindo called the $|F^2|$ method (but which was soon called the Patterson method). They soon sent the data and in the meantime Lindo computed the $|F^2|$-series of hexachlorobenzene from Kathleen Lonsdale's paper.[6]

It must be remembered that at this time large calculators were used for all computation, which was done by hand. Bragg had also told Lindo that 'Lipson and Beevers had had a method for computing series by using "strips".'[6] Lindo had been computing by a method suggested by George Kimball 'which involved multiplication of every term in the series by its appropriate sine or cosine.'[6] Since this entailed a lot of repetition, Lindo collaborating with George Tunell set up a 'strip' method where you read the strips through holes in a matrix negative or positive (see Chapter 25, this volume). That summer at the New Jersey shore Betty's father rented an electric calculator (with hand shifting) on which Lindo and Betty calculated $CuSO_4$ using the strips and matrices (Betty's mother strongly disapproved of this activity which kept her daughter off the beach and tennis courts).

It seems clear from Lindo's letters that it was not until he struggled with the $CuSO_4$ results that he realized the power of his method. He wrote in another letter: 'The program this week is to redo $CuSO_4$ with the origin taken out and with the "bugger factor" [a sharpening factor] applied. I plotted the map last week as I told you and found that the Cu and S located themselves immediately, but that the oxygens were not so good. I could find four of the nine pretty well, but the others could not be located'.[7]

Things were now really coming together. In a letter in November 1934, referring to writing papers for the *Zeitschrift für Kristallographie*, Lindo wrote, 'I finished the first draft this morning and felt that it was pretty lousy. I then went into the lab and had a long talk with Bert. Some things he said reminded me of the fact that I had set out to write a cookbook paper, *i.e.*, "How to use the method and get results," with theory in words of one syllable. I had certainly lost sight of that ideal in the writing, so I came home and started the rewriting. And instead of it being just a brushing up on the machine of what I had already written, it is a real rewriting except that I have most of the facts. The order of presentation has been completely changed. Actually, however, I think that what I have rewritten is about 100% improved. By that I mean that perhaps 6 people will read the paper instead of 3.'[11]

In 1962 Lindo wrote: 'In retrospect it is a source of satisfaction to me that I did so much work on the $|F^2|$-series method before publishing the second paper. This made it possible for me to draw attention to many of the difficulties which were likely to and did arise. I must say that I was very annoyed at myself for missing the beautiful extension of the method made by Harker. I guess I really could not get out of the plane.'[6]

The depression years provided great hardships for Lindo and Betty. In later years he sadly remarked, 'I didn't really understand about depressions and did not contemplate three years out of a job'. It was not until the spring of 1936 that Lindo was offered a job in which he could continue as a crystallographer.[1] Other offers required him to change his field. 'It was therefore very gratifying ··· that I was offered an assistant professorship in Physics at Bryn Mawr College, with the express purpose of developing X-ray analyses in parallel with the wider interests of Walter Michels in the solid state.'[6]

Lindo and Betty were married in September 1935, feeling demeaned by having to depend on their families for support. Betty had given up her job at the Rockefeller Institute in 1934 to start graduate work at Cornell Medical College. In 1935-1936 they lived in Wellesley, and Betty, who had been a chemistry major, caught up on biology courses at Wellesley while continuing a research program started at Cornell. One night, well remembered afterwards, the doorbell rang late in the evening, and the physicists Jim Fisk and Ralph Johnson appeared carrying a bottle of rum and a can of grapefruit juice. They said, 'We heard you too had something to celebrate. We just passed our final Ph.D. exams.' The hilarious evening culminated by Lindo and Ralph Johnson writing a somewhat off-color telegram in Latin to Keffer Hartline who had just gotten married. (Lindo had known Keffer well at the Johnson Foundation.) Lindo's dictation of this telegram to Western Union had everyone close to hysterics. Of course, Betty was delighted with Lindo's Bryn Mawr job because she knew from friends that she could find just what she wanted in Ph.D. study there.[1]

The years 1936-1949 at Bryn Mawr were happy ones but not productive as far as Lindo's research was concerned. He had some excellent graduate Ph.D. students such as Selma Brody and Beatrice Magdoff but all free time was taken up with individual student conferences, by committee work or summer projects such as sorting surplus war equipment, a Naval Ordnance Laboratory research project, using manual labor to rebuild the laboratories after a disastrous fire at Bryn Mawr and writing a physics textbook *Elements of Modern Physics* with Walter Michels. As you might guess, they were affectionately called Pat and Mike by their Bryn Mawr friends. Funds for X-ray equipment were hard to obtain. Pat had a Research Corporation grant for $500 for a transformer. He found a second-hand one and made the rest of his equipment in the shop. He did finish two particle-size papers started at MIT.[4] His work on homometric structures was greatly aided by collaboration with the mathematicians, notably John Oxtoby and his wife Jean (see Chapter 40, this volume).[1] From this collaboration he derived much pleasure as he did also from association with colleagues in the many fields of research, scientific and humanistic, existing at Bryn Mawr. It was a stimulating atmosphere.

Betty received her Ph.D. in 1940 in cytology and physical chemistry and spent the next year in research in ultramicroenzyme chemistry with William Doyle who had come to Bryn Mawr after working in the Carlsberg Laboratories in Denmark with Heinz Holter. It was felt still at that post-depression time that a woman should not take a job that a man could hold. In October 1940, Lindo and Betty took into their small apartment a young English cousin of Lindo's, David Pilkington, aged 10. He stayed until April 1944, just before D-day (June 1944) when he was almost 14. Lindo felt, after his experience with English public schools, that 6 foot-tall David could wait no longer to start his English education. He was a charming child, caused no trouble and rapidly adapted to American life.

After Pearl Harbor, Michels embarked on a research project for the Navy which later led to his being commissioned as Commander and leaving for Washington. Pat was still a British subject and worked on the project at Bryn Mawr until leaving for Washington for a 6-month stay as a civilian in the fall of 1944. Actually the project involved a pressure mine for large vessels and thus it was necessary to analyze ocean waves. Therefore, Lindo's expertise in Fourier series came into play. He organized a project for students on an hourly basis to carry out the calculations based on data supplied by the Navy. Betty supervised the group teaching them how to use the matrices, strips and electric hand calculators, and plot the results. Actually, no one was told they were analyzing ocean waves.[1]

Many professors (including William Doyle and Joseph Berry in Biology) began to leave Bryn Mawr for war work so when the Navy calculation project was completed in the fall of 1943, Betty accepted a job as 'Demonstrator' in biology which entailed running the lab in bacteriology and biochemistry while other female recent Ph.D.s lectured. At that same time Glenn Richards of the University of Pennsylvania had a war project involving the action of DDT. He supplied to Mary Dumm and Betty 100 frozen dried 'honeybee brains' (supraœsophagial ganglia) on which they were to use ultramicrochemical techniques to analyze their differential lipid content. Dr. Dumm soon left for Harvard and Betty was left with the project. Although her Bryn Mawr students were excellent, she resented the time they took away from research and realized that teaching was not for her. When Lindo left for Washington and their rent-controlled apartment was much in demand, Betty decided to stay on in Bryn Mawr and accept the job offered by Dr. Stanley Reimann, Director of the Lankenau Hospital Research Institute (LHRI) to work with Jack Schultz on enzymes in the genetically controlled fruit fly. She started in November 1944 after she finished the paper on the lipid content of honey bee brains. Actually, Lindo returned to Bryn Mawr in June 1945, so it was just as well that they kept their apartment.

While in Washington at the Naval Ordnance Laboratory, Lindo's Naval colleagues used the fact that Lindo was not a citizen to get him to pressure higher official personnel to force through some projects. They begged Lindo to stay a British subject until after the war. Actually the project was finished in 1945 and Lindo received the Meritorious Civilian Service Award for his work on submarine warfare. He became a U.S. citizen in the same year.

After the war he continued his main interests which were particle-size broadening and the study of 'homometric structures'. Lindo was also a good citizen in the crystallographic community. He served the ACA (American Crystallographic Association) with great dedication throughout its history and played an important role in its formation in 1950 when the American Society for X-ray and Electron Diffraction (ASXRED) and the Crystallographic Society of America (CSA) merged. He was President of ASXRED in 1949 and served on the ACA Publications Committee in 1955-1957 and 1959-1961. He was responsible for initiating a clause in the new society's constitution barring from office anyone who had previously been an officer in one of the merging societies.

Another great service that Lindo did for the crystallographic community was his masterfully lucid section on 'Fundamental Mathematics' in Volume II of *International Tables.*

Lindo contributed to international crystallography by his service on the U.S. National Committee for Crystallography in 1951-1955, 1957-1962 and 1964-1966; he was Chairman from 1948–1950. He played a major part in the writing of the by-laws of the International Union of Crystallography. [14] He served as a U.S. Delegate to several I.U.Cr. meetings, for instance the meeting in Paris in 1954.

Betty loved her job at the LHRI and found the discussions with her new colleagues very stimulating. In the meantime Lindo was less happy at Bryn Mawr College; many of his colleagues had been promoted to full Professor while he remained an Associate Professor. His crystallographic friends told him he must move to a place where he had time for research. So in 1948-49 he took a long overdue sabbatical leave and looked around for another job. Two definite possibilities arose and Betty told Jack Schultz that Lindo was leaving Bryn Mawr and that she would also have to go. Jack knew Lindo and protested and said, 'We're building a new building with laboratory space reserved for a physicist, Lindo would be well suited to the job.'

It so happened that for a long time Dr. Reimann, founder and first Director of the Institute for Cancer Research at Fox Chase in Philadelphia, had wanted a physicist on the staff. As early as 1932 he wrote to a friend:[13] 'Stereochemistry is what we need. I would like an X-ray spectrometer in the lab with two or three experts to run it. Also we need pure mathematicians to develop equations for the apparatus and its complications.'[13] So Reimann called Lindo, interviewed him and was very enthusiastic and offered him the position. For one reason or another the other places Lindo might move to had disadvantages so he took Reimann's offer seriously. He submitted a proposal for a Physics department that would not only help other people with pertinent problems but would be set up for X-ray diffraction studies of the three-dimensional properties of single crystals of 'biological interest.' Still, Lindo wondered if it was worthwhile to gamble on a place with little endowment, a shaky future and where his wife worked, just for an opportunity to do full time research in the crystallographic structure of compounds of biological interest. With quite some trepidation, Lindo and Betty took the gamble on the basis of the people there and of the excellent research already in progress in the LHRI and the Institute for Cancer Research.[1] The Institute for Cancer Research (ICR) was incorporated in 1945 and housed in the same building as the LHRI. In 1958, the name LHRI was dropped and the ICR continued.

It took some time and struggle before Lindo had his laboratory set up for experimental research. He was one of the first people to move in the fall of 1949 into the brand new Fox Chase laboratories of the (LHRI) and (ICR). He immediately made friends with the maintenance crew and shop personnel and Dr Reimann somehow provided money for X-ray equipment. The Institute was just beginning to apply for and receive government grants. Lindo rewrote an application that Dr Reimann wrote for funds for his work and obtained a grant. He had an excellent Research Assistant, Joan Clark, who supervised all the purchasing, filing, and setting up of the laboratory. He soon acquired another bright Research Assistant, Alice Weldon, a classmate of Betty's at Wellesley. In the fall of 1949 the Women's Auxiliary of the Institute provided a Research Fellowship for Cecily Darwin (later Littleton) the great-granddaughter of Charles Darwin. Cecily had studied crystallography at Oxford under Dorothy Crowfoot Hodgkin, and in Fox Chase completed a structure determination of the dimer of *para*-bromonitrosobenzene started at Oxford. Cecily went on to determine the structure of potassium gluconate after starting with the isomorphous rubidium salt.

In succeeding years Lindo had a series of excellent Postdoctoral Research Fellows and Associates who have all now become well-known crystallographers: Christer E. Nordman; 1953–1955, Warner E. Love, 1955–1957; Jenny Pickworth Glusker, 1956–1966; Dick van der Helm, 1959–1962; Carroll K. Johnson, 1959–1962; Jean A. Minkin, 1960–1968; Eric J. Gabe, 1963–1966; Max Taylor, 1964–1965. Marilyn Dornberg was invaluable as a Research Assistant.

The work of Lindo at ICR covered many themes of which the principal one was the study of citrate conformation. He became interested in this as a result of discussions with Dr Sidney Weinhouse at ICR. Weinhouse was interested in metabolism and the role of citrate in the Krebs cycle. Otto Warburg had suggested that some defect in the Krebs cycle might be important in cancer cells. But this idea has not survived the test of time. However it did lead Lindo to start working on citrates, particularly rubidium citrate and citric acid when Chris Nordman was working with him. Rubidium citrate crystallized with a 'pseudocell' of one third the true cell, *i.e.*, there was a slight disorder in the true unit cell from averaged positions in the pseudocell. The averaged structure in the smaller cell was studied by Lindo and the other members of his lab.

Joan Clark assisted with the work on phenylpropionic acid and cinnamic acid derivatives. This was an extension of the work of his good friend Gerhard Schmidt who was interested in solid-state reactions. Another crystal of great interest was that of an unstable modification of tetraacetylribofuranose. Dr George Brown of the Sloan-Kettering Institute for Cancer Research suggested that Lindo's group investigate by X-ray diffraction the solid state transformation of this compound and a crystal was obtained from Dr Irving Goodman of the Wellcome Laboratories. Since more and more labs found their unstable modification had been converted to the stable modification and they could no longer grow crystals of the unstable form, Lindo began to correspond by telegram, rather than contaminate his lab with the stable form. He collected some preliminary data and then found that to change the crystals he actually had to sprinkle a powder of the stable form over the crystals before they converted; it was not enough to leave dishes with the two forms lying side by side.[14] George Brown visited the lab to watch these experiments.

Then Warner Love and later Jenny Glusker and other members of the laboratory studied sodium citrate and the isomorphous lithium citrate. Unfortunately the two crystals were not sufficiently isomorphous for the data to be used in direct phasing. Warner Love, and later, Dick van der Helm and Carroll Johnson, had a crystal farm and studied many of the citrates. Carroll spent most of his energies on magnesium citrate.

The work on citrate led to interesting interactions with biochemists, particularly Vickery at the Agricultural Experimental Station in Yale. This led to work not only on citrates but also on isocitrates, first as their lactones and then as the potassium salt. The isocitric lactone structure established the relative configuration of the two carbon atoms and the work on potassium isocitrate established the absolute configuration. This structure was solved by Dick van der Helm with the aid of the P_S (sine Patterson) function (see Chapter 29, this volume); the positive and negative vectors immediately indicated the absolute configuration.

Lindo then became interested in absolute configuration. As a result of a question posed to him in conversation at a meeting, he analyzed the problem of computing electron density maps in noncentrosymmetric structures when anomalous scattering occurs and the structure factors are more complex. He felt particularly pleased when Patterson methods came up with structures and Dick van der Helm was especially adept at that, and he was even more pleased when new methods gave interesting results. There are many excellent examples of those from his laboratory.

All of this took careful experimental measurements of intensity. This had always intrigued Lindo and he worked on an integrating precession camera with Chris Nordman and Charles Supper and on various other aspects of data collection. By the late 1960's the lab had acquired a General Electric XRD-5 diffractometer which was used to collect intensity data manually. Before that, integrated Weissenberg and precession photographs were scanned by a densitometer belonging to the ICR geneticists (George Rudkin and Jerry Freed).

In the later 1960's Lindo spent considerable time on 'function spaces between crystal space and Fourier-transform space.' This problem was concerned with the continuous sequences of transforms that can be set up between function space and Fourier-transform space. This was of particular interest to him as a crystallographer because most phase-determining properties such as positivity, atomicity and resolution are properties of crystal space while the Fourier coefficients whose phases are to be determined are defined in reciprocal space, *i.e.*, Fourier-transform space. The question he asked was whether there is a function space between the crystal space and reciprocal space that is simply related to both of these. He limited himself initially to the one-dimensional Fourier transform, but discussed how his results (using Hermite functions for the kernels of integral transforms) could be expanded to two and three dimensions.

Of course, the insight gained from his scientific discoveries tells us little of the kind of person Lindo Patterson was. How would he appear to each of you at a meeting? He was a tall, handsome, good-looking red-head with freckles and a ready smile, generous-hearted and yet he could also be firm. Cecily Littleton wrote that he was a gentle wise man, a moral support for people lucky enough to know him.

Joan Clark wrote of him,[15] 'He never missed an opportunity to encourage young scientists, not only by a word of praise or an interest in their work, but also by treating those with whom he was in contact as personal friends, giving them opportunities to meet older scientists and participate in discussions at a level that might otherwise not have been possible to them for many years. Although the temper that went naturally with his red hair became calmer in later years, it still appeared on occasion, usually sparked by some injustice or unnecessary stupidity. He was always a valiant proponent of the ability of women in science, and he did what he could to prevent any kind of illogical discrimination, retaining only that based on true intellectual capability.'

'A social person, he thoroughly enjoyed conversation, good food preceded at dinnertime by a shot of bourbon, and good music. He could sing most of the Gilbert and Sullivan arias, having acted in productions of these operettas in his earlier years, and he delighted in jokes, especially scientific ones. One of his favorites was a poem on entropy, which he would sing to the tune of *Rock of Ages*, ending with the lines, 'ΔS is always plus, No matter how you fume or fuss.'

Unfortunately, although he was extremely good-looking, not many good photographs of him remain. He tended to 'freeze' in front of a camera and his dynamic character was lost. He had a delightful sense of humor and said that the best crystallographic information was obtained by informal discussion, preferably at the bar. That was in the days before poster sessions. He had an endless store of funny stories. In his textbook with Michels there are some out-of-the-ordinary diagrams, including a schmoo, and a Charles Addams child on a see-saw.

In his files were found the words of a song, to the tune of 'My bonnie lies over the ocean'. It was written in August, 1932, for (or during) a solar eclipse expedition (August 31, 03h. 27m. 29.6s). The verse for Lindo ran

> But when the morale of our party
> Is getting exceedingly low,
> Our spirits will rise as we look at
> Our jolly stake-driver Lindo.
> Lindo, Lindo,
> As a diplomat you are so smart, so smart,
> Lindo, Lindo,
> We hate to have you depart.

Lindo also delighted in amusing phrases. Most of those he used were not original to him, but he was a master in choosing the right moment to use each. One favorite of his was:

> Proof by the 'Method of Rapid Speech.'

Another, that he gave to Larry Anderson, the photographer at the Institute, who often had to keep impossible deadlines, was:

> The only person who got everything done by Friday was Robinson Crusoe.

No article about Lindo would be complete without mention of his article on 'Wheaks' (white radiation streaks). He was tired of reading lots of acronyms and yet realized that jargon was taking over crystallography and therefore wrote an article, signed 'A. L. Pon' in which he tried to save space. It was accepted by *Acta Crystallographica* since it was good science, but, since the journal was in the red the editors asked people to shorten their papers. At this point Lindo withdrew his paper because he did not want to use up space that properly should have gone to a scientific paper. However he got offprints

Not reprinted from *Acta Crystallographica*

PRINTED IN DENMARK

On the symmetry of the wheaks produced by the Bucessera. By A. L. Pon, *Senics, Incanearch, Philpa, U. A.*

(*Received* 5 *December* 1952)

Recent publications on the reory (some workers prefer the term rekhaning) of diffions of X-ray (Raster, 1951*a*, *b*; Helister, 1951; Hoester, 1952*a*, *b*) have suggested the mathiques used in the prote.

It is clear that in all discussions of diflems, the radius of the Ewere can be set at unity.* If then a given relp is associated with monion of a given wangth, then a relve joining this relp with the relor corresponds to the whion which is necessarily pruced with the monion in the X-rube (see any took on the quary). Thus any difromena pruced by the Bucessera possesses planretry about a sylane through the relp, the relor and the difray.

Rences

Helister, N. H. A. (1951). *Intergraphs*. London: Macan.
Hoester, J. A. (1952*a*). *Experientia*, **8**, 297.
Hoester, J. A. (1952*b*). *Acta Cryst.* **5**, 626.
Raster, G. A. (1951*a*). *Acta Cryst.* **4**, 335.
Raster, G. A. (1951*b*). *Acta Cryst.* **4**, 431.

* Vernished.

FIG. 41.1. The paper on 'wheaks' by A. L. Pon, unpublished in *Acta Crystallographica*.

marked 'Not published in *Acta Crystallographica*'. The paper is appended as Fig. 41.1; a translation is in order.

On the symmetry of white radiation streaks produced by the Buerger precession camera, by A. L. Patterson, Senior Scientist, The Institute for Cancer Research, Philadelphia, Pennsylvania, U.S.A. Recent publications on the reciprocal lattice theory of diffracted reflections of X-rays have suggested the mathematical techniques used in the preliminary note. It is clear that in all discussions of diffraction, the radius of the Ewald sphere can be set at unity (Verner Schomaker, unpublished). If then a given reciprocal lattice point plane is associated with monochromatic radiation of a given wavelength, then a reciprocal vector, joining this reciprocal lattice plane with the reciprocal lattice origin corresponds to the white radiation which is necessarily produced with the monochromatic radiation in the X-ray tube (see any textbook on quantum theory). Thus any diffraction phenomenon produced by the Buerger precession camera possesses plane symmetry about a symmetry plane through the reciprocal lattice plane, the reciprocal lattice vector and the diffracted ray.

Lindo agonized over letters of recommendation. He explained to Betty that he did not want to be responsible for ruining the career of a young person. Yet he felt he must be honest in all fairness to the individual requesting the letter. It often took him hours to complete such a letter.

Lindo never gave any time for caring for his own physical fitness. He had been a good golfer and a skier in his Montreal days, but such exercise was not readily available to the impecunious in New York or Philadelphia. He was a heavy smoker of the first inch of Raleigh cigarettes and accumulated large ashtrays full of partially smoked and smoldering cigarettes. He hated hospitals and doctors and Betty could never persuade him to have a physical exam. He was rarely ill, even with a cold, and saw no need for the type of periodic exam that she indulged in.

On November 3, 1966, Betty was driving Lindo to lunch with three friends from the lab. Lindo, who so rarely complained about anything physical, suddenly said he had a terrible pain in the back of his head. When they arrived at the restaurant Lindo started to try to get out of the car but had such difficulty that Betty told him to get back in. He said, 'You had better take me to the ... ' and collapsed over her in the car. He never regained consciousness and did not know that he had had a massive cerebral hemorrhage. He died three days later. For him it was the best way to die; his friends and relatives would remember him only as a vigorous, elegant and charming person.

The funeral was private but a memorial service for him was held at the Institute for Cancer Research. It was a unique and beautiful service and the numbers of people who came gave testimony to the deep friendships and admiration he commanded. His impact on his chosen field was inestimable and his influence on his colleagues is described with love in many of the reminiscences that now follow.

References

1. Personal communication from A. L. Patterson and personal reminiscence, Betty Knight Patterson.
2. Personal communication from Nellie Tweedale Patterson.
3. Unpublished diary of Nellie Tweedale Patterson of trip from England to Nelson, New Zealand.
4. Personal communication from A. L. Patterson to Jenny P. Glusker.
5. Letters of N. T. Patterson to his family from Army camps in Belgium in 1916 (WW I).
6. Patterson, A. L., Experiences in Crystallography — 1924 to Date. In *Fifty Years of X-ray Diffraction.* (P. P. Ewald, ed.) pp. 612-622. International Union of Crystallography: Oosthoeks' Uitgeversmaatshapp, Utrecht, The Netherlands, (1962).
7. Bernal, J. D. (ibid.) pp. 522-525.
8. Courant, R. and Hilbert, D. *Methoden der Mathematischen Physik* Springer: Berlin, Germany (1931). Translated as *Methods of Mathematical Physics.* **I** (1953); **II** (1962). Wiley-Interscience: New York, NY.
9. Diary of A. L. Patterson of his year in Berlin (1927).
10. White, T. N. *The X-ray investigation of certain substitution products of cyclohexane.* Ph.D. Thesis. McGill University: Montreal, Quebec, Canada (1929).
11. Letters of A. L. Patterson to his fiancée, Betty Knight.
12. Letter of A. L. Patterson to his fiancée, Betty Knight (2/27/34).
13. Reimann, S., letter to J. J. Durett (17 May 1932).
14. Patterson, A. L. and Groshens, B. A solid state transformation in the tetraacetyl-D-ribofuranose. *Nature (London)* **173**, 398 (1954).
15. Glusker, J. P. A. L. Patterson. In *Crystallography in North America.* (D. McLachlan, Jr. and J. P. Glusker, ed.) pp. 103-107, American Crystallographic Association: New York, NY (1983).
16. Clark, J. R. *Am. Mineral.* **53**, 576–586 (1968).

Betty K. Patterson
Jenny P. Glusker

Curriculum vitae

NAME: Arthur Lindo Patterson
DATES: July 23, 1902 – November 6, 1966
PLACE OF BIRTH: Nelson, New Zealand
Naturalized U.S. Citizen, 1945
MARITAL STATUS: Married (Elizabeth Knight Patterson) 1935
PARENTS: Arthur Henry Patterson and
Nellie Tweeddale Slack Patterson

A. EDUCATION:

Montreal High School, Montreal, Canada
Tonbridge School, Kent, England
McGill University, Montreal, Canada

B.Sc.	1923	Mathematics and Physics
M.Sc.	1924	Physics
Ph.D.	1928	Physics and Mathematics

B. SCIENTIFIC POSITIONS AND EXPERIENCE:

McGill University, Montreal —Molson Scholar	1921–1923
National Research Council, McGill University, Canada — Bursar	1923–1924
The Davy Faraday Research Laboratory of the Royal Institution, London, England Moyse Traveling Fellow (McGill)	1924–1926
National Research Council, Canada, Scholar, at Kaiser Wilhelm Institute für Faserstoffchemie, Berlin, Dahlem, Germany	1926–1927
National Research Council, Canada, Fellow at McGill University	1927–1928
McGill University — Lecturer in Physics	1928–1929
Rockfeller Institute, New York City — Associate, Division of Biophysics	1929–1931
Johnson Foundation, University of Pennsylvania — Fellow, Medical Physics, Lecturer in Biophysics	1931–1933
Massachusetts Institute of Technology —	1933–1936

Scientific Guest, Physics Department, research in crystal analysis	
Bryn Mawr College	
Assistant Professor	1936–1940
Associate Professor	1940–1949
Professor Elect	1949
Physicist, Research contracts (U.S. Navy Bu. Ord. & B.M.C.)	1942–1943
Naval Ordnance Laboratory, Washington, DC	1944–1945
Research Physicist (on leave from Bryn Mawr College) Meritorious Civilian Service Award (1945) for work on mine warfare	
Alabama Polytechnical Institute	1948
(on leave from Bryn Mawr College) Auburn Research Foundation and Visiting Professor of Physics — Research Associate	
The Institute for Cancer Research, Philadelphia, PA	1949–1966
Senior Member, Head of Department of Physics and University of Pennsylvania, School of Medicine	
Professor of Biophysics (adjunct appointment)	1966

C. SCIENTIFIC AND PROFESSIONAL SOCIETIES:

Society of the Sigma Xi		1924
Institute of Physics of Great Britain	Associate Fellow	1925
	Fellow	1929
Physical Society of London	Fellow	1926
American Physical Society	Member	1932
	Fellow	1939
Physics Club of Philadelphia		1936
American Association of University Professors		1937
American Association of Physics Teachers		1939
The Franklin Institute	Member	1939
American Society for X-ray and Electron Diffraction (ASXRED)	Member	1941–1949

	Vice President	1947
	President	1948
American Crystallographic Association		1949
Mineralogical Society of America	Member	1944
	Fellow	1947
New York Academy of Sciences	Member	1944
	Fellow	1947
Societé Française de Minéralogie et de Cristallographie	Member	1949

D. COMMITTEES:

Provisional Commission on *The International Tables for the Determination of Crystal Structures*	Chairman	1946–1948
U.S. National Committee for Crystallography	Chairman	1948–1950
	Member	1948–1955
	Member	1957–1962
	Member	1963
Editorial Committee on International Tables (resigned)	Member	1948–1949
International Union of Crystallography		
U.S. Delegation to.		
1st Congress, Cambridge, MA	Chairman	1948
2nd Congress, Stockholm, Sweden	Member	1951
3rd Congress, Paris, France	Member	1954
4th Congress, Montreal, Canada	Member	1957
5th Congress (resigned)	Elected Delegate	1960
6th Congress, Rome, Italy	Member	1963
Executive Committee	Member	1948–1954
Special Subcommittee on Revision of Statutes and By-laws	Chairman	1955–1957
U.S.A. National Research Council	Member	1956–1959
Executive Committee, Division of Physical Sciences		
American Crystallographic Association	Representative	1956–1959
	Reappointed	1959–1962

PUBLICATIONS: A. L. PATTERSON

1. Patterson, A. L. An X-ray examination of the lower ω-phenyl normal saturated fatty acids. *Phil. Mag.* **3**, 1252–1262 (1927).
2. Patterson, A. L. The scattering of electrons from single crystals of nickel. *Nature (London)* **120**, 46–47 (1927).
3. Patterson, A. L. Über das Gibbs-Ewaldsche reziproke Gitter und den dazugehörigen Raum. *Z. Physik* **44**, 596–599 (1927).
4. Patterson, A. L. Über die Messung der Grösse von Kristallteilchen mittels Röntgenstrahlung. *Z. Krist.* **66**, 637–650 (1928).
5. Patterson, A. L. The Gibbs-Ewald reciprocal lattice. *Nature (London)* **125**, 238 and **447** (1930).
6. Patterson, A. L. Methods in crystal analysis. I. Fourier series and the interpretation of X-ray data. *Z. Krist.* **76**, 177–186 (1930).
7. Patterson, A. L. Methods in crystal analysis. II. The enhancement principle and the Fourier series of certain types of function. *Z. Krist.* **76**, 187–200 (1930).
8. Patterson, A. L. Glucose and the structure of the cycloses. *Nature (London)* **126**, 880–881 (1930).
9. Patterson, A. L. and White, T. N. The X-ray investigation of certain derivatives of cyclohexane. I. General survey. *Z. Krist.* **78**, 76–85 (1931).
10. Patterson, A. L. and White, T. N. The X-ray investigation of certain derivatives of cyclohexane. II. Quebrachitol. *Z. Krist.* **78**, 86–90 (1931).
11. Patterson, A. L. A Fourier series method for the determination of the components of interatomic distances in crystals. *Phys. Rev.* **46**, 372–376 (1934).
12. Patterson, A. L. A direct method for the determination of the components of interatomic distances in crystals. *Z. Krist.* **90**, 517–542 (1935).
13. Patterson, A. L. Tabulated data for the seventeen plane groups. *Z. Krist.* **90**, 543–554 (1935).
14. Patterson, A. L. A note on the synthesis of Fourier series. *Phil. Mag.* **22**, 753–754 (1936).
15. Cameron, G. H. and Patterson, A. L. The X-ray determination of particle size. Symposium on Radiography and X-ray Diffraction Methods, Philadelphia, PA, 1936. pp. 324–338. Am. Soc. Testing Materials: Philadelphia, PA (1937).

16. Patterson, A. L. Homometric structures. *Nature (London)* **143**, 939–940 (1939).
17. Patterson, A. L. The use of an MKS system of units in a first course in electricity. *Am. Phys. Teacher* **7**, 335–336 (1939).
18. Patterson, A. L. The diffraction of X-rays by small crystalline particles. *Phys. Rev.* **56**, 972–977 (1939).
19. Patterson, A. L. The Scherrer formula for X-ray particle size determination. *Phys. Rev.* **56**, 978–982 (1939).
20. Michels, W. C. and Patterson, A. L. The remodeled physics laboratory at Bryn Mawr College. *Am. J. Phys.* **8**, 117–119 (1940).
21. Patterson, A. L. Crystal lattice models based on the close packing of spheres. *Rev. Sci. Inst.* **12**, 206–211 (1941).
22. Michels, W. C. and Patterson, A. L. Special relativity in refracting media. *Phys. Rev.* **60**, 589–592 (1941).
23. Patterson, A. L. and Tunell, G. A method for the summation of the Fourier series used in the X-ray analysis of crystal structures. *Am. Mineral.* **27**, 655–679 (1942).
24. Patterson, A. L. Ambiguities in the X-ray analysis of crystal structures. *Phys. Rev.* **65**, 195–201 (1944).
25. Patterson, A. L. Review: D. Wrinch. *Fourier Transforms and Structure Factors.* ASXRED Monogr. No. 2, 1946. *J. Am. Chem. Soc.* **69**, 2252 (1947).
26. Patterson, A. L. Review: A. D. Booth. *Fourier Technique in X-ray Organic Structure Analysis.* Cambridge University Press, 1948. *Am. J. Phys.* **17**, 322–323 (1949).
27. Patterson, A. L. Review: R. W. James. *The Optical Principles of the Diffraction of X-rays.* Macmillan: New York, 1948. *Rev. Sci. Inst.* **20**, 449 (1949).
28. Patterson, A. L. An alternative interpretation for vector maps. *Acta Cryst.* **2**, 339–340 (1949).
29. Patterson, A. L. Review: K. Lonsdale. *Crystals and X-rays.* D. Van Nostrand: New York, 1949. *J. Optical Soc. Am.* **40**, 181 (1950).
30. Michels, W. C. and Patterson, A. L. *Elements of Modern Physics.* D. Van Nostrand: New York. 659 + x (1951).
31. Patterson, A. L. Approximate formulae for triclinic calculations. *Am. Mineral.* **37**, 207–210 (1952).
32. Patterson, A. L. Symmetry maps derived from the $|F|^2$-series. In *Computing Methods and the Phase Problem in X-ray Crystal Analysis.* (R. Pepinsky, ed.) pp. 29–42. Pennsylvania State College: College

Park, Penna. (1952).

33. Patterson, A. L. and Clark, J. R. Crystal structures of two *para*-substituted phenylpropionic acids. *Nature (London)* **169**, 1008–1009 (1952).
34. Bijvoet, J. M., Bernal, J. D. and Patterson, A. L. Forty years of X-ray diffraction. *Nature (London)* **169**, 949–950 (1952).
35. Patterson, A. L. An orthogonal unit vector triplet associated with a general lattice. *Acta Cryst.* **5**, 829–833 (1952).
36. Patterson, A. L. Review: *Structure Reports*. Vol. **12**. (A. J. C. Wilson, gen. ed.) 1949. *Chem. Eng. News* **31**, 88 (1952).
37. Patterson, A. L. Review: W. Nowacki. *Fouriersynthese von Kristallen und ihre Anwendung in der Chemie.* Berkhauser: Basel, 1952. *Biochim. Biophys. Acta* **10**, 201–202 (1953).
38. Patterson, A. L. Review: J. M. Robertson. *Organic Crystals and Molecules Theory of X-ray Structure Analysis with Applications to Organic Chemistry.* Cornell University Press: Ithaca, New York, 1953. *J. Am. Chem. Soc.* **75**, 6089 (1953).
39. Patterson, A. L. and Groshens, B. P. A solid state transformation in tetra-acetyl-D-ribofuranose. *Nature (London)* **173**, 398–399 (1954).
40. Patterson, A. L. Review: *Structure Reports*, Vol. **10**. (A. J. C. Wilson, gen. ed.) 1945–1946. *Rev. Sci. Inst.* **25**, 818 (1954).
41. Patterson, A. L. Review: J. D. H. Donnay and W. Nowacki. *Crystal Data: Classification of Substances by Space Groups and their Identification from Cell Dimensions.* Geol. Soc. Am., Mem. 60: New York. *Science* **120**, 836–837 (1954).
42. Patterson, A. L. Review: H. P. Klug and L. E. Alexander. *X-ray Diffraction Procedures for Polycrystalline and Amorphous Materials.* John Wiley: New York, 1954. *J. Am. Chem. Soc.* **77**, 2030–2031 (1955).
43. Nordman, C. E., Patterson, A. L., Weldon, A. S. and Supper, C. E. Integrating mechanism for the Buerger precession camera. *Rev. Sci. Inst.* **26**, 690–692 (1955).
44. Patterson, A. L. and Love, W. E. Remarks on the Delaunay reduction. *Acta Cryst.* **10**, 111–116 (1957).
45. Nordman, C. E. and Patterson, A. L. Integrating attachment for the Weissenberg camera. *Rev. Sci. Inst.* **28**, 384–385 (1957).
46. Patterson, A. L. Review: B. D. Cullity. *Elements of X-ray Diffraction.* Addison-Wesley: Reading, MA 1956. *Rev. Sci. Inst.* **28**, 660 (1957).
47. Glusker, J. P., Patterson, A. L., Love, W. E. and Dornberg, M. L.

Crystallographic evidence for the relative configuration of naturally occurring isocitric acid. *J. Am. Chem. Soc.* **80**, 4426–4427 (1958).

48. Patterson, A. L. Function spaces between crystal space and Fourier-transform space. *Z. Krist.* **112**, 22–32 (1959).
49. Patterson, A. L. Fundamental mathematics. Section 2. In *International Tables for X-ray Crystallography.* Vol. **2**, Mathematical Tables. (J. S. Kasper and K. Lonsdale, eds.) pp. 5–83.Kynoch Press: Birmingham, England (1959).
50. Nordman, C. E., Weldon, A. S. and Patterson, A. L. X-ray crystal analysis of the substrates of aconitase. I. Rubidium dihydrogen citrate. *Acta Cryst.* **13**, 414–417 (1960).
51. Nordman, C. E., Weldon, A. S. and Patterson, A. L. X-ray crystal analysis of the substrates of aconitase. II. Anhydrous citric acid. *Acta Cryst.* **13**, 418–426 (1960).
52. Love, W. E. and Patterson, A. L. X-ray crystal analysis of the substrates of aconitase. III. Crystallization, cell constants, and space groups of some alkali citrates. *Acta Cryst.* **13**, 426–428 (1960).
53. Glusker, J. P., van der Helm, D., Love, W. E. Dornberg, M. L. and Patterson, A. L. The state of ionization of crystalline sodium dihydrogen citrate. *J. Am. Chem. Soc.* **82**, 2964–2965 (1960).
54. Patterson, A. L. and Love, W. E. Error analysis for the Buerger precession camera. *Am. Mineral.* **45**, 325–333 (1960).
55. Patterson, A. L. Review: A. Holden and P. Singer. *Crystals and Crystal Growing.* Doubleday: Garden City, New York, 1960. *Am. J. Phys.* **29**, 127 (1961).
56. Patterson, A. L. Review: M. J. Buerger. *Crystal Structure Analysis.* John Wiley: New York, 1960. *Chem. Eng. News* **39**, 132 (1961).
57. Patterson, A. L., Johnson, C. K., van der Helm, D. and Minkin, J. A. The absolute configuration of naturally occurring isocitric acid. *J. Am. Chem. Soc.* **84**, 309–310 (1962).
58. Patterson, A. L. Experiences in crystallography — 1924 to date. Chapter VII. In *Fifty Years of X-ray Diffraction.* (P. P. Ewald, ed.) pp. 612–622. N.V.A. Oosthoek's Uitgeversmaatschappij: Utrecht, The Netherlands (1962).
59. Reichard, G. A., Jr., Moury, N. F., Jr., Hochella, N. J., Patterson, A. L. and Weinhouse, S. Quantitative estimation of the Cori cycle in the human. *J. Biol. Chem.* **238**, 495–501 (1963).
60. Glusker, J. P., Patterson, A. L., Love, W. E. and Dornberg, M. L. X-ray analysis of the substrates of aconitase. IV. The configuration

of the naturally occurring isocitric acid as determined from potassium and rubidium salts of its lactone. *Acta Cryst.* **16**, 1102–1107 (1963).

61. Patterson, A. L. Treatment of anomalous dispersion in X-ray diffraction data. *Acta Cryst.* **16**, 1255–1256 (1963).
62. Patterson, A. L. Mathematical problems in crystallography. In *Proc. IBM Scientific Computing Symposium on Combinatorial Problems.* pp. 53–70. Thomas J. Watson Research Center: Yorktown Heights, New York (1964).
63. Glusker, J. P., van der Helm, D., Love, W. E., Dornberg, M. L., Minkin, J. A., Johnson, C. K. and Patterson, A. L. X-ray crystal analysis of the substrates of aconitase. VI. The structures of sodium and lithium dihydrogen citrates. *Acta Cryst.* **19**, 561–572 (1965).
64. Taylor, M. R., Gabe, E. J., Glusker, J. P., Minkin, J. A. and Patterson, A. L. The crystal structure of compounds with antitumor activity. 2-Keto-3-ethoxy-butyraldehyde bis(thiosemicarbazone) and its cupric complex. *J. Am. Chem. Soc.* **88**, 1845–1846 (1966).
65. Patterson, A. L. Review: M. A. Jaswon. *An Introduction to Mathematical Crystallography.* American Elsevier: New York, 1965. *J. Am. Chem. Soc.* **88**, 3183–3184 (1966).
66. Gabe, E. J., Glusker, J. P., Minkin, J. A. and Patterson, A. L. X-ray crystal analysis of the substrates of aconitase. VII. The structure of lithium ammonium hydrogen citrate monohydrate. *Acta Cryst.* **22**, 366–375 (1967).
67. Glusker, J. P., van der Helm, D., Love, W. E., Minkin, J. A. and Patterson, A. L. The molecular structure of an azidopurine. *Acta Cryst.* **24B**, 359–366 (1968).
68. van der Helm, D., Glusker, J. P., Johnson, C. K., Minkin, J. A., Burow, N. E. and Patterson, A. L. X-ray crystal analysis of the substrates of aconitase. VIII. The structure and absolute configuration of potassium dihydrogen isocitrate isolated from *Bryophyllum calycinum. Acta Cryst.* **B24**, 578–592 (1968).
69. Glusker, J. P., Minkin, J. A. and Patterson, A. L. X-ray crystal analysis of the substrates of aconitase. IX. A refinement of the structure of anhydrous citric acid. *Acta Cryst.* **B25**, 1066–1072 (1969).
70. Gabe, E. J., Taylor, M. R., Glusker, J. P., Minkin, J. A. and Patterson, A. L. The crystal structure of 2–keto–3–ethoxybutyraldehyde–bis(thiosemicarbazone). *Acta Cryst.* **B25**, 1620–1631 (1969).

ABSTRACTS

71. Patterson, A. L. A Fourier series representation of the average distribution of scattering power in crystals. Am. Phys. Soc., Washington, April 1934. *Phys. Rev.* **45**, 763 (1934).
72. Patterson, A. L. The determination of the components of interatomic distances in crystals. Am. Phys. Soc., Pittsburgh, December 1934. *Phys. Rev.* **47**, 330 (1935).
73. Patterson, A. L. The determination of the size and shape of crystal particles by X-rays. Am. Phys. Soc., Washington, May 1936. *Phys. Rev.* **49**, 884 (1936).
74. Patterson, A. L. The uniqueness of an X-ray crystal analysis. Am. Phys. Soc., New York, February 1939. *Phys. Rev.* **55**, 682 (1939).
75. Patterson, A. L. Ambiguities in X-ray analysis. ASXRED, Ann Arbor, June 1943. *Phys. Rev.* **64**, 314 (1939).
76. Patterson, A. L. Ambiguities in X-ray and electron diffraction analysis. Am. Phys. Soc., New York, January 1946. *Phys. Rev.* **69**, 256 (1946).
77. Patterson, A. L. Ambiguities in diffraction analysis. ASXRED, Pittsburgh, PA December 1946 (1946).
78. Patterson, A. L. Ambiguities in the diffraction analysis of structure. First Congress, IUCr, Cambridge, MA, July (1948).
79. Patterson, A. L. The information contained in a vector map. Second Congress, IUCr, Stockholm, Sweden, July (1951).
80. Patterson, A. L. (with J. R. Clark). Structures of p-chlor- and p-brom-phenyl-propionic acids. Am. Cryst. Assoc., Tamiment, PA, June (1952).
81. Patterson, A. L. (with J. R. Clark). Structure of p-chlor-phenyl-acrylic acid. Am. Cryst. Assoc., Tamiment, PA, June (1952).
82. Patterson, A. L. Phase operators and homometric structures. Am. Cryst. Assoc., Tamiment, PA, June (1952).
83. Patterson, A. L. An orthogonal unit vector triplet associated with a triclinic lattice. Am. Cryst. Assoc., Tamiment, PA, June (1952).
84. Patterson, A. L. A transform space 'midway' between crystal and reciprocal space. Am. Cryst. Assoc., Ann Arbor, MI, June (1953).
85. Patterson, A. L. (with C. E. Nordman, A. S. Weldon and C. E. Supper). An integrating mechanism for the Buerger precession camera. Third Congress IUCr, Paris, July 1954. *Acta Cryst.* **7**, 619 (1954).
86. Patterson, A. L. (with C. E. Nordman, and A. S. Weldon). Crystal structures of rubidium dihydrogen citrate and citric acid. Paper 49.

Am. Cryst. Assoc., Pasadena, CA July (1955).

87. Patterson, A. L. (with J. P. Glusker, W. Love and M. L. Dornberg). The crystal structures of the monopotassium and monorubidium salts of the lactone of naturally occurring isocitric acid. Paper 26. Am. Cryst. Assoc., Pittsburgh, PA (1957).
88. Patterson, A. L. (with J. P. Glusker, D. van der Helm, W. E. Love and M. Dornberg). The structures of sodium and lithium dihydrogen citrates. Paper C-5. Am. Cryst. Assoc., Washington, DC (1960).
89. Patterson, A. L. (with C. K. Johnson, D. van der Helm and J. A. Minkin). The absolute configuration of naturally occurring isocitric acid. Paper L-3. Am. Cryst. Assoc., Boulder, CO (1961).
90. Patterson, A. L. (with J. P. Glusker, D. van der Helm, W. E. Love and M. L. Dornberg). Paper L-4. Am. Cryst. Assoc., Boulder, CO (1961)
91. Patterson, A. L. (with E. J. Gabe, M. R. Taylor, J. P. Glusker and J. A. Minkin). The crystal structure of KTS (2-keto-3-ethoxybutyraldehyde bis(thiosemicarbazone)). Paper K-4. Am. Cryst. Assoc., Atlanta, GA (1967)

42. Personal reminiscences and history of crystallography

42.1 Timothy R. Talbot, Jr., M.D.

Dr Talbot was Director of the Institute for Cancer Research from December 1957, while Lindo Patterson was here until 1977. He also became President of the Fox Chase Cancer Center in 1974. He retired in 1981, but has still maintained an active interest in the affairs of the Institute and Center.

I am extremely grateful because of the person that we are thinking about. Between Stanley Reimann and myself, starting in 1927 until about three years ago, the two of us covered close to 55 years of directorship of this Institution. Stanley Reimann set the stage for this place in a way that only those of us who recall the history, and who knew him, and who knew the people that he chose, can possibly appreciate. That, I think, is extremely important for us to recall. Because the goals, objectives, and somehow the inner spirit and driving force of Stanley Reimann made him and me ultimately strangely compatible as I look at his history and mine. I don't suppose that we realized that we were that compatible when we first met. We came almost from a different generation, and therefore it took us a little while to get the language together to understand that we really were talking about the same thing, and had the same goals and the same objectives, and somehow or other the same in-built inner driving forces. Although this Institution was in trouble when I came here, Stanley had built what I have previously called the launching pad for this rocket. Now among the people who made this place what it is, were a remarkably small cadre of people. When I got here you could count them on one hand. All of them had been chosen by Stanley Reimann before I got here. Among those was Lindo.

Lindo Patterson had a great and beneficial influence upon this institution at a time when it was very much needed. The objective that Stanley Reimann had as he carried the institution from its creation to its initial development was clearly to serve and foster creative endeavors in biomedical research, with an underlying driving force of an interest in cancer. Lindo was

one of the small group of people who were critically important in helping to preserve and strengthen our adherence to those goals. And he remained a very strong pillar of influence and support during the first nine years of my directorship at the Institute.

As I wrote many years ago in our Scientific Report,

> It has been one of the great privileges of my life to have been associated with him. I know what he stood for as a man and as a human being. Quite apart from his very high standing as a scientist, he possessed qualities that will always leave their mark upon those of us who knew him and upon this institution. He had a passionate belief in the freedom of the individual human spirit. He knew how to breathe excitement and, as Linus Pauling has so beautifully said today, optimism and joy into his efforts. He had no tolerance whatever for inaccuracy or evasion, whatever it may have been concerned with. He had an abundance of the most pleasing of all human qualities — integrity, humor, responsibility, loyalty and affection. Therefore he was one of those terribly rare and wonderful men who exemplify a balance of loyalties — loyalty to the creativity and morality of the individual, and loyalty to the community of mankind within which he worked and lived.

On a lighter side, of course, there were his salty commentaries, his mimicry of the cockney and other facets of British life, his launching into Gilbert and Sullivan at apt moments. In about 1965, when we were building new additions to the laboratory, I discovered that the architect had failed to design the rather attractive sun screens on our southern windows so that they would really function, which annoyed me a little bit. So I asked Lindo if he would calculate the angles of these so that they would be effective and at the same time attractive. A small thing, but important in my memory. He listened and ran out in a few seconds. And the next day he was back, smiling, with his usual simple, beautiful and correct solution, as he had done with every question of which there were many, that I had presented to him. Those questions included science, values, everything you can think of. I often walk out to that side of the Institute, usually about noon, and look at the patterns of light, and as I do I remember that Lindo was here when it mattered so much to me, and to the world of science and to this institution.

42.2 Bertram Warren

Lindo worked out the Patterson function while he was a guest in Bert Warren's laboratory at MIT. Born in 1902, Warren obtained a Ph.D. from MIT in physics in 1929. He worked with W. L. Bragg on the structure of diopside. Professor Bragg calculated a two-dimensional Fourier map, the first one so computed, at the suggestion of his father, W. H. Bragg; for this he used the structure factors for diopside. Warren then went on to study silicates and alloys. This led to his studies with Gingrich, which influenced Patterson. These studies were an extension of the work of Zernike and Prins on the interpretation of liquid diffraction patterns. Dr Warren's interests include amorphous materials, imperfections and order-disorder phenomena. Dr Warren was President of the American Society for X-ray and Electron Diffraction in 1942 and Chairman of the U.S. National Committee for Crystallography in 1958–1960. The American Crystallographic Association has honored him by giving a Warren Award every three years.

Lindo spent two years, back around 1934, as a guest in my lab. At several different times during that two year period, I remember distinctly his saying, 'I have a strong feeling that more can be done in developing methods for working out crystal structures.' I also remember, even more strikingly his saying again and again, 'I think that more use can be made of Fourier Series methods than has been done up till now.' I think it was that very intense interest in Fourier series which was at least partly responsible for his coming to MIT, so that he could have discussions with Norbert Wiener in the Mathematics Department. Of course we know that his two hunches worked out. There was a better way, and there was a better use of Fourier series, and it all ended up in his Patterson function, F^2 function. Now I'm going to say something that I'm not dead sure of — it's just a hunch on my part. I have the feeling that Lindo had his F^2 series clearly in mind quite a while before we ever heard about it. He must have been completely familiar with the convolution operation on a Fourier series which leads to a series with squared coefficients. I have a feeling that he had the formalism of the series, but at that point he hadn't completely realized the significance of the series and therefore he wasn't

saying anything about it. That's just my hunch. Another reason why I think this way came from contact with some work that I was doing. At that time I was interested in amorphous materials — structures of glass, and liquids. We worked them out always by applying the Zernike–Prins sine transform to the intensity, and from that getting a radial distribution function. I was also interested in intermediate materials — things like carbon black, whose structure at that time was in question. Was it an amorphous form of carbon, or was it small graphite crystals, or what was it? I remember distinctly one weekend I was really all steamed up about this problem, wondering what can you do with intermediate materials like this to work out something about them without making any a priori assumptions about crystallinity. Of course, the obvious answer was that you can apply the Zernike-Prins sine transform and get a radial distribution function from that. It was only one short step to saying that you could do the same thing to the powder pattern of a crystalline material, apply the sine transform and get a radial distribution function.

I was all steamed up about that because that sort of thing had not been done, and I thought it would be interesting to try it. In those days there was a restaurant across the street from MIT, Walton's Lunch. Lindo and I used to have lunch there noons, and the Monday noon following this busy weekend, I was talking to Lindo and telling him what I'd been doing over the weekend. I remember saying that you can take the intensity function represented by the powder pattern of a crystalline material, perform the Fourier sine transform on it, and get a radial distribution function. And I pointed out that this is interesting because unlike the problems we've always had about applying Fourier methods this doesn't involve any phase. This works directly with intensities, so you can do it, but what you get out of it is not a structure, it is a distance of atoms from other atoms. At that point Lindo practically rose right up off his chair. 'Yes,' he said 'and in my series what you get is the displacement of atoms from other atoms.'

I think this response was the indication that he had the F^2 series all worked out. He was completely familiar with it, but he hadn't completely seen the significance of his function. You

might say that that point marked the birth of the Patterson function, because at that instant he saw the significance of the whole thing.

That would have been around January or February of 1934, I'm not sure just when it was; Betty maybe could figure that out better than I can. There was due to be a meeting of the Physical Society in Washington in April of that year. The time was awfully short and I remember that Lindo went frantically to work to crank out some material to send in an abstract for the Washington meeting. This was the first public presentation of the Patterson function. It was a typical Physical Society ten minute paper. The title of that paper was 'A Fourier series representation of the average distribution of scattering power in crystals.' I can't imagine any title which was so far from telling you that this is introducing the Patterson function. I suppose that you might call this the birth of the Patterson function since it was the first public presentation.

It happened that Sir Lawrence Bragg (W.L.) was in the country that spring. He was giving the Baker Lectures at Cornell, and he was there at the Washington meeting. And, of course, he saw right away the significance and the importance of this new function and, as Lindo said afterwards, 'Sir Lawrence asked all the right questions to bring out the importance of the new function.' Well, I'll leave you one last thought. If you want to say that from Sir Lawrence Bragg asking all the right questions to bring out the importance, and if you want to say that that first public presentation was really the birth of the Patterson function, then you'd have to say that Sir Lawrence was chief midwife at the birth of the Patterson function.

42.3 Dorothy Hodgkin

I first heard of Lindo Patterson when I heard of his first publication of the F^2 series in Cambridge, where it caused something of a stir and a great deal of serious discussion, which I will deal with on a later occasion. I first heard from him when he wrote me a letter asking for a reprint of the paper I wrote, in which I had calculated the Patterson in three dimensions from air-dried crystals of insulin. I asked my granddaughter the other day if

she could put her hands on his letters because she had been sorting my files, and she said 'Oh, yes,' and there they were, all the letters that I had ever had from Patterson, beautifully arranged. And I said 'Oh I think I'll take this first one with me to Philadelphia,' but Kathy, who had a historian's training, said 'Oh no, you don't. I'll give you a xerox to take with you.' So I have the xerox, two xeroxes, one for Betty Patterson and one to use.

I first met him when I came first to this country in 1947. I was invited by the Rockefeller Foundation to go round and make different visits, and naturally a visit to A. L. Patterson was one of the first on my list. Well, you know how it is. You read papers and letters and you form a picture of a very distinguished scientist in a beautiful laboratory with many people working for him and having every kind of apparatus that he would like. So it was a surprise for me to come to Bryn Mawr, to the very small physics department, and to find Patterson with two students of physics, whom he talked to kindly for a minute or two and then sent away for the afternoon, so that he could have a long conversation with me about the determination of the structure of penicillin. Of course it was very exciting for me to tell him everything that had happened, because we had used the Patterson calculations to find our heavy atom positions, and they helped to solve the structure, and Lindo was very delighted with everything that had happened. Then we went home in the evening to find Betty Patterson and a lovely and quiet home, and I began to hear the story of Lindo's life and of all the different things that had brought him where he was at that moment.

Other people will tell you many of the stories he told me then. I remember particularly the story of how he was a student at McGill and got a fellowship to go to London to work with W. H. Bragg. He had had a nice life at McGill, enjoying his time there playing a lot of bridge and other such games, and his professor sent for him before he went and said 'Now it's important, you are going abroad, you must represent this country and show how well people work in this country, and perhaps make changes in your way of life.' When he got to the Royal Institution, he went, of course, to call on W. H. Bragg,

and W. H. Bragg, after a certain amount of conversation with him, said 'Now you haven't come only to this country to work. Be sure that you have a good time and enjoy yourself.'

Well, this was typical of Lindo. He did enjoy his life, and insisted that those around him should enjoy theirs, and particularly, of course, he enjoyed science, scientific discovery, and he must have been, in a curious way, one of the most successful scientists anyone can imagine. He set out as a young man to find a way by which the structure analysis of complicated molecules could be achieved; and he found it, and could see everybody using his methods everywhere. All the same, I think it was remarkably foresighted, perspicacious of the Trustees of this Institute for Cancer Research to adopt Lindo Patterson, and set him up in the proper kind of laboratory that he needed, with research assistants and research facilities with which he could continue to work in the fields he cared about for the rest of his life, even though he must have seemed to them sometimes to be wandering off into fields that they could little understand. And we too, in the same way, owe an enormous debt to the Trustees of this Institute who have brought us here.

I would like to tell you more — I could now go on to my long series of letters which would take you through a lot of the history of the International Union of Crystallography and the early troubles we had with constructing the constitution of this institution, but I think I would like to end just with my last Christmas card from A. L. Patterson, which had scribbled on the back of it something you will all understand 'Every day of my life I bless you for Jenny.'

42.4 How I First Learned of the Patterson Function and Bragg's Second Law

E. W. Hughes

In the late winter of 1933-34, Professor W. L. Bragg (later Sir Lawrence Bragg) arrived at Cornell University to begin a semester as Baker Non-resident Lecturer in Chemistry. I was then a graduate student and had the good luck to be appointed as his Baker assistant. For a person beginning a career in X-ray

Crystallography the opportunity to work closely with Bragg for a full term was almost unbelievable good fortune.

His lectureship kept him in Ithaca only Tuesdays through Thursdays and he traveled a lot of weekends to neighboring institutions. One wintery weekend he spent at MIT and there he learned, prior to publication, about the Patterson method from A. L. Patterson himself.

He arrived back in Ithaca early on a very bad winter's day; it had sleeted all night, then frozen and snowed slightly and the streets were sheets of ice. He arrived at the laboratory safely by taxi and asked if I could deliver him to a local appointment. My Model A Ford had sat out all night in a nearby parking lot and would not start. I suggested that he get in and steer while I pushed it down a nearby hill. He objected that he was not familiar with American cars and insisted that I operate it while he pushed. Fortunately it started and we set off south on East Avenue. The University B & G trucks had spread ashes on the east side of our road but not yet on our side.

He then started enthusiastically to tell me about the Patterson method, which he pronounced to be the most important advance since his father, Sir William Bragg, had introduced the use of Fourier series. He used his fingers of one hand to represent atomic position vectors and those of the other hand to represent their differences, and I soon forgot all about the ice. Until suddenly, approaching Central Avenue, I became aware of a red traffic light and a heavy truck rushing west on Central to make his green light. Ashes had been spread on both sides of Central. We skidded about badly but by pure good luck lurched to the left onto the ashes at the last instant and scrunched to a stop, well into the intersection. The truck managed to swerve and miss us by about a foot. Our light turned green and I continued, very cautiously. Through all this Bragg continued to wave his hands and lecture on Patterson's vectors, but to a deaf audience! When he finished there was a brief pause and he suddenly said, 'I say! we skidded a bit back there, didn't we?' Later, in the security of his office I got a repeat of the lecture.

The following year I was in England with proofs for the figures in his Baker monograph. Once when we were driving (in his car) to his office we passed the scene of a recent bad road accident, which reminded me of what could have happened in Ithaca. I asked if he remembered the incident and when he said that indeed he did, I remarked upon how impressed I had been by his calm nonchalance in the face of very real danger. He then stated what I have come to call

Bragg's Second Law:

> When travelling in a foreign country I make it a point of personal honor not to show fear, anger, or mirth, or surprise at any happening that does not seem to be unusual to the natives.

'And,' he added, 'you didn't seem to be frightened so I was jolly well determined not to be frightened either.'

I have considered this 'law' to be excellent advice and have tried to follow it, but one thing was clear; on that winter morning in Ithaca he was not very good at estimating the reaction of the native.

42.5 David P. Shoemaker and Clara B. Shoemaker

David Shoemaker received his Ph.D. degree at the California Institute of Technology in 1947, and served on the faculty of the Massachusetts Institute of Technology for 19 years before accepting the post of Chairman of the Chemistry Department of Oregon State University in 1970. He served as Chairman of the US National Committee for Crystallography 1967–70, and as member of the Executive Committee of the International Union of Crystallography 1972–1978. The former Clara Brink received her Ph.D. at the University of Leiden in 1950, and came to MIT in 1953 to work with David; they were married in 1955, and have continued their collaboration on the structures of tetrahedrally close packed metals and alloys ever since. Both are now Professors Emeriti at Oregon State University, having retired in 1984.

Clara and I are very sorry that we will miss the Fox Chase Symposium celebrating the Patterson function, for many reasons. An important one is the missed opportunity to see many old friends including ones we have not seen for a great many years. Another is my own involvement with Verner Schomaker in producing and interpreting the first full three-dimensional Patterson function ever to be calculated and solved, to our knowledge. We also regret missing what promises to be a very interesting symposium program. Finally, we would have liked to be present to pay due homage to Lindo Patterson, of whom we have vivid and fond memories. We arrived home in October from an extensive and exhausting stay abroad and we promised ourselves that we would stay home for a while.

I am glad that Verner will be there to represent our 37-year-old work with the Patterson function for L_s-threonine. After 37 years my own memory of the circumstances has dimmed somewhat, but Verner, in a recent telephone conversation, was very helpful in reminding me of details I had forgotton about that enterprise. Here I offer my Verner–reinforced reminiscences about that three-dimensional Patterson function, trusting that Verner will not hesitate to correct any errors of fact that I let slip through.

I was at the time a graduate student at Cal Tech working on the crystal structure of L_s-threonine under Professor Robert B. Corey. With an old Weissenberg camera and a cold-cathode copper-anode X-ray tube continuously pumped with a noisy Hyvac mechanical pump, with multiple-film technique and exposures lasting several days, recording as many layer lines as possible around all three orthorhombic axes (the space group was $P2_12_12_1$), and estimating the intensities visually with interfilm comparisons, I acquired a full three-dimensional set of intensity data for 636 non-equivalent reflections. This was one of the first such three-dimensional data sets to be measured at Cal Tech; many more were to follow there and elsewhere, although one really has to be something of an old timer these days to remember having obtained data sets in this way.

As this was a more-than-averagely complicated structural problem for those days (eight non-hydrogen atoms per asymmetric unit!) we knew we were in trouble after use of the traditional two-dimensional structure factor maps and the then much more in vogue Harker sections failed to yield clues to the structure. Interpretation of the Harker peaks was frustrated by the presence of the more numerous non-Harker ones. (In retrospect, we might have succeeded had we realized how those non-Harker peaks might have been identified and interpreted with the aid of non-Harker peaks on the special sections $0kl, h0l$, and $hk0$, no more difficult to calculate than the Harker sections themselves.) It was at this point that a three-dimensional Patterson function was considered — urged upon us first by Verner Schomaker, who had been looking over our shoulders. To anyone at the time who had labored for days doing a mere two-dimensional Patterson or Fourier calculation with Beevers–Lipson strips, the prospect of doing anything in three dimension had to be awesome. However, the stage was all set; sometime earlier Phil Shaffer, Verner Schomaker, and Linus Pauling had shown how accounting equipment manufactured by International Business Machines could be used to facilitate Fourier summations. Then, at the suggestion of Eddie Hughes, Verner had put the contents of the Beevers–Lipson strips on 80-column IBM cards, creating the 'L' cards which made the summations easier and faster. (Much of the real work was done by Judy Schomaker, who actually hand-keypunched the cards, while Verner wired up the complicated plugboard, no mean job, and the equivalent today of writing a fairly sophisticated computer program). Related developments had taken place elsewhere, and a few three-dimensional electron density syntheses had actually been calculated.

So Verner and I set about calculating the Patterson function for threonine with an IBM 400-series electromechanical tabulator, a sorter, and a large bank of drawers of 'L' cards. For each dimension the cards were pulled for the required amplitudes and frequencies and summed on the tabulator, one reciprocal lattice line at a time, until all points in the reciprocal lattice (for the first dimension, or in the combination lattice in the other two dimensions) had been covered, and the cards were sorted and laboriously refiled in the drawers for the next use. The output sheets from the first dimension furnished the coefficients for the second, and so on. Verner and I put in two 15-hour days on the machines, followed by several days plotting contours.

We didn't have well established ideas on how to interpret the three-dimensional pattern, beyond looking for clear vector relationships; the idea of superposition developed too slowly

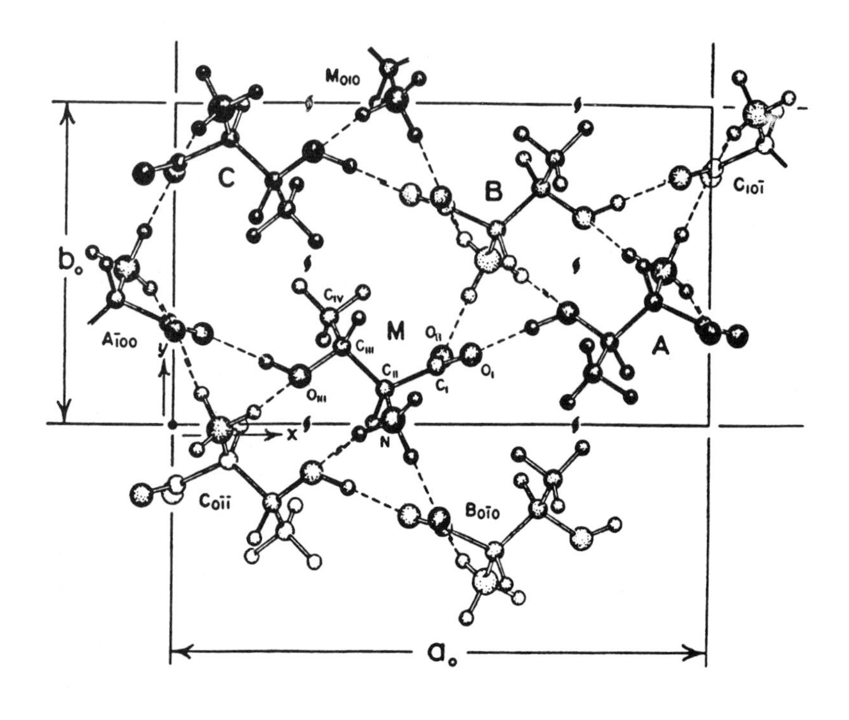

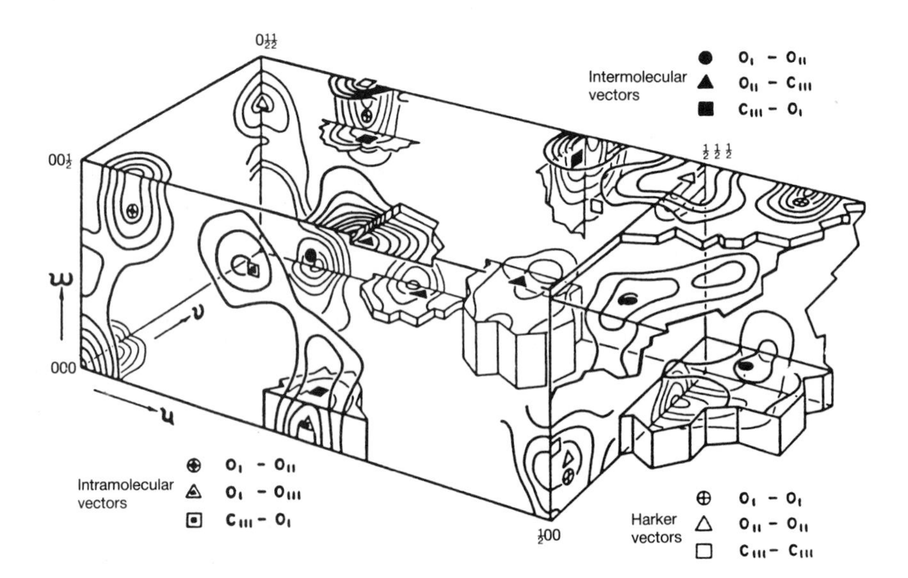

FIG. 42.5.1. Threonine and its Patterson map.

and too late to do us much good at the time. We were lucky in having one strong and easily identifiable vector, of length 2.25 Å, assignable on the basis of previous structural knowledge to the two carboxyl oxygens. We were then able, with the Harker sections, to assign tentative atomic coordinates to these atoms. Plodding onward we eventually found all of the atomic positions. However, the intensity agreement for the trial structure fell short of expectations. It was then realized that Carl Nieman, Professor of Organic Chemistry, had allowed me to walk out of his office with the wrong relative stereochemical configuration at the α- and β-carbon atoms; our model was not threonine but ***allothreonine***. It will never be known whether the error was due to misinformation from him or (perhaps more likely) confusion on my part. Be that as it may, on switching the hydroxyl oxygen with the terminal carbon we converted the model to threonine, which is fortunate because methyl hydrogens don't usually form very good hydrogen bonds. The corrected model refined satisfactorily to at least the standards of those days (I think rather better!) in the hands of Verner Schomaker and Jerry Donohue with four three-dimensional Fourier electron density syntheses, and also with a cycle of full-matrix least squares, which is a story in itself.

News of the successful solving of a three-dimensional Patterson function travelled fast to certain quarters. Not long after the first trial structure was obtained I received an envelope, postmarked from somewhere in Pennsylvania with no sender identification inside or out, containing only a photograph of a highway road sign — the familar shield, inscribed 'Penna 309'. The best I could think of to do in response was to send Lindo Patterson a thank you in the form of an 80-column IBM card, punched in alphanumeric code. It was not long before we learned that our identification of the sender had been correct. It had been no secret to us that Lindo knew of our enterprise at the outset and was following its progress with great interest.

To know Lindo Patterson was a cherished privilege of our scientific and personal lives; he will forever have a very special place in our hearts and memories. Some years ago I had the honor, and a sad honor it was, of co-authoring with Dick Marsh an obituary of Lindo Patterson for *Acta Crystallographica.* I read it again the other day; it still expresses, about as well as mere words can express, my own feelings about Lindo Patterson. 'Lindo Patterson's high standing among crystallographers, particularly in America, goes far beyond his published scientific work. Large segments of two generations of crystallographers — numerically all out of proportion to the number he trained in any formal sense — were inspired by Lindo Patterson and proud of even a casual acquaintanceship with him ...' Our own acquaintanceship was somewhat more than casual, and he certainly inspired both Clara and myself.

42.6 Verner Schomaker

Verner Schomaker obtained a Ph.D. from Caltech in 1938 and taught there in the chemistry department for many years. After some time with Union Carbide Company he moved to the University of Washington in Seattle, where he is now. He has had a profound influence on those aspects of science that Lindo valued most — the development and understanding of the basics of the discipline. He was the person who was 'vernished' in the A. L. Pon paper. There's a Shoemaker, Schomaker and Shoemaker; they're different people and two of them worked on the structure of a rather interesting amino acid, threonine, and it was a famous paper, published in the *Journal of the American Chemical Society*,[1] which first made use of a three-dimensional Patterson map to solve a structure containing no heavy atoms.

We were proud to have solved the structure of threonine, with its eight atoms, by calculating the Patterson function in three dimensions, in the way told so well by David and Clara Shoemaker in this volume. My recollection of Eddie Hughes's suggestion is that he remarked one day 'Look, you'll never get it done with those cards that give you thirteen decimal places'. (Eddie had it wrong, or I misquoted him. The instant design that Dr Pauling made when the possibility of using punched cards was brought to his attention, and that was described by Shaffer *et al.*, had only six-digit entries). 'Put a whole Beevers-Lipson strip on a card.' We found a way to do it, and Judy typed them up. Then I had the notion that when you came right down to it, *resolution* was the important thing. I didn't know that it was in principle possible, except for ambiguities, to solve a Patterson straightaway (Wrinch), but nonetheless it was my conviction that getting greater resolution of the individual peaks was the important thing, and that if we looked in the three-dimensional body of the function, we would be spared looking at Dave Harker's special peaks, the special sections. So after being accused by Bob Corey of discouraging his student by saying 'Come on, let's make a three-dimensional function,' we finally calculated it — David [Shoemaker] says in two 15-hour stints — and managed to blunder through one of those first cases of 'applying Patterson techniques' to solve the structure. When we were together in England later, Dorothy was very nice to say fine words about how we had done this, and we were very proud. And the paper eventually got published.

I'm glad that Isabella has confirmed tonight that Ewald was right when he said one time, some thirty years ago, 'These things [*i.e.*, whether the structure should be sought in the Patterson or in the phase relations] ought to be looked at in both ways.' By now I'd given up thinking that on hard jobs the Patterson would be the way to get the answer — that instead the only way was the way of the direct methods. But if it's true that on a hard job, as Isabella told us tonight, it *is* useful to look both ways, I'm glad. I am also ashamed that, those thirty years ago, I could hardly see that there was another way.

I want to say something about when I came to Caltech in 1935, when those wonderful seminars mentioned by Dave Harker were going on. It was true, I was bewildered. The first seminar I ever heard there was on the structure of ice; Professor Pauling talked about that [his theory of the residual entropy]. And the second or the third one was by Harker, about proustite and pyrargyrite, and the Harker sections. I think I half-way understood it. But most of the time, as I recall, was given to telling how the structures of proustite and pyrargyrite can be related to other structures that were already well known, I suppose, to everyone else in the group but me. Do you remember, David? (***David Harker: No, I don't remember***). I was impressed by the feat. I didn't understand it, but I think that was the true crystallographer in you. We still have to be interested in such relationships.

I didn't get to know Lindo very well in a personal way. I had one feeling about him that I guess has not been expressed directly here tonight. He was always helpful in the ACA business meetings, trying to make the discussion sensible when sometimes it wasn't sensible. It often helped.

It's been wonderful to be here. One thing that hasn't been taken up enough, perhaps, is the question of homometric structures. I wish that I could say what this young mathematician at Caltech did, and find out from you whether you think it is important. He found ways of getting lots of homometric structures, of creating them, that I think Lindo would have enjoyed, but maybe Lindo already understood them. His paper appeared in *Acta Crystallographica* about ten years ago,[2] and I thought it was quite wonderful. Did you see it, Jerry?

And about those theses, I was very much taken by what you could see in the Library at Caltech, and it was not quite the way David has described it here. Instead, it was a mark of distinction to have a thesis of only a few pages. I noticed there (not just then, but a few years later) that Pauling, E. Bright Wilson, Ira Sprague Bowen, and David Harker — and I don't remember who the few others were — have theses that were very thin like that. They were the ones that counted. It was not getting away with something; it was showing how good you really were.

References

1. Shoemaker, D. P., Donohue, J., Schomaker, V. and Corey, R. B. *J. Am. Chem. Soc.* **72**, 2328–2349 (1950).
2. Franklin, J. *Acta Cryst.* **A30**, 698–702 (1974).

42.7 John H. Robertson

The Patterson function has played a central role in crystallography, and crystallography, meantime, has come to have a central place in physical science. Crystallography was still a relatively small, pioneering science, relative to chemistry, physics and biology, when I grew up into its ranks, in the late 1940's. In those days the Patterson function, still barely a decade old, was recognised by all of us as a wonderful, almost indispensible key for the initial unlocking of crystal structures generally. My own personal involvement then was to make one of the first fully 3-dimensional summations (done with strips in those days!) and use it for solving the structure of strychnine, for my Ph.D. (1949). That work used heavy atom vectors (Br–C, etc.) but we also realised, in solving sucrose, with no heavy atoms ($\approx$ 1950) that specific patterns of vectors characteristic of known groupings (glucose rings) could be seen within the Patterson and used to solve otherwise intractable problems.

Although these past events now seem quite childish, I might give some detail of these episodes. First, strychnine (1948–49). This was when Arnold Beevers and I invented the *Vector Convergence Method* (VCD). Using the Beevers-Lipson strips, and many long hours of mental arithmetic, we calculated the full 3-D Patterson of the HBr salt (albeit, with a restricted set of 750 intensities, out to 0.7 Å), and then, having got the Br position from the Harker peaks, we superimposed the full Patterson on itself four times, with its origin successively on the four Br positions of the space group ($P2_12_12_1$). The resulting 3-D plot, on glass plates, allowed us to see the molecule quite well, solving the structure. Then, sucrose (1949–50). Intensities had been gathered in 1943-44 but this structure remained stubbornly unsolved until Beevers was able to compute the 3-D Patterson on X-RAC (courtesy of Ray Pepinsky) at Penn State. At that time the structure of glucose had been solved in Beevers' laboratory, and we knew, both from first principles, and from seeing that Patterson map, that the pyranose ring gave a fine, characteristic rosette of vector peaks. So, we plotted the sucrose sections out on cellophane sheets, to study them in three dimensions, and lo, we saw that rosette standing there, in space, enabling us to place the glucose moiety and hence to solve that structure. I was personally at the front line in both these crusades but I would like to emphasise that Arnold Beevers was always an enthusiastic protagonist of Patterson vectors and certainly taught me all I ever knew about them. Also, it should be acknowledged that these powers of the Patterson function received their clearest treatment at that time in the papers of Martin Buerger.

Those were the early days, when the total number of crystal structures determined amounted to a few hundred, and when a molecule of 30 non-hydrogen atoms, without symmetry, could appear to be a heroic objective.

Now, several decades on from those days, crystallography has become one of the major girders of the great building which is modern physical science, responsible for huge tracts of well-established knowledge, underlying many successful excursions of theory and even, in effect, part-creator of several active new sciences (such as molecular biology or crystal physics). Today, in the organic/organometallic field alone, some 40,000 structures are known and catalogued; in the inorganic/mineralogical fields there are probably an equal number. The prodigious volume of detailed documentation of atomic dimensions, bonding arrangements, molecular connectivities and conformations, intermolecular contacts and spatial dispositions is basic to chemistry and physics; so much so that it is nowadays taken almost for granted. Such is the groundwork that crystallography has done, and is still doing. Yet this is not all, for crystallography has also supplied and is still supplying imaginative initiatives in the best tradition of innovative scientific enquiry — such imaginative leaps of intuition as those of Pauling, and of Watson and Crick (each based on a wealth of background structural knowledge) which have given us our modern insights into the design of protein molecules, the architecture of DNA, and the most central features of the processes of life. Crystallography is nowadays relevant to discussions of enzyme/substrate binding, chemical reactivities, altered redox potentials of cations, synthetic metals, and drug-design ... not to mention innumerable aspects of solid state physics which, in their turn, are relevant, these days, to industrial applications of considerable economic importance.

It is against this background that Lindo Patterson's contribution of the 'Patterson function' should be seen. It was, of course, not the only key that unlocked the doors to progress, but it was, at the beginning, almost the only one and, for many years (until the development of direct methods), it continued to be the major one. Now, it remains today still an extremely valuable tool, employed often in modified forms, applicable even to problems as gigantic as the structures of big viruses — still invaluable for its ability to get an initial grip on a crystal structure through its capacity for location either of heavy atoms or of specific molecular components.

Patterson's contribution to present day science was, in some ways, a simple thing, and it was very fruitful. He must have been very happy to have been able to see so much use made of it, in the 30 years or so left of his career before he was taken from us. For us, remembering him now, and celebrating his function as it reaches its 50 year anniversary, it is a particular joy to recall what a lovable, warm, irrepressibly good-humoured and altogether delightfully human person it was that gave crystallography this function, half a century ago!

42.8 Raymond Pepinsky

Raymond Pepinsky was educated at the University of Minnesota and obtained a Ph.D. in physics from the University of Chicago in 1940. He worked for the U.S. Rubber Company and then went to Alabama Polytechnic Institute in 1941 where he remained until 1949 when he moved to Pennsylvania State University (then College). He is now at the University of Florida in Gainesville. His interactions with Lindo involved his analogue computer X-RAC which was started in Alabama in 1949 and finished at Penn State. This was used by many crystallographers to solve structures. He also, with Okaya, worked out details of analyses of structures by the Patterson function with anomalous scattering data.

It is a large pleasure for me to be among so many old and good friends again. I bring with me greetings from Louise [Mrs Pepinsky], who has shared in these friendships.

And it is a consummate pleasure to join with you in thinking about Lindo Patterson: what he was as a person and as an influence on us all. Lindo was a very special man. With him as the focal point, this commemorative event is more than a celebration of a great step forward in science, a seminal idea revealed, one which had tremendous practical value immediately and which deepened our understanding of diffraction theory and brought the mathematics of Fourier transforms onto center stage in crystallography. Despite all that, *in re* Lindo Patterson there is a supervening concept, having to do with the human aspect of science. What qualities in a person permit the development of such insight and orderly integration of components of a profound and long-standing puzzle? Why, when we speak or hear or read his name, as a metaphor for a fundamental process in crystal analysis, do we feel an inner warmth and a smile? How did his discovery, and the admiration and honor it brought him from all of us, affect him? What sort of fellow did he reveal himself to be?

We all know the answers. I can only echo them. He enjoyed knowledge, science, and a life in a humane society, and he participated in each with gentle wisdom mixed with delightful good humor. Nothing made him happier than helping young people toward understanding and intellectual strength. He was entirely unselfish, un-self-centered. He radiated gentlemanliness. It was fun, and refreshing, and illuminating, to be in Lindo's presence or to think about him.

I cannot be sure, at this moment, where it was that I first met Lindo. It was probably at the Washington meeting of the American Physical Society in 1938. Willie Zachariasen had arranged for his retinue of graduate students to make the trip from Chicago (in Arthur Compton's Cosmic Ray Laboratory station wagon — a wild bit of travelling which I still remember with alarm). Those springtime meetings were held on the grounds of the old campus of the National Bureau of Standards, on Connecticut Avenue and Upton Street. We sat on the grass in the warm sunshine (Chicago was still snow-covered) and I remember Bert Warren and Sterling Hendricks and Rose Mooney joining us. Several geophysicists and mineralogists from the Geophysical Institute would be there as well, eager to talk with Bert and Sterling and Willie. There were few physicists involved in X-ray structure analysis in those days; but there were a good number involved in X-ray physics and spectroscopy. Most structure analysis was in the hands of mineralogists and chemists. Lindo was another among the limited number of physicists, and all physicists made the annual pilgrimage to the Washington meeting. Lindo was generally in the company of his bearded physics colleague from Bryn Mawr, Walter Michels. I remember the beard as being a flaming red;

but perhaps I am just dreaming in colors. I cannot remember such a meeting when I did *not* encounter and enjoy Lindo. It seems to me now that I have always known him.

The first meeting which Louise and I attended together was the ASXRED affair at Gibson Island in 1943. It was held as one of the summer Gordon Conferences. It was particularly memorable because crusty old Dr Gordon himself appeared, in his pajamas, at midnight, to castigate Betty Wood, Fankuchen, and a number of other reprobates for singing Christmas carols in harmony at that hour in the bar (in July). Whether Lindo was among the sinners, Betty Wood will remember. I suspect not since, had he been, the songs would probably have been from Gilbert and Sullivan rather than carols (Louise and I were not in the bar at *that* hour since we were still on our honeymoon). The consequence of the old boy's reprimand, repeated by him during the next morning's lecture session, was that the crystallographers elected never again to assemble at a Gordon Conference as long as Dr G. still controlled them.

The first meeting of the crystallographers after the war ended, in 1945, was at Lake Geneva, Wisconsin, near Yerkes Observatory. Willie Zachariasen organized it. It was notable for several reasons, which I briefly enumerate. (1) It ran very smoothly, despite the fact that the lectures were held in a hall which in horse-racing season served as a (highly-illegal) bookie joint. The walls were covered with tally-sheets, odds information, and the like. More people had come than pre-registration had indicated, and Willie had been forced to find a larger room than he had been led to believe would be necessary. Willie was justifiably proud of the fine arrangements he had made, and he let everyone know it thereafter. Some of you will remember his mischievous grin when screw-ups would occur in later meetings (like in that old broken-down inn where the Lake George meeting was held, and where the chef decamped the day before the crystallographers arrived). Willie would grin, and call out 'bad planning, bad planning!', and everyone's exasperation would subside. (2) Wisconsin being famous for its cheeses, and with cheese just having been removed from war-time rationing but supplies being almost non-existent elsewhere in the country, everyone loaded up with as much cheese as he (and/or she) could carry. Martin Buerger staggered back to Cambridge with a fifty-pound carton. (3) Crystallographers were increasingly concerned with attempts to reduce the dreadfully-limiting tedium of computing. It was obviously a tremendous bottleneck in our field. It was at the Lake Geneva meeting that Dan McLachlan first discussed his sand machine for analogue summation of two-dimensional Fourier series. Dan McLachlan was another of the dearest, warmest, most amusing persons I have ever known, and he was unquestionably one of the most ingenious. He brightened the lives of many people in the time he was on this earth, and he made some significant contributions to our field. The sand machine is well described in his 1957 book *X-Ray Crystal Structure* (McGraw-Hill). As he spoke about that device in the bookie room at Lake Geneva, I decided I should make some use of the electronics I had learned during the war, and design an analogue to Dan's analogue.

So much for Lake Geneva. I rode the train from Chicago back to Auburn, sitting up all night in the club car because I could not sleep with a flood of electronic schemes racing through my mind. I spent most of a year, then, working out and discarding schemes for generating the frequencies which would be required for a two-dimensional series with indices H, K for $F_{HK} = A_{HK} + iB_{HK}$ over the range $0 \leq H \leq 20$, $-20 \leq K \leq 20$. Jim Vann, one of my graduate students, and I argued over these, and Jim wrote his master's thesis on the various schemes.

A travelling representative, Lawson McKenzie, from the Office of Naval Research, visited Auburn near the end of that year. He was looking for research projects to support (those were the days!). I asked whether he thought the ONR would be interested in the proposed computer. He encouraged me to submit a proposal, and I wrote out what I had in mind. When asked for suggestions about referees, I suggested Willie Zach, Lindo Patterson, and Martin Buerger.

It was then I discovered, full force, what sort of fellow Lindo was!

Willie, whom I saw every few months, one way or another, had known of my planning; but he was interested in crystal chemistry and structure determination, and he had no interest in electronics. Besides, his brain was equipped with an inbuilt lightening calculator. What he said to me, in effect, was, 'Well, it's probably all right; but think of how many structures you could solve in the time you would need to build the machine.' But he evidently was supportive. Martin Buerger was extremely supportive and encouraging, and said that if anyone could make progress in the matter, probably I could.

But Lindo! He read every word of the proposal, thought through every circuit suggestion I had made, and then got on a train for Washington and marched into the office of the Director of the Division of Mathematical Sciences of the ONR. The director was a great lady, Mina Rees. Lindo never told me this, but Mina did. He praised the discussion I had prepared concerning the phase problem, the review of existing computers and methods, and explained why he thought the electronics was entirely feasible as well as sensible. There were at least two negative reviews, from crystallographers whom most of you know but whom I shall not mention. Lindo's comments, his reputation, his obvious knowledgeability, simply overrode the doubters.

We built it at Auburn. Paul Jarmotz, a dour electronics genius, who had been assigned to me as a technician in the Radiation Lab at MIT during the war — assigned to me because in those days I was reputed to have a sunny disposition and could handle difficult people (how times have changed!) — Paul came down to Auburn and did a magnificant job on the electrical circuit engineering. Phil Hemily came from his position in mathematics at the University of Michigan, and took over the mechanical engineering. There he is, in that enlarged photograph on the wall, standing alongside Lindo and another fine student, Hugh Long, looking at the gear-box system which Phil designed (Dave Harker is hiding in the background). That is reproduced as Figure 112 on page 280 of the 1952 article on X-RAC and S-FAC, a copy of which some of you may still have around.

The other piece of great good fortune was my success in inviting Dave Sayre to join us. Dave and I had become friends in a biophysics lecture series at MIT. We spent a delightful two years together in Auburn while he helped with the fundamental design problems on X-RAC and wrote his master's thesis on Fourier transforms (which is still one of the best treatments of the subject ever printed). It is reprinted as the last Appendix in the 1952 report. I have since corrected all the typos in that reprint, which errors were *not* in Dave's thesis.

Caroline MacGillavry spent a year with us in Auburn also, primarily to work on ferroelectric transitions with me, but she contributed many good ideas on the uses of X-RAC as well.

Lindo! It is heartwarming to think of him. When we were far enough along with X-RAC for someone to look it all over, Dave Sayre and I decided we would try to get Lindo to come for the summer of 1948. He came, and it was a marvelous time for us. José Donnay and Gai Hamburger both wrote, independently (I had assumed), asking whether they might come also; so we had quite a colony. Dave, Lindo, Gai and José had breakfast together each morning. It became necessary to adopt and enforce a rule that there would be *no* crystallography discussed in those breakfast sessions until everyone had enjoyed at least one cup of coffee. Thereafter, all controls were off. The talk went on all day long between Gai and José. After dinner, everyone assembled in the parlor of our rambling old house in Auburn, and we listened to music (more or less) and talked much more. We had many close friends in town, and they would drop in — in waves, thoughout the evening — to talk and share in Lindo's radiance. Eventually the talk always turned to crystallography. That was rather an imposition on Louise, who had learned to utter a few terms like 'orthorhombic' or 'birefringence,' mostly as expletives to suggest that she had endured a surfeit of crystallography and would someone please change the subject. Such hints being of no avail, she would finally rise from her chair and remark resolutely: 'Sines and cosines, cosines and sines. Oh, my aching back!' And thereupon she would bid all a rather forgiving Good Night, and retire.

It was on an evening some months after those summer visitors had left that the telephone rang, and I answered and heard Lindo's voice. 'Put Louise on, would you please, Ray. I have a story for her.'

It was about two housewives in Tibet, who had put their

evening meals on to cook and who had nothing more pressing to do for a while but to meet beside the fence which separated their respective back yards and exchange gossip. The conversation was going along at a great rate when one of the two stopped suddenly, and sniffed the air. 'I smell something burning,' she said. 'I wonder what it is.' 'Oh,' cried the other lady, 'Oh, my baking yak!'

There are many other stories about Lindo, some of which he told on himself. They have brightened our lives, as he did in person, and as our memories of him will continue to do.

It is his spirit, and himself in the persons of Betty Patterson and Jenny Glusker, which have provided me with the opportunity to return in vivid memory to the times we enjoyed together, and to do so in the company of so many old friends. I am indeed grateful.

42.9 A. Lindo Patterson, Impressions and Memories

Isabella L. and Jerome Karle

Jerome Karle was educated at the City College of New York, Harvard University and The University of Michigan where he obtained a Ph.D. degree in 1944. He worked on the Manhattan Project (1943–44) and then went to work for the US Naval Research Laboratory where he is head of the Laboratory for the Structure of Matter. For many years, he was professorial lecturer in the University College of the University of Maryland. He was President of the American Crystallographic Association in 1972, Chairman of the US National Committee for Crystallography (1973–75) and is a member of the National Academy of Sciences. He was also President of the International Union of Crystallography (1981–84). Together with Isabella Karle he developed the quantitative aspects of gas electron diffraction analysis. Together with Herbert Hauptman, now of the Medical Foundation of Buffalo, he established the foundation mathematics and procedural concepts for modern phase determination using mainly mathematical inequalities, probability theory and a combination of group theory and invariant and seminvariant theory. The methods that developed from this, largely owing to the pioneering work of Isabella Karle, are the methods of choice for the solution of structures containing up to about 200 or somewhat more independent nonhydrogen atoms. In 1985 Jerome won the Nobel Prize in chemistry, shared with Herbert Hauptman.

We met Lindo Patterson at a meeting of the American Society for X-ray and Electron Diffraction (ASXRED) in June of 1943 at the University of Michigan in Ann Arbor. It was a meeting that brought together many of the leading crystallographers in the United States. Our doctoral degrees were soon to be awarded and after a sojourn at the University of Chicago we returned to Ann Arbor, Isabella as instructor in the Chemistry Department and Jerome as an employee of the Naval Research Laboratory engaged in a wartime project in the laboratory of Lawrence Brockway, who had been our thesis adviser in gas electron diffraction. The ASXRED meeting is quite vague in our minds, but we probably attended parts of the meeting on the suggestion of Lawrence Brockway and did so as observers, rather than as participants.

We especially remember making the acquaintance of Lindo Patterson at the ASXRED meeting. His attraction to young people was immediately evident and this led to pleasant conversations that included among other matters a discussion of a subject that was of considerable interest to him at the time, cyclotomic sets.[1] The cyclotomic sets provided a formal mechanism for determining the ambiguities that may arise in atomic arrangments from information that solely concerns values for interatomic vectors. The question, of course, arose from the fact that such information was, in principle, obtainable from the Fourier series that Lindo Patterson discovered and bears his name. It appealed to Jerome as an interesting problem in mathematics. The context in which Lindo was interested in the sets had not been worked out in any detail. Jerome corresponded with Lindo on the subject, but Lindo's interest in pursuing it waned. The difficulty of determining the nature of the ambiguities increases rapidly with complexity and, at the same time, increasing complexity soon makes it impossible to obtain a completely resolved set of interatomic vectors from the Patterson function. In addition, much, if not all, of the ambiguity inherent in a set of values for the interatomic vectors can probably be resolved by use of knowledge of the chemistry of the substance of interest.

As the years went by, we became more acquainted and met at a variety of meetings, for example, the Crystallographic Society of America, the Mineralogical Society of America, the American Society of X-ray and Electron Diffraction, Gordon Conferences, the International Union of Crystallography, the US National Committee for Crystallography and the American Crystallographic Association. We also met at such gatherings as the inauguration in 1949 of X-RAC by Ray Pepinsky at Auburn University, Alabama.[2] X-RAC was an analogue Fourier series synthesizer that played a useful role in crystallographic computing until it was superseded by the development of digital computers.

The first meeting of the newly formed International Union of Crystallography was held in Cambridge, Massachusetts in 1948. We attended only part of the meeting and missed some of the peripheral activities. We were there in time for the dinner, however. When Lindo saw us entering the dining room he very kindly asked us to join him and the other persons at his table. The other persons turned out to be Paul P. Ewald and Max von Laue. It was a distinct pleasure to meet them for the first time and we appreciated the opportunity to converse with them. The conversations were substantive and quite pleasant. Paul Ewald and Max von Laue reflected the same kindness and friendliness toward young people that was so evident with Lindo.

Lindo played a key role in merging the two organizations, the Crystallographic Society of America and the American Society for X-ray and Electron Diffraction to form the American Crystallographic Association. This had the expected beneficial effect of bringing together people of broadly common research activities who had been holding meetings of overlapping interests. It is usually not possible to determine how such things happen but it is likely that Jerome became the first treasurer of the new American Crystallographic Association in 1950 and a member of the US National Committee for Crystallography in 1953 as a consequence of strong support from Lindo. This would be consistent with his commitment to encouraging young people to have a broad participation in the activities of the crystallographic community.

Jerome recalls meeting with Lindo in Philadelphia and driving to a meeting of the ASXRED in June 1949 at Cornell

University during which Lindo spoke extensively of his stay at the Massachusetts Institute of Technology in the laboratory of Bert Warren and his respect and affection for him. It was during that stay that Lindo was stimulated to develop the Fourier series that bears his name.[3,4] He indicated that Warren's studies of sulfur structures motivated him to consider the possibility of finding a function that could yield values for interatomic vectors from the measured intensities of scattering. Specifically, Warren's work in obtaining a radial distribution of the interatomic distances in a crystal from powder diffraction data provided it. This was reported in a paper in the ***Physical Review*** [5] that immediately precedes the one by Lindo on the interatomic vectors.

Lindo also spoke of another matter during the drive. In the late forties there was a spate of publications and presentations at meetings that emphasized the limitations of the Patterson function and minimized its virtues. It caused him the usual distress that derives from a sense of unfairness and the observation of the regression in competence and understanding that so often accompanies valuable developments in complex areas. Fortunately for science, such aberrations are normally overcome in time, although they may persist for years and sometimes for decades. At the present time, the very extensive development of computer technology and its applications to science have a tendency to foster such regressions. Concommitant with the great saving of time and effort and the facility that computers provide for performing calculations that would otherwise be impossible are the program packages that replace thought and understanding and are often quite mistakenly presumed to represent the ultimate in analytical competence. How often have we read in the literature that some method for structure determination was concluded to fail because of the fact that a computer program based on the method did not produce an answer? In many such cases, the method was not tested, only the computer program was. And sometimes there is the question of whether the computer program was properly tested. For example, there is the recent case of two different laboratories publishing an investigation of the same material quite independently of one another in different journals. One group stated that they could not solve the structure by direct methods (computer program specified). The other group stated that they did solve the structure by direct methods with the use of the same version of the computer program as was specified by the first group.

We continued our research in gas electron diffraction at the Naval Research Laboratory in Washington, D.C. when we went there in 1946 and developed an interest in the investigation of crystals only after some years. Jerome became interested in the mathematical consequences of the non-negativity of the electron density distributions in crystals and investigated this with Herbert Hauptman.[6,7] This led to further studies which resulted in procedures for phase determination. Isabella entered the field about 1956, initially to facilitate the testing and further development of the procedures. Important substances of interest soon entered the program.

The applications of the Patterson function continue to be very broad and are disproportionately appropriate when there are particularly heavy atoms in a structure. Analyses of macromolecular structures as well as those with fewer atoms benefit from use of the Patterson function. Such matters should be extensively documented in this volume. We would like to mention some implications and applications of the function that are not usually discussed. The Patterson function has implications concerning the solvability of the phase problem from use of the intensities alone. Atomic coordinates can be readily deduced from values for resolved interatomic vectors obtainable from a Patterson map for simple structures. Phase values can be computed from the atomic coordinates, thus showing that they are obtainable from the measured intensities. As a practical matter, the usefulness of this conclusion is realized when phases are directly derived from the intensities and then used to obtain the atomic coordinates, rather than the other way around. This, of course, characterizes direct methods of phase determination.

Another property of the Patterson function is its possible use in a procedure to correct faulty intensities such as those seriously affected by extinction. In making calculations for their paper entitled 'Positivity, Point Atoms and Patterson',[8] Herbert Hauptman and Jerome found that a Patterson function computed from data having extinction errors could be used to correct those errors. This can be achieved by making the function positive by, for example, setting all negative values of the function equal to zero and then performing a Fourier inversion. It was found in a test calculation that much of the error could be eliminated by the inversion. This observation was not investigated further at the time, but could, however, provide the basis for a useful method for improving intensities having major errors and merits additional study.

One of the common uses of the Patterson function is to locate the positions of heavy atoms in a structure. Comparatively light atoms favored by small thermal motion can also be so located. In an investigation of a magnesium dihexapeptide[9] in which there were one magnesium atom and two perchlorate groups per asymmetric unit, the highest peaks in the Patterson function arose from the magnesium atoms in the unit cell. Peaks associated with the chlorine atoms were essentially obliterated by the effects of large thermal factors. Although the magnesium atoms contributed only about 3% of the total scattering, they were readily found in the Patterson function and provided the starting point for partial structure development to the complete structure by use of the tangent formula.[10]

There are a number of other special contexts in which the Patterson function can be usefully coupled with direct methods for phase determination. Partial structures from E-maps can be tested against the Patterson function for consistency and a translation function derived on the basis of concepts associated with the Patterson function[11] can be used to shift the partial structure if it is incorrectly located in the unit cell.

The Patterson function is useful for determining whether there are some heavier atoms in substances of unknown composition or not. We have had experiences in both respects. We have had heavier atoms turn up in a structure, brought in as part of initially unexpected solvent molecules. We have also been told of the presence of heavier atoms that did not appear in the analyses. For example, a pyrophosphate group was supposed to be contained in a structure, but did not appear. The absence of an appropriate phosphorus-phosphorus peak in the Patterson function was easily verified. On the other hand, the Patterson function was consistent with the presence of orthophosphate groups which did appear in the structure.

It may be considered superfluous to involve the Patterson function in such matters since it would usually be possible ultimately to resolve such questions and find the proper solution to the structure by direct methods. The response to this point of view is that often the analysis can be greatly facilitated by use of the Patterson function.

In our last example of the coupling of the Patterson function with direct methods, we note that, in the course of the determination of the structure enkephalin[12] the Patterson function played a valuable role in verifying the presence of a β-sheet. A notable aspect of this use of the Patterson function is that it was computed from only a small subset of the strongest reflections. Only 55 of those reflections with $|E| > 2.4$ were used. Certain regularities in the data suggested that it would be worthwhile to compute the map from only these special reflections. As it turned out, the predominant contributions to the strongest reflections came from the β-sheet. When the Patterson function was computed from large numbers of data, contributions from the side chains obscured those peaks which were particularly associated with the β-sheet part of the structure.

The very valuable contributions that the work of Lindo Patterson has made and continues to make to crystallographic science in the analysis of structures large and small brought him few distinctions beyond the most appropriate attachment of his name to the F^2 series, as he called it. He was not offered membership in the National Academy of Sciences and he received no significant awards of which we are aware. The history of science is replete with such oversights. In the case of Lindo Patterson, it may be more flagrant than most.

Contacts with Lindo Patterson (see Fig. 42.9.1) were stimulating and filled with humor. There was a strong sense of his commitment to his profession and his generosity of spirit, particularly in his dealings with young people.

It was a distinct pleasure to have known Lindo Patterson

and it is an honor and privilege to have the opportunity to reflect upon our association with him.

FIG. 42.9.1. Two resolved Pattersons: Betty and Lindo at the IUCr excursion on an island in the archipelago, Sweden, 1951.

References

1. Patterson, A. L. *Phys. Rev.* **65**, 195–201 (1944).
2. Pepinsky, R. (ed.) *Computing Methods and the Phase Problem in X-ray Crystal Analysis.* X-ray Crystal Analysis Laboratory.Pennsylvania State College: College Park, Penna. (1952).
3. Patterson, A. L. *Phys. Rev.* **46**, 372–376 (1934)
4. Patterson, A. L. *Z. Krist.* **90**, 517–542 (1935).
5. Warren, B. and Gingrich, N. *Phys. Rev.* **46**, 368–372 (1934).
6. Karle, J. and Hauptman, H. *Am. Mineral.* **35**, 123 (1950). Abstracts of presentations at a meeting of the Crystallographic Society of America in Ann Arbor, Michigan, April 1949.
7. Karle, J. and Hauptman, H. *Acta Cryst.* **3**, 181–187 (1950).

8. Karle, J. and Hauptman, H. *Acta Cryst.* **17**, 392–396 (1964).
9. Karle, J. and Karle, I. L. *Proc. Natl. Acad. Sci. USA* **78**, 681–685 (1981).
10. Karle, J. *Acta Cryst.* **B24**, 182–186 (1968).
11. Karle, J. *Acta Cryst.* **B28**, 820–824 (1972).
12. Karle, I. L., Karle, J., Mastropaolo, D., Camerman, A. and Camerman, N. *Acta Cryst.* **B39**, 625–637 (1983).

42.10 Gabrielle Donnay

Gabrielle Donnay was educated at UCLA and obtained a Ph.D. at MIT in 1948. She went to the Johns Hopkins University, then to the Geophysical Laboratory of the Carnegie Institute before going, in 1970, to McGill University where she is Professor of Crystallography. Her interests lie in the area of crystal chemistry and especially hydrogen bonding, with an emphasis on the chemical bonding present in mineral structures.

I belong to Jerry Karle's generation, and so when I first met Lindo (Pat) he was very much the senior to whom one looked up. He was very shy to talk with, and yet as you've all heard so many times, it was very easy to talk even shop with him. He would very often tell me that I was jumping to conclusions and I mustn't do that; and he was absolutely right, and I still remember every time I come to a conclusion, I ought to be thinking of Lindo's advice not to jump to it. So he gave me very good advice very early.

The summer that we spent with him in Auburn, Alabama, was 1948. I'd just gotten my Ph.D. I had gone through a lot of troubles with my Professor, Martin Buerger, attempting to unravel the Patterson function. You haven't heard very much about that because it sounded as if we all get the structure immediately after we have got all these interatomic vectors from the (F^2) Patterson function. But you know all these interatomic vectors get translated back to the origin, and the problem really is to find where in space these interatomic vectors are situated. So Buerger then had a procedure that he called the 'implication method,' and while he was doing a lot of book writing in Brazil, I was supposed to be using the implication method to solve the structure of tourmaline. That, at the time, appeared to be a complicated mineral on which a lot of students had already tried to determine the structure and had given up; I was considered very much of a fool because I was still trying to tackle it. Well I got an answer for the structure from the implication method. At the meeting where Buerger and I then gave a paper, Pat discussed the structure a little bit, and he made me somewhat dubious about it. Very fortunately for me, I was allowed to visit Bert Warren's office occasionally (I had taken his introductory course on X-ray diffraction). He looked at my abstract, and he told me that if *he* were tourmaline that certainly wasn't the way *he* would choose to crystallize. He took my structure home with him, and asked me to come back after I found the correct structure. So, using Dr Pauling's procedure of balance of charge and polyhedral edge sharing, etc. to get the right interatomic distances, I finally solved that structure and went back to Bert Warren. The first thing he said to me was, 'The right structure of tourmaline is locked up in this drawer of mine, and when you think you've found it come back, we'll check it out'. I produced my structure, he pulled his out — they were identical! Pat was very much pleased with my now-much-more-reasonable structure and encouraged me to write it up. The end of the story is that I couldn't get that structure published. *Acta* had just come out. It was 1948.

Paul Ewald was the Chief Editor, Fan (Fankuchen) was the American Editor (their offices were side by side). Fan accepted the structure but not the title of the manuscript! Paul Ewald, luckily for me, picked up the paper (he later told me) and looked at all the calculations I had done.This was a noncentrosymmetric structure, and, as you have heard by now, these were the precomputer days, so it had taken me, not weeks or months, but years to do all the calculations of structure factors. Ewald wrote me a note saying, 'I've gone over the head of my American Editor. The paper is accepted, and it will get published. And it doesn't matter about the title' (As it turned out, Fan's proposed title won out in the end). So that's how it happened, and again Pat was wonderful, encouraging, in all these trials and tribulations. I'm quite sure that if that structure hadn't appeared in print, I would never have remained a crystallographer. So for my career, Pat's influence helped from the very start.

I should perhaps add a few words to the story of the 'Translation Group' at Johns Hopkins, where I moved in 1949 as Mrs Donnay. We had a lot of wonderful meetings. The 'translation' was that of people coming from Philadelphia and from Washington to Baltimore once a month for crystallographic discussions. Without Pat's continuous support and the many papers he gave us at these meetings, the group would have died out rather quickly, I think, but with his support, and with all the nice stories he could tell, we had very good times. We would go to the Faculty Club afterwards to have dinner before our visitors undertook that long trip back to Philadelphia.

Finally, I will tell you a little about Pat's early career at McGill. As Pat himself described it in his contribution to ***Fifty Years of X-ray Crystallography***, McGill Physics Department had had Rutherford, who was the great man, as a Professor, and anybody (like Patterson) who would want to go to England to study with someone else than Rutherford (with Bragg in this case) was considered out-of-line. Pat, however, stuck to his guns and went to work with Bragg. When we came to McGill in 1970, we thought we could change the prevailing attitude a little bit. We decided right away that we would put a nice plaque outside of our X-ray lab, in memory of Patterson, on which I inscribed the dates of Pat's bachelor's degree, Ph.D. degree, and the year he spent at McGill as a lecturer (he built the first X-ray diffraction lab in Canada), and then, of course, the F^2 function, just the way it was written in his first paper. The plaque also contains the emblem of the F^2 map of potassium dihydrogen phosphate. We had a nice opening ceremony. Betty Patterson came to unveil the plaque. Anna McPherson, who had been one of Pat's classmates also came. She had obtained her Ph.D. at the University of Chicago under Compton and then came back to be a Professor of Physics at McGill, where she spent her whole career. She told us what a wonderful classmate Patterson had been; she had a lot of nice stories to tell about him. This, I think, was the beginning of an attempt to let McGill appreciate the very important and famous men they had produced. But, after fourteen years at McGill now,

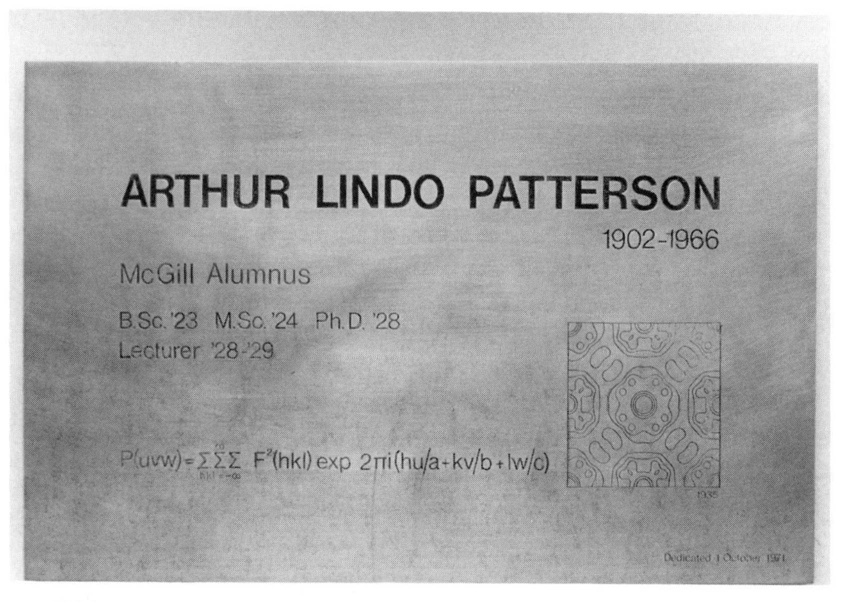

FIG. 42.10.1. Plaque commemorating Lindo Patterson's time at McGill University. This plaque is now outside the room that was Lindo's office in the Institute for Cancer Research.

I am sorry to say I haven't really succeeded. They could have made a lot of publicity about this sucessful McGill graduate in Physics at this time on this occasion. We did all we could to stir a little bit of enthusiasm but we haven't succeeded, so maybe this is a sign that the people that you raise yourself are the ones you appreciate the least. I think that sooner or later McGill will wake up and find out they had a really outstanding alumnus in Lindo Patterson. We all agree he should have received the Nobel Prize for his contribution.

42.11 Cecily Darwin Littleton

Cecily Littleton obtained her B.Sc. degree from Somerville College in Oxford with Dorothy Hodgkin as her tutor, and then she came in 1949 to work in Patterson's lab in Philadelphia. Her Fellowship there was financed by the Northern Branch of the Woman's Auxiliary of the Institute for Cancer Research. She worked on the structure of the gluconate ion, and used the X-RAC calculator in Pepinsky's lab at Penn State. Cecily's father, C. G.

Darwin, made great contributions to the understanding of the interpretation of crystal diffraction patterns, particularly dynamical diffraction and extinction theory, and her great-grandfather, Charles Darwin, wrote *The Origin of Species.*

Lindo was marvelous, but I'm afraid that when I first arrived in '49, I was young and ignorant, and I can only give you a worm's eye view of the great man. But first I want to say that if it weren't for A. L. Patterson I wouldn't be here today, and I wouldn't be here at all if I hadn't become a crystallographer 35 years ago, in Oxford, under the care of Dorothy Hodgkin. Bless you Dorothy!

Then here in Philadelphia the ICR was just beginning. It was all very exciting! We were the new Physics department and Lindo had the pleasure of creating it. And he certainly enjoyed it! He was the Chief. I was fresh from Oxford with my B.Sc. and Joan Clark was our third member. She was young then too and only later she got her Ph.D.

Pat tried to teach Joan and me some of the basic math of crystallography, but, I'm sorry, we couldn't rise to the occasion. Before coming to the ICR he'd been teaching physics at Bryn Mawr College and I think he missed the teaching when he came here. In those days our lab was brand new, starkly grey, and empty, with concrete floors. By now, with 35 years of good work and Jenny's care, it has become comfortable and human.

In our conversations about solving crystal structures, we talked of the Patterson function, but naturally around here we couldn't call it that, it would have been embarrassing to that good man. He always called it the F^2 synthesis, which was fine. And he was always doing more mathematics around it to try and find new meanings in his great discovery.

I solved my crystal, potassium gluconate — it was full of hydrogen bonds — with the constant advice and encouragement of Lindo. He never failed to be patient with my idiocies, and to help me understand what was going on. And in a wider sense too, he explained things about America to me. He had lived in England for a while as a boy and as a young man, and knew where I was coming from. Work was a pleasure, largely due to his good nature, and to his fund of anecdotes and funny stories, and also to the friendly welcome given to me by Betty Patterson, Dr Reimann, the ICR and all its people.

Back then he was starting a new department. He had a lovely time reading the catalogs and ordering suitable things. Now here is a really trivial tale. As well as buying all the expectable equipment, we had to buy a desk calculator. Several salesmen came and showed us their wares and, in the end, we chose a Friden machine. But we had to decide which color, soft-orange-beige or blue-grey. That was when I realized that Pat's clothes were always one of those two colors. His suit was either soft-orange-beige to match his hair, or blue-grey to match his eyes. So we asked him to choose the color. And *he* chose the one to match his red hair.

I told you that things were exciting! There was a mini-meeting of crystallographers at the Brooklyn Polytechnic, where Professor Isidor Fankuchen was in charge. I had been in these United States exactly one month when I met someone there who was an even newer immigrant than me. He was P. P. Ewald, from Germany and England. He had arrived only 10 days before, and I believe was at the Brooklyn Poly thereafter. He was charming and pleased to meet *me*, because he knew my father, Charles Galton Darwin. Together, or separately, they had worked out how crystals reflect X-rays. That was in 1920 or so. I also remember a conversation that evening during dinner at a restaurant. Dave Harker said to me, 'I have just found someone to give me a million dollars, and I am going to solve proteins.' This was in 1949. *I* knew that proteins were impossible, and that a million dollars was impossible. But *he* knew better.

Another time we drove to State College, Pennsylvania, to visit Ray Pepinsky and to see his early computer, X-RAC. We put the data from my Oxford crystal on it — and wonder of wonders — immediately there was the electron density map. Nothing laborious about that! Since then computers have revolutionized crystallography, and everything else. You lucky people!

Then there was an occasion when Max Perutz of Cambridge, England, visited Philadelphia and the University of Pennsylvania. He, too, was going to solve proteins. It's just that I have read the very recent issue of that periodical *Nature* where there's a long paper about DNA and its nucleosome core particle, and I saw that the whole article is dedicated to Max Perutz because he is 70 years old. Seventy years deserves celebrating, doesn't it? You know, it's 70 years since X-ray crystallography began. Max von Laue took that first spotty picture in 1912, remember? None of us was there then, but I hope you won't mind if I say that I feel that I'm here at this great international gathering of creative crystallographers, representing the early pioneers, von Laue, W. H. Bragg, P. P. Ewald, C. G. Darwin and W. L. Bragg. I'm sure they too admire Patterson and his work.

Think of the long, long way we've come since then. And think what a tremendous difference Patterson himself, and the Patterson function, have made — a quantum leap! — and how he has opened the door to the solution of all kinds of new structures, and enriched the lives of all who knew him in his own incomparable way. I'm so happy to be one of them, and to rejoice with all of you.

42.12 Joan R. Clark

Joan Clark was a Research Assistant in Patterson's laboratory from 1949 (when he first moved to the Institute for Cancer Research) to 1953. She then went on to get a Ph.D. and worked at the U.S. Geological Survey in Washington DC and at Menlo Park.

When I started working for Lindo Patterson in September, 1949, I didn't realize that I was entering a new and fascinating world and that my whole life would be changed as a result. During my four years at ICR, Lindo not only taught me about crystallography, as he had promised when I was hired, but also graciously opened the door to people and events of crystallography. His laboratory was, of course, a focal point for all scientists involved with crystallography. The thought of banishing assistants from visitors and outside activities never even occurred to him. This generous spirit, supported wholeheartedly by his dearly loved wife, Betty, brought me so many unusual and memorable contacts that it would take pages to list all. Among those I cherish particularly, I will mention three.

One was during a party at Betty and Lindo's home when Peter Debye and Isidor Fankuchen took off gloves and sparred happily together verbally, each recognizing the other's strengths and weaknesses, and each thoroughly enjoying the combat. Another was the entire last meeting of ASXRED at the Franklin Institute, Philadelphia, December 1949, particularly the intense 'no holds barred' discussions that followed papers by Dorothy Wrinch and Isabella Karle. Clearly women crystallographers were viewed as peers, fully capable of defending their theories, an attitude that Lindo himself encouraged at every opportunity. Last but not least were Lindo's own reminiscences of his early days in crystallography — his experiences at the Royal Institution with the Braggs, in Germany where he found himself in a debate with von Laue, and the developments that led to his discovery of the F^2-series which everyone else soon referred to as the 'Patterson series.' His delightful sense of humor and his often wicked wit pungently spiced these reminiscences. What a loss it is that tape recorders were not then commonly available!

With such a remarkable initiation by Lindo into crystallography and its world, I continued on to a successful and satisfying 30-year career in mineralogical crystallography. In retrospect those first few years were among the best of all, thanks to the unique, vibrant, and joyous personality of Lindo Patterson.

42.13 Alice Weldon

Alice Sword Weldon, B.A. 1930 was a student of physics and mathematics and a classmate of Betty Knight Patterson at Wellesley College. She was a Research Assistant in Patterson's Laboratory from 1951-1955.

My memories of Lindo Patterson are of what a great teacher he was to me, a complete novice, when I came to the crystallography lab at Fox Chase. Joan Clark, Christer Nordman and Warner Love were beginning their careers as researchers while I was there.

Lindo had total recall of all the things he had heard and seen in his life time and was generous in sharing his experiences and insights. One couldn't help learning a relaxed and joyful approach to life from him.

42.14 Caroline MacGillavry

Caroline MacGillavry was born in Amsterdam, the Netherlands and studied at the University of Amsterdam where she obtained a Ph.D. in 1937. She continued as a research assistant at the University of Amsterdam and was Professor of Chemical Crystallography there from 1950 until her retirement in 1972. She was a member of the Executive Board of the International Union of Crystallography (1954–1960) and, with G. D. Rieck, an Editor of *International Tables* Volume III. She has published many crystal structures and, in 1965, published a book entitled *Symmetry Aspects of M. C. Escher's Periodic Drawings*; the second edition is entitled *Fantasy and Symmetry*.

I think it must have been in 1936 that my friend and colleague, H. J. Verweel and I, both of us assistants of J. M. Bijvoet, carried out the first Patterson synthesis in the Netherlands. Our tools were a table of goniometric functions and a slide rule; additions were made by hand. It was a projection, of course, and also not about an axis that would have given the most relevant information. Moreover, our use of visually estimated intensities was looked upon with great scepticism by our 'boss' Bijvoet. He was convinced you should measure them with the classical, beautifully machined, but extremely slow and tedious Bragg-spectrometer. Nevertheless, the result of this first effort gave enough information about the position of the molecule in the unit cell to encourage us to try to go ahead. Verweel's tragic and sudden death put me off for some time, but later I took the investigation up again, going over to Fourier projections and solving the structure of succinic acid, my first organic molecular structure, published just before World War II started. It led to a life-long friendship with Kathleen Lonsdale, who worked on the same problem without success because the insufficient resolution of the Bragg spectrometer had led her to the wrong space group!

This first effort was followed by many Patterson syntheses, as a first step to extract structure information from the wealth of experimental data. Even now, after fifty years in which the so-called 'direct' methods have gained so much ground, I still feel that the Patterson method can be a useful counterpart to the use of phase relations between reflections. The first method, working in direct space, gives direct information about internal regularities in the atomic arrangement. The second approach, operating in reciprocal space, starts from the assumption of random arrangement of atoms; large fluctuations in the intensity distribution then give indirect clues to more or less systematic deviations from such random arrangement. When correctly phased, the E-map does give direct information about the structure!

At the Cambridge, Mass. Conference in 1948, where the International Union of Crystallography was created, I was the only representative from The Netherlands. For me, it was the first full-size international meeting. I was quite overwhelmed by seeing all the celebrities! Then, to my surprise, a kind, red-haired giant, spotting an unknown and rather shy person, came up to meet me. This was the famous Patterson himself! He took me under his wing and introduced me to all these VIP's. This was for me the beginning of a full year's stay in the U.S.A. During that year I attended all the meetings of both the CSA and the ASXRED (old-timers will remember these two groups), notably their turbulent common meetings during which the two were supposed to fuse. I remember how Patterson's humor, wisdom and tact smoothed out the storms of this courtship, so that finally the ACA could be born. Since then I met Pat at many scientific meetings. I visited him and Betty in Bryn Mawr and later at the Institute for Cancer Research. Also, he came to see me in Amsterdam, where I showed him how to sip a drink from a brim-full glass on the counter of a very old pub. Apart from such diversions, we worked under Kathleen Lonsdale's strict leadership at the *International Tables for Crystallography*, Patterson as the author of the section 'Fundamental Mathematics' in Volume II, I as co-editor of Volume III. Kathleen would tell Patterson that he should not put in things 'purely for the sake of mathematical elegance'. He would just grin, and to the benefit of all of us he produced a marvellous, elegant and very clearly and concisely written Section. He must have worked extrèmely hard on it, but he never boasted about it.

In these first, say, fifteen years of the International Union of Crystallography, the crystallographers from all over the world seemed to know each other personally. We were an exceptionally close-knit group within the realm of science and many warm friendships were formed. It was a great shock when this group was broken by the loss of Fan and Pat within such a short interval. They were great contrasts, *Homo faber* and *Homo ludens*, but both of them personalities characteristic of the crystallography of those days.

I, among many others, kept the bonds of friendship with Dina Fankuchen and with Betty Patterson. I regret that it was not possible for me to attend the 'Fifty Years of the Patterson Function' in Philadelphia. I thought of all of you, old friends who were there. But mostly I thought about Lindo, and of course about Betty, to whom this meeting must have meant much.

42.15 John S. Kasper

Both the Patterson function and Lindo Patterson personally have played a major role in my career as a crystallographer. Actually, my first real experience with the Patterson function was somewhat traumatic. In attempting to solve the structure of decaborane, with my wife Charlys and with David Harker, we relied totally on interpretation of vector maps. After many months of intensive study of the maps with no success, I was inclined to believe the skeptics who maintained that the Patterson Function was severely limited and that one had to possess a special, rare gift to have any success.

The result of this impasse was that we were led to the discovery of a method for obtaining phase information from inequalities relating structure factors and their squares. So, while the Patterson function did not contribute directly to our discovery, we may not have made it without this particular experience.

At the time of our work on decaborane, the systematic sophisticated methods of extracting the maximum information from Pattersons had not been developed. With their advent, I was pleased to see the greater utilization of the Patterson

Function, and subsequently, I myself had only successful results with its use.

Perhaps my best appreciation of the Patterson Function has come about in the last few years. I am associated now with a routine structure determination effort employing a state-of-the-art automated diffractometer and up-to-date sophisticated programs for utilization of 'direct methods.' I am impressed with the power of these methods and their success in the great majority of cases. I am more delighted, however, that when there are failures, the use of Patterson maps has provided the solution.

It is well to remember that the Patterson Function is unique and exact and any plausible model must conform to it. It is my belief that the Patterson is here to stay and that fifty years hence it may be in order to have a centenary celebration.

My personal acquaintance with Lindo Patterson is something that I shall cherish all my life. I had the good fortune to have many lengthy sessions with him in connection with the production of Volume II of *International Tables of X-ray Crystallography*. Lindo was the first selection for Editor but in declining, agreed to produce the section on Basic Mathematics and to be a consultant and advisor, in which role his contributions were very substantial. While so much of the work on the Tables was arduous and sometimes painful, there was the pleasurable side of the meetings with Lindo. These were lengthy, not only because of the work at hand, but also because of the delightful diversions covering many things — basic problems of crystallography, general problems of science, philosophy and wonderful anecdotes from his considerable experience. I had ample opportunity to develop a great admiration for his warm personality, as well as for his intellect.

It is only fitting that in celebrating the Function, we also recognize the great man responsible for it and how fortunate many of us are to have known him.

42.16 J. D. H. Donnay

José Donnay, born in 1902, was educated at the University of Liège and obtained a Ph.D. in geology from Stanford University in 1929 where he remained until 1931. Then he went to The Johns Hopkins University where he was Professor of Crystallography and Mineralogy. He is now working at McGill University, Montreal. His interests lie in the relationship between crystal morphology and structure and he has been a leading figure in mineralogical crystallography. He was President of the Crystallographic Society of America just before it merged with ASXRED to form the American Crystallographic Association.

First of all, I want to tell this Institute, its directors, present and past, my appreciation for the idea of this symposium which is one of the most beautiful ideas I have met in my career. I am enjoying it to the limit.

Pat became my friend very early in the game, in fact we were born in the same year, 1902. So you see we are not chickens. The job that we tackled together was to collect the errata in the *International Tables* first edition (IT 1935). Many scholars around the world had been sending whatever errata they had found, so we had quite a long list. Of course, pottering with that list day after day, we found a few more errata, which we added to the list, which thus became quite impressive. Pat told me one day, 'You know, José, what we are doing here turns out to be, for my career at Bryn Mawr, my most important contribution of the year'. Well, we all know that this is not a scientific contribution, it is a labor of love, and one that is very poorly rewarded. But there are two kinds of people that do that kind of work, the very gifted people who do extremely good work and do it over and above whatever else they do, just to help the profession, and then the ones who are not so very gifted and have to spend their time doing something useful; that's the other kind. Well, Pat was obviously in the first category. He was very generous with his time, trying to make a good job of that lousy task. Unwittingly he had the opportunity to show his modesty. His name does not appear on the first page of the reprinting of 1935, which was published by Edwards Brothers in Michigan. The only place where it is mentioned is at the end of an appendix, explaining how the errata were gathered; this is signed A. L. Patterson and J.D.H. Donnay. He was extremely conscientious, never accepted any shoddy work or any quick conclusions (as my wife told you she found out from him). I too learned a lot from Pat.

Another remark I'd like to make is something that happened during the 'Translation Group' period. You know that in England there was a 'Space Group', and then when Fankuchen came back to the U. S. A. he created the 'Point Group'. Well, when we jumped into the fray at Hopkins with yet another group, the name 'Translation Group' was still available, and that's what we called it. Patterson was very devoted, came from Pennsylvania to Baltimore once a month and kept that thing going. Many, many people have enjoyed it and benefitted by it. One day the mathematician Wintner told me 'Do you know what I see in mineralogical journals and books, about the so-called Patterson theorem? That's not a theorem, I have a counter-example right now'. Well, I wasn't very much aware of what was going on, but it soon turned out the natural scientists had confused physical observations of Patterson with the mathematical theorems that can be proved rigorously. And the mathematicians were quite right. I mean there was no *Patterson theorem*. So, at the next 'Translation Group' meeting, Pat came from Bryn Mawr and clarified the thing to the satisfaction of mathematicians. He said, 'No, no, no, don't blame me, I never called it a theorem'. And he explained that it was a set of observations made physically, which turned out to be true most of the time and therefore were what in the middle ages would have been called a law of nature. In early modern times the expression 'law of nature' was redefined as a valid generalization based on many observations. This process is known as induction, as opposed to deduction which draws conclusions from a postulated principle. I'd like to stress this point, because so many scientists, to this day, confuse such terms as theorem, principle, rule, law. It would be nice if we could get these names straight. Anyway, so much for this Patterson 'theorem' that was not a theorem, as Patterson himself well knew. Since then we haven't seen the word theorem used in the literature any more in connection with the Patterson function.

I was so overwhelmed by the solemnity of this occasion that I forgot to mention one remark of Patterson which is one of the gems in my collection of Pattersoniana. It is the remark he made when he first heard about the Harker sections. He said, and this is expressed in authentic Pattersonese 'I could have kicked myself in the pants for not thinking of it myself'. This expression shows his scientific humility. It also shows that he was a big enough man to salute a beautiful discovery coming from someone else.

42.17 Sidney C. Abrahams

Sidney Abrahams was educated at the University of Glasgow, Scotland working with J. Monteath Robertson. He did postdoctoral work at the University of Minnesota, MIT and Glasgow University before going to Bell Laboratories in 1957. His interests lie in the area of solid state physics and the relationship of properties, such as dielectric, elastic, optical and magnetic properties, to the atomic arrangement of the compound. Dr Abrahams was President of the American Crystallographic Association in 1968, Chairman of the US National Committee for Crystallography 1970-1972 and is Editor of the major international journal *Acta Crystallographica*.

I first met Pat while spending the summer of 1948 at MIT. A rather mature group of World War II veterans on their return to MIT had raised enough funds to bring one or more promising young graduate students from almost every European country to Cambridge for their participation in MIT's summer program. Their objective was to develop strong ties between potential leaders from these countries and America, both to decrease the possibility of future wars between them and to help stimulate the resurgence of peaceful science in their war-torn countries. The Iron Curtain had yet to cut off a complete exchange of young scientists with the eastern countries. It was my privilege that summer to be working in Bert Warren's laboratory, where a major interest was the X-ray study of order-disorder in simple systems such as Cu_3Au. The summer of 1948 was also the period, of course, in which the International Union of Crystallography held its first Congress, at Harvard University. I was then one of Monteath Robertson's graduate students and, although most of my thesis research was complete, it was not yet written up. Robertson was unable to come to that first Congress and had instead suggested that I present a joint paper.

A total of 83 papers at nine main sessions, with an attendance of 310 crystallographers, leaves a very different impression from that given by our most recent Congress in Hamburg with its 1519 participants and 1398 papers. Further, most of the participants at that first Congress were already well-known. My paper, for example, was immediately followed by Dorothy Hodgkin's. At the end of my session, I was sought out by the already eminent Lindo Patterson who said he had enjoyed my talk and who thereupon launched into a perceptive discussion of my paper. He followed this by telling me that he had a check to give me, in his capacity as Chairman of the U.S. National Committee for Crystallography. He also gave a check to each of the other overseas visitors who had presented papers at the Congress. So far as I know this generosity was unmatched by the organizers of subsequent Congresses and it particularly reflected the warm interest Pat clearly felt for his fellow scientists as people.

I came to appreciate his personality more fully the following year. I had returned to Scotland in the meantime and finished my thesis, then had travelled to Minnesota as Bill Lipscomb's first postdoc. We took the long train trip together to Ithaca in June 1949 for the penultimate meeting of the ASXRED. Lindo Patterson turned out to be the chairman of my session that morning, which had Eddie Hughes, Caroline MacGillavry and Dan McLachlan among the other authors. Multiple sessions had not yet become necessary at crystallographic meetings. Pat not only sparked lively discussion after each paper in his session but, as often as not, he kindled similar discussion after many others. His skill in reaching the heart of so many contributions, and his willingness to debate the other 'elder statesmen' present such as Fan (I. Fankuchen), Dave Harker, Willie Zachariasen and José Donnay with insight and wit, enlivened those sesssions and led many early participants to sit metaphorically and sometimes literally at his feet after the sessions ended as they basked in his erudite good humor.

I felt most fortunate from that time on to be one of Pat's friends. He continued for many years to encourage newcomers at ACA meetings, stimulating thoughtful discussion in all areas of our field whenever possible. It is fitting to close my remarks with the famous aphorism attributed to Sir Isaac Newton: 'If I have seen further, it is by standing on the shoulders of giants.'

Lindo Patterson was truly one of our giants and we have all been able to see further because of his great stature.

42.18 Christer E. Nordman

I, too, spent some time here at the Institute for Cancer Research, a couple of years before Dick van der Helm. This was from 1953 to 1955. During this period, Lindo was heavily involved in the writing of the mathematical section of the second volume of the *International Tables.* This was, I can tell you, an enormous task, and on which he spent a huge amount of effort, and on which he did a masterful job. It is also, as was the case with many other things he did, a task for which he received very little professional recognition or scientific fame. But any user of this material can appreciate the value of his contribution to the crystallographic profession.

It has also been pointed out how nice it was for young graduate students and young postdocs, who gave their first paper at a meeting, to be approached by Lindo afterwards with some encouraging remarks or questions about their paper. I can also say how beneficial it was to a fresh postdoc like me to be introduced to crystallographic meetings by being taken in tow by Lindo Patterson and being introduced to everybody who was anybody in the field at that time. And how good it was to be included in informal gatherings and small dinner groups, and to get to know many people, essentially all the colleagues a bit older than I. I was introduced to some of them at get-togethers such as evening 'hospitality hours.' It was also at such an occasion that I first encountered Gilbert and Sullivan, thanks to Lindo's talents and ability to inspire spontaneous participation.

It has also been pointed out how Lindo took his own presentations seriously, how he was a bit tense before he would give a talk and showed the same reactions that most of us, I think, have experienced. But let me comment on another aspect of his contribution to crystallographic meetings. There was a period that I'm sure some of you remember, a period when there were very heated sessions at the ACA meetings, particularly sessions dealing with methods of crystal structure solution. In retrospect, this may be an 'in' joke among the older people here. At the time I was the only postdoctoral in the lab, and Lindo on one or two occasions shared with me his reflections on the task that invariably fell on his shoulders, namely, the task of serving as chairman of every such methods session. I would of course be very curious and I listened very carefully to everything he said, and I would ask what he thought of this and what he thought of that. Now when I recall it, his characterization of his colleagues was always such a genuine expression of respect and friendship, that although the privacy of these conversations would for most humans have allowed the slipping of a few snide remarks about one side or the other, there never was the slightest hint of that. I also recall a remark Warner Love made to me about the same time; Warner was a frequent visitor in the lab and he learned much of his crystallography from Lindo. Warner once made the observation — he was trained as a biologist as you know — he said, 'I think crystallographers are so much nicer people than biologists.' What he meant was that there seemed to be so much mutual support, helpfulness and absence of cutthroat competition among crystallographers, in contrast to what he felt he had observed among scientists in other fields. This was, I'm sure, exactly what Warner had perceived in his contacts with Lindo. I also think that there is a general validity to this observation and that it stems, at least within the North American crystallographic community, in large measure from Lindo's influence and example.

42.19 Dick van der Helm

My first two years (1957-1959) in the U.S. I worked with Lynne Merritt at Indiana University. This stay was arranged by Edy Eichhorn, an earlier student than myself of Caroline MacGillavry in Amsterdam. These two years were very fruitful in allowing me to work independently on the solution of several structures using the Patterson synthesis and superposition methods but also to become adept in analyzing crystallographic computations and to program the IBM 650 which at that time was a middle-sized computer and the main resource of many university computing centers. It was possible to do three dimensional calculations with the 650, however these used all of the computer peripherals. Thus, I spent many nights as the sole occupant of the computer center catching up at the same time with American literature.

Being young and naïve, I assumed that Fourier and Patterson were contemporaries who lived in the 18/19 th century, which was true for one but not for the other. I therefore, wrote Lindo Patterson to ask if I could work with him. He invited me to Philadelphia for an interview. My first surprise was the open Chrysler New Yorker powered with many horses in which he picked me up from the station. The second surprise was the ease of meeting with him. This was done on his part by being interested in the other person and, of course, by telling stories, while maneuvering the powerful car, being more interested in the flow of the story than in that of the traffic. Lindo apparently approved of me, and offered me a job in the lab, which I began in April of 1959. The subsequent $3\frac{1}{2}$ years were for me the most pleasant of my life, both from a personal as well as a scientific point of view. During that time the lab consisted of Jenny Glusker who had been with Lindo a number of years. A lot of the structural work I did was done in cooperation with Jenny, who already in those days was a marvel at organizing her time around her family while being an excellent crystallographer. Carroll Johnson also joined the lab in 1959. For Carroll the difficulty in representing both the structure and thermal motion of the complicated magnesium citrate structure, which he solved, formed the basis for his famous ORTEP program, which he later wrote at Oak Ridge. Also in the lab at that time were Marilyn Dornberg and Jean Minkin who were both valuable contributors to the crystallographic work which was done in the lab. Betty Patterson, who worked independently in the Institute, was a very important member of the social and intellectual life of the crystallography lab, and she still is. I will be forever thankful for the many evenings of conversation I spent with Betty and Lindo during those years and for the grace and friendship I received from them.

Initially, one troublesome structure was sodium citrate. The only computing equipment in the lab at the time was an IBM 602A, which did not allow anything other than two-dimensional calculations. I, therefore, used 650's at both Princeton and Indiana University to calculate a three dimensional Patterson. In Princeton I worked with my cousin U. v. Wijk, for whom later a crater on the moon was named. It was, of course, Lindo who pointed out that his was the ideal mathematical name, in subscript. In unhurried peace, which was normal for the lab, Jenny and I were able to solve the complete structure from these maps. I believe that this gave real enjoyment to Lindo, indicating to him the unexpected power of his vector method when applied in three dimensions. He therefore, decided that we needed better computing facilities and I was put in charge of deciding which of three or four computers could be leased with the available funds. Analyzing the requirements for a three-dimensional Fourier calculation and a general SFLS computation, using the methods of Bill Busing and Mills, the choice became quite clear and was the IBM 1620, even though this new computer was not yet available. Once they were produced the lab received one of the first of these computers and I spent most of a year writing the Fourier and SFLS programs, which together with other programs written by Lindo, Carroll, Eric Gabe and Max Taylor were sent to more than 100 labs located on every continent, and at no charge. Lindo thoroughly enjoyed the computer and its capabilities, spending more and more time using it for his long-time interest which was the investigation of homometric structures. It truly is a loss that the many notes he left behind about this problem can not be collected and analyzed. At this time he also received the Franklin Medal. He was always shy about his own accomplishments and not eager to assess the advances which the vector method had made possible. However, he was very pleased with the recognition which this honor reflected.

During the years in the lab I learned a tremendous amount of crystallography and mathematics from Lindo. He was an excellent teacher who would start from first principles, writing carefully with pencil on blocklined paper, and developing the answer at the listener's speed, without ever using a proof by 'the method of rapid speech'. I kept the notes and even now sometimes use them. His lecture notes for courses, such as those on space group theory used at Bryn Mawr, were similar in nature. This same care was used in the preparation of his publications, and this quality of writing, with great care of interpretation and presentation of results is obvious in all his publications. Examples can be seen in the 1934 paper or in the 'Fundamental Mathematics' section of Volume II of the *International Tables*. These and other characteristics, easily could be learned from Lindo. He was open, direct and generous. His honesty was the only approach he used to tackle his function in the Institute or in other areas. It was due to Lindo's influence that there was more emphasis in the lab on doing things correctly rather than doing many things superficially. At least once a day there was a sound and in-depth discussion of a scientific problem, some of immediate significance but often of a general nature, in which the whole lab participated. Again Lindo was the teacher with patience, completely involved, and at the cost of a number of barely lit cigarettes, using, in that case the large green blackboard behind my desk. For me, the atmosphere in the lab was easy.

The good computing permitted more structures to be done while various methods besides vector methods were applied for the structure solutions such as inequalities, isomorphous replacement and the Okaya and Pepinsky P_S method. The latter method was used to solve the structure of potassium isocitrate giving at the same time the absolute configuration, which confirmed the earlier result on the isocitric lactone. This settled a long-standing dispute among chemists. This also required that Jenny learn to translate Japanese in a couple of weeks in order to understand one of the key (and correct) contributions to this problem.

It also showed me an example of Lindo's temper which was immediate but very short lived. The reason was that I had talked with a reporter from *C & E News* and had responded to his question about the results before the paper on the absolute configuration was actually published, unaware of the possible breach of scientific ethics. But I definitely became aware of the proper procedures, later, after I had mentioned the telephone conversation to Lindo. The results on the absolute configuration formed the basis for the elegant mechanism proposed later by Jenny Glusker for the enzymatic conversion of citrate to isocitrate.

Although the structures of the citrates remained the major effort, many other compounds became part of the program, especially after Carroll and I left for other jobs, and Max Taylor and Eric Gabe joined the lab. Lindo was a strong and active supporter, in the early days, of the work of Jerry Karle and Herb Hauptman, and direct methods, as well as F^2 series, became a normal method used in the lab.

Lindo was, of course, known by everyone in the crystallographic community. His presence was striking; rather tall, a large head with a ready smile and fun in his eyes. At ACA meetings he made it a point, quite naturally for him, to meet all new and young crystallographers and have them meet and talk with other established members of the community. I am sure that it amazed all newcomers how easy it was to get to know Lindo and to be made to feel a part of the crystallographic community. On the other hand it was surprising to see how hesitant or even nervous Lindo was to ask questions in scientific sessions. Probably this uneasiness was due to his adherence to scientific correctness, or his hesitancy to imply possible criticism of people, along with the fact that a question could be misunderstood or misinterpreted. To him science was at its best in the written form. On the other hand, Lindo was famous for his hundreds of stories and remembrances. They were told at lunch, after and during work, travelling, at meetings and at parties. I can not repeat them and possibly no one can, because, more often than not half the fun of the story was the way it was told by him. The aspect of the stories which I did not really notice at the time but now amazes me, later in life, was his tremendous generosity to other people. This same characteristic was visible in all his encounters both with friends and strangers, and in his scientific and personal life. His whole demeanor showed enthusiasm for all he did and at all times.

42.20 Jean A. Minkin

Jean Minkin obtained a B.A. in physics from Bryn Mawr College in 1947, with Lindo Patterson as one of her Professors. She was Research Assistant, then Research Associate in his laboratory at the Institute for Cancer

Research from 1960-1968. Since 1968 she has worked at the U.S. Geological Survey in Washington, DC and Reston, VA, studying meteoritic crater materials, moon rocks, coal and shale.

This brief article is intended to convey some of my fond memories, not with regard to the derivation of the Patterson function, but with regard to Patterson the person. It was my unique good fortune to be Lindo's student and later to work in his laboratory.

I arrived at Bryn Mawr College in the fall of 1943 as a wide-eyed member of the Freshman class, fortified only with the general notion that science or mathematics were my preferred areas of study. In my orientation interview, the Dean of Freshmen suggested that I make first year physics my area of concentration in science for that year. Thus came my first encounter with the man who provided me with the background, enthusiasm and encouragement to aim at a career in physics.

The Department of Physics at Bryn Mawr was then headed by A. L. Patterson and Walter C. Michels, better known to the campus community as 'Pat and Mike'. In 1943 'Mike' was off helping in the war effort as a commander in the U.S. Navy, and so it was 'Pat' who introduced my class to the realm of matter and energy. Patterson was a tall, lean, freckled, carrot-topped individual who brought his inexhaustible energy and marvelous wit to enliven each lecture with colorful demonstrations, songs, poems and anecdotes. I shall always remember Lindo standing and spinning on a rotating turntable with a dumbbell in each hand, alternately raising his extended arms to shoulder height and bringing them back to his sides. We were left with no doubt that 'in an isolated system the angular velocity is inversely proportional to the moment of inertia!' Our introduction to relativity was reinforced with Lindo's recital of a poem:

> There was a young lady named Bright
> Who traveled much faster than light;
> She left home one day in her relative way,
> And returned home the previous night!

Upon protest by a student, after a quiz early in the year, that not all of the data furnished at the bottom of the page were relevant to solving the problems, the Patterson quip in response was, 'Hell's bells, lady, I can write that the moon is made of green cheese if I so choose.'

Although he loved fun at all times, Patterson, along with Michels, was intense in his desire to provide undergraduate students with a strong foundation in physics, whether for a career in the field or as a part of their general education. The culmination of their efforts was ***Elements of Modern Physics***,[1] affectionately referred to as 'The Elephants' by students of the forties who were the 'guinea pigs' for the preliminary mimeographed editions of the text. Whenever a high rate of failure on a quiz indicated lack of sufficient understanding of the contents of a given chapter, the authors patiently went back to the drawing board, and shortly emerged with another version! To quote Lindo Patterson,[2] 'Walter Michels and I ended my stay at Bryn Mawr in the full glory of a book on elementary physics.'

My post-Bryn Mawr years led me into marriage and employment, first at the Franklin Institute Laboratories for Research and Development in Philadelphia, and then at the National Bureau of Standards in Washington, DC. A 'time-out' for rearing two young sons and my husband's job transfer had us back in the Philadelphia area and me ready to return to study and/or research work in the summer of 1960. Though reading the 'want-ads' was not a habit of mine, one lazy Sunday afternoon I found I had read most of what interested me in the local newspaper, and so I leafed rather aimlessly through the classified section. One advertisement caught my eye with its wording to the effect that there was an opportunity to work in X-ray crystallography, learning modern computer methods, at a research institute in northeast Philadelphia. 'Bells began to ring' as I recalled that A. L. Patterson had left Bryn Mawr some years before and had joined the Institute for Cancer Research, which was somewhere in northeast Philadelphia! I quickly got the feeling that a reply to this advertisement would ultimately be received by A. L. Patterson. With some misgivings because of my several-year hiatus from professional activity and my need to limit my working hours while my family responsibilities still presented heavy demands, I submitted a letter of application to the listed P. O. box address. Three days later the phone rang, and it was Lindo with his usual enthusiasm, wanting to know how soon I could get in to talk about the job! Thus began my eight years of happy association with the Department of Molecular Structure at ICR.

Tribute must be paid here to Lindo especially and to the Institute for their early advocacy of allowing women to pursue careers in research part-time during their child-rearing years. Permitting, let alone encouraging, women to maintain part-time work schedules was not a widespread practice in the early 1960's! Lindo is also warmly remembered, I am sure, by many of today's prominent crystallographers for the words of encouragement and praise he took the time to express after they delivered their crucial first papers at meetings of the American Crystallographic Association.

I shall be eternally grateful for all the exciting adventures in learning afforded me under the guidance of Lindo in both the academic and laboratory environment. Moreover, it has been a distinct honor and privilege for me and my family to have known him as a true friend.

References

1. Michels, W. C. and Patterson, A. L. *Elements of Modern Physics*. D. van Nostrand: New York, Toronto (1951).
2. Patterson, A. L. In *Fifty Years of X-ray Diffraction*. (P. P. Ewald, ed.) pp. 612–622. International Union of Crystallography: N. V. A. Oosthoek's Uitgeversmaatschappij, Utrecht, Netherlands (1962).

42.21 Max R. Taylor

I first met Lindo Patterson in 1963 in the North Philadelphia railway station. He, with Eric Gabe, had driven down from Fox Chase to welcome me as I arrived to take up a post-doctoral position in his laboratory at ICR. It was typical of him that he took the time to do this, for not only was he a great scientist, but he was also a friendly, kind, generous and caring person. Joan Clark, who was working in Hans Freeman's laboratory in Sydney where I was a Ph.D. student, had suggested that I consider applying to come to ICR. When I discovered that

Patterson and I were born in the same locality in the South Island of New Zealand, it semed a good omen and I needed no more persuading. I've never regretted this decision because not only did it result in a valuable two-year post-doc, but it was the beginning of an association with Jenny Glusker and the Institute that has continued throughout the years.

I found that ICR was a stimulating place to work in, and that Patterson's lab had a relaxed and friendly atmosphere. Patterson allowed a good deal of freedom and was always ready to discuss, teach and help whenever the opportunity arose. He was never too busy to share his deep knowledge and understanding of the fundamentals of crystallography with others. I soon learnt that Patterson preferred the term 'the F^2 series', and for my part it was better to use this in the lab than to be heard referring to 'that bloody Patterson.' He insisted that we attend as many meetings as possible and I had not been in the lab many months before I was on my way to the ACA meeting in Bozeman, Montana. At this meeting and many others he always took time to introduce me to his many friends and colleagues and I will always be grateful to him for initiating these contacts for me.

Lindo had a great sense of humour and the ability to laugh with others. For example, I remember him accepting good-naturedly some teasing about his 'part-time women' — a reference to the number of part-timers that were working in the lab — five while I was there, and I was always amused by his use of 'spice' as the plural of 'spouse.'

I had been working on transition metal complexes of peptides for my Ph.D. and when I came to ICR I intended to continue with transition metal complexes of biological interest. I attempted to crystallize complexes with citrate but I soon moved on to investigate the structure of the copper complex of KTS which was among the early metal complexes to show antineoplastic activity. Of course, the crystal structure of the complex was solved by the application of the Patterson method. When it came to determining the structure of the ligand itself, we wanted to take the opportunity to gain some experience in using direct methods (Symbolic Addition Procedure, SAP). I felt a little apprehensive about doing this but, of course, there was no need to worry for Lindo enthusiastically embraced any development that facilitated the study of crystallography. He was a great supporter of Karle and Hauptman's work. We wrote some computer programs for the IBM 1620 to automate the SAP to some extent, although a great deal of manual intervention was still required. Eric and I motored down to Washington to get help from the Karles about many aspects of the procedure and after that it wasn't long before the solution was obtained.

An interesting project that I was involved in at this time was the determination of the absolute configuration of a molecule in which the chiral centre involved one hydrogen and one deuterium atom. X-ray diffraction could not distinguish between H and D and the space group was found to be $P2_1/n$ which is centrosymmetric. The neutron diffraction experiment indicated the noncentrosymmetric space group $P2_1$; a good reminder that the space group is not an invariant property of a crystal but is also a function of the method of observation. The absolute configuration was determined by utilizing the anomalous scattering of the Li^6 salt.[1] It was Carroll Johnson's idea to do this experiment, but Lindo kept a close watch on things and supervised proceedings along the way.

A good deal of cameraderie existed in the lab. The chore of data collection was always shared among us, each taking turns to work a two-hour shift on the manual diffractometer no matter whose structure was under investigation. It was customary for Lindo to provide a bottle of champagne to share every time a structure was considered solved. This custom I have continued in my lab at Flinders, although the frequency is now restricted to a bottle each time a person (usually a student) solves his or her first structure. It gives me the opportunity and the pleasure to recount some of the anecdotes associated with Patterson and Pattersons.

It was my privilege to have worked with him.

Reference

1. Johnson, C. K., Gabe, E. J., Taylor, M. R. and Rose, I. A. *J. Am. Chem. Soc.* **87**, 1802–1804 (1965).

42.22 A. Tulinsky

My first exposure to A. L. Patterson was in 1951 when I was a junior undergraduate chemistry major at Temple University in Philadelphia. Crystals had held some of my interests up to that time because of their unusually orderly and well-defined morphological appearances. Lindo Patterson presented a seminar in the Department of Chemistry that year which I attended. I understood little of the seminar but one thing I still remember about the talk was a slide showing the projected electron density of (I think) *p*-chlorobenzoic acid. The slide startled me because the molecule could be plainly visualized from the electron density and this experience was probably the beginning which led me to ultimately pursue the field of X-ray crystallography as a life work.

My first personal encounter with Lindo was as a graduate student at Princeton University while I was working with John G. White. Pat and John were good friends, so when Linus Pauling came to Princeton to present the Vanuxem Lectures, Pat came up from Philadelphia to visit with Pauling and John and to attend the Lectures. I met him at that time and was surprised to find such a notable and renowned scientist so friendly and personally interested in me and my work. It was only later in my life that I realized that this was the mark of Lindo Patterson and that he was the same with all new and young crystallographers and people in general. This initial acquaintance grew into a much deeper friendship during my four years as a postdoctoral research associate at The Protein Structure Project with Dave Harker. Pat was a member of a 'trustee' team which visited and reported on the Project annually.

Our friendship continued and grew over the years until Pat's passing away. We met at many ACA and IUCr meetings, Pat was partially instrumental in placing me at Yale University and he also arranged a position for me at the Institute for Cancer Research as I was leaving Yale. Making a decision about the appointment was one of the most difficult of my life, since the Institute not only had Pat but Pat had such bright young crystallographers like Jenny Glusker, Carroll Johnson and Warner Love. While we were at Yale, Pat visited us in the early 60's and presented a classic seminar on anomalous scattering studies with citrates where he obtained quantitative agreement between observed and calculated Bijvoet differences. Surprising to me was the fact that the audience was not particularly impressed by such a notable achievement. This was but another example of Pat being ahead of his time as was his introduction in the 60's of an in-house IBM 1620 computer which only recently has become standard procedure in larger crystallographic laboratories.

There is no doubt but that except for John White and Dave Harker, Lindo Patterson was the most influential scientist in my life, not so much for the science learned from him, but more for the personal example he presented to follow which is practically impossible to copy. We have to feel that we are a better person because of our brief and occasional associations with him. Pat's picture taken at the Stockholm Meeting, which reminds us so much of the way we knew him, hangs in our laboratory.

42.23 Leonard Muldawer

Len Muldawer is a Professor in the Physics Department at Temple University. He also worked in the laboratory of Bert Warren. His interests are alloys, and phase transformations, X-ray and neutron scattering, and small angle scattering.

My first acquaintance with the work of A. L. Patterson came in the course 8.28 at MIT given by Professor Warren in the spring of '45. It was a second course in X-ray diffraction and covered advanced topics of research interest to him. We were lectured on this ingenious technique, which had been derived by someone called A. L. Patterson in the year 1934. To make sure that we really understood Patterson plots, we were given a problem. Find the projected Patterson-Harker plot $P(xy)$ for rutile, TiO_2, from intensity data supplied with the problem. This was in the precomputer era, and we used Lipson–Beevers strips, a table of sines and cosines in 'convenient' form, to obtain the two dimensional map. Well, it took quite a bit of time. The result was very pretty and I think I enjoyed it. We had nice long problems in those days.

When Warren learned that I was going to Philadelphia upon graduation from MIT in 1948 to teach physics at Temple University, he suggested that I look up Patterson who was then teaching Physics at Bryn Mawr College. I'd been one of Warren's students and had just completed a dissertation in electron diffraction. I was told that Patterson was doing X-ray diffraction, and would be receptive to helping out newcomers to the area. I cannot quite remember my first meeting with Lindo Patterson, but it must have been early in 1949, and we maintained some kind of contact on an intermittant basis.

In 1949 I heard that Lindo was leaving Bryn Mawr College to go to a place called the Institute for Cancer Research. I didn't understand then what a crystallographer would do at such a place but I do now.

In 1955 I performed some neutron small-angle scattering experiments at the Brookhaven National Laboratory. I needed to carry out Stokes corrections to remove instrumental broadening, a technique I had learned in connection with Warren's methods of particle size and strain determination. Lindo said 'Sure,' that he could help. So I traveled out to the Institute for Cancer Research, where I sat for the entire day using Patterson-Tunell strips, which were similar to Lipson-Beevers strips. Lunch was a cheery affair with Lindo, Betty and others at the Institute. There was an incident related about a daddy-long-legs crossing over a small car, and I didn't quite understand, and it had to be explained to me. That's something I do remember and that's the vagaries of one's memory.

In 1957 Lindo Patterson felt that there was a need for a Philadelphia area group devoted to X-ray diffraction and crystallography. He contacted some local people, T. H. Doyne, R. E. Hughes, L. R. Levine, J. G. White, H. G. Wyckoff and myself and we sent out a call for an organizational meeting at this Institute. Some of us wanted to name the group, the 'Patterson Group', but Lindo would have none of that. Thus was born the Delaware Valley Association of Crystallographers, DELVAC (see a notice in Fig. 42.23.1). It served a useful function in the area for about ten years, bringing together the university and

Crystallographers and Diffractionists in the Philadelphia Area

A number of crystallographers and diffractionists living and working in the neighborhood of Philadelphia have felt the need of occasional meetings for informal discussions and talks, from local talent and visitors, on the field which interests them most. The subjects would range through all the applications of diffraction techniques (x-ray, electron, or neutron) to the investigation of the structure of matter, crystal structure, molecular structure, polymer structure, texture, order-disorder, strain, etc. It is suggested that six or so meetings a year would be appropriate during the fall, winter and spring months and that the place of meetings would be variable to distribute the travel load evenly unless a truly central and congenial location can be found.

In order to facilitate the arrangements for such meetings, the Institute for Cancer Research invites you and any of your colleagues or students who might be interested to attend an organization meeting at the:

Institute for Cancer Research
7701 Burholme Avenue
Fox Chase, Philadelphia 11, Pa.

on

Friday, December 6, 1957, at 8:00 p.m.

A map is enclosed for your guidance. Although no speaker is arranged for this first meeting, it is hoped that there can be considerable discussion of the range of work going on in the area.

We hope that you will be able to attend the organization meeting but we realize that some will not. In any case, please answer and return the copy of the enclosed questionnaire which bears your name and address and hand on the other copies to anyone who is interested.

Thomas H. Doyne, Villanova University, Villanova, Pa.
Robert E. Hughes, University of Pennsylvania, Philadelphia, Pa.
Louis R. Lavine, Remington-Rand Corp., Philadelphia, Pa.
Leonard Muldawer, Temple University, Philadelphia, Pa.
A. L. Patterson, Institute for Cancer Research, Philadelphia, Pa.
John G. White, RCA Labs, Princeton, N. J.
Harold G. Wyckoff, American Viscose Corporation, Marcus Hook, Pa.

FIG. 42.23.1. Notice of a DELVAC meeting.

industrial researchers, pure crystallographers and applied and electron diffractionists. Some of our well-attended meetings were held at the old General Electric sales office on Hunting Park Avenue with Tom Workman as host.

When I was writing the article on the Warren School of X-ray Diffraction at MIT for *Crystallography in North America*, I checked references in the 1930's. By this time I knew that Patterson had been at MIT when he made his discovery, and that he was an unpaid member of Warren's group. Only when I read his (Patterson's) statement in his 1934 *Physical Review* article, did I realize that some credit should go to the milieu in which he was located. Patterson wrote 'The result obtained here is an extension of the application to crystals of the theory of scattering of X-rays reported by Gingrich and Warren at the Washington meeting of the APS and arose in a discussion of that work.' I believe that the Patterson method was certainly original, and it is to Patterson's credit that he was appreciative of the stimulation he received. It is also to Warren's credit that he never even suggested,when we covered the material in class, that any of the credit would come in any way from his previous work. These were honorable men.

There's one paper that I recall Patterson giving at a meeting. I don't recall whether it was ACA or American Physical Society. The title was something like 'On a space midway between direct and reciprocal.' I had learned about exploring reciprocal space in studying short range order. After hearing Patterson's paper, I had nightmares about getting stuck between these two spaces. A terrible predicament. But I always knew that Lindo Patterson was a kind person and a true gentleman, and I'm sure that he would have rescued me.

42.24 Elizabeth Wood

Betty Wood obtained a Ph.D. in Geology from Bryn Mawr College and then went to Bell Labs where she spent her professional life. Her interests were in ferroelectrics, epitaxy, and solid state properties in general. She was a very active member of the crystallographic community, as President of the American Crystallographic Association in 1957 and Chairman of the U.S. National Committee for Crystallography in 1956-1957.

If Len Muldawer is doubly honored to be here, I'm triply honored to be here. But I should say to Dr Pauling that there is 'something funny' about me. I got my Ph.D. in Geology. I first knew Lindo Patterson when I was at Bryn Mawr College where

I was in the geology department and he was in the physics department. The other major member of the physics department was Walter Michels, whose nickname was Mike, and so when Lindo Patterson arrived in the department, obviously he was Pat. Pat and Mike ran the physics department at Bryn Mawr, when I was there in the geology department.

I think the way I interacted most closely with Lindo Patterson was in the writing of the constitution for the American Crystallographic Association. We spent a number of hours together trying to make sure that the constitution said what we thought it ought to say, and didn't say what we thought it ought not to say. And I can remember a gesture of his, and many of you who remember Pat will remember this. He would run his hand down his face from the forehead down to the chin in a kind of a wiping-the-slate-clean sort of a gesture, and then he'd come forward with a simple, direct, straightforward statement that was so clear that it indeed wiped the slate clear of all the confusion that we had had about how to write a constitution.

I remember also that he had a sense of humor. This has been mentioned before. One of the things that bothered him was the fad that there was for people abbreviating things. It made them practically unintelligible. So he sat himself down and wrote a paper which was full of abbreviations. I can't remember exactly the title of the paper and someone can correct me perhaps, but I think it was 'The wheaks in the relp of the Buccessera.' Obviously, this means 'the white streaks in the reciprocal lattice patterns of the Buerger precession camera.' The paper occupied only one page. It was signed by A. L. Pon, and I do remember the one footnote. You remember it, don't you, Verner? Yes. It was 'Vernished' which meant 'Verner Schomaker, unpublished.' This paper was set up in type in the proper way for *Acta Crystallographica* and then at the last minute, it was decided that people might think that this was a frivolous way to spend money for publishing *Acta Crystallographica*, and so it was never published. But many reprints of it were available, and I am the proud owner of one. It's a shame to know that sometimes people don't do things because of what somebody might say.

One thing I remember about Pat was something very nice that he did for young people who gave papers at Crystallographic meetings. Whenever some young person was giving a paper (perhaps the first paper that he or she had given, but certainly a paper for which the person was apparently a little reticent, a little nervous) after the paper had been given, Pat would look up the young person and go and talk to him or her about some aspect of the paper that had interested him. Not 'That was a nice paper, you gave' or 'You did a very good job,' but showing that he really had listened, had cared about what was being said, had some thoughts about it. Think how much that must have meant to young people giving their papers for the first time, or nearly the first time, to have Lindo Patterson interested in what they were doing.

I was impressed by another aspect of Pat. One time when he was about to give a paper himself, and he was sitting down in the front row of the auditorium, I had some question that I wanted to ask him and I went down and sat beside him. I said 'Pat I want to ask you something.' He said 'Betty, don't speak to me now, I'm about to give a paper.' I was amazed because I had thought that giving a paper was something he did as easily as rolling off a log. But he had a very high standard that he set for himself. If he was going to give a paper, it had to be done right. That's what I remember about Lindo Patterson.

42.25 Isabella Karle

Isabella Karle obtained her Ph.D. from the University of Michigan with Lawrence Brockway, working on electron diffraction problems. In 1946 she went to the Naval Research Laboratory in Washington where she has remained ever since. She was President of the American Crystallographic Association in 1976, won the Garvan Medal of the American Chemical Society that year, and is a member of the National Academy of Sciences. Her interests are in peptides, enkephalins, frog poisons and structures of unknown compounds and she, of course, is extremely well known in the crystallographic community for having developed the practical applications of the methods that her husband and Herbert Hauptman have proposed.

There has been a great void in my life in that I have not been a student of Lindo Patterson, or have even been closely asso-

ciated with him in my professional life. As Jerome had mentioned, we first met Lindo in the 40's. At that time my interests lay mainly in the vapor phase, and they weren't quite as solid as my later interest in crystals. During those years, and the following years, I had a most agreeable social relationship with Lindo. He was always extremely friendly. I remember him best, as do my children, for the amount of attention that he lavished upon them. In those days when both of us were coming to meetings, we also brought our family.

I entered X-ray crystallography by the back door, at the urgings of Jerome that someone should apply some of the direct phasing formulas, and there were so many of them, that he and Herb had developed. Someone should show that they do indeed work. So I was elected, which meant that I got a copy of Buerger's book, I used a borrowed X-ray generator and a borrowed camera. The first problem I encountered concerned which way to mount a monoclinic crystal, because I got a different pattern if I mounted it at one end or the other. But after a while I learned that it didn't really make any difference. I was, of course, quite grateful to Buerger for having written a sufficiently lucid book that I could follow to make the X-ray photographs, to index the patterns, and so forth.

I am very, very pleased that I was pushed into this field because I have found it extremely exciting, and it becomes more and more exciting as the years go by because, of course, we are able to look at larger and more important and more fascinating structures. Lindo's approach to the problem and the direct phase approach to the problem are really just different facets of the same sort of thing and, of course, it was inevitable that the two different approaches should merge at one time or another. I think the time has come now. There have been quite a number of examples in which the Patterson function and direct methods have come together very neatly, especially for the larger and more ambiguous types of structures. Let me just give you two short examples.

The first one is a structure of a magnesium complex of a peptide with one magnesium ion and one hundred carbon, nitrogen and oxygen atoms. The magnesium contributed only to the extent of 3% of the total scattering. Nevertheless, a

Patterson function was computed with (E^2 − 1) coefficients rather than F^2 (where the E's are normalized structure factors), and in the Harker sections, there were found three enormous peaks that corresponded to the position of the magnesium ion. Obviously the magnesium ion is rather stiff, rather rigid in the crystal, whereas the peptide part is floppy, so that the contribution of the magnesium ion is emphasized. Knowing the location of the magnesium ion, then, by using phase extension methods (part of direct methods procedures) the structure came out quite readily. And so, instead of calling this the 'heavy atom method' perhaps we should call it the 'light atom method.' However, the magnesium was sufficiently heavier and more rigid than the carbon, nitrogen and oxygen atoms, that it functioned as a heavy atom.

In another example in which there were only light atoms, well over 200 of them, a Patterson function was computed with only about 50 terms. Those 50 terms were chosen by using the largest E values, the largest normalized structure factors, and immediately the resulting map showed a very beautiful β-sheet. Fortunately it showed only the β-sheet, and not the vectors associated with the side chains. The side chains contributed to the smaller E values since they had larger thermal factors. Leaving out all the other E values, aside from the very, very strong ones, emphasized the β-sheet and its orientation, which was of considerable help in phase extension by the tangent formula.

In addition to these two examples there are others that are now appearing in the literature. Olga Kennard just told me on the way down the steps here, that she had solved a very nice porphyrin structure recently using a combination of Patterson methods and direct methods in a program that had come out of Sheldrick's laboratory. I am sure that Lindo would have been very excited about these new applications of his procedures. I am sorry that he isn't around to enjoy these developments. It was really a pleasure to have known Lindo before I really had used his method, and I appreciate his work even more so now.

42.26 Thomas J. King

Tom King worked at the Institute for Cancer Research from 1950 to 1967. He and Bob Briggs conducted the classical experiment of initiating embryonic development by transplanting a nucleus from a somatic cell in successive stages of cellular differentiation back into an egg, the nucleus of which had been removed.

As the previous speakers have noted, I too have fond memories of Lindo Patterson. When I came to the Institute for Cancer Research in 1950 as a graduate student, he encouraged me to persist, despite repeated failures, in the new and unorthodox research problem that Bob Briggs and I had begun. That encouragement led to the success that allowed me to experience the true joy of research, namely, to accomplish something that no one else had done before and in doing so to open up a whole new way of looking at the stability of cell differentiation in embryogenesis. Once having initiated the development of a reconstituted egg, Bob and I would come back to the lab after dinner to observe the day's results and set up for the next day's experiment. On most evenings Betty and Lindo were also in their laboratories. After finishing up, we would all stop by the Patterson's apartment for a nightcap. Those evenings became a time that all of us treasured as an opportunity to discuss the exciting areas of research that were going on throughout the Institute by the new and energetic staff that Dr Reimann had assembled and to contrast it with the political machinations that were going on at the level of the Board of Trustees that threatened to jeopardize the future of the Institute. Lindo was a stabilizing force in the crises of those days. He became the conscience of the Institute and, as Tim Talbot said earlier, it was he, and only two or three others, who forced the resolution of issues and made superior science survive in this house. Without that shoring up of the decaying foundation of the Institute for Cancer Research, Dr Talbot would never have been able to develop the Fox Chase Cancer Center as we know it today.

It is a pleasure to come back to where my professional career began and to see so many of Lindo's friends and colleagues, many of whom knew him in quite diverse circumstances, but all of whom are as one in recognizing the added dimension that he brought to everything that he did.

43. Reprint of article entitled 'Symmetry maps derived from the $|F|^2$ series'

by A. L. Patterson

Published with permission.

SYMMETRY MAPS DERIVED FROM THE $|F|^2$ SERIES

A. L. Patterson

Institute for Cancer Research and
Lankenau Hospital Research Institute
Philadelphia, Pennsylvania

The purpose of this paper is to show that the "vector maps" in common use in X-ray crystallography can also be interpreted as "symmetry maps", i.e. as maps which indicate the degree to which any point, line, or plane in the crystal cell possesses any one of the symmetry operations of the space group of the crystal. As a first step in this process, we shall show that the $|F|^2$-series can be obtained from the density distribution in the crystal by as many different Faltung integrals as there are symmetry operations of the space group. In the second step, we shall derive the necessary transformation which must be applied to the basic $|F|^2$-series in order that it may be interpreted as a symmetry map for any specific symmetry element. These transformations are essentially those used by Buerger (1946) in deriving his implication diagrams, but the interpretation suggested here differs from his, and is in addition applicable to the whole cell of the $|F|^2$-series, rather than to the Harker sections alone.

We now proceed to show that, if $\underline{A}$ is any operation of the space group of the periodic function $f(\underline{x})$, the generalized Faltung integral*.

$$\phi(\underline{u}) = \iiint_{-1/2}^{1/2} f(\underline{x})\, f(\underline{Ax} + \underline{Bu})\, dx_1 dx_2 dx_3 \tag{1}$$

depends only on the coefficients $|F(h)|^2$, where the Fourier series for $f(\underline{x})$ is written in the form:

$$f(\underline{x}) = \sum_{\underline{h}} F(\underline{h})\, e^{-2\pi i \underline{hx}} , \tag{2}$$

in which $\underline{h}$ is an integral vector in the reciprocal lattice and the summation is to be taken over all values of this vector.

Any symmetry operation $\underline{A}$ of the space group $f(\underline{x})$ can be written explicitly in the form:

$$\underline{Ax} = \underline{A}_0\underline{x} + \underline{a} , \tag{3}$$

in which $\underline{A}_0$ is a point operation (rotation or rotatory inversion) which does not shift the origin, and $\underline{a}$ is a translation. The inverse of this operation is clearly:

$$\underline{A}^{-1}\underline{x} = \underline{A}_0^{-1}\underline{x} - \underline{A}_0^{-1}\underline{a} . \tag{4}$$

---

* The argument of the present paper is developed for three-dimensional periodic distributions only. The extension of the discussion to one, two, or more than three dimensions is quite clear. In a similar way the application to non-periodic distributions, in which the Fourier integral is used instead of (2), can readily be carried out.

If $\underline{A}$ is a symmetry operation of $f(\underline{x})$, we must have

$$f(\underline{x}) = f(\underline{Ax}) = \sum_{\underline{k}} F(\underline{k})\, e^{-2\pi i \underline{ka}}\, e^{-2\pi i \underline{kA}_0 \underline{x}}$$

Setting $\underline{h} = \underline{k}\,\underline{A}_0$, it follows that $\underline{k} = \underline{hA}_0^{-1}$ and we have

$$f(\underline{x}) = f(\underline{Ax}) = \sum_{h} F(\underline{hA}_0^{-1})\, e^{-2\pi i \underline{hA}_0^{-1} \underline{a}}\, e^{-2\pi i \underline{hx}} .$$

Since this result must hold for all values of $\underline{x}$, comparison with (2) shows that

$$F(\underline{hA}_0^{-1}) = e^{2\pi i \underline{hA}_0^{-1} \underline{a}}\, F(\underline{h}) . \tag{5a}$$

By consideration of $f(\underline{A}^{-1}\underline{x})$ it may be shown that

$$F(\underline{hA}_0) = e^{-2\pi i \underline{ha}}\, F(\underline{h}) . \tag{5b}$$

These are of course the familiar expressions for the relationships between Fourier coefficients introduced by symmetry operations.

We may now rewrite the integral (1) in the form

$$\phi(\underline{u}) = \sum_{\underline{h}_1} \sum_{\underline{h}_2} F(\underline{h}_1)\, F(\underline{h}_2)\, e^{-2\pi i \underline{h}_2 \underline{a}}\, I(\underline{h}_1 \underline{h}_2)\, e^{-2\pi i \underline{h}_2 \underline{Bu}} , \tag{6}$$

in which

$$I(\underline{h}_1 \underline{h}_2) = \iiint_{-1/2}^{1/2} e^{-2\pi i (\underline{h}_1 + \underline{h}_2 \underline{A}_0)\underline{x}}\, dx_1 dx_2 dx_3 . \tag{6a}$$

This integral has the value unity when

$$\underline{h}_1 + \underline{h}_2 \underline{A}_0 = 0 , \tag{6b}$$

and has the value zero otherwise. Thus (6) reduces to a single summation in which we set $\underline{h}_2 = \underline{h}$ and $\underline{h}_1 = -\underline{hA}_0$, i.e.

$$\phi(\underline{u}) = \sum_{\underline{h}} F(-\underline{hA}_0)\, F(\underline{h})\, e^{-2\pi i \underline{ha}}\, e^{-2\pi i \underline{hBu}} , \tag{7}$$

which, with the aid of (5b), may immediately be written

$$\phi(\underline{u}) = \sum_{\underline{h}} |F(\underline{h})|^2\, e^{-2\pi i \underline{hBu}} , \tag{8}$$

which differs from the usual form of the $|F|^2$-series (Patterson, 1934, 1935) only in the presence of the operator $\underline{B}$, whose significance will be discussed later.

We shall also be interested in an alternative form of the integral (1) which we write in the form

$$\phi_0(\underline{u}) = \iiint_{-1/2}^{1/2} f(\underline{x})\ f(\underline{A}_0 x + \underline{Bu})\ dx_1 dx_2 dx_3 \ , \tag{9}$$

in which $\underline{A}_0$ is the pure rotational part (proper or improper) of $\underline{A}$ which is an operation of the space group of $f(\underline{x})$. The argument given above, when applied to the integral (9), leads to

$$\phi_0(\underline{u}) = \sum_{\underline{h}} F(-\underline{hA}_0)\ F(\underline{h})\ e^{-2\pi i \underline{hBu}} \ , \tag{10}$$

instead of (7) and the use of (5b) then leads to

$$\phi_0(\underline{u}) = \sum_{\underline{h}} |F(\underline{h})|^2\ e^{2\pi i h\underline{a}}\ e^{-2\pi i h\underline{Bu}} \ , \tag{11}$$

which is simply the series (8) with the origin translated a distance $\underline{a}$ in the scale of $\underline{Bu}$. We shall return to the discussion of the operators $\underline{B}$ after we have written specific expressions for the operators $\underline{A}$ of crystallographic interest.

The Combination of Rotations and Translations

To simplify our discussion we shall assume that the point, line, or plane, which defines the symmetry element under discussion, includes the origin. The argument can be developed for other locations of the element, but the simplification does not reduce the generality. In a symmetry element which possesses a definite direction, we shall choose $\underline{a}_3$ for that direction. In such cases $\underline{a}_1$ and $\underline{a}_2$ will be chosen normal to $\underline{a}_3$ and their lengths and the angle α_3 between them will if necessary be specified to conform with symmetry requirements. Such a choice of axes can always be made. We shall use matrix representations (Seitz, 1934, 1935, 1936; Burckhardt, 1947) in terms of such a choice of axes.

The screw-rotation $\underline{n}_m$ is represented in operational symbolism by the equation

$$\underline{n}_m \underline{x} = \underline{n}_0 \underline{x} + m \quad \tau/n \tag{12}$$

where τ is the translation parallel to the axis and $\underline{n}_0$ is the pure $(2\pi/n)$ rotation operator. The rotatory inversions (in the origin) take the simpler form $\bar{\underline{n}}_0 \underline{x}$. The matrices for the operators $\underline{n}_0$ and $\bar{\underline{n}}_0$ are listed in Table 1.

We shall use the symbol $\underline{\alpha}$ for the operator which transforms a vector into its components parallel to a given axis, thus $\underline{\alpha}\ \underline{w}$ is the component of the vector $\underline{w}$ parallel to the axis in question. The operator $\underline{\nu}$ takes the normal component, i.e. $\underline{\nu}\ \underline{w}$ is the component of $\underline{w}$ normal to the axis. These operators are represented by the matrices

$$\underline{\alpha} \equiv \begin{pmatrix} 000 \\ 000 \\ 001 \end{pmatrix} \quad \text{and} \quad \underline{\nu} \equiv \begin{pmatrix} 100 \\ 010 \\ 000 \end{pmatrix} \tag{13}$$

Table 1. Matrix Operators for Rotations

<table>
<tr><th colspan="3">Axes</th><th rowspan="2">$\phi = \frac{2\pi}{n}$</th><th colspan="3">Proper Rotations</th><th colspan="3">Improper Rotations</th></tr>
<tr><th>$a_1 a_2$</th><th>$\alpha_1\ \alpha_2$</th><th>α_3</th><th>$\underline{n}$</th><th>$\underline{n}_0$</th><th>$\underline{B}$</th><th>$\bar{\underline{n}}$</th><th>$\bar{\underline{n}}_0$</th><th>$\underline{B}$</th></tr>
<tr><td>$a_1 a_2$</td><td>$\alpha_1\ \alpha_2$</td><td>α_3</td><td>2π</td><td>1</td><td>$\begin{pmatrix}100\\010\\001\end{pmatrix}$</td><td>$\begin{pmatrix}100\\010\\001\end{pmatrix}$</td><td>$\bar{1}$</td><td>$\begin{pmatrix}\bar{1}00\\0\bar{1}0\\00\bar{1}\end{pmatrix}$</td><td>$\begin{pmatrix}200\\020\\002\end{pmatrix}$</td></tr>
<tr><td></td><td></td><td>α_3</td><td>π</td><td>2</td><td>$\begin{pmatrix}\bar{1}00\\0\bar{1}0\\00\bar{1}\end{pmatrix}$</td><td>$\begin{pmatrix}200\\020\\001\end{pmatrix}$</td><td>$\bar{2}$</td><td>$\begin{pmatrix}100\\010\\00\bar{1}\end{pmatrix}$</td><td>$\begin{pmatrix}100\\010\\002\end{pmatrix}$</td></tr>
<tr><td rowspan="6">$a_1 = a_2$</td><td rowspan="6">$\alpha_1 = \alpha_2 = \frac{\pi}{2}$</td><td rowspan="2">$\frac{2\pi}{3}$</td><td rowspan="2">$\frac{2\pi}{3}$</td><td>3</td><td>$\begin{matrix}0\bar{1}0\\1\bar{1}0\\001\end{matrix}$</td><td>$\begin{pmatrix}110\\\bar{1}20\\001\end{pmatrix}$</td><td>$\bar{3}$</td><td>$\begin{pmatrix}010\\\bar{1}10\\00\bar{1}\end{pmatrix}$</td><td>$\begin{pmatrix}1\bar{1}0\\100\\002\end{pmatrix}$</td></tr>
<tr><td>3^{-1}</td><td>$\begin{pmatrix}\bar{1}10\\\bar{1}00\\001\end{pmatrix}$</td><td>$\begin{pmatrix}2\bar{1}0\\110\\001\end{pmatrix}$</td><td>$\bar{3}^{-1}$</td><td>$\begin{pmatrix}1\bar{1}0\\100\\00\bar{1}\end{pmatrix}$</td><td>$\begin{pmatrix}010\\\bar{1}10\\002\end{pmatrix}$</td></tr>
<tr><td rowspan="2">$\frac{\pi}{2}$</td><td rowspan="2">$\frac{\pi}{2}$</td><td>4</td><td>$\begin{pmatrix}0\bar{1}0\\100\\001\end{pmatrix}$</td><td>$\begin{pmatrix}110\\\bar{1}10\\001\end{pmatrix}$</td><td>$\bar{4}$</td><td>$\begin{pmatrix}010\\\bar{1}00\\00\bar{1}\end{pmatrix}$</td><td>$\begin{pmatrix}1\bar{1}0\\110\\002\end{pmatrix}$</td></tr>
<tr><td>4^{-1}</td><td>$\begin{pmatrix}010\\\bar{1}10\\001\end{pmatrix}$</td><td>$\begin{pmatrix}1\bar{1}0\\110\\001\end{pmatrix}$</td><td>$\bar{4}^{-1}$</td><td>$\begin{pmatrix}0\bar{1}0\\100\\00\bar{1}\end{pmatrix}$</td><td>$\begin{pmatrix}110\\\bar{1}10\\002\end{pmatrix}$</td></tr>
<tr><td rowspan="2">$\frac{2\pi}{3}$</td><td rowspan="2">$\frac{\pi}{3}$</td><td>6</td><td>$\begin{pmatrix}1\bar{1}0\\100\\001\end{pmatrix}$</td><td>$\begin{pmatrix}010\\1\bar{1}0\\001\end{pmatrix}$</td><td>$\bar{6}$</td><td>$\begin{pmatrix}\bar{1}10\\\bar{1}00\\00\bar{1}\end{pmatrix}$</td><td>$\begin{pmatrix}2\bar{1}0\\110\\002\end{pmatrix}$</td></tr>
<tr><td>6^{-1}</td><td>$\begin{pmatrix}010\\\bar{1}10\\001\end{pmatrix}$</td><td>$\begin{pmatrix}1\bar{1}0\\100\\001\end{pmatrix}$</td><td>$\bar{6}^{-1}$</td><td>$\begin{pmatrix}0\bar{1}0\\110\\00\bar{1}\end{pmatrix}$</td><td>$\begin{pmatrix}110\\\bar{1}20\\002\end{pmatrix}$</td></tr>
</table>

Note 1: Every power and every product of the above operators is equivalent to one of the above operators.
Note 2: Proper Rotations: $\underline{B} = (\underline{1} - \underline{\nu}\ \underline{n}_0)$, $(n \neq 1)$; $\underline{B} = \underline{1}$, $(n = 1)$.
Improper Rotations: $\underline{B} = (\underline{1} - \bar{\underline{n}}_0)$, $(n \neq 2)$; $\underline{B} = (\underline{1} - \underline{\alpha}\ \bar{\underline{n}}_0)$, $(n = 2)$.

From Fig. 1, it is clear that the operation $\underline{n}_0\underline{x} + \underline{\nu}\,\underline{w}$ is equivalent to a rotation through $2\pi/n$ about the point P, such that OP is the vector $\underline{\nu}\,\underline{u}$ defined by the equation

$$\underline{\nu}\,\underline{w} = \underline{\nu}\,\underline{u} - \underline{\nu}\,\underline{n}_0 u = \underline{\nu}(\underline{1} - \underline{n}_0)\,\underline{u} \qquad (14)$$

This equation specifies $\underline{\nu u}$ in terms of $\underline{w}$ but leaves $\underline{\alpha}\,\underline{u}$ undetermined. The operation

$$\underline{n}_m\underline{x} + \underline{w} = (\underline{n}_0\underline{x} + \underline{\nu w}) + (m\,\underline{\tau}/n + \underline{\alpha}\,\underline{w}) \qquad (15)$$

is thus a screw rotation about P through an angle $2\pi/n$ with a translation $(m\,\underline{\tau}/n + \underline{\alpha}\,\underline{w})$. If we specify

$$\underline{\alpha}\,\underline{w} = \underline{\alpha}\,\underline{u} \qquad (16)$$

then the vector $\underline{u}$ is specified in terms of $\underline{w}$ by the equation

$$\underline{w} = \underline{u} - \underline{\nu}\,\underline{n}_0\underline{u} = (\underline{1} - \underline{\nu\,n_0})\,\underline{u} \qquad (17)$$

Thus if we choose $\underline{Bu}$ in (1) equal to $\underline{w}$, i.e.

$$\underline{B} = \underline{1} - \underline{\nu\,n_0}\ , \qquad (18)$$

there will be a contribution to the integral (1) if two atoms are related by a screw axis of rotation $2\pi/n$ displaced from the origin by a translation $\underline{\nu\;u}$ and having a screw translation $(m\,\underline{\tau}/n + \underline{\alpha}\,\underline{u})$. Thus $\underline{\alpha}$ $\underline{u}$ measures the departure of the screw translation from $m\,\underline{\tau}/n$, the translation associated with the space group operation $\underline{n}_m$. Values of $\underline{B}$ for the screw axes are included in Table 1.

Fig. 1 indicates that for $n = 1$, $\underline{u}$ will be infinite so that the argument leading (14) is invalid. For this case we choose $\underline{B} = \underline{1}$ and obtain the standard form for the $|F|^2$-series.

From Fig. 2, we can see that the operation

$$\bar{\underline{n}}_0\underline{x} + \underline{w}$$

is equivalent to an operation $\bar{\underline{n}}$ located at $\underline{u}$ provided that

$$\underline{\alpha}\,\underline{w} = \underline{\alpha}\,\underline{u} - \underline{\alpha}\,\bar{\underline{n}}_0\underline{u} \qquad (19a)$$

and

$$\underline{\nu}\,\underline{w} = \underline{\nu}\,\underline{u} - \underline{\nu}\,\underline{n}_0\underline{u}\ , \qquad (19b)$$

i.e. that

$$\underline{w} = \underline{u} - \bar{\underline{n}}_0\underline{u} = (\underline{1} - \bar{\underline{n}}_0)\,\underline{u}\ ,$$

and

$$\bar{\underline{n}}_0\underline{x} + \underline{w} = \bar{\underline{n}}_0\underline{x} + (\underline{1} - \bar{\underline{n}}_0)\,\underline{u}$$

Thus the value of $\underline{B}$ for operations $\bar{\underline{n}}$ is clearly

$$\underline{B} = \underline{1} - \bar{\underline{n}}_0\ . \qquad (20)$$

Using this operation in the integral (1), it is clear that the vector $\underline{u}$ then gives the location of the center for the operation $\underline{\bar{n}}$. Thus, if two atoms are related by the operation $\underline{\bar{n}}$ with respect to the point $\underline{u}$ there will be a contribution to the function $\phi(\underline{u})$.

For the case $\bar{1}$, the transformation (20) clarifies the discussion, already given (Patterson, 1949), of the integral (1) with $\underline{A} = \bar{1}$ and $\underline{B} = \underline{1}$. With $\underline{B} = \underline{1} - \bar{1} = 2.\underline{1}$ instead of $\underline{1}$, the interpretation as a "symmetry center map" is clearer as will be seen from the examples given below.

For the operation $\bar{2}$ ($\equiv$ m), the argument leading to (20) breaks down, since $\underline{u}$ becomes infinite and (19b) is invalid. We consider a space group having the reflection or glide reflection operation

$$\underline{G}\,\underline{x} = \bar{\underline{2}}_0\underline{x} + 1/2\,\underline{g} \tag{21}$$

where $\underline{g}$ is the lattice translation in the glide direction for a glide plane, or is zero for a pure reflection. The operation

$$\underline{G}\,\underline{x} + \underline{w} = (\bar{\underline{2}}_0\underline{x} + \underline{w}) + (1/2\underline{g} + \underline{w}) \tag{22}$$

is then a reflection in a plane of height $\underline{u}$ given by (19a) with glide translation $(1/2\underline{g} + \underline{\nu}\,\underline{w})$. We are thus led to write

$$\underline{\nu}\,\underline{w} = \underline{\nu}\,\underline{u} \tag{19b}$$

and obtain (using 19a)

$$\underline{w} = \underline{u} - \underline{\alpha}\,\bar{\underline{2}}_0\underline{u} = (\underline{1} - \underline{\alpha}\,\bar{\underline{2}}_0)\,\underline{u}\,. \tag{23}$$

Thus for a glide plane, we write

$$\underline{B} = (\underline{1} - \underline{\alpha}\,\bar{\underline{2}}_0)\,, \tag{24}$$

and in interpreting the integral (1) and its series expansion (8), a peak at $\underline{u}$ indicates density peaks related by a glide plane at height $\underline{\nu}\,\underline{u}$ with glide translation $1/2\underline{g} + \underline{\alpha}\,\underline{u}$. Thus $\underline{\alpha}\,\underline{u}$ indicates the departure of the glide from that of the space group, as in the case of the screw axis translations.

In many ways, the "displaced and transformed" $|F|^2$-series (11) is easier to interpret than the "transformed" $|F|^2$-series (8). It is not necessary to repeat the discussions of the present section to see that the screw or glide translations indicated by the peaks of the series (8) will correspond directly to the appropriate component of $\underline{u}$, since in this case the operation used in the Faltung is the point operation $\underline{A}_0$ instead of the more general operation $\underline{A}$. The latter will usually contain a translation. If it does not the two series (8) and (11) will of course be identical.

Examples

The new interpretation of the $|F|^2$-series is illustrated by the four possible interpretations for the familiar space group $P2_1/n$, for which a set of general equivalent points, and the symmetry elements are illustrated in the upper diagram of Fig. 3. The second diagram of the figure exhibits the usual $|F|^2$-series for this space group, which is now to be interpreted as ϕ or ϕ_0, with $\underline{A} = \underline{B} = \underline{1}$. The third diagram exhibits ϕ (and ϕ_0) for $\underline{A} = \bar{1}$ and $\underline{B} = 2.\underline{1}$. Its interpretation as a "center of symmetry" map becomes clear by considering a few of the peaks in the ϕ map. The peaks of weight 4 at 000; 1/2, 00; 1/2, 1/2, 0; 1/2, 1/2, 1/2 etc. indicate that all the equivalent points are related by centers at these points. The peak at xyz indicates that there is a point at xyz which is itself centro-symmetrical. That at 1/2 - x, y, -z relates the point at 1/2 - x, 1/2 + y, 1/2 - z to its translation equivalent 1/2 - x, - 1/2 + y, - 1/2 - z. The weight 1 indicates that the particular situation is met with only once in integrating with respect to x over the cell. The peak of weight 2 at 1/4, 1/4 - y, 1/4 arises from the central relation at this point between 1/2 + x, 1/2 - y, 1/2 + z and -x, -y, -z. The weight arises from the fact that the central relation is reciprocal, in the same way that two peaks centrally related arise from a single distance in the usual form of the series. A second interpretation can be based on the fact that the pair is met twice in integrating over the cell. Similar interpretations can be given to all peaks in this map.

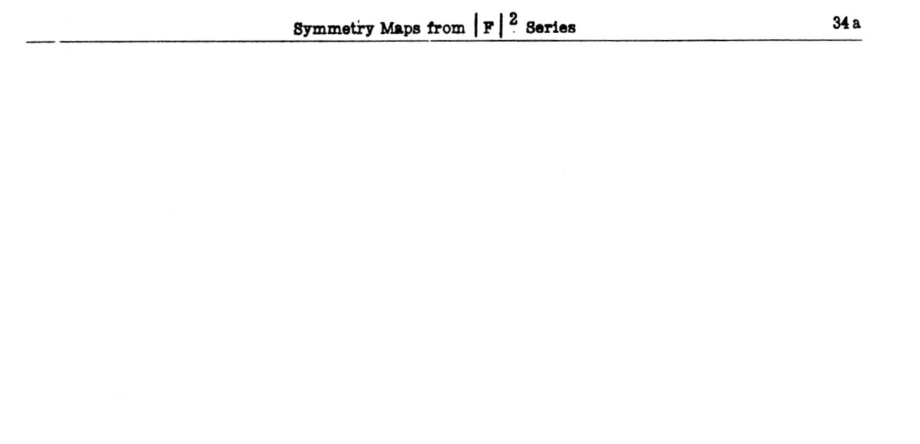

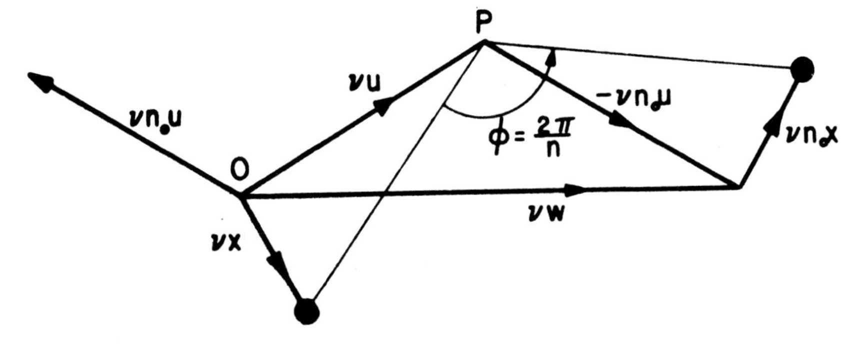

Fig. 1 - Combination of rotation and translation

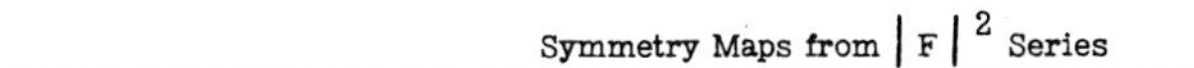

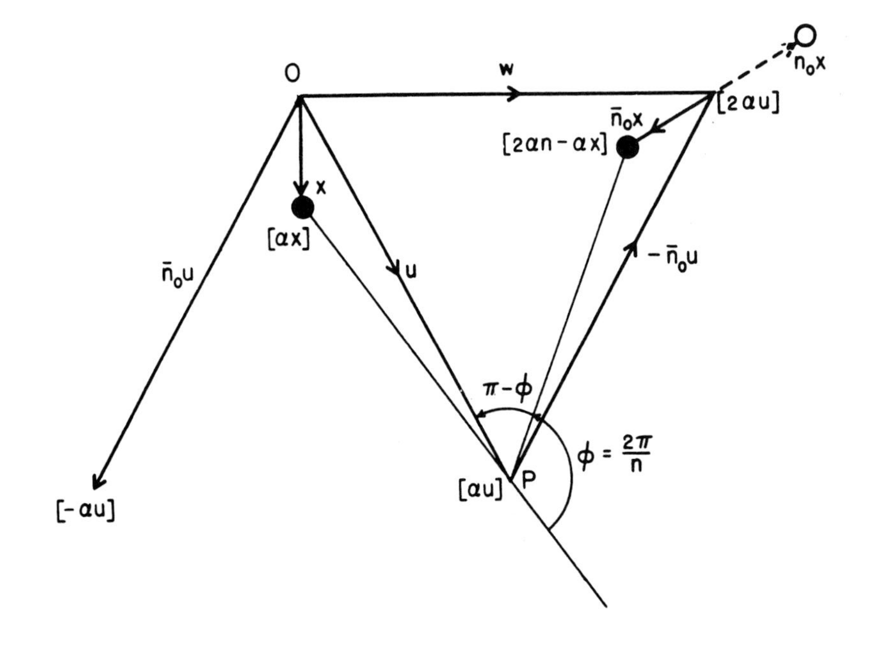

Fig. 2 - Combination of rotation-inversion and translation
The heights are indicated in brackets

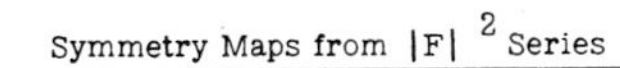

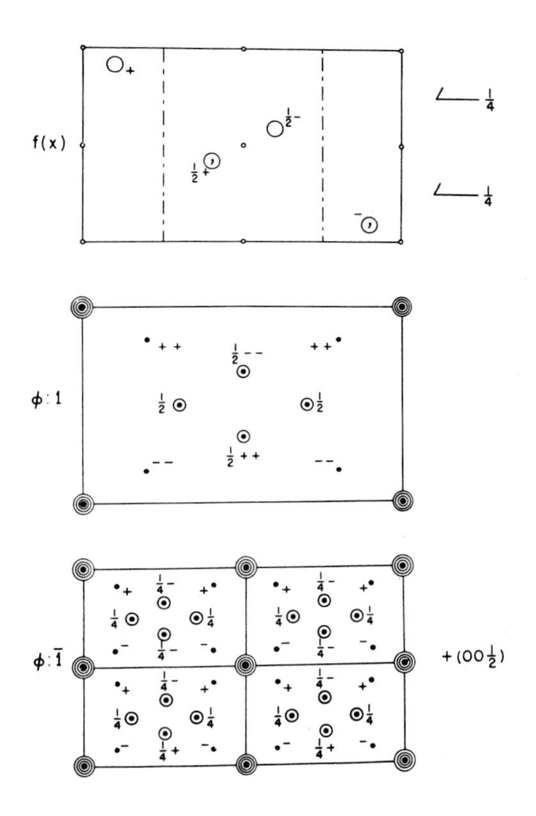

Fig. 3 - Symmetry maps for $P2_1/n$: Identity 1 and inversion $\bar{1}$

In Fig. 4, the glide plane n is investigated. In the upper diagram the origin of the space group is shifted to the intersection of the 2_1-axis and the n-plane, since in our theoretical discussion we have assumed that the symmetry element, on which $\underline{A}$ is based, contains the origin. For this operation we have from (21)

$$\underline{A}\underline{x} = \bar{\underline{2}}_0\underline{x} + 1/2\underline{g}$$

which with the aid of Table 1 may be written in the form

$$\underline{A}\underline{x} = \begin{pmatrix} 100 \\ 010 \\ 00\bar{1} \end{pmatrix}\begin{pmatrix} x_1 \\ x_2 \\ x_3 \end{pmatrix} + 1/2\begin{pmatrix} 1 \\ 1 \\ 0 \end{pmatrix} = \begin{pmatrix} x_1 + 1/2 \\ x_2 + 1/2 \\ -x_3 \end{pmatrix}$$

For the operation $\underline{B}$ we have from (24)

$$\underline{B}\underline{u} = (\underline{1} - \underline{\alpha}\,\bar{\underline{2}}_0)\,\underline{u}$$

$$= \left\{\begin{pmatrix} 100 \\ 010 \\ 001 \end{pmatrix} - \begin{pmatrix} 000 \\ 000 \\ 001 \end{pmatrix}\begin{pmatrix} 100 \\ 010 \\ 00\bar{1} \end{pmatrix}\right\}\begin{pmatrix} u_1 \\ u_2 \\ u_3 \end{pmatrix}$$

$$= \begin{pmatrix} 100 \\ 010 \\ 002 \end{pmatrix}\begin{pmatrix} u_1 \\ u_2 \\ u_3 \end{pmatrix} = \begin{pmatrix} u_1 \\ u_2 \\ 2u_3 \end{pmatrix}.$$

We now may insert these values in (1) and write ϕ (u) in the form

$$\phi\,(uvw) = \iiint f(xyz)\ f(u + x + 1/2,\ 2v - y,\ w + z + 1/2)\ dxdydz \tag{25}$$

in which we have written xyz respectively for x_2, x_3, x_1 and uvw for u_2, u_3, u_1. The series (8) then takes the form

$$\phi\,(uvw) = \sum_{\underline{h}} |F(\underline{h})|^2\, e^{-2\pi i(hu + 2kv + lw)} \tag{26}$$

while the series (11) for ϕ_0 is clearly

$$\phi_0(uvw) = \sum_{\underline{h}} e^{-\pi i(h + l)}\, |F(\underline{h})|^2\, e^{-2\pi i(hu + 2kv + lw)}\,. \tag{27}$$

These two series are thus contracted by a factor 1/2 in the y direction as shown in Fig. 4. In this case the peaks of the ϕ series are generated by the glide plane, and those of weight 4 at 000 and 0, 1/2, 0 indicate this operation. The singly weighted peaks now no longer arise from a lattice operation, but rather from the screw axes. Thus the peak at 2x, y, 2z implies a glide plane at the level y with a translation 1/2 + 2x, 0 , 1/2 + 2z relating the point at -1/4 - x, -1/4 + y, -1/4 - z with that at 1/4 + x, 1/4 + y, 1/4 + z. On the other hand, the double peaks now arise from a lattice identity operation, for example that at 1/2, 1/4 - y, 1/2 indicates a glide plane at the level 1/4 - y with translation 1/2 + 1/2, 0, 1/2 + 1/2, i.e. 000, that is to say a plane of symmetry through the point 1/4 - x, 1/4 - y, 1/4 - z. This plane also relates 3/4 + x, 3/4 - y, 3/4 + z with its translation equivalent 3/4 + x, -1/4 - y, 3/4 + z, and thus has double weight.

The series ϕ which has had its origin peaks shifted to 1/2, 0, 1/2, and 1/2, 1/2, 1/2 indicates the glide plane translations of the group directly by these peaks. Similarly the doubly weighted peak, which we referred to above is now at 0, 1/4 - y, 0 and indicates a plane of symmetry at the level 1/4 - y directly.

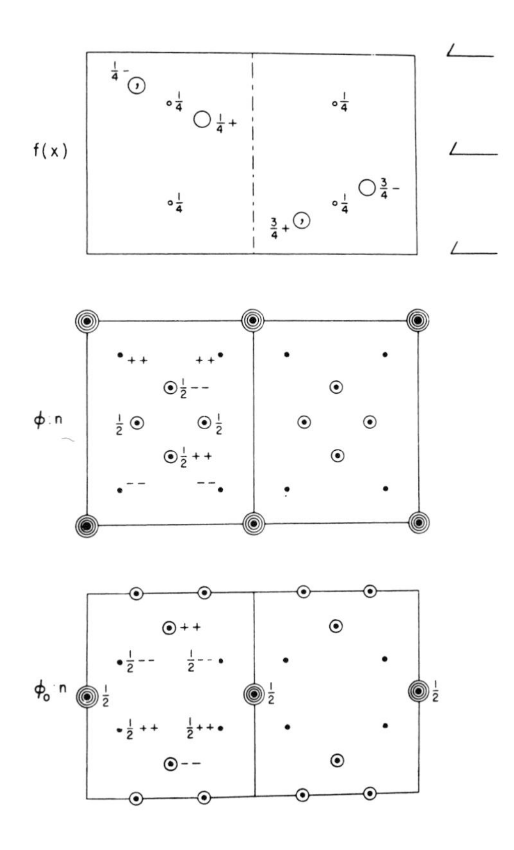

Fig. 4 - Symmetry maps for $P2_1/n$: Glide plane n

In Fig. 5, the peaks derived from the ϕ and ϕ_0 series based on the operation 2 are illustrated. In these series the operation $\underline{B}$ (Table 1) indicates that the $|F|^2$-series must be contracted by a factor 1/2 in both the x and z directions. The interpretation follows directly the lines used in discussing Figs. 3 and 4.

The discussion of the series ϕ and ϕ_0 for the other space groups follows exactly the line of argument developed here for $P2_1/n$. The integrals (1) or (9) need never be written down, although they can be with the aid of Table 1. The operator $\underline{B}$, taken from Table 1, indicates the transformation which the usual form of the $|F|^2$-series must undergo to produce a symmetry map. Such transformations may be identical, may involve contractions in specific directions, rotations, or both (Buerger, 1946).

Discussion

It is clear from the theory and from the examples discussed above that any two atoms, related by symmetry or not, can be related by an appropriately located symmetry element with appropriate glide or screw translation, provided that only two atoms are studied and the remainder of the structure is ignored. In this sense the new interpretation has theoretical interest in that it does give a more general approach to the study of the Faltung integrals and a more general interpretation to the Buerger implication diagrams.

The new point of view has practical interest in the interpretation of the $|F|^2$-series when the molecule (asymmetric unit) possesses exact or approximate symmetry of the same type as that possessed by the space group. In the example of Fig. 6 an imaginary molecule, which possesses a plane of symmetry, is laid down on the space group Pa. The usual form of the $|F|^2$-series is shown in the second diagram in which the large peaks (weight $5Z^2$) can be interpreted in terms of equal parallel vectors. In the third diagram, which is a symmetry map derived from the operation $\underline{a}$, the large peaks have immediate interpretation in terms of the "planes of symmetry" which pass through the molecule and between two molecules related by a translation. It is of course recognized that it is only chance which makes a symmetry element in a part of the "asymmetric" unit have the same or similar orientation to a corresponding element in the space group. This situation does however occur and when it does a number of peaks from light atoms can combine to simulate a peak from a single pair of heavy atoms.

The occurence of a center of symmetry within the asymmetric unit is by no means infrequent and has already been partially discussed (Patterson, 1949). Fig. 7 illustrates an artificial case in which two benzene-like rings are laid down on the space group $P\bar{1}$. In the symmetry map $\phi : \bar{1}$ the large peaks give the implication for the centers of the rings. This benzene effect is still further enhanced if the benzene ring is para-substituted. Unless the two atoms in the para-position are very different, their interaction will add to that of the ring and a very large peak will result. It has been pointed out that the central peak of a benzene ring alone will have a weight 6 x 6 = 216 which already approaches that of a peak due to two chlorines ($17^2 = 289$).

Remarks on Further Generalization

In the present paper we have considered only those operations $\underline{A}$ in the Faltung (1) which were operations of the space group or of the lattice of the function $f(\underline{x})$. In this case the vector

$$\underline{h}_1 + \underline{h}_2 \underline{A}_0 = \underline{H} \tag{28}$$

is a whole-numbered vector in the reciprocal lattice, and the double lattice summation of (6) reduces to the single summation of (7). If $\underline{H}$ is not a whole-numbered vector, i.e. if A_0 is a point operation which is not an operation of the lattice, the integral (6a) can be evaluated, but it will no longer have the simple values which we have discussed, and the specific values will have to be inserted as indicated in the series (6). Such series represent one form which the generalized series referred to by Karle and Hauptmann (1950) may take.* In special cases in which the phases can be determined by inequalities this more general Faltung may be of interest. It will be worth while, perhaps, to consider the transformations which take the lattice into a sub-lattice or a super-lattice of itself.

* The writer is indebted to Dr. J. Karle for the privilege of reading this paper in manuscript and for several valuable discussions. It is probable that his presentation, at the time of the opening exercises for X-RAC at Auburn, suggested the line of thought which led to the present work.

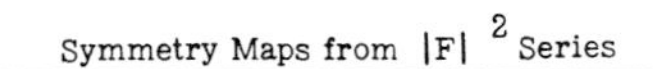

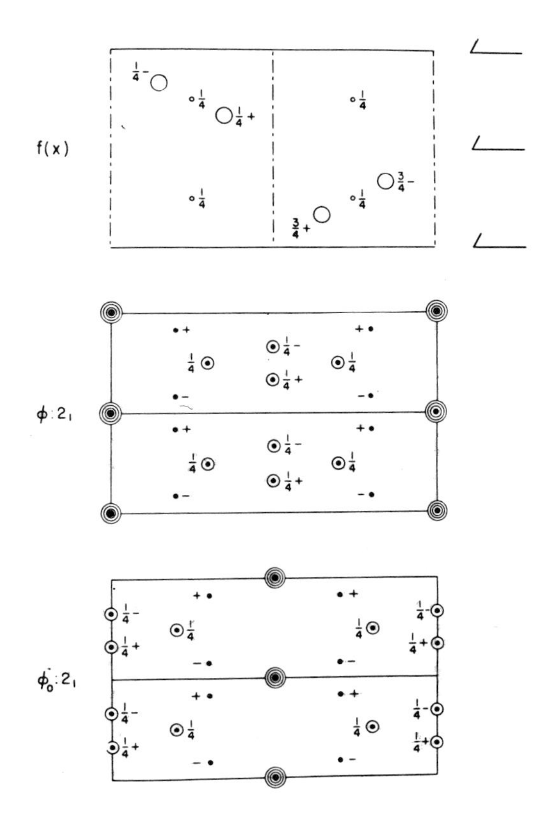

Fig. 5 - Symmetry maps for $P2_1/n$: Screw axis 2_1

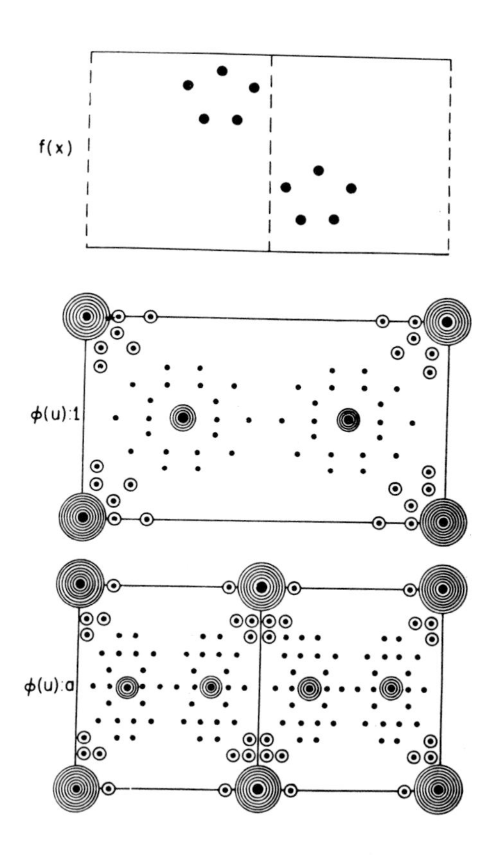

Fig. 6 - Molecule with plane of symmetry on space group Pa

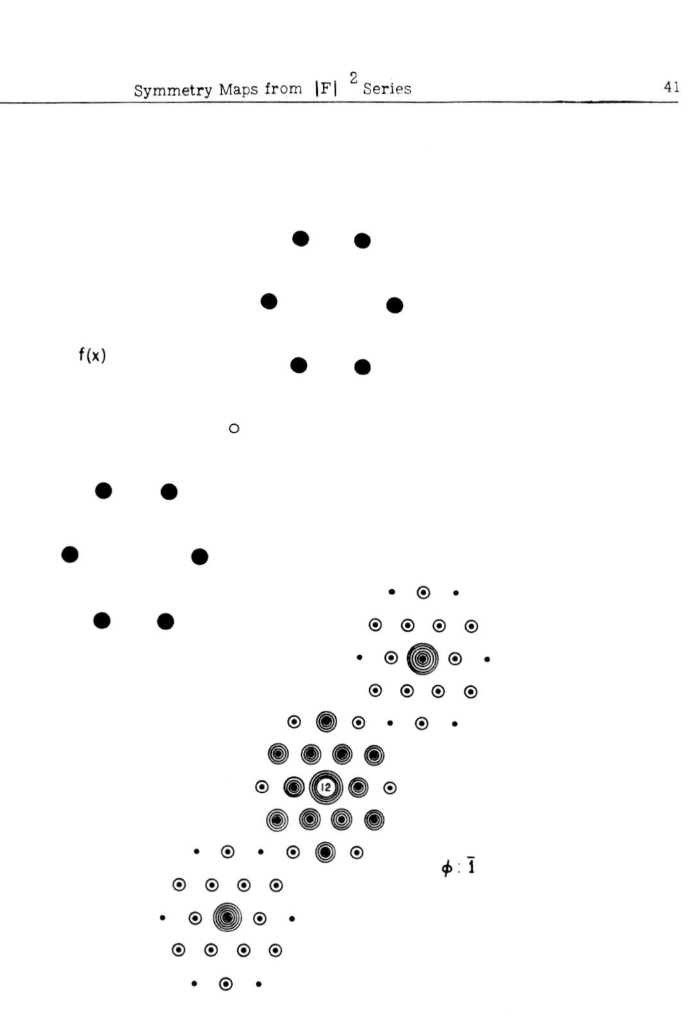

Fig. 7 - Molecule with center of symmetry on space group $P\bar{1}$

As a further generalization of the integral (1), a weight function may be introduced as an additional factor under the integral sign. In its simplest form, such a weight function could have the value zero in certain regions of the variables u and x and unity elsewhere and would lead to the "slices" of Booth (1945) and Schomaker (unpublished) and possibly to other computable functions of interest. More complicated weight functions (Hughes and Wilson, 1949) may well lead to useful series which are easier to compute.

Acknowledgements

The work reported here was commenced during a sabbatical leave granted by the Trustees of Bryn Mawr College. While at Bryn Mawr College, the writer's work was supported by a Frederick Gardner Cottrell Grant of the Research Corporation and at present is supported by a research grant from the National Cancer Institute, of the National Institute of Health, Public Health Service. The writer wishes to express his gratitude to these three bodies for their very generous assistance.

References

Booth, A. D. (1945). Trans. Faraday Soc. 41, 434.

Buerger, M. J. (1946). J. Appl. Phys. 17, 579.

Burckhardt, J. J. (1947). Die Bewegungsgruppen der Kristallographie, Basel, Verlag Burkhauser.

Hughes, E. W. and Wilson, J. (1949). Abstracts: Cornell Meeting ASXRED.

Karle, J. and Hauptman, H. (1950). Acta Cryst. 3, 181.

Patterson, A. L. (1934). Phys. Rev. 46, 372.

(1935). Z. Krist. A90, 517.

(1949). Acta Cryst. 2, 339.

Seitz, F. (1934). Z. Krist. A88, 433.

(1935a). Z. Krist. A90, 289.

(1935b). Z. Krist. A91, 336.

(1936). Z. Krist. A94, 100.

INDEX